T0206120

CAMBRIDGE LIBRARY COLLECTION

Books of enduring scholarly value

Botany and Horticulture

Until the nineteenth century, the investigation of natural phenomena, plants and animals was considered either the preserve of elite scholars or a pastime for the leisured upper classes. As increasing academic rigour and systematisation was brought to the study of 'natural history', its subdisciplines were adopted into university curricula, and learned societies (such as the Royal Horticultural Society, founded in 1804) were established to support research in these areas. A related development was strong enthusiasm for exotic garden plants, which resulted in plant collecting expeditions to every corner of the globe, sometimes with tragic consequences. This series includes accounts of some of those expeditions, detailed reference works on the flora of different regions, and practical advice for amateur and professional gardeners.

A Dictionary of the Economic Products of India

A Scottish doctor and botanist, George Watt (1851–1930) had studied the flora of India for more than a decade before he took on the task of compiling this monumental work. Assisted by numerous contributors, he set about organising vast amounts of information on India's commercial plants and produce, including scientific and vernacular names, properties, domestic and medical uses, trade statistics, and published sources. Watt hoped that the dictionary, 'though not a strictly scientific publication', would be found 'sufficiently accurate in its scientific details for all practical and commercial purposes'. First published in six volumes between 1889 and 1893, with an index volume completed in 1896, the whole work is now reissued in nine separate parts. Volume 6, Part 3 (1893) contains entries from *silk* to *tea*, two of India's most important economic products.

Cambridge University Press has long been a pioneer in the reissuing of out-of-print titles from its own backlist, producing digital reprints of books that are still sought after by scholars and students but could not be reprinted economically using traditional technology. The Cambridge Library Collection extends this activity to a wider range of books which are still of importance to researchers and professionals, either for the source material they contain, or as landmarks in the history of their academic discipline.

Drawing from the world-renowned collections in the Cambridge University Library and other partner libraries, and guided by the advice of experts in each subject area, Cambridge University Press is using state-of-the-art scanning machines in its own Printing House to capture the content of each book selected for inclusion. The files are processed to give a consistently clear, crisp image, and the books finished to the high quality standard for which the Press is recognised around the world. The latest print-on-demand technology ensures that the books will remain available indefinitely, and that orders for single or multiple copies can quickly be supplied.

The Cambridge Library Collection brings back to life books of enduring scholarly value (including out-of-copyright works originally issued by other publishers) across a wide range of disciplines in the humanities and social sciences and in science and technology.

A Dictionary of the Economic Products of India

Volume 6 – Part 3: Silk to Tea

George Watt

CAMBRIDGE
UNIVERSITY PRESS

University Printing House, Cambridge, CB2 8BS, United Kingdom

Published in the United States of America by Cambridge University Press, New York

Cambridge University Press is part of the University of Cambridge.
It furthers the University's mission by disseminating knowledge in the pursuit of
education, learning and research at the highest international levels of excellence.

www.cambridge.org
Information on this title: www.cambridge.org/9781108068802

This edition first published 1893
This digitally printed version 2014

ISBN 978-1-108-06880-2 Paperback

A

DICTIONARY

OF

THE ECONOMIC PRODUCTS OF INDIA.

BY

GEORGE WATT, M.B., C.M., C.I.E.,

REPORTER ON ECONOMIC PRODUCTS WITH THE GOVERNMENT OF INDIA.
OFFICIER D'ACADEMIE; FELLOW OF THE LINNEAN SOCIETY; CORRESPONDING MEMBER OF THE ROYAL HORTICULTURAL SOCIETY, ETC., ETC.

(ASSISTED BY NUMEROUS CONTRIBUTORS.)
IN SIX VOLUMES.

VOLUME VI, PART III.

[Silk to Tea.]

Published under the Authority of the Government of India,
Department of Revenue and Agriculture.

LONDON:
W. H. ALLEN & Co., 13, WATERLOO PLACE, S.W., PUBLISHERS TO INDIA OFFICE.

CALCUTTA:
OFFICE OF THE SUPERINTENDENT OF GOVERNMENT PRINTING, INDIA,
8, HASTINGS STREET.

1893.

CALCUTTA
GOVERNMENT OF INDIA CENTRAL PRINTING OFFICE
8, HASTINGS STREET.

DICTIONARY

OF

THE ECONOMIC PRODUCTS OF INDIA.

1726

SILK AND SILKWORMS OF INDIA.

In perhaps no other country of the world does the necessity exist so pressingly as in India to treat the subject of its silk industries under two important and distinct sections, *viz.*, the DOMESTICATED or MULBERRY-FEEDING and the WILD or NON-MULBERRY-FEEDING worms. The term "wild" may at the outset be explained as here used in a commercial more than a literal sense, for, of the insects so designated, some have for centuries been completely domesticated both in India and China. **Mr. Cobb** (*British Manufacturing Industries*) may be quoted in support of the very frequent acceptation of this classification. He writes: "The silks now generally recognised as *tussahs,* and already mentioned as produced principally in China and India, are a description of wild silk hitherto neglected by Europeans, but now deservedly attracting much notice. In the East they have been long appreciated, from their cheapness and durability." Thus the isolation into two sections is not only in conformity with the accepted European commercial view, but it approximates to the Indian conditions of the trade. To the inhabitants of this country, more especially to the uncultured races who live in the forests and hill tracts, the wild insects are of the greatest value, and the interests of these people are perfectly distinct from those of the growers, reelers, and weavers of, and traders in, Mulberry silk. Although, of course, the total value of the trade in domesticated silkworms is considerably greater than that in the wild insects, it has been held by many writers that India's best interests are, or should be, concentrated in the development of one or more of the wild insects, since these are more suited to the climatic conditions of the country. Whether or not that view be correct it is hoped the reader will be able to judge for himself from the review of facts and opinions which the author has thrown together in the pages that follow. The comparatively large space which has been devoted to the wild silks has been partly due to the necessity that seemed to exist for a full and exhaustive statement of the subject and to the fact of the greater number of species of wild silkworms, each of which, to the people of India at least, has its own peculiar merits and uses, and therefore, in a *Dictionary of the Economic Products of India*, had to find its own place, even though, to European silk experts, it may at present be regarded as devoid of interest. It is further anticipated that a detailed statement of the chief facts known regarding the wild silks may in a manner save the recurrence of false hopes which, in everything connected with the undeveloped resources of India, are spasmodically pressed on public consideration as new and invaluable discoveries. Of our so-called wild silks, the *Tasar, Eri* and *Muga* are those which can alone be regarded as of commercial value, and of the remainder it may be said that only a very few have the least chance of ever proving of much use. The classification which the writer has thus adopted may be said to be into the two great Tribes of silk-yielding insects, *viz.*, the BOMBYCIDÆ, and the SATURNIIDÆ. Under each of these sections will, therefore, be found, *first*, a brief alpabetical enumeration of the more important species that fall into these tribes, and, *second*, a detailed statement of such as are thought of present commercial value. The Editor desires to here acknowledge the very great obligations he is under to several distinguished entomologists for generously reading the proofs of this sketch of the literature of the Indian Silk Moths. More especially is he indebted to **G. F. Hampson, Esq.**, for many useful suggestions.

Unfortunately, the proofs had advanced too far, before being shown to Mr. Hampson to admit of all his synonymy being fully incorporated. This defect, it is hoped, may, however, be mitigated by Mr. Hampson's *Complete List of the Indian* Bombycidæ and Saturniidæ (furnished as an Appendix, *p. 235*), which has been utilized as an index to the chief passages in this article, that deal with the insects, while, at the same time, it has been made a bibliographical register to the scientific publications that treat of the various species. The Editor has had to keep two main objects in view: to prepare a commercial rather than a scientific treatise; and to furnish a fairly complete review of the writings of Indian authors. Scientific precision, under these conditions, was not always possible, since opinions had to be quoted, side by side, even although at times these might be conflicting.

THE BOMBYCIDÆ.

A.—ENUMERATION OF THE CHIEF INDIAN BOMBYCIDÆ.

(For B. **Saturniidæ**, see pp. 67 to 96.)

I. BOMBYX.

I.—BOMBYX.

The genus **Bombyx** includes six species or forms which have been enumerated by Mr. F. Moore in a list published in Mr. Thos. Wardle's *Hand-book of the Wild Silks of India.* They are as follows:—

1727

1. Bombyx arracanensis, *Hutton.*

The Burmese silk-worm, domesticated in Arracan; said to have been introduced from China. It yields several broods annually, and the cocoons are larger than the Bengal monthly species. (*For fuller particulars, see p. 7.*)

1728

2. B. crœsi, *Hutton.*

The *Nistrí* or *Madrasí* of Bengal was probably introduced from China. Has been domesticated in Bengal perhaps for centuries. It yields seven or eight broods of golden yellow cocoons in the year, and these are of larger size than those of **B. sinensis.** (*See also p. 10.*)

1729

3. B. fortunatus, *Hutton.*

The *Desí* (commonly spelt *dasee*) of Bengal. This yields several broods annually, and spins a very small cocoon, which is of a golden yellow colour. (*See p. 12.*)

1730

4. B. mori, *Linn.*

The common silk-worm, domesticated in China, Bokhara, Afghánistán, Kashmír (? the Panjáb and Dehra Dún), Persia, South Russia, Turkey, Egypt, and Algeria, Italy, France, and Spain, in all of which countries it produces but one crop annually, and spins the largest cocoon and the best silk, which is of a golden yellow or white. (*See p. 15.*)

1731

5. B. sinensis, *Hutton.*

The *Siná, China,*—the small Chinese monthly worm of Bengal. This has been domesticated in Bengal, where it is said to have been introduced from China. It produces several broods in the year, and the cocoons are white or yellow. (*See p. 18.*)

1732

6. B. textor, *Hutton.*

The *Boro polo* of Bengal where it is domesticated. It is an annual, and produces a white (sometimes yellow) cocoon, of a different texture and more flossy than that of **B. mori.** (*See p. 19.*)

The other three genera are commonly classed amongst *wild silkworms* though they belong to the **Bombycidæ**; they are as follows:—

II.—OCINARA.

II. OCINARA.

It will be seen from the Appendix (*p. 235*) that **Hampson** reduces the three species of this genus (as here dealt with) to two forms, but adds a third. Nos. 7 and 8 he regards as one species, *viz.*, **O. signifera,** and No. 9 he refers to **O. apicalis.** For **Hampson's** third species of this genus see the remarks under **Trilocha,** p. 5.

7. Ocinara diaphana, *Moore.*

1733

This insect (a native of the Khasia hills) was described originally by **Moore** (*Lep. Atk., I., 83, 1879*), but very little is known regarding it: it is probably only a form of **O. lactea.**

8. O. lactea, *Hutton.*

1734

Mussurie and the North-West Himálaya, generally up to Kulu. It feeds on **Ficus infectoria,** spins a small, yellow cocoon, and yields several broods during the summer. **Hutton** (*Notes on Indian Bombycidæ, Jour. Agri.-Hort. Soc., Ind., New Series, III., 133*) speaks of this insect as yielding too little silk to be of any practical value. It was found feeding along with **O. lida** on **Ficus infectoria,** *Roxb.*, near Mussurie.

9. O. moorei, *Hutton* (*Conf. with p. 235*).

1735

Mussurie, North-West Himálaya, as, for example, at Dehra Dun. This also feeds on **Ficus infectoria,** as well as on several other wild figs. It spins a small, white cocoon, is multivoltine, and probably doubtfully distinct from the Javanese, **O. lida.** This view was taken by **Hutton** himself, subsequent to his having published it under the above name.

III.—THEOPHILA.

III. THEOPHILA.

The character which **Moore** regards as separating the insects (now placed in this genus) from **Bombyx** (=the true mulberry-feeding silkworms) is the presence of rows of spines on the larvæ. Although thus closely allied to the species of **Bombyx,** it seems problematic if it would ever pay to cultivate any of the insects that belong to this genus, since the worms mostly feed on the mulberry, which might be more profitably given to some of the forms of **Bombyx.** In the Appendix (*p. 236*) it will be seen that **Hampson** views Nos. 10, 11, 13, and 14 as synonyms of No. 12.

10. Theophila affinis, *Hutton* (*see p 236*).

1736

Modern entomologists place this as intermediate between **T. bengalensis** and **T. huttoni.** It occurs plentifully in Chutia Nagpur and may be reared either on the mulberry or on **Artocarpus Lacoocha.** The Rev. **A. Campbell** of Govindpur attempted its domestication but with doubtful results. The insect appeared spontaneously on mulberry bushes in his garden. (*Conf.* with **Hutton's** *Notes on Indian Bombycidæ, Jour. Agri.-Hort. Soc., 1871, III., 131.*)

11. T. bengalensis, *Hutton, Trans. Ent. Soc. Lond., 3,* [*II., 322, pl. XIX., f. 5* (*see p. 236*).

1737

The wild silkworm of Lower Bengal. This insect has been discovered in the neighbourhood of Calcutta, feeding on **Artocarpus Lacoocha** and **A. Chaplasha,** as also at Ranchi in Chutia Nagpur and in Sikkim up to an altitude of 2,000 feet. **Hutton** tells us (*Jour. Agri.-Hort. Soc., Ind., XIV.,*

THEOPHILA.

Sel. 95) that it was first discovered by the late Mr. Frith in 1849 near Moorshedabad, but adds that he thought there might be some mistake, since Frith spoke of it as feeding on the common mulberry. Subsequently Mr. Grose furnished Hutton with a carefully coloured drawing of the larva, and stated that it was found on the bread-fruit tree (**Artocarpus**). See Hutton's *Notes on Indian Bombycidæ* (*Jour. l. c., III., 130*).

1738

12. **Theophila huttoni,** *Westwood* (*Conf. with p 236*).

This, by some writers, is spoken of as the wild silkworm of the North-West Himálaya, where it occurs between 2,000 and 8,000 feet. It was discovered by Hutton at Simla in 1837 and again in Mussoorie in 1842. Its distribution is from Kumáon westward. The worms are found abundantly feeding on the common mulberry of the mountain forests and hedgerows near cultivation.

Acting under the orders of Government, Captain Hutton performed a series of experiments in 1858, with the object of ascertaining whether it was possible to domesticate this insect, and, when domesticated, whether it could be reared profitably. The result was unfavourable, and the experiment on Captain Hutton's own recommendation was abandoned. The insect has so many enemies that it cannot be profitably reared in the open, the worm, in fact, proved too restless for domestication.

It is, however, a strong and hardy species, which affords a beautiful soft whitish silk. Its cocoon is encased in a hard polished exterior layer and embraced by leaves. It yields two crops in the year. The caterpillars of the first generation appear about April, and the cocoons are formed in May and again in September. Hutton tells us that in 1849 he succeeded in crossing this insect with the Kashmír silkworm, the female **T. huttoni** being coupled with the male **Bombyx mori**, but the progeny to all intents and purposes was **T. huttoni.**

For further information see *Jour. Agri.-Hort. Soc., Ind., VI. (1848), 178-179; XI., 403-414; XIII., 58; XIV., Sel., 91; New Series, III., 129; Moore, Cat. Lepid. Insects, II., 379; Rondot, L' art de la soie, II., 6.*

1739

13. **T. religiosa,** *Helfer.*

This, by most writers, is supposed to be the *jori* insect of Assam (according to Helfer) and the *deo-muga* of Cachar (according to Hugon). Hutton says it was first discovered by Hugon in 1834 in Cachar, and that Helfer's specimen was supplied by Jenkins from Assam. Hutton regarded these insects as belonging to **Ocinara** (*Jour. Agri.-Hort. Soc., Ind., XIV., Sel., 97*). The insect here dealt with feeds on the *bar* tree (**Ficus bengalensis**) and the *pipal* (**F. religiosa**), but it may be added that Moore thinks it is one and the same with **T. huttoni.**

The greatest possible confusion exists regarding the *deo-muga* worms of Assam and Cachar. Thus, for example, the late Mr. Stack says that that name is given to a special form of *tasar* (which feeds on the *phutuka*), and to the present (the true *deo-muga*) insect. The following is Mr. Stack's account:—

"The *deo-muga* silk-worm is so called from its size. It is the largest of all the worms, attaining a length of 6½ inches,* and it is also the handsomest. Mr. Buckingham writes:—

"'This worm appears at times on *sum*-trees with the common *muga*, but it is of rare occurrence. The worm in its second and third stage is particularly handsome, with rows of turquoise spots on each side. When

* Hampson hinks this must refer to the *Tasar*.

the worm enters upon its fourth stage, the turquoise spots vanish, and spots of gold appear in their place, and on each side of the body stripes having all the colours of the rainbow tend to make this worm by far the most beautiful of its tribe.'*

DEO-MUGA. 1740 *Conf. with p. 139.*

"The *deo-muga* worm is said to live 30 days, and to spend three days in spinning its cocoon; the period of the chrysalis is 15 days in the hot and 30 days in the cold season, and the life of the moth lasts about four days. "The cocoon is ($3'' \times 1\frac{1}{3}''$) large, and gives a large quantity of strong but coarse and dark-coloured silk. The hardness of the cocoon renders it difficult to reel, and the silk easily gets into knots. The thread of the *deo-muga* is said to be used for fishing lines in Bengal. In Cachar the *deo-muga* feeds on the *banyan* (**Ficus bengalensis**) and *pipal* (**Ficus religiosa**). The worm occurs generally in the Assam Valley."

Conf. with Helfer in *Jour. Asiatic Soc. Beng., VI., 41 (1837)*: Hugon, *Jour. l. c.*; Hutton, *Notes on Indian Bombycidæ, Jour. Agri.-Hort. Soc., Ind., New Series, III., 132*; Moore, *Cat. Lep. Insects, II., 381.*

14. Theophila sherwilli, *Moore, Trans. Ento. Soc. Lond. (3), [II., 423, pl. XXII., f. 1.*

1741

This is the wild silkworm of the Eastern Himálaya.

Hutton says of it (*Notes on Indian Bombycidæ, l. c., 132*) that it was first seen in a collection made at Darjíling by the late Major I. L. Sherwill, but whether captured in the plains or at Darjíling itself no one knows. Moore remarks that it is allied to **T. huttoni**, but differs in being larger, and in the abdomen being tipped with black.

IV.—TRILOCHA.

IV TRILOCHA.

Hampson regards the genus **Trilocha** as possessing no structural difference from **Ocinara**; the latter being the older generic name he therefore transfers Nos. 15, 16, and 17 as synonyms for one species under **O. varians**, *Wlk.* See Appendix, p. 235.

15. Trilocha albicollis, *Walker.*

1742

This insect was found by Forsayeth at Mhow feeding on **Ficus religiosa**. It is doubtfully distinct from **T. varians.**

16. T. cervina, *Walker.*

1743

This is an imperfectly known species, but it is supposed to have been collected in India.

17. T. varians, *Walker.*

1744

This insect is found over the greater part of India, Ceylon, and at the Cape.

Hutton says (*Notes on Indian Bombycidæ*) that it had been met with at Kanara and again by Mr. Grote in Calcutta, but that as a silk-yielder it is useless. Although subsequent collectors have recorded its occurrence in many localities, it has nowhere been found in such abundance as to justify its collection commercially. It spins a small yellow compact cocoon which is embraced by the leaves of the tree on which it feeds. The worm has been found feeding on **Streblus aspera, Ficus indica, F. religiosa, Artocarpus integrifolia, Mimusops Elengi,** &c.

* Hampson regards this passage as a description of the larva of an **Antheræa**, possibly **A. assama.**

GUNDA. 1745

CONCLUDING NOTE ON BOMBYCIDÆ.—The above enumeration embraces all the better known genera and species of this tribe. A few imperfectly known insects have been alluded to by entomological writers, but these cannot be regarded as silk-yielding. Amongst these may be mentioned the species of **Gunda** (**G. sikkima,** *Moore,* and **G. apicalis,** *Hampson*). (*See pp.* 235-6.)

THE DOMESTICATED OR MULBERRY-FEEDING SILKWORMS.

References.—*The number of works, reports, and newspaper articles which have appeared regarding Indian Silk during the past 100 years is so very great that an enumeration of even the principal ones would fill many pages. Those of a general nature will therefore be quoted in the text of this section (below), and the more technical publications will be found mentioned under the headings of various sections, as, for example, the specific names of the species or races of silkworms, the provinces to which the passages or works refer, as also such special subjects as diseases, under the chapters devoted to these.*

VARIETIES. 1746

SPECIES, VARIETIES, OR RACES.

The following information on this subject is taken from the Indian Museum Notes, published by **Mr. E. C. Cotes,** which will be found to very closely follow on the lines established by **Hutton** in his *Revision and Restoration of the Silk-worm* (*Jour. Agri.-Hort. Soc., Ind., XIV., Selections, pp.* 8-37 *&* 67-103, being a reprint of his paper which appeared in the *Transactions of the Entomological Society of London*) :—

"The classification of the domesticated mulberry-feeding silkworms, which are reared in different parts of the world, has long been a puzzle to entomologists; the fact being that, while the extreme forms of each variety are well marked and distinct, both in habits and appearance, they are connected by so many intermediate forms that in most cases it is impossible to fix any line of demarcation which shall separate the varieties into groups having distinct characteristics; added to this, so far as has at present been observed, even the most distinct forms are subject to the same diseases, and interbreed readily, when allowed to do so, producing fertile offspring which present characteristics intermediate between those of their parents. On the whole, therefore, it seems best to look upon all domesticated mulberry-feeding silkworms as belonging to the one species **Bombyx mori,** the innumerable varieties being considered as merely Sub-species or races, though for convenience we may retain their old nomenclature, which accords them the rank of Species. Of these races, or sub-species, we may notice the common annual silkworm (**Bombyx mori**) which is reared in Japan, Central Asia, Southern Europe, and, indeed, throughout the whole of the temperate zone. It comprises innumerable local varieties which agree, more or less absolutely, in being *univoltine* (that is to say, in going through but one generation in the course of the year); in the cocoons being of a firm and close consistency, so that the silk can be readily reeled off them; and in the eggs requiring to be exposed to a certain degree of cold to enable them to hatch out regularly and healthily. Connected with this race are *bivoltine*[1] varieties, which produce two crops in the course of the year, the eggs of the second generation only being kept for the next year's crop, as those of the first generation hatch soon after being laid; also *trevoltines*[2], which pass through three generations in the year, and *quadrivoltines*[3], which pass through four. There is an annual, or *univoltine* silkworm, **Bombyx textor,** known in Bengal as the *Boro polo,* which produces cocoons of loose texture,

[1] **Rondot,** in his *L'Art de la soie,* writes that the Genoese were the first to introduce the bivoltine worms of China into Europe. The cocoons, which are white in colour were reared at Novi Ligure with the greatest care and gave good results, the silk becoming well known under the name of "*candide di novi.*"

[2] **Rondot,** *l. c.,* writes that there is a constant variety of *trevoltine* silkworms to be found at Pistoria and other places in Tuscany.

[3] Riley, U. S. Department of Agriculture Bull. No. 9.

VARIETIES.

which are therefore more difficult to reel than the firm cocoons made by the univoltine silk-worms of the temperate zone. But the most important varieties in India are the *desi* (**Bombyx fortunatus**), *madrassi* (**Bombyx cræsi**), *chota pat* (**Bombyx sinensis**), and *nya paw* (**Bombyx arracanensis**), all four of which pass through a succession of generations, sometimes amounting to as many as eight, in the course of the year, the eggs hatching out healthily without exposure to cold, and the cocoons containing comparatively little silk, and that so loosely wound upon the cocoon as to be difficult to reel off without entanglement. It is these small *multivoltines* which yield the bulk of the silk produced in India; but three,[1] or at most four, of the generations produced in the year being raised in sufficient numbers to yield cocoons for the production of silk, and the intermediate generations being only reared in comparatively small quantities by men who devote themselves to the work of raising *seed* (eggs) from which the regular cocoon crops or *bunds*, are reared.

Neglecting varieties which are not reared in large quantities, we may say that the general silk crop of Europe is produced by a variety of silkworm which thrives in a temperate climate, requires cold for the hatching of its eggs, and produces but one crop of cocoons in the year, these cocoons, however, containing a large amount of silk which can be easily reeled. In India, on the contrary, the general silk crop is produced by smaller varieties, which thrive in sub-tropical climates, do not require cold for the hatching of their eggs, and produce each year a series of crops of cocoons which contain, comparatively, a small amount of silk, itself perhaps equal in quality to that produced by the European variety, but so loosely wound upon the cocoon that it is almost impossible, in reeling it off, to prevent entanglement and thus to produce a thread equal in value to that easily obtained from the European variety.

In the steamy plains of Bengal, where the silk industry is chiefly carried on, the mulberry will yield several crops of leaves in the year. A *multivoltine* silkworm, therefore, which can be raised several times in the course of the year, suits the requirements of the country. The superiority, however, of the cocoons of the European variety is so obvious that many attempts have been made to introduce this form into India, or to cultivate a cross between it and the native races. Except, however, in Kashmir, which has an almost European climate, and upon a small scale in Dehra Dun and the Punjab, where the eggs are sent up into the Himalayas annually for the necessary cold, the introduction of the European variety has not been successful; while crosses between it and the *multivoltine* varieties, though at first often producing cocoons superior to *multivoltine* ones, rapidly deteriorate, and are considered unsatisfactory[2]. It is possible that a further attempt will be made to introduce the European variety into Bengal, special arrangements being made for cooling the eggs before the hatching, which it is anticipated can be arranged to take place at two different times in the year, so as to give two crops, as has already been done in Italy. It remains, however, to be seen to what extent this attempt will prove successful."

The various forms of **Bombyx mori**, briefly alluded to above, in these introductory remarks, require to be more fully dealt with. They shall, therefore, be taken up in alphabetical order (as is customary in this work), and the information given by Hutton, Moore, &c., as also in the *Museum Notes*, briefly reviewed and amplified, where thought desirable, from the writings of more recent entomological and sericultural authors, as also from the official Proceedings of the Government of India:—

BURMESE SILKWORM. 1747

I. Bombyx arracanensis, *Hutton*. (*Conf. with pp.* 2, 235.)

Vern.—*Nyapaw*, BURM.

[1] In Bengal, where most of the silk-rearing is done, the regular crops or *bands* are known as the November *band*, the March *band*, and the July *band*; a fourth *band* being only attempted on a small scale, after the close of the July *band*, by such rearers as happen to have leaf to spare; the fourth *band* therefore is of but little importance.

[2] See Bashford's Experiments (Geoghegan's *Account of Silk in India*; 1872, p. 21) in crossing French, Italian, and China annuals with *madrassi* and *desi* multivoltines. The results, though promising at first, were not considered satisfactory as the stock rapidly deteriorated and generally reverted to inferior annuals. See also an account of a similar experiment carried out in the Indian Museum (*Indian Museum Notes*, Vol. I, No. 2, p. 123).

SILK: Mulberry.

The Domesticated or Mulberry-feeding

BURMESE SILKWORM.

References.—*Fytche, Jour. Agri.-Hort. Soc. Ind., VII., 281—285 (1850); Hutton, Trans. Ento. Soc. London (3), II., 313 (1864); Jour. Agri.-Hort. Soc., Ind., XIV. (Selections, 1865), 85; III., New Series, 129 (1871); Burma Gazetteer, I., 412 (1880); Moore in Wardle's Wild Silks of India, 3; Manuel, Jour. Agri.-Hort. Soc., Ind., N. S., VII., 291, also as separate pamphlet at Rangoon, 1884; Report of the Lyons Laboratoire, D'etudes de la Soie, 10 (1886); Rondot, L'art de la Soie, II., 483 (1887).*

Habitat and Description.—This is a multivoltine silkworm, which may be regarded as doubtfully distinct from the *desí* and *madrassí* races of Bengal. It is the peculiar form bred in Burma, hence the specific name given to it by **Captain Hutton**.

BURMA. 1748

Burma.—One of the earliest notices regarding it is that which will be found in the *Journal of the Agri.-Horticultural Society of India for 1850*. **Lieutenant Fytche** forwarded samples, together with an interesting note on the methods of breeding and manufacture pursued at Sandoway. He alluded to a former sample which had been communicated in 1847. These samples seem to have excited some attention, owing to the cocoons being larger than those of Bengal, but interest in the subject appears to have died out until **Hutton** re-awakened it. The best account of this insect, as well as the most recent, is that given by **Mr. R. A. Manuel** in a pamphlet on the Silks of Burma, certain portions of which will be found quoted *in extenso* in connection with two wild silks, the **Atlas** and the **Cricula**. **Mr. Manuel's** remarks regarding the Burma mulberry silkworm may, therefore, be here reproduced:—

"The domesticated silkworm of British Burma is technically known as the **Bombyx arracanensis**, a name given it by **Captain Hutton**, who believed the worm to be a species distinct from any of those domesticated in Bengal. His belief was based, he says, on two facts, which, as they are not exact, are given here in order that the identification of the insect may be settled authoritatively. The *first* is *the deposit by the moth of its ova in rings*. Now this circular deposit of ova is due to the laying moths being confined within palm-leaf circlets. Hundreds and thousands of moths have laid their eggs in our nurseries, but they have never deposited their ova in circles unless confined under circular glasses, or by means of the palm-leaf circlets used by the natives. The *second* is *the change of colour of the worm from bluish to pinky-white* just before spinning. This is also incorrect. In not a single case has a worm been noticed to change into pinky-white, although thousands of them have been under observation. There are two kinds of cocoons in the country,—one *white*, the other *yellow*,—and what has been invariably observed is that the white spinners change from a bluish to pearly, semi-transparent white, and the yellow ones from bluish to amber colour. And it is owing to these changes that it is possible to keep distinct the white from the yellow cocoons without loss of much floss. Then the white and yellow cannot be said to be the produce of distinct species, or even distinct varieties, for yellow spinners have been known to give sometimes five per cent. of white cocoons and *vice versâ*. It would be interesting to know what it is that causes such changes.

Change in colour. 1749 *Conf. with pp. 14, 15, 20-1.*

"*Districts in which the Industry occurs.*—Silk-growing is a profitable occupation in this province; nevertheless it is not followed largely. The industry can be rendered much more profitable than it is by the introduction of better methods (*a*) of rearing the worm, (*b*) of reeling the silk, and (*c*) of cultivating the mulberry plant. As at present followed, the industry has its chief seats in Tharrawaddy, Prome, Thayetmyo, and Toungoo. Spinners and weavers of silk are found in other places as in Henzada, Shwegyin, Tavoy, and Mergui; but the occupation of breeding, with few exceptions, is confined to the higher latitudes of the country, on the slopes of the Pegu and Arakan Yomas.

Districts where found. 1750

"*Food.*—The chief food here, as in other parts of the world, is the mulberry, of which there are many varieties. The principal in Burma are the red and white, the latter being preferred. Lately the Agricultural Department has introduced the Philippine variety. There is no doubt, if breeders can be induced to take to it, that the silk industry will be largely benefited by such a step. The Philippine mulberry is a hardy plant and will suit the native breeders admirably, because it not only stands a good deal of rough treatment but it is adapted to a variety of soils, is not inclined to grow into trees, and flushes earlier than other kinds. It puts forth

Food. 1751

BURMESE SILKWORM.

Food.

many flexible shoots with a rapid growth of both stalk and leaves, making the gathering of the latter easy to women and children. In times of scarcity the natives use the leaves of the **Brousonnetia papyrifera,** a plant of the same natural order as the mulberry, just as in Europe they use the lettuce, and in America the Osage-orange. But the silk of worms fed on these substitutes is never so good as that of those fed on the mulberry.

Crops. 1752

"*Crops.*—The local worm is multivoltine, that is to say, it spins a number of times during the year, and it is the most prolific of known varieties, as it completes a cycle of its existence in from 35 to 44 days, thus:—

	Days.
In the egg state	8
In the worm state	15 to 23
In the cocoon state	8 to 10
In the moth state	2 to 3
TOTAL .	35 to 44

"The length of the cycle, however, depends on the season of the year; it is longer in the cool weather and shorter during the warm; the silk-yield of the cool weather is, however, finer than at other times. The average weight of a single fresh full cocoon is 12 grains, while an empty one weighs from 2 to 2½ grains. The 'seeds,' as the ova are commercially termed, are sold in the bazar on cloths containing eight circles of eggs for one rupee.

Cultivation. 1753

"*Cultivation.*—It takes the female moth about two days to deposit all her eggs, which average from 200 to 250 in number. The pieces of cloth on which the eggs are laid are put away till the sixth day, when they are taken out and inspected. By this time the worms have matured in the ova which has changed colour from white to dark slate. On the eighth day the worms begin to appear,—tiny, black specks. The ova cloths are then covered with tender mulberry leaves, to which the worms speedily crawl. The earliest risers are considered the best worms, and those which do not crawl at all are considered too weak and worthless and are usually thrown away. The selected ones are then kept in large circular trays, being fed in them without any change of bed and without being disturbed in the least. In these trays, during all their life, they moult and defecate, and here the refuse of their food accumulates till the mass attains to almost the level of the border of the trays. By that time the worms show, by their restlessness and attempts to spin, that they are 'ripe,' they are then picked out by the hand and deposited in the cocooning trays. These are of large size, from 3 to 6 feet in diameter, and within them is a long ribbon of plaited bamboo a couple of inches broad wound round, with the edge on the flat of the tray, in a helix or spiral. The worms are scattered over these trays by the handfuls without any care or regularity; and, left to themselves, they soon begin to spin. They would form much better cocoons if a little care were taken to provide each with a separate place for cocooning. They finish the 'cradle' in about six hours, in eight or ten hours the worms have disappeared from view, and in from 24 to 36 hours the cocoon is completed. In from 48 to 50 hours the last transformation is effected and then the insect sleeps for eight, sometimes and especially in the cool weather, for ten days, and eventually emerges a moth. If male, he is active and restless, seeking a mate; if female, remaining quiet till found by a male, whom she at once admits. The males become violently active if enough of the other sex are not provided, and in such case it is not unusual to find two males attached to one female.

"The whole treatment of the worm, as locally pursued, from its first entrance into world to the time it disappears from sight within its silken enclosure, is careless, slovenly, and dirty. No separate place is provided for it, except it be that a portion of the family sleeping-room is screened off with a *kalaga.* The trays are never changed, the excreta never removed, the refuse of the food never cleared out, and all these, with the sloughs of the moults together with the silk the worm makes at all times, form a dense matting of stinking, fermenting materials, which must be deleterious to the healthy growth of an insect so sensitive as the silk-worm. Under such circumstances it would be surprising if the worms were not subject to disease. Enquiries amongst the breeders of Kynegyi and Shwele proved the truth of such suspicions. As a matter of fact, the mortality amongst the worms is said to be always great, and sickness often sweeps away large broods. Hence the men are anxious to secure

The Domesticated or Mulberry-feeding

BURMESE SILK-WORM.

'seed' of the Bengal worm; and, as another matter of fact, breeders rarely depended on their own seed to any great extent but purchased fresh stock annually from the itinerant Shan traders who bring in quantities of ova from the Karen States beyond the frontier.

Manufactures. 1754

Reeling. 1755

"*Manufactures.*—The operations connected with obtaining silk from the cocoons are as careless and crude as they are wasteful. After the cocoons have matured, and before the exit of the moths, they are prepared for reeling. Torn away from the cocooning-trays by handfuls they are thrown into baskets, when the women and children of the family divest the 'pods' of all their 'waste' or floss; then, without sorting or selection of any kind, except that the yellow and white 'pods' are kept apart, the cocoons are put into a *chattie*, or earthen pot, of water and slowly simmered over a fire. The reeler, generally a woman. who makes it her sole business to reel silk, tries the pods after they have simmered for a while, and as soon as she finds the fibre come away easily, she picks up a handful of cocoons each by a thread (*brin*) of silk,—the number usually being from 18 to 25,—shakes them well to a sufficient length, and then runs them through a loop of brass wire on to a reel fixed to a pair of cross-sticks of bamboo. From the reel the filaments are given a slight twist and carried on to a cylinder of wood with a handle and turning on a trestle: one woman manages the whole operation. She sits beside the fire opposite the pot over which the cross-sticks with the loop and reel are supported, in her right hand she holds an iron fork, with which she regulates the outcome of the threads from the pot, and with her left she turns the handle of the cylinder of wood on which the silk is reeled. Some practice is necessary to attend to and carry out operations with both hands so as to produce a tolerably even and fine thread; and good reelers generally command good wages in their villages, so that it is difficult to get one to leave her home.

Waste. 1756

"As much silk having been obtained from the cocoons as it is possible by the crude method used, the pods are taken out of the pot and, while still moist and warm, are stretched into a kind of coarse, knubby thread, which finds a sale in the market for coarse work. The chrysalis, now divested entirely of its silken covering, is taken up by the children and eaten either fried in oil or unfried; and not unfrequently is sun-dried and sold as a sort of condiment in the bazár.

Fibre. 1757

Trade. 1758

"The silk thus obtained is coarse and unfitted for export, though it answers very well for the well-known fabrics *tamaings, lungyis, pasos* worn by the Burmese. The extent of European enterprise is simply *nil*, and the same answer must also be given to the query 'quantity and value of silk exported.' At the same time imports of raw silk by sea and land during the three official years ending 1879-80 averaged annually nearly twelve lakhs of rupees, while the value of the imports of manufactured silks during the same time was about fifty lakhs of rupees. A short consideration of these figures will show what room there is for capital and enterprise in this direction—either in reeling or in the manufacture of piece-goods. And a still further consideration of the produce of the worm and the vile treatment to which it is subjected will show what room there is for improvement in nearly every department of the industry. The usual reeled thread represents from 18 to 25 filamenst taken together: it is no wonder the produce is pronounced coarse and harsh. Very different would be the result if 5 instead of 25 filaments were used and better machinery brought into play. Samples of thread reeled with only five brin have been pronounced equal to the best Chinese crops. With better treatment of the worms, better cultivation of the trees, and better apparatus for reeling, the produce of the domesticated worm of Burma is capable of very much better use than that to which it is now put."

Measurements. 1759

M. **Rondot** gives five generations as peculiar to this insect, no rearing taking place between the 16th January and the 22nd June. The fibre, he says, is about 22·9 thousands of a millimetre in thickness, its tenacity 6·56 grammes, and its elasticity 17 per cent.

THE RAINY or HOT-WEATHER BAND. 1760

II. Bombyx crœsi, *Hutton.* (*Conf. with pp. 2, 235.*)

Vern.—*Nistrí* or *Madrassi*, The Rainy and Hot Weather *Band.*

References.—*Hutton, Trans. Ento. Soc., London (3), II., 312 (1864) Jour. Agri.-Hort. Soc., Ind., XIV., 84; III., (N. S.), 127 (1871); Moore; Proc. Zool. Soc., Lond. (1867), 683; Wardle, Wild Silks of India, 3; Stack, Silk in Assam (1884); Cleghorn, Letter dated 9th March 1888.*

NISTRI SILKWORM.

Habitat and Description.—**Cotes** says of this insect that it "is a small multivoltine variety of the mulberry silkworm; it is reared chiefly in the March and rains *bands* in Bengal, where it is second in importance to the *desí* variety. It is also reared in Assam. Like the *desí* it produces a succession of crops throughout the year, most of which are only reared for seed. Unlike the *desí*, however, it thrives best in the hot weather. The cocoons, which are generally yellow, at least externally, are somewhat larger than *desí* cocoons, but the fibre has less elasticity and brilliancy. The moth is milky white in colour, the caterpillar is milky white with two black spots on each segment; it is reared like the *desí*, of which it appears to be little more than an artificial variety." In **Hutton's** account of this insect he gives the further fact that in the hot weather it goes through all its changes from egg to the cocoon in 25 days, but in the colder months it occupies 35 days. He speaks of it as "the largest of the monthly worms."

BENGAL & N.-W. P. 1761

BENGAL AND THE NORTH-WEST PROVINCES.—**Hutton** says: "This species is cultivated in several parts of India, and thrives well at Mussoorie. It is to be particularly remarked, however, that none of the Chinese species, whether annual or monthly, have hitherto succeeded in the North-West Provinces. Dr. **Royle** long since remarked that none of the Old Company's Filatures extended higher up the country than about 32° of north latitude, owing to the dry hot weather of the North-Western climates." It is worthy of remark that one of the earliest accounts of this insect is that given by **Mr. D. W. H. Speed** (*Trans. Agri.-Hort. Soc., Ind., III., 20 (1836)*, in which it is spoken of as the *madrasí* or *china pulu*. It would thus seem possible that the **B. sinensis** or *sina* or *china* insect may be but a degenerate state of the *madrasí*, and that the two names were in Bengal, sixty years ago, accepted as synonymous. **Speed** tells us that the *madrassí* "was introduced into this country, though by whom not ascertained, about the year 1780 or 1781; but degenerating, by reason of carelessness and improper management of the worms, a fresh supply of eggs was, two or three years after, obtained by a **Mr. Frushard**; which, again falling off immediately his direct superintendence was withdrawn, a third supply was brought by the late **Colonel Kyd** in 1788, from Canton; which, from the vast increase he effected by personal attention for a considerable time, forms the origin of the present stock of this description of worm,—cocoons monthly between November and June, if attention be paid; but more generally from January to May." **Hutton**, in his numerous reports on silk cultivation on the Himálaya (Mussoorie), alludes to his success with this insect as also with a cross between it and the Kashmír form of the true **B. mori.** In consequence of these experiments he formed the opinion that the neighbourhood of Mussoorie would "prove to be a good silk-yielding district." He contended that **Count Dandolo** was in error when he affirmed that the silk-worm was a native of the Southern Provinces of China and therefore required "a high temperature to bring it to perfection." It was found, says **Hutton**, that the silk reared in Sweden was far superior to that of the best Italian produce. Hence he argues the superiority of Kashmír over Bengal silk. He adds: "Next to that of Kashmír, therefore (if indeed ours be not superior), the climate of the lower ranges of the Himálaya lying between the Sutlej and the Ganges will probably become the most productive, provided always that proper care and attention are bestowed upon the management of the insects; for even the much belauded Panjáb, although in my opinion, infinitely better than Bengal for silkworm cultivation, is yet far too hot to preserve it long in perfect health, and with a better system of cultivation than that which is practised in Kashmír, or indeed than any Native cultivators are likely to bestow,

Bengal. 1762

Superiority of Kashmir. 1763

Conf. with pp. 18, 35, 49.

The Domesticated or Mulberry-feeding

NISTRI SILKWORM.

Conf. with pp. 17-18, 35-36, 61-62, 215.

the probabilities are greatly in favour of our hill districts eventually being made to 'bear the bill' in India, and even to rival the far-famed Italian produce."

It will be found that the failure that has, since **Captain Hutton** wrote, attended the extensive experiments in the Panjáb, bears out his theory so far, and the greater success up to date, in the migration of the enterprise to Derha Dun, would seem also to strongly support his position.

The **Rev. A. Campbell**, of the Santal Mission, Chutia Nagpur, in a letter to **Mr. N. G. Mookerji**, gives the following information regarding his experiments with this insect:—"My *madrasís* have been spinning cocoons for the last eight days, and about half of them are now in the cocoon. I have, I think, succeeded very well this time. I had plenty good leaves, and the worms were very healthy and vigorous. I send you a few cocoons to be examined for *pebrine*. I must give you an account as to how I rear my worms. I have little bamboo trays, about two feet square, in which they are kept. The trays are then placed in a stand, which contains eight trays one above the other. The trays are open, and have no protection whatever from flying insects, and are kept in a room, the doors of which stand open from morning to night. So if ventilation is what is required, they get plenty of it. I tried to get nets, but did not succeed, so I have to change the insects at intervals from one tray to another to clean them. But after the first moult it is rare to see a dead insect.

"My opinion is that they have improved during the time they have bred here. There are very few small worms among them, and some excessively large. I was surprised to see from your printed report that from 40 to 80-fold was regarded as a good crop of cocoons. From my first *bund* I had upwards of 100-fold, and from my second nearly 200. This *bund* will not turn out so well, as during the first week I was away from home, and they were not taken care of; but the cocoons are the best I have yet obtained. However, no worms died after the first moult."

COLD WEATHER BAND. 1764

III. Bombyx fortunatus, *Hutton.* (*Conf. with pp. 2, 235.*)

Vern.—*Desi* or *chota polo.* The November or Cold Weather *Band.*

References.—*Hutton, Trans. Ent. Soc. London (3), II., 312 (1864); Jour. Agri.-Hort. Soc. Ind., XIV., 84; (N. S.), III., 128 (1871); Moore, Proc. Zool. Soc. London, 1867, 683; Wardle, Wild Silk of India, 3; Rondot, L'Art de la Soie, I., 312 (1885); Mukerji's Report, dated 6th April, 1888; Geoghegan, Silk in India, 15, 16; Liotard, Silk in India, Pt. I. 3; Cleghorn, in letters dated 18th November 1887, 9th April, 1888 and 12th December, 1889, also in Agri.-Horti. Soc. Jour.*

Habitat and Description.—This insect is by most writers accepted as more especially peculiar to Bengal. **Speed** (*Trans. Agri.-Hort. Soc., Ind., III., 20*) speaks of it as of a small size, but as yielding cocoons five times a year, at periods of from 40 to 110 days. The long period, he remarks, produces the best silk. Commenting on this subject **Hutton** points out that the longest period is found to be the November *band*, and hence he maintains that the heat of Bengal is inimical to good sericulture (*Jour. Agri.-Hort. Soc. Ind., XI., 223*).

Cotes gives the following abstract of the main ideas regarding the *desi* silkworm:—

BENGAL. 1765

"BENGAL.—This is a small multivoltine variety of the mulberry silk-worm; it is largely reared in Bengal, where it yields the principal cold-weather crop of cocoons. The cocoons are generally golden yellow in colour, and, compared with the European annual variety, they are small and of loose consistency. **Cleghorn** observed that the moths are dusky in colour, the worm being bluish-white without

DESI or CHOTA POLO SILKWORMS. Bengal.

distinctive marking; while **Hutton** noticed that the variety can be distinguished from all other varieties by the fact that when near to maturity the caterpillar becomes of a dull leaden blue colour.

"The following is an abstract of **Mookerji's** report upon the species:—

"The rearers prefer a south aspect for the rearing-house, but all rearing-houses do not face the south; they are covered with specially thick thatch, and generally have but one small window and a door. The window is always kept shut at night, and during the cold season in the day time also; the door is always kept shut at night and in the cold weather all chinks are carefully filled up, the fermenting refuse from the trays being often piled up inside the rearing-house to further raise the temperature.* In one rearing-house in the cold weather **Mookerji** found as many as thirty-two trays, each tray containing about 2,500 worms, besides a man with his wife and children and a cow. He does not give the dimensions of the house, but notices that this was rather an exceptional case, the rearing-house being generally set apart for the worms, though one or two men usually sleep in it with the idea of protecting the worms from supernatural influence. These men, both by the warmth of their own bodies and by letting in the air when the room becomes too hot and by stuffing up the cracks and crannies when it is too cold, for their own comfort, no doubt, as **Cleghorn** suggests, unconsciously tend to keep the conditions of the atmosphere suitable for the worms; they thus furnish an example of the practical utility of a custom which is followed on account of the supernatural benefits supposed to be derived from it.

"The moths emerge from their cocoons within eight or ten days after the spinning has been completed. The caterpillars moult four times within a period of about 25 days, at the end of which time they are ready to spin. As they gradually grow bigger they are distributed over a larger and larger number of trays. The worms are fed three times a day in the cold weather and four times a day in the hot. The trays are cleaned about once every five days; and as the worms are moved by hand, the cleaning is often very imperfect, and by the time it takes place the tray is covered with a fermenting mass of leaves, excreta and dead worms, upon the top of which the live worms are feeding.

"When full fed, the worms are removed by hand to cocooning trays, which are fitted with a coiled strip of basket-work, about two inches broad, along which the cocoons are spun. The spinning takes about two days to accomplish, so that the whole period which elapses between the bringing in of the seed and the selling of the cocoons is about 35 to 40 days. Some seven or eight crops of cocoons could be raised in the course of the year, if it were desirable to do so, but, as a matter of fact, rearing is only done in the regular *bunds*, both because a continuous supply of mulberry leaf cannot usually be obtained, and because the presence of the silkworm fly, *Trycolyga bombycis* (*see p. 61 below*), renders continuous rearing inadvisable."

Efforts put forth by the East India Company to improve Indian Silk.
1766
Conf. with pp. 57, 62, 184-6, 190, 193, etc.

Many writers, accepting the name *desí* to mean literally indigenous, speak of this insect as a native of India, more especially of Bengal and South India. **Hutton** does not venture any opinion as to its origin, and **Speed** is also silent on this subject. **Geoghegan** says it was doubtless the first domesticated species naturalised in India. He adds that "in Hurripal the *desi* worm had in 1819 only just been introduced, and it was then unknown in the Radnagore circle." According to **Atkinson** the *desi* worm had also degenerated. He describes only three forms—the large annual (**B. textor**), the *desí* (**B. fortunatus**), and the *chína* or *madrassí*. He classes them in the above order as regards quality of silk. From the history he gives of these insects from 1781 to 1788, **Geoghegan** points out that the third species must be the *china* not the *nistri*. "Under careful breeding, he says, he obtained good silk from it, but in the hands of the Natives, who crossed it with the country breed, it had so degenerated that he had tried to persuade them to give it up. But they clung to it

* The necessity of keeping up the temperature in the rearing-houses, during the cold weather, is shown by the failure of **Mookerji's** attempt to rear worms in some old barracks, without following the usual method of keeping the place warm.

SILK: Mulberry.

The Domesticated or Mulberry-feeding

DESI or CHOTA POLO SILKWORMS.

Dusky-coloured cocoons and worms. 1767

Adverse opinion. 1768

Microscopic Selection. 1769

Conf. with p. 59.

Remedy for Fly-pest. 1770

Change of colour. 1771

Conf. with pp. 8, 15.

for three reasons: (1) it was more rapid in its evolutions; (2) it was hardier; and (3) it was less squeamish as to its food than the worms of country stock." **Cleghorn** urges that to produce a pure *desí* stock of silkworms only the most dusky moths should be bred from, and by following this principle, he further tells us, he had succeeded in producing a dusky coloured worm. This result he regards as one of the most valuable advances hitherto attained, against the calamity of the silkworm fly—a pest which he views as of far greater moment than *pebrine.* "Microscopic selection," adds **Cleghorn**, "to me appears simply a flash advertisement, and I fancy that it has seen its day in Europe, for there is no evidence that microscopic selection has succeeded but in the hands of professional breeders of the stocks. There is also evidence that it always fails in the hands of the inexperienced, however scientific they may be. There is, in fact, evidence that the experienced breeder of stocks produces highly successful results without its help. And lastly, and which is of very great importance, those breeders who have the capacity for keeping up very large stocks continuously and without interruption can alone hope to effect much improvement." Large stocks, **Cleghorn** contends, cannot, however, be kept up in Bengal owing to the fly-pest. That pest has made it compulsory for every breeder, belonging to a certain locality (with well defined natural boundaries), to dispose of his entire silkworm stocks every four months, or after every third generation, as the only means of being able to produce any economical result. After thus keeping their establishments closed for about two months, during which the fly-pest disappears from their neighbourhood, the breeders are able to recommence operations with seed cocoons purchased at a distance. **Cleghorn** speaks of this Bengal system of treating the *desí* breed and its protection from fly as merely a defensive measure as compared to the offensive position that his dusky worm enables the rearers to assume, since his worms enjoy practically an immunity from the pest in question. The discovery that I have made, he continues, is that the dark skinned worm, which in Europe is considered to be peppered with pebrine disease, a dusky moth, also in Europe viewed as pebrinized, and its lustrous white cocoon, considered in Europe to be the product of a weakly worm, is the animal best suited for feeding on the bush mulberry in a hot climate, whereas the milk-white worm and milk-white moth, which spins a yellow cocoon and feeds on the tree mulberry, appears more suited for Europe. He cites in support of this opinion a passage from **Wallace's** *Darwinism,* to the effect that in tropical countries it has been observed that white or pale coloured cattle are much more troubled by flies than are those of a black or brown colour. "The same law," adds **Wallace** "even extends to insects: for it is found that silkworms which produce white cocoons resist the fungus disease much better than do those which produce yellow cocoons." (*Second Ed., 1889 p. 171.*)

This observation regarding dark coloured insects and worms will be found to have a peculiar bearing on **Hutton's** opinion regarding the original form of **Bombyx mori**, more especially when it is added that **Mr. Nitya Gopal Mookerji**, while experimenting with certain cocoons furnished by **Cleghorn**, arrived at the opinion that they "were a cross between the country breeds and **B. mori** of Europe." "There is no doubt," continues **Mr. Mookerji**, "the cocoons are very superior: not quite so large or firm as the European cocoons, but very much superior to those of the *desí* or *madrassí*. The stock, however, being diseased, the quality is not at all uniform, and a large number was inferior to good *desí* or *madrassí* cocoons." In certain experiments performed by the **Rev. A. Campbell** in Chutia Nagpur it was found the *desí* worm did not do so well as

DESI SILKWORM.

the *madrassi*. Mr. Campbell, in fact, arrived at the opinion that they were weaker and more liable to die than the *madrassi*.

ASSAM. 1772

Conf. with pp. 35, 55, 142.

ASSAM AND MANIPUR.—Samples of the Manipur mulberry silk were communicated to the Agri.-Horticultural Society of India in 1863. These were examined and reported on in the *Proceedings*. They were identified as belonging to this race. The system of rearing is extreme y crude. Some years ago the writer (while botanising in Manipur) was, after much trouble, permitted to visit one of the chief silk-producing villages. He was not allowed to approach any of the houses nearer than to within a few yards, in case his intrusion should be avenged by the death of the entire stock. The apparatus used in reeling was, however, shown to him and also the process pursued. There was little of an exceptional character in the methods of unwinding the silk from what could be seen in Bengal. The rearing, however, was remarkably primitive. The mulberry was grown as a small jungle bush around the villages, and a good deal of the cocoons were obtained from the open air, the worms being as far as possible protected against their common enemies, insects and birds. To a certain extent also the worms were fed in houses, but these the writer was not shown, but on some of the bushes in open air hundreds of worms were to be seen.

CHINESE DOMESTICATED. 1773

IV. Bombyx mori, *Fabr.* (*Conf. with pp.* 2, 235.)

THE TEMPERATE ANNUAL MULBERRY SILKWORM. The Chinese Domesticated Silkworm now largely reared in Europe.

Syn.—PHALÆNA MORI, *Linn.*

References.—*Pasteur, Etudes sur la Maladie des Vers à Soie (1870); Maillot, Leçons sur les Vers à Soie du Murier; Rondot, L'Art de la Soie (1885—87); Riley, U. S. Dept. Agri. Div. Ent. Bull. No. 9 (1886), Trans. Agri.-Hort. Soc., Ind., III., 57; Hutton in Jour., XIV., Selections, 76-81.*

Habitat and Description.—Of this insect Hutton writes that it was originally a native of "the northern mountainous provinces of China, especially that of Tchekiang; now domesticated in China generally, in Kashmír, Afghánistán, Bokhara, Persia, Syria, France, Italy, Spain, Sweden, Russia," and, it may be added, in Turkey, Egypt, Algeria, America, and Australia. He also remarks that it had recently been introduced into Oudh and the Panjáb. It everywhere thrives best, he says, where the temperature is moderate. "It is the largest and strongest of the domesticated species, and is an annual, which produces naturally but one crop of silk in the year, although in certain temperatures it is possible to make it produce a second crop." The appearance of dark-coloured or black worms after the first moult is by no means unusual, and this Hutton regarded as the type of the original species.

Dark-coloured worms. 1774

Cocoon and Silk. 1775

In the *Museum Notes* Cotes, remarks regarding this insect: "It requires cold for the uniform hatching of its eggs and produces a close-grained cocoon containing a large amount of silk, of a golden yellow or white colour, that can readily be reeled. This silkworm is essentially suited to the conditions of a temperate climate, and is not generally cultivated in India, though it has been grown on a small scale in Dehra Dun (North-Western Provinces), and also in some parts of the Panjáb. In Dehra Dun (2,300 feet above sea level) the eggs are hatched in February, and the cocoons are ready by the end of March. The eggs which are retained as seed for the next year's crop are then sent up to Mussurie (7,400 feet above sea level), where the comparatively low temperature

SILK: Mulberry.

The Domesticated or Mulberry-feeding

TEMPERATE ANNUAL SILKWORMS.

Origin in China.

1776

prevents their hatching out until brought down again to Dehra in the February of the following year."

It is perhaps scarcely necessary to republish here the voluminous correspondence and discussions that have taken place in India regarding the annual silkworms of this country. Suffice it to say that **Hutton** and other early writers clearly established the existence of two forms or species, *viz.*, the true **Bombyx mori** and **B. textor**. After reviewing the historic facts and fancies regarding the origin of silk-culture **Hutton** says :—

"This after all is, I think, the most probable and rational account of the matter, and hence we appear to gather from these several narratives that at all events *the silk-worm* in the time of **Justinian** was not cultivated out of China, but was then carried westward from the northern provinces of that empire, and it appears that even in the present day the silks which are exported from that country are brought down from the interior in bales, and sold at Canton to the British, Dutch, and French merchants.

"The particular species thus derived from the northern parts of China, and to which the distinctive appellation of **Bombyx mori** has been assigned, is evidently, from the various accounts given of its habits by entomologists and cultivators of silk, an annual worm yielding but one crop.

"In his remarks on the rearing of silk-worms, **Dewhurst** clearly shows that his description refers to an annual when he loosely observed that it '*dies annually;*' by which, however, he merely means that the operation of making silk is only performed once in the year, since he afterwards proceeds to say that after the eggs have been deposited they are kept in a cool place until the following season.

"**M. Boitard**, likewise, describes the same species; while that it was an annual which **Kirby & Spence** regarded as the true **B. mori**, is proved beyond a doubt, by their saying, after some remarks upon it 'other species as may be inferred from an extract of a letter given in **Young's** *Annals of Agriculture*, are known in China, and have been recently introduced into India.' 'We have obtained' says the writer, 'a monthly silkworm from China, which I have reared with my own hands, and in twenty-five days have had the cocoons in my basins, and by the twenty-ninth or thirty-first day a new progeny feeding in my trays.'

"This at once establishes the fact that the monthly worms were not known in India earlier than between thirty and forty years ago; for that they were, at least, not cultivated there is shown when the writer adds 'this makes it a mine to whoever *would undertake* the cultivation of it.'

"Here then is direct proof that previous to the time here indicated, an annual worm and it only was the species under cultivation both in India and in Europe, and consequently that it is the true **Bombyx mori**, *Linn.*

"But now a question arises as to what annual the name applies, since there are very strong reasons for believing that there are two annual worms, unless (which I hardly think will prove to be the case) the one should turn out to be a degenerated variety of the other. These are, first, the annual worm, with white cocoon, known in Bengal as the *boro pulo*, and is said to be domesticated in China, Bengal, France, and Italy; while the second is the Kashmír, Afghán (and Persian (?)) worm.

"The first of these appears to be fast disappearing from Bengal, if we can depend on **Mr. Bashford's** remarks already published in a previous volume of this journal, and judging from a list of desiderata received not long since from **M. Guerin de Meneville**, I am inclined to think that the species is not much known in France, since he says *in epistolá*, 'les diverses especes que je desire sont: 1° Le Grand cocoon Annuel, qui n'est élevé qù une fois par au et qu' on recolte en Mars.'

"It is, however, quite possible that he may have been misled by the flaming Report entitled '*A Synopsis on Bengal Raw Silk*,' by **M. L. Nerac**, and drawn up, it is said, by order of the East India Company, for the cocoon of the Bengal annual is not to be compared in point of size to that of the Kashmír worm, and which I strongly suspect is the species domesticated in France, and whose cocoons so much astonished **Mr. Bashford**. In the event then of the Bengal annual being little known in Europe, the Kashmír species will be entitled to the name of **Bombyx mori**, while the former will either prove to be a distinct and unnamed species, or a degenerated variety of the other.

TEMPERATE ANNUAL SILKWORM.

"That the Kashmír worm will in all probability prove to be the true **Bombyx mori** need not surprise us if the story of its introduction into Europe be correct, since having once been established in Kashmír from Northern China it would easily find its way through Afghánistán into Persia, and may be the species which the monks are said previously to have introduced into Constantinople; for the present, however, the question must remain unsolved.

"The reason why the **Bombyx mori** is said to be finer in the countries of Europe than in Asia is said to be attributed to the fact of their climates more nearly approximating in temperature to that of the natural habitat of the insect, than does the fervid climate of the plains of India, and indeed this may be said of all the species now domesticated in Bengal where the climate being totally different from that of the provinces from whence the worms were originally procured may have a most debilitating and deteriorating effect upon the constitutions of the insects, showing itself in the inferior size and quality of the cocoons as compared with those of Europe."

N.-W. PROVINCES & PANJAB. 1777

NORTH-WEST PROVINCES AND THE PANJÁB.—Reference has already been made to the fact that within the past few years the experiments which were conducted with the view of introducing the true mulberry silkworm into the Panjab have been pronounced a failure. **Mr. Lister**, who, for some years past, has prosecuted the endeavour with great energy and perseverance, has had to admit that the result was a complete failure. Years ago the subject of the Panjáb becoming a possible rich field for silk culture was prominently brought before the public in a series of articles mostly by **Mr. Henry Cope.** The following may be given as an enumeration of the more important papers which appeared on this subject:—*Jour. Agri.-Horticultural Society of India, IX.,* (*1854*), *140-142; X., 117-138; XI., 194-201,* (*Proceedings, 1859*), *63-65,* (*Proceedings, 1860*), *53-59; XII., 129-141, 286-299; New Series, II,* (*Proceedings, 1870*), *42-43*. Both **Captain Hollings** and **Mr. Cope** speak of the seed that had been introduced into the Panjáb and which was being experimentally grown there, as having come from Kashmír, a circumstance which may be admitted as justifying the inference that it was **B. mori.** During the time that the Panjáb was being thus strongly upheld as the region which would likely prove India's best silk-producing area, **Captain Hutton** contributed most of his papers on Indian silk culture. The chief refrain of these was a powerful protest against the continuance of the effort to produce silk in Bengal. He contended that both the annual and the monthly worms there grown had been obtained from colder countries than Bengal, and that, while for a time it might be possible as in the past to produce silk, the climatic conditions were inimical to the insect, and that consequently, disease and degeneration were the natural results, to guard against which involved an unnecessary expense and a constant renewal of stock. He pointed out that this fact was fully appreciated by the East India Company, by their having endeavoured to establish the industry in the coldest or most northern portions of the dominions over which they then exercised control. He further urged that the time had long since transpired when the migration should have taken place to the still colder regions which had fallen under British domination. **Captain Hutton,** in fact, held that while the Panjáb, which **Mr. Cope** so strongly recommended, would be preferable to Bengal, that even colder areas might be found where the climatic extremes between summer and winter were not so great as on the plains of the five rivers. His contention hinged mainly on the fact that since Kashmír had produced silk of a quality far superior to that of any other part of India, the effort should be made to establish the silk industry in localities that, in climate and soil, approached as near as possible to those of Kashmír. In his opinion Dehra Dun would possibly be found far more productive of good results than the Panjáb. From the official and other papers which will be found below it will be seen that

Tendency to Degeneration. 1778

Panjab. 1779

SILK: Mulberry.

The Domesticated or Mulberry-feeding

TEMPERATE ANNUAL SILKWORMS.

Conf. with pp. 12, 17, 35-36, 61-62, 215.

after the loss of much time and money **Mr. Lister** and others interested in silk production in Northern India have now practically abandoned their experiments in the Panjáb and have even already obtained results in Dehra Dun which would appear to confirm the accuracy of **Captain Hutton's** suggestions which were made fully a quarter of a century ago. It remains, however, to be seen whether even Dehra Dun can produce silk that might compete in the same markets with that of France and Italy. It, indeed, seems highly probable that Kashmír (or perhaps also Manipur), of all the silk-producing areas of India, is likely to ever turn out silk of a superior quality.

Kashmir. 1780

Conf. with pp. 11, 35, 49.

KASHMIR.—In 1853 **Captain Lowther** (*Jour. Agri.-Horticultural Society of India, VIII., 209*) wrote: "In a Mussulman village, on my road, I found a silk establishment; the worms were just being hatched (May 20th) by wrapping the eggs in a woollen cloth, and putting it in the bosom of a man: the young brood are put with a feather on the new shoots of mulberry, and these are gradually changed to leaves, with the growth of the worm, which may be said to attain cocoon-ship, or maturity, in two months. The species struck me as being unusually large, and the silk of extra fine quality. Certainly no country in the world has greater natural resources of silk-growing—fineness of climate, cheapness of labour, abundance of food, and excellent markets at hand (on the Indus in our territories);—but none of these under the present *regimé* of unscrupulous exaction, appear to be of any value to the growers, who are thereby much reduced in number; indeed, the 'Lion' himself seems to have 'put a strong paw' on the whole concern, together with everything else of any value."

In the passage quoted above from **Hutton's** *Notes on the Silkworms of India*, it will be seen that he regards the annual insect found in Kashmír as the true **Bombyx mori**; indeed, that insect is in India often designated the Kashmír or Bokhara worm.

V. Bombyx sinensis, *Hutton*. (*Conf. with pp. 2, 235.*)

SINA OR CHOTA PAT. 1781

Vern.—*Sína, Chína* or *Chota Pát.*

References.—*Hutton, Ent. Soc. London (3), II., 313, (1864); Jour. Agri.-Hort. Soc., Ind., XIV., 75, 85; III., N. S., 128 (1871); Moore, in Wardle's Wild Silks of India, 3.*

Habitat and Description.—**Cotes** says of this insect that it is a small multivoltine mulberry silkworm, which produces cocoons inferior to those of the *desí* or *madrasí*, and that, accordingly, its cultivation in Bengal has been almost entirely abandoned. He also suggests that **B. meridionalis** (described by **Wood-Mason** in the Indian Museum Report for 1886) may be this insect. The samples so named by **Wood-Mason** were obtained from the Cuddapah and Coimbatore districts of Madras. **Hutton** (*Jour. Agri.-Hort. Soc., l.c.*) gives the following account of *chota pát*:—"This is known as the *sína* of Bengal, but, like the others, it originally came from China; it is very prolific, and even at Mussoorie goes on yielding crop after crop, up to the middle of December. The cocoons vary in colour, some being white and others yellow, while others even have a beautiful faint, greenish hue. There is a peculiarity about these also which may enable the tyro to distinguish them from any of the others; while all the other species hatch slowly during the morning, from six to twelve o'clock, the *sína* worms come forth all in a batch, or continue hatching all day and all night."

Hutton informs us that a Japanese insect, he had bred, had much the appearance of this silkworm, but that with its passage through the various phases of its life, it manifested at the same time certain traits of **Bombyx mori.** He therefore thought that it was a hybrid between these two insects,

CHINA or CHOTA PA SILKWORM

but a hybrid or cross which rapidly degenerated towards **B. sinensis.** Atkinson (*Conf.* with remarks above, p. 13) speaks of a cross between this and the *desí* worm, which was very hardy, but yielded inferior silk.

ASSAM. 1782

ASSAM.—Mr. Stack furnishes the following particulars regarding this insect:—

"The smaller kind of the *pát* silkworm gives a white silk, which is reeled into a coarser and less valuable thread than that of the larger; but as the worm is multi-voltine, yielding four broods in the year, it finds greater favour with the cultivators, and is, perhaps, supplanting the univoltine variety. It lays its eggs chiefly in December, January, March, April, June, July, September, and October. The eggs are hatched on trays woven from slips of split bamboo. An experiment made by Mr. Buckingham, a tea-planter in the district of Sibsagár, during the month of June, with an average temperature of 82°F. by day and 76° by night, gave the following results:—

June 2nd	Cocoons obtained.
„ 4th	Moths appeared.
„ 5th	Laid eggs.
„ 9th and 10th	Moths died.
„ 14th	Worms hatched.
„ 17th	1st moulting.
„ 21st	2nd „
„ 24th	3rd „
„ 28th	4th „
July 6th	Spinning began.

"In this experiment the life of the worm lasted 23 days.

"The number of fresh cocoons that weighed one pound was found to be 720, and of whole dried cocoons 2,048. It is estimated that 7,200 whole cocoons would yield one pound of reeled silk, but this calculation seems open to question.

The results of an experiment made by Krishna Kanta Ghugua, a Native gentleman of the Jorhát sub-division, were as follows:—

August	9th	Cocoons obtained.
„	16th	Moths emerged.
„	18th and 19th	Laid eggs.
„	25th	Worms hatched.
„	30th	1st moulting.
September	4th	2nd „
„	9th	3rd „
„	16th	4th „
„	20th	Spinning began.
October	6th	Moths emerged.

"Here the life of the worm lasted 27 days, and 16 days elapsed between the commencement of spinning and the emergence of the moth."

BORO PAT. 1783

VI. Bombyx textor, *Hutton.* (*Conf. with pp. 2, 235.*)

Vern.—*Boro Polo* or Large *Pát.*

References.—*Hutton, Trans. Ent. Soc., London (3), II., 309 (1864); Jour., Agri.-Hort. Soc., Ind., XIV., Sel., 81—84; N. S., III., 126 (1871); Moore, Proc. Zool. Soc. Lond., 1867, 683; in Wardle's Wild Silks of India, 2; Louis, A few words on Sericulture in Bengal (1880), 20; Stack, Silk in Assam (1884); Rondot, L'Art de la Soie, I., 320 (1885); Mukharji, Report dated 6th January (1888); Cleghorn, Letter dated 9th March (1888).*

Habitat and Description.—Cotes, in the *Indian Museum Notes,* gives the following brief statement regarding this insect: "This is an annual mulberry silkworm, larger than either the *desí* or *madrasí.* It produces a considerable amount of good silk, and is occasionally reared in Assam and Bengal; owing, however, to the fact that it produces but one crop of cocoons in the year, and that its eggs do not hatch simultaneously, its cultivation has now been generally abandoned. Rondot, in his *L'Art de la Soie,* writes that this variety spins a white cocoon smaller than that of **Bombyx mori,** and differing from it both in form and structure, being generally pointed at each end, a little soft, the silk not closely wound,

BORO PAT SILKWORM.

and containing comparatively little gum. He notes that in the earlier part of the century this variety was reared almost everywhere in the Kasimbazar circle and other places in Bengal, and that it has also been found in Ceylon."

One of the earliest direct reference to this insect is that by **Dr. Speed** (*Trans. Agri.-Hort. Soc., Ind., III., 19-20*), who in 1836 wrote:—

"The *Burra*, or large annual *púlú*, supposed to be the Italian, and seems to have been introduced into this country about 120 years ago, but in what precise year, or by whom, is not ascertained; cocoons in March or April; 1 *puhun*, or 80 yielding about 3½ *kouhuns* (a *kouhun*=16 *puhuns*) or 4,480 cocoons, which is a proportion of 56 to 1,—that is,—as allowing, by the Natives, for the chance of irregularity of *pairing* and failure, or destruction of eggs, during the long keep from March or April to January or February;—but 80 males and females, carefully selected, that is to say, 40 *true* pairs, will produce upwards of 10 *kouhuns*, or 12,800 eggs, being a proportion of 160 to 1. In Italy, according to the best accounts, the proportion is 192 to 1. The colour of this egg is, first yellowish white, changing to a slate colour in the course of 36 hours; about the fifth day from the change, the centre of the egg contracts, leaving the circumference full, as if the worm was actually formed; and in this state it remains for about 10 months. The life of this worm is from 42 to 50 days; and the cocoon lasts from 10 to 15 days."

"**Hutton** says, 'This species, hitherto confounded with **B. mori**, is said to have been introduced from China, where it is still cultivated, under the name of the *white cocoon*, but the time of its introduction into India appears to have been forgotten. In Bengal, as well as in its native country, it is an annual, hatching early in the spring, usually in January, yielding generally pure white cocoons, far inferior in size to those of **B. mori**, and altogether of a different shape, character, and texture, having an inclination to become pointed at each and, and with the silk not closely interwoven, but externally somewhat flossy and loose, whereas the cocoons of **B. mori** are closely woven, compact, hard and smooth, ovate in shape and four or five times larger; some that I have received from France being little inferior in size to those of the *tasar* moth.'

"In **Dr. Bonavia's** report on Sericulture in Oudh for 1864, he remarks of **B. textor**: 'I cannot find any reason to believe that this worm belongs to a different stock from the Kashmír and Bokhara worms'; others, continues **Hutton**, have said the same thing, which only proves to me that they have never looked beyond the worm itself, since had they done so they might have found, as I have done, abundant proofs of specific distinctness."

Dr. Bonavia proceeds to inform us that "**Captain Hutton** favoured me with a small quantity of eggs, of his selected dark-coloured worms. According to his views the dark-coloured variety approaches more to the wild kind, and therefore has more healthy blood in it than the white variety, which he considers as a degeneration of the original worm. It is strange though, that the *Boro púlú*, which has been reared in Bengal for a long time, contains a large number of the dark variety. One would be inclined to think that, considering the bad mode of rearing and the climate of Bengal, it would have degenerated into the white variety by this time, according to **Captain Hutton's** theory.

"The writer, however, shows by the admission that 'a large number of the dark variety' occurs among the *Boro púlú* worms 'that **Captain Hutton's** theory' actually does hold good. Just as with the originally dark-coloured worms of **B. mori**, so also the originally dark-coloured worms of **B. textor**, 'have degenerated into the white variety.' The occurrence of these dark worms, as I have previously pointed out. is due to an effort on the part of nature to return to the original stock, from the sickly degenerated state into which the species have fallen.

Dark-coloured worms. 1784

"Again, we are told that, 'the selected dark-coloured worm of Mussoorie did very well, but I could not detect any difference between the cocoons of these and those of the white Kashmíri ones. I selected many of the black ones of the Kashmiri, Bokhara and *Boro púlú* and kept them separate, but did not find that they produced better cocoons than the rest, and they all had one disadvantage, that is, on account of their colour, it was not easy to discover when they were ready to spin.'

"Be it observed, however, that in furnishing these dark-coloured worms, I did not guarantee the same in Oudh as are obtainable in the European climate of Mussoorie. I should not have been at all surprised to hear, considering the heat of Oudh and the inexperience of the conductor of the experiments, that every worm had returned to a state of sickly whiteness. As to the difficulty of discovering when they

BORO PAT SILKWORM.

were ready to spin, this could only have occurred to an unpractised eye, since there is always a semi-transparent yellowish waxy hue about a mature worm that is quite unmistakable to an experienced eye.

"According to **Mr. C. Blechynden** and **Mr. Bashford,** this species is the one that in Bengal is recognised as '*The Italian stock;*' in which case it would appear to be identical with that which in France is termed '*The Milanese worm,*' though if such be the case, how are we to account for its only undergoing three moults in France and Italy, while in India it invariably has four, like all the others? I incline very strongly to the belief that this alleged peculiarity is altogether fabulous.

"In Bengal, according to **Mr. C. Blechynden** and others, the worm is also sometimes dark-coloured like those of **B. mori,** thus showing clearly that it is not in its original healthy state; the worms attain lengths varying from 2 to 2½ inches, as is the case also at Mussoorie.

"As regards the colour of the silk, nothing could more strongly support my view that white is a sign of weakness and degeneracy. In Italy, we are informed, there are generally nine white cocoons in every ten,—but when cultivated in France bright golden yellow is the predominant colour; this is undoubtedly an effect of climate, showing that the warmth of Italy is less adapted to the health of the insect than the cooler temperature of France, which in some districts is nearly the same as that of Mussoorie, where precisely similar results have been observed. The eggs of this species, hatched in March 1864, from the deposit of May 1863, gave seventy-eight black to thirty-one white worms, in a batch of 109, whereas in 1863, eggs produced from Bengal produced white worms without a single exception. The cocoons spun in 1863 by the Bengal worms were all white, with the exception of about half a dozen, whereas in 1864 there was not one white cocoon, all being of a bright golden yellow. In China, as in Bengal, the usual colour is white, with an exceptional sprinkling of yellow cocoons. Here we have the effect of climate distinctly marked, and showing that while a high temperature produces both white worms and white silk, a temperate climate, by imparting strength, produces dark worms and yellow cocoons.

Dark & light-coloured cocoons. 1785 *Conf. with* pp. *8, 14, 15.*

"The worm which in France gives permanently a white cocoon, and which was imported from China into the '*arrondissement d'Alais,*' would appear to be distinct both from **B. textor** and the other two varieties; so that if No. 1, or the Milanese worm, be our *Boropúlú* as I suspect is the case, and Nos. 2 and 4 are true **B. mori,** then No. 3, with the permanent white silk, is in all probability a distinct species."

BENGAL. 1786

BENGAL.—The above passage, taken from **Hutton's** *Revision and Restoration of the Silkworm* incidentally alludes to the Bengal *bara palú* insect, and **Speed** fixes the date of its introduction. **Geoghegan** says that this insect once predominated in the Cossimbazar Circle, where it yielded the greater part of the March crop of silk, but was found also in Huripal, Jungypore, Radnagore, and Sonamúkí. The Jungypore Resident complained, however, that in 1819 the cultivation of this insect had become "extremely precarious and uncertain," and he attributed this to "degeneracy in the stock."

ASSAM. 1787

ASSAM.—**Mr. Stack** furnishes the following account of this insect as met with in Assam:—

"The *pát* worm is a **Bombyx,** akin to the silkworm of Europe. Under this name are included two distinct species,—the univoltine **Bombyx textor,** called *bor polu,* or large worm, and the multivoltine *horu polu,* or small worm, of which the scientific name is **Bombyx crœsi.** Both kinds are reared indoors, on the leaves of the mulberry **Morus indica,** called *nuni* in Upper and *meshkuri* in Lower Assam.

"The peculiarity of the *bor polu,* or large *pát* silkworm, is that the period of batching lasts ten months. To this circumstance it owes its name of *lehemia* or slow. During this time the eggs are kept in a piece of cloth deposited in a wicker basket (*japá*), which is carefully placed out of the reach of rats and insects. The cultivators look for the appearance of the young worms about the time of the festival of the first day of *Magh,* that is, towards the middle of January, when the mulberry is putting forth green shoots.

Process of Breeding.—"The worms are fed at first on young mulberry leaves cut into pieces and shred over them. They change their skin four times. After the second moulting, they are able to feed on entire leaves. A hundred worms in this stage will eat about one seer of leaves in a day. The tending of the worms usually devolves upon the women and infirm members of the family. The life of the worm lasts thirty to forty days, of which ten or twelve days elapse between the final

SILK: Mulberry.

Cultivation or Rearing of

TROPICAL ANNUAL SILKWORM.

moulting and maturity. The mature worms are removed to a basket divided into compartments each allotted to two or three worms. Here the cocoons are spun.

"The cocoon is completed in about six days. Those selected for breeding are placed on a sieve. The moths emerge in about a fortnight (the time is also stated as ten to twenty days, according to the heat of the weather), and remain in pairs on the sieve for three days, when the females are taken away and placed on a cloth suspended in some quiet corner, where they deposit their eggs, and die a day or two later.

"About seven per cent. of the cocoons are reserved for breeding. Their price for this purpose runs as high as one rupee per hundred."

Method of Reeling.—"The cocoons intended for use are placed in the sun, to destroy the life of the chrysalis. This having been effected, a score of coooons are thrown into a pot of scalding water, and stirred with a splinter of bamboo; the fibres attack themselves to the bamboo, and a thread is thus carried to the reel and reeled off. Sometimes the bamboo fails to pick up the filaments, and a twig of the *makudi* creeper with the leaves on has to be employed.

"The cocoon is of a bright yellow colour, but the silk, when boiled in potash water, becomes perfectly white. About 320 cocoons yield a tola of thread, hence 25,000 to 30,000 will yield a seer.

"From the breeding cocoons after the escape of the moth, and also from the refuse of reeled cocoons, a coarser thread, called *lát*, is made by spinning. One thousand such cocoons weigh about 4½ *tolas*, and yield a thread about one quarter as valuable as the same weight of reeled yarn.

Assam. 1788

"The *pát* silk is a much rarer and more valuable article than either *eri* or *muga*. The thread sells for R16 to R24 per seer, and the cloth for R3 to R4 per square yard. Like the *mezankuri* variety of *muga*, the *pát* silk is rather an article of luxury than of ordinary trade. If a piece is wanted, it usually has to be made to order. Nothing like a market for *pát* thread or cloth can be said to exist. The breeding of the worms is restricted by custom to the Jugi caste, who used to supply the requirements of the Ahom kings and their court, and the industry is hardly known out of the district of Sibságar, the ancient centre of Ahom rule. The Jugis still make a profound mystery of the business, refusing to let a stranger see the worms, and answering enquiries in a manner calculated to mislead. They say, for instance, that the worm takes nine months to spin its cocoon. There can be little doubt that the production of *pát* silk has greatly declined since the annexation of Assam, nor is there any prospect of its revival. Writing to the Government of India in 1877, Colonel Keatinge observed that the question of extending the *pát* silk industry need not be seriously discussed."

N.-W. PROVINCES & OUDH. 1789

NORTH-WEST PROVINCES & OUDH.—In the passage quoted above Hutton incidentally alludes to Dr. Bonavia's experiments in Oudh, with this insect. In a later publication he returns to the subject of these provinces in relation to **Bombyx textor**. "This species," he writes, "is cultivated sparingly in several parts of India, but its constitution is thoroughly worn out, and it ought to be sent to the hill climate. At Mussurie it thrives well, and although like **B. mori**, an annual everywhere else, here it yields a second or autumnal crop also. It was originally brought from China, near Nankin, in north latitude 32°, but is fast fading away from Bengal. It is cultivated in France and Italy and in China, as well as in Bengal, and in these countries generally produces a pure white silk; in Italy there are more white that yellow cocoons, but in France more yellow than white; this is dependent upon climate, as is well shewn at Mussoorie, where worms introduced from Bengal produce *white* cocoons for the first crop, but almost all yellow in the second crop. The worm being northern is impatient of heat and suffers accordingly in constitution, the silk in consequence becoming white, which, as I have elsewhere pointed out, is generally a sign of loss of constitution, not only among silkworms, but among animals still higher in the scale of nature; the natural colour of the worm of **Bombyx mori** is nearly black-brindle, whereas the worms under domestication are of a sickly creamy white. So, then, the climate of France being more temperate than that of Italy, produces more yellow than white cocoons. The species is often termed the Milanese or Italian stock, and in Bengal is known as the *barra palu*, because its cocoon is larger than those of the so-called *desi* worms or polyvoltines."

CULTIVATION OR REARING OF MULBERRY SILK-WORMS IN INDIA.

CULTIVATION 1790

In drawing up the present article on the *Silk and Silkworms of India* the author has deemed it the preferable course to devote the space which could be allotted to the subject, more especially to certain headings in preference to others. This article does not, therefore, profess to be a complete monograph. Methods of Cultivation or Rearing, as also the Systems of Reeling the Cocoons, were thought to be almost sufficiently dealt with by the remark that in both these sections of the subject the information available might be summed up in a very few words, namely, that it denoted differences from the methods or systems pursued in Europe only in being crude, careless, and wasteful. The information available regarding the various BREEDS or RACES of Mulberry worms having now been reviewed, in which will be found particulars of the regions in India where these occur, it need only here be remarked that separate sections will be found below on special subjects such as the WILD or SEMI-DOMESTICATED INSECTS; the DISEASE and PESTS; the SPECIES or VARIETIES, of the Mulberry Plant grown for feeding the worms; and the Indian Trade in Silk, both Internal and External. It remains, therefore, to give in this place a few brief notes regarding the sericulture of the Provinces of India; but before doing so it may be as well to give the main Indian historic facts regarding Silk.

HISTORY OF SERICULTURE IN INDIA.

HISTORY. 1791

So much has been written on this subject, without any satisfactory conclusion having been arrived at, that the subject might be dismissed with the remark that no practical good can be attained by joining issue with either the one school of thinkers or the other. Hutton, than whom no modern writer can be cited who has as yet devoted an equal amount of careful study to the numerous questions connected with India Silk, was of opinion that all the domesticated forms of the mulberry-feeding insects came to India from China. He even went further and affirmed that they originated in the northern colder tracts of that vast country and were, therefore, purely exotics in the tropical districts of Bengal, where the industry had for some time taken root. That in consequence they were constantly liable to degeneration and disease, so much so that instead of advocating the necessity of periodic renewal of stock he urged that the only solution of the Indian Silk difficulty was for the industry to migrate into more northern and colder tracts. How far that idea has succeeded the reader will be able to judge for himself through the extensive republication here given of the reports on the Panjáb and North-West experiments.

On the other hand **Mr. N. G. Mookerji** holds that all the domesticated mulberry insects originated somewhere on the slopes of the Himálaya. Indeed, he fixes the home of the silk moths (BOMBYCIDÆ) as centreing around a limited tract on the skirts of Mount Everest. Becoming diffused to "the sunnier regions of the north and south," the diversifications of form became "entrapped as slaves to the Turanians and the Aryans." Poetic and happy though **Mr. Mookerji's** theory is, and possessed of many points that can be linked on to certain historic facts, it cannot be said that he has done more than advance a hypothesis which possesses some elements of probability. Long ago one or two writers went so far as to advocate that India was the *serica regio* of classic Rome.* The agency of man in the causation and diffusion of certain forms of life is doubtless great, but it does not follow as a necessity that he need have

Serica, Regio. 1792 *Conf. with pp.* 25, 100.

* *Conf.* with the remarks under "Sheep and Goats" more especially the Silingia Sheep, *Vol. VI., Pt. II.*, p. 573.

SILK: Mulberry.

Cultivation or Rearing of

HISTORY.

carried to remote sections of the tribal area the genera or species of moths which, out of a long list of useless forms, came to be selected for human purposes. Long before necessity arose for the utilisation of silk, the useful genera or species might have been diffused to remote corners of the theoretical home of the series and only matured, so to speak, their useful properties in the countries of their adoption. Thus even after having accepted one of **Mr. Mookerji's** strongest points, *viz.*, that India possesses a very complete and extensive series of forms of the SATURNIDÆ and BOMBYCIDÆ, it does not follow that the typical genus of the latter tribe need be a native of India, still less that its domestication was first accomplished anywhere on the Himálaya or in Hindustan.

The Mulberry. 1793 *Conf. with pp. 37, 48, 66.*

Perhaps a more trustworthy method of treatment, of this perplexing subject would be to follow up the indications of truly wild states of the species of **Morus** [the food of the most interesting silk moths, such as the species of **Bombyx** (the true silkworms)]. The writer is aware that modern botanists have come to regard certain forms of **Morus** as truly wild in India. Personally, he is disposed to refer all these forms to one species, and to view some of its varieties as manifesting a greater degree of acclimatisation than others; in other words, to regard all the forms as constituting an introduced and highly variable species. Whether that view be correct or not the main point at issue may be briefly stated. Can it be said that any form of **Morus** is indigenous to the Bengal silk districts? No. There is of course a vast difference between the popular word "wild" and the more accurate synonym "indigenous." The yellow Mexican poppy (**Argemone mexicana**) is sufficiently plentiful in the hotter parts of India, from the mouths of the Ganges and Indus up the slopes of the Himálaya to 5,000 or 6,000 feet, as to deserve the popular designation of being wild. It is, however, not indigenous to India. This illustration is perhaps needless, but if the species of **Morus** are not truly indigenous in India or rather to the silk districts of India, the whole argument about the **Bombyx** moths being so, goes to the wall. The writer has certainly never seen jungles or forests in India, far from human dwellings, that possessed here and there a trace of **Morus**, still less forests that mainly consisted of mulberry bushes. The plant springs up naturally in the scrubby jungles around human dwellings, but it is met with chiefly on roadsides or around the margins of fields. The fruit is greedily eaten by birds and many wild animals. Had it been indigenous, or did it even take freely to India, the mulberry might have been expected (from a wide distribution of its seeds through the agency of birds), to be plentiful in the forests; can this be said to be the case? Then, again, the fact that none of the so-called wild forms can with certainty be pronounced as well marked and characteristic or exclusively Indian types, we obtain a negative indication that favours the improbability of any of them being Indian plants. Added to all this, has any entomologist discovered the ancestral state of the **Bombyx mori** group of silkworms in a truly wild state in India? It is very generally admitted that the annual and polyvoltine conditions of that insect, met with in India and China, are all forms of one species, which have been brought into existence through man's agency and the conditions of the country in which they have for centuries been reared. The occurrence of a truly wild species of **Bombyx** in a country possessed of an unmistakably indigenous species of **Morus**, might be accepted as affording the best claim to being the original birth-place of sericulture. So far as the writer can discover these conditions stand a better chance of being established for China than for India. But we have another line of evidence which must be viewed as important, *viz.*, modern direct historic records. The experiments conducted by the East India

HISTORY.

Company leave no room for doubt that the early European pioneers introduced many of the common silkworms now met with in Bengal. The remarks, in the Proceedings of the Company, regarding these introduced insects narrow very much the list of possible indigenous species.

Serica regio. 1794 *Conf. with pp. 23, 100.*

But the main argument raised by writers on this subject may be said to localise the area of contention. The idea as to whether some part of India was the original home of the mulberry moths is less important than the point urged by **Hutton**, namely, that the insect is an exotic in Bengal. In the writer's opinion the earliest records of Bengal silk very probably denote the *tasar* worm not the mulberry. It is from this point of view, therefore, of little moment whether we regard the people spoken of in the Mahabharata as the "Chinas" as having been inhabitants of a portion of the Himálaya or of the China of modern writers. The silken goods and silk which they brought to Yudhisthira were not produced in Hindustan. We have, in fact, to pass over the chief classic period of this country before we obtain any positive evidence of the cultivation of silk on the plains of India. As in Hebrew so in Sanskrit, the greatest possible confusion exists as to the early names and synonyms that should be viewed as denoting silk. Reference to this subject will be found in connection with the chief commercial fibres of India such as Hemp (**Cannabis**), Jute (**Corchorus**), Sunn-hemp (**Crotalaria**), Cotton (**Gossypium**), Wool (SHEEP & GOATS), etc.. The reader will find, from the information given in connection with these fibres, that the greatest difficulty prevails in fixing comparatively the date at which a specific signification was given to such names as *Urna*, *Pat* or *Patta*, *Kshauma*, and the like. In the writer's opinion all the undoubted references to mulberry silk, in early Hindu literature, speak of it as an imported article, and it is not, until comparatively modern times, that we have direct indications of silk culture in India. This view is not, however, held by **Mr. Mookerji**, and since he reviews some of the opinions that have been published on this subject, the present chapter may fitly be concluded by the following passage from one of **Mr. Mookerji's** reports, the remark only being added that the author of this paper does not of course commit himself to the accuracy of **Mr. Mookerji's** review :—

"The ancient literature of the Hindus and the Sanskrit language afford valuable historical arguments in support of our thesis. China is not mentioned in the oldest Sanskrit works, although silk is taken for granted as the proper article of wear for ceremonial purposes. In the chivalrous age that followed the patriarchal, warlike connection with China is mentioned, but even this China was probably in the Himálayas. In the dramatic and mythological age (which, according to the best authorities, must be placed subsequent to the Christian era), China-silk is mentioned as an imported article or of commerce [*c.f. Chinangshukamibaket orniya manang pratibatasya.—Sakuntalá of Kalidásha*]. But it was indigenous fabrics alone, such as were used in the old Vedic days, that could be used in these later days, and in fact, even in the modern Hindu society, for ceremonial purposes, the Hindus, from very ancient times, have regarded all foreigners and foreign articles as impure. Not only was there an indigenous silk fabric, but there was also an indigenous art of bleaching, and an indigenous art of embroidery, unborrowed from foreign sources. Silk dress has been always used among the Hindus for the most ancient of all religious ceremonies, *viz.*, that of marriage [*Kshoume basane basáná bágnábáhayátáne.—Vedas*]. The ancient method of bleaching silk had come to be looked upon, even in the days of Manu the codifier of Hindu laws, as so venerable that it had already been associated with a vile superstition that lingers to our days. '*Kouceyábikayo rushiah*' (silk and shawl are to be purified with liquid excrements of animals and water) was no doubt meant by Manu as a ceremonious method of purification by sprinkling, against ceremonious defilement, but it points to those days when the method of bleaching, which is employed even in modern Europe was employed in Old India.

Bleaching Silk. 1795 *Conf. with remarks, pp. 57 and 153 on bleaching of Tasar.*

"Now, if our contention that the old Aryan Hindus obtained their knowledge of silk in their land of adoption independently of the Turanian Chinese, who obtained their knowledge in the ultra-Himálayan regions, and that the Semetics also probably

SILK: Mulberry.

Cultivation or Rearing of

HISTORY.

received their knowledge independently of either of the other two races in the western Himálayan regions, there would be no common word for silk in these languages, and the facts bear out this supposition. The Sanskrit synonyms for silk are very numerous. They are *Kouceya Patta, Patta-pattaja* (Mohabharata 2, 51, 26), *Krimija, Krimija Sutra, Kitatantu, Kitasutra, Kitaja, Bardara, Urna, Kshouma, Dúkula* and *Dugúla*. The word for silkworm in Sanskrit is *Pundarika,* and the present Hindu caste of silkworm rearers are called *Pundarikakshas* or *Pundas.* None of these words can be traced in the Sanskrit signification to the common Aryan source. The word *Koca* is a generic name for shell or any swelling, and is no doubt allied to *cocoon*. But it is a mere accident that led the French and the English and the Indians to adopt the same idea of swelling or shell to denote cocoon. The Italians use a different word, *Bozzoli,* evidently derived from Latin *Bocas,* box, to signify cocoon. Nor would Grimm's Law allow us to trace any connection between the Sanskrit word *Patta* and the Italian word *Bozzoli,* or the Anglo-Saxon word *Botsian* (=swelling). But of all these words *Urna* appears to be the oldest (Rig Veda V. 52, 9; IV., 22, 2; X, 75, 8), and this has not even an apparent connection with any Aryan word for silk, cocoon, or silkworm. Much less is there any connection between any of these Sanskrit words and the Chinese word *Tsau* (=cocoon) or *Tsi* (=silkworm) to which most of the later words for silk in other languages can be traced. All the European countries evidently derived their knowledge of silk latterly from the Chinese or perhaps the Mongols. The Mongol, *Sirkeh,* the Corean, *Sir,* the Greek, *Σήρ,* the Latin *Sericum,* the German, *Seiden,* the French, *Soie,* the Russian, *Sheolk,* the Anglo-Saxon, *Seolc,* the Icelandic, *Silke,* the Burman, *Tsa,* have all an apparent family likeness, and they point to one common source for silk in the countries concerned, *viz.*, China or Mongolia. The Assamese word *Pat* meaning cocoons, and the Tamil word *Pattu* meaning silk, are also apparently derived from the Sanskrit *Patta,* to which the word *Pat,* used in the silk districts of Bengal as equivalent both to silk and cocoon, is also related. In the language of Kashmír, *Pat* means silk, and *Krimkas* means silkworm-rearer, both words being apparently Sanskrit in origin. The Assamese and the Tamils, though they ethnologically belong to the Chinese race, have no doubt derived their knowledge of silk from, or along with, their Hindu neighbours, although their easterly neighbours, the Burmese, seem to have derived their knowledge of silk from the Chinese. What we have said for the Aryan Hindus and the Chinese holds also with reference to the Semetics. Although no mention is made of China in the Old Testament, the knowledge of silk is taken for granted. The traditions regarding the existence of silk and the mulberry tree in countries west of India also go back beyond the Christian era. The prejudice in the minds of scholars regarding the universal origin of silk from China is so strong that the well known classical name of silk not occurring in the Hebrew (Proverbs xxxi., 22; and Ezekiel XVI., 10 and 13), a doubt has been entertained *solely* on this philological reason whether the Hebrew *Meshi* and *Demeshek* (Amos iii., 12) and the Arabic *Dimakso,* refer to silk. But the reference to silk in Revelation xviii, 12, tracing the substance to Babylon, and the reference to the mulberry in I. Macchabees vi., 34, have been allowed by scholars to be undoubted. So strong, indeed, is the prejudice that there must be a philological connection of silk with China that *Sherikoth* of Isaiah xix., 9, has been understood by some scholars to mean silk, although from the context it appears clearly that it cannot mean silk, and that the authorised translation is correct. The philological difficulty, however, would vanish at once if we supposed that as the Aryan Hindus discovered the silkworm in the sub-Himalayan regions, as the Chinese and other Turanians discovered it in the ultra-Himálayan regions, so did the Semetics discover it in the western Himálayan regions bayond Kashmír; and the name of silk ought to be, and is, quite different in these languages, and it is therefore also that there is no reference to China, either historical or philological in this connection, in the oldest records we have of silk. Under such a supposition we can easily understand why the Hebrew words *Meshi* and *Demeshek,* the Arabic words *Dimakso* and *Kus,* and the Persian words *Abresham* and *Resham,* though related to one another, have no affinity whatever with the Chinese or the Indian words for silk. These differences in fact in the three classes of languages is an argument in support of our contention that the Himálayan regions were the original home of the silkworm. None of the references to silk in the classical literature of the west referring it to China, are older than those we have mentioned already, and they do not therefore affect our argument either way. We are, in fact, inclined to believe that the silkworm was known in India and in the land of *Uz*, which was somewhere in the north of Arabia, before the time of Moses, and that the reference to the moth that "buildeth his house" (Job, xxvii., 18) in the most ancient of the Old Testament books, *viz.*, that of Job, is not to a clothes-

HISTORY.

destroying **Tinea**, as is usually supposed, but to a cocoon-forming **Bombyx**. The Vedas, though claimed by Hindus to be coeval with creation, is easily allowed by European scholars to be at least as old as 500 (B.C.). If any reliance is to be placed on native Chinese dates, the invention of the *magnanerie*, the *filature moulinage tissage*, and a great many other things ascribed to the **Empress Silingchi** must be fixed in that country about 2700 (B.C.). To the **Emperor Fo-hi**, the first Chinese Emperor, the husband of **Silingchi**, is ascribed the invention of numbers, music, &c. But modern historians regard all the Chinese stories dating earlier than the third century (B.C.) as chiefly fabulous. The **Emperor Chi-Hoangti**, who built the great wall, also burnt nearly the whole of the classical literature of the Chinese, and the old histories were subsequently re-written from memory. The existence of silk in China before the third century (B.C.) is undoubted, but beyond this we are not warranted to go much further, although from the fables extant this much can be gathered that the Emperors and the Nobles of China have *always* taken a prominent interest in sericulture; and that China stands at this time at the forefront in the trade both in soft silk and in *tusser* is to be attributed chiefly to this and not to any natural superiority China has ever enjoyed above other countries. The notion that silkworms spread originally from China is in fact due chiefly to the fact that Europe received them directly from China about the sixth century of our era, when, according to **Procopino**, the **Emperor Justinian** induced certain Persian monks to revisit China and bring back with them eggs of the superior classes of silkworms. The commercial spread of sericulture in Europe dates from this epoch. It is in fact an exception to the rule of the gradual spread of silk industry in other countries from one common focus. The most ancient Greek and Roman traditions extant, regarding silk, have indeed no reference to China. **Pliny** (who lived in the beginning of the Christian era) mentions the *Seris* (probably the modern Chinese), but it is not in connection with silk, but with wool, that he mentions them. The same **Pliny**, however, mentions the produce of the Assyrian silkworm, and also relates the tradition of southern Europe of the native discovery of the silkworm, and the art of weaving silk. **Pamphile**, the daughter of **Plotes**, according to **Pliny** (who followed **Aristotle's** account), first discovered the art of spinning cocoons and weaving silk in the island of Cos. The celebrated *coan-vest* of European classics must therefore have been silk."

The reader who desires further particulars of the historic facts that bear on the slik question should consult the chapter on TRADE, pp. 182—196, and of special works the following:—**Dr. Royle's** *Productive Resources of India*; **Sir George Birdwood's** *Indian Arts*; **The Very Rev. Dr. D. Rock's** *Textile Fabrics; The Encyclopædia Britannica*; **Milburn's** *Oriental Commerce*; **Ure's** *Dictionary of Arts, Manufactures, &c.*; **Balfour's** *Cyclopædia of India*; *The Indian Art Journal, &c., &c.*

BENGAL. 1796

I.—BENGAL.

Among the works of special interest on the subject of Bengal silk the following may be mentioned: **Geoghegan's** *Silk in India* (Ed. 1880); **Louis'** *Present State and Future Prospects of Sericulture in Bengal* (1882); **Liotard's** *Memorandum on Silk in India* (1883); the *Annual Reports of the River borne Traffic of the Lower Provinces and Inland Trade of Calcutta*; and the reports which have from time to time appeared in connection with the investigations presently being conducted by **Mr. N. G. Mookerji**. The annual official publication on the trade of Bengal (above cited) contains from year to year a review of the various district reports on the subject of sericulture, and the remarks there given are often extremely valuable as manifesting the opinions of district officers on the prospects of the enterprise. The pessimist view, taken by **Mr. Louis** and other writers whose opinions carried alarm some eight or ten years ago into the hearts of all persons interested in Bengal silk, might be dismissed with the remark that the decline which they saw, and profess to still see, was, and is, far less real than imaginary. That a decline has taken place, in certain features of the trade, there can be no doubt, but in some directions it has greatly expanded. **Mr. Louis** says: "It is

Cultivation or Rearing of

CULTIVATION in Bengal.

Disturbing elements in the Silk Trade.

1797

clear that the production of silk is decreasing at an alarming ratio, and if this is the result, as I shall endeavour to show, of a steady deterioration in the breed of the Bengal **Bombyx** and in the quality of the cocoons, the total disappearance of this wealth-producing industry is a question of a few years only." **Mr. Louis** then proceeds to state his case in the strongest terms possible. He maintains that deterioration of the breed is the sole and only cause. It has transpired, however, that another element of disturbance was operating, and which has since manifested itself in an irresistible form, namely, a change of fashion in Europe. To meet the necessities of the altered trade new machinery rapidly came into existence. A demand arose for wild and waste silks, to be carded and spun, instead of reeled. For the new purposes fashion had created, cheap inferior silks and cocoons found relatively a more remunerative market, than the results of the most patient endeavours to produce, in Bengal, silks of a superior nature. The exports of reeled silks accordingly declined, but there sprang into existence a compensating export of wild and waste silks, and of cocoons, so that if the totals of all raw Indian-produced silks be compared, the decline will be found to be serious only in value not in quantity. **Mr. Louis** seems to have overlooked entirely this feature of the modern trade. He gave the returns from 1870 to 1878, but disposed of the question of wild and waste silks by the remark that the exports from the whole of India may, roughly speaking, be accepted as the exports from Bengal since "other silk manufactures of India, such as kincobs, tussahs, &c., are made from silks grown in other provinces, and are almost entirely absorbed for consumption in India." There are several errors involved and implied by **Mr. Louis** in the passage cited, but the most serious is the contrast made between Bengal mulberry silk (in the expression given), and the waste and wild silks of India as a whole. His words might, indeed, be read as referring to the manufactures from these silks, but if so, why contrast them with the decline of the Bengal mulberry raw silk? But would it be correct to say that kincobs (*i.e.*, silk brocades) are not made of Bengal mulberry silk? Would it be wise to accept "*tussahs*" as clothes made of the silk of **Antheræa paphia**? At the present day *tasar* or *kutní* cloth almost invariably means a fabric woven of silk and cotton, the silk not necessarily being *tasar* in its restricted and correct signification, *viz.*, the fibre of **Antheræa paphia** (=**A. mylitta**). But in addition to these misconceptions implied, **Mr. Louis'** contention involves a direct error, since he gives the returns of the exports of Indian-reared silk *plus* the re-exports of foreign silk, as the exports of Indian silk. The remarks on this subject in the section on TRADE below should be consulted. It will thus be seen that **Mr. Louis** starts his argument of the decline of the Bengal silk trade from several distinct errors. While the writer is unable to distribute the figures given by **Mr. Louis** into Raw Mulberry Native and European Filature-reeled Silks; Reeled Tasar and other Wild Silks; Mulberry Chasam; Tasar and other Wild Chasams; Mulberry Cocoons; and Tasar and other Wild Cocoons, still these were the items that made up the totals of raw silk shown by **Mr. Louis**, and there can be no doubt that the proportion of the waste raw silk in 1870 (the first year of the series he deals with) was very probably close on half the total recorded exports. The author would further venture the suggestion that the decline of reeled mulberry silk from India may have affected more seriously the Native or inferior, than the European or superior filature silks. In other words, whether produced by European or Native methods, the superior reeled silks would appear to have held very nearly their own, while the inferior have given way to the comparatively more remunerative prices offered for waste and wild silks. This, indeed, might be regarded as a result that

Re-exports, foreign silk.

1798

Conf. with pp. 196, 201-2.

CULTIVATION in Bengal.

Decline of Trade in Reeled Silk.

1799

could have been foretold. But while it is probably the correct view, it has to be added that opinions differ greatly; some writers even maintain that the European branches of the trade have felt the change that has taken place more seriously than the Native. That a general decline in the foreign exports of reeled silk has taken place, the writer does not for a moment wish to doubt. He desires only to point out the necessity of carrying to its final issue the investigation of the chief direction of the admitted degree of depreciation. The more inferior European-reeled silks would doubtless have shared the same fate as the Native article, and it is possible that the additional profit obtained for the "waste" may, in the case of the Native artisans, have encouraged the reeling of more silk, rather than have retarded its production, the inferior reeled article going into consumption in India itself, but disappearing from the returns of exports. As opposed to this idea we have, however, the all but universal outcry that the chief and serious depreciation of India's Silk trade is the collapse of the demand for Native manufactures, *i.e.*, Native-made silk piece goods. In Mr. Liotard's memorandum the following facts are given regarding the Bengal silk trade. The Memorandum, it may be added, was compiled in 1882-83:—

"The cultivation of the mulberry and the rearing of the worms are conducted by the peasantry themselves by two different classes of people who are under no obligation but their own interests. The destination of the cocoons is two-fold: they are as a rule either sent to small Native filatures where the silk is roughly wound and usually consumed in the hand-looms of the country or consigned to Madras or to the Bombay mills; or they are brought to the great European factories in Bengal, where, after being reeled by steam machinery, the silk is consigned direct to Europe. The chief silk-producing districts are Rajshahye, Murshedabad, Maldah, Birbhum, and Midnapur, with Nuddea, Bankura, Bogra, and Rungpore of less importance.

1800

"I.—In Rajshahye Messrs. Watson & Co., of Calcutta, and Messrs. L. Payen & Co., of Lyons, still have large filatures at Surdah and Klrahja respectively, with smaller establishments scattered throughout the district. The annual outturn (of raw silk?) is given by Dr. Hunter at about 2,00,000 seers * of raw silk, and the area under mulberry at 80,000 acres. The rearing of silkworms is carried on by the small cultivators, who formerly used to get their cocoons reeled by Natives for the European markets and for the great cities of the North-West; but at present they are almost entirely in the hands of the above European filatures, who purchase the cocoons from them at prices governed by the silk markets of Europe. Efforts were recently made to reduce the rates of rent on mulberry lands, as the high rates prevailing pressed heavily on the silk trade: the result is not known to me.

1801

"II.—In the Murshedabad district there are several filatures owned by Europeans: some by Messrs. L. Payen & Co., and the others by Messrs. Lyall & Co. The system adopted by the firms is to fix the price of cocoons so as to manufacture silk at the market rate prevailing at the commencement of the season. These cocoons are bought from brokers generally under a pecuniary advance from the filature: the brokers procure them from the Murshedabad and neighbouring districts, and are bound to buy them from the rearers at a still lower price than they have to get from the filatures. The rearers thus suffer the heaviest loss owing to being compelled to sell at any price or see their cocoons spoilt; but they are not, as those of Rajshahye, entirely in the hands of the European filatures of the district: a large proportion of their cocoons is bought by Native reelers and reeled into common silk for export to Madras and Bombay, and for the silk called *khangru*. This latter class of silk is purchased by Native weavers who manufacture *cora* cloths in many villages under contract and on advances received from wholesale dealers.

1802

"III.—Regarding the Maldah district fuller information is available in reports submitted by the Collector (Mr. R. Porch) in June 1880 and February 1883: the

* The figures given above do not appear to occur in Sir W. W. Hunter's statistical account of Rajshahye, though a long and interesting statement of the silk industry of that district is given in Vol. VIII., 82—86. Sir William gives the outturn (Native and European) at 5,000 maunds, which would be 410,000lb. That figure is very probably considerably above what the outturn actually was for the year in which the Gazetteer (1876) was published.—*Ed., Dict. Econ. Prod.*

SILK: Mulberry.

Cultivation or Rearing of

CULTIVATION in Bengal.

chief seat of the industry is in the centre of the district; the total area under mulberry cultivation is about 35 square miles; rents for mulberry lands, sub-*ryoti*, are so high as from R16 to R25 per *bigha*, and press heavily on the industry: the annual outturn of cocoons is variously stated at from 12,000 to 60,000 maunds; the number of people to whom silk gives employment either by mulberry culture, worm-rearing, silk-spinning, or silk-weaving, or silk with cotton weaving, is said to be 300,000; the rearing of the worms is done by small cultivators from whom the cocoons are purchased under special arrangements by the numerous Native filatures and the two European firms established in the district (Messrs. Watson & Co., and Messrs. L. Payen & Co.). At this stage the industry divides itself into two classes: (1) under the European system of reeling and supervision; (2) under the Native system of reeling. The *European* part of the industry is stated by the Collector to use variously from one-third to one-seventh part of the cocoons reared in the district which are bought by them from the rearers and reeled into good silk for the Calcutta and European markets: the business is not prospering; in the words of the Collector 'it is at best a very precarious industry.' The *native side* of the industry on the contrary is prospering: 'in its agricultural aspects, and as regards theleasy profits made by the natives, the mulberry silk industry must be considered as brisk, prosperous, and flourishing.' The greater portion of the cocoons reared in the district is either bought by Native manufactures or are reeled by the rearers themselves, who most of them have one or two reels: the silk reeled by the natives and called *khangru* is partly bought by silk-piece manufacturers of Bombay*, Benares, Delhi, Mirzapur, &c., and partly is used in home manufacture for *corahs*, *masru*, and other kinds of cloth. These cloths are mostly exported to Calcutta, Bombay, Madras, Nagpore, Allahabad, Benares, and Delhi (the largest exports being to Calcutta and Bombay), and partly worn in the district. A very large proportion of the cocoons reared in the district are reeled into this silk (*khangru*) in the small Native filatures; it commands a uniform rate of between R11 and R13 per seer of 80 sicca weight, and the price of cocoons is regulated by the market rate for the native silk.

1803

"IV.—In Birbhum, silk is now produced in the eastern part of the district only. One European factory owned by Lyall & Co., says Dr. Hunter, gives employment to about 15,000 people, and turns out silk to the value of £160,000 annually, which is usually sold in the raw state, and finds its way to the Calcutta and European markets. The number of people employed and the value of the outturn are probably both exaggerated; at any rate the figures do not represent the present condition, for the industry has declined considerably. There are also numbers of small native filatures whose reeled silk is either consumed in the local manufacture of piece goods, or sent to Murshedabad and the silk-consuming towns of the North-Western Provinces and the Panjab, and to Bombay. But the native reelers and weavers (says the Commissioner in a recent letter) are labouring under the disadvantages arising from want of capital and the competition of foreign goods.

1804

"V.—In Midnapur Messrs. Watson & Co. have four factories, and Natives have three; but the business does not apparently pay as well as it did before, and the industry has in this district declined more than elsewhere; the cocoons, too, are more inferior than those of the other districts, although some say that they are superior.

1805

"VI.—In Nuddea the industry is reported by the Commissioner to have become almost extinct; there are one European and several small Native filatures, but the business is not a very flourishing one.

1806

"VII.—In Bankura, cocoons are reared on a very small scale by petty capitalists, but the quantity of silk produced is not sufficient for local demand, which is mostly met from imports from Singbhum. A stuff called *kutra* (a texture of cotton and silk) is made to some extent.

1807

"VIII.—Bogra, once famous for its silk, now possesses a lingering industry; the few cocoons now reared are mostly exported to Rajshahye to be worked up there; the worm-rearers and mulberry-growers are distressed. The Rungpore silk industry is confined to the south of this district, whence about 300 cwts. of cocoons and 50 cwts. of raw silk are annually exported.

1808

"IX.—In Cuttack, sericulture has been carried out as an experiment since 1877, at Government expense, under the supervision of the Executive Engineer of the

* The constant reference to Bombay as a market to which Bengal silk was consigned is interesting. (*Conf. with p. 52.*) It will be seen in the chapter on Internal Trade below (*pp. 205, 209, 210-13*) that no such trade now exists—Bombay is supplied mainly by China.—*Ed., Dict. Econ. Prod.*

Mulberry Silkworms in India. (*G. Watt.*)

SILK: **Mulberry.**

CULTIVATION in Bengal.

Mahanudi division, and the result after four years' trial is shown in the following table:—

Amount expended.	Quantity of outturn.	Value of outturn.	Number of mulberry trees planted.
	Sr. Ch.	R a. p.	
Government outlay, including Superintendent's charge for two years, R3,065¼.	Silk . . 6 15 Chussum . 20 14 Cocoons . 1 0	318 11 6	2,376 in 1877-78 2,667 " 1878-79 10,973 " 1879-80
		TOTAL .	16,016

"It is not known whether further prosecution of the trial has been abandoned or not.

"Different opinions have been advanced to account for the decline in the silk industry of Bengal. The extensive importation from Japan and China to Europe since the opening of the Suez Canal,—the larger yield of recent seasons in Italy and France, helped by regular supplies of silk-worm eggs from Japan,—the indifferent quality of the Bengal silk in that it wants strength and elasticity,—and the probable fact that the demand for silk goods has not kept pace proportionately with the increased supply thrown upon the market,—have all been brought forward as so many causes of the stagnation and gradual decline of the Bengal silk industry; and perhaps there is some truth in each and all of these opinions.

"But there seems to be evidence to lead to the belief that of the European and Native sections of the industry, it is the European * that has suffered the more seriously. In regard to the Maldah district, for instance, this has already been shown to be the case in the preceding paragraphs. In other districts it is more or less the same (except in Rajshahye, where the Native is entirely in the hands of the European section); and the European section complains of the obstinacy with which the Native workers in silk demand high prices in the face of the active competition with Bengal silk which has set in from Europe, China, and Japan. The Native section, however, does not seem ready to lower its prices or accept any radical change of custom. The Bengal worm suits its circumstances; it eats little comparatively and thrives on the immature or shrub mulberry leaf which is renewed at every cutting; it is considered less troublesome in rearing, and spins often, being multivoltine, except in Murshedabad and Midnapur, where the annual worm (*boro-poloo*) is reared. The silk thread obtained is generally wanting in wiriness, and a bad system of reeling makes the threads endy and an abomination to the European silk-thrower. The European firms who have so great a stake in the Indian industry have repeatedly made efforts to bring improvements in the Native system of rearing and reeling. But the Natives care little about that so long as their Industry goes on according to custom and they can raise and dispose of their produce by reeling it off themselves for despatch to other parts of India; and European firms find themselves compelled to buy the native grown cocoon or close their filatures. These facts may lead to the inference that the native section can go on prospering whatever may happen to the European section; but the Collector of Maldah (Mr. R. Porch) is of a contrary opinion. He writes:—

"'If the European-supervised silk filatures were closed the Native silk industry would still thrive for a long time, but undoubtedly such collapse would recoil upon it and be disastrous to the Native silk industry which is so largely subsidised and indirectly guided by European capitalists. Without that capital and guidance and support the Native silk industry would, it is believed, become very precarious and collapse after a time.'

"The Native section is not without its vicissitudes: sometimes the worms fail to spin from extremes of heat and cold, from too much rain and cloudy weather in their last stage, and from the want of opportune showers for the mulberries. Sometimes when the rainy season is good, mulberry leaves are abundant, and then the crop of

* More recent information would seem to throw doubt on this opinion. Certain Europeans have doubtless suffered, but the facts regarding the trade would seem to justify the idea that there has been the usual effect of capital and enterprise prospering over poverty and indifference. The fittest have survived in an altered condition of the trade which lowered the value of inferior goods.

SILK: Mulberry.

Cultivation or Rearing of

CULTIVATION in Bengal.

cocoons is fine and there is a glut in the Native silk market which brings down the price of cocoons and of reeled silk; sometimes, again, the silk crops in France, Italy, and China are very good, and the market for Indian silk is then very bad. Lastly, the rents of land under mulberry cultivation are excessively high, and this, which is not the least of the drawbacks, enhances the cost of producing silk, tempts the rearers to give the very least quantity of leaf required, and causes, by a semi-state of starvation of the worms, the weakness in the silk which renders it difficult to reel without breakage."

Total Trade. 1809 *Conf. with pp. 186, 189, 191, 195, 196, 198-9, 200-1.*

In order to ascertain how far the views advanced by **Mr. Liotard**, in the above passage, are borne out by the more recently published facts regarding the trade, the following review may be given of the annual returns of the internal trade of Bengal and of the reports of the Committee of merchants recently formed with the view of investigating the cause of the decline of the trade. The imports into Calcutta may be taken in the first place as a fairly good indication of the fluctuations of the whole trade, since Calcutta is the chief, if not the only, exporting emporium; but it should be recollected that the figures given below denote all kinds of raw silk collectively. During the past fourteen years these were as follows:—

	Maunds.		Maunds.
1876-77	24,290	1883-84	24,308
1877-78	27,398	1884-85	29,689
1878-79	22,844	1885-86	24,147
1879-80	22,745	1886-87	28,234
1880-81	23,140	1887-88	27,806
1881-82	20,874	1888-89	31,475
1882-83	25,421	1889-90	28,864

It cannot, therefore, be said that the total quantity of silk imported by Calcutta has fallen off, but a critical examination of the share of these imports,taken in reeled silk, manifests the fact that since 1858 there has been a steady increase in the trade of waste and wild silks, and for the past 20 years at least a considerable decline in quantity, but a marked improvement in the quality, of reeled mulberry. The average exports in reeled silk have, during the past five years, been about 400,000lb. The Bengal silk district reports speak, however, of decline as if it had been far more serious than the figures of the trade justify, and it seems probable that these remarks refer more to local manufactures than to reeled silk. It would, in fact, appear that the changes that have taken place, the ruin of a filature here and the closing of looms there, were viewed as indications of a general and wholesale shrinking of the trade throughout the Province, instead of a redistribution. It will, perhaps, suffice for the present purpose to amplify the information conveyed by the above extract from **Mr. Liotard's** *Memorandum* to quote a few passages which bring the facts regarding the silk-producing districts down to the year 1890. These it had perhaps be best to deal with in the same order as the paragraphs given by **Mr. Liotard**.

Changes in the Trade. 1810 *Conf. with pp. 59, 62, 72, 106, 118, 121, 148, 156, 187, 193, 201, etc.*

Rajshahye. 1811

I. RAJSHAHYE.—In the Internal Trade Reports of Bengal it is stated of Rajshahye Division for 1884-85 that "It is to be regretted that silk manufacture is steadily falling off in Rajshahye and Bogra. In the former district, whence the article is chiefly exported, the two principal European manufacturing firms turned out only 1,628 maunds against 2,097 maunds in the previous year." In the report for 1885-86 a further decline is notified in the produce of the European firms to 113,400lb. In 1886-87 decline is still again recorded, the falling off having amounted to 4,640lb (or 58 maunds), but it is remarked that "the causes were the same as in the former year—drought in March and April and heavy rains in July and August, the result being that in both years the outturn was only half of the November or cold-weather crop. The condition of the industry,

CULTIVATION in Bengal.

Rajshahye.

particularly as regards disease and deterioration of the worm, occupied the attention of the Government during the year, and a Silk Conference was held in Calcutta in March 1887." We next read of improvement. In the report for 1889-90 it is stated that "the manufactured outturn of silk in Rajshahye shows an increase, being 1,380 maunds, as compared with 1,112 maunds in 1888-89 and 1,244 maunds in 1887-88." It is thus safe to assume that in Rajshahye the ordinary fluctuations have been more marked than in other parts of Bengal, but that no serious decline has occurred more than could be accounted for by the idea advanced, namely, a radical change in the conditions and location of the Bengal silk trade. In Bogra, on the other hand, the silk industry has been almost driven out by competition. "The families who formerly wove silk have almost all taken to agriculture." But here again the decline complained of is in manufactures not in the production of cocoons or of reeling.

Murshedabad. 1812

II. MURSHEDABAD.—This is the only silk-producing district of the Presidency Division. In the report of Internal Trade of Bengal for 1884-85 it is said that the silk trade of Murshedabad had declined, but in the following year more details are given. "The number of silk factories," we are told, "which worked in Murshedabad district during the year was 72. They produced 200,911lb of silk, valued at R13,03,914. The silk industry was, it is said, not profitable. The depression is attributed to a strong competition in the European market of Italian silk." In the report for 1886-87 it is stated that the largest filature proprietors are the European firms of Messrs. Luis, Payen & Co., the Bengal Silk Company, and Messrs. Watson & Co.; "but this last firm has, during the past two years, closed many of its filatures which were not paying. The French firm, some two or three years ago, we are told commenced making *tasar* at Bagorpara, which is now a great success, and is carried on there on a large scale. Prices of Bengal silk are said to have risen somewhat lately, but the trade is still much depressed." The report goes on to say, however, that during the year there were 91 filatures and factories working in the district (*e.g.*, 19 more than in the previous year), and that the outturn was 231,120lb, valued at R17,22,765. The report then adds that "it thus appears that the past year was a favourable one." For 1888-89 the outturn was given as at 285,567lb of silk, valued at R23,43,341. "It may be noted that during the year six filatures were closed and three new ones started. The value of outturn from three of the filatures at work during the year was not furnished." The report for 1889-90 states that eight filatures were stopped, and three new ones started during the year. The outturn was 352,866lb, valued at R27,53,981.

It will thus be seen that Murshedabad, at all events, has manifested no decline during the past five or six years, although serious changes have taken place.

Maldah. 1813

III. MALDAH.—It seems almost unnecessary to say anything regarding this district. The silk interests of the Bhagulpore Division as a whole are mainly concerned in *tasar* or in silk manufactures. The reports which have appeared on these branches of the trade are unanimous in holding that a serious decline has taken place. There are five silk factories in Maldah, under European management, besides numerous filatures. In the report for 1885-86 it is stated the season for mulberry was unfavourable, the plants having suffered severely owing to the drought and great heat. Of the succeeding year it is remarked that there are two large factories, *viz.*, at Bholahat and Baraghore. "The large manufacturers export direct, but the produce of the small filatures is taken to the neighbouring markets and there disposed of, if possible. The outturn of raw

Cultivation or Rearing of

CULTIVATION in Bengal.

Maldah.

silk from the Native filatures, during the last two years, has been about the same, and the following table shows the prices :—

	1885-86.		1886-87.	
	R a.		R a.	
Fine silk . . .	16 0	per seer.	15 0	per seer.
Coarse silk . .	11 0	" "	10 8	" "
Jatam or silk refuse .	1 8	" "	1 12	" "
Cocoons . . .	36 0	per maund.	40 0	per maund.

"The production of, and demand for, silk fabrics have considerably declined. The Collector believes this decline to be due to better and cheaper qualities being manufactured in Europe. Three hundred and twenty maunds of raw silk were produced at the European factories during the year, as compared with 288 maunds in the preceding year." The outturn of the two European filatures of Maldah for the year 1888-89 is given at 30,424lb (say, 379 maunds), so that it cannot be said that the trade in superior reeled silk is declining so far as the European filatures are concerned.

Burdwan.
1814

IV. BURDWAN DIVISION.—"The manufacture of mulberry silk is on the decline in almost all places where this industry is carried out. The *kutni*, a mixed fabric of silk and cotton, produced in the towns of Bisoonpore and Soonamukhi, in the district of Bankoora, is still in demand in the North-Western Provinces and among Europeans. Rupees 70,000 worth was turned out during the year. Gonotia is the principal seat of the silk industry of Beerbhoom. The Bengal Silk Company has its principal place of business there. The only available statistics of silk-spinning are those relating to the filatures of this Company. The outturn of the year was 378⅝ maunds, worth R1,91,372 against 620 maunds, worth R3,86,650 of the previous year. The falling off is attributed to some extent to a corresponding decrease in the supply of cocoons, which varies according as the mulberry crop turns out well or ill. The Company buy their cocoons from local growers, and the supply fell short of the demand. The importation of foreign silk also prejudiciously affected this branch of the industry in all places where it is carried on. Silk fabrics of various kinds are made in parts of the Ghattal, Tumlook, and Sudder Sub-Division of Midnapore. Those manufactured at Radhanugger in Ghattal are much in requisition for the use of ladies of the upper classes of the Panjáb and Rájputána" (*Report for 1883-84*).

In the above extract, and in all the other notices of the Burdwan silk interests, the tone of decline which pervades these remarks appears to proceed from a certain loss of Native manufactures rather than from any positive falling off in the filature produce. Thus the report for 1886-87 records the fact that "silk cocoons are reared in the Sudder, Tamlook, and Ghattal Sub-Divisions of Midnapore. It is stated that 268 maunds of silk, valued at R10,65,674, were sold in the Tumlook Sub-Division alone during the year. In the Ghattal Sub-Division 595 maunds of silk, worth R3,99,401, are reported to have been manufactured during the year, against 252 maunds, worth R1,31,550 in the previous year." So, again, in the report for 1888-89 there occurs the following: "In Beerbhoom the value of the silk turned out is reported to have improved 10 per cent." And lastly, in the report for 1889-90 we are told that "The Bengal Silk Company, which has a factory at Ganotia, in Beerbhoom, sold R2,64,000 worth of silk, as against R3,22,885 in 1888." So far, then, for the Burdwan Division, silk production has been marked, during the past eight or ten years, with great fluctuations, but no positive or serious decline.

A change in Silk Trade, but no serious decline.
1815

Sufficient has been shown to justify, or at least to partially justify, the opinion which the author holds, *viz.*, that a material change has and is

CULTIVATION in Bengal. Burdwan.

taking place in the Indian silk trade. The superior article turned out by the better class filatures is finding, if anything, a better market, but the inferior produce has fallen under the new demand in Europe for waste and wild cocoons in preference to badly wound silk. This change has upset the conditions of old and well-established marts, and caused the reports from these to be characterised by complaints of a serious falling off in the trade. The production of silk has been directed into new channels and has come under new conditions. The outturn of the filatures has been affected by the change to a greater extent than might have been anticipated, but it seems probable that year by year they will be found to benefit. The closing of Native defective filatures should in time place a larger supply of cocoons at the disposal of the European filatures of Bengal, and, alive to their best interests, if the owners of these put forth a streneous effort to improve their produce, the silk industry of India may, and doubtless will, be placed on a sounder basis than it has ever before attained. Improvements must, however, be directed not to systems of reeling alone. The preservation of the quality of the breed (a much hackneyed subject) must not be entirely overlooked, but what is of even greater importance, the interests of the rearers must never be ignored. So long as the growing of mulberry bushes or trees is remunerative to the peasant, so long as the feeding of the worms and the production of cocoons gives a fairly good return, the supply will continue. But if these conditions fail, the entire industry must be taken in hand by the owners of filatures, or abandoned. It will be seen from a letter quoted below that Mr. Lister thinks the time has even now come when the capitalist should directly control every branch of the enterprise. The rearer is certainly too poor and too ignorant to contend against degeneration in the breed, or loss from disease and pests. So long as India was allowed to move at an oriental pace the trade was, in its multifarious branches, more prosperous than it is to-day. It was once on a time able to compete with and drive Turkey out of the market. It has now more formidable rivals. The struggle with Europe, China, and Japan is a reality that has to be faced, or the battle given up without an effort against impending ruin. But that ruin is imminent cannot for a moment be entertained.

Conf. with pp. 61-2.

II.—NORTH-WEST PROVINCES & OUDH.

N.-W. PROVINCES & OUDH. 1816

These Provinces have never taken any very pronounced position in the production of silk. Their connection with that subject derives interest from the important experiments presently being performed by Mr. Lister at Dehra Dún. The history of all past experiments in these Provinces will be learned from Mr. Geoghegan's work and from Mr. Liotard's *Memorandum.* It does not seem necessary to reprint some 10 or 12 pages to record only failure. Some of these experiments will at the same time be found briefly alluded to in the present article such as those performed by Captain Hutton. To Mr. H. G. Ross, Superintendent of the Dún, belongs the credit of having, with great courage, skill, and untiring energy, prosecuted the attempt to produce silk in the Dún. The reports of his numerous experiments can be readily procured, but it may, even with perfect justice to Mr. Ross, be said that if success be ultimately attained it will be due entirely to Mr. Lister. Until the results of that gentleman's efforts have been made public, however, it would perhaps be premature to hazard opinions, but the writer, supported in his view of the case by the mature convictions formed by Hutton and other authors, does not believe in the possibility of producing **Bombyx mori** (proper) on a commercial scale anywhere in India, except perhaps in Kashmír or Manipur. In many respects the last-mentioned country resembles more closely Japan or China than it does India, and speaking from botanical evidence, sericulture might

Conf. with pp. 14, 49, 55, 142,

CULTIVATION in N.-W. P. & Oudh.

be expected to prosper better in Manipur than in any part of India. But the writer is aware that his view of the possibilites of **B. mori** is opposed by that of some of the best modern bservers. **Mr. Cleghorn** has much faith in a hybrid between the Bengal insects and **B. mori. Mr. Mookerji** goes even further. He writes: "The important step to be undertaken, is the introduction of the superior **B. mori** cocoons. These are now employed in every other silk-producing country, and I fully believe they can be introduced with success in Bengal." Yes, perhaps so, if fresh seed be imported each year, if the opposition of the Natives to a univoltine insect can be got over, and if the imported stock can be induced to thrive on the Bengal mulberry. The whole question is one of expediency.

But to return more immediately to the subject of the North-West Provinces, the present remarks may be fitly concluded by quoting **Mr. Mookerji's** report—the most recent statement which has appeared on the prospects of sericulture in the Dún:—

"*System of Rearing that Ought to be Introduced.*—Cocoon-rearing industry does not exist among the Native peasantry in the North-Western Provinces as it does in the Panjáb, and it is still in the pioneering or experimental stage. That sericulture is suited to the Dún or any other part of the North-Western Provinces yet remains to be proved, and no proof will be conclusive until the Native peasantry take to it of their own accord as a means of livelihood, and 'it is in this direction that the enterprise has most interest' (to quote the words of the Director of Agriculture from a letter on this subject addressed to the Secretary to the Government of the North-Western Provinces and Oudh on the 14th September 1878). An industry like cocoon-rearing, which is simple and inexpensive enough to be carried on by the peasantry, cannot in India be carried on by capitalists, for the simple reason that it can be managed almost entirely by women, who, on account of the *zenana* system (more or less rigidly prevalent in every part of the North-Western Provinces), would have almost nothing to do if they had no such employment. Besides, the Indian peasant, however lazy he might appear working for a master, is the hardest-working individual working for himself. Labour, therefore, does not count in cottage industries in most parts of India, although the same industries, if they are carried on in an organized state by capitalists, will cost an enormous amount in labour. The *grainage* and the reeling will never be conducted satisfactorily by peasants, but the rearing of silk-worms, if it succeeds at all, will be taken up by them, and they are sure to keep capitalists out of this department of the silk industry. Not only will labour cost the peasantry of the North-Western Provinces almost nothing in doing silkworm-rearing but the leaf also they will get almost without price for a long time to come. In the North-Western Provinces and the Panjáb road and canal sides abound in large mulberry trees which cost nothing now to Government or anybody else to keep them up. Thousands of trees can be now had from Government for the rearing season for merely a rupee. In this way the peasantry of these provinces have an advantage over the Bengal silk-rearers, as the shrub mulberry costs a great deal in cultivating and manuring. Each maund of leaf costs to the Bengal peasant 4 to 8 annas. A thousand trees in the North-Western Provinces will yield as much as 2,000 maunds of leaf which represents 4,000 seers of green cocoons, or about R4,000 worth of silk. In pioneering this industry these two advantages must be borne in mind, and attempt should be made from the first to interest villagers who have large trees before their houses in the rearing of silkworms. The education of the *rayat* in rearing silkworms in a healthy manner should, therefore, be the chief end of the experiments. The enterprise is a generous one and properly belongs to the State, but **Messrs. Lister & Co.** offered to carry on what **Hutton** and **Ross** had begun under Government patronage, and it is well known what enormous sums of money have been spent by the Company since 1879 in these experiments.

"*Results of the Early Experiments.*—The results of the experiments in the North-Western Provinces have hitherto proved nothing. The most carefully conducted of these experiments, *viz.*, those undertaken by **Captain Hutton**, gave unsatisfactory results at the end. **Captain Hutton**, who began so hopefully and in a locality (Mussoorie) more naturally suited for sericulture than the Dún, writes thus discouragingly towards the end of his sericultural career: "If the cultivation of silk in India is to be extended in earnest, it is with these wild worms that the game must be played, for we already know that nothing can be gained by introducing the **Bombyces** of China into localities where experience has shown us that they cannot thrive."

CULTIVATION in N.-W. P. & Oudh.

'I should be very much tempted to begin my experiments again; with **Bombyx** I will have no more to do.' 'In the hills of the North-West no doubt suitable localities and elevation may be found, but I confess I do not consider Mussoorie either sufficiently elevated or far enough north to enable any one to work the worm with full success, or extract from it all that it is capable of yielding.' 'Where the speculator possesses more money than brains, the best possible way of equallising the two will be to attempt silk cultivation with Chinese worms in the North-Western Provinces of India."

The Mulberry. 1817 *Conf. with pp. 24, 48, 66.*

"*Fresh Endeavours.*—These discouraging statements of **Captain Hutton's**, however, have satisfied no one with his eyes open that the mulberry silkworm is an exotic and unsuitable for the Indian climate. Judging from several of his experiments, one can see that **Captain Hutton** jumped to conclusions too quick, and that he worked too much on nature's principles to achieve any practical success. That the rearing of silkworms succeeded occasionally was enough to convince subsequent experimentalists that there were certain conditions under which success could be achieved, and that if the conditions under which failures sometimes took place were understood, they could be avoided in future. The presence of mulberry trees in the wild state on the hills and the submontane tracts was a standing inducement to try further experiments. So late as 1882, the Government of the North-Western Provinces took in hand the distribution of mulberry cuttings and silkworm eggs all over the country through the Forest School at Dehra Dún and the Saharanpur Botanical Garden; and it was in 1883 that the grant of a large tract of land was made by the Government to **Messrs. Lister & Co.** with a view to organise sericultural operations on an extensive scale.

"The Company had already large experiences in sericulture acquired in Assam and in the Panjáb. The sericultural experiments in Assam were abandoned, as labour was found to cost too much; and when **Mr. Lepper** began operations in the Dún in 1879, most of the rearing he used to get done through independent parties (chiefly villagers), giving them seed and buying cocoons from them. This was the right method of work, and it would have been very fortunate if this method was adopted and persevered in in the grant. No better examples illustrating the French proverb 'Petites magnauries, grandes filatures' are needed than what **Mr. Lepper's** experiments in 1880 have to offer. The following table will show that the European experience, that each rearer should rear only from about an ounce of eggs, is worth following in practice wherever the industry is introduced:—

Names of growers.	Quantities of seed used.	Weights of green cocoons obtained.		
	Ounces.	M.	s.	c.
1. Sipahi Singh	20	3	29	8
2. Suk Lal	4	0	33	13½
3. Shib Ram	4	1	1	7
4. Laljee	8½	0	35	3½
5. Indor Singh	2	1	17	6
6. Kishen Singh	3	0	36	14
7. Kishna Nund	6	2	7	33½
8. Mahunta	5	1	4	1½
9. Bishun Singh	2	1	11	10½
10. Bhana Mull	1	0	31	12½
11. Mr. Bell	1	0	36	9½
12. Mr. Macnaughton	24	6	2	4
13. M. Nagona	14	1	35	0
14. Mr. Lepper	Not recorded.	7	14	1½

"*Advantage of rearing small Quantities.*—Where one ounce of seed was used in the above experiments the result came up almost to the European average. **Captain Murray**, who succeeded **Mr. Lepper** in the management, had also good result when he had rearings on a small scale. The chief evil of the Kashmir system of rearing is that the rearers like a large quantity of seed. In the Panjáb the same evil exists among the Native rearers, and it is important that better example should be set from

SILK: Mulberry.

Cultivation or Rearing of

CULTIVATION in N.-W. P. & Oudh.

the very first in introducing the industry in the North-Western Provinces. Even the most successful rearings in Kashmir and the Panjáb on the ordinary Native method are poor compared to rearings conducted on the proper method with small quantities of seed. But as leaf and labour cost the up-country rearer very little, even the poor result obtained by him hitherto has brought him profits. The famous Jaffer Ali of Gurdaspur used an average quantity of 32 ounces of seed per year, and his highest expectation was 21 seers of silk which he had in his best years, whereas he ought to have had at least 2 maunds and a half if he worked on the proper system. Yet Jafir made very large profits, and not only did he himself make silk-rearing his profession for upwards of 20 years, but all his family followed him. I will quote here some figures from Baden Powell's *Punjab Products*, Volume I., p. 173, which show that even with the poor results obtained by the Native rearer, his clear profit was over 100 per cent. There is not a single example furnished by the hundreds of successful and unsuccessful experiments conducted by European capitalists or officials, with hired labour, of pecuniary profits, and it is vain to expect it in any future undertaking of this character:—

	R	a.	p.
Cost of one seer of silkworm eggs	11	0	0
Do. mulberry leaves	10	0	0
Do. coolies for gathering and bringing in	25	0	0
House rent	5	0	0
Own wages, calculated at R5 per month, for tending worms and reeling	25	0	0
Wages of an assistant at R2-8	12	8	0
Cleaning silk at 12 annas per seer	10	8	0
Interest on money borrowed	0	3	0
TOTAL	99	3	0

Net proceeds of 14 seers of silk at R16-8=204-8-0, leaving a net profit on the season's operations of R102-8-0, or more than 108 per cent. per annum.

"The economical management which is represented in the above table is possible only with a person who works himself with his family and has little to pay for mulberry leaf. To raise 14 seers of silk a Bengal silkworm-rearer would have to spend R70 or R75, supposing he grew his own leaf. If he had to buy leaf, he would have to pay twice the amount. The advantage of cottage cultivation will be still more apparent when it is considered that the result obtained, *viz.*, 14 seers of silk from one seer of seed, is an extremely poor one.

"*Subsequent method of work.*—Mr. Cunliffe Lister came out to this country in 1881. He was so disgusted with the unhealthy methods of rearing in vogue in the Panjáb that he was convinced that nothing but skilful and rational management of silk worms could secure the permanency of the new industry in the Punjab and its introduction in the North-Western Provinces. Direct management of silkworms was, therefore, undertaken both in the Panjáb and the North-Western Provinces. Better result in quantity and quality has been generally obtained by this system, but the system has not been found to pay. It will be well to go into the details of the management of the grant from this period, especially as Liotard's report does not go beyond this.

"*The history of the Grant.*—On the 9th March 1883 a deed of grant was executed by which 3,218 acres of forest land known as *majri* was made over by Government to Messrs. Lister & Co. on condition that '*bonâ fide* silk operations shall be commenced within five years from the date of signing the deed of grant. *Bonâ-fide* silk operations are understood to mean the having not less than 50 acres of land under mulberry and the having spent R4,000 on buildings and on irrigation channels.' Another condition of the grant is that 'at the end of ten years, provided the grantees have 10 per cent. of the assessable area under three-year-old mulberry trees, they may retain the whole grant on condition that they turn out not less than 100 maunds of green cocoons annually.' Again, the allowance of 16 mulberry trees to the acre was declared by the Director of the Department of Agriculture and Commerce in his letter No. 1181A. of the 23rd August 1882 as 'amply sufficient to satisfy the terms on which Messrs. Lister & Co. hold the grant.' All these conditions have been satisfied with the exception of that referring to the 100 maunds of cocoons. It must be remembered that, although the deed of grant was executed in 1883, Messrs. Lister & Co. had begun operations in the Dún in 1879, and even at Majri in 1882. But, notwithstanding strenuous efforts on the part of the

CULTIVATION in N.-W. P. & Oudh.

managers, the 100 maunds of cocoons are still to come. If the proper method of work had been followed up for the last ten years, a great deal more might have been achieved.

"*Difficulties Experienced.*—I will describe here the various difficulties with which the managers of the Majri grant have had to contend:—

"(1) Majri is not naturally adapted for silk-rearing any more than Bengal is. The exact conditions necessary for silk-rearing must be undertood before these can be artificially introduced. The proximity of the mountains naturally suited for sericulture is, however, an advantage, and hibernation of the proper kind so necessary in effecting even and complete hatching of eggs and in producing a strong brood of silk worms naturally capable of resisting flacherie can take place without difficulty.

"(2) A second disadvantage from which Majri suffers at present is the ravages of the deer on the mulberry plantation. If the mulberry plants were allowed to grow to trees or standards, there would have been nothing to fear; but the trees being closely planted in most places and being pruned down low, nearly all the plants are shrubby or brambly, and therefore easily attacked by animals. Extension of the plantation on the right method will do away with this difficulty.

"(3) Another disadvantage the plantation suffers from is its unhealthiness after the rains, which is one reason for the unwillingess on the part of *rayats* settling in the grant. But so long as the plantation remains jungly, and so long as there are no *rayats* to keep the lands occupied with crops instead of allowing grass to rot on all sides, the plantation must continue to be malarious. If a large colony can be got to settle in the grant, say in December, and bring a large tract of it under cultivation before next August, I believe the malaria will disappear.

"(4) A fourth disadvantage is the distance of the grant from Hurdwar or Dehra Dún, it being about 14 miles from either town. During the rains all communication with market towns is practically cut off. The Company have, however, always afforded facilities to settlers to get provisions from *bunias* or from the plantation itself at reasonable rates. If this system is extended, there need be no difficulty on this score for *rayats* to settle. Even if rearing is done in future directly by the Company, to have resident labourers will be found very economical. At present the supply of labour is uncertain and unsatisfactory, a large staff of labourers being kept up for six months from December to May to ensure their attendance at two or three urgent occasions, when the full staff is found barely sufficient for the work.

"(5) The dryness of Majri at the rearing season is also a disadvantage. Apart from other considerations, a dry atmosphere is an advantage for silk-rearing. But dry atmosphere means great variation between the day and night temperatures, which, during the rearing season at Majri, is about 25°F., both when it is very cold, as at the beginning of the rearing, when the temperature varies from 40° to 65°F., and also when it is very hot, when the temperature varies from 85° to 100°F. When I left the grant on the 10th of April, the temperature ranged from 80° to 96°F. If rearing-houses are well shaded with trees, a local moisture is induced in them which helps to keep the houses cooler in the daytime and warmer at night. The rearing-houses both in the North-Western Provinces and in the Panjáb are, however, too open. This is one of the principal disadvantages, as, next to healthy seed, the condition of the evenness of temperature is of first consideration in securing successful cocoon crops. I was told the rearing-houses were purposely made so open, the idea being that silkworms wanted plenty of air. But in reality silkworms need very little beyond bare ventilation. If the rearing-houses were made in the present Kashmír style, *i.e.*, properly shut up with glass windows, fire-places, and aloft with ventilators, they would be all that could be desired. Under such arrangements it would be quite easy to get a month to 35 days with the temperature of the rearing-houses always standidg at about 75°F. The Dún, however, is better situated in this respect than most other places in the North-Western Provinces and the Panjáb. The Siwalik Hills, only five miles to the south of the grant, protects it from the scorching winds from the plains.

"(6) The large rearing-houses will be always found a drawback. They can be easily divided up each into two houses; and to introduce the system of

SILK: Mulberry.

CULTIVATION in N.-W. P. & Oudh.

Cultivation or Rearing of

cottage cultivation at once the houses can be let to the rearers at present employed, and the rearing can be done by them under supervision on a system of contracts. The rearing will go on much better if the rearers and their families live on the premises.

"(7) Of the 900 acres planted with mulberry, about 600 acres consist of a late variety which is practically useless for the silkworms, as they begin to put forth leaf in April, when it is so hot that all the rearing should be over before that. The remaining 300 acres also yield leaf which is not so nourishing as the properly elaborated leaf from trees; the planting being generally too thick (the object being to keep down the jungle), the trees do not get the chance of growing vigorously and producing large and full-formed leaves.

"*The Management.*—The practical difficulties in the way of introducing sericulture on a commercial basis have been found so great in the midst of all the disadvantages mentioned that one manager after another has left the task as impracticable or impossible. Mr. Lepper began by building a wall to protect the future plantation from the deer. He wrote a pamphlet on the brilliant prospects of sericulture in the Dun, began the work on the proper system of getting cocoons reared by villagers, but he soon retired from the sphere of practice. Captain Murray, who succeeded in the management, effected great improvements. Besides getting 250 acres of land under mulberry cultivation, building three large rearing-houses (each 100 feet x 20 feet), and constructing a canal and irrigation channels which now beautifully ramify the plantation, he got three villages settled. From what has been already said, the settling of the villagers was a very good thing for the plantation. Captain Murray's plan was to make advances to tenants for breaking up land and cultivating it. He advanced them R2 per every *bigha* (one-sixth of an acre) brought under cultivation and took a fourth share of their crop; and when the land was cleared by the tenants, he used to plant mulberry in it. In 1884 Captain Murray obtained a crop of 1,076℔ of green cocoons out of 15½ ounces of seed, *i.e.*, about 33 kilos per ounce, which has been the best result obtained at Majri. Eight acres of three years' old mulberry plants were used for this crop. Mulberry plants should not, however, be touched for about five years. But the Company was apparently impatient to see result, and the Captain was replaced in the management by Mr. Herdon, as he had spent an enormous sum of money and had reared but one shed of silk worms. Mr. Herdon thought he could rear 20 sheds with the leaf there was in the grant. He persevered in the attempt for three years, and each time the leaf fell short and had to be carted from forests and roadsides, sometimes from a distance of 20 to 25 miles. The result was poor each time. In 1885 he used 20℔ of seed and he got 31 maunds of green cocoons, which cost him R30 per maund in labour alone. During the rains Mr. Herdon used 1½ seers of Bengal eggs and he obtained 23 maunds of green cocoons. The next year he again used 20℔ of seed and got 18 maunds of cocoons, which cost him R20 per maund in labour. A portion of this crop failed through muscardine, and as it happened to be thundery, the worms were supposed to have been struck by lightning. A second crop of seven sheds was attempted this year (1886) with Bengal seeds, but it failed entirely through muscardine. In the last year of his management (1887) he again used 20℔ of seed and obtained 55 maunds of cocoons, which cost him R73 per maund in labour. In thus attempting to push on and get the maximum result, he had not only to cart leaf from long distances, but to strip every leaf from the young plantation and cut the plants down very low to encourage the growth of layers. This has injured the plantation permanently. It is to be borne in mind that the most important element of plant food, that which goes to form the woody portions of a plant, is obtained through the leaf. Plants that are stripped of leaf when very young and never get rest from stripping afterwards continue sickly and stunted. That is the condition of most of the plants in the grant now, and it will be found difficult to make decent standards of the existing plants.

"To push sericulture in the Dún, Mr. Herdon also erected a large filature—the most expensively equipped filature in India. When Mr. Herdon left, he had to show a plantation of about 800 acres, 27 large rearing-sheds, and a filature which to all outward appearance seemed a complete and satisfactory get-up and went almost to satisfy the requirements of the deed of grant. But the 100 maunds of green cocoons were still to come. To satisfy the conditions of the deed of grant before March 1893, working with the existing plantation, will require all the energy and vigilance of the present manager, Mr. Farrant, who succeeded Mr. Herdon in 1887. Besides ruining the plantation Mr. Herdon made another unfortunate

CULTIVATION in N.-W. P. & Oudh.

mistake: ne dispersed the three villages set up by **Captain Murray** to make room for his plantation. It has thus become doubly difficult to get men to settle in the grant.

"**Mr. Farrant**, the present manager, reared, in 1888, 65 maunds of green cocoons out of 112 ounces of seed at the cost of R9 per maund for labour. In 1889 he used 150 ounces of seed and obtained again about the same quantity of green cocoons. Pebrine killing a large proportion of the worms at the last stage, the cost came to about R14 per maund for labour.

"*The Present Year's Crop.*—This year the stock was renewed with imported seed. The seed from Europe arrived hatching, and it could not be hibernated in this condition: so it went on hatching for about two months, and the whole quantity did not hatch. From Smyrna also a consignment of eggs arrived which had hatched on the way. A second consignment of Smyrna seed, however, arrived safe, and it was sent up to a mountain for hibernation, and it hatched properly. As there was too little seed to use up the leaf, it was thought advisable to use some Bengal seed also, although the Bengal cocoons are not worth growing if it can be helped. The European seed arriving and hatching out at improper times, the rearing went on from the third week of January to the end of April. It was too cold at the beginning and too hot at the end, the earliest worms being fed on lettuce-leaf in the absence of mulberry. When I left the grant on the 10th of April, the spinning ot the cocoons was proceeding most satisfactorily and a crop of fully 100 maunds of green cocoons was expected. I have to report the appearance of muscardine and flacherie in some of the sheds, but they were checked ás soon as they were noticed. The appearance of these diseases served, in fact, as a means of demonstration as to the methods that are to be adopted in preventing their spread. The following figures as to the outturn came to me, after I had left the grant, from **Mr. Herdon**, the officiating manager; **Mr. Farrant** having, according to my advice, just left for Europe to study the sericultural practices that prevailed in France and Italy, as my recommendation was to adopt them bodily in the Dún. The weights representing the outturn must be those of smoothered and dried cocoons, or else the outturn would come to be very poor, which was not the case. The details, however, of the crop I could not get, as with the figures came a memorandum from the officiating manager to say 'As I have just taken over charge, I am sorry I am unable to give you any particulars.' If the figures for the outturn represent dry weights, they ought to be multiplied three times over and the weights of the green cocoons will be obtained approximately. Under such a supposition the present year's crop weighed over a hundred maunds:—

Kind of seeds used.	Weight of seed.	Weight of crop.
	Ounces.	M. s. c.
French yellow . . .	26	16 31 9¾
Italian white	20	5 13 14
Smyrna	10	1 34 12
Bengal	69	16 7 8
TOTAL .	125	40 7 11¾

A good average result from 125 ounces of seed would be about 120 maunds of *green* cocoons, or about 40 maunds of dry. In the Kashmír experiments also similiar results were obtained in the State nurseries, where 37½ ounces of seed produced 35 maunds 14 seers and 14 chatacks of *green* cocoons. As most of the Kashmír cocoons were used for seeding, they had to be weighed green.

Forms of Mulberry Plant and methods of cultivation. 1818 *Conf. with pp. 66-67.*

"*Recommendations as to Planting and Rearing.*—The system of mulberry plantation in **Messrs. Lister & Co.**'s establishment in the Dún and the Panjáb is neither the Bengal system of quick return for capital spent nor the European system of true economy. The Bengal system is good only for the Bengal worms, especially the most degenerate of all species, *viz.*, the *Deshi* or *Choto palu* worms. The European system of cultivation is good for the annual worms. **Messrs. Lister & Co.**'s system has answered fairly well for both the Bengal and the annual silkworms. It is more economical, however, in the long run to take only one crop of the superior annual cocoons than two or three crops of the polyvoltine cocoons in addition to an annual one; and if the annual cocoons alone

SILK: Mulberry.

Cultivation or Rearing of

CULTIVATION in N.-W. P. & Oudh.

are reared, there is no question but that the European system of planting is the best. It is possible in the Dún to take three or four crops of the poor Bengal cocoons, and they have been taken; but they should not be taken. Complete rest should be given to rearing operations for at least eight months to protect the annual crop from infection. There is another advantage in taking one cutting of leaf from a tree where it is possible to take three. It strengthens the tree to leave it unstripped. I would therefore recommend all future extension of the plantation on the European plan. Let the forest land beyond the present plantation be leased out to tenants, the Company reserving the right to plant mulberry trees, 66 feet by 33 feet apart, on the tenant's fields, to prune the plants with a view to get good standards, and to gather leaf from them. In this way there will be 20 trees per acre, there being at present thousands of shrubs per acre, though the conditions of the grant require only 16 trees per acre. The trees will be thus further apart than in Europe (where 40 trees per acre is the usual allowance), and for the first five years there will be so little shade that the *rayats* will have no objection to the trees being on their fields, especially if they are told that the mulberry trees are meant to benefit them eventually, when they will take to rearing silkworms themselves. If these plants are not touched when in leaf, and pruned only when the leaves have done their duty in nourishing them, *i.e.*, about November, they ought to grow into fine trees in five years, each tree yielding in one cutting 12 to 15 seers of leaf. After 10 years they ought to yield on an average one maund of leaf per tree. No manuring will be required to keep the plants up in condition, except what is done by the *rayats* immediately in the interest of their crops; and the cultivation by the *rayats* in between the plants will benefit them, especially by preventing the growth of the tall grass that now so often gets the better of the mulberry by chocking them to death. The maximum yield of a mulberry tree is not a maund of leaf, but three to four maunds; and a tree, if it is properly grown, will yield this in 20 years. But to compare the relative merits of the two systems, *viz.*, the existing system and the one I am recommending, I will reckon only middle-sized trees of ten years' growth. Although about 900 acres have been planted, it is only 300 acres in the grant which yield leaf at the proper time, and which have reached almost the maximum of yield, that I will reckon for the purpose of this comparison. Three hundred acres under the system of planting recommended will yield 300 x 20 maunds of leaf after 10 years. Let us see what the 300 acres of mulberry now used yield. Mr. Farrant's experience is that it is not possible to grow more than 30 sheds of silkworms with the present stock of leaf. From what I saw of the plantation I thought the same. Now 30 sheds of worms represent about 200 maunds of green cocoons, a quantity that has never been grown yet, although the leaf is pretty well used up every year. Now 200 maunds of green cocoons can be raised on 200 x 20 maunds of leaf. The total capacity of the present plantation is, therefore, much less than that of a plantation with only 20 middle-sized trees per acre. The combination of the planting and the '*rayati*' system will thus only appear to compromise the object of the grant, but it will really be beneficial to the sericultural enterprise in every way. In a word, I would recommend not only for the Dún, but for every place where the introduction of silk is contemplated, the adoption of the European systems of mulberry cultivation and of rearing bodily, it being perfectly safe to grow the superior European cocoons with artificial methods of hibernation and incubation which are practised even in France and Italy now to get better results.

"*Why the Bengal system should not be followed.*—The temptation to take to the Bengal system of planting and rearing is so great that I would conclude this report by pointing out the evils of this system and emphasizing the introduction of the European method. The Bengal system of planting utilises every inch of land for the silkworm, and seems at first sight extremely economical and admirably suited to the object and requirements of the grant. The Bengal silkworm again breeds several times in the year and the seed requires no artificial treatment to make it hatch. This also may seem at first sight to be a great advantage. The Bengal system is invaluable for training rearers in a locality where silk-rearing does not exist, and in the new spheres where silk-rearing has just commenced, such as Gwalior, the Sonthal Pergunnahs, and the Darjeeling Division, I have advised the rearing of Bengal silkworms, which keep men continually in training. But when the mere pioneering is over, the Bengal method should be abandoned and all temptation to take a second and a third crop avoided:—

"(1) It is found troublesome guarding mulberry plantations from the attack of cattle. If there were standard trees at Lister's grant, the deer could have done no harm.

Mulberry Silkworms in India. (*G. Watt.*)

SILK: Mulberry.

CULTIVATION in N.-W. P. & Oudh.

"(2) The tree-leaf yields a larger proportion of silk. When mulberry is kept down to the height of beans or lucerne while the natural habit of the plant is to grow to a tree, the leaf is not properly elaborated and is wanting in the proper nutritive properties which go to form silk. The *Deshi* silkworms of Bengal, however, are too weak in their jaws to thrive on the tree leaf, and for this the most degenerated species only is the Bengal mulberry more suitable.

"(3) The frost of winter will be too severe for the Bengal mulberry in cold localities.

"(4) The Bengal mulberry, if it is allowed to grow tall, dies out, being closely planted. They have thus to be kept cut down three or four times in the year. This entails a twofold disadvantage. First, it exhausts the soil, and it requires heavy manuring to keep the plants in condition. Secondly, three or four cuttings of mulberry necessitated the growing of three or four crops of cocoons in the year. If selected seed is used there is not much chance of pebrine ruining the crops under such treatment. But the germs of flacherie and muscardine which propagate readily do great damage if long rest is not given to the rearing-houses: and as for pebrine, even when it does not ruin the crop, it does some harm.

"(5) The Bengal mulberry gives rise to grasserie. Grasserie is of no consequence where the tree mulberry is used, but in Bengal it is epidemic in character. When young worms are given nutritious leaf, and when they grow old if they are given more watery leaf, gresserie is the result. So in dry season a shower of rain at the last stage of the worms often means total loss from grasserie. The consistency of the leaf growing on a large and deep-rooted tree is not changed so readily, and grasserie is, therefore, of rare occurrence in countries where the tree mulberry is used.

"(6) The outturn from an acre of Bengal mulberry is about 150 maunds. The outturn from an acre of mulberry trees which have attained their maximum size, supposing there are 40 trees to the acre (I have recommended 20 trees to the acre for Lister's grant for special reasons), is about 120 maunds of leaf in one cutting. A second and a third cutting are possible, but, as I have said before, more than one cutting should not be taken. An acre of land with 40 trees will yield one or two short crops of vegetables or areals which will bring an additional outturn.

"(7) The Bengal system of cultivation is costly, trees requiring little or no attention when they have once grown beyond pruning every second year. It may look stupid to allow a tree to grow for five years without making use of the leaf, but this is the proper treatment and it is most economical in the long run. In Bengal, if sericulture fails for a number of years, mulberry fields get neglected and they cease to be mulberry fields in no time. A mulberry tree when it has once grown is there, whether silk-growing fails or succeeds, and it always invites fresh efforts in rearing. The potential sericultural wealth of countries like Kashmír, the Panjáb, and the North-Western Provinces, where large mulberry trees abound, is always very great and stable. In Bengal two or three years of total failure of cocoon crops may permanently ruin the industry. There will be nothing to tempt the *rayats* to take to it again.

"(8) The Bengal cocoons alone are suitable for the Bengal mulberry, and Bengal cocoons are not worth growing if the superior annual cocoons can be grown. The annual European worms will eat the Bengal mulberry, but as it is wanting in nutrition they consume more of this than of this tree mulberry, and they spin comparatively poorer cocoons. The superior annual cocoons produce a high-class silk, the demand for which in European markets is great stable.

"*Conclusion.*—Messrs. Lister & Co. have already introduced Pasteur's system of grainage in their establishment in the Dún, and when Mr. Farrant comes back from Europe, I hope he will see the benefit of introducing the European system bodily and adopt exact methods of hibernation, incubation, housing, feeding, and delitage, and teach this system only to the *rayats*, who alone will be able to do the rearing part of the industry so as to keep the debit and credit sides of their account in a satisfactory condition. The rearing of silkworms may go on successfully under the system of direct management, but, unless financial success is attained, the industry will not be established."

SILK: Mulberry.

Cultivation or Rearing of

CULTIVATION in N.-W. P. & Oudh.

The above somewhat detailed report has been here given because of its linking together many facts that will be found scattered throughout the present article. Mr. Mookerji's report is, therefore, not alone applicable to the North-West Provinces. It will be found to convey opinions on many of the all-important aspects of sericulture, such as the manner in which disease should be coped with, the advantages of this system and of that of growing the mulberry, and the views held by Mr. Mookerji as to sericulture being more successful as a peasant industry than when under European Companies. Below will be found Mr. Lister's view on this subject which is opposed in its entirety to that advanced by Mr. Mookerji.

Conf. with pp. 49, 59-62.

The reports issued by the Director of Land Records and Agriculture contain full particulars of the Government experiments in sericulture up to the date of the transfer of the farms to Messrs. Lister & Co. The following passages may be specially consulted:—Reports, 1877-78, 11-12; 1878-79, 21-22; 1880-81, 10-11; 1881-82, 21-23; 1882-83, 22; and 1883-84, 13.

III.—THE PANJAB.

PANJAB. 1819

The earlier experiments in the Panjáb will be found fully dealt with by Mr. Geoghegan and by Mr. Liotard. Recently an instructive monograph has been published on the subject of the Panjáb silk trade by Mr. H. C. Cookson from which the following passages may be here republished:—

"The silk industry of the Province has undoubtedly declined since the establishment of the British rule. Silk fabrics are much less worn by the well-to-do classes than they were in Sikh times; European broadcloth and cotton goods having to a large extent taken their place.

"At no time does the cultivation of the silkworm seem to have been carried on systematically by the Natives of the Province. Where sericulture has existed at all, it has been taken up rather as a means of adding a few rupees to the income of the household without incurring much outlay or involving much trouble, than as an earnest means of livelihood. Efforts have been made from time to time by European gentlemen to establish the industry on a firmer footing; but the success achieved has been very small.

"The silkworm is supposed not to be indigenous in the Panjáb, but to have been imported from Kashmír, or perhaps from Central Asia. However this may be, silkworm rearing has been carried on in a desultory way for many years in the low hills of Kángra, Hoshiárpur, and Gurdáspur. The Hazára district would, doubtless, be equally favourable, as climates of almost any degree of heat or cold can be found in the district. The experiment has, indeed, been tried with some success by Colonel Waterfield when Deputy Commissioner of Hazára, but it was dropped after he left the district, and has never taken root.

"The worms when first hatched require plenty of fresh young mulberry leaves. One of the difficulties that has to be dealt with at the very outset is in procuring such a supply. It is found that the country mulberry does not put forth young leaves till some 15 days after the eggs first disclose. To remedy this, the China and Philippine varieties of the mulberry, which come into leaf earlier, have been imported. It is found that the Philippine variety comes into leaf in the beginning of February; the China variety 15 days later, and the country variety at the beginning of March.

"These varieties have been planted in considerable quantities by the District Boards in Gurdáspur and Kángra, and by Messrs. Lister & Co., at Madhopur and Gurdáspur, and by the late Captain Bartlett, in Nurpur, in the Kángra district. The rearing of the silkworm requires considerable care and constant attention. Airy sheds, perfect cleanliness, fresh food, and a tolerably even temperature, are among the chief requirements, and it is owing to the want of these that so small a measure of success is obtained among Native rearers. The worms, unless kept perfectly clean, are liable to epidemics which carry them off in large numbers; excessive heat also developes in them a kind of apoplexy which is invariably fatal. It is probably owing to the excessive heat of Changa Manga that the attempt by the Forest Department to create a silk-rearing industry there has not proved a success. A temperate climate appears to be essential. The improvement in the character of the silk that can be effected by even an ordinary amount of care is well illustrated by the report made on some cocoons reared under considerable difficuties in the Kángra district, by Mr. James Montgomery, which were said to have struck

CULTIVATION in Panjab.

competent authorities as of a quality extraordinarily good to have come from India. This was in 1876; since then **Messrs. Lister & Co.** established their silk filature at Madhopur, in the Gurdáspur district. Considerable capital was put into the business; large rearing sheds were raised at Gurdáspur, and plantations of the various kinds of mulberries were made. The experiment at one time promised considerable success, the quality of silk produced was much superior to that ordinarily produced by the Native growers, and gained first prizes at the Nurpur and Pathánkot exhibitions, held yearly; notwithstanding this, it has now been closed. This fact, and the death of **Captain Bartlett**, who took much interest in the industry in the neighbouring district of Kángra, is a blow to sericulture in the Panjáb, from which it is not likely to recover.

"Sericulture on a large scale, requiring considerable capital and the use of machinery, does not so far seem to have proved a súccess. The only chance seems to be to try and improve the already existing cottage industry, by showing the people the advantage of cleanliness, of better ventilated sheds, and more constant attention. At present the rearing is almost entirely carried on by Kashmíri colonists, and notwithstanding the efforts made by Government to encourage the industry by giving prizes for the best cocoons exhibited annually at the Nurpur-Pathánkot Cocoon Exhibition, the industry seems to be flagging more and more every year.

"The species usually reared is supposed, as said above, to have come from Kashmír. Foreign eggs can, however, now be procured. **Mr. Coldstream**, in his excellent report, gives a history of the introduction of foreign varieties of the worm into the Gurdáspur district, about twelve years ago, by **Mr. F. Halsey**. Since then the French, Italian, Chinese and Japanese várieties have been cultivated by **Messrs. Lister & Co.** It is said that the French variety is most suited to the climate of the Panjáb, though some of the Tahsíldárs and others, who have made experiments, have given the preference to the Japan variety. There seems no doubt that foreign eggs require more care than country seed; the people, however, readily admit that the silk produced from the foreign species is both better in quality and more in quantity than that obtained from the country cocoon. The number of people employed, or who give a portion of their time to silk-rearing in the two districts of Kángra and Gurdáspur, is said to be 551; this is probably somewhat under the mark.

"The above remarks refer entirely to the silkworm proper. Experiments were made some years ago by **Mr. Coldstream**, as Deputy Commissioner, Hoshiárpur, in domesticating the *tasar* (**Anthereæa siwalika***). The report furnished by him in 1884 is printed as an appendix to this monograph. The result of his experiments seems to show that with patience considerable success in the production of Tasar silk is possible. **Messrs. Lister & Co.** are also believed to be making experiments, the result of which is not at present known.

"From the above account of the silk-rearing industry in the Panjáb, it will be obvious that a sufficient quantity is not produced to meet its requirements. The deficiency is made up by importation. The chief manufacturing centres are Amritsar, Lahore, Mooltan, and Jullundur. The silk grown in Kángra and Gurdáspur is, very little of it, wound in those districts; the cocoons are sent to the market and fetch about R1-4 a seer according to the figures given in the Kángra report; and a price varying from R250 to R40 per maund according to quality as given in the Amritsar report.

"The first process is the unwinding of the silk from the cocoon, and separating the different qualities of fibre, and winding them on different reels or *uras*. The cocoons are boiled in water, and then two or more fibres are taken, according to the fineness or coarseness of the strand required, and reeled on to a small wheel or spindle (*uri*); the silk is then made up into skeins. Imported silk is found to be so badly wound, as to require this process to be done over again before dyeing. The quality of the thread is tested by passing it through the finger and thumb, and the reelers, or *patpheras* † as they are called, become wonderfully adept in distinguishing the various qualities. A maund of dry cocoons will yield from 8 to 10 seers of fine silk, with a value of from R4 to R20 per seer according to quality, besides a certain amount of wastage, which is used in embroidery and for stringing ornaments, and is worth from R3 to R5 a seer.

† *Pat* is the common word for 'silk.'

"The raw silk that is imported into the Panjáb comes principally from three sources—(1) from Bengal, (2) from China by Bombay, (3) from Yarkand and Bokhára. The last is the most extensively imported, and is the kind chiefly used in

* A local variety at most of **A. paphia**. *Conf. with pp. 79, 237.—Ed., Dict. Econ. Prod.*

SILK: Mulberry.

Cultivation or Rearing of

CULTIVATION in Panjab.

Amritsar, the chief centre of the trade. It is imported in skeins, yellow or white, according to the natural colour of the silk, and in appearance is quite dull and glossless, full of dirt, and has apparently been steeped in some form of size or starch, giving it a coarse appearance like horse-hair.

"These skeins are handed over to the *patpheras* to clean and wind. The illustration accompanying the Amritsar report will show the process, and the machinery used. Both will be found described on page 58 of Powell's *Panjab Manufactures.* The skeins are first stretched on a *charkha,* and then wound off on to different *uras,* or *parailas,* according to the quality. Should the thread break, it is joined with the tongue. After dyeing the silk is frequently roughly wound again. As it passes through the winder's hands, impurities are separated, and a gloss is given to the silk.

"The total reported number of *patpheras* or winders in the Province is 2,755; this is probably an under-estimate. They earn on an average from 3 to 6 annas a day, getting from 8 annas to R2-8 per seer of silk wound according to the degree of fineness of the thread reeled. They are, as a rule, paid by the piece and not by the day. According to the Deputy Commissioner, Lahore, there are three qualities of thread, the *táni,* the *peta,* and the *kachar.* The *tani* is used for forming the strands of the warp, and frequently undergoes another process, that of twisting two threads into one at the hand of the *tauzi* before being woven. The Deputy Commissioner, Gurdáspur, gives five different kinds of thread, or rather threads of five different qualities, which are often wound off from the same skein. They are—(1) *Wana,* used for the woof; (2) *Tani;* (3) *Makhtul;* (4) *Khota;* (5) *Shish Mahal.* Nos. 3 and 5 are the coarsest. There are other local names in other districts, which denote rather difference of quality than difference of origin. For instance, the Deputy Commissioner, Ludhiana, gives the following:—(1) *Singal,* which sells at from 3½ to 5 tolas the rupee; (2) *Abu,* a coarser quality, sold at from R7 to R12 per seer, and is used in weaving the ordinary cloth; (3) *Pat,* used extensively in embroidery.

"By far the greater part of the silk manufactured in the Panjáb is imported from Bokhára, the best quality of which is known as *Wardan.*"

Mr. Cookson furnishes certain particulars of the silk manufactures of the Panjáb. From the impossibility of discovering particulars regarding all the Provinces of India the author has thought it the better course to practically omit from the present paper the Indian manufactures of silk. A considerable amount of information on that subject will, however, be found in the *Journal of Indian Art* and in Sir George Birdwood's work—*Industrial Arts of India.* The seats of the Panjáb manufactures are Amritsar, Jullundur, and Mooltan.

Mr. N. G. Mookerji furnished (*August 1890*) the following review of the position of sericulture in the Panjáb:—

"*History and Traditions.*—The history of sericulture in the Panjáb is wrapped up in still greater obscurity than that of Kashmír. The modern Native tradition traces the silk industry of the Province to Bokhára and Kashmír. A belief, however, is prevalent among the Native peasantry that the hill districts of the Panjáb are suitable for raising silk. It is more than probable that sericulture has been only revived in the Panjáb within the last 50 years—that the industry had existed in very ancient times, but had gradually decayed. The manufacturing industry is still considerable, although, before the introduction of the British rule, it used to be in a still more flourishing condition (*vide* Cookson's *Monograph on the Silk Industry of the Panjáb,* 1886-1887). That the Panjáb is a good field for sericulture will be apparent from the fact that even now 25 to 30 lakhs of rupees' worth of silk is manufactured in this Province, chiefly in the neighbourhood of Amritsar. Ancient traditions favour the idea that silkworm-rearing was practised in Northern India by mountain tribes. They are mentioned by Manu and in the *Mahabharata.* I am, in fact, inclined to believe in the ancient tradition that it was from Sirhind (accepting Captain Hutton's etymology *Sir-i-Hind,* or northern part of India, if that is necessary) that silk was first imported into the western countries of Asia and Europe. M. Boitard, a French writer on the subject, believes, that it was from Sirhind and not from China that the monks in the reign of the Emperor Justinian obtained the eggs of silkworms. The same writer, however, follows the popular belief that the silkworm and the mulberry tree were introduced into India as exotics, and gives this imaginary description of their introduction from China into the Gangetic plains,

CULTIVATION in Panjab.

which the writer erroneously believes to be particularly suitable for silk industry, overlooking the fact that all the natural advantages for sericulture exist in the Himálayas:—'The Chinese not finding a sufficient consumption of the article at home, sallied forth from behind the Great Wall in order to exchange their silk for the perfumes and spices of the Hindus. The superb climate of the Gangetic Provinces was far too favourable for the culture of the silkworms for the people to remain long indifferent to the advantages to be derived from the introduction of this new species of commerce, and accordingly, in spite of all the precautions of the Chinese, the mulberry tree and the silkworm were speedily acquired and introduced to increase the riches of the fertile plains of Hindustan.' That the Hindus carried on traffic in silk with the Persians long before the Christian era is allowed by this writer. Captain Hutton alludes to another incident in his notes on the silkworms of India (p. 7):—'The town of Turfan in Little Bokhária was for a long time the rendezvous of the caravans coming from the west, and the principal entrepôt for the silks from China.' [Captain Hutton here assumes that the ancient land of the Seres was the modern China.] 'This town was the capital of the Seres of Upper Asia, or of the Serica of Ptolemy. Driven from their country by the Huns, these people established themselves in Great Bokhária and in India, where they founded among other colonies that of Sirhind, in the ancient kingdom of Delhi, where they applied themselves assiduously to the rearing of silkworms.' The Chinas, Hunas, and Pamdrakas (the last word meaning silkworm-rearer) are mentioned in the Sanskrit literature as mountain tribes of the Himálayas who reared silkworms. The Hindu silkworm-rearing caste of Bengal is still known as Pundarikaksha, and there is no doubt that silk-rearing was a recognised indigenous industry in the Panjáb, chiefly in days gone by.

"*Recent Establishment of the Industry.*—Be that as it may, the present infant industry of the Panjáb is wholly the creation of the British rule. Civil, political, and military officers have, since the commencement of British influence in the Province (1836), taken a lively interest with a view to the introduction of this industry in suitable localities. These efforts have been described by Baden Powell, Geoghegan, and Liotard in their excellent reports. Following in the wake of British enterprise, silk-rearers from Kashmír came and settled near Gurdáspur, which is now the centre of the silk-rearing industry of the Panjáb. This has had great effect in introducing the industry among the villages. The industry has been decaying for a number of years on account of the silkworm plague, aided by the corrupt system of rearing in vogue, introduced no doubt from Kashmír. The number of silk-rearers has now dwindled down to only about 600, and they are confined to the districts of Kángra, Gurdáspur, and Hoshiárpur. Messrs. Lister & Co. have had to close their filature at Madhupur in consequence of this decay, and the cocoons reared in their own establishments at Gurdáspur, Gulpur, and Sujanpur are sent to the Dún for reeling. A great many of the village rearers still continue to do the rearing simply with a view to getting prizes in the Pathánkot Exhibition, which, I believe, will not be held from next year. But I look upon this decay as a temporary one, destined to revive shortly, as Messrs. Lister & Co. have now made arrangements for the supply of seed selected by Pasteur's system. The industry having been taken up by the people, it is bound to spread when once they get healthy seed. Instruction should be given to villagers how to do the rearing in a healthy manner, and they should be also told that it is bad economy rearing too much. One or two ounces of seed is ample for one rearer. Now the practice is in the Panjáb, as in Kashmír, to rear from several seers. But when the seed is good it stands a good deal of rough treatment, and, whether any special instructions are given or not, the industry must be considered as standing on a firm basis in the Panjáb, provided, of course, that villagers use healthy seed. That the industry has achieved such a footing within 50 years is not to be considered as a trifling matter when it is borne in mind how difficult it is to establish a new industry among the peasantry in a country like India. The attempts at the introduction of the industry were mostly of a spasmodic character, and they were made chiefly by Government officers whose hands must have been very full with their ordinary work. The success achieved, therefore, in the Panjáb is not by any means to be ignored.

"*Difficulties of Experiments.*—It will be well to describe here the principal difficulties experienced by the promoters of this industry, so that they may be avoided in future undertakings. These difficulties could have been all overcome, and the failure that took place in certain cases need not have taken place:—

"(1) Eggs in some of these experiments hatched before mulberry was out. This, for instance, took place this year in the Dún, where the young

CULTIVATION in Panjab.

worms were fed for ten days in January on lettuce leaf. If the experimenters knew how to retard or accelerate the hatching of eggs according to will, they would not nave experienced this difficulty.

"(2) Eggs went on hatching for a long time in nearly all these experiments. Even and complete hatching, however, can be always secured by adopting proper methods of hibernation and incubation, as illustrated in the present year's experiment in Kashmír.

"(3) Diseases spoilt many of the rearings; but if their nature had been understood, the worst of them, pebrine and muscardine, could have been avoided altogether, and the others kept within bounds.

"(4) No arrangement was made in most of the experiments to secure even temperature (about 75°F.) throughout the rearing, next to healthy seed this being the most essential condition for a successful rearing. It is cheap enough keeping the temperature of a room at about 75°F. in the cold weather, but it is almost impossible to keep a rearing-house at this temperature when the outside temperature in the shade varies from 90° to 110°F. Not hastening on the rearing when the temperature was 40°, 50°, 60°, or 73°F., it generally took two or three months for worms to spin cocoons, whereas they ought to do it in about 35 days. The consequence was that towards the termination of a rearing the heat became too intense for a successful crop. There is scarcely a place in India, however, where a temperature of 75°F. cannot be obtained by adopting simple methods of heating up the room for 35 to 40 days, whether it is called summer or winter at the particular locality. In one district this temperature may be easily secured in January, in another in March, while in a third in July, but it is very essential that the proper season should be chosen for each locality for the rearing.

The Mulberry. 1820 *Conf. with pp. 24, 37, 66.*

"(5) Some experimentalist found the indigenous mulberry more suitable than the China or Philippine mulberry; others found the latter more useful. Here, again, a knowledge of what the silkworms require at the first three stages and what at the last, would have saved Government and private parties importation of foreign species of mulberry which are in no way superior to the indigenous variety of the **Morus alba,** which grows abundantly in the Himálayan and Sub-Himálayan districts. The split-leaf kind of the indigenous mulberry is suitable for young worms and the thick whole-leaf kind for the worms in the last stage. Wherever silkworms are reared, there should be *three trees of the latter kind to one of the former kind,* if success is desired.

"(6) The experiments were too soon abandoned in some cases where the experimentalists themselves were just learning, by failures and successes, the proper conditions of rearing. A sustained effort on a small scale is much better than an experiment conducted on a large scale for two or three years ending in big failures.

"(7) Experiments were too often conducted with hired labour and found impracticable, being expensive. If they were conducted through *rayats,* this objection would not have appeared.

"*Commercial Speculations.*—The local consumption of silk in the Panjáb is so enormous, and the general result of all the experiments in rearing cocoons locally has been so satisfactory, that European capitalists have of late years appreciated the importance of raising the raw materials in the Panjáb. Of these recent commercial speculations **Mr. Halsey's** seems to have been the most successful. **Mr. Halsey** worked on sounder principles than his successors (**Messrs. Lister & Co.**) **Mr. Halsey** gave the rearers seed, secured for them trees belonging to Government on road and canal sides, made also small advances of money, and then bought their cocoons at reasonable prices and reeled them in a filature which he had erected. **Mr. Halsey** lived only for three years to carry on his experiments, which have been since taken in hand by **Messrs. Lister & Co.** During this time he purchased 250 to 400 maunds of green cocoons annually from the *rayats.*

"**Messrs. Lister & Co.** have now carried on sericultural enterprise in the Panjáb for ten years, but without attaining commercial success. The abandonment of **Mr. Halsey's** principle of keeping the distribution of seed and the reeling of cocoons in his own hands and getting cocoons reared by villagers working under a certain amount of supervision is, in my opinion, the cause of the failure.

Mulberry Silkworms in India. (*G. Watt.*)

SILK: Mulberry.

CULTIVATION in Panjab.

"Messrs. Lister & Co.'s *doings in the Panjáb.*—"The rearing in the Panjáb and in the North-Western Provinces is now conducted on similar lines, and as I had already spent a month at the Company's establishment in the Dún, I simply paid a visit to their establishment at Gurdáspur to see if there was any peculiarities to notice with regard to the enterprise in the Panjáb. These I detail below:—

"(1) The rearing-houses are situated at inconvenient distances for the supervision by one manager. They are separated from one another by distances of several miles, being located in three different towns, *viz.*, Gurdáspur, Gulpur, and Sujanpur.

"(2) There is a want of proper supply of mulberry close to the rearing-houses. There are plantations no doubt, but the planting has been done so thick that the leaf is not suitable for worms in the last and voracious stage. As a matter of fact the manager prefers using leaf from the trees growing on road and canal sides. But carting it from a distance favours the generation of flacherie in the nurseries.

"(3) There is no arrangement in the rearing-houses, either here or in the Dún, for keeping the rooms warm at the early stages of the worms nor cool at the last stage of the worms. Large rearing-houses, inequality of temperature, great heat at the last stage, and leaf coming heated in carts—all these mean unavoidable loss from flacherie. The inequality of temperature, however, is not so great in these Panjáb rearing-houses as in those in the Dún.

"(4) There is no necessity in the Panjáb to carry on rearing directly under the Company's supervision as there is in the Dún where the industry is unknown among Natives. It is enough to rear a small quantity of worms only for seed purposes, and this rearing will serve as an example to the Natives for rational treatment of the worms.

"*The present year's crop.*—This year fresh seed was imported for the rearing both in the Panjáb and in the Dún, as last year the crop suffered very much from pebrine. In the Panjáb 24 ounces of Smyrna seed and 26 ounces of Bengal seed were used. No disease was noticed in the crop from the Bengal seed. With regard to the Smyrna seed, it was remarked by the manager, in his letter dated the 5th June, that six ounces of it did not hatch, and that the last lot of worms died of flacherie. From the Bengal seed 6 maunds 24 seers of cocoons were obtained, and from the Smyrna seed 5 maunds and 7 seers. The weight was evidently taken after the cocoons were killed and dried, or else the result would be extremely poor. When I visited Gurdáspur the rearing of the Bengal had finished and had proved very satisfactory. The Smyrna worms were also spinning their cocoons and the whole lot looked very healthy. In searching the litter through and through, I discovered a few worms with flacherie. On enquiring of the manager, I was informed that cleaning was done only once during each month, and that if he had to do the cleaning every other day when the worms were small, and daily when they were large, he could not do the rearing at less than R20 per maund for labour alone, whereas his instructions were to spend only about R9 per maund for labour. My suggestion was that he could do the cleaning as cheaply if he introduced the nets I used in Bengal. The first cost would not be prohibitive for any but the poor *rayats*. I am not surprised, therefore, that since I came away from Gurdáspur flacherie did some damage to the crop, as the germ that causes this disease is the ordinary germ of fermentation of mulberry.

Microscopic Selection. 1821 *Conf. with pp. 14 44, 59, 62.*

"*Concluding remarks.*—In conclusion, let me repeat that Messrs. Lister & Co.'s efforts in the Panjáb should be confined to the distribution of cellular and industrial seed among villagers, encouraging the rearing industry among them, keeping up a few model rearing-houses for rearing cocoons for seed purposes only, and reeling the cocoons purchased from the villagers. The Company ought to be satisfied by this time that cocoon-rearing, except as a cottage industry, can never pay."

Kashmir. 1822 *Conf. with pp. 11, 18, 35.*

Kashmir.

Incidentally reference has been made above to the silk of Kashmír, and the author has repeatedly expressed his conviction that if India is ever likely to produce **Bombyx mori** silk in competition with the produce of France and Italy, it will be by the extension and improvement of sericulture of Kashmír more than by effort being expended with that object in view within the limits of British India. Mr. Liotard's *Memorandum* will be found to contain many details regarding Kashmír silk. Mr. N. G. Mookerji, the most recent writer on the subject of Kashmír silk (*Report of Tour of Inspection, August 1890*) furnishes much useful information an

SILK: Mulberry.

Cultivation or Rearing of

CULTIVATION in Kashmír.

offers many practical suggestions for future guidance. Space can only be afforded for one or two passages of his report, however, of which the following facts and theories regarding the history of sericulture in Kashmír may be given :—

"Before 1869 the silk industry of Kashmír had existed in the unorganised crude state in which it had probably existed for centuries, from the days when Bactrian silk was exported to Damascus and other centres of manufacture. There is little doubt that Kashmír formed a part of the ancient kingdom of Bactria before the Christian era, and that some of the raw silk that found its way to the west came from Kashmír also. Nothing, however, is known in Kashmír about the origin of its silk industry beyond the fact that it is very ancient, and that it is intimately connected with that of Bokhára, with which it has always had interchange of seed and silk.

"In 1869, under the orders of the Maharaja, Babu Nilambar Mukerji undertook the organisation of this industry on a systematic basis, and within eight years the industry was put in a most flourishing condition. The income of the State gradually rose to five lakhs of rupees in one year according to the report of Babu Rishibar Mukerji, the present Director of Kashmír Sericulture, the gross outturn, however, of the best year (1876) being about one lakh of rupees according to the table given by Mr. Liotard (page 54). I was very much struck with the number and magnificence of the rearing-houses and factories erected by the State at this time of the prosperity of the industry. The number of native rearers increased enormously, and the industry being a State monopoly, the whole was put under a centralized system of control. This was both the source of weakness and of strength of this industry. Under this centralized system bad seed resulted in general calamity, which would not have been possible if the industry had remained in the old unorganized condition (each rearer choosing his own seed). Pebrine was introduced into the country with the highly domesticated though superior foreign cocoons from Europe, China, and Japan. The artificial system of selection introduced by Pasteur to avoid pebrine became a necessity when the foreign breeds were introduced, and, owing to the absence of this system and the evil methods of rearing which were in practice, the source of seed supply became also the source of the silkworm plague, and the downfall of the Kashmír silk industry was, therefore, so precipitate and complete. The industry, which was put on a new basis in 1870, was nearly gone in 1878, and in 1879 it was virtually abandoned by the State. The foreign cocoons were no doubt richer in silk than the indigenous cocoons, but the latter had a native vitality of their own belonging to a country that was peculiarly adopted for sericulture, and the introduction of foreign cocoons without the system of grainage proved to be the ruin of the Native industry.

"In 1888, when the Government of Bengal had laid the foundation of resuscitating a decaying industry, the Kashmír Government also determined to do likewise, and in December of that year Babu Rishibar Mukerji, Chief Judge of Kashmír, and brother of Babu Nilambar, visited Berhampur with the object of enquiring how the sericulture of Kashmír could be revived. From his description of the plague it seemed to have been nothing else than pebrine; and as the system was centralised, I assured him that it would take but two seasons to revise the silk industry of Kashmír completely. The question of pebrine was as easy of solution in Kashmír as it was in Europe, and I assured Mr. Mukerji that the difficulties experienced in Bengal would not be experienced in Kashmír.

"A few words of explanation are here necessary to show how it is that Pasteur's system answers so well in countries where the European or annual silkworms are reared, and that it does not seem to answer so well in Bengal. In Bengal, nowhere in the silk districts is a longer rest than that of three months allowed, though a rest of seven or eight months is absolutely necessary to kill the germs of pebrine. This rest is naturally obtained where the silkworm eggs take about nine months to hatch, and if the seed used is free from the germs of pebrine (a condition which is easily secured by cellular selection), there is nothing to fear from the disease producing germs that might have abounded in the previous year. In Bengal, however, the germs of infection are always being renewed, and they are always present in their nascent and virulent condition. So in Bengal it is not a question of mere heredity, of which Pasteur's system is a complete selection, but of infection, which a gale or a bundle of leaf may bring into a rearing-house, where, notwithstanding the use of pure seed, loss from pebrine may take place.

Mulberry Silkworms in India. (*G. Watt.*)

SILK: Mulberry.

CULTIVATION in Kashmir.

"There was thus no infection to fight with in Kashmír, and Pasteur's method was as easily applicable there as in Europe. There were no disturbing elements, such as the fly pest, or unsuitable climate, or improper leaf, to throw one out of calculation, and the question of improving Kashmír silk simply meant that of using seed selected by Pasteur's method.

"To make sure that it was pebrine that was the cause of the sudden decline of the silk-trade in Kashmír, I requested Mr. Mukerji to send me a little seed on his return to Kashmír. This seed arrived in February 1889 and hatched well, but by the middle of March the worms nearly all died of pebrine. In my letter dated the 25th March 1889, which appeared in the *Englishman* dated the 29th March, I said about the rearing from the Kashmír seed: 'I have also finished the rearing of about 10 ounces of eggs sent from Kashmír, where the silkworm plague was most intense last year, and, as will no doubt be heard two months hence, will be more so this year.' In my letter dated the 16th April, which appeared in the *Englishman* of the 19th, I said: 'When the rearing in Kashmír this season will be over, he will find that my forecast regarding this crop will also come true. Here more than 95 per cent. of the worms will die before spinning, in spite of the careful arrangements undertaken this season by the Government of Kashmír in the rearing of the worms.'

"Mr. R. Mukerji, in his report of the rearing in May from the same seed, writes: 'In April 1889 I collected three seers of eggs from the rearers with the object of rearing them under my own supervision in the State nurseries, and for this purpose I selected three nurseries in three different places. My object in rearing them in the State nurseries was to do away with the objections regarding improper ventilation, light and other inconveniences usually found in the rearers' houses here, so that it might not be said that the worms died more for not being properly looked after and treated than from any disease. The worms in the above nurseries appeared to be in very satisfactory condition up to the third moulting, but, unfortunately, soon after it they showed signs of decay and a deplorable loss took place at the fourth stage.' The rearing was done with special care in Kashmír, as the authorities were warned of the failure, and the failure was as pronounced in villages as in the State nurseries.

"The collapse of the industry being now certain, the State agreed to try Pasteur's method the very next season, that is, the one that is just over. I wrote to Signor Susani of Italy to send 100 ounces of purely cellular seed and of the three best classes of cocoons, *viz.*, round, white Chinese, which was recently getting into such favour in Italy and France, and Italian white and yellow cocoons specially reared and selected by Signor Susani. I advised Signor Susani to send the seed by post and not as cargo, to pack it up in a particular manner, and to see that it left Europe before the cold weather set in, for, if once hibernation took place, the development of the embryo must go on unchecked until the hatching or death of the worms. How important these precautions were would be apparent from the fact that out of the three consignments of eggs received from Europe this year for the Kashmír experiments only this one consignment of 100 ounces arrived safely and gave perfect result, while of the other two the eggs in the one had hatched on the way and in the other got smothered and arrived dead.

"In fact, the success of the Kashmír experiments this year was wholly due to the 100 ounces of seed obtained from Susani's. The State also got a consignment of seed from Bokhára and another from China for this year's experiment. Both of these I found badly pebrinised, and I advised that they should be reared 20 or 30 miles away from Susani's seed and promised failure from these seeds. The China seed failed almost completely, 2 seers and 3 chattacks of cocoons being the result of a quantity of seed which was not measured, the eggs being attached to pieces of paper. From the Bokhára seed (4 seers and 12 chattacks) only 8 maunds 12 seers and 12 chattacks of cocoons were obtained when there ought to have been about 150 maunds, if similar result to what was obtained with the seed from Susani's had been obtained with the Bokhára seed also. From the 100 ounces (about 3 seers) of seed from Susani's 64 maunds and 26 seers of green cocoons were obtained; of this 1 seer 2 chattacks and 4 tolas of seed, or about 37½ ounces that was reared in the State nurseries gave 35 maunds 14 seers and 14 chattacks of cocoons, and the 1 seer 15 chattacks and 1 tola or about 62½ ounces reared by villagers gave 29 maunds 11 seers of cocoons. The seed used was the same, and yet the result obtained in the State nurseries is far superior both in quantity and in quality and equal to what is obtained in Europe.

SILK: Mulberry.

Cultivation or Rearing of

CULTIVATION in Kashmír.

"The following table gives a comparative view of the relation of the produce to the seed used in Kashmír and Jammu this year and in some other years:—

Place.	Year.	Quantity of eggs used.	Quantity of green cocoons produced.	Proportion of increase.	REMARKS.
		S. C. T.	M. S. C.		
Jammu	1873 .	6 8 0	42 20 0	1·261	
	1874 .	78 0 0	232 10 0	1·119	
	1875 .	128 0 0	258 0 0	1·80	
	1876 .	135 0 0	308 30 0	1·91	
Kashmír	1876 .	2,200 0 0	8,500 0 0	1·154	The figure given in Liotard's report for 1876 is 10,025½ maunds.
	1888 .	52 12 2½	8 37 13	1·7	
	1889 .	17 13 0	7 16 4	1·6	
	1890 .	3 0 0 (The total quantity reared in villages and in the State nurserries.)	64 26 0	1·860	For 1890 I have taken into consideration only the seed that came from Susani's, as I was not responsible for the importation of other seeds that failed almost completely.
	1890 .	1 2 4 (The quantity used in the State nurseries.)	35 14 14	1·1204	

The improvements introduced in the system of rearing in the State nurseries this year will be obvious from the above table. What the defects of the Kashmír system are, and what improvements I found successfully introduced by the Silk Superintendent, a man who was trained for over three years at Berhampur, will be found in my preliminary report, of which I enclose a copy. The introduction of the clearing nets, of chopping up of leaf, and regulation of temperature day and night throughout the rearing season at 75°F., were the principal improvements. The method of hibernation adopted was also very exact, and the method of incubation was not the ordinary European method, but that invented recently by Signor Chrici, and which was described in the *Balletin des Soies et des Soieries* dated the 1st June 1889. The method is much simpler than the ordinary European method, but quite as effectual. Instead of raising the temperature gradually from 32°F. to 75°F., the eggs as soon as they were brought out of the refrigerator were put in a room which was kept uniformly day and night at the temperature of 75°F. Daily to raise ¼° or ½° is a long and a tedious process, and the hatching under Signor Chrici's method was found to be as complete and uniform as wit, the ordinary European method. This is an excellent confirmation of the view put forth by Chrici, and there should be no hesitation in future experiments to adopt this as the proper method of incubation instead of the one still practised in Europe.

"I stayed only thirteen days in Kashmír—seven days being spent in villages and six at Srinagar. The arrangements for the rearing were so good that I assured the Resident there was no use my staying there to see to the end of the crop, and that the rearing was perfectly safe in the hands of the Superintendent and controlled by Rishibar Babu. The results have, happily, borne out the pormises of success."

BOMBAY & SIND. 1823

IV.—BOMBAY & SIND.

The various attempts which have been made to introduce the rearing of mulberry silk in Western India may be said to have established the hopelessness of the enterprise. The reader should consult Section II. of

CULTIVATION in Bombay & Sind.

Mr. Geoghegan's work, and pp. 56-58 of Mr. Liotard's *Memorandum*. It may at once be admitted that the chief Bombay silk interests are connected with the power loom mills which run almost exclusively for the Burma market. In the review of the trade below will be found particulars regarding the consumption of silk in the Bombay City Mills and by the local manufactures of the Presidency.

The Gazetteers of Bombay contain numerous articles on the silk manufactures, but Thana, Nasik, and Poona are the most important. The following references to instructive articles on Bombay Silk may be consulted:—*Surat, II., 179; Kaira, III., 54; Ahmedabad, IV., 135; Cutch, V., 126; Kolaba, XI., 131; Khandesh, XII., 186; Thana, XIII., Pt. I., 379, 384; Nasik, XVI., 155; Dharwar, XXII., 375; Poona, XVIII., 65, 185; Bijapur, XXIII., 371.*

Bombay supply. 1824 *Conf. with pp. 30, 205, 209, 210-13.*

Though silk culture has made less progress in Bombay than perhaps in any other Province, Western India can boast of having given the subject a more searching and patient trial than has been the case with the provinces of India generally. So early as 1838 **Signor G. Mutti** was Superintendent of Silk Culture in the Deccan. In his *Guide to Silk Culture*, **Mutti** furnishes much practical information of great value. That work will be found in the *Transactions of the Agri.-Horticultural Society of India*, Vol. VI., pp. 149-197. **Tavernier** devotes a special chapter of his highly instructive book of travels in India (published in the seventeenth century) to the Bombay silk manufactures. Speaking of the rearing of silk in "Kásimbazar" in Bengal he remarks of the surplus stock that, after supplying the Dutch merchants and those of "Tartary," "all these silks are brought to the kingdom of Gujarát, and the greater part come to Ahmedabad and Surat where they are woven into fabrics."

V.—MADRAS.

MADRAS. 1825

It may be said that **Mr. Geoghegan's** report on the experiments conducted in Southern India and Mysore, with a view to introducing mulberry silk culture, fully exhausts the subject. **Mr. Liotard** gives the following further particulars:—

"The eggs procured from **Messrs. Lister & Co.** and those from Kashmír and China were distributed by the Madras Government for experiment to the Superintendent of the Government Farm at Sydapet, to the Superintendent of the Jails at Coimbatore and Rajamundry, and to **Dr. Shortt** at Yarkand in Salem.

"The report of the trial at Sydapet does not make any mention of the quantity of eggs tried, or of the date of hatching, mode of rearing, or quantity and quality of cocoons produced. The trial was made some time in the month of February 1883. The only information given is the following:—

"In connection with **Lister & Co.'s** eggs, it is said that 'the results seem to show that it is possible here in the colder months to rear silkworms;' but that very grave doubts exist 'whether any satisfactory commercial results can be obtained in conducting sericultural operations under such adverse circumstances as must be encountered in this vicinity.' The average maximum temperature during the trial was 88·3°. The cocoons obtained averaged 1,800 to a ℔, and ½oz. eggs were produced which have been kept for further trials.

"In connection with the Kashmír and China eggs, it is stated that the whole of the worms died from the effects of the high temperature, which ranged during the trials between 93·2° and 94·8°; and that possibly the condition in which the eggs arrived, combined with the fact that the eggs were from climates differing widely from that of Madras, may also have contributed to the unfavourable result.

"On the question of the food-supply for the worms, it is said that there was 'a number of healthy acclimatised bushes,' and that an abundance of mulberry leaves enabled the worms to be well fed during the whole experiment.

The Superintendent of the Coimbatore Jail reports more details, and the following is a summary thereof:—

"'**Lister & Co's** eggs: 2 ozs. received on 23rd January. Many of the eggs had hatched. Five thousand two hundred and thirty worms obtained in all, none died;

SILK: Mulberry.

Cultivation or Rearing of

CULTIVATION in Madras.

cocoons finished by the 27th February; the food was of course mulberry leaves, but the variety used is not mentioned. Temperature of the room during the rearing ranged from 75° in the morning to 85° in the afternoon. Five thousand one hundred and thirty of the cocoons were reeled off, giving 10½ ozs. of silk; the remaining 100 cocoons were set aside for breeding. The first cocoon opened on the 5th March, and the laying of the eggs was completed by the 8th idem. The eggs of this brood have been kept for the continuance of the experiment. The eggs of one of the moths were found to be 30.

"'Kashmír eggs: 3 ozs. received on 2nd March. Hatched between 4th and 10th idem. Variety of the mulberry leaf used is not stated, but the supply was ample. Temperature of the rearing-room was from 80° to 90°. The worms lived on through the different stages of their growth till they commenced to spin on the 1st April, when an epidemic broke out amongst them and large numbers died daily, the number of deaths decreasing as the number of cocoons increased. One peculiarity in the disease was that the worm turned quite black after its death. By the 11th April the last cocoon was formed. During the eleven days 3,918 worms died and 7,944 cocoons were spun. Many of the worms died in the act of spinning, and numbers of the chrysalis died in their cocoons. Fifty-one empty cocoons weighed 1 oz. Three hundred cocoons were retained for breeding and 7,644 cocoons were reeled off, yielding 1℔ 2 oz. of silk, which was of two tints—a creamy white and a pale yellow. The breeding cocoons began to open on the 19th April and ceased on the 26th idem: seventy-eight only opened, and the chrysalis of the remainder was found dead. Of the 78 moths, 24 females only copulated with the males, some having died before the males were available. The 24 females laid ½ oz. eggs, which have been preserved for further trial.

"'China eggs: 'a quantity' received on 2nd March. On the 7th idem a few of the eggs hatched out, and by the 1st Aptil 36 commenced to spin. By the 9th idem 90 cocoons formed. The cocoons obtained were very poor and of a white tint. They all opened by the 18th idem (with the exception of 4). Only 25 female moths, however, eventually laid eggs, as several of the females died before the males emerged from their cocoons. One hundred and fifty-one empty cocoons weighed 1 oz. On 27th April a few of the new brood eggs hatched, being multivoltine, and by the 3rd May 168 worms were out. These worms were doing well when the report was written. The remainder of the new brood eggs seemed good, and not having hatched, lead to the inference that they are univoltine.

"In the *Rajamundry* Jail the majority of the egg received were found to have hatched on the way and the worms had died. From the residue, 49 worms were obtained between the 7th and 13th February; these did well, and between the 26th idem and 6th March they finished spinning their cocoons. Most of the cocoons produced were retained for reproduction, and 34 moths appeared between the 8th and 17th March, and gave over 1,500 eggs. Eight days after, the fresh breed of worms began to appear.

From *Salem* Dr. Shortt's report furnishes the following particulars: Lister & Co.'s *eggs* were received on the 23rd February; a portion had hatched on the way; the remainder was successfully reared and spun their cocoons about the 21st March. The cocoons were 1,080 in number, and were small and oblong; the majority was of a creamy white colour, 75 were pinkish and a few were yellowish; 200 of the cocoons were reserved for reproduction, and the rest were steamed for reeling. From the former, moths issued in due course and eggs were produced which, at the time the report was written (12th July), had not hatched out. The *China eggs* were received in March, and are said not to have been fertilised: a few worms hatched, but died almost immediately. The *Kashmír eggs* hatched we ll, the worms grew healthy and strong, till, within a few days of the spinning, 'they died off in two days in the most mysterious manner.'"

In the reports issued by the Director of Land Records and Agriculture frequent mention is made of the subject of silk, more especially *tasar*. The report for 1883-84, however, contains the details of an extensive experiment (pp. 54-59) in which the failure experienced may be regarded as due to starvation of the worms. A considerably greater interest may be said to exist in mulberry silk in Mysore than in any other part of South India. Very little progress has, however, been made since the time of Dr. Buchanan-Hamilton's visit to that province: his report should be consulted.

CENTRAL PROVINCES. 1826

VI.—CENTRAL PROVINCES.

These Provinces, along with Chutia Nagpur and the adjoining portions

CULTIVATION in Central Provinces.

of Orissa and Madras, may be said to constitute the *tasar* area of India. In local reports, therefore, which speak of silk without specialising the class of silk meant, it may safely be admitted that *tasar* silk is being dealt with. Some few years ago, however, **Mr. J. B. Fuller** obtained seed of mulberry silk from **Messrs. Lister & Co.** **Mr. Fuller** reported that the climate of the Satpura District most nearly resembled that of Dehra Dún, and that the experiment appeared to have been fairly successful at Chhindwara. **Mr. Fuller's** report will be found in **Mr. Liotard's** *Memorandum*.

VII.—ASSAM.

ASSAM. 1827

While mulberry silk culture is pursued here and there to a limited extent in some of the temperate tracts of this Province, as, for example, in Manipur, it may safely be said that Assam silk consists of *eri* or *muga*, two of the so-called wild forms. Some 200 years ago Tavernier wrote of the Assam "wild silks" that "the stuffs made of them were very brilliant." He speaks of the cocoons as being round and remaining on the trees for a whole year. Perhaps his reference was to *tasar* silk and confused Assam with Bengal generally. In an article which has recently appeared in the *Nineteenth Century*, **Sir J. Johnston** informs us that the tradition prevails in Manipur that silk culture and manufacture was introduced into that little State from China. The author had the opportunity, while on a visit to Manipur, to witness the system followed of rearing the worms and reeling the cocoons. The worms were to a large extent allowed to run wild on a scrub of mulberry bushes, and absolutely no care was bestowed on them. Yet the silk obtained was of superior quality and the manufactures highly creditable. The people were, however, so superstitious on the subject that it was impossible to learn more than the most ordinary facts. Judging from the rampant growth of the mulberry bushes and the prevailing climatic features of the State, Manipur, next to Kashmír, would appear to offer the best prospects of a future extension of sericulture in India. Labour could be had plentifully, and large expanses of rich land, perfectly level, would be available, which for centuries have not been cultivated, and which bear a wild vegetation that in many respects resembles that of China and Japan. The average height of the northern portions of the valley is about 3,000 feet, but much land could be got at even 5,000 feet, in which the humidity and temperature closely resemble that of France or Italy. Perhaps no better country exists for the oak-feeding **Antheræa pernyi** than Manipur, so that both "Chinese *tasar*" and mulberry worms might be reared. Manipur might, in fact, be described as a land of oaks, and in many respects it possesses the characteristic features of Shantung, the home of **Antheræa pernyi,** which might be characterised as the best of all the so-called wild silkworms.

Manipur. 1828

Conf. with pp. 15, 35, 142.

VIII.—BURMA.

BURMA. 1829

Mr. Geoghegan informs us that at the time he wrote, sericulture was pursued mainly in Prome, Thayetmyo, Henzada, Toungu, and the northern portion of Rangoon, by a class of people who lived in villages by themselves. The silk produced was, however, of an inferior kind, being derived from the insect already dealt with under the name of **Bombyx arracanensis.** **Mr. Liotard** gives us the following particulars regarding a sample of Burmese silk:—

"From samples of raw silk of the Prome District received by the Government of India in the early part of 1882, there was certainly no improvement apparent. The samples were transmitted to the Secretary of State, and were reported upon by **Messrs. Durant & Co.,** silk brokers of London, thus:—

"We are not able to give you a very encouraging report on the samples: (1) the thread is very imperfect, being uneven, gouty, and knobby; (2) the length of reel

SILK: Mulberry.

Properties & Uses

CULTIVATION in Burma.

(*i.e.*, the circumference of the skein) is too large for any machinery now in use in this country; and if such did exist, it is doubtful if any employment could be found for silk so defective in all respects. There being no market here for such silk, it is difficult to give anything like an accurate quotation of the value; but if a market could be found in Marseilles, where there is an outlet for 'dappioni' of Italy and France, with which it might compete (at a distance), the value would be about seven shillings per lb.

"'So far as we can judge, we are inclined to think favourably of the nature and quality of the cocoon; and if technical skill was brought to bear so that the temperature of the water, the laying of the thread on the reel, and some minor points in the manipulation of the cocoons could be accurately determined, the silk produced from such cocoons as the samples indicate would, we think, compete successfully with that of Bengal, as also with that produced in some of the north-west provinces of China.'"

USES. 1830

PROPERTIES AND USES OF BENGAL SILK.

Incidental references have been made to this subject here and there throughout the foregoing pages, but it may serve a useful purpose to bring together into one place the main ideas. The following passage from Mr. N. G. Mookerji's 'Note on the Decline of the Silk Trade in Bengal' (*Published October 1888*, may be accepted as conveying the chief facts of importance:—

"But it is very doubtful whether Bengal silk of the present quality will be used at all in Europe in future, except for certain special purposes. I do not deny Bengal silk has some special qualities, for which it will probably always command a small amount of demand. It is the best silk for gloss, elasticity, and for taking the black dye, and no other silk will make better Sunday hats than Bengal. But taking it all in all, it is the worst silk in the market, inferior to European, Japan, and China silks. When these silks sell at 45 francs per kilo., Bengal silk sells at 32 francs. The reason for this is, as Mr. Wardle pointed out before, that Bengal silk is the worst reeled silk in the market, and it will be impossible ever to reel Bengal cocoons to produce silk like the European, Chinese or Japanese silk, however the reeling machines may be improved. A European or Chinese or Japanese cocoon of a good class contains thread four times longer than a Bengal cocoon, the former containing 800 to 1,000 metres of reelable *bave*, the latter only 200 to 250 metres. This must make our silk at least four times more uneven than the other silks, whatever reeling machines may be employed. It is necessary to add Bengal cocoons at the time of reeling at least four times as often. I say *at least*, because the *bave* of the Bengal silk is the weakest of all and thus very liable to break. Another cause of the unevenness of Bengal silk, due to the character of the cocoons, is the lightness of the chrysalid causing the *telette* to jump up and cause *duvets* or *paqeuts*, whereas the chrysalid of the **Bombyx mori** cocoons by its own weight helps to break the thread whenever it becomes very fine towards the *telette*. The effect of unevenness is enhanced by the practice of trying to hide this unevenness prevalent in Bengal filatures. When the skein is taken from the reel, it looks more uneven than when it is packed up. Picking and clipping help to make it look more even, but this makes the continuity of the *grege* to be severed in several places. Skeins, in their rough state, are really better than skeins prepared in this manner; but the former look so dirty that in the European market the latter fetches a higher price. This is the excuse for the practice of picking the skeins prevalent in Bengal filatures. When the Bengal grege is made into organzine and trame, it breaks so continually, that one girl can manage only 6 to 30 skeins at a time during the process of milling, whereas 100 skeins of European or superior Japanese and Chinese grege can work at the same time under the supervision of one girl. This makes it so costly to turn the Bengal silk into account, and where labour is dear as in America, they will have nothing to do with Bengal. In France and Italy also wages of labourers have been steadily increasing. This I believe to be another cause, besides the diminished production of cocoons in Bengal, of the fall of the Bengal silk trade. With increase of production of superior cocoons in China and Japan, and with the enhancement of the manufacturing industries of silk in Europe, a still further diminution in the price of silk may be reasonably expected. On the other hand, of course, an increased production of cocoons in Bengal, due to the introduction of Pasteur's system, will tend to lower the price at which Bengal grege can be produced, by giving full work to filatures. But the strain will continue just the same and even with the introduction of Pasteur's system, if that alone is done and nothing else, no permanent amelioration can be expected."

of Bengal Silk. (*G. Watt.*)

SILK: Mulberry.

USES OF BENGAL SILK.

Defects of Bengal Silk.
1831
Conf. with pp. 62, 184, 185, 186, 187, 188, 190, 191.

The East India Company Paid £1,000,000 for their Experiments to improve the Bengal Silk.
1832
Conf. with pp. 13, 184-186, 190.

The great defect of Native-reeled silk is the want of thread uniformity. The fibre in being drawn from the cocoon every now and then passes off in loops or folds. This constitutes the *duvet* of French writers, which is removed or minimised by the *croisure* effected by the superior and modern reeling apparatus. That is to say, a certain strain is given to the *bave* (or individual cocoon thread) by passing it round carefully-adjusted pulleys and by the *bave* being several times crossed around itself. The process of *croisette* causes the *baves* of two or more cocoons to lie together into a uniform thread. All these and many other delicate adjustments have, in Native reeling, to be accomplished by the hand, and added to the imperfections unavoidable in crude, as compared with refined methods, the cocoon possesses intrinsic defects that increase the inferiority of Native reeling. As mentioned by **Mr. Mookerji** the shortness of the *bave* is a serious drawback, but the cocoon has a greater amount of gum, *grés*, than the European one, and it is withal lighter, smaller, and softer, defects which increase the amount of waste and make the *bave* shorter than it need be. The presence of "knibs," "slubs," etc., lower greatly the value of Bengal silk. It is characterised as "foul," but when reeled at the European filatures all these defects are combated, more or less successfully, and an article is produced that commands a good market, and which will always hold its own for certain purposes. The discovery of methods of utilising waste silk can, however, be seen to naturally operate unfavourably on the inferior Native-reeled silk. Instances are even on record where a manufacturer has found it, within recent years, desirable to cut, not only Bengal mulberry silk, but native reeled *tasar*, into short lengths, and to card and spin these as if they had been "waste" instead of reeled silk.

Bleaching Bengal Silk.
1833
Conf. with pp. 25, 153, 156.

Some 200 years ago **Tavernier** made a somewhat interesting observation regarding the silk of Kásimbázar. "As for crude silks," he says, "it should be remarked that none of them are naturally white except that of Palestine, of which even the merchants of Aleppo and Tripoli have difficulty in obtaining a small quantity. The silk of Kásimbázar is yellow, as are all the crude silks which come from Persia and Sicily. But the people of Kásimbázar know how to whiten them with a lye made of the ashes of a tree which is called **Adam's** fig (= the plantain), which makes it as white as the silk of Palestine."

DISEASES.
1834

DISEASES & PESTS.

In the *Indian Museum Notes* (*Vol. I., 133—135*) the following brief account is given of the **Diseases** to which the Mulberry and several of the other silkworms are liable :—

"The Mulberry silkworms of the temperate zone have long been known to suffer from a number of diseases. These were studied exhaustively by **Pasteur** between the years 1865 and 1870, and the remedial measures which he recommended have since been widely adopted in Europe, where they have proved the salvation of the silk industry. In Bengal, mulberry silkworms suffer from similar diseases, which have been found by **Wood-Mason** and **Mookerji** to be identical with the diseases known in Europe. Within the last few years accordingly experiments have been carried on in Berhampore, with the support of the Government, with a view to introducing **Pasteur's** system into India. **Mr. N. G. Mookerji**, who conducted the investigation, has found some practical difficulties in the way of applying to the multivoltine insects of Bengal, the remedies that were devised for the univoltine silkworms of Europe; there would seem, however, to be considerable reason to hope that **Pasteur's** methods may ultimately be adapted to the requirements of the Bengal silk industry."

The following summary is given by Cotes (*Museum Notes*) of the chief diseases to which the domesticated worm is liable—the original will

DISEASES.

be found in Riley's Manual, U.-S. Department of Agriculture, Bull No. 9 (1886):—

1835

a. Pebrine,* which is known in Bengal as *kata*, or when in an aggravated form, as *tali*, is characterised by the presence of microscopic corpuscles† of oval shape which are found in the tissues of the silkworm, and also in the pupa, moth, and eggs. The disease is not always fatal, but when it does not kill the worm it damages the quality of the cocoon. Besides being contagious, the spores preserving their powers of communicating the disease for considerable periods, it is also hereditary, the eggs laid by a pebrinized female tending to produce pebrinized worms. The remedy therefore consists in general sanitary precautions to prevent infection, and in breeding only from eggs laid by such females as are found, on miscroscopical examination, after they have laid their eggs, to be free from corpuscles; the eggs laid by moths which prove to be pebrinized being carefully rejected.

1836

"*b. Flacherie*, which is known in Bengal as *kala shira*, is characterised by the presence of a chain ferment ‡ in the digestive tract of the silkworm and pupa; the disease is contagious, and to a certain extent hereditary in that the larvæ of moths which show symptoms of flacherie have a predisposition to take the disease. The remedy therefore consists in general sanitary precautions to prevent infection, and in the rejection, for breeding purposes, of all eggs obtained from batches of cocoons, which, on microscopical examination of the digestive tracts in a percentage of the pupæ, show signs of the chain ferment.

1837

"*c. Muscardine*, which is known in Bengal as *chuna*, is caused by a fungus, **Botrytis bassiana**, which appears as a white efflorescence on the body of the worm some hours after it has died of the disease. The disease is contagious, but not hereditary, for though the worm may be so slightly affected that it is able to spin, it invariably dies before it becomes a moth, while healthy pupæ. being protected by their cocoons, are not liable to be affected. The disease is spread by the spores, which only appear several hours after the death of the diseased worm; in the speedy removal therefore of all dead worms from the breeding trays, is found an efficient preventive.

1838

"*d. Grasserie*, which is probably the same as the disease that is known in Bengal as *rasa*, is of but little importance, and is never hereditary. Little seems to be known about it.

"Besides being subject to the above diseases silkworms suffer from the attacks of various parasitic and other enemies. In Bengal considerable loss is occasioned by the Tachinid fly§ **Trycolyga bombycis**, which lays its eggs upon the body of the worm. It grubs, on emerging from the eggs, bore into the tissues of the worm, and remain there until they are full grown; they then cut their way out and betake themselves to the ground, where they pupate. When attacked by this pest the caterpillar generally continues to live and feed as before, and in some cases spins a cocoon, but it invariably perishes when the grubs cut their way out. The cocoon made by a parasitised worm is generally a poor one to begin with, and is rendered unfit for reeling by the hole made by the grub in escaping from it. This pest can be to a great extent kept under by the removal of all rubbish in which the grubs pupate, the speedy suffocation of all cocoons that are known to harbour fly, the establishment of well marked intervals of time between the *bunds*, so that the flies that are bred with the worms of

---

* Mookerji reported in January 1888 that, while *Flacherie*, *Grasserie*, and *Muscardine* have always been known in Bengal, *Pebrine* has only appeared within the last ten or twelve years, becoming each year more destructive, and causing fears of a total collapse of the silk trade.

† These are the corpuscles of **Panhistophyton ovatum** (**Nosema bombycis**, *Micrococcus ovatus corpuscles du ver à soie*), shining oval cocci 2-3 μ. long, 2 μ. wide occurring singly and in pairs, or masses, or in rods 2·5 μ. thick and 5 μ. long. They multiply by subdivision. They have been experimentally proved to be the cause of *pebrine, gattine, maladie des corpuscles*, or *flecksucht*; and are found in the organs of diseased silkworms, as well as in the pupæ, moths, and eggs.—(*Crookshank's Introduction to practical Bacteriology*: London, 1887.

‡ This ferment is **Streptococcus bombycis** of Bechamp (**Microzyma bombycis**); oval cocci, ·5 μ. in diameter, occurring singly or in pairs or chains. They are found in the contents of the alimentary canal, and in the gastric juice of silkworms suffering from *flacherie* (*maladie de morts blance, flaccidezza*, or *schlaffsucht*).—(Crookshank's *Introduction to Practical Bacteriology*; London 1886.

§ For a detailed account of this pest see *Indian Museum Notes*, Vol. I., No. 2. (Also reprinted below, pp. 62—65. *Ed., Dict. Econ. Prod.*)

DISEASES.

one *bund* die out before the next bund commences, and by precautions to prevent the propagation of the fly in the intermediate generations of silkworms that are reared for the production of seed. A somewhat similar parasite attacks the mulberry silkworms of China, while in Japan the *oudji* fly (**Udschymia sericaria***), which, however, has a somewhat different life-history, affects silkworms much in the same way. Some loss is also occasioned in silk-rearing establishments by rats, mice and ants, also by **Dermestes** and **Anthrenus** beetles, particularly by **Dermestes vulpinus**, which penetrates the cocoons, and thus renders them unfit for reeling. The damage however that is done by these pests is generally of only secondary importance."

It may briefly be said that in no essential feature are the mulberry worms of India different (as far as liability to disease is concerned) from those of Europe, America, or the Colonies. The reader who may, therefore, desire detailed information on the subject should find no difficulty in procuring a fairly extensive library of books and reports that treat on the diseases of the worm.

Decline of Bengal Silk Industry. 1869 *Conf. with pp. 32, 62, 75, 106, 118, 121, 148, 156, 158, 187, 193, 197, 201.*

The Calcutta Silk Committee, in one of its communications to Government, affirm that there seems to be "some misapprehension as to the causes of the decline in the silk industry of Bengal, the true solution of which was apparently set forth by **Mr. Reynolds**, *i.e.*, that the whole question hinges upon the supply of cocoons." Thus the Committee appears to view with greater alarm a falling off in the annual yield than in any supposed or real decline in quality. Disease and the fly pest are the chief causes of the diminished yield and the Committee is, therefore, fully alive to the desirability of combating these evils. "The Committee," writes **Sir A. Wilson**, "have watched with interest the experiments which have been made from time to time in introducing foreign breeds of silkworms, none of which have resulted in success, and they are sufficiently alive not only to their own interests, but to 'the benefit of their fellow-subjects in India,' and also of the silk industry in England, to take full advantage of any improvement which may be discovered by those working in this direction." "It may be interesting," he adds, "to the members of the Silk Association of Great Britain and Ireland to learn that with eradication of disease considerable improvement in the yielding capabilities of the cocoons has been already obtained." But it was subsequently found by no means an easy matter to carry out in Bengal the microscopic system of selection of healthy stock from which to distribute seed. Indeed, many of the experts hold that the selective method had proved a complete failure. **Mr. Mookerji** at the same time, in contending with the perplexing difficulties with which he found himself beset, changed his plans and scene of operations (or desired to do so), and thus laid himself open to the charge of vacillation. **Pasteur's** system of dealing with the disease was even characterised as having given origin in Europe to an intermediary class of persons who constituted a drain on sericulture. That the past thirty years of microscopic selection had not sufficed to eradicate the disease from Europe, nor to lessen the necessity for microscopic selectors of seed. It was pointed out that results nearly as bad as those normally attained by the *rayats* were got from the second generation of selected seed, so that a continuous selection and a constant supply of fresh seed became a necessity, and in a country of poor and uneducated people such as Bengal this was viewed as an impossibility. **Mr. Mookerji** recommended the formation of a laboratory in which the improvement of the breed of silkworms, as also the estimation of the value of silk, etc., might be conjoined with that of the endeavour to eradicate the disease. But the Committee disapproved; they desired the one object alone to be dealt with, for the present at least, namely, the investigation of the prevention of disease and pest. The Committee accordingly wrote that it could not "accept **Mr. Mookerji's** plan as being at all reasonable, or believe

Microscopic Selection. 1840 *Conf. with pp. 14, 49.*

* Ugimyia sericariæ. *Ed., Dict. Econ. Prod.*

SILK: Mulberry.

Diseases and Pests to which

DISEASES.

that any good can result from his abandoning the campaign against pebrine, to do nothing at Seebpore but educate a few *kahans* of cocoons. They believe that if the true cause of failure were sought out, that success must be attained. Having consulted **Mr. Wood-Mason** and **Mr. Blechynden**, they are advised that no progress can be made so long as Natives are not compelled to cremate all their diseased worms, *debris* of leaf, and excreta of the worms. These items are now, they understand, universally used as manure for the mulberry fields, so that all leaf in these districts must be more or less infected with pebrine spores. However pure the seed cocoons may be, that **Mr. Mookerji** can supply, the feeding with tainted leaf is certain to cause the infection of the worms in the rearers' hands."

Cremation of Diseased Worms and of *Debris* of Leaves, &c. 1841

Below will be found a complete review of the facts which have been brought to light regarding the silkworm fly pest. In this place it need only be remarked that many writers hold that fly is in India even more destructive than pebrine. Intimately connected with the question of the eradication of disease or the prevention of the fly pest is the question of whether the future of sericulture would progress more rapidly by the capitalists—the owners of filatures—taking over the cultivation of the mulberry and the rearing of their own cocoons. **Mr. N. G. Mookerji** has advocated that sericulture in India depends upon perfecting it as a peasant enterprise. He says, "it is impossible for foreign capitalists, with hired Native labour, to succeed in this industry." Why it should be so is difficult to see more than that each villager in Assam should grow a few tea bushes and produce tea much after the same fashion as pursued in China, and thus supplant "foreign capital and hired labour." The two industries admit of an exactly parallel comparison. That tea does pay under European management and pays better than as a peasant industry, may perhaps be admitted as proved through the greater success of India over China or by the higher price obtained for the Company-prepared indigo of Bengal as compared with the peasant produce of Madras. It would probably not pay a capitalist to enter on rice cultivation, on a large scale: it pays the peasant to grow not only rice but indigo, silk, sugar, or anything else, so long as he brings intelligence and a requisite amount of capital to bear on his undertakings. The point at which large capital finds it profitable to supersede peasant enterprise may be said to be when the article produced is of high value, thus admitting of liberal profits through luxurious perfections; or when expensive appliances or the removal or mitigation of the liability to disease, lower the cost of production or improve the quality, and thus produce results not attainable by the peasant or artisan. In few industries could greater and more urgently needed improvements be discovered than in that of sericulture. That the peasantry of India can and do produce silk goes without saying; but a century of education, of village to village lecturing, on the part of Government experts, and of prizes at local exhibitions will effect no appreciable reform. Ignorance and deep-seated inimical religious traditions have to be conquered and the social habits of the people raised to a far higher platform than they have as yet attained. In a discussion of the possibilities of improvement, it proves nothing that they do produce silk. Their best endeavours, when actuated by the prospects of reward or enhanced returns, fall far short of the possibilities. And it is not the fault of the *rayats* or peasants of India that they are unable to give the necessary ventilation, cleanliness, and food to their worms. They do their best with their limited resources. The fault is poverty. But so long as such a state of affairs prevails capital and capital alone can effect improvement. Whether improvement would be remunerative is quite another matter. **Mr. Lister** tried the encouragement of the village system in the Panjab. He admits that he failed signally. In spite of the rewards paid, in

Sericulture—a village Industry. *Conf. with p. 49.* 1842

Reforms Difficult. 1843

DISEASES.

Conf. with pp. 12, 17-18, 35-36, 61 62, 215.

spite of the patient efforts to instruct the cultivators, in spite of the expenditure of large sums in the supply of good seed, disease spread, and dirt and carelessness prevailed. The Panjab experiment was abandoned, and Mr. Lister removed his scene of operations to Dehra Dún, where he has since conducted every stage of the experiment under his own management and with hired labour. In a letter on this subject dated January 2nd 1889 Mr. Lister writes:—

"You have two diametrically opposite opinions, and opposite systems; and time and experience can alone show which is right. I have tried both systems, and have paid dearly for my learning, and therefore can speak with some authority; and I am fully persuaded that this great problem is now completely mastered, and that the future of sericulture in India is thereby assured. Nothing, therefore, could give me greater pleasure than that Mr. Mookerji, or any other official, should go and see for himself what is being done at Lister's Grant, and examine and test everything relating to the cost and the quality of the silk produced. If his report is, as I believe it will be, satisfactory, then another year I should propose that the Silk Association should send some one along with a Government official, and should again examine and test everything relating to the cost and quality of the silk produced, and so prepare the way for its being carried out on a much larger scale by British and Native capital. The time for the silk crop is close at hand, and, from its commencement in the first or second week in February, only requires from 30 to 40 days to complete it, so that either Mr. M. or some other expert appointed by Government, might easily devote a month to testing the results.

"A considerable portion of the mulberry plantations are now in fair bearing, and surround the rearing-houses, and we are provided with seed of the first quality, being the produce of Italian and French breeds, reared on the estate, so that there should be and can be no difficulty in testing everything. For this I have patiently worked year after year, and at last the time has come. I have said, give me a fulcrum, and I can move the sericultural world. Give me labour, sufficiently trained; seed free from disease; plantations of sufficient age; and rearing-houses adapted for the purpose (and without this it is all labour in vain); and then there can be no difficulty in obtaining the most positive, accurate, and undeniable results.

"There are certain things of great importance that we have already tested and proved. First, that disease, when the worms are properly fed and attended to, is unknown to us; second, that the seed of the Italian and French Bombyx mori, reared in the Dún, gives just as good cocoons as in Europe, and, so far, does not appear to degenerate. Last year we compared the two, and found that the cocoons raised from our own seed were quite as good as those from imported European seed. We have therefore ceased to import any, and rely altogether upon our own; and last year, Mr. Ferrant, the Manager, to whom much of our success is due, in a small experimental way, raised four crops of the ordinary polyvoltine Bengal sort, without so much as losing a worm. Such have been the results of intelligent and careful cultivation, and I am perfectly satisfied that disease, about which we hear so much, is only another name for ignorance, neglect, dirt, and rearing-houses altogether unsuited for the purpose.

"I am not proposing to write a treatise on sericulture (I must leave that to those who have more time at their disposal); but the whole art and mystery may be expressed and enforced in three or four simple rules. *First*, sound seed; *second* air, space, and cleanliness; *third*, regular feeding; *fourth*, suitable rearing-houses. And where do you find these conditions in the native cottage? I have never seen anything of the kind; they may exist, but, I again say, I have never seen them. Air, space, and cleanliness the worms must have, or disease is certain. Then comes regular feeding, and at night, if possible, as we find that the worms are healthier, spin sooner, and make much finer cocoons, with night feeding.

Conf. with pp. 48-49.

"Mr. Bose, Secretary of the Gurdaspur Board, is right, when he says (and he has evidently taken infinite pains), 'My own impression is that the prevalence of disease was far more owing to the want of care, the negligence and general ignorance which prevail amongst rearers than anything else.' At last, the Government officials are beginning to find out what I have long since discovered, that it is impossible to rear silkworms in dirt accompanied with neglect. And he says: 'They never care to make the rearing-houses airy, and to keep them clean.' Under such conditions I am clear sericulture is utterly impossible. Give what prizes you will, it is all money thrown away. Mr. Dane, Deputy Commissioner, Gurdaspur, says: 'The first prize for foreign seed cocoons fell to Lister & Co.,' and further on he

SILK: Mulberry.

Diseases and Pests to which

DISEASES.

says: 'It seems somewhat absurd' (I should think it does) 'to award over R1,000 worth of prizes for a total outturn of silk of such trifling value, *viz.*, R6,415.' And what is more, if they gave every shilling in the Indian Treasury they would not be one bit nearer. All the wealth of India can never make silkworms thrive in the hands of dirty, careless, ignorant native rearers. I have paid for my learning, as for several years I joined the Government in giving prizes; but I soon saw that it was a perfect waste of time and money. Then it was that I determined to try what could be done by having everything carried out in a proper, business-like manner; and I am now, as I think, on the point of having a great success, after years of trouble and expense.

"Just a word with regard to cottage cultivation, and then I have done. Where mulberry trees abound and the climate is suitable, cottage cultivation should be possible, provided the native rearers are supplied with sound seed, and, above all, are taught how to use it. A certain number of intelligent, trained rearers, going from house to house, might soon bring about abundant success; but it is quite useless to offer prizes to men who know nothing of sericulture, and who are totally ignorant of the fundamental fact that silkworms cannot be reared excepting with sufficient air, space, cleanliness, and regular feeding."

Government Rewards. 1844

Mr. Lister's argument has much to be said for it, but it may be pointed out that it is one thing for Government to offer annually small sums as rewards to be given at local exhibitions; it is quite another matter to combat national defects. It neither argues ignorance of these defects, nor wilful neglect, for the Government to extend a helping hand in a direction which may do good and can certainly do no harm. It is for the capitalist—for private enterprise—to take such industries as silk under more direct control where essential reforms can be made a condition of employment.

But it may be pointed out that the whole of Mr. Lister's efforts have practically been directed towards establishing the rearing of **Bombyx mori** in India. His agents have accordingly laboured in the Upper Provinces, that is to say, many hundred miles to the north and north-west of the chief Indian area of mulberry silk production. Mr. Mookerji, on the other hand, has directed his attention almost solely towards the improvement of the multivoltine insects—the cultivation of which is undoubtedly accomplished by the peasantry. The two objects aimed at and the conditions of the countries severally dealt with are to a large extent independent. It would have to be considered whether such improvements as are possible in the rearing of the Bengal insects would prove remunerative. Although Mr. Lister, in the letter quoted above, was sanguine of success, it has since transpired that neither do his insects enjoy an immunity from disease, nor are his results so completely encouraging as to warrant the inference of an immediate and large extension of sericulture (of the superior silk producer) in India.

Improvements. 1845

It has still, in fact, to be proved that such radical improvements are possible in this country as would lead to the supersession of the village industry by the capitalist. But it seems safe to conclude these remarks by the opinion held by many writers, that if India is found to actually offer in its physical aspects a formidable barrier to profitable improvements sufficient to discourage an extension of European capital and enterprise, little progression need be looked for from the peasantry. The only measures that can and are likely to be taken up by Government and the merchants interested in the trade (such as it is) are those of a palliative or remedial nature against the spread of disease or the mitigation of the other evils to which the worm is subject. Long years ago the Honourable the East India Company effected a satisfactory improvement in the substitution of a superior mode of reeling the cocoons to that which then prevailed. It seems probable that the continuance of that reform will year by year assure more and more the displacement of the village reeling business by the establishment of filatures. The cheapest appliances as yet brought out are pecuniarily beyond the reach of the peasant, even were he not actuated

Conf. with pp. 13, 57, 184, 186-190, 192, 193.

DISEASES.

by a deep-seated opposition to reform and improvement. The silk he is at present able to produce finds a ready enough market in India, and the waste obtained, as also the cocoons, however indifferently reared, are greedily bought up. Indeed, the tendency of the Indian trade is to encourage the production of a cheap article in large quantities rather than to turn out a small amount of silk or cocoons intended for competition with Europe, China, and Japan.

The silkworm fly is by many writers considered quite as serious a drawback to sericulture in India as are the diseases above briefly dealt with. A short notice of the fly will be found under **Pests** (Vol. VI., p. 151), but the following article by **Cotes**—taken from the *Indian Museum Notes*—furnishes an abstract of all that has been written on the subject of the chief pest, *viz.*:—

PESTS. 1846

Trycolyga bombycis.
The Bengal Silkworm Fly.

"*Sources of information.*—A technical description of **Trycolyga bombycis** by the late **Dr. E. Becher** is given on page 77 of these *Notes*. The economic side of the subject has been dealt with in the following papers:—

1. A few notes on Sericulture in Bengal by **J. A. H. Louis** (1880).
2. Notes on the natural history of the Bengal silkworm fly in the Rajshahye district by **James Cleghorn**, 1887 (published by the Government of Bengal).
3. A letter by **Nitya Gopal Mookerji**, dated 7th October 1887 (published by the Government of Bengal).
4. Report of a meeting held in Berhampore on 12th November 1887, by **Nitya Gopal Mookerji** (published by the Government of Bengal).
5. A letter by **C. W. Marshall**, dated Berhampore, 20th July 1888 (published in *The Asian*).

In the present note, which is a summary of what has been ascertained, much has been taken from **Cleghorn's** paper, which is a valuable record of a very complete series of observations on the life-history of the insect.

"*Extent of the Injury.*—The Bengal silkworm fly causes considerable loss to silk-rearers in Bengal, but somewhat contradictory statements have been made as to the extent of the evil.

"**Marshall** notices that the fly causes 'enormous loss every year.' **Louis** estimates this loss in Bengal at between £200,000 and £300,000 annually. **Cleghorn** notices a loss of five lakhs of rupees in a single crop, as indirectly due to fly. **Mookerji** recounts how the fly destroyed 90 per cent. of a lot of silkworms he attempted to rear in Berhampere, while his two village nurseries, which might have been expected to yield 40 *khaons* each if the fly could have been kept off, produced only 8½ and 3½ *khaons* respectively. In this case, however, the loss does not appear to have been caused by fly alone, for muscardine was also present.

"On the other hand, **Mookerji** states that the regular cocoon bunds in the villages do not suffer much from flies; and if this is the case some of the preceding estimates must be excessive. Where so many causes are at work to affect the outturn of a crop, exact estimates of the actual loss occasioned by the fly are no doubt difficult to obtain, and estimates, based on the amount of injury, done by the fly to small experimental rearings, are likely to be excessive, as it is just the small rearings which suffer most. Taking everything into consideration, however, we may conclude that the pest is a real and serious evil, though perhaps not such a deadly one as some have been inclined to suppose.

"*Life-History.*—The Bengal silkworm fly is much like a big house-fly, but its great wing power makes its capture particularly difficult. In Rajshahye it is found about rearing rooms and silk stores all the year round, but it greatly increases in number during the months of July, August, and September. It attacks the common Bengal mulberry worm (**Bombyx**), the eri worm (**Attacus ricini**), and is also supposed to be the species which was found to be parasitic upon the caterpillar **Dasychira thwaitesii**, a pest which has been known to do much injury in the Doars by defoliating *tea* and *sál*.

"Copulation takes place in the air, one male fertilizing several females. The impregnated female is very active and persistent in her efforts to get at the worms,

SILK: Mulberry.

Diseases and Pests to which

PESTS.

and once having reached a rearing tray, she will wander all over it, depositing an egg here and an egg there indiscriminately on the worms. The act of oviposition is rapid, and the fly simply glues her minute egg into the worm's skin with her ovipositor without making an incision. She moves under and about the leaves and stems on the tray, sometimes even ovipositing from below on to the ventral surface of a worm. When she suspects dan.er she conceals herself amongst the leaves and escapes from one part of the tray to another, so that an hour may sometimes be spent in endeavouring to dislodge her from a single tray. According to **Louis** two or three dozen flies are sufficient to destroy a whole roomful of worms. A silkworm is only capable of nourishing about four maggots, but if there is a scarcity of worms the flies will lay many more than this number of eggs upon one caterpillar. **Cleghorn** found that the fly survived for about four or five days in confinement, but he is of opinion that it usually lives about eight days.

"When freshly laid, the egg has a hard white shell that can just be seen with the naked eye. About fifteen hours after the egg is deposited, it hatches, and the maggot penetrates the skin of the worm. **Cleghorn** found that worms, which moulted within fourteen hours of having eggs deposited upon them, escaped puncture by the maggots; while those which moulted later were more or less punctured. The opening made by the maggot, in penetrating into the body of the worm, does not close up, but becomes the entrance of a tube lined with a hard dark-coloured substance. This tube is small at the entrance, where it perforates the skin of the worm, but increases in size internally to correspond with the gradual growth of the maggot. It is curved a short distance from the entrance, and expands like a bell where it is filled up with the posterior truncate extremity of the maggot. Beyond the bell-shaped expansion, the hard lining of the tube is continued by a soft gristly substance, which envelopes the body of the maggot, so that it is only the maggot's head which projects into the actual tissues of the silkworm. The stigmata of the maggot are situated, as is usual in dipterous larvæ, at the posterior extremity of the body; and the maggot breathes air, which enters by the bell-shaped passage. When full fed, the maggot is much like a large grain of boiled rice, except that it is pointed at the head, and truncate at the posterior extremity, which is of about the same thickness as the middle of the body. It is armed with a long pair of thin mandibles, which are slightly hooked at the ends, and are hard and powerful. The body is ten-jointed, the kin remarkably tough. After living for about seven days inside the silkworm the maggot becomes full fed and cuts its way out. The presence of the grub is indicated on the silkworm by a black spot, waich increases in size day by day, though the silkworm continues to feed as usual, and even spins a cocoon, if it was not parasitised before the last days of the fifth month. A fly-blown worm spins about two days earlier than it would do when healthy, and it produces a very inferior cocoon. In cases where a cocoon has been spun the maggot forces its way through both chrysalis and cocoon, destroying the chrysalis and rendering the cocoon useless for reeling. After freeing itself from its host the maggot crawls away, and tries to reach the ground; if it succeeds in its endeavours and can find a soft place, it buries itself about an inch deep in the ground, and transforms into a pupa enclosed in the larval skin, which dries and hardens so as to form a case. About six hours elapse between its leaving the body of its host and transforming into a pupa; and if in this time it fails to find a place to bury itself, it pupates in the open.

"The pupæ that were observed by **Cleghorn** rested about 12 days before producing imagos. The imago breaks through the anterior end of the puparium, by expanding the membrane of the front part of its head, exactly as **Sasaki** describee to be the case with the Uji Fly of Japan, and emerges as a fly ready to commence the cycle of existence that has been traced above.

"The whole life of the insect, from the laying of the egg to the death of the fly, after depositing its own eggs, may thus be comprised within about twenty-eight days.

"These periods, however, vary with temperature and other conditions, for maggots have been known to become pupæ as early as the fourth day of their existence, and to produce small flies: while **Cleghorn** believes that a large number of pupæ hybernate from November to February.

Remedies. 1847

Remedies.—According to **Louis** the flies bred from the rains bund emerge so long before there are any worms of the November bund, upon which to oviposit, that they die without producing offspring, and hence at this time of the year the pest almost completely disappears. Reproduction, however, is so rapid amongst the few individuals which do manage to survive that by the following rains bund, or even by the end of the March bund, there are sufficient flies to occasion much injury. **Louis**

Mulberry Silkworms are Subject. (*G. Watt.*)

SILK: Mulberry.

PESTS.

recommends that a persistent effort should be made in silk factories, in the early part of each year, to destroy the pupæ of the fly. At present little is done, except by keeping the rearing-houses in almost total darkness, and closed in with thick *purdahs*, and this purely defensive attitude, he considers, is insufficient to keep out the fly; while it deprives the worms of much of the light and air they require.

"Mookerji notices that flies are specially injurious in the August-September bund, which is not one of the principal bunds, rearing being only done by those who have leaf left after the immediately preceding July bund. The reason why this bund is specially injured is, he believes, because the flies, which have emerged from the worms of the July bund, appear just in time to oviposit in the worms of the August-September bund. He, therefore, suggests that the practice of withholding rearing every alternate bund in the same neighbourhood, should be made compulsory through the agency of village *panchaiats*. This practice, being already very generally followed, the loss of the silk that is reared between the July and November bunds would not be considerable. If this action were taken, he thinks that the only remaining source of flies would be the seed cocoons, which could be obtained free of fly by rearing them under wire netting in properly managed central establishments. With regard to the advisability, however, of making regulations which must necessarily occasion at least temporary loss to those who now find, in spite of the fly, that it pays to raise an autumn-September bund, the writer would suggest that investigations should first be made to ascertain to what extent the reduction, which undoubtedly occurs, in the numbers of the fly at the end of the rains, is due to causes other than the scarcity of silkworms; for it is a well known fact that very many insects in India increase inordinately in numbers during the rains and die down again naturally with the advent of the cold weather.

"The suffocation by heat to which cocoons intended to yield silk are subjected, in all cases, destroys any maggots they may contain. It is obvious, therefore, that the speedy suffocation of all cocoons obtained from fly-blown worms must tend to reduce the evil; while general cleanliness in the rearing-rooms must also be useful in preventing the accumulation of dust and dirt in which the maggots conceal themselves when about to pupate.

Parasites. 1848

Parasites.—Cleghorn notices that spiders destroy numbers of the flies, and that a small beetle (probably **Dermestes vulpinus**), which also attacks silkworms and chrysalids, will eat the maggot of the fly; but that the greatest enemy of the fly is a smaller fly. He names this smaller fly "the midge,"* and supposes it to be parasitic upon the silkworm fly. In proof of this he confined twelve fly-blown worms with some midges in a muslin-covered glass with earth at the bottom. The fly-blown worms had a total of twenty-one black marks upon them, each mark showing the presence of a maggot, about four days old, capable of developing into a fly. The worms were carefully tended, but produced only two silkworm flies together with over one hundred midges, indicating that the midges had in some way caused the death of the silkworm flies.

"The midge proves to be a dipterous insect, belonging to the family MUSCIDÆ and is probably a new species.† Specimens, obtained by Marshall of Berhampore, have been sent to Europe for examination; and it is hoped that practical entomologists in the silk districts, where specimens can readily be obtained, will set themselves to elucidate the interesting question of the connection which exists between this insect

* The following note has been furnished by Mr. E. E. Austen of the British Museum: "This cannot be a true midge, as these insects belong to a family quite distinct from the **Muscidæ**. The flies usually known as midges are gnat-like creatures belonging to the family **Chiropomidæ**. Their larval stage is passed under bark, in water or in decaying vegetable matter and none of them are parasitic upon other insects. It is just possible that the fly called "the midge" by Cleghorn is a species of **Phora** a dipterous insect belonging to the family **Phoridæ**. The larvæ of **Phora** are usually found in decaying animal and vegetable matter, but the flies have been bread from Lepidopterous and Coleopterous papæ. If the "midge" is really a **Phora**, it is unlikely that its larvæ are, in the natural state, directly parasitic upon those of the silk-worm fly: it is more probable that the eggs of the **Phora** are laid upon a diseased silk-worm, and that the larvæ which are produced from them, incidentally destroy the larvæ of the silk-worm fly in feeding within the body of the caterpillar." *Ed., Dict. Econ. Prod.*

† It has been described as **Phora Cleghorni**, *Bigot*, in the Indian Museum Notes, Vol. I., pp. 191-92. *Ed., Dict. Econ. Prod.*

Food of the Mulberry Silkworms.

PESTS.

and the silkworm fly. It is by no means improbable that the midge may be found to be parasitic upon the maggot of the silkworm fly, much as the latter is parasitic upon the silkworm; and this supposition is supported by **Marshall's** observation of the deposition of the eggs of the midge upon the grub of the silkworm fly, within a few hours of the latter's cutting its way out of a silkworm cocoon.

"In order to breed the midge in sufficient numbers to keep down the fly, **Marshall** suggests that in silk establishments a practice should be made of always putting any maggots that can be found of the silkworm fly into a close worm basket, with a perforated lid, the perforations being sufficiently large to allow the entrance and egress of the midge while not allowing the fly to escape. He claims that this would serve two purposes; first, every maggot secured would be a fly put out of the way of doing mischief; and, second, from its being unable to shelter itself in the ground it would become a certain victim to the midge. This is no doubt an excellent suggestion for experiment, but too little is yet known of the life history of the midge to make it possible to judge whether or no the suggestion is likely to be of use. It is worth noticing, however, that if the midge were suited for extensive multiplication in the plains of Bengal, where it already exists, in all probability it would have become vastly numerous without artificial aid; for the fly upon which it is said to feed must offer, in the silk districts, an unlimited supply of food, and we know that under favourable circumstances the rate of multiplication amongst insects is excessively rapid. If, therefore, as would seem not improbable, the midge is only able to attack and thrive upon maggots which happen to be particularly exposed,—as, for example, on those collected in a basket,—it is evidently unlikely to be of much use, even if the measures suggested by **Marshall** have the effect of raising it in vast numbers; for the maggots which the midge is wanted to destroy are just the ones that are not caught and confined for its benefit."

FOOD. 1849 *Conf. with pp. 41-43.*

FOOD OF THE MULBERRY-FEEDING SILKWORMS.

The greatest possible difference of opinion prevails on the subject of the advantages of this and that form of **Morus**, and this and that method of cultivating the plant, as being best suited to India. The Journals of the Agri.-Horticultural Society of India contain numerous papers on the subject, which the interested reader should consult. One of the earliest articles of this kind is that by **Signor G. Mutti** (*Trans., 1837-38, VI., 149-182*), in which he strongly urges the European standard (tree) system of growing the mulberry as superior to the Bengal bush method. In the *Journal* (*Vol. XII.*), 1861, pp. 136-141, **Mr. H. Cope** furnishes particulars regarding four kinds of mulberry, *viz.*, the common standard tree of the Panjáb; the Philippine Island form (**M. multicaulis**); the Shah or Royal mulberry of Kashmír; and the Chinese mulberry. The Philippine mulberry was, however, strongly recommended by **Colonel Sykes** as a desirable one for Indian sericulture many years before the date of **Cope's** paper (see *Trans. Agri.-Horti. Soc., Vol. VIII, 340*). **Hutton**, in his *Reversion and Restoration of the Silkworm* (*Jour., Vol. XIV.* (*Sel.*), *25-36*), furnishes much useful information on the subject of the species of mulberry, in which he shows that the black, red, and white-fruited forms are scarcely entitled to be regarded as varieties, still less species. He found, for example, that the white-fruited form of Kashmír, after a few years' cultivation in India, yielded purple fruits. He views the white fruit accordingly as only an albino condition of the purple, and is of opinion that that condition may be met with in any of the species of **Morus** recognised by botanists. **Hutton** thus regards the whole question of the plant as governed by the same laws as the insect itself. That is to say, that, just as there are numerous races of silkworms, some peculiarly adapted to the climatic and other conditions of the regions where found, so in a like manner there are local conditions of the mulberry, many of which experience has proved are better suited to the insect or insects of that region than would be the mulberries of other countries, though both may botanically be the same species.

Conf. with pp. 24, 37, 48.

In his paper on the Cultivation of Silk in the Colonies, **Hutton** gives

FOOD.

some additional information (*Jour. Agri.-Horti. Soc., New Series*, 1870' 155-158), in which he urges that in new regions of sericulture all forms of mulberry should be experimentally tried till it is found which answers best. He repudiates the statement often advanced that only the white-fruited mulberry should be selected, on the ground that the colour of the fruit, as already stated, is not a constant character; it is, therefore, in his opinion, of no value in determining the nature of the plant.

The reader should consult the opinions given above by **Mr. N. G. Mookerji** in the passage from his report on sericulture of the North-West Provinces. See also the remarks on the species of **Morus** in Vol. V., pp. 279-284, of this work.

THE WILD SILKS OF INDIA.

WILD SILKS. 1850

The silkworms which, by popular usage, are spoken of as the "wild" insects, may be said to be those which the entomologist (for the most part) refers to the family SATURNIIDÆ, in contradistinction to the domesticated or mulberry silk-worms which belong to the BOMBYCIDÆ. Some of the SATURNIIDÆ exist, however, in a state of domestication very nearly as complete as that of the mulberry worms. Others are only half domesticated. But of the SATURNIIDÆ, out of a long list of species actually met with in India, only three can be said to be of present commercial importance. In dealing with this subject, therefore, it seems desirable to place the information of more immediate importance by itself and apart from what, in the present state of knowledge, may be regarded as of interest mainly to the entomologist. The pages which follow will, accordingly, be found to practically assume two distinct sections: (*a*) an alphabetical enumeration of the better known SATURNIIDÆ with brief notes under each, more especially those which may be called of secondary commercial value; and (*b*) a detailed chapter on each of the commercial SATURNIIDÆ, *viz.*, the *Tasar*, *Eri*, and *Muga* silks. The reader will find it instructive, however, to consult **Mr. Hampson's** complete list of the Indian SATURNIIDÆ, which is given as an appendix at the end of this article (pp. *236-238*). **Mr. Hampson** there gives the most recent classification and a reference to the more important entomological publications on the species.

B.—AN ENUMERATION OF THE CHIEF INDIAN SATURNIIDÆ.

THE SATURNIIDÆ.

(For A, **Bombycidæ**, see pp. 1-6.)

I.—ACTIAS.

I. ACTIAS. 1851

This genus is generally regarded as embracing some three or four forms, all of which, like the species of ANTHERÆA, form hard entire cocoons, from which the imago could not escape but for the fact of possessing spurs on the shoulder of the anterior wings. These spurs are employed to distribute the solvent and to mechanically separate the fibres when once the cement has been dissolved. **Hutton** remarks that the cocoons of this genus are not so full of silk as those of ANTHERÆA, but that what there is has been well spoken of, being "strong, tenacious, elastic, and brilliant." **Actias** differs from ANTHERÆA chiefly in the posterior wings being tailed.

1. **Actias ignescens**, *Moore* (*Proc. Lot. Soc., 1877, 602*).

1852

This insect occurs in the Andaman Islands, but is doubtfully distinct from **A. mænas**.

2. **A. mænas**, *Doubl.* (*See p. 236.*)

1853

Habitat.—The lower slopes of the Eastern Himálaya and in Assam.

According to **Moller** it is bivoltine, the moths appearing in spring and autumn. The worm is said to feed on **Turpinia pomifera**. It is a large insect closely allied to **A. selene**, but it is scarce and but imperfectly known even to entomologists. Its chief peculiarity is the great difference in the size and coloration of the sexes, which are so dissimilar as to have led to their being viewed as two species—the male as **A. leto** and the female as **A. mænas**.

ACTIAS SELENE. 1854

3. Actias selene, *Hübn.; Moore, Cat. Lepid. Insects, II., 400.* (*See p. 236.*)

THE SWALLOW-TAILED SILK-MOTH : SELENE SILK-WORM.

Syn.—TROPÆA SELENE, *Hübn.*

Habitat.—Recorded as found at Simla, Mussoorie, Sikkim, the Khásia Hills, Shillong, Sylhet, Sibsagar, Chutia Nagpur, Calcutta, Bombay, and Madras, etc. It ascends the Himálaya to altitudes of 3,000 to 8,000 feet. Distributed to China.

Food. 1855

Food of the Worm.—The following may be given as the plants reported to be suitable for this silk-worm :—

1. **Bischofia javanica,** *Blume.*
   A euphorbiaceous tree of sub-tropical regions.
2. **Cedrela Toona,** *Roxb.*, and **C. serrata,** *Royle.*
3. **Coriaria nepalensis,** *Wall.*
   A shrub of the Himálaya, found from Murree to Sikkim and Bhután.
4. **Juglans regia,** *Linn.*
   THE WALNUT.
   Wild in the North-West and Sikkim Himálaya, often also cultivated.
5. **Odina Wodier,** *Roxb.*
   A deciduous tree met with throughout the hotter parts of India, and along the foot of the Himálaya to Assam, Burma, and the Andamans.
6. **Pieris ovalifolia,** *D. Don.*
   A tree or shrub common in the Khásia Mountains and in the Himálaya from Bhután to Kashmír, also in British Burma and Manipur.
7. **Prunus Puddum,** *Roxb.*
   THE WILD CHERRY.
   One of the most plentiful trees of the North-West Himálaya.
8. **Pyrus Pashia,** *Ham.*
   A small, very common tree which is often called by Himálayan writers the wild pear.
9. **Terminalia tomentosa,** *W. & A.*
   Rev. Mr. Campbell says that in Chutia Nagpur this tree and **Odina** are those most frequented by **A. salene.**
10. **Zanthoxylum acanthopodium,** *DC.*, and **Z. alatum,** *Roxb.*

A. selene passes the winter inside the cocoon, and goes through two generations during the year. The first set of moths appear in early spring and the second in midsummer. The cocoon is said to be coarse in texture and to afford but a small amount of silk.

NOTES Himalaya. 1856

Notes regarding this Silk arranged mainly according to Provinces.

HIMÁLAYA.—**Hutton** says that it is very common in a wild state at Mussoorie, also at Almorah, Darjíling, Assam, Cachar, Saugor, and has been even collected at Serampore near Calcutta. He further tells us that **Mr. C. Turnbull** failed to reel silk from the cocoons sent down from Mussoorie, but that they had been reeled by others, though the amount of silk obtained was not very great. In another of his publications **Hutton** says that as the insect may be fed on the cherry, pear, and walnut, there should be no difficulty in rearing it in France or England. It yields, he remarks, three or four crops a year, hence an abundant supply of silk might be obtained from it, since the worm is as easily reared within doors as on trees. He then adds : "In India it has a wide range, occurring along the coast line from Pondicherry eastward, along the base of the Himálaya even to the Sutlej in the north-west, and it has been found also in Central India, so that there can be no question of its being rendered otherwise than productive in this country." (See *Hutton, Trans. Ent. Soc., IV., 221 ; V., 85 ; Jour., Agri.-Hort. Soc. Ind., IX., 167-169.*)

ACTIAS SELENE.

Assam. 1857

ASSAM.—The late Mr. Stack wrote that it occurs though rarely in Cachar. "The cocoon yields but little silk and no attempt is ever made to use it." In the appendix to the Assam Government Resolution, dated 6th June 1879, it is enumerated in a list of Assam wild silkworms regarding which no information is obtained. It apparently bears no vernacular name.

Bombay. 1858

BOMBAY.—Major Coussmaker simply alludes to the Selene insects as one which occurs in Bombay and which he has had under observation.

Madras. 1859

MADRAS.—Mr. Morgan, in a report on the wild silks of Malabar, gives the following brief notice regarding this silk:—

"This lovely moth is not common anywhere.* The larvæ are very handsome, of a bright brass-green, with yellow spots on the head. In Malabar I have only found it feeding on two species of trees, **Bischoffia javanica**, *Bl.*, and **Odina Wodier**, *Roxb.* The cocoon is of a yellowish-brown colour, and wiry in texture. The cocoon assumes the form of the leaves that the worm cements round it to conceal it from its enemies. I have tried to manufacture fishing gut from it, and obtained very fine long strands; but owing to some fault in the manufacture, it lacked strength. As a silk-producer, it does not seem to me to be worthy of comparison with the *tasar* or atlas worms." Dr. Bidie, in a report on the wild silkworms of Madras (*published March 1877*) says that this moth is widely distributed in the Carnatic, but is less common than the *tasar*. He remarks that it could easily be domesticated, and that he has learned it is reeled in Pondicherry.

Conf. with pp. 76, 134.

Bengal. 1860

BENGAL.—Hutton, in the notes above, says that this insect has been found near Calcutta. The writer has seen one or two specimens collected in Lower Bengal, and it is fairly plentiful in Chutia Nagpur. The Rev. A. Campbell, for example, showed him a considerable number which he was breeding at his Mission station of Govindpur in Manbhum. Mr. Campbell is, however, of opinion that the habit the worm has of leaving the tree and spinning its cocoon on or near the ground makes it less valuable than it might be, as the cocoons are both injured and are difficult to find.

Nature of the Silk, &c.—Mr. Wardle, in his *Wild Silks of India*, says: "The silk does not appear to be windable, but it is of a coarseish kind and might be spun if it could be obtained in sufficient quantity." "The diameter of the cocoon fibre is $\frac{1}{1000}$ of an inch, and it has about an average strength and tension, and thus presents no obstacle to its use. The cocoon is rather large, being 3 inches long and $1\frac{1}{4}$ inches in diameter."

For further information see the remarks below under **Antheræa perotteti**, pp. 75-76.

II.—ANTHERÆA.

II. ANTHERÆA. 1861

India possesses some eight or nine species which, by entomologists, have been referred to this genus. Of these by far the most important are the various forms collectively spoken of as the *Tasar;* next in value stands the *Muga* and *its* allies. These two types may, in fact, be viewed as representing two important sections in the genus, *viz.*, those with cocoons attached to the twigs of the food-tree by means of a peduncle, and those devoid of a peduncle. **Antheræa paphia, A. frithi, A. helferi,** and **A. knyvetti,** are names for pedunculated forms, with (in addition) several local varieties of the first mentioned. **Antheræa assama, A. roylei, A. pernyi, A. yamamai, &c.**, are names of forms which do not possess a peduncle. Hutton, while discussing the importance of ascertaining the nature of the gum or cement used in consolidating the cocoon and of the solvent secreted by the imago, in order to soften

* Hampson informs the Editor that it occurs on the Nilgiri hills up to an altitude of 8,000 feet.

II. ANTHERÆA.

the gum and thus allow of its escape, remarks that in **Antheræa** and **Actias** "in which the cocoons are extremely hard and tough and without the least opening, the solvent from the mouth of the moth would, if unassisted by other means, be wholely insufficient to enable the insect to come forth, and these insects are consequently furnished, in addition, with a strong sharp-pointed horny spine, or spur situated on the shoulder of each anterior wing, close to the *patagia* or tippets, where it is concealed by the long down, and with which the moistened mass of threads is then divided, and an orifice formed through which the moth is enabled to escape. How difficult, not to say how impossible, it would be for the insects of these genera, if unarmed with the wing spur, to escape from the cocoon, will be readily understood when I explain that the true silk-yielding cocoon in the genus **Actias,** and also in **Antheræa roylei,** is encased within another which is so tough and closely woven as to be impervious to water, and is partially glazed externally for the protection of the pupa from inclement weather; while in **Antheræa paphia, A. frithi, A. assama,** and others of the true *tasar* group, the threads are so massed and agglutinated as to render it impossible for the solvent to effect more than the dissolution of the gum." "A good and well formed *tasar* cocoon is so extremely hard and resisting, that it is difficult and often impossible by the strongest pressure of the fingers to make the least indentation in it, and indeed so hard and durable are those of **A. paphia** that the Natives are in the habit of using them when cut into transverse rings, for binding the barrels of their matchlocks to the stocks."

1862 *Conf. with pp. 72, 73, 77, 98, 157.*

4. Antheræa andamana, *Moore.* (*See p. 237*)

An allied species to the common *tasar,* which inhabits the South Andaman Island. Could it possibly be the form described by **Rumphius** as met with in Amboyna? **Mr. Cotes** thinks not, since the fauna of Amboyna differs greatly from that of the Andaman islands.

1863

5. A. assama, *Westw.; Moore, Cat. Lepid. Insects, II., 398.*

MUGA. 1864 *Conf. with pp. 174-183.*

This is the *muga* silkworm which occurs in Assam, Eastern Bengal, and Darjiling. It is often met with in a state of complete domestication, but there are also numerous wild forms, some of which have, by certain entomologists, been accepted as separate species. **Hutton** mentions that he had found a specimen of this insect in Dehra Dún, so that it is probably much more widely distributed than is generally supposed. A greater difficulty exists in extending its cultivation than is the case with the *tasar,* owing to its being much more particular in the selection of food. In the pages below will be found a detailed account of this insect and of its silk, and at page 175 is given a list of the plants on which the caterpillar feeds. The majority of these will be seen to be species of Laurel:—plants which inhabit sub-tropical regions, their area of distribution being towards the temperate rather than the tropical zones. (*See p. 237.*)

1865

6. A. frithi, *Moore.* (*See p. 237.*)

Habitat.—A species said to be found in the hot valleys below Darjíling, and, to a certain extent, along the outer ranges of the Himálaya, ascending to 2,000 feet. **Mr. F. Moore** remarks that "The cocoon is stated to be similar to that of *tasar* in form, but of finer silk" (*Wardle, Wild Silks of India, 6*). **Cotes** has reared it in captivity and figured the cocoon in the *Museum Notes, plate VI.* According to the late **Mr. Otto Moller** (*Museum Notes, Vol. I., 201*) this insect is common at low elevations in Sikkim. It is bivoltine. The moths appear in March and again in August. It hybernates as a pupa, and the worm feeds on the *sál*—**Shorea robusta. Rondot** (*L'art de la Soie, II., 117*) says that **Fallon** reared it in France. The cocoons, he remarks, are yellowish white in colour and very similar to those of **A. roylei,** but less silky on the exterior. He regards them as containing a considerable amount of silk.

1866

In the *Museum Notes* an incidental remark is made regarding a single-

ANTHERÆA FRITH II.

walled pedunculate cocoon found by Mr. A. V. Knyvett upon wild cherry trees in Sikkim, in which it is said that it was much like the cocoons of **A. frithi.** This is the form which Hampson recognises as an undescribed species, and to which in the Appendix below he assigns the name **A. knyvetti** (*p. 237*). There would appear to be no doubt that it is a good species. It is said to feed on the "Wild Cherry." Could this be accepted as **Prunus Puddum**, a warm-temperate tree, on which **Actias selene** is said to be sometimes found? (*Conf. with p. 68.*) The writer pointed out to Mr. Hampson that there was apparently some confusion in the reportts on the food of **A. roylei & A. knyvetti.** These two species seem to have been confused with **A. frithi**, where it is stated they feed on the *Mauha* tree.

Conf with pp. 76, 77.

7. Antheræa helferi, *Moore.* (*See p. 236.*)

1867

Met with, along with **A. frithi**, in the sub tropical East Himálayan valleys. Although many collectors have specimens of this and the last-mentioned moths, the cocoons of the insects are rarely seen in collections. This fact was deplored by Hutton, so long ago as 1871, and in a note issued by the Indian Museum only very recently it is stated that the "cocoons are unknown." It is somewhat curious, however, that Hutton should make the almost self-same statement regarding the cocoons of this species as Moore offers regarding the cocoons of **A. frithi**, *viz.*, that they resemble those of the common *tasar*. It would, in fact, seem just possible that Moore's statement is derived from Hutton's but assigned to the wrong species. This idea is all the more likely to be correct, since Moore makes no such remark regarding the cocoons of **A. helferi.** Accepting this explanation of the confliction, in the writings of authors, regarding these cocoons, **A. helferi** may for the present be accepted as belonging to the pedunculate series of **Antheræa.**

8. A. mezankuri, *Moore.* (*See p. 237.*)

1868

This yields the *Mezankuri* silk of Assam, a fibre nearly white, and valued at about 50 per cent. above that of *Múga.*

Most writers appear to agree with Hutton that the "existence of this species distinct from **A. assama** is altogether apocryphal." On another page the reader will find a separate paragraph on this subject, in which the late Mr. Stack's opinion is quoted, an opinion which goes to show that the special conditions which led Moore to establish the above name as that of an undescribed species are purely the result of the food on which the caterpillar is reared. A long and instructive account (and the earliest on record) of this insect will be found in Vol. VII., *Trans. Agri.-Horti. Soc. Ind.* (*1839*), *pp. 97 to 100*, which was contributed by Jenkins, being a translation of a paper by Muniram Bur Bandari Barrua. The insect is said to be known as *Mezanguri*, and the plant on which it feeds as *adahkori*. It is stated that "the *khoteah* crop yields less silk, but is the finest in colour and most valuable, from not being exposed to the rains. The crops of the rainy season are plentiful, but the silk is far inferior in colour and strength. "The produce from 1,000 cocoons is generally estimated at 20 tolahs of silk, nearly ½ lb. The price of common *muga* silk in thread is from R3 to 4 per seer of 80 tolahs, but the *mezanguri* fetches from R6 to 8 per seer, whilst the silk of the true mulberry worm is only valued at R5 to 7."

Conf. with pp. 175-176 & 183.

9. A. nebulosa, *Hutton.** (*See p. 237.*)

1869

This worm is reported to be met with in the jungles of Colgong, Sing-

* Mr. Cotes informs the Editor that "what appears to be Hutton's type specimen of this form is preserved in the Indian Museum. It is obviously only a dark coloured individual of the common tasar."

ANTHERÆA NEBULOSA.

bhum, Chutia Nagpur, and Central India. It differs from the ordinary *tasar* in being darker-coloured with cloudy bands on the wings. **Hutton** thought that the silk from this form would probably be found to rival that of the ordinary *tasar*. It seems at least probable that the Southern India form, figured by **Shortt** in the *Madras Monthly Journal of Medical Science*, has a stronger claim to being accepted as a separate species than has **A. nebulosa**. **Shortt** shows the caterpillar as devoid of the spinose processes and the moth as greyish yellow with a deep purple transverse line across the wings and round the talc-like spots. In the remarks below, p. 98, it will be seen the writer draws attention to the similarity of the Madras insect figured by **Shortt** with that exhibited and described by **Rumphius** 120 years ago. (*Conf. with p. 97.*) (*See p. 237.*)

1870

10. **Antheræa paphia,** *Linn.; Moore, Cat. Lepid. Insects, II., 385.*

It is unnecessary to do more in this place than to refer the reader to the special chapter below which treats of the insect now is recognised by the most recent entomological writers under the above name, but which was formerly more generally known as **A. mylitta**—The Tasar—,pp. 96-166.

It may be said to be met with throughout India, some of its local forms being viewed as varieties, others as mere accidental variations or the result of food. Much has been written on the hybrids which may be produced from it, but in the majority of cases these disappear from consideration if the forms are accepted as but local manifestations of one common species. In that view of the case they are not hybrids. One writer, however, maintains that he has crossed the Indian *tasar* with the Japanese *yamamai*—an insect which appears to be more nearly allied to the *muga* than to the *tasar*. This discovery excited no small interest in the public newspapers, since it was claimed that the hybrid yielded a superior quality of silk, which was more easily reeled than that of the *tasar*, and that the insect withal was more amenable to domestication. It is to be feared, however, a false interest was created, since nothing more has been heard of the hybrid. For information regarding the source of much of the so-called Chinese *tasar* silk in the trade returns of Europe consult the article **A. pernyi** below. [*f. 1.*

Hybrids.
1871
Conf. with pp. 71, 78, 79, 101.

1872

11. **A. pernyi,** *Guér., Méne., Rev. et Mag. de Zool. (1855), 297, Pl. 6,*

Habitat.—This insect is a native of China, where it has been semi-domesticated for centuries as a source of silk. The cocoons obtained from it are believed to constitute by far the major portion of the material imported into Europe as "Tasar Silk." The worm, like that of **A. yamamai** of Japan, and **A. roylei** of the Himálaya, feeds on a species of oak. These three insects attracted considerable attention in Europe (some 25 or 30 years ago) during the then great alarm of the ruin of the mulberry industry, through the spread of disease. Shortly before that date (1850), the **Abbe Perny** made this insect known to entomological science, and **M. Guerin Meneville** associated it with the name of its discoverer. Not long after, and apparently in ignorance of what **Perny** had done, **Surgeon Maingay**, during a botanical excursion to the Shantung promontory of Northern China (September 1862), made the discovery of extensive plantations of oak, and on enquiry as to the use of these, found that they were grown in order to feed the worms of a silk moth which **Hutton** pointed out was nothing more than the then recently described **Antheræa pernyi**.

Conf. with p. 159.

But **Maingay** added greatly to our knowledge of the nature of the country where produced, the systems of oak culture pursued, and the methods of utilising the silk of this, even to-day, little known insect. He found it in a region which consisted of irregularly disposed ranges of granitic hills, of an elevation of from 1,000 to 3,000 feet. The oaks

ANTHERÆA PERNYI.

on which it fed were associated with **Pinus sinensis,** *Lamb.* The rivers of the promontory are broad and sandy, but the streams insignificant, since little rain or snow falls, though the winter frosts are most intense. "The climate," wrote **Maingay,** "in its essential characters, seems to appoximate closely to that of the northern slopes of the Himálaya." "The country is admirably cultivated and produces leguminous and other crops in great abundance." These brief indications of one of the chief regions of the domestication of this silkworm have been here given in order to allow the reader to form his own opinion as to the possibility of acclimatising **A. pernyi** in India. The idea that the insular though sparsely rainy promontory, which has an average breadth of only 30 miles and is washed on its northern shores by the Gulf of Pechili and on its southern by the Yellow Sea, resembles climatically the slopes of the Himálaya (which are, perhaps, 1,000 miles from marine influence, and on the average have a high though periodic rainfall with alternating severe seasons of drought) is likely to be doubted by most persons who have studied the results of the sericultural experiments made in India.

Maingay says that "The cocoons are very slightly pedunculated, but by no means resemble those of **A. paphia.** They are generally surrounded by leaves. I imagine the Chinese breed the young worms in confinement and then place them on the plantations, which are not used indiscriminately but are allowed regular periods of rest, or they may adopt the plan of placing the pregnant females on the trees selected as the future food of their progeny." Here again the imperfectly pedunculated cocoon embraced by leaves suggests the idea, urged on more than one occasion, that entomologists may be in error in referring **Rumphius'** Amboyna insect to **A. paphia.** But **Maingay** next deals with the subject of the independence of this insect from the *jaru* of Assam, and he adds that the Shantung silkworm is also quite distinct from that of Central China "from which the main supplies of commerce are derived." There can be no doubt on that point, since the silkworm of Soochow and of Central China feeds on the mulberry. The author's object in these controversial remarks is, however, to suggest the improbability of the *tasar* silk of Chinese commerce being mainly (if at all) derived from **A. paphia.** Indeed, he is disposed to go further and to think the so-called *tasar* of China, like that of Amboyna, has, by many writers, been incorrectly referred to that species. It seems highly probable, in fact, that future investigations will show that **A. paphia** does not occur in China at all.

1873
Conf. with pp. 70, 72, 77, 98-9, 157.

The tyranny of the Taiping rebels, **Maingay** remarks, was so great as to have in his day threatened the stability and permanence of the mulberry silk of Soochow. He accordingly urged that a greater future was open for India in which it might be desirable to ascertain whether the Shantung worm could be acclimatised in the drier warm temperate tracts where species of oak abound. But that China has not since lost its hold nor India gained a stronger footing in the *tasar* silk supply of the world, the reader will perhaps be satisfied who may take sufficient interest in the subject to peruse the review of the facts of the case given in this article. **Maingay** was, however, sanguine of success with the Shuntung insect, and in all fairness it should be added the experiment has never been honestly tried. He writes, "I now invite attention to the ease with which the Shantung silkworm may be introduced into India, as being more hardy than the Indian varieties. It has, moreover, the advantage of feeding on the various species of oak, all hardy in their growth, otherwise they would perish in the rigorous winter climate of North China, and in other respects admirably adapted for growth on the dry slopes of the Northern Himálaya or in Kangra and other North-Western valleys, in which already cognate species

SILK:
Saturniidæ.

The Wild Silks of India.

ANTHERÆA PERNYI.

are indigenous." We have now learned that it is just the absence from the possible regions of Indian cultivation, of those conditions which Maingay characterises as "hardy" that has proved the stumbling-block of all sericultural experiments of India. With the mulberry silkworm, for example, it has been found imperative that the eggs should be conveyed far above the altitude of the plantations to a climate sufficiently cold to effect that check on germination (so to speak) which has been proved necessary to the preservation of the good qualities of the worm. And the same difficulty would doubtless be experienced with the oak-feeding insects. At the lower altitudes of oak production, where land, labour, and facilities of export exist, the winter cold is by no means severe enough, and the summer heat far too great to justify the entertainment of high expectations with **A. pernyi.** Hutton believed in a possible future for **A. roylei** or a hybrid between that species and **A. pernyi,** but it should be recollected (what is true of all such experiments), that while in the hands of an enthusiast it may be possible to rear a limited number of these or of almost any other silkworm, it is quite another matter to obtain all the conditions necessary for commercial success. Solution of the questions of labour and facilities of export are quite as pressing as the possibility of cultivation. The conditions of success with **A. pernyi** may, indeed, be said to exist more completely in Europe than in India. Indeed, the alarm of complete failure which became so pressing with the mulberry-growers of Europe, some few years ago, caused (as already indicated) attention to be directed to India and China as greater sources of silk supply in the future than they had been for many years past. One of the natural directions of this new aspect of the silk trade was the effort to discover an insect free from disease which might be introduced into Europe, so as to take the place of the mulberry-feeding worm. Amongst others **A. pernyi** was experimented with, especially in France, and the habits of that insect both in China and in the countries to which it had been conveyed, became the subject of careful study. **Cotes** gives the following passage from M. Rondot's account of this insect:—"**Antheræa pernyi** is a bivoltine species, found both wild and also in a semi-domesticated state upon oak trees in many parts of China, where it is reared extensively for the production of silk; the amount of fresh cocoons annually reared having been estimated at twenty-two millions of kilogrammes, of which a considerable amount is imported into Europe. Two yields of silk are obtained in the year, one in the spring and another in the autumn; the spring rearing occupying about sixty days and the autumn rearing about a hundred. The insect is generally reared indoors, but to a certain extent also upon trees or bushes in the open air. The worm is also common in a wild state in the forests and copses of oak trees on the mountain sides. In Koueitcheou (according to Father Perny) there is an annual variety of the worm which is less esteemed than the bivoltine one, a fact which is noticeable in connection with the tendency to become annual, which has been observed in species when reared in France. The cocoon is enveloped in two or three oak leaves drawn together by a network of silken strands, and is further attached at one end to some small branch or leafstalk by a flat silken cord. The cocoon of the spring rearing contains only about half as much silk as that of the autumn rearing, but the silk itself is far more brilliant, that of the autumn rearing being somewhat dull and lustreless. The cocoons are either reeled or spun. The reeling is done in two ways—either wet or dry. In the dry process the cocoons, after having been dipped in a mordant made from oakwood ashes, are washed in clean water and then reeled dry, the basket containing them, however, being sometimes steamed over a vessel of boiling water. In the wet process, the

Hybrids.
1874
Conf. with pp. 72, 78, 79.

ANTHERÆA PERNYI.

cocoons are simply reeled as they lie in the iron boiler, which contains either a solution of raw soda or strong mordant made out of oak ashes, the liquid being but just sufficient to cover the cocoons; the wet process therefore differs mechanically from that of mulberry silk filatures, where deep basins of water are used for holding the cocoons while they are in process of reeling. Of the wet and dry processes, the dry one is preferred and gives the most satisfactory results. A large portion of the autumn crop and of the pierced cocoons is spun, the spinning being done either by hand or with a jenny worked by the foot."

Tasar Trade of China. 1875

In concluding this brief notice of **A. pernyi** the suggestion already offered may be here repeated, that the cocoons of this species constitute almost exclusively the so-called "Chinese Tasar" of European commerce. Hence to a considerable extent the success of China in the so-called *tasar* trade as compared with that attained by India. A domesticated insect that can be reared with ease on a plantation in an accessible locality, can hardly help proving more profitable than the collection of wild cocoons over an extensive inhospitable tract, the more so since the insect of the latter, after the most careful and exhaustive experiments, has proved intractable of domestication. It is clearly an unwarrantable conclusion to urge that India has failed in the past to compete with China, in the *tasar* market, through the indifference of her people or of their rulers.

Comparison with that of India. 1876

The capabilities of **A. pernyi** in China are as different from those of **A. paphia** in India as any two subjects of inquiry could possibly be. The one is a denizen of a salubrious and invigorating richly cultivated temperate country, the other of enervating tropical jungles, infested with disease and animals inimical to human life. The inhabitants of the one are industrious and energetic, of the other apathetic and enslaved by religious restrictions and obligations that make the collection of *tasar* cocoons distasteful and unpopular. The whole question of the present greater importance of China than of India, and of the future of the *tasar* supply, would thus seem to turn on a more precise and honest comparison of the facts regarding **A. paphia**—the *tasar* proper—than has been made in the past, and a more thorough appreciation of the difficulties that have in India to be contended against.

Extent of Indian Supply. 1877 *Conf. with pp. 59, 106, 115-118, 121, 132, 148, 149, 151, 156.*

It is one thing to say an insect is "found in the forests in all parts of the Indian continent," and that therefore "there is no reason why it should continue" in a more backward state than the *tasar* of China (*Pioneer, June 11th, 1891*); but it is quite another matter to make that wild insect tractable to the necessities of commerce, or even to overcome the religious prejudices of the agents that have to be primarily employed in the development of the trade. But in stating the case thus briefly, the popular view has been admitted (a by no means established one) that a portion of the *tasar* of China is obtained from the self same insect as the *tasar* of the wild hilly tracts of the Central table land of India. It is highly probable that were seed of the Indian *tasar* to be required by China, the failure of that country to produce the Indian article commercially would be as complete as the failure hitherto experienced in India. (*Conf.* with the detailed information on *tasar* below, as also that under **Attacus cynthia.**)

12. Antheræa perotteti, *Guér. Mén.* (*See p. 237.*)

1878

Habitat.—A silkworm, said to be a native of Pondicherry, which has been reported to produce four broods a year.

Hutton says he suspects this to be only one of the numerous forms of the *tasar* (*Jour., Agri.-Hort. Soc., N. S., III., 137*). Geoghegan remarks, however, that M. Perottet of Pondicherry (the discoverer of the insect) never could induce it to eat anything but **Odina Wodier**—a tree which the

Conf. with pp. 69, 100.

ANTHERÆA PEROTTETI.

ordinary *tasar* is not reported to feed on, but which, most writers say, is the natural food of **Actias selene.** This fact may be accepted as favouring the idea that either some mistake has been made, or that **A. perotteti** is a distinct species from **A. paphia.** Geoghegan adds, "The silk is reported to be strong, wiry, and brilliant, but had to be carded. The worm breeds in captivitiy, undergoes four moults, and yields four crops in the year. The chrysalis of the fourth generation remains in cocoon till the tree it feeds on (which is deciduous) is again in leaf." In a foot-note to the above Mr. Geoghegan says, "Fishing lines are said to be made in Dinagepore from the silk of a worm which feeds on **Odina Wodier.** The food of the worm and the quality of the silk point to the identity of the insect with **A. perotteti.**" Mr. Wardle accepts (*p. 69*) the suggestion of the above foot-note as an actual fact regarding the silk of **A. perotteti**, but it is highly probable that whatever M. Perottet's insect may have been, the Dinagepore fishing lines and the silk referred to by Dr. Bidie (*p. 69 ; also p. 134*) and other writers as obtained from an insect, the caterpillars of which feed on the leaves of **Odina Wodier**, should be transferred to **Actias selene.** Since the above was written the author has had the pleasure to see a proof copy of Mr. E. C. Cotes' forthcoming paper in the Indian Museum Notes on the *Wild Silk Insects of India.* In that paper the following passage occurs regarding this insect: "**Antheræa perotetti**, *Guérin Méneville*, from Pondicherry: this insect, according to Walker (*B. M. Cat. Lep., VI., p. 1379*), is very nearly allied to **A. assama***, and Guerin Meneville's description of the cocoon (*Mag. de Zool., VI., pl. 123, 1844*) also answers to that of **A. assama**, and it is not improbable that it may be a variety of that species, though Guerin Meneville's figure of the moth presents some peculiarities. According to Rondot (*L'art de la soie, II., p. 117*), it feeds upon **Eugenia Jambolana, Zizyphus sp.**, and other trees."

It will thus be seen that the greatest confusion prevails regarding this species, if indeed, it has any existence. If it feeds on the plants named by Rondot it is very probably only a form of the common *tasar*, or if structurally it be actually allied to **A. assama**, the plants mentioned by Geoghegan and Rondot are not likely to be its food. The writer is more disposed, therefore, to regard M. Perottet's insect as hypothetical, resulting from a combination of the peculiarities of **A. paphia** and **Actias selene.** How this error came into the literature of the Indian silkworms will probably always remain obscure until original investigations are instituted and the insect recollected.

1879

13. Antheræa roylei, *Moore.* (*See p. 237.*)

Habitat.—This is the oak-feeding silkworm of the Himálaya, which may be said to be found throughout the entire length of these mountains.

Much confusion still, however, exists regarding this insect, though it has been studied by Hutton under domestication, and in spite of the fact that detailed scientific descriptions of it have appeared (see *Jour. Agri.-Hort. Soc. Ind., XIII., 147;* also in the publications of the *Entomological Society of London*). Geoghegan (*Silk in India, p. 160*) remarks that "it is properly an annual, but can be made to yield two or three crops. Its silk is favourably spoken of, but it is not abundant. The true cocoon is contained in a large closely-woven glazed case and enveloped on all sides by the leaves

* Mr. Hampson, in the List given as an Appendix (p. 237), quotes **A. perotteti** as a synonym for **A. assama.** If this be correct the difficulty is immensely increased, since that species has never been recorded as feeding on **Odina Wodier**, nor, indeed, on any member of the Natural Order to which that tree belongs—*Ed. Dict. Econ. Prod.*

ANTHERÆA ROYLEI.

of the tree, the impression of the nervures being deeply imprinted on the glazed surface. It is like the *tasar* moth, but smaller. It can be domesticated." **Hutton** tells us that it is necessary to remove the external glazed coating before the silk-yielding cocoon can be arrived at (*Jour. Agri.-Hort. Soc. Ind., XIII., 57*). It may be pointed out that the peculiarity of causing the leaves of the tree to adhere to the cocoon recalls **Rumphius'** drawing of the Amboyna insect, which he says in that respect differs from the *tasar* of Bengal. In the *Indian Museum Notes* (*Proofs of Vol. II., No. II.*), which the writer has had the advantage of consulting, a brief account of this species is given, in which doubt is thrown on the idea of the cocoon always consisting of two layers. It would also appear from the "*Notes*" that, in some cases, the cocoon is pedunculated, in others not. These differences of opinion, regarding so common an insect, are most perplexing, and it seems, therefore, more than probable that the descriptions of two widely different moths have been confused one with the other. * * The liberty may be here taken to publish the following passage from a paper by **Cotes** in the Museum Notes, although professedly submitted to the author in rough proof, and, therefore, subject to revision before being published in the "Notes":—

Conf. with pp. 70, 72, 73, 98-9, 157.

1880

"Little has hitherto been recorded about this insect, though it has been bred by several people both in India and in Europe (see *Entomologist, XIV., p. 246*, and *Bull. Soc. Ent. France (5), IV., p. 154*). **Hutton** writes in the *Jour. Agri.-Hort. Soc. Ind., III., p. 125, 1871*: '**Antheræa roylei** is common at Simla, Mussoorie, Almorah, and I think Darjíling. It feeds upon the common hill oak, spinning a large but thin cocoon between three or four leaves . . . . The outer coating is very strong, and I do not think it could be reeled, but within this case is the true cocoon, of an oval form and yielding good silk. The worms are easily reared and sometimes give two or three crops, but this is when treated in the house.' **Major Harford** also, writing recently from the North-West Himálayas, notices the peculiar double-walled cocoons which he has found upon *ilex* (hill oak?) and the khaki-coloured males and pinkish females he has bred from them. Some observations also on the habits of **A. roylei**, by the late **Mr. Otto Moller**, appeared on page 201 of Vol. I. of these *Notes*; it is thought, however, that these observations referred to some other species, the cocoon forwarded with them being single-walled and pedunculate, instead of having the double-walled structure characteristic of typical **A. roylei** cocoons.* A double-walled cocoon found by **Mr. A. V. Knyvett** * * on a chestnut tree in Sikkim was sent to the Indian Museum in May 1890, and produced, in the early part of the rains, a female moth of the typical pinkish colour, and **Mr. Knyvett** also writes that he has found **A. roylei** cocoons on *mohwa* trees, oak, and birch in Sikkim. The above is all that we at present know about **A. roylei** proper. The figures o the caterpillar and cocoon are taken from specimens obtained by **Mr. A V. Knyvett** in Sikkim; those of the moths from **Hutton's** type specimens which are in the Indian Museum collection. Closely connected with **A. roylei** proper is an insect with a firm single-walled pedunculate cocoon* * which has been

Conf. with pp. 70-71.

* **Mr. Otto Moller** states that the insect he had forwarded fed on **Evodia fraxinifolia** —a Rutaceous, or on **Daphniphllum himalense,**—a Euphorbiaceous plant. The fact of the food thus differing so materially from that on which **A. roylei** is generally known to feed, confirms the suspicion that **Mr. Moller** may not have found **A. roylei** but some other and perhaps an undescribed species of **Antheræa.**

* * The suspicion indicated by the above footnote has been abundantly confirmed by **Hampson**, for he has isolated at least the specimens collected by **Knyvett** as a distinct species, *viz.*, **A. knyvetti.** See the Appendix, p. 237.

SILK:
Saturniidæ.

The Wild Silks of India.

ANTHERÆA ROYLEI.

found by Mr. Knyvett upon wild cherry trees in Sikkim. The cocoons of this form, which have been sent to the Indian Museum, are much like the cocoons of **Antheræa frithi,** but moths reared from them by Mr. Knyvett are almost indistinguishable from moths reared from the typical double-walled cocoons of **A. roylei,** the chief distinction consisting in the greater pinkness of those reared from the pedunculate cocoons. As, however, the females reared from double-walled cocoons show every variety of colour from pink to greenish brown, and are in some cases altogether indistinguishable from females reared from pedunculate cocoons, it would seem most probable that the difference in the structure of the cocoon is to be attributed more to the difference in the food plant than to any specific distinctness in the insects. The differences observed by Mr. Knyvett between the caterpillars of the two forms, though very remarkable, are not of a sufficiently radical nature to warrant the creation of a new species."*

Mauha Tree. 1881 *Conf. with p. 71.*

It will be observed that the suggestion of a possible error is covered by the above passage. The *mohwa* (=*mauha*, or **Bassia latifolia**) tree does not grow in the Darjíling district much above 4,000 feet in altitude, whereas the birch tree (also mentioned as one of those on which the worm feeds) might more correctly be said to occur at altitudes from 7,000 to 14,000 feet. The writer has seen this insect feeding near Simla on the leaves of **Quercus incana,** and he believes that or some alied species of Himálayan oak, such as **Q. dilatata, Q. semecarpi-folia,** or **Q. Ilex** to be its natural food, and is therefore disposed to doubt the accuracy of Major Harford's observation that he found the worm feeding on the **Ilex,** still more so of its having been on the mauha tree.

Hybrids. 1882 *Conf. with pp. 72, 79, 101.*

Hutton informs us that he had succeeded to produce a hybrid between this and the ordinary *tasar*, but that the produce never came to anything. **A. roylei** is, however, closely allied to the Chinese insect, **A. pernyi,** a semi-domesticated species, which for centuries has been reared on a species of oak. It, in fact, affords one of the commercially recognised forms of Chinese "Tasar" silk. M. Rondot mentions the fact that a hybrid has been produced between **A. roylei** and **A. pernyi,** but he does not tell us whether or not the cross possesses any advantages over either of the pure species. Hutton mentions the fact that he had succeeded to send live cocoons of **A. roylei** to France, and as **A. pernyi** had some time previously been introduced into that country, a hybrid might have originated from Hutton's contribution to French sericulture. It seems probable that an industry, similar to that of China, dealt with under **A. pernyi** might be established on the Himálaya in rearing either **A. roylei, A pernyi,** or the hybrid between these. Hutton says of **A. roylei**:—

"This species is very easily reared in the house, but the most effectual way of securing a brood is to tie the female out at night upon some tree, when a wild male will almost always be found with her next morning, remaining until sunset, when he again departs. It is, strictly speaking, an annual, but it may be found in some stage or other from April even to September. When first hatched the caterpillar is entirely black with a rufous brown head; the tubercles, of which there are six longitudinal rows giving out tufts of whitish hairs, are rather prominent. At this stage the body is much broader in front than towards the rear, the anterior lateral tubercles projecting forwards on each side of the head. The second stage after the first moult, the colour becomes green, the rows of tubercles crowned with orange, and the apex giving out generally one and sometimes two long stiff hairs which are clubbed or buttoned at the tips; a central

* This view has not been upheld by Hampson—an entomologist not given to unnecessarily multiplication of species.—*Ed., Dict. Econ. Prod.*

ANTHERÆA ROYLEI.

dorsal tubercle on the penultimate segment; head and anterior segments pale greenish; legs faint brown; the second and third segments are the most prominent; a faint flesh-coloured stripe above the stigmata ending in a triangular brown patch on each side of the anal foot; a row of small tubercles beneath the stigmata of an ultramarine blue; two black spots obliquely transverse on the back of the first segments; anal shield bordered with ultramarine blue."

After the second moult the colour is still green, with the tubercles of the back tipped with pale yellow, and in the third and fourth moults the colour becomes pale green with a faint greyish tinge, but the hairs from the dorsal tubercles are long and pale-coloured and no longer clotted at the end. Though the worms attain a large size, so much, says **Hutton**, does their colour approximate to that of the back of the oak leaf upon which they feed, that although the ordure beneath betrays their presence, they are by no means readily seen.

14. Antheræa sivalika, *Moore.* (*pp. 45, 122, 237.*)

1883

This is the *tasar* worm met with on the submontane districts of the Panjáb. It feeds upon the **Zizyphus Jujuba** (*ber* or *beri*), and is plentiful in the Hoshiarpur District. There would seem no doubt but that this should be accepted as a specially pale-coloured local form of the ordinary *tasar*. For further particulars see p. 120.

15. A. yamamai, *Guér. Méne.*

1884

This may be designated the Japanese form of *muga*, which has been several times introduced into India without apparently any permanent result. **Hutton** says that it "is well thought of both in England and in France, where great efforts have been made to introduce it, but as yet with very indifferent success." He then adds that it may be reared on the Himálayan spiny oak (**Quercus incana**), but that it requires to be carefully protected from its insect enemies. Reference has already been made (p. 71) to the reported discovery of a valuable hybrid between this insect and the ordinary *tasar*. This discovery was made by a **Mr. Mowis**, who, for some short time, was employed in connection with the Poona experiments with *tasar*. Some idea of the very high, and it is feared misplaced, expectations of the value of **Mr. Mowis'** discovery may be learned from the following passage from an official publication: "**Mr. Mowis**, in spite of repeated calls from Government, has forwarded no report, and it has been ascertained indirectly that he declines to make known the results of his process, preferring to use his knowledge for his own benefit. He has, however, shewn the result of his experiments both in the shape of the hybridised worms and their produce to many private individuals, and there can be no doubt as to their complete success. They undoubtedly, in the opinion of this Government, open out a prospect of the rise of a most valuable industry in India, where the food on which the worm lives, *viz.*, the *nandrúk* (**Ficus Benjamina**) and other varieties of the **Ficus** are indigenous and abundant, and it is strongly recommended that a supply of eggs of the **yamamai** be procured from Japan, and distributed to the different model farms in this country with a view to the experiment becoming widely known and encouraged."

Hybrids.
1885
Conf. with pp. 72, 78, 101.

It perhaps need only be added to what has been said above that nothing more has been heard of this valuable hybrid, and that, consequently, neither **Mr. Mowis** nor the people of Bombay have as yet reaped the vast fortunes that they were supposed to be on the eve of obtaining. It may also be added that, like **A pernyi**, this species is assumed to contribute to some extent to the consignments of cocoons returned in the trade statistics of imports into Europe under the name of "Chinese Tasar."

III. ATTACUS. 1886

III.—ATTACUS.

Entomologists have referred some eight or nine species to this genus. One, the Eri or Castor-oil-feeding insect, is of considerable importance to India, and its cocoons are even exported to some extent (see the detailed article pp. 162-174). Two others may be said to have attracted a greater degree of consideration than is generally the case with the truly wild silk-producers. These are the Atlas moth and **Attacus cynthia,** an insect which by some writers has been called the Ailanthus silkworm, because of its feeding on the leaves of the tree of that name.

Hutton speaks of the genus **Attacus** as being characterised by the moths being devoid of the power to secrete a solvent fluid, and consequently the egress of the insect is provided for by the worm leaving the threads at the head of the cocoon unagglutinated, and only loosely drawn together, so as to be easily pushed aside by the head and forefeet of the imago.

(*See p. 236.*)

1887

16. Attacus atlas, *Linn.; Moore, Cat. Lepid. Insects, II., 405.*

Habitat.—Found in China, Burma, India, Ceylon, and Java. In India its habitat extends from Sylhet and Cachar to Sibsagor, Johore, Sikkim, Mussoorie, and Almora; also Bengal, Burma, Madras and Ceylon. According to **Gosse** it ranges in South-Eastern Asia over 35° of lat. and 55° of long., being abundant in India and China as well as throughout the Malay Archipelago (*Entomol., XII., 25*).

Food. 1888

Food of the Worm.

1. **Ardisia sp.**

Mr. **Manuel** mentions a common species of this genus, found all over Pegu and Martaban, as being eaten by the worm.

2. **Artemisia vulgaris,** *Linn.*

Vern.—*Dona,* HIND., BENG.; *Titapat,* NEPAL; *Nagdana,* CACHAR.

A gregarious shrub found in Sikkim and many other parts of the Himálaya, also in Bengal and Assam. It is mentioned by **Brownlow** as a tree on which the **Atlas** may be found.

3. **Bischofia javanica,** *Bl.*

4. **Cedrela Toona,** *Roxb.*

This is known in Cachar as the *Lud.*

5. **Clerodendron infortunatum,** *Gærtn.*

Mentioned in the Burmese list of foods.

6. **Dillenia indica,** *Linn.*

7. **D. pentagynia,** *Roxb.*

8. **Glochidion lanceolarium,** *Dalz.*

A shrub on which the insect is said to feed in Mussoorie.

9. **G. velutinum,** *Wight.*

10. **Lagerstrœmia Flos-Reginæ,** *Retz.*

Mentioned by **Mr. Manuel.**

11. **Leucosceptrum canum,** *Sm.*

Said to be eaten in Sikkim.

12. **Melastoma malabathricum,** *Linn.*

Vern.—*Choulisy,* NEPAL; *Tungbram,* LEPCHA; *Shapti,* MECHI; *Lutki,* CACHAR; *Yetpyai,* BURM.

A large bush met with throughout India up to an altitude of 6,000 feet, chiefly near water-courses. The Atlas worm, when fed on the leaves of this tree, is said to give a very white silk.

13. **Ocimum sp.**

Mentioned by **Cameron** as eaten by the worm in Bangalore.

14. **Phyllanthus Emblica,** *Linn.*

ATTACUS ATLAS.

This is mentioned by **Hutton, Manuel,** and other writers.

15. P. lanceolaria.

Important Practical Observation. 1889

The Rev. **Mr. Campbell** says the atlas insect is very plentiful in Chutia Nagpur, where it feeds chiefly on this low shrub. **Mr. Campbell** succeeded, however, to rear it on the leaves of **Dalbergia Sissoo,** and he adds that he has seen the worm feeding on **Eugenia caryophylla** and **Symplocos racemosa.** The cocoons of the worms fed on *Sissoo,* **Mr. Campbell** remarks, in no way resemble the normal type. They are smaller and appear more easily reeled.

16. Sapium insigne, *Benth.*

Vern.—*Khinna,* HIND.; *Dudla,* PB.

A tree of the Sub-Himálayan tract, also found in Chittagong, Burma, and Western Gháts.

17. Schleichera trijuga, *Willd.*

Mentioned by **Manuel** as one of the trees on which the insect feeds in Burma.

18. Stephegyne diversifolia, *Hook. f.*

This is mentioned by several writers as the plant on which the insect feeds in Chittagong and Martaban.

19. Symplocos cratægoides, *Ham.*

Vern.—*Lood,* CACHAR; *Lodh,* KUMAON; *Loja,* SUTLEJ; *Loj,* PB.

A small tree inhabiting the Himálaya at an altitude from 3,000 to 8,000 feet; also Assam, Khásia, and Martaban Hills.

20. Vangueria spinosa, *Roxb.*

A plant reported to be eaten by the worm in Assam.

21. Wendlandia Notonia, *Wall.*

The worm is said to eat the leaves of this plant in the Madras Presidency.

Brownlow gives a list of Cachar plants (most of which it is believed are included in the above), but as he furnishes only local names for them it has not been thought desirable to make the attempt to determine what they may be.

Provincial Notices Regarding this Silk.

Bengal. 1890 Assam. 1891

BENGAL & ASSAM.—Very little has been written regarding this insect in the Provinces named. It occurs, however, fairly abundantly, and has been practically domesticated by **Mr. C. Brownlow** in Cachar, who urged that this insect possessed even stronger claims to consideration than any of the other wild silkworms. His papers will be found in *Vol. XIII (1865), 392-415* of the *Jour. Agri.-Hort. Soc. Ind., viz.,* "On the feasibility of a mixed system of open air silk-culture," and *Vol. III., 1872:* "A few notes on the Atlas worm." **Mr. Brownlow's** paper is given in the appendix to **Geoghegan's** *Silk in India,* and is quoted by **Mr. Wardle** as "from a Government Report" (*Wild Silks of India, p. 63*). **Stack** says this is the *petogore muga* of Assam. In an official communication **Mr. W. C. Taylor,** Settlement Officer, Khurda, forwarded to the Collector of Purí in 1879 samples of this insect together with a note on the subject, from which it would appear to be fairly plentiful in the forests of that portion of Bengal. **Hutton** in a paper (*Agri.-Hort. Soc. Ind., Vol. II. (1870), 146-176*) on "the Cultivation of Silk in the Australian Colonies" recommends this species to consideration, and **Gosse** (in the *Entomologist, 1879*) gave an account of his endeavours to acclimatise it in England.

Himalaya. 1892

HIMALAYAN DISTRICTS.—One of the earliest accounts of the Altas moth is that by **Lady Isabella Rose Gilbert** (1825), but **Hutton** and other writers

SILK: Saturniidæ.

The Wild Silks of India.

ATTACUS ATLAS.

speak of it as found on the lower hills and as feeding more especially on the species of **Sepium** and **Glochidion**. Although found, therefore, on the Himálaya, it is not very well known to the people, and no use is made of its cocoons.

Madras. 1893

MADRAS*.—The following notice regarding this moth appeared in Mr. Morgan's Report of the Silkworms of Malabar:—"The Atlas moth (**Attacus atlas**) is found throughout the Province, but cannot be said to be abundant anywhere. I found the larvæ feeding in Wynaad on **Bischofia javanica, Wendlandia Notonia, Mallotus subpeltatus, Eurya japonica,** on three species of **Tetranthera,** and on a host of others too numerous to mention here. The cocoons are of a very dark-brown colour, and are invariably hidden by leaves, which the larvæ gums round itself when it commences to spin its cocoon. I have carefully watched those I have collected and brought in from the forests, and when about to spin they invariably commence by drawing together the edges of the largest leaves they can find, and attaching them to one another by silken threads. Within this leafy screen, the cocoon is rapidly perfected; and if the leaves are afterwards removed, it will be found that the cocoon has exactly taken the shape of the leaves, their nervation even being clearly impressed on the silk. The cocoons of this species are never collected or utilised in the district.

It is easily bred under shelter; for I successfully hatched two eggs under a glass shade in my house. The larvæ were fed on **Bischofia** leaves, and spun a fine cocoon. The imagos came out in due time and were fine specimens. The larvæ of **Antheræs paphia** will not stand such confinement, but die when about three-fourths grown."

Burma. 1894

BURMA.—By far the most complete local account of this insect is that given by Mr. R. A. Manuel (*Jour. Agri.-Hort. Soc. Ind., 1881-82, VII., 303-307*) in a paper on the silks of British Burma. Omitting the paragraphs on the plants on which the worm feeds (which have been amalgamated in the remarks above on that subject), the information furnished by Mr. Manuel may be here republished:—"The great Atlas moth of Further Asia, the vastest of all known lepidoptera, has been described by every naturalist of note from Linnæus downwards, so that few of the family are better known than '*Le Geant des Papillons.*' It is a widespread species, ranging over the south and east half of continental and insular Asia, common on the slopes of the Himálaya and all through India to the points of both peninsulas, abundant in China, scattered over the isles of the Archipelago, from Java to Molucca, to Borneo, and to the Philippines, a range of 35° of latitude and 55° of longitude. As is often the case with animals of extensive habitat, this magnificent insect is subject to considerable variety. Here in Burma it has its home in all the lower moist forests, and in some of the upper mixed ones too. In the sunlight its strong, vigorous flight gives one more the idea of a small bird than of a moth. On account of its great size,—the expanse of its wings at times exceeds 9† inches,—and from its vigorous flight the natives call it the *Seen-lypea,* or elephant moth. The late Mr. Blyth, of the Asiatic Society's Museum, Calcutta, writing of some specimens of it sent from here, says: 'A well-known moth from Burma' . . . . 'a splendid species, common in collections from China, Assam, Sylhet, and Aracan.' And the late Dr. Mason mentions it in his work on Burma as abounding in the province."

"*Crops.*—The Atlas spins once a year, but it can in domestication be got to spin twice,—once towards the close of the rains and again just before its commencement.

*Mr. Hampson informs the editor that this insect occurs on the Nilgiri hills up to altitudes of 6000 feet.

† Sometimes over 11 inches.—*Ed.*

ATTACUS ATLAS.

The cocoons on the latter occasion are never so good as those spun during September to November. The cycle of its existence is from 81 days to 8 or 9 months:—

	Days.	
In the ova state	8 to 11	
In the larval state	36 to 40	
In the cocoon state	30	to 8 or 9 months.
In the moth state	7 to 15	
TOTAL .	81	to 8 or 9 months.

"The cocoons are solitary, enclosed in leaves, in colour they are a light umber or drab, the surface (independently of the impress of leaves) roughly granular, scarcely at all silky or floccose except at the mouth; its substance is thin, parchmenty, very firm; the interior very smooth and even subglossy. The upper extremity forms a natural orifice for the exit of the moth, but the great number of silk fibres which are here left ungummed and are loose, soft, and flossy, effectually prevent the ingress of intruders. The cocoon is not closed like those of **Bombyx mori** and **Antheræa mylitta.** As a result of this structure the exit of the moth leaves no disturbance behind, no disarrangement of the fibres. The cocoon is from 2 to 3 inches long and about 1 inch broad in its widest part; it weighs, without the chrysalis, about 6 or 7 grains.

"*Cultivation.*—The Atlas is an indolent worm, seldom travelling far from its birthplace unless compelled to go in search of food. The ova are laid in groups, or strings, on the under-surface of leaves which are the food of the worm: this the instinct of the female leads her to accomplish. They are all strongly gummed to the surface, so that they not only are not shaken off by the wind but the crawling out on the young larvæ is facilitated. They are as large, or perhaps slightly larger, than *tusser* ova, broadly ovate, and granular on the surface; in colour white, clouded with purple brown, which tint centres in an irregular mass of intense depth. All this colour is readily washed off by a few moments' immersion in water, the tinge being communicated to the water, leaving the egg of a delicate greenish white. The young larvæ are about three lines long in repose and five lines when crawling. Their general colour is black with a broad-based band of light grey running down the back for the whole length and crossed at the side of each segment by two white lines. The tubercles are tall cylinders of pure white, tallest in front; all of them have white hairs which, uniting laterally, form conspicuous transverse bars of white, one on every segment. From each tubercle proceed several very slender black hairs of great length. Head glossy black, unspotted; the clypeus grey; anal region white, feet black, prolegs grey. The little worms do not congregate like the larvæ of **Cricula** but sit on the under-side of a leaf almost always in a double position with the head bent round towards the tail. They moult on an average once a week, and go through six moults. The first moult, however, takes place sometimes after seven days. Very soon there is an entire change in the whole appearance of the worm, which, for the greater part of its existence, seems as if thickly powdered with very fine white dust-like flour. Full grown, they become greenish white, the skin all studded with minute oval darker specks, which give the impression of translucent cells in the substance. The orange clouds on the sides are nearly all obsolete, especially the posterior ones; last segment azure with the oval specks dark blue. A rondo-triangular ring of rich pale orange is conspicuous on the outside of each hindmost proleg; the face is wholly pale green, leg and clypeus margined by a black line; thoracic tubercles shorter and blunter; the rest much increased in length and become soft spines, lying nearly flat, pointing backwards and overlapping, the lowest row, dark-iron grey. Feet and prolegs iron grey, the latter crossed by a band of greenish white, the whole thickly covered by a snow-white farina. Just before spinning the worm is nearly 5 inches long and 1 inch high.

"In spinning, the worms generally use one leaf, but they attach the cocoon by the means of their silk not only to the peduncle of the leaf but to a great extent to the branch to which the leaf is attached, so that in the event of the leaf withering the cocoon does not fall but remains pendant from its tip.

"*Manufacture.*—Although the silk of the Atlas can be reeled, this is not done on account of the difficulty experienced in freeing the filaments of silk from their position in the cocoon. **Captain Hutton**, the entomologist of the Himálayas, pronounced the silk as decidedly good, and **Dr. Chavannes**, of Lausanne, considered its introduction into France desirable. In appearance the fibre is very much like *tusser*; if anything, it is in hue a little darker, but to the feel it is much finer

ATTACUS ATLAS.

and softer. The 'waste,' or floss, is easily prepared and the dark brown umber or drab colour is easily discharged. Like *tusser*, however, the fibres are flat and striated and will not take all dyes. The uses to which *tusser* is put are suited to Atlas silk, and China annually exports large quantities of Atlas-waste to the European markets."

COCOON. 1895 SILK. 1896

The Atlas Cocoon & Silk.

Mr. Gosse, in the paper already quoted, says of the cocoons that they are in colour "a light umber, or drab; its surface (independently of the impress of leaves) roughly granular, scarcely at all silky or floccose, except at the mouth; its substance thin, parchmenty, very firm; the interior very smooth, and even sub-glossy. The upper extremity forms a natural orifice for the exit of the moth, made by the conveyance of a great number of silk-fibres, which are left ungummed, and are thus soft and flossy, the gummed, stiff silk passing up on one side and contracting into the cord. Thus the cocoon is not closed, like those of **Bombyx mori,** of **Telea,** of the **Antheræa,** but open, like those of **A. cynthia,** of the **Samiæ,** of the **Saturniæ.** As a result of this structure, the exit of the imago leaves no disturbance behind, no witness, no disarrangement of these soft fibres, such as is the case with **yamamai, pernyi,** and **paphia (Antheræa mylitta).**"

Hutton goes into great detail on the subject of this cocoon (*Jour. Agri.-Hort. Soc. Ind., XIII, 82-83-87*). "After repeated trials in various solvents," he says, "I have hitherto proved unsuccessful in doing more than reeling to a certain extent, with the certainty of the thread snapping every few feet. The intractability seems more attributable to the peculiar manner in which the cocoon is formed than to the insolubility of the agglutinating substance, for the cocoon appears to be composed of separate layers or silken bags one within the other, and the thread is sure to snap at the end of each of these. There seems no reason, however, why the silk should not be carded and spun after the method pursued with the Assamese *eri* (**A. ricini**), and if this be accomplished, the silk may be brought into the market, as the insect is easily reared, and is abundant in some of the Sub-Himálayan tracts." Hutton says that he found a cheap coarse vinegar the best solvent for the cocoons.

Mr. Wardle informs us that "the diameter of the external fibres of the cocoon is very variable, averaging about $\frac{1}{1320}$ inch, whilst that of the internal fibres is more uniform, and about $\frac{1}{1000}$ inch. The outer fibres are capable of supporting an average weight of $2\frac{1}{8}$ drams, and the inner $2\frac{1}{3}$ drams. The tension of the outer fibres is one inch to the foot, and the inner $1\frac{1}{4}$ inches. The fibres are flat and longitudinally **striated,** and united in pairs by their edges."

1897

17. Attacus canningi, *Hutton.* (*See p. 236.*)

Habitat.—North-West Himálaya; common in a wild state. It produces annually hard, compactly-woven cocoons of a rusty orange or grey colour. A writer in the *Jour. Agri.-Hort. Soc. Ind., XI. (1859)* speaks of this as the Himálayan *eri*. It feeds on the leaves of the following bushes:—**Coriaria nepalensis,** *Wall*—a shrub of the Himálaya from Murree to Sikkim, Bhután, and Manipur; and **Zanthoxylum alatum,** *Roxb.*, an equally plentiful bush on the lower slopes of the Himálaya.

In the *Jour. Agri.-Hort. Soc., Ind., XI. Proc., 1859), 84,* Mr. Turnbull announced that he had been unsuccessful in reeling these cocoons, though he had tried many different methods, such as warm water, solution of borax, soda, cow-dung, etc. This species Hutton subsequently found was one and the same as the next. Hampson will be seen (*p. 36*) to quote the name here given as a synonym for **A. cynthia.** The species is retained for the present, however, since the name **A. canningi** is so very frequently

ATTACUS CYNTHIA.

mentioned by Indian writers, and is even given by **Moore** in the list furnished to **Wardle's** *Wild Silks of India.* (*Conf.* with remark under **A. guerini** below, p. 86.) [*pp. 162, 263.*)

1898

18. Attacus cynthia, *Drury; Moore, Cat. Lepid. Insects, II., 407.* (*See*

Habitat.—This moth inhabits the regions from Sylhet, Cachar, and Shillong to Sibsagor, Sikkim (common up to 5,000 feet), Mussoorie, and Simla; it is also said to be found in South Andaman, and, **Mr. Hampson** adds, on the Nilghiri hills. It is distributed to China and the Malay Islands.

Food. 1899

Food.—Much confusion still exists regarding this insect and its separation from the *eri* (**A. ricini**). **Moore** says that it is the *eri* of **Royle** and of **Hugon**, as also the *arrindy* or *arrundi* of **Roxburgh**. **Hutton** remarks (*Jour. Agri.-Hort. Soc. Ind., XII., 63 & 71*) that, like the domesticated *eri*, it will eat the castor oil plant, but in Mussoorie it is generally found on **Coriaria nepalensis, Zanthoxylum alatum,** and other such bushes In an essay (*Jour. Agri.-Hort. Soc. Ind., New Series, II., 173-74*) on the "Cultivation of Silk in the Australian Colonies," **Hutton** gives some useful hints regarding this insect. Alluding to **Mr. Brady's** successful introduction of it into the colonies, he would seem to have thought that good results might be anticipated. He repudiates the idea that it is the wild state of **Attacus ricini,** on the ground that it cannot be successfully reared on the castor-oil plant. **Hutton** also says: "I have even reared it easily enough on the common Cape Woodbine, so that its introduction into England seems feasible enough, especially as the **Ailanthus glandulosa** is said to thrive there." In China this insect has been domesticated for centuries and fed on **Ailanthus,** hence, by European writers, it is often spoken of as the Ailanthus silk-moth. In the *Proceedings of the Agri.-Horti. Soc. Ind. for 1861* (*p. 135*) occurs a correspondence on the subject of **Dr. Bonavia's** attempt to domesticate this worm in Oudh. **Hutton,** in referring to this subject, commends the zeal of **Mr. Brownlow** of Cachar in his experiments with silk-moths, and announces that that gentleman had recently found the **cynthia** insect there. The late **Mr. Stack,** however, informs us that the people of Assam believe this to be the *eri* in a wild state. He adds "it appears to be commonest in Cachar, but it is also known in Kamrup."

It had been found by **Captain** (now **General**) **Strachey** in Kumáon, so that **Hutton** was constrained to add "I have no doubt that the insect may be found along the entire line of the Sub-Himálaya from the Sutlej to the sea." In a report on Sericulture in Kangra the late **Captain Bartlet** wrote in 1883: "The **Ailanthus** tree (native name *tirimul*) grows wild about this district, and the wild cocoons are those of the Ailanthus silk-worm (**Attacus cynthia**), which I found in the larva state on the bushes on the estate. I have no knowledge of any other wild silkworm, and am under the impression that either attempting to rear or collect the cocoons of this kind would not pay, particularly when the **Bombyx mori** flourishes so well here." Commenting on **Dr. Bonavia's** experiments **Hutton** thought the heat of Oudh would, however, prove inimical. "There is nothing like trying, however, and as we have the real Simon Pure at Mussoorie, you need not bewail the absence of **Mr. Fortune** from China at this moment. The fact of our possessing the **Attacus cynthia** in India affords another proof of the necessity of our setting our own shoulders to the wheel if ever we wish to ascertain the resources of the country, and it says little for our own enterprise and knowledge, when we allow ourselves to be sent to seek in China what, without knowing it we already abundantly possess in India."

The Ailanthus silkworm has been successfully introduced into England, Europe, Australia, etc., but it may be said that interest in the

The Wild Silks of India.

ATTACUS CYNTHIA.

subject has practically died out, and that no further action appears to have taken place in India than the futile experiments in Oudh. While this result is extremely disappointing, it seems likely that attention may once more turn to this insect, for, if it has been successfully domesticated in China, there would seem no good reason why it might not be also reared in some part of India.

In the forthcoming number of the *Indian Museum Notes*, the following passage occurs regarding this insect and its silk:—

"Its larval stages do not appear to have been observed in India. The cocoons, though much smaller in size, are very similar in structure, coloration, or general appearance to the cocoons of **Attacus atlas**; they are smaller and more compact than cocoons of **A. ricini**, but appear, nevertheless, to contain a considerable amount of silk. This silk would no doubt be valuable for carding purposes if it could be obtained in any considerable quantities, and of late they have appeared in the Calcutta market where they are known as *junglies*. Mr. G. C. Hodson writes that they are brought to market by Muhammadans, while the *eri* and *muga* trade is entirely in the hands of *Marwarries*. According to **Hutton**, the insect also is identical with a semi-domesticated species, which is reared on a small scale in some parts of China upon **Ailanthus glandulosa** for the production of silk, and which has also been raised experimentally upon the same plant in Europe."

It is not known to what extent the cocoons and silk of the **Ailanthus**-feeding worm are exported from China and Japan, but, as already pointed out under **Antheræa pernyi** and **A. yamamai**, the so-called "Tasar Silk of China" would appear to embrace more than the produce of **Antheræa paphia** (if indeed any of the true *tasar* is exported from these countries), and it is probable therefore that a considerable amount of Ailanthus silk finds its way to Europe. If this suggestion proves correct, it would give the comparison with the Indian trade more correctly to class **Antheræa pernyi** as *muga* and **Attacus cynthia** as *eri*. Some such classification would not only allow a correct comparison between the Chinese and Indian trade, but it would facilitate an appreciation being formed of the value of the true *tasar* in the European markets. The writer has been unable to discover the amounts of silk and cocoons of **Antheræa pernyi** and **Attacus cynthia** imported into India, but this much can be said, the entire Chinese imports are certainly not mulberry silk. The writer when visiting a silk mill in Bombay was shown the material imported. It proved to be merely waste cynthia silk. (*Conf. with pp. 210, 213.*)

1900

19. Attacus edwardsii, *White.* (*See p. 236.*)

Habitat.—Sikkim, Khásia Hills, Mussoorie, the Western Gháts, and Mysore, but is nowhere abundant. According to the late Otto Moller it occurs at an altitude of from 6,000 to 7,000 feet.

Hutton, in his essay on "The Cultivation of Silk in the Australian Colonies," says that although the Chinese are said to have domesticated this insect as also **Attacus atlas**, "it would be a difficult thing to rear them profitably." In his "Notes on the Indian Bombycidæ (*Jour. Agri.-Hort. Soc. Ind., Vol. XIII., 87, also New Series, III., 139*), he remarks that it was found originally at Darjíling, is much darker in colour and rather smaller in size than **Attacus atlas**. His remarks regarding it are, however, apparently derived from Moore's descriptive notes and drawings, (see *Cat. Lepid.-Insects, II., 486*). In the *Museum Notes* (*Vol. I., 201*) Mr. Otto Moller says that this insect is an annual: the moth appears in the rains (July to September) and it hybernates as a pupa. The cocoon is much like that of **A. atlas**, except that it is denser in structure.

20. Attacus guerini, *Moore.* (*See p. 237.*)

ATTACUS GUERINI. 1901

Habitat.—It inhabits Eastern Bengal.

Hutton, however, says this insect has only been seen in the Museums in England, and he is accordingly disposed to regard it as little more than an ill-fed condition of **A. ricini**; he cites a case known to him where under-fed **Actias selene** only attained about one quarter their natural size (*Jour. Agri.-Hort., Soc. Ind., III.* (*New Series*) *140*). In the *Museum Notes* (Vol. II., Pt. II., proofs of which have just been seen by the author) it is stated that "**A. guerini, A. obscurus, A. canningii,** and **Saturnia iole** have at different times been described as distinct species. From the descriptions and figures that have been given of them, however, they appear to be so nearly related to **A. cynthia** and **A. ricini,** that until cause is shown for their separation, it seems best to look upon them as synonyms of one or other of these two species." Hampson treats **A. guerini** and **A. obscurus** as synonyms for **A. ricini,** and places **A. canningii** under **A. cynthia.**

21. A. obscurus, *Butler.* (*See p. 237.*)

1902

Habitat.—Occasionally found in Cahar; the worm feeds on a plant called in the vernacular *Lood* (= **Cedrela Toona**). See remark under **A. guerini** above.

22. A. ricini, *Boisd.; Moore, Cat. Lepid. Insects, II., 407.* (*See p. 237.*)

1903

Habitat.—This insect inhabits Assam and parts of Bengal, and extends from Darjíling to Nepál and Kumáon. It is the *eri* of Assam and *arindi* of Bengal. **Hutton** offers the suggestion (Essay on "The Cultivation of Silk in the Australian Colonies") that this insect may be "an importation from that 'old curiosity shop' termed China." He repudiates the statement that its wild condition is the insect known as **A. cynthia** on structural as well as practical considerations, chief among the latter being the fact that the natural food of **cynthia,** in the wild state, is not **Ricinus communis.** He lays stress on the presence of rows of black spots between the tubercles of the larvæ in **A. cynthia** throughout all its stages, while **A. ricini** is devoid of these in its perfect condition (*Jour. Agri.-Hort. Soc. Ind., XIII., 71-82*). A fuller account will be found at the end of this enumeration of the species p. 162.

Eri. 1904 *Conf. with pp. 162-164.*

IV.—CALIGULA.

IV. CALIGULA. 1905

Hampson does not regard this genus as separable from the SATURNIA. [*See p. 238.*

23. Caligula simla, *Westwood; Moore, Cat. Lep. Insects, II., 309.*

Food. 1906

Habitat.—Simla, Kumáon, Mussoorie, and the Khásia Hills.

This insect forms an open, net-like cocoon and feeds on—

1. **Juglans regia,** *Linn.* (**A selene** feeds on this tree *see p. 68 also p. 95.*)
2. **Pyrus communis,** *Linn.*

THE COMMON PEAR.

Vern.—*Naspati, tang, sunkeiut,* PB.

Cultivated and sometimes wild in the North-West Himálaya and Kashmír. The "wild pear," alluded to by **Hutton**, instead of being the pear, may have been **Pyrus Pashia,** which is often called a pear.

3. **Salix babylonica,** *Linn.*

Vern.—*Bisa, bada,* PB.; *Giur,* KASHMIR; *Tissi,* NEPAL.

A tree commonly cultivated in North India.

The above are the plants mentioned by **Hutton,** but they are so utterly diversified, both in habitat and nature, that it is difficult to believe the same insect could feed on all of them. The willow is more likely to be **Salix tetrasperma** (*Conf. with Jour. Agri.-Hort. Soc. Ind., III.* (*New Series*) *141*).

CALIGULA CACHARA.

1907

24. Caligula cachara, *Moore, Proc. Zool. Soc., 1812, 578.* (*See p. 238.*)
Syn.—RINACA EXTENSA, *Butl.*
Habitat.—Cachar.

1908

25. C. thibeta, *Westwood.* (*See p. 237.*)
Syn.—RINACA EXTENSA, *Butl.*
Habitat.—North-West Himálaya, Mussoorie, and Sikkim.
It forms a light, open, net-like cocoon and feeds on the—

1. **Cydonia vulgaris,** *Pers.*
THE QUINCE.
Vern.—*Bihi,* HIND.; *Bamsunt,* KASHMIR.
Cultivated in North-West India and up to 5,500 feet in the Himálaya.

2. **Pieris ovalifolia,** *D. Don.*
3. **Pyrus communis,** *Linn.*

Hutton says it "occurs also in Kumáon, but the specific name is a misnomer; the insect never approaches Thibet. Specimens were taken out of a collection made in Kumáon, but because the collector travelled into Thibet it was ridiculously enough called a Thibetan collection and the specimen named accordingly. The cocoon is a coarse open net-work through which the larva is visible, but there is no available silk."

V. CRICULA.

V.—CRICULA.

1909

26. Cricula drepanoides, *Moore.* (*See p. 238.*)
Habitat.—Sikkim.
This is presumed to be a very rare insect, but it differs sufficiently from **C. trifenestrata** to justify its retention as a distinct species.

1910

27. C. trifenestrata, *Helfer; Moore, Cat. Lepid. Insects, II., 384.* (*See p. 238.*)
Syn.—SATURNIA TRIFENESTRATA, *Helfer, Jour. Asiatic Soc. Beng. (1887), VI., 45; Herr Schoffer, Lep. Exot. Spec., Nov. Ser., I., pl. 17, f. 80;* CRICULA TRIFENESTRATA, *Walker, List Lep. Het. Brit. Mus.;* EUPHRANOR TRIFENESTRATA, *Herr Schoffer, &c., &c.*
Habitat.—An insect which is probably much more widely distributed than is generally supposed. Records of its collection occur from Assam, Bengal, Burma, Madras, Ceylon, the Andaman Islands, Bombay, and Central India. It is also distributed to Java.
It is described as a gregarious species, the worms spinning large agglutinated masses of cocoons. Each cocoon is composed of a reticulated tissue of the colour of gold. The insect is regarded as going through various generations, each occupying about two months, but of these it is reported that the rains brood is the best.

Food.

1911

Food of the Cricula Silkworm.

1. **Anacardium occidentale,** *Linn.*
THE CASHEW NUT TREE.
A small, evergreen tree in the coast forests of Chittagong, Tenasserim, Andaman Isles, and South India. Although an American tree now acclimatised in India, many writers affirm that this is a favourite food of Cricula silkworms.

2. **Camellia theifera,** *Griff.*
Hutton says that the larva is reported to sometimes do considerable injury to the tea plants.

3. **Careya arborea,** *Roxb.* The chief wild tree on which the larvæ feed.
4. **Eugenia fruticosa,** *Roxb.;* and
5. **E. Jambolana,** *Lam.*

These trees are mentioned by Manuel as two of those on which the insect feeds in Burma.

CRICULA TRIFENESTRATA.

Food.

6. Grewia Microcos, *Linn.*

Manuel mentions this tree as one of those on which the la væ feed.

7. Litsœa sebifera, *Pers.*

Manuel mentions this tree (the *ong-tong*) as one of those on which the worms feed.

8. Machilus odoratissima, *Nees.*

Vern.—*Súm*, Ass.; *Dingpingwait*, KHASIA; *Phamlet*, LEPCHA; *Kawala*, NEPAL, HIND.; *Dalchini*, PB.

A large tree of the outer Himálaya, extending to the Khásia Hills, Assam, and Burma.

9. Mangifera indica, *Linn.*

Hutton says that the larvæ sometimes appear in such numbers as to denude the mango trees of every leaf. The mango seems indeed to be one of its most favourite foods.

10. Pterospermum semisagittatum, *Ham.*

Mentioned by Manuel as one of the trees on which this silkworm feeds.

Notes. Assam. 1912

Provincial Notices Regarding Cricula Silk.

ASSAM.—**Mr. Stack**, in his Note on the Silks of this Province, furnished the following particulars:—

"The *ámluri* or *ampotoni*, so called from the mango or *ám* tree on which it feeds, is one of the commonest wild silkworms of Assam. It occurs in the Assam Valley under both the northern and the southern hills, and likewise in Cachar, where the wild tea-plant often supplies it with food. It is also frequently found on *sum*-trees. Its favourite tree, however, is the mango, whether the wild mango of the forest, or the cultivated trees in the vicinity of villages. The *amluri* spins a bright yellow cocoon, in clusters so closely interwoven that they cannot be separated for reeling, which, indeed, their very texture prohibits. It is said that a single tree will sometimes furnish as much as a maund of cocoons. In the number of broods and times of breeding this worm is said to correspond with the *muga*. The *katia* brood (October-November) is the most plentiful in Kámrúp. Subjoined is the record of an experiment made with some cocoons of this worm by Krishna Kanta Ghugua:—

September	15th	Cocoons obtained (from a *sum*-tree).
,,	26th	Moth emerged.
,,	27th, 28th, 29th	Laid eggs.
October	4th	Worms hatched.
,,	10th	First moulting.
,,	16th	Second ,,
,,	23rd	Third ,,
,,	29th	Fourth ,,
November	15th	Spinning began.
December	6th	Moth emerged.

"The silk of the *ámluri* is almost worthless. The cocoon is one of a thin and open texture, yielding very little silk. It cannot be reeled. The worm is covered with hairs, which produce irritation of the skin, and for this reason it is regarded as unclean by the Hindus; but Kacháris, Rábhas, and Meches occasionally mix the silk with *eri*, where it reveals its presence by the itching it causes. This irritating property of the worm is said to protect it against crows and bats. The chrysalis, however, is eaten by Kacháris, Rábhas, Meches, and Mikirs. A smaller variety of the *ámluri*, called *bisha*, and feeding, like the *ámluri*, on the mango tree, is found in small numbers in the sub-Himálayan jungles of Kámrúp. The name expresses the irritating quality of the worm."

It seems likely that the discoverer of this insect in Assam was **Captain Jenkins**, an officer who took an enlightened interest in the development of

SILK: Saturniidæ.

The Wild Silks of India.

CRICULA TRIFENESTRATA.

the resources of that Province. Dr. Helfer (*Jour. As. Soc, Bengal (1837), VI., 43*) furnished a fairly accurate and complete account of the cocoon. Mr. Brownlow speaks of the worm as feeding on **Anacardium**, a plant which might impart the stinging property assigned above to the worms (*Conf.* with the concluding paragraph under the heading Cocoons, silk, &c.).

Bengal. 1913

BENGAL.—Several writers allude to this silkworm as occurring in Bengal, more especially in Chutia Nagpur. Thus, for example, in the Proceedings of the Agri.-Horticultural Society of India for 1868 mention is made of a sample of the cocoons having been submitted to the Society by Mr. R. W. King. These were reported on by Mr. Lotteri as follows:—

"As I have had the pleasure of informing you personally, the cocoons of which you gave me a box of samples, although it is not possible to reel them, yet by the means of chemical and mechanical processes they can be made to give a good result, as you will judge from the six cocoons which I have reduced to pure silk.

"Several firms to whom I offered this produce, as prepared by me, have given me considerable orders, and you will help me very much if you could refer me to those that can furnish large quantities either here or at Moulmein, for it would be new riches to India and an encouragement to increase the production."

Madras. 1914 *Conf. with p. 134.*

MADRAS.—In an official report dated 1877 Dr. George Bidie, then Superintendent, Government Central Museum, wrote of this moth that "there was not a specimen in the museum." "I have not seen this insect in the Carnatic, but it is very common in Southern Coorg, and possibly extends into North Wynaad. In Coorg the larva feeds on the leaves of **Careya arborea**, and are gregarious, constructing their cocoons in clusters closely interwoven. I am doubtful if the silk of this insect could ever be turned to any practical purpose, as it would be impossible to reel it. Captain Hutton, in his Memoir on the Indian Bombycediæ, says that in some parts of the country the cocoons are carded, so it is just possible that some manufacturers might be able to make something of them if sent to them in the raw state."

Mr. R. Morgan, in *Note on the Wild Silks of Malabar*, furnishes the following particulars:—

"The sociable silkworm, **Cricula trifenestrata**, is undoubtedly the most abundant species of silk-producer found in Malabar. It is common enough in the plains; but in the Wynaad plateau I have seen every tree on 5 or 6 acres of ground entirely denuded of leaves by the larvæ of the species. It particularly affects the **Careya arborea**, which grows well on the grassy downs in Wynaad. Curiously enough the caterpillars, which are grey, and covered with stinging hairs, never denude the same tree two years successively, thus allowing them a whole season to recover from the check received during the previous one. During the larval stage, it may be said that they have no enemies, for they are able to protect themselves with their poisonous spines; but when they are helpless pupæ, then they are destroyed wholesale, principally by ants.

"The cocoons, which are of a brilliant yellow colour, are open and reticulated, and the chrysalis is easily visible inside. As a rule, the worms spin in company, and great masses of cocoons, upwards of 20lb in weight sometimes, may be seen pendant from the branches of the trees they have denuded of leaves.

"The silk is very strong and lustrous, but is cemented together very firmly in the cocoon. Doubtless it could be treated in the same way as *tassa*.

"The difficulty would be to get rid of the chrysalis, and this could only be done by thorough dessication, to render them inoffensive (for when crushed, they putrify), or by picking the pupæ out, which would be not only laborious, but would break the fibre of the silk.

"In Wynaad alone, tons of this silk might be produced without the slightest trouble, if the precaution were taken of collecting the cocoons the moment they are spun and storing them safely, until the imagos emerge and lay their eggs.

"The eggs should be hung in cocoanut shells on the branches of careya and mango

CRICULA TRIFENESTRATA.

trees the day they are expected to hatch. No trouble whatever need be taken with the worms afterwards, and nothing has to be done but collect the cocoons when spun and render them marketable.

"I have experimented largely with this moth, and have attempted to obtain hybrids between it and **Bombyx mori & Antheræa mylitta**, but failed. I manufactured a considerable quantity of fishing gut from it, which was very strong, but proved to be too opaque for the purpose it was wanted."

Burma. 1915

Burma.—One of the earliest notices of this silk in Burma is that of **Captain J. C. Haughton**, who forwarded samples of the cocoons to the Agri.-Horticultural Society of India in 1856. They were then regarded as worthless. **Helfer** had, of course, described the insect some time before, in the *Journal of the Asiatic Society of Bengal*, but the existence of the insect in Burma was not definitely established till the receipt of **Captain Haughton's** samples. **Mr. R. A. Manuel** furnished the following useful account of this silkworm as met with in Burma at the present day:—

"Among undomesticated, or *wild silk* spinners, although a number are to be met with in all three divisions of the Province, the most notorious as well as the most extensively spread, is the **Cricula trifenestrata**, or the mango silkworm, called by the Natives the *thayet-po*, because it is usually met with on the mango or *thayet* trees. So great is the ravage committed on this fruit tree by these insects that they have been known to denude a large tree of its leaves in a single night. The insect is also met with in India, where it is known as the Malda silkworm, but there its cocoon is so much reticulated, the silk so scanty and loaded with gum, that it is only sparsely used, and that to adulterate *eri* silk. **Dr. Mason**, in his *Burma*, gives this most common worm only a passing notice, saying, 'The Malda silkworm, the cocoons of which are mixed with those of the eri silkworm, is found on the mango tree.'

"**Captain Haughton**, however, mentions it as being found in Moulmein feeding on the cashew nut, and **Mr. Moore** has identified it as the **Cricula trifenestrata**. The *raison d'être* of the specific name may be gathered from the following accurate description of the wings of the moth by a writer in one of the earlier numbers of the *Journal of the Asiatic Society*: *Alis superioribus ad marginem externam fenestris tribus transparentibus.* As a matter of fact, the worm is to be met with all over the country, and consumes the leaves of many more plants than the mango.

"*Crops.*—The worm is multivoltine, working all the year round, though most extensively and vigorously during the rains. The cycle of its existence is about from 55 to 62 days:—

	Days.
In the egg state	8 to 10
In the worm state	25 to 27
In the cocoon state	20 to 22
In the moth state	2 to 3
Total	55 to 62

"The length of the cycle depends on the time of the year, the worms attaining a larger size and spinning a better cocoon during the rains. The cocoon is a little over 1⅓ inches long and ¾ inch in its broadest part. It tapers to a point at both ends, where it is open, hence the moth escapes without leaving any trace behind of its exit as with the *tusser*. But, just as with the *tusser*, the approach of the time for its exit may be known by the appearance of a dark moist spot at one end of the cocoon. The weight of an ordinary cocoon is 20 grains with the chrysalis and 3·5 grains without it.

The Wild Silks of India.

CRICULA TRIFENESTRATA.

Burma.

"*Cultivation.*—The worm is of a wild nature and never does well in the house, however slight the restraint. The best way to rear it is to leave it alone on one of its many food-plants. It is gregarious and does not wander much, and neither sun, wind, nor rain affects it in the least; on the contrary the more open it is kept—the more its condition approaches its wild state—the better it thrives and spins. Under this head, therefore, all that will be related is the result of observations of the insect's habits and ways in its larval state in semi-domestication.

"The ova are laid in irregular patches and are minute; in colour white and in shape like a miniature hen's egg, having one end broader than the other. The female lays from 200 to 250 eggs, and 32 of the fertilized ova weigh one grain. Shortly after laying the ova pit in the centre, and as the time for emergence of the worm approaches, the ova become first yellow, then light orange, then dark slate, until, a day previous to the cracking of the shell, a dark spot appears, oftener nearer the broader than the narrower end. Next the shell cracks and breaks, and the shiny, dark brown head of the young worm appears. Soon it wriggles out of its shell and begins life by nibbling at it. It then remains quiet for a few hours and then attacks the leaf. During all their larval state the worms are gregarious,—they feed in rows or clusters, move in batches from place to place in search of food, and, when the time comes, they spin their cocoons in masses, frequently overlaying one another and invariably overlapping each other at the ends.

"On emergence the worms are one-eighth of an inch in length, of a light yellow or ochrish colour, with two rows of dark-brown spots (tubercles) ending in brown hairs, two spots on each segment of the body, and one big dark brown spot on the penultimate segment, that is, the one just before the anal segment. The head is dark-polished brown, almost black with a fringe of yellow advancing hairs just behind it like a lady's lace collar of Elizabethan days. Behind this collar two brown crescent-like markings appear concave towards the head. The first moult occurs in five or six days, after that there are four other moults every five days. At the end of the moults the worms, if healthy, are fully four inches long, hairy, gaudy creatures, their prevailing colour being a rich maroon with bands of black and yellow with tubercles having erect hairs surrounding their base and one occupying the centre; and the prolegs have brown markings at their joints. The cocoons are completed in from 36 to 48 hours, and in two or three days, after disappearing from sight in their silken envelopes, the last metamorphosis is completed, and the insect becomes a light-brown chrysalis, in which condition it rests from 20 to 22 days before leaving its confinement as a brown moth having an expanse of wing of from 2½ to 3 inches, with three glass-windows on its superior, and one small one on each of its lower, wings. They pair the very night of their emergence, though some eight or ten hours after. The males desert the females at early dawn, and then the females remain quiet all day through, beginning to lay their eggs almost as soon as the males have quitted them.

"*Manufacture.*—In consequence of the very irregular manner in which the insect lays the fibres of its cocoon, and also on account of the quality of the gum with which it loads its silk, the cocoons, though very handsome in appearance, looking like beaten gold, are very difficult to reel. In fact all attempts to reel them have hitherto proved failures. The only use they can be put to is the manufacture of 'waste,' or floss, from which silk may be spun. And very fine floss the cocoon does yield, so fine that English and Continental manufacturers are ready to purchase

CRICULA TRIFENESTRATA, Burma.

it for all purposes of spun silk. The threads are glossy, long, and fine, about one-half finer than *tusser*, and about half as coarse again as the local mulberry silk. A great future awaits this silk, as, in consequence of improvements in modern spinning machinery, a large demand exists for all kinds of silk which can be carded and spun. The *eri* of Assam is woven into cloths of great strength and durability from yarn spun out of its waste, and there really is no reason why the silk of the cricula of Burma should not be utilised in the same way as *eria*. The fibres under a microscope of low power appear beautifully transparent and round, unlike *tusser*, and thus capable of taking the dyes which *tusser* refuses. Made into waste, the colour of the silk is of a light yellow ochre, which, however, can be easily discharged and the fibre made quite white."

Captain J. O. Haughton (*Jour. Agri.-Hort. Soc. Ind., X., 101 (1858)*) found this species at Moulmein feeding on **Anacardium**.

The Cocoons, Silk, &c.

Cocoons. 1916

Hutton says, "The cocoons are very irritating, from a number of minute bristly hairs from the caterpillars. I am inclined to think there are two species now standing under this name, as some cocoons are very thick, much reticulated, while those from other localities are far more closely woven and scarcely reticulated at all.* This will never prove productive as a silk-yielder, unless the cocoons can be reduced to a gummy pulp, and used for some other purposes." Geoghegan, on what authority the writer cannot discover, says of this cocoon: "The cocoon is of a transparent yellow, with an opening at one end. The silk is capable of being spun like *eria* silk, but not used, because it excites severe itching." This may be but a more powerful mode of expressing Hutton's remark above, but if the silk does actually cause itching, the prospects of a trade being developed in it seems remote. Mr. Wardle, however, appears either to have never observed the above references to the irritation caused by this silk or to have disregarded that opinion, for he says, "Very abundant in British Burma, where the cocoons rot in the jungles for want of gathering. The silk of this species promises to be most useful, and only waits importation to Europe and utilisation." Mr. Manuel also, in the passage quoted above, makes no mention of any report of the fibre producing itchiness, and, like Mr. Wardle, is very hopeful of the future of this silk. He describes the Burma cocoon as more closely spun than that of other parts of India, and it is thus probable that Hutton was right in thinking there were two insects commonly treated as one under the name given to this species. If this suggestion proves correct it is probable that the objectionable form may be the insect seen by Hutton, Geoghegan, Stack, and Morgan, and that the Burma form does not possess the stinging hairs which Stack and Morgan allude to as covering the worm found in Assam and Malabar. Manuel's account is so minute in its details that it is difficult to believe he overlooked the stinging hairs of the worm. His description of the worm also differs from that given by Morgan. This feature of Cricula silk seems well worthy the consideration of Entomologists.

Mr. Wardle in his table of measurements gives the diameter of the fibre in fractions of an inch, as $\frac{1}{880}$th taken from the inner part of the cocoon, while the Tasar is $\frac{1}{770}$ th and Eri $\frac{1}{1500}$th. It is thus intermediate in thickness between the *tasar* and *eri*.

* Swinhoe described the insect that forms a solid, that is, non-reticulated cocoon, under the name of **C. burmana.**

VI. LOEPA.

VI.—LOEPA. (*See p. 238.*)

1917

28. Loepa katinka, *Westwood; Cat. Or.Ent., p. 25, pl. 1., f. 2 (1847).*

Habitat.—Sikkim, Shillong, North Khásia Hills, Sibsagar, Assam, Upper Burma, Nilgiri Hills.

This species is found mainly between altitudes of 5,000 and 7,000 feet, and in the higher altitude in which it occurs, it is quite distinct from the other species of the genus. The worm is said by **Horsfield** to feed on the *galing* or *girang*, a plant thought to be a species of **Cissus** or **Leea.** In Nepál and Sikkim **Leea robusta,** *Roxb.*, and some three or four other species are known as *galeni*, and they are therefore likely to be the plants on which **Horsfield** found the worm. No species of **Vitis** (**Cissus**) bears any such name. The insect is said to be "abundant," and to appear from "December to February." **Hutton** (*Jour. Agri.-Hort. Soc. Ind., III., (New Series), 141*) decribes this as "a beautiful yellow moth, found originally in Assam, occurring also according to my ideas, in Mussoorie. **Mr. Moore**, however, considers mine as distinct. I am not quite satisfied that the cocoon will not yield silk, but there is very little of it." **Hampson** supports **Hutton's** hesitation regarding **Moore's** two additional species by reducing **L. miranda** and **L. sikkima** to synonyms for this species; but in the Appendix he will be seen to mention two other species. (*Conf.* with **Moore**, *Cat. Lep. Insects, II., 399.*)

1918

29. L. miranda, *Moore, Trans. Ent. Soc. London (3), II., 424.*

Habitat.—Sikkim Himálaya (*Conf.* with **Hutton**, *Jour. Agri.-Hort. Soc. Ind., III. (New Series), 142*).

1919

30. L. sikkima, *Moore, Proc. Zool. Soc. (1865), 818.*

Habitat.—Hot valleys of Sikkim.

1920

31. L. sivatica, *Hutton.**

Habitat.—Mussoorie.

It forms a long-pointed cocoon of a dark greenish-grey colour (*Conf.* with **Hutton**, *Jour. Agri.-Hort. Soc. Ind. III. (New Series), 141-142*).

VII. NEORIS.

VII.—NEORIS.

This genus is probably doubtfully distinct from **Saturnia.**

1921

32. Neoris huttoni, *Moore.* (*See p. 238.*)

Habitat.—Mussoorie, North-West Himálaya.

The worms appear in April; feed on the wild pear-tree; and spin a thin, open, worthless, silken cocoon (*Jour. Agri.-Hort. Soc. Ind., III. (New Series), 141*).

1922

33. N. stoliczkana, *Felder.* (*See p. 238.*)

Syn.—SATURNIA STOLICZKANA, *Feld.*; N. SHADULLA, *Moore.*

Habitat.—Ladak.

VIII. RINACA.

VIII.—RINACA. (*See p. 237.*)

1923

34. Rinaca zuleika, *Hope, Trans. Linn. Soc., XIX., 132, pl. 11., f. 5.*

Syn.—SATURNIA ZULEIKA, *Hope.*

Habitat.—Sylhet, Sikkim, and Simla.

Mr. Otto Moeller informs us that in Sikkim the wormf eeds on a laurel (**Actinodaphne sikkimensis**) and a species of maple (**Acer caudatum** or **A. campbellii**). **Rondot** (*L'art. de la soie II., 205*) regards the silk as worthless. The cocoon is reticulated. In the *Museum Notes* **Mr. A. V. Knyvett** is reported to say that the caterpillar is found at an elevation of about 7,000 feet during October and November. It spins, he remarks, a

* "This is a synonym for **Antheræa paphia** and not a **Loepa** at all." *G. F. Hampson.*

rough cocoon on the ground at the foot of the food-plant; the moth appears in the following August.

RINACA ZULEIKA.

IX.—RHODIA.

IX. RHODIA. 1924

35. **Rhodia newara,** *Moore.*

Habitat.—Sikkim and Nepál. The author collected several cocoons of this species in Manipur. The insect spins a brilliant green cocoon and feeds on the walnut (*Conf. with p. 87*) and a species of willow, occurring at elevations between 4000 and 7000 feet. **Moeller** says that it goes through but one generation in the year, and hibernates in the egg. The larvæ emerge early in spring, and finish spinning their cocoons by the end of May. The moth appears in November. It will be seen from the Appendix (*p. 238*), that **Hampson** places this species in **Loepa.**

X.—SALASSA.

X. SALASSA. 1925

36. **Salassa lola,** *Westwood, Cat. Or. Ent., p. 25, pl. 12, f. 3.* (*See p. 238.*)

Habitat.—Sylhet to Sikkim. In the proof copy of the forthcoming number of the *Museum Notes*, **Mr. A V. Knyvett** is said to have observed that in Sikkim the caterpillar, previous to transformation into the chrysalis, "spins a few leaves and chips together into a sort of rough covering in the ground exactly as is done" by **Salassa royi.**

37. **S. royi,** *Elwes, Proc. Zool. Soc. (1887), 447.* (*See p. 238.*)

1926

Habitat.—A species found in Sikkim at an altitude of 10,000 feet. Its habits have been carefully studied by **Mr. A. V. Knyvett.**

XI.—SATURNIA.

XI. SATURNIA. 1927

Formerly this genus was made to embrace the major portion of the insects here described under the various genera of the tribe SATURNIIDÆ. With the advance of Entomological Science the species of that tribe were broken into distinct genera, but it is perhaps a matter for some doubt if that process has not been carried too far. It will, for example, be seen that **Hampson** (in the enumeration given in the Appendix) has returned the species of **Caligula, Neoris,** and **Rinaca** to **Saturnia.**

38. **Saturnia anna,** *Moore.* (*See p. 237.*)

1928

Habitat.—Sikkim Himálaya.

39. **S. cidosa,** *Moore, Trans. Ent. Soc. Lond. (3) II., 423, pl. 22, f. 2.*

1929

Habitat.—Hot valleys of Sikkim Himálaya. The specimen seen by **Hutton** (*Jour. Agri.-Hort. Soc. Ind., III. (New Series), 141*) was from **Captain Sherwill's** collection. In his remarks regarding that insect **Hutton** says it is very nearly allied to **S. pyretorum,** and **Hampson** reduces it to that species (*see p. 238.*)

40. **S. grotei,** *Moore, Proc. Zool. Soc. (1859), 265, pl. 62, f. 2.* (*See p. 237.*)

1930

Habitat.—Kulu and Sikkim.

In the proof copy of the forthcoming part of the *Museum Notes* (before the author) it is stated that this, as also **S. lindia, S. anna,** and **S. hockingii** are so closely allied to each other, that it seems convenient to consider them together, since very little is known about any of them. Of **S. hockingii,** *Moore,* it is stated that it was described from a Kulu specimen and is noticeable from the fact that **Moore** records that the cocoon, which is formed "under stones," is "pyriform, dark brown, hard, pointed, and lax at the upper end."

41. **S. lindia,** *Moore.* (*See p. 238.*)

1931

Syn. S. HOCKINGII, *Moore.*

Habitat.- Found in Kulu.

2. **S. pyretorum,** *Westwood, Cat. Or. Ent., p. 49, pl. 24, f. 2.* (*See p. 238.*)

1932

Habitat.—This insect, supposed to be a native of China, is stated by

SILK: Tasar.

The Home of the Tasar Insect.

SATURNIA PYRETORUM.

Hutton (*Jour. Agri.-Hort. Soc. Ind., III. (New Series), 140*) to occur also in Cachar and Sikkim. In the forthcoming number of the *Museum Notes* that idea seems to be questioned. Then follows the remark that— "In China, according to the *resumé* given by Rondot (*L'art de la soie II., 1887, p. 205*), the caterpillar is of medium size, longitudinally streaked with bright turquoise blue, alternating with canary yellow, and covered with bristling yellow hairs; they feed chiefly upon the Liquidambar formosana and camphor trees. The cocoon is ovoid, much produced in length, pointed, open at one end, and surrounded with floss which is deep brown in colour; the silk is silver grey or brownish grey in colour, coarse and very tough; it is used on a considerable scale commercially; and the silk glands of the worm are also largely used for the preparation of gut for fishing lines."

XII. BRAHMÆA. 1933

XII.—BRAHMÆA.

This genus has been incorrectly referred to the SATURNIIDÆ by certain writers on Indian silk-worms. It has been retained here in order to prevent future misconceptions. There are two Indian species, *viz.*, **B. wallichii**, *Gray* (=**conchifera** and **rufescens**, *Butler*) and **B. hearseyi**, *White* (=**B. whitei**, *Butler*). **B. certhia** is a Chinese not an Indian species. (*See p. 238.*)

The BRAHMŒIDÆ do not spin cocoons, and they are therefore not silk-worms.

THE TASAR. 1934

THE TASAR SILK-WORM.

(For Eri see pp. 162—174; for Muga pp. 174—183.)

THE TUSSER OF TASAR SILK MOTH, *Eng.*; TUSSORE, *Fr.*; BOMBYX SEIDE, *Germ.* (ANTHERÆA PAPHIA).

Vern.—*Tasar* or *bughy* and *jarvo tasar*, BIRBHUM; *Chattisghari tasar*, SANTHAL PARGANAS; *Guti tasar*, BANKURA; *Dasa, daba, ampath, ampatia tasar*, MANBHUM; *Lumam, lumang*, SANTALI; *Jarú* (described by Buchanan-Hamilton in the districts of Bhagulpore and Dinagepore), BENG.; *Katkura* or *tussar, kutkuri konkuri mung, gori deomunga*, ASS.; *Tusuru*, HIND.; *Kolissura*, DEC.

Habitat.—The Tasar silkworm is, perhaps, the most abundant, as it is the most important, of the so-called Indian wild silks. It seems to occur in a wild state in the forests of the lower undulations of the plains of India, but is apparently absent from North India, Burma, and Ceylon. It is occasional, however, in the lower and hotter forests of Manipur, and often ascends to 2,000 or even 4,000 feet in altitude. This fact extends the apparent eastern line of habitat of the insect to the mountain slopes of North Burma.

Mr. Hugon regarded the insect called in Bengal *bughy*, which is met with feeding upon the *bér* (**Zizyphus**) to be a different species from the *jarvo*, found on the *asan* (**Terminalia**), but Moore and other Entomologists think this is not the case. As economic products, however, they differ considerably from each other, and the worm is of a different colour. Although not distinct species, these forms illustrate, in a marked degree, the effect of different food in changing many of the characters of an insect. The Tasar of the Panjáb is also a very different looking insect, but it must be admitted that these and many other forms have been named when absolutely nothing has been done towards advancing the knowledge of the races of *tasar*.

Synonyms. 1935

FORMS OF, AND SYNONYMS FOR, THE TASAR INSECT.—Perhaps few insects are more variable in size, markings, and colour than the *tasar*. This has led to much difference of opinion as to their classification. Some writers regard these variations as mere sexual peculiarities, others as domesticated races, while a third set of authors speak of them as varieties of one species or different species with innumerable hybrid domesticated

SYNONYMS.

stocks. Such a state of conflicting opinion has naturally lead to a multiplication of names. In a preliminary list of the insects contributed by the Indian Museum to the Colonial and Indian Exhibition, Mr. J. Wood-Mason enumerates eight specimens grouped under the designations of **Antheræa mylitta,** *Drury;* **A. mylitta,** *Drury,* **var.?;** and **A. mylitta,** *Drury,* **var. nebulosa,** *Hutton, &c.* It is difficult to discover whether Mr. Wood-Mason regards all the eight forms as referable to the type of the species, and to three or four varieties, of which **nebulosa** alone had received a name, or whether he viewed them as all constituting one form except **nebulosa.** In the more recent publication issued from the Museum, Mr. E. C. Cotes speaks of **nebulosa** as only a dark-coloured individual of the common *tasar* which is common in Chutia Nagpur and Central India. Hutton, on the other hand, not only regarded the species as embracing several distinct varieties characteristic of certain localities or altitudes, but maintained that in the state of semi-domestication followed in the chief *tasar*-producing districts, it was impossible to obtain pure stock, all or nearly all being hybrids.

Conf. with p. 72.

The following may be given as the better known synonyms for this insect:—**Antheræa mylitta** (*of Hübn., Walker, Moore, Aurivillius, Wardle, Rondot*), **Phalæna (Attacus) mylitta** (*Drury*); **Attacus mylitta** (*Blanch.*); **Bombyx mylitta** (*Fabr., Oliv.*); **Phalæna paphia** (*Cramer, Roxburgh*); **Antheræa paphia** (*Moore, Beavan, Hmpson, &c.*); **Saturnia mylitta** (*Westw.*); **Saturnia paphia** (*Helfer*); **A. cingalesa,** *Moore;* **A. sivalika,** *Moore;* **A nebulosa,** *Hutton.*

Home of Tasar. 1936

THE HOME OF THE TASAR.—It may be said to be a denizen of the highland forests inhabited by the Santhal, the Kol, the Khond, and the Gond, extending west and south-west of the Gangetic alluvial basin. Commencing at Rajmehal the region of the *tasar* silk-worm may be said to stretch away south through the Rajmehal and Kurackpore hills to the table-land of Chutia Nagpur, thence to the mountain tracts of Orissa, of the Central Provinces, and the Northern Circars to Hyderabad. Taking the westerly direction this region may be said to pass from Rajmehal and Bhagulpur, through Behar, to the Kaimor mountains and Bandelkhand, thence to the Central Provinces and Berar. Practically speaking, this region may be said to have the Ganges for its northern boundary and the Godavari for its southern, with the coast ranges from Midnapur in Orissa to Ramgar in Hyderabad, as its south-eastern, and the Narbada river and the Kaimor mountains as its north-western boundary. Of course the *tasar* insect crosses these limitations to a certain extent, being met with on the north of the Ganges along the foot of the Himálaya from Nepal to Sikkim, Assam, and the Khásia hills, the Naga hills, and the Lushai country to Chittagong, the Sandarbans, and, sparingly, in the neighbourhood of Calcutta. It also crosses the Godavari and extends into the mountains of the Madras Presidency, and is even reported as met with in Mysore. Beyond the region which has been defined, however, it can only be said to occur to a small extent, and in a wild and neglected condition, for, with the exception of a small corner of the North-West Provinces at Mirzapur, the cocoons are not even collected. The name *tasar* has unfortunately been applied to all fawn-coloured indigenous silks, and in the North-West Provinces at Mirzapur a mixed cotton and silk fabric bears that name. It is exceedingly doubtful therefore if the *tasar* worm proper occurs anywhere beyond the region defined, and it is incorrect to regard it as met with throughout the entire Peninsula of India. Mr. Wardle gives a map of the region inhabited by the *tasar* insect in which he colours the whole of India, except Kashmir, Rajputana, Bhutan, Burma, and Ceylon. This is very probably a mistake. It seems, practically speaking, to be

HISTORY.

absent from the Panjáb, Rajputana, the North-West Provinces and Oudh. It nowhere occurs upon the Himálaya proper and never ascends above 4,000 feet in altitude, and is rarely, if ever, found on the alluvial plains except where these are limited and confined by hilly undulations.

1937

HISTORY OF TASAR SILK.

The influence of the French method of spelling Indian vernacular names has, in few cases only, led to greater confusion than the present. The word *tasar* is, by universal admission, supposed to come from the Hindi *tasar*, "a shuttle," being derived from *tasara*, or *trasara* in the Sanskrit. *Tasara* is, however, not given by any Sanskrit author as a name for silk or any special silk. Indeed, the Sanskrit authors appear to have been ignorant of this substance, so that the adaptation of the name *tasar* must be viewed as of modern, or comparatively modern origin. By most French writers it is spelt *tussore*, but it assumes numerous forms, and many writers regard the terminal "r" as a vulgarism. The most general form of the word however, and the one in use with writers in Europe, is *tussur*, but it is also given as *tusur* (=*toosoor*), *tusser, tussah, tusseh, tusha*, &c.

By some writers, on the other hand, the word denotes a peculiar kind of cloth, mostly made of cotton with strips of silk or embroidered with *tasar, muga*, or any other wild silk. When, therefore, the word alone occurs (that is to say, when it is not accompanied with a description sufficient to lead to the conclusion that this particular silk is indicated), it is safer to assume that one or other of the peculiar garments that commonly bear that name is meant. Thus, for example, in the *Áin-i-Akbarí* (published originally in 1590) we are told *tasar* sold at R$1\frac{1}{2}$ to R2 per piece. But, fortunately, we have an unmistakable allusion to this silk which bears the date of April 1691. **Rumphius**, the indefatigable botanist of Amboyna, while describing **Sonneratia acida**—a small tree which inhabits the swamps on the coast of Bengal and the Malay, etc.—took the opportunity to figure a silk-worm which he found feeding on its leaves. His description goes into the greatest detail, and even directs attention to the points in which the Mollucas insect differs from the "tesser" of Bengal. These points of difference are very remarkable, since certain Indian writers have held that what appears to be the exact same insect is found in similar localities of India, and that it is quite distinct from the *tasar* of the more interior tracts of the country. Modern entomological writers, however, regard **Antheræa paphia** as a protean insect that varies to a remarkable extent according to the climatic conditions under which it is found or the nature of the food on which it has been fed. **Dr. Shortt's** Madras and Mysore form, which he met with near the coast, agrees remarkably with **Rumphius'** figure and description, and it seems likely that the Sandarband *tasar* may be found to feed, like the Amboyna worm, on **Sonneratia.** In passing it need only, therefore, be necessary to add that **Rumphius** lays stress on the fact that the cocoon of the Amboyna *tasar* is larger, paler-coloured, and has a shorter peduncle than the Bengal, and, moreover, that the leaves of the tree are also attached to the cocoon of the Amboyna insect. Another feature of some interest is the fact that **Rumphius** urges that the moth to which he had given the name of **Eruca amboinensis** has always a deep purple transverse line along its wings and around the transparent eye-like spots. Whether Entomologists are likely to reconsider their position and to accept **Rumphius'** insect as distinct from that of India, or may even admit the existence of the Amboyna *tasar* on the coast of India, are points of far less importance to the light **Rumphius** throws on the history and character of the

Conf. with pp. 70, 72, 73, 77.

HISTORY.

tasar trade, past and present. The following may be given as a translation of the concluding sentences of **Rumphius'** description.

"The moth had only escaped from the cocoon for the space of an hour when it began to deposit eggs. These were mostly of a dirty white colour, of the size of mustard seeds, round and flattened above. The cocoon is as white as ordinary silk, and there can be reeled from it a filament of some 20 or 30 yards in length. The cocoon of the Bengal insect is dirty yellow, hence it is evident why it has been said that Bengal silk is called *gingang* and that the silk is prepared from the bark of a tree, since the cocoons are so neatly constructed, fixed with a pedicel and suspended from the branches of the tree on which the worm fed in such a way as to appear like the fruits of the tree. Indeed, a Surgeon, who came here from Bengal, maintained that they *were* the fruits of the tree, an opinion which I shared until I had dissected one and found the chrysalis within.

"The cocoons from which the moths have escaped are not so good for reeling, since the filament breaks frequently. The Bengal cocoons, as already explained, are much darker coloured than are those of Amboyna, and have an earthy appearance, yet I have never heard of the Amboynas preparing a texture from them. From the *tesser* as just described, *mogta*, or, as some people call it, *moock*, differs; it affords very bad silk being reeled from cocoons from which the moths have escaped, and this in Tonkin is by the Chinese called *samungi* or *samœgi*, of which an inferior fabric called *baas* is prepared."

From these remarks one might be almost justified in supposing that **Rumphius** was of opinion that all the silk of Bengal was of this nature. But setting speculation on one side there is no room for doubt that the *tasar* silk of Bengal was in 1691 relatively of far greater importance than at the present day. The particulars of the trade were at least so well known, that a distant traveller was enabled to produce an account of it which might easily enough be transcribed to a modern gazetteer without its being detected that the sentences were originally penned 200 years ago. **Rumphius** gives us a new meaning for the much-disputed word "Gingham" of modern commerce, which would confirm the idea advanced by many writers that, like "Chintz" and "Calico," Gingham was of Indian, not French, origin. It was, according to **Rumphius**, the name for the dirty yellow silk cloths of Bengal. So again the inferior nature of pierced, as compared with unpierced, cocoons, and even the name for these was known to **Rumphius**. But the concluding words regarding China may be viewed as of peculiar interest. It is often asserted that the Chinese have only recently been made acquainted with *tasar* silk. The caution may, however, be here offered that the *tasar* of Tonkin (to which **Rumphius** alludes) is distinct from the modern so-called *tasar* of Chinese trade.

The first mention of *tasar* in English official reports (1796) is said to be that by **Michael Atkinson** of Jangipur. This fact is alluded to by **Roxburgh**, but in his account of *tasar*, **Roxburgh** was shortly after followed by **Buchanan-Hamilton** (in 1809), who furnished much useful information regarding that silk at Bhagulpur. It does not, however, seem necessary to allude here, in chronological order, to all the subsequent nineteenth century writers. The reader will find these freely consulted in the pages which follow and extensive passages taken from such of them as are thought less likely to be accessible. The historic facts may, therefore, be viewed as completed by the chapter on TRADE, but, in conclusion, it may be said that, although no direct proof exists in support of the statement, it is probably safe to assume that in Bengal at least *tasar* silk was known and

SILK: Tasar.

History of the Tasar Silk.

HISTORY.

much appreciated long before mulberry silk was heard of, except in the form of imported piece goods from China and other foreign countries. The new interest awakened in it within the past twenty or thirty years may, in fact, be regarded (as far as India is concerned) as an attempt to revive an industry which by modern facilities and necessities had scarcely room for existence, and which, far from giving any indications of revival, continues year after year along its downward course, even in spite of the European demand for the silk. It is more profitable to grow rice, jute, indigo, and a thousand other articles of modern commerce than to either collect the cocoons or attempt to produce the coarse *tasar* cloths which once upon a time were the natural garments of the rich and intellectual leaders of the community.

The important part taken by the Agri.-Horticultural Society of India in the effort to make known the *tasar* silk and, if possible, to establish a trade in it may be admitted as sufficient ground for giving here a brief note regarding the chief papers which appeared in its Journals on this subject, in addition to those which will be found quoted below in the Provincial Sections of this article.

Serica regio. 1938 *Conf. with pp. 23, 25, 35.*

1848.—**Messrs. Hodgson & Frith** published conjointly a series of notes on the wild silks of Assam, and gave a coloured illustration of the *tasar* insect, caterpillar, and pierced cocoon In two of his letters **Mr. Hodgson** suggests that *tasar* silk may have been the article exported from India during the classic period of Rome. **Mr. Taylor** (*Jour. Asiatic Soc. Beng.*) even contended that Assam was the *Serica regio* of the classics and that *tasar* silk was probably used as the awnings of the great Roman theatres.

1856.—**Hutton** discusses the process by which the insect opens its cocoon.

Conf. with pp. 69, 75-76.

1859, Vol. XI.—**M. Perottet** gave an account of the *tasar* and other wild silks of Pondicherry. He speaks of a species unknown to him as feeding on **Odina Wodier.** It has already been pointed out that the insect in question (though it has received various names by subsequent writers) is very possibly only **Actias selene.** In the same volume **Dr. G Smith,** Secretary to the Hyderabad Exhibition Local Committee, forwarded to the Madras Exhibition of 1859 a most instructive paper on "the way the *tasar* silk-worm is reared and the silk dyed by the Telegus of the Hyderabad country together with receipts for making the dyes."

1865.—**Captain Thomas Hutton** contributed to the Journal a long paper on "the Indian Bombycidæ, in which he incidentally discussed many features of the *tasar* worm. By far the most curious paper in that volume, however, is one from the pen of **Dr. Maingay,** the late distinguished Chinese botanist During a journey through the Shantung Province of North China, **Dr. Maingay** was struck with the number of oak plantations he saw. On enquiry as to their use he found that the leaves were employed to feed a silkworm which closely resembled, though in many respects it differed from, the *tasar* of India" (*Conf.* with remarks under **Antheræa pernyi,** *pp. 72-75*).

1867.—**Captain J. Mitchel** drew attention to the fact that **Captain Hutton** was in error when he described the two filaments given out by the *tasar* worm as twisted together to form the thread of which the cocoon was built up. The filaments, **Captain Mitchel** said, were parellel and bound together by the peculiar gum; that the filaments, unlike those of the ordinary silk, were not round but flat, and that the finest filaments he had examined measured $\frac{1}{35,000}$th of an inch (*Conf.* with **Mr. Wardle's remarks** on this subject in the paragraph at pp. 145-146).

HISTORY.

1872.—**Captain Hutton** continued his *Notes* on the Indian Bombycidæ and gave particulars of the hybrids known to him between the *tasar* and other allied forms of **Antheræa.** (*Conf. with pp. 72, 74, 78-9.*)

1885.—**M. C. J. Dumaine** gave full particulars of his experiments with *tasar* silk-worm in Hazaribagh, treating the subject under the following sections: Seed cocoons, breeding, eggs, the field, the rearing, the cocoons, the enemies, the native superstitions, and the trees (23 in number) on which the worm feeds. In a later paper in the same volume he gave further particulars, and added that **Actias selene, Attacus atlas, Attacus ricini, Theophilia huttoni, Cricula trifenestrata,** and **Bombyx religiosa** were also procurable in the district of Hazaribagh.

Besides the above (which are the chief papers) incidental allusion is made to *tasar* in nearly every one of the numerous papers on Indian silk which occur in the Journals. In the Asiatic Society of Bengal an almost equally comprehensive series of papers will be found on this form of silk, but since these are very frequently mentioned in the succeeding pages, it is not considered necessary to specialise them any further. Perhaps the most useful papers on *tasar* which appeared in the Asiatic Society's Journal were, however, those by **Hugon** (*VI., 32 (1837)*), **Helfer** (*Vol. VI., 43*), **Taylor**, &c.

Food of the Worm.

Food. 1939

The following are the principal trees on which the *tasar* (or as it is often written the *tusser*) caterpillar feeds:—

1, **Anogeissus latifolia,** *Wall.*

Vern.—*Dháwa, dháura, bakil,* HIND.; *Góbra, dhokridan,* RAJPUTANA.

Captain Brooke mentions the wild worms in the Seoní forests as being met with on this tree.

2. **Bassia latifolia,** *Roxb.*

The **Rev. A. Campbell** reports that in Chutia Nagpur the *tasar* is sometimes seen to feed on the *mahuá.*

3. **Bauhinia variegata,** *Linn.*

The *kanchan* is mentioned by **Major Coussmaker.**

4. **Bombax malabaricum,** *DC.* (*Conf. with p. 125.*)

Vern.—*Semul, shembal, semur, bouro,* HIND., BENG., ASS.; *Illavam,* TAM; *Búrga, buraga,* TEL.

Mr. Hugon alludes to this tree as being one of the chief trees in Assam upon which it feeds.

5. **Careya arborea,** *Roxb.*

Vern.—*Kumbi,* HIND.; *Komba,* BOMB.; *Boktok,* LEPCHA; *Dambel,* GARO; *Ayma,* TAM.

6. **Carissa Carandas,** *Linn.* (*Conf. with pp. 125, 131*).

Vern.—*Karaunda, kurunda,* HIND.; *Kurumia,* BENG.; *Kalaka,* TAM.

7. **Celastrus paniculata,** *Willd.*

The *mal kangoni* of Bombay.

8. **Chloroxylon Swietenia,** *DC.*

The *billú* or *halda* alluded to by **Major Coussmaker.**

9. **Dodonæa viscosa,** *Linn.*

This bush **Mr. J. Cameron** (Report, Bangalore Gardens, 1887-88) says is suitable for the purpose.

10. **Eugenia Jambolana,** *Lam.* (*Conf. with p. 127*).

Described by **Major Coussmaker** as being a good plant to feed this worm on.

SILK: Tasar.

Food of the Tasar Silkworm.

FOOD OF THE WORM.

11. Ficus Benjamina, *Linn.* (*Conf. with pp. 129, 131.*)
The *nandruk* of Poona, the leaves of which were used by Major Coussmaker for in-door feeding.

12. F. religiosa, *Linn.*
THE PEEPUL; *Aswát, asud,* BENG.
Mr. Gamble says the *gori* or *deomuga* silk-worm feeds upon this plant in Assam. It is difficult to decide as to what insect is meant, but it may be guessed at as the *tasar,* though of course it may probably be **Bombyx religiosa,** *Helfer.*

13. F. retusa, *Linn.*
Vern.—*Kamrup, zir,* BENG.; *Jamu,* NEP.; *Situyok,* LEPCHA; *Yerra juri,* TEL.; *Pilála,* KAN.; *Nyoungop,* BURM.
A large, elegant tree, often cultivated in avenues in India.

14. F. Tsiela, *Roxb.* (*Conf. with p. 129.*)
The *jari* or *pimpri* used by Major Coussmaker.

15. Lagerstrœmia indica, *Linn.*
Vern.—*Telinga-china,* HIND.; *Daiyeti,* SIND & PB.
A small bush, much cultivated in Indian gardens on account of its rose-pink flowers. (*Conf. with pp. 125, 127-28, 132, 137-38.*)

16. L. parviflora, *Hook.*
Vern.—*Lendya, dhaura,* HIND.
A small tree or large bush, wild in Bengal, Central and South India.

17. Ricinus communis, *Linn.*
Vern.—*Rand, arund, arendi,* HIND.

18. Shorea robusta, *Gærtn.* (*Conf. with p. 122.*)
Vern.—*Sál, sála, sálwa,* HIND.; *Koroh,* OUDH; *Gúgal,* TEL.
Mr. B. H. Hodgson (*Jour. Agri.-Hort. Soc. of Ind.*) says the *tasar* feeds chiefly on this tree in the Mechi forests at the foot of the Sikkim Himálaya. Dr. Helfer mentions this same fact, and it is also reported to be the tree in Midnapur upon which the worms feed.

19. Tectona grandis, *Linn.*
Vern.—*Ságun,* HIND.; *Tekku,* TAM.; *Kyum,* BURM.
Colonel Sykes states that the *kolisurra* (or *tasar*) worm feeds upon this tree in the Deccan.

20. Terminalia Arjuna, *Bedd.* (*Conf. with p. 131.*)
Vern.—*Anjan, arjun,* HIND., BENG.; *Vella marda,* TAM.

21. T. balerica, *Roxb.*
The Rev. Mr. Campbell says it sometimes feeds on this species.

22. T. Catappa, *Linn.*
Vern.—*Badam,* BENG.; *Tari,* KAN.; *Vedam,* TAM.; *Catappa,* MAL.
Mr. Hugon mentions that the *tasar* feeds largely in Assam upon this tree.

23. T. tomentosa, *W. & A.* (*Conf. with pp. 115, 119, 131, &c.*)
Vern.—*Saj, seni, asan,* HIND.; *Piasal,* BENG.; *Amari,* ASS.
This is one of the most favoured *tasar* trees.

24. Zizyphus Jujuba, *Lam.* (*Conf. with pp. 125, 129, 131.*)
Vern.—*Kúl,* HIND.; *Bér,* BENG.; *Bhór,* MAR.; *Blair,* BOMB.

25. Z. xylopyra.
The *Gúti.*
Captain Brooke, writing of the *tasar* silk, in Seoní, states that it is not known to feed upon this tree, but in other parts of India it is reported as doing so.

FOOD OF THE WORM.

Much has been written regarding the improvement of the *tasar* cocoon, and the question of food has naturally taken a first place in the controversy. **Lagerstrœmia indica** and **L. parviflora,** since they possess the property of rapid growth, and would seem to luxuriate under severe pruning or plucking of the leaves, are unquestionably the most successful bushes for this purpose. The **Zizyphus** (or *Bér*) is also a favourite, and of the remainder perhaps **Treminalia tomentosa** and **Shorea robusta** are the most important. The *tasar* worm may, however, be described as an indifferent feeder, and it is possible the above list could be more than doubled. In Bhagulpur Dr. Buchanan states that the tree chiefly used in his time was the **Terminalia** just mentioned, the worms being lifted within baskets on to the trees and changed from tree to tree as the leaves were consumed. They were only applied to the same tree once in two years.

REARING OR PRODUCTION OF TASAR.

REARING. 1940

In the year 1880, Mr. J. Geoghegan issued from the Revenue and Agricultural Department of the Government of India a publication entitled *Silk in India.* In that little work was brought together not only the substance of all the official correspondence which had taken place regarding *tasar* silk up to the date of publication, but a brief review was furnished of the writings of the earlier authors. It does not therefore seem necessary to republish the main facts there brought out, since the reader should find little difficulty in procuring a copy of Mr. Geoghegan's book. The present article aims accordingly at supplementing, where necessary, and carrying down to the present date, the record of official and commercial proceedings connected with this silk, rather than at being a complete compilation of the subject.

Briefly stated, the most important regions of supply are the Bhagulpur, Chutia Nagpur, and Orissa divisions of Bengal, and the Chattisghur, Nagpur, Narbudda, and Jubbalpur divisions of the Central Provinces. The experiments at semi-domestication, performed some few years ago at Poona, would appear to have proved that, so far at least as Western India is concerned, the hopes of greater success from cultivation over a restricted area, as compared with collection from wild sources, have proved misleading and the experiment has accordingly been abandoned. The modern improvements in reeling, spinning, and dyeing *tasar* silk have surmounted all the difficulties that were alone thought to stand in the way of a greatly enhanced Indian *tasar* industry, but the low price paid for the cocoons in Europe would appear to be now, as it has always been, of a far greater moment. Many of the modern reports speak of the collection of *tasar*, beyond that for local demands, as proving year by year less remunerative, and accordingly the trade with foreign countries from India may be said to be on the decline or rather to have manifested no tendency to expand. It has not, perhaps, been satisfactorily proved, however, that an extension of the process of semi-domestication in regions more suitable for the worm than Bombay would not greatly reduce the cost of production and transport, and thus favour the development of the Indian *tasar* industry. As matters stand, China affords a more trustworthy European supply than India of what is designated "Tasar Silk" (*Conf.* with **Antheræa pernyi**), and the modern extensive utilization of waste silk for purposes and by processes not hitherto thought of, has probably tended to lower the price offered for *tasar*. That *tasar* silk has meantime attained a recognised position in commerce and has met demands scarcely dreamt of a few years ago, are now, however, established facts. At the same time also indications are not wanting which tend to show that India may yet come to assume a more honoured position in the world's supply

SILK:
Tasar.

Rearing or Production of Tasar.

REARING.

of this, one of her most abundant wild silks. In the vast improvements that have been effected in reeling the cocoons, Indian enterprise and capital has, by no means, taken a backward place. Most of the filatures of this country and many private individuals hold patents for processes of their own, and the results obtained have called forth the highest commendations from the European manufacturers. A great advance also has been accomplished in the utilization of *tasar* waste and pierced and other *tasar* cocoons in the preparation of special textiles.

The energetic action of the Government of India in responding to applications from merchants who called for details of the Italian and French new methods of reeling and for more precise information on the Indian supply is manifested by a voluminous correspondence, the final conclusions of which were from time to time made public. Foremost among these official records may be cited a "collection of papers regarding *tasar* silk" which appeared in the Proceedings of the Revenue and Agricultural Department for 1879, and may be said to have culminated by the production of **Mr. T. Wardle's** Report on the *Wild Silks of India.* Prior to these proceedings, however, **Mr. J. Geoghegan,** then Under-Secretary in the Department, issued (1872) a most valuable report which has since passed through several editions in the form of the small book designated *Silk in India* to which reference has been made above. **Mr. Geoghegan** devotes several pages to the subject of *tasar* silk and republishes *in extenso* most of the more valuable papers on the subject which had appeared before the time at which he wrote. His book was, however, more especially devoted to the mulberry silk-worms, and the subsequent official proceedings on *tasar* silk are of considerably greater interest than those to which he had access. In 1875 the Government of India published a Resolution which set forth the extent of the knowledge which then existed and invited Her Majesty's Secretary of State for India to endeavour to procure certain information desired by Indian merchants who were even then showing considerable interest in the subject. It may serve a useful purpose, at least in marking the rapidity and extent of the development of the subject, to refer to a few of the points set forth in the Resolution. It is announced that **Major Coussmaker** had commenced his experiments which were designed to ascertain how far a more extended system of semi-domestication would bring the insect into the field of constant and cheap supply. But it is added that "at present there is practically no demand in the European market for this silk, except in the shape of fabrics prepared in this country, and for these the demand is limited." It is noted in the Resolution that Her Majesty's Secretary of State for India had, through the Vice-Consul at Lyons and the Consul at Genoa, ascertained a few particulars regarding the new methods of reeling and dyeing *tasar* silk which had been discovered at these great centres of commercial enterprise. It is also announced that **Mr. Wardle** had discovered certain methods of "dyeing *tasar* silk in brilliant colours and of giving it the lustre of Chinese silk." It is thus acceded that the two great obstacles hitherto supposed to impede the path of progression in the *tasar* industry had been practically removed by the discoveries made in Italy and France. These obstacles are expressed thus:—

"*1st*—Defective reeling in connection with the difficulty of properly dissolving the natural gum exuded in spinning by the worm.

"*2nd*—Difficulties in dyeing."

"As regards the first, owing to the defective way in which the silk is reeled, the thread is not continuous and retains more or less of the peculiar cement (compared to plaster-of-paris by **Major Coussmaker**) which the worm exudes, the presence of this cement detracting from the appearance

REARING.

of the fibre rendering it unfit for fine fabrics, and preventing it from taking fine colours. It has, however, it is understood, been proved by actual experiments made within the last year, that by the use of a simple alkaline solvent and by keeping the basins of water in which the cocoons are plunged, when being reeled off, at a temperature of about 200°F., or a little higher, all difficulties of reeling, so far as the mere unwinding of the silk is concerned, disappear." "In respect to the second, there seems to be little doubt that no real effort to dye the silk has been thoroughly made, and that if attention were turned to the subject by competent persons, the difficulties in question could easily be overcome." The Resolution concludes in the following words:—

"The other and more serious difficulties which will require to be overcome appear to be: *1st, the manner in which the cocoons are naturally distributed.* It is true that the supply is inexhaustible, but owing to its distribution over a vast area a thousand square miles will, in the natural condition of things, seldom probably yield as many cocoons as a single hamlet in Italy produces of the domesticated worm, and although the cocoons can be obtained for nothing, yet, as the search for them has to be made over enormous areas, if large quantities, such as a filature would require, are to be obtained (and it is only during one brief period of the year when the trees, the **Antheræa** chiefly haunt, are shedding their foliage that any successful search can be made), it appears doubtful whether, under these conditions, the wild cocoons would not cost more than cocoons obtained from domesticated worms.

"It is true that the *tasar* worm can be entirely domesticated, but if regularly domesticated like the **Bombyx mori**, the produce obtained would perhaps be more costly than that of the **Bombyx mori**, inasmuch as many more worms would be required to produce an equal value of silk, and if the manufacture of silk from the **Bombyx** worm is not remunerative, as seems sufficiently proved by the state of the industry in Bengal, *à fortiori*, the *tasar* worm will not yield any profit.

"Again, the *tasar* may be half-domesticated, a certain number of moths being kept for laying purposes yearly, the eggs hatched, and the young worms turned out to feed themselves, thus avoiding the heavy expense (especially during the latter stages) of constantly supplying fresh leaves to the worms, but here also it appears doubtful how far it will be possible to concentrate the worms or protect them from birds or other enemies if they are at all abnormally numerous on any group of trees. Under these circumstances it appears probable that it is only in a nearly wild condition that the *tasar* can prove remunerative.

"*The second difficulty (which is even greater than the first) depends upon an inherent defect in the filaments spun by the worm.* It must be remembered that the thread of the *tasar* silk-worm is spun from a double spinnaret, and that these filaments are not parallel, lying close side by side, but are spirals touching each other only at the exterior points of their curves, but united by the natural gum in and with which they are exuded, and it is not on this spirality that the elasticity of the silk depends. Now, in reeling the silk, it is necessary that the spirals should be ground well into each other so as to form an even round thread, but it is doubtful whether the filaments can be brought to bear the amount of croissure necessary to produce the round thread, and without this it will be impossible to provide an article of export which will be acceptable in the European market.

"Granting that this difficulty may be surmounted, it appears certain that it can only be done under skilled European supervision, aided by the best mechanical appliances in properly appointed filatures. It will be

SILK: Tasar.

Rearing of the Tasar Silkworm

REARING.

hopeless to expect that such reeling, as is required to fit *tasar* for manufacture into superior fabrics for the European market, can be done by natives working in their own homes. If success is to be expected in the manufacture of *tasar* silk, the operations of villagers must be confined to the production or collection of cocoons. The reeling processes, if manufacture is to be attempted at all in India, must be carried out in properly organised filatures, possessing means and appliances, machinery and systematic supervision, such as are wholly unattainable by villagers in their own homes. Thus, for the proper reeling of the *tasar*, where the basin must be kept at a heat of from 200° to 205° Fahr., nothing but steam can keep them uninterruptedly at precisely that temperature which is essential not only to enable the silk to unwind, but to keep the gum still retained by the filaments at just such a temperature when they reach the *croissure* as to be soft and yielding, but not so soft as to be worked out.

"The conditions of successful manufacture being such, there dees not appear to be any prospect of reviving the reeling of silk as a village industry, whether the silk is produced for local consumption or export.

"Under no circumstances would there appear to be any reasonable prospect of any proximate material enlargement of the local demand. If, therefore, anything is to be done for the country in silk, whether for the domesticated or the *tasar* worm, it must, it seems to the Government of India, be in the way of increased exports, either in the shape of cocoons or as raw silk so reeled as to be acceptable to the European purchaser.

"In regard to *tasar*, many of the most important data necessary towards forming a satisfactory conclusion in the matter are altogether wanting, and the Government of India are of opinion that the subject should be systematically investigated, so as to set at rest all doubts which now exist.

"Towards the attainment of this end, the first thing in regard to which it is requisite to obtain definite information is the exact cost at which the raw material can be collected or produced in commercial quantities, both in its wild and semi-domesticated state. The next points on which further information is requisite are the cost of reeling off the silk, the amount of silk there is in proportion to cocoon, the degree in which the filaments will bear *croissure*, and the consequent ultimate value of the silk in the market."

Expectations of a Future for Tasars.
1941
Conf. with pp. 59, 148, 156, 159, 161, 193.

It will thus be seen that long before the appearance of any of the recent publications, the authors of whom have severally claimed to have discovered the true nature of *tasar* and to have, through their energies, placed that silk on a sound commercial basis, the Government of India were not only in possession of the main facts, but fully appreciated the issues which alone could, and have since, governed the growth of an Indian foreign trade in the product. That the obstacles offered by the nature of the silk to its utilization could and would be removed was a matter on which all past experience justified the most positive convictions. That India possessed a practically inexhaustible supply was upheld by a consensus of local opinion. That the worm could be domesticated or rather semi-domesticated, experiment had demonstrated beyond doubt (*see pp. 108 et seq.*; also *115, 124, 129, 132*). The question was reduced, therefore, to one of profit or loss (*see pp. 132, 149*). Would the price paid at the European markets cover the expense of collection and transport? The product was wild. The cocoons could be had for the trouble of collection. But a thousand square miles would usually not afford more than an Italian hamlet could produce of mulberry cocoons. If the insect had to be domesticated or semi-domesticated, the expense of collection would be greatly reduced; but would the produce even then pay better than might be obtained by rearing the more tractable mulberry-feeding worm? These were the issues which the Government of India

Extent of Indian supply
1942
Conf. with pp. 59, 75, 115-18, 126.

REARING in Bengal.

realised were of far greater moment to India than the possible European utilizations of the fibre. It was seen at once that while for local demand a few maunds might be collected here and there and conveyed to the reelers and manufacturers to be made into a coarse textile for which there was always a ready and remunerative sale, it was quite another question to expect the ignorant and poverty-stricken peasantry of India to become reelers of the silk for the foreign markets. If, therefore, *tasar* silk had to be reeled in India, the enterprise would have to pass into large filatures where the necessary appliances could be set up and the trade carried out with European capital and under trained supervision. India's interest primarily was, therefore, in the production of the cocoons. In order to ascertain the present location and extent of supply and to discover the expectations entertained as to a possible expansion of the trade throughout the country, the Government of India issued to Local Governments the Resolution from which the above passages have been taken. The correspondence that ensued would, if published *in extenso*, run to a large volume. A brief review of the main facts, therefore, is all that can be attempted in this work. For the earlier facts and opinions, the reader should consult **Mr. Geoghegan's** *Silk in India*, since in the following notes, grouped provincially, the attempt is only made to give the more recent aspects of the subject.

I.—BENGAL.

IN BENGAL. 1943

Roxburgh described the *tasar* worm as a species that could not be domesticated, and he furnished much useful information regarding it (*Trans. Linnean Soc. for 1804*). He viewed it as a native of Bengal, Behar, and Assam, and added that it was the *bughy* insect of other parts of India. **Dr. Buchanan-Hamilton** gave a detailed account of it and its silk as met with in Bhágulpur and Dinagepur. Indeed, it may be said that his paper is one of the most complete statements that exists even at the present day on the *tasar* silk of Eastern Bengal. **Mr. Charles Blechynden** furnished, in the *Journal of the Agri.-Horticultural Society of India*, much valuable statistical and other information regarding the *tasar* of Midnapur. He was then Superintendent of the Radnagore Silk Factories and was thus an expert writer. About the same time (1836) **Mr. Homfray** wrote of the collection of *tasar* cocoons in the Sandarbands (*Trans. Agri.-Hort. Soc.*). It will thus be seen that, during the first half of the present century, the subject of Bengal *tasar* silk was brought prominently before the public. Not the least important of these publications were the numerous papers and official documents regarding the *tasar* of the mountainous tract of Chutia Nagpur which, with the continuance of similar physical peculiarities into the Central Provinces, constitutes the chief producing area. As a minor wild forest product, it has, to the savage tribes who inhabit the great central, western, and southern table-land of India, always been of the utmost importance in supplementing their scanty sources of sustenance. As was customary with many of the Indian writers of the 3rd to the 6th decades of the present century, the Journals of the Agri.-Horticultural Society of India were made the channel through which publicity was given to the discoveries in economic products. Papers bearing on the subject of the *tasar* of Chutia Nagpur appear not infrequently in these Journals. One of the most instructive will be found in the volume for 1861 together with the controversy raised in connection with some of the statements there made. **Captain Sherwell** also furnished in 1854 much useful information on the system of breeding and semi-domestication then prosecuted in Hazaríbagh.

In its reply to the Resolution of the Government of India, the Govern-

SILK : Tasar.

Rearing of the Tasar Silkworm

REARING in Bengal.

ment of Bengal, in March 1877, published the substance of the particulars which had been collected. The subject was dealt with fully, and much valuable information was brought to light by the inquiry instituted by the Bengal Government. The Commissioners of Burdwan, Chutia Nagpur, Orissa, and Bhágulpur were desired to circulate to all district officers the following series of questions:—

(1) Is *tasar* silk (under any name) usually collected in the state of cocoons, or are eggs collected and hatched in so-called gardens (or enclosures in the jungles), and are the worms tended with care by the growers?

(2) Are *tasar* eggs produced by moths in captivity ever fruitful, and are they kept for a future crop of worms?

(3) On what trees do the worms feed?

(4) If *tasar* cocoons are collected in the jungle, can any estimate be formed of the weight annually collected in the districts? Is the right to collect them farmed out by the *zemindars*, and what is approximately the area on which they are collected?

(5) What caste collects the cocoons, who rears the *tasar* worms, and what caste sells the *tasar* silk? How does such silk sell?

(6) Are there any, and (if any) what, varieties of insect producing silk beside the ordinary silk worm and *tasar*-worm?

The reports received by the Bengal Government in reply to these searching questions deal mainly with the points connected with the methods of collection, degree of domesticity, castes of the people who collect, reel, and weave the silk, and the varieties or classes of cocoons recognised in local trade. Practically speaking, none of the district officers ventured an opinion as to the annual production, extent of traffic (import or export), and price; yet these are the very points regarding which more precise information was an indispensable condition of possible future expansion. One or two passages having a bearing on the Bengal sources of supply may be here quoted, however, from the Government summary of the local reports. "The result of the enquiries that were made shows that the insect (or insects) known as the *tasar* moth is found in the jungles of Bankúra, Midnápur (west), in the northern part of Balasore, Bhágulpur (south), Sontal Parganas (chiefly Dúmka, Godda, and Rajmehal), Hazaríbagh (practically in the *gaddies* of Gawan and Satgawan, and the *parganas* of Gola, Ootara, Kunda, and to some extent in the *gaddies* of Palgunge and Seerampore), Lohardugga (Palamow, Bilounjah, Tamar, part of Yúrpa Palkote, and Bira), Singbhúm, and Manbhúm. It is also found throughout the Orissa hills, though it is strange that the replies of the Chutia Nagpur tributary chiefs report little or no *tasar*.

"It is probable that, though the cocoons may not be collected, the insect is found in those tracts. On the Orissa side, Keonjhar is expressly mentioned as one of the sources of the *tasar* worked in Burdwan. Roughly speaking, it appears that the *tasar* insect will be found throughout all tree and scrub jungle lying south and west of a line marking the southern and western limits of the continuous cultivation of the Gangetic alluvial and the Orissa littoral tracts of country." In the various volumes of Sir W. W. Hunter's *Statistical Account of Bengal*, occasional brief notices occur regarding *tasar* silk. Thus, for example, in the volume on Birbhúm (*IV., 377*) it is stated that cocoons are kept over from one season to furnish a breed for the next. If this really be the case, a nearer approach to the domestication of the worm has been accomplished in Birbhúm than is on record regarding the other districts of Bengal, since in all other districts the nucleus of each year's breed is sought afresh from the forests. The volume on Bhágulpur (*XIV., 35 & 180-181*) gives the main facts originally made known in 1810 by **Buchanan-Hamilton**. There were then,

Domestication in Birbhum.
1944
Conf. with pp. 106, 124.

REARING in Bengal.

Dr. Hamilton tells us, 3,275 looms employed on *tasar* silk alone in the Bhágulpur district. In the same volume (p. 338) particulars are given regarding the Santal Parganas. Volume XVI. (pp. 168-169) deals with the *tasar* silk of Hazaríbagh, and in a further portion of that volume details are furnished (pp. 346-349) regarding Lohardugga. Lastly, in the volume on Manbhúm (*XVII., 314-315*) much useful information will be found.

In the review, furnished by the Government of Bengal, of the reports it had received, much space is devoted to a discussion of the classification of *tasar* cocoons pursued in each district. Geoghegan forms three sections: (1) *bonbunda*, or cocoons found in the natural state in the forest; (2) (*a*) *mooga*, female, and (*b*) *teerah*, male, cocoons formed by worms reared in captivity, the worms having begot from moths escaping in July or August from *bonbunda* cocoons; and (3) (*a*) *dabba*, (*b*) *buggoy*, and (*c*) *tarroy*—names given to cocoons according to size, colour, or season at which bred. The term *buggoy* (used by Roxburgh nearly a century ago) seems to denote a peculiar colour of cocoon and not a form recognisable in the insect. Dr. Hamilton found the people of Bhágulpur classifying *tasar* into four kinds, *viz.*, *dabba*, *sarihan*, *jarhan*, and *laugga*.

It does not seem necessary in this place to give the multifarious classifications furnished by the local officers in their replies to the Bengal Government. Suffice it to say that none of these justify the belief that they denote varieties entomologically and perhaps not even domesticated races. They mostly convey the idea of the season of the year when spun by the worm, the fact of their being wild cocoons, or of one or more seasons' domestication, or, more generally still, the size and colour of the cocoons on being assorted by the dealers. Speaking of Bengal as a whole, it may be said that the reports under consideration concur in the statement that the seed is not reared from year to year in domestication, but is annually renewed from the jungles.

The information regarding Bengal *tasar* in the possession of Government, up to the date of the reports here dealt with, may be concluded by the following passages from the Bengal Government's review:—

"In Bankoora, the first jungle cocoons seem to be collected in *Asharh*, *Sraban*, or *Bhadra* (*i.e.*, from June to August), and the eggs are laid in *Bhadra* (August). In Midnapore the season of breeding is not given. In Bhagulpore, *Jyaishtha* and *Asharh* (May and June) are said to be the collecting months. In the Sonthal Pergunnahs, January and February, September and October. In Hazareebagh, the season varies; in some parts there are three crops, in others only one. In Manbhúm the jungle cocoons must be collected about March; in Singbhoom, about April.

"The mode of breeding seems everywhere nearly the same. As soon as the moths emerge from the cocoons kept for seed, the females are exposed and left to form their own connexions, whether with the males emerging from the same batch of cocoons, or with jungle moths. Impregnation having taken place, the eggs, which are immediately laid, are collected in baskets, trays, pots, or on tufts of grass; and when, in some eight days, the worms begin to emerge, they are placed on the *asan* or *sál* trees growing in sites defended from their enemies (birds and insects), and moved from tree to tree as they devour the foliage of their first location, till finally, in about fifty days and after five moults, they spin their cocoon upon the tree.

"How many complete cycles there are in the year cannot be easily stated. In Midnapore it is not apparent how often the process is repeated in the year, nor is the Collector's account for Bankoora very intelligible. It is meant apparently that the jungle cocoon is collected in June, that its eggs are laid at once, and that moths from these eggs lay eggs in August, which produce cocoons in September; that the eggs laid by moths emerging from these cocoons produce cocoons in January, while from January to June they are left to Nature. There must be a cycle in that interval; so in Bankoora the insect would appear to be quadrivoltine. The same would appear to be the case in Bhagulpore and in Singbhoom and Manbhoom. In the last, three out of the four bunds are formed in captivity, *i.e.*, on the trees of the breeder's plantation. Mr. Geoghegan adds that he is not sure whether, between the *dabba*

SILK: Tasar.

Rearing of the Tasar Silkworm

REARING in Bengal.

cocoon formed in August, and the jungle cocoons which must be collected about February, *two* cycles do not intervene. This would make it quinquevoltine.

"Mr. Forbes, the Sub-Divisional Officer of Palamow, gives a different account. According to him the domesticated tusser is reduced to a bivoltine. Cocoons are formed once in August and once in November. From November to the next June 'the cocoons kept for seed are carefully packed in *kodo*-straw, just as seed-dhan is packed, and stored in some dry place till June following.' Not only then, if Mr. Forbes' account be strictly accurate, has the insect been thoroughly domesticated, but it has been so completely changed in habit as to become a bivoltine; the cocoons formed in November not opening to let out the moth till the following June. The whole subject requires more careful consideration. In the first place, it appears open to question whether the facts are ascertained beyond all doubt. If they are correct, it might be possible to domesticate the worm without reducing the rapidity of reproduction. If without degeneracy of the breed a tusser domesticated, yet quadrivoltine, could be raised, we should have a silk-producer of twice the value of this Palamow worm.

"With regard to outturn, whether of cocoons collected from the jungle, or cocoons regularly reared, it was not to be expected that any very accurate estimate could be framed. The Collectors have (wisely) not even hazarded guesses on the subject. Perhaps some estimate might be got of the amount of cocoons sold at the marts in Burdwan, Beerbhoom, Bhagulpore, Bankoora, and in Gya or Patna, and possibly of the exports to Mirzapore; but we should still be unable to reckon the amount absorbed by the local village weavers.

"As to any cess levied by zamindars, it is said that in Midnapore the collectors pay 8 annas or 12 annas per head for the right to collect; in Balasore they pay as much as a rupee or a rupee and a half; in Bankoora no cess is said to be levied; in Bhagulpore the rate is 5 annas a head; in the Sonthal Pergunnahs the charge is either so much a tree, or 12 annas to R1-8 per patch of jungle used as feeding ground; in Hazareebagh each rearer pays 6 annas to 8 annas for the privilege; in Palamow a 'koa' revenue appears to be paid to Government. In consideration of this, the proprietor levies a rate which ultimately assumes the shape of a hansua-tax, or tax on the bill-hook, of 4 annas per annum.

"The chief collectors and breeders of the tusser are the aboriginal tribes—Sonthals, Paharias, Kols, Uraons, Cheros, Kherwars, &c.; and the lower Hindu or semi-Hindu castes—Bagdis, Bowris, Tewirs, Bhuiyas, Bhogtas, Samantas, Chamars; Dosads, Koormees, and Gowalas are also mentioned. The breeders seem to collect their own wild cocoons from the jungle, a different practice from that which prevails in the Central Provinces. The breeders sometimes deal direct with the weavers, but generally receive advances from a baniya or mahajun, who thus receives the cocoons and makes them over to the local weavers, or exports them to some of the marts outside the district. The belief in the necessity of ceremonial purity, and the exclusion of women from the operation of rearing tusser on this ground, seem universal.

Prices & cost of Production.
1945
Conf. with p. 115.

"The price of cocoons seems to vary a good deal. Mr. Geoghegan expresses his doubts of the accuracy of much of the information on this point. In Midnapore prices run from R6 to R10 per kahan (1 kahan=16 pans of 80 cocoons each.∴= 1,280). At Kalipore, near Soory, the Sonthals are said to get R8 per kahan; yet the Deputy Commissioner puts the price in the Sonthal Pergunnahs at 350 to 450 per rupee, or, say, R3 to R4 per *kahan*. In Bankoora the Tantis are said to pay R6 or R7 per *kahan*.

"The Burdwan Collector's report gives prices varying from 120 to 450 per rupee, according to the eight kinds furnished; this would be from about R3 to R10 per *kahan*. Yet it is said that the cheapest sort (*ampatia*) fetches R16 per *kahan*, if brought for breeding. This is not very intelligible, for it is not clear how the purpose of the buyer can affect the demand of the seller. The information is hearsay. In Hazareebagh the rearers get indebted to the *baniya*, and sell at R5 to R6 the *kahan* of 1,680, or about R4 to R5 the *kahan*. The *baniya* gets about 7½ to 8½ per *kahan*. In Pergunnah Golah the rearers seem to get a trifle better price. In Palamow, after repaying advances at 50 per cent., the breeder gets from R4 to R6 per *kahan*; in Singbhoom the price is R8 to R10 per kahan; in Manbhoom the price averages R6 per kahan. The cocoons of Chota Nagpore seem almost entirely exported to Bhagulpore, Gya, Burdwan, Benares, Azimgurh, and Mirzapur. They are not even wound off on the spot.

"An alkaline lye is always used in reeling off; the process is of the rudest, and the thread produced foul and uneven.

"Details as to prices of raw tusser or tusser-cloth are meagre. In Midnapore raw tusser is said to fetch 6 tolahs to 8 tolahs the rupee. At Soory a piece of stuff

REARING in Bengal.

containing 15 chittacks of raw tusser fetches R12. In Bankoora a kahan of cocoons will make two 'pairs of dhotis,' fetching R14 to R18; a thánof twilled tusser fetches R9 or R10, but the weight of silk is not given. At Soory a kahan of cocoons is said to yield 15 chittacks of silk. Prices of the Burdwan stuffs of tusser or tusser mixed with cotton are not given. At Bhagulpore the raw silk fetches R10 to R13 per seer. Width of stuff is not mentioned, so details on that point are useless. In the Sonthal Pergunnahs a kahan of cocoons is said to yield 1½ to 2 seers silk, which fetches R9 to R13-8 per seer."

The Government of India (in August 1878) received a communication on the subject of Bengal *tasar* from **Mr. J. Deveria**, in which that gentleman made the suggestion that the Forest Department should be asked to undertake the supervision of an experiment to rear *tasar* worms in Chutia Nagpur. While the Government of India did not see its way to authorise that experiment—an experiment which might have been viewed as a direct interference with private enterprise—**Mr. Deveria's** communication furnished certain facts of interest which were very justly viewed as having considerably advanced the available information on the subject. It may, however, be here remarked that in giving some of **Mr. Deveria's** estimates and figures the writer does not wish to commit himself to an unreserved concurrence in their accuracy. Indeed, he views them as, in some respects perhaps, sanguine, though they are at least suggestive, and are perhaps directed on the only likely lines upon which success may be attained. **Mr. Deveria** estimated that three men could look after at least 300 trees and protect the worms from their natural enemies—birds and insects. He affirmed that each tree would yield 100 cocoons, so that the crop from 300 trees would be 30,000 cocoons, which, at R4 per 1,000, would give the following return:—

	R	a.	p.
Creditor—			
Salary of three men for 45 days	18	0	0
Seed for 30,000 caterpillars at, say, 200 eggs per moth or 150 cocoons, at R4-8 per 1,000	0	12	0
	18	12	0
Debtor—			
Value of 30,000 cocoons at R4 per 1,000	120	0	0
Less half share to be given to the rearers	60	0	0
	60	0	0
Refund of salary paid to men or rearers	18	0	0
Balance in favour of account .	78	0	0

Mr. Deveria suggested, upon the above estimate of profit, that Government should direct the Forest Department to cultivate the worm over 20 acres of the land around each of some 800 villages, that is to say, over 16,000 acres bearing 30 trees per acre or 480,000 trees. He assumed that if each of these trees yielded only 50 cocoons, the produce would be 24,000,000 cocoons per annum. Applying his estimate of a profit of R78 per every 30,000 cocoons, his total estimate came to R62,400 a year as the revenue from the cultivation. To work the experiment he proposed the staff of a Superintendent on R500 a month, a Moonshee, 20 peons, house-rent, &c., &c., or, say, R10,400, a year expenditure, which would thus show a net annual return of R52,000.

If these estimates are anywhere near the mark, it would seem probable he undertaking might be made remunerative. But the danger of too sanguine expectations is perhaps a greater evil than the want of direct encouragement. Thus, for example, **Mr. Deveria** in one of his early communications (August 1872) expressed the opinion that a *bíghá* of land

SILK: Tasar.

Rearing of the Tasar Silkworm

REARING in Bengal.

in Chutia Nagpur might easily bear 50 trees, and that each would annually yield 250 cocoons. That opinion was at once reported on by the district officer in Palamow (**Mr. L. R. Forbes**), whose able report on the *tasar* worm of that district will be found placed under free contribution below. In dismissing **Mr. Deveria's** recommendations for the present, it may, however, be remarked that with the extension of railway communication, tapping, as it is now doing, large tracts of Chutia Nagpur hitherto closed to commerce through their inaccessibility, it seems probable a new interest may be aroused in this subject. Large portions of the land on which the trees grow which are chiefly employed to feed the worms are practically unsuited for agricultural purposes. The chief difficulty is to induce the people to live in these tracts and to take to *tasar* cultivation as a regular industry. Though agricultural pursuits have greatly expanded and emigration to the tea-growing districts have also drawn on the labour market of the province, still land and labour are cheap and plentiful. To consumers of *tasar* silk the possibility of profitable rearing in Chutia Nagpur is a subject that should therefore commend itself as at least worthy of definite solution. If *tasar*-rearing can be made profitable anywhere in India, it seems likely the industry will find its natural home in the mountainous tracts where the insect and its food are both plentiful and capable of great expansion without any encroachment on the area of agricultural possibilities. The necessary land could very likely be had easily, and in no other part of India could labour be obtained so cheaply and plentifully as in Chutia Nagpur. Before passing away, however, from **Mr. Deveria's** recommendations, it may be stated that he gives 4,000 cocoons as the quantity equivalent to a maund, and that he points out that for many months of the year the Indian filatures have little or no work for their reelers. The *tasar* cocoons, if carefully killed and properly dried, he adds, may be kept for a long time, so that they could be worked up at the Indian filatures during the period of least activity. To a certain extent the Indian filatures are using *tasar* cocoons, but judging from the returns of foreign exports, the trade in *tasar* cannot be said to have progressed during the past five years; indeed, it has, if anything, retrogressed. The amounts used therefore by the Indian filatures are probably easily enough procured, so that it would be incorrect to affirm that the Indian demand has as yet justified any very pronounced step in the direction of expanding the supply.

Cocoons to Maund. 1946 *Conf. with pp. 113, 120, 121, 126*

As manifesting more completely the salient features of the Chutia Nagpur *tasar* industry and its capabilities, the following passages may be republished from **Mr. Forbes'** able statement regarding Palamow. Reference has already been made to his report on that subject as refuting some of the opinions advanced by **Mr. Deveria**, but the reader will be better able to form a correct opinion by having placed before him both sides of the controversy:—

Production 1947

PRODUCTION.—"In dealing with this subject in my settlement report I stated that the *hansua* really means the producer, or *assami*, so called from the *hansua*, or reaping-hook used by him in pruning the trees and collecting the cocoons. I have tried in vain to arrive at some definite decision as to the real or average number of *kharies* contained in a *hansua*, that is to say, the number of cocoons produced by one person, and by this means to get at the area; but I have found it quite impossible to do so with any degree of exactitude. A *hansua* means, as I said before, a producer, *i.e.*, each man or family who is found guarding a plantation on which cocoons are feeding is counted as a *hansua*, and pays to the farmer a tax of R4, whether his cocoons are feeding on five trees or five hundred; and as therefore this is really the case that while one poor man is working (say) three trees, another is working one hundred it is manifestly impossible to arrive even at an approximate estimate of *hansua*. I have therefore adopted other means. On enquiry from 20 producers I find that on an average, and a very outside average too, the *hansua* may be considered as five *kharies* of 1,100 cocoons each, or 5,500 cocoons per *hansua*, but this, as I have said,

REARING in Bengal.

is a very high average, as taking the average market value of a *khary* to be R6, we find the average money value produced by such *hansua* or *assami* to be R30,—a very large sum when we consider the class of people who are engaged in the work. The average amount produced by the 20 growers mentioned above was below R20; but allowing room for concealment of the true amount, I prefer to assume the *hansu* to be as above. On enquiry from the present farmers of the *koakuth* revenue, I find that the average number of *hansuas* of the last three years on which they have collected tax or rent is 861. To this I would add, as outside figures for the four exempted estates, 750 *hansuas*, and for the Belounja and Jupla 400 *hansuas*, or (say) a total for the entire sub-division of 2,000 *hansuas*, or 10,000 *kharies* of 11,000,000 cocoons, which, assuming Mr. Deveria's estimate of the number going to the bazár maund to be correct, would give a total of 3,168, or, in round numbers, 3,170 maunds of cocoons for the whole of Palamow, including Pergunnahs Belounja and Jupla. Now as the Palamow sub-division measures 4,260 square miles, or one-fourth of Mr. Deveria's area, it is clear—always supposing that, as a cocoon-producing district, Palamow fairly represents the whole, and which I have very little doubt it does—that the whole of the area included in Mr. Deveria's calculation as cocoon-producing yields at present at the very outside 12,680 maunds of cocoons,—very different from the marvellous figures given by Mr. Deveria in his memorandum. Of course the figures I have given above appertain only to the quantity now produced. There cannot be a doubt that with European capital and supervision the amount produced could be very largely increased. Supposing Mr. Deveria's calculations to be correct as to the number of trees per *bigha*, and the average number of cocoons per tree to be 250—a rather high estimate—we find, assuming the number of cocoons produced to be 11,000,000, that there are 44,000 × 2 trees now being worked contained in an area of 1,760 *bigas*, or 586 acres, less than one square mile.

Cocoons to Maund. **1948** *Conf. with p. 121 as to loss of weight.*

Decrease in Production. **1949** *Conf. with p. 121.*

DECREASE IN PRODUCTION.—"From enquiries I have made I find that the production of *tusser* cocoons has greatly diminished within the last twenty years; some say about one-third. This falling off is ascribed to (1) increase of cultivation and consequent destruction of *asan* plantations; (2) greater demand for agricultural labour and higher wages, making the production of cocoons a less profitable employment.

"As I have before said the *asan* plantations are of indigenous growth, and are of use only for about 10 to 15 years. At present the Government farmers of *koa* revenue, having the right over about 1,800 villages, show by their accounts and registers that cocoons are bred in 105 only. In a large number of villages there is no *asan* at all, but it is to be found in most village areas, and is destroyed in large numbers yearly in *jooming* operations; and though the figures given by Mr. Deveria are absurd and impossible, still the area of cocoon-producing trees would be found considerable and quite capable of supplying a very extensive trade. I should say that a truer estimate of the area would be one-sixteenth of the whole, but I should prefer to fix it at one-half of that, or even at one *bigha* in 50. The limit, therefore, in the extent to which the trade could be developed, depends rather upon the number of *hansusas*, or who could be persuaded to undertake the production of cocoons; and for this reason I should say that a multiplication by eight of the number of cocoons now produced would fairly represent the limit to which the trade could be expected to reach.

Cocoons. **1950**

CASTES.—"The rearing of *tusser* cocoons is not confined to any particular class. I have found the following castes occupied in the work: Cheros, Kherwars, Nraouis. Bhunias, Chamars, Dosadhs, Mullahs and Ghasis, but no Musulmans. Their *modus operandi* may be thus described: Very few cocoons are kept for seed at the *Kartic* harvest, and these only by well-to-do producers, who can afford to breed during the *Bhadro* harvest for the great harvest of *Kartic*. In *Bhadro*, therefore, the poorer producers apply to the *mahajuns* for advances as *byjhan* for the purchase of seed-cocoons, and as *kyhan* for the means of supporting them during the breeding season. This advance is generally given at *derha*, or 50 per cent., *i.e.*, for every rupee given R1-8 is returned. When the cocoons are ready, the *bania* or *mahajun* appears on the ground and collects his due; the remainder is sold by the producer either to the *mahajun* or in the *hauts* at prices varying from R5 to 7 per *khary*.

"Most of the *tusser* cocoons are exported to Benares, Mirzapore, and Patna. There are a few Patwas (Gosains) in the sub-districts who reel off a few cocoons for making the *zerbunds, dundahs,* &c., worn by most of the respectable castes, worn as waistbands or for fastening ornaments to, but no cloth of any kind is made. Some cloth is made, I believe, in the Gya district, but I have never seen it.

SILK: Tasar.

Rearing of the Tasar Silkworm

REARING in Bengal.

Form of Cocoons. 1951

Forms of Cocoons.—"I have never met with any other variety of silk-producing insect in this part of the country but the ordinary *tusser* silkworm, but I believe some such do exist in a wild state.

Domestication. 1952

Degree of Domestication.—"It will be seen from the foregoiug that in this part of the country the *tusser* silkworm is almost altogether domesticated. The only thing left to render it completely so is to feed it under cover instead of in open plantations. This is the only difficulty, but one which, I am persuaded, can be easily overcome. Before, however, capital could be safely embarked in the trade, it is absolutely necessary that it should be overcome as at present, in open plantations. Not only are the worms exposed to certain pests, such as I have detailed, but rain falling upon them in the '*hatia*' is almost complete destruction; and as rain falls during the *hatia* every other year, and sometimes even for two or three years in succession, it follows that under the circumstances the trade would be a very precarious one.

Method. 1953

Methods of Rearing.—"The habits of the worm are so different from those of the ordinary silkworm, that none of the means employed in rearing the former would be applicable to the latter. The system adopted by Captain Hutton might answer very well for the purposes of a limitted experiment; but I am inclined to doubt its capability of being applied on a large scale, chiefly on the score of expense. I should be inclined to try a simpler method. I would erect a shed on the plantation itself, a simple bamboo frame-work roof covered with grass and supported on green poles cut from the neighbouring jungle, would suffice; the walls to consist of stout bamboo-matting, sufficiently open to admit air and light, but not sufficiently so as to allow of the moths escaping; the flooring I would raise a foot or so with gravel or soil well beaten down and cover it with date palm-matting, which is not only absurdly cheap, but possesses the invaluable quality that no white-ants or insects will touch it. In the centre of the shed I would have a rough frame-work running nearly the whole length, and resting on the matting, but not reaching the roof, so as to cut off communication within this frame-work which may be made of bamboos and rough poles. I could have shelves or *machans* of date palm-matting; the first being at least six inches from the ground, and the next two or three feet above it. On these shelves I would place or heap dry leafless twigs or branches of any bush or tree, interlacing each other and running the whole length of the shelf. To one of these twigs I would fasten the hatching cradle, placing lightly over it a few green leaves from the plantation, and then daily, as the young worms began to wander, I would lightly sprinkle freshly-gathered leaves over the twigs, until they had been induced to spread themselves over the whole length of the shelf. When the time came on for spinning, they would attach themselves to the twigs, which of course should be sufficiently far apart as to freely permit the operation of spinning to go on. Should a little moisture be found necessary, a little rain water lightly sprinkled over them with a grass broom would answer very well.

Future Prospects. 1954 *Conf. with pp. 126, 129.*

Future Prospects.—"For my part, I should be very glad, indeed, to see the production of *tusser* cocoons in these districts in the hands of enterprising European capitalists working on fair and honest principles. Apart from the creation of a valuable industry, the capital yearly thrown amongst the very poorest classes at a time when they most need it, would go very far to relieve their necessities and keep them from the door of the usurer. I would not recommend the initiative being taken by the Government, as these kinds of experiments are not generally successful when undertaken by public officers having other and more important calls upon them but I would suggest that private parties be invited to take the matter up, and, by way of assistance, that the Government allow such persons the free and unconditional use of any one or more of the *asan* plantations lying within the Government farms."

In bringing these remarks regarding the *tasar* silk of Chutia Nagpur down to more recent times, it may be said that Mr. J. F. K. Hewitt, in a letter to Mr. Wardle (*dated September 1880, printed in Wild Silks of India, pp. 22-24*), views the *tasar* supply from the mountainous tracts of the central tableland of India as possible of great development. "With a little effort," he says, "a large quantity of *tusser* can be produced in the division under my charge, and the very great likelihood of establishing a trade which will conduce to the permanent increase of the wealth of the country, if means can be found to increase the production of *tusser* to meet the present growing demand. I must say, however, that Government can never stimulate a trade so quickly as private capital. A much quicker and more easy solution of the difficulty could be worked out by

REARING in Bengal.

the large silk-houses if they once determined to turn their earnest attention to the subject. Before *tusser* silk can be produced at all approaching to that for China silk, the methods of the trade must be entirely revolutionised." "The country over which the *tusser* is found," says **Mr. Hewitt**, "includes an area larger than France; and in almost the whole of the hills and forest country, which cover fully three-fifths of its surface, *tusser* silk could be profitably produced, if the requisite arrangements were made, as the *asan* (**Terminalia tomentosa**), on which the worm thrives best, is found abundantly everywhere. The numbers, however, of the people who cultivate are small, and these numbers are yearly decreasing as the jungles are cleared and the distance between the village sites and the jungles increases. The cultivator requires that the trees on which the worms are fed should be constantly watched, and superstition adds to the difficulty of the cultivator by insisting that no one can hope to cultivate *tusser* worms and reap profit from the cultivation, while maintaining his own health and that of his family, without submitting to a long series of ascetic ordinances during the whole time of watching the worms. If he fails in doing this, the anger of the gods will inevitably destroy him and all belonging to him. As very few of these people will ever consent to live in houses separated by any great distance from their nearest neighbours, it invariably happens that the cultivators find out, as cultivation extends and the area of cleared land between the village and the forest increases, that cultivation is easier and quite as profitable work as *tusser* growing with its concomitant annoyances, and so he gives up *tusser* and takes to cultivation instead. Of course, if they kept a large number of the *tusser*-feeding trees near the village, they might contrive both; but this they do not do, as their great object is to get rid of every cover near the village within which wild beasts can hide, and the universal practice, therefore, is to cut down every tree and leave the cultivated land totally unshaded and open. The increase of population, and the yearly increasing area of forest cleared away, therefore, tend to lessen the cultivation of *tusser*, and it is not possible to increase the production to meet a large demand except by inducing the cultivators to betake themselves to uncleared forests, or else to utilise the partially cleared lands for the cultivation by planting fresh trees and preventing the destruction of those still left."

Conf. with p. 102.

Difficulties to Expansion.

1955

Conf. with pp. 75, 106, 121, 149, 156, 158.

The difficulties which thus beset a possible expansion of *tasar* production in Chutia Nagpur are by no means insignificant, and it need only be necessary to add that the decline above alluded to has by all subsequent writers, been abundantly confirmed. Thus in the Administration Report for 1882-83 it is affirmed that the collection of the cocoons is no longer profitable, and that this state of affairs will continue till certain changes in the administration of the forests have been effected. Hopeful results are likely, therefore, to be attained alone by an improvement in the price paid, and when capital and an amount of enterprise not easily daunted by initial failures is brought to bear on the undertaking, in conjunction with liberal co-operation from Government. That success may, and is even likely to, be attained, however, the writer has little doubt, but until Europeans engage in the rearing of the worm and prove it both more remunerative and less irksome than at present, the decline in production, which all writers deplore, must steadily take place.

From year to year (since 1882-83) in the annual publication on internal trade, the Bengal Government has continued to deal with *tasar* as distinct from mulberry silk, and it may, therefore, be found instructive to give here a selection of the more important passages that have appeared in these reports. These will be observed to denote the chief regions where the

REARING in Bengal.

cocoons are found, the more important towns where *tasar* silk is manufactured, and to discuss both the production and price of the crude and manufactured articles. It has, in fact, been deemed the better course, in furnishing the particulars on these all-important topics, to republish the passages as they stand, rather than to attempt a review of them even although a certain amount of repetition is necessarily involved:—

1956

Report for 1883-84—BHAGULPORE DIVISION.—"The manufacture of *tusser* and *bafta* silk pieces in Bhagulpore, and of silk fabrics of Maldah, is gradually decaying. As regards the former, it is reported that formerly the quality was superior, but the *tusser* made in Bankura and Midnapore is now better than the *tusser* made here. The manufacture is confined to the town of Bhagulpore only, where there were during the year under report 880 looms. Last year only 300 looms were reported to have been in existence,—I believe on uncertain data. The outturn is said to be 46,726 pieces, valued at R2,01,778. The industry is decaying in Maldah, because there is not a sufficient demand for the fabrics. It suffers particularly during the year under report owing to absence of rain and the extreme heat of the weather. It is said that in the late *Choitra bunda* the outturn was less than 4 annas on an average, entailing a loss of several lakhs of rupees. The bulk of the silkworms reared died, and those that survived yielded an inferior quality of cocoons. In the Sonthal Pergunnahs and Monghyr *tusser* silkworms are reared to a limited extent, and the cocoons are mostly exported to Beerbhoom and Bhagulpore. In only a few places in the district cloth is manufactured."

1957

1884-85—BURDWAN DIVISION.—"*Tusser* silk cloth is manufactured in all the three districts of Burdwan, Bankura, and Beerbhoom, more or less, but the principal seats of the industry are, as stated in last year's report, at Kuddea and Tantipara in Beerbhoom. The outturn of cloth manufactured in these two places during the year under report is estimated at 17,460 pieces, each 10 yards long, valued at R96,030. The value of the previous year's outturn was estimated at R46,675. The demand for *tusser* cloth appears for some reasons to have been greater than usual during the past year."

1958

BHAGULPORE DIVISION.—"In my last year's report I said that the manufacture of *tusser* and *bafta* pieces in Bhagulpore and of silk fabrics in Maldah was gradually decaying. There has been no change during the year worth noticing. The manufacture of the former is confined to the weavers in Chumpanagur, the western part of the town of Bhagulpore. There are now 880 looms, and the outturn is said to have been 44,555 pieces, valued at R1,91,500, against 46,726 pieces, valued at R2,01,778 in the preceding year.

"The Sonthal Pergunnahs export *tusser* cocoons to the neighbouring district of Beerbhoom, where the silk is manufactured. About 3,500 maunds of cocoons are said to have been exported during the year. The industry in this district, therefore, is confined to rearing of the worms. There is no local manufacture of the *tusser*."

1959

1885-86—BURDWAN DIVISION.—"*Tusser* silk is manufactured in all the three districts of Burdwan, Bankura, and Beerbhoom. In Midnapore also *tusser* cocoons are obtained from the jungle mehals, and are reeled and manufactured into cloths in parts of the Sudder and Ghatal Sub-divisions. The principal seat of the industry in Burdwan is at Bagtikari, Moosthal, and Ghoranash, in the Cutwa Sub-division. The cocoons are imported from the districts of Moorshedabad and Chota Nagpore. Over 200 persons are employed in this manufacture. The outturn is estimated at 30,010 yards, valued at R40,242. In Bankura the principal seats of the industry are Bankura, Beersinghapore, Sonamukhi, and Bishenpore. The supply of *tusser* cocoons in the district is not adequate for the requirements of the manufacture. Large importations are therefore made from Chutia Nagpur. The *tusser* cloth finds a ready market in Calcutta and North-Western Provinces. The value of the fabrics produced is estimated at about R20,000. In Beerbhoom 21,500 pieces of *tusser*, valued at R1,40,000, against 17,460 pieces to the value of R96,030, were exported from the weaving centres of Kaddya, Beersinghapore, Purulia, and Tantipara. The demand for this article from Europe is on the increase."

1960

BHAGULPORE DIVISION.—"The manufacture of *tusser* and *bafta* pieces by the weaver in Champanagore, Nathnagore, and Khanjarpore of Bhagulpore, is still maintained. The outturn during the year under report is estimated at 22,417 pieces, valued at R1,13,088 as compared with 44,555 pieces worth R1,91,500 in the previous year; the figures show a sad falling off, and I am afraid this too must be written down as a decaying form of industry."

1961

CHUTIA NAGPUR DIVISION.—"In Hazaribagh the manufacture of *tusser* silk is said to have received an impetus owing to the establishment of a filature in

REARING in Bengal.

the town of Hazaribagh by a French gentleman, who is working in cocoons on an improved method. The same industry is said to have been very much depressed in Manbhoom in consequence of the large export of raw silk to Europe and the consequent rise in the price of the raw product."

1962

1886-87—BURDWAN DIVISION.—"Excepting Hooghly and Howrah, *tusser* silk is manufactured in all the districts. In Burdwan, Mankor is the chief centre of this industry. The outturn of last year is estimated at 50.000 maunds against 56,910 maunds in the previous year. It is reported that this industry, notwithstanding some spasmodic revivals, due to changes of fashion in Europe, is waning. In Bankura the seats of *tusser* industry are the towns of Bankura and Birsingpur in the Sudder, and Bishenpur, Sonomookhi, and Joypur in the Bissenpur Sub-divisions. *Tusser* cocoons are imported from the districts of the Chota Nagpore Division to supplement home production. It is reported that the business of the *tusser* weavers is at a low ebb in consequence of the high price of cocoons consequent on the large purchases of several European firms. This has had the effect of driving many a weaver to betake himself to cotton-spinning, a less lucrative occupation. *Tusser* fabrics of Bankura, though coarse, are much liked in Calcutta and the North-Western Provinces; and the ruin of the industry is much to be regretted. *Tusser* is also manufactured in Beerbhoom. Five thousand eight hundred *kahans* (a *kahan* is equal to 1,280 cocoons) of cocoons were imported from Chybassa and Calcutta by native weavers. Here also competition with the European traders has much reduced the margin of profit earned by a native weaver. It is reported that, notwithstanding the reduction in the profit, 11,000 pieces, each 10 yards, were manufactured, worth about R88,000. *Tusser* silk is also manufactured in the Sudder and Ghatal Sub-divisions of Midnapore. The cocoons are procured from the jungle mehals and the adjoining districts of Singhboom, Manbhoom, etc. A new factory has recently been opened at Ghattal by a European firm, where *tusser* silk is manufactured. It is said that a discovery facilitating the reeling of a continuous thread from the cocoon, has imparted a fresh impulse to this industry. The cocoons are now fetching nearly twice their old price."

1963

BHAGULPORE DIVISION.—"The manufacture of *tusser* in Bhagulpore is maintained, but the industry is a decaying one. During the year 20,544 pieces of the cloth, valued at R1,23,264, were exported. This shows a decrease in the amount manufactured as compared with last year's figures, but there was an improvement in the prices obtained. The industry remains in the same hands. During the past season, however, agents of **Messrs. R. Watson & Co.** had been buying up the *tusser* silk cocoons in this district to the value of R20,000, and the rise in prices resulting from these operations is likely to increase the production of cocoons."

1964

CHUTIA NAGPUR DIVISION—"The price of *tasar* cocoons was doubled, and in some parts of the division trebled, during the year owing to the keen and extensive competition of several European firms in making purchases. The trade in cocoons has thus been stimulated in a remarkable degree all over the division, and has penetrated to the more accessible Native States. So far the advantage has been very great, but one unfortunate result has been to check, and almost annihilate for a time, the local manufacture of *tasar* silk. The weavers cannot afford to compete with the wealthy speculators for the raw material. The Deputy Commissioner of Manbhum says that he has ascertained by enquiry that numbers of silk-weavers at Raghunathpur and other places have taken to cloth weaving, finding it impossible to make a livelihood out of silk. He thinks that the silk-weaving industry on its present basis is moribund, and cannot hope to revive unless filatures are established within the district under skilled supervision. In Hazaribagh two new filatures have been established at Hasir and Sarun, two large villages in the south-east of the district, and in Lohardugga a filature is being erected in Palamow, and is expected to commence working during the current year."

1965

1888-89.—BURDWAN DIVISION—"*Tasar* silk is manufactured in all districts except Howrah. In Bankura the principal centres are Bankura, Rajgram, Beersingpur, Sonamukhi, and Bishenpur. Business is reported to have been bad during the past year, cocoons having risen in price owing to a deficient supply from Chota Nagpore and increased demand for Europe. It is reported that 20,000 *kahuns* of cocoons, valued at R2,80,000, were imported, and cloths, valued at R15,000 manufactured during the year and mostly exported to Calcutta. In Burdwan, Mankor is the chief manufacturing centre. The total outturn during the year is reported to have been 335,325 yards value(d at R1,61,671), against 322,505 yards (worth R1,52, 603) in 1887-88. In Beerbhoom the value of *tasar* turned out improved nearly 25 per cent."

1966

BHAGULPORE DIVISION.—"The *tasar* silk industry in Bhagulpore still strug-

SILK: Tasar.

Rearing of the Tasar Silkworm

REARING in Bengal.

gles on, but the price of cocoons was high, and silk thread sold at 5 tolahs per rupee against 8 tolahs in the previous year. Twenty-four thousand three hundred and eleven pieces, valued at R80,197, are said to have been exported this year, against 39,948 pieces, valued at R1,00, 356 in the previous year."

1967

CHUTIA NAGPUR DIVISION.—"In Manbhum the total va lue of the outturn of *tasar* silk amounted to R5,500, against R9,000 in 1887-88, and R10,762 in 1886-87. In Palamow the attempt of Messrs. Hodges & Radford to start a filature for reeling *tasar* cocoons has proved a failure. In Hazaribagh *tasar* silk reeling is now carried on only by native methods in some half-a-dozen villages by a class of men called *Patwas*."

1968

1889-90.—BURDWAN DIVISION.—"In the Burdwan district *tasar* silk is manufactured at Mankor and *garad* at Memari. The total value of the outturn in 1889 was approximately R1,67,333, against R1,61,671 in 1888." The manufacture in Bankura is again spoken of as declining. The district not being able to meet its own demands for cocoons it drew its supplies for Midnapur and Chutia Nagpur. "About R25,000 worth of *tasar* silk is said to have been exported to Calcutta and the North-Western Provinces during the year under report. In Beerbhum R9,450 worth of *tasar* silk was sold in 1889-90 at unusually low prices, against R13,840 in the preceding year."

1969

BHAGULPORE DIVISION.—"The *tasar* manufacture in Bhagulpore shows some signs of improvement, attributable to the cheaper rate at which cocoons were sold. *Tusser* worms are reared and a small amount of *tasar* silk is made in several parts of the Sonthal Pergunnahs district, but the chief seat of the industry is in the Pakour Sub-division. The rearers sell mixed cocoons to the *goladars*, who separate the different qualities and sell them to European firms. The chief sale centres are Amrapara and Hiranpur in the Damin-i-Koh and Dangapara in pergunnah Ambar."

1970

CHUTIA NAGPUR DIVISION.—*Tasar* silk still continues to be manufactured on a small scale in Hazaribagh, Lohardugga, and Manbhum.

Decline of Tasar Trade.
1971
Conf. with pp. 59, 75, 156-7, etc.

It is believed the above passages convey a faithful conception of the present position of the Bengal *tasar* silk industry, an industry which appears to be declining in the inverse ratio to the foreign demand for the cocoons and the success of the European filatures in India. The Native manufacturers are, year by year as it would seem, feeling it to be more and more difficult to carry on their trade. This loss, however, cannot in any way be said to be compensated to India as a whole, by the greater prosperity of the rearing or collection of cocoons, nor, indeed, by the expansion of the *tasar* transactions carried out at the European filatures of the country. Indeed, it would probably be more correct to ascribe the decline to the successful supply by China of a cocoon suitable for the same purposes as *tasar* and to the cheaper manufactures of Europe.

IN CENTRAL PROVINCES.
1972

II.—CENTRAL PROVINCES.

The system of collection and semi-domesticaiton (or to be more correct, temporary domestication) which has already been detailed in connection with Chutia Nagpur, prevails in these Provinces also. The chief districts are Raipur, Bilaspur, Sambulpur, Upper Godavery, Chanda, Bhundara, Nagpur, Seoní, Balaghat, Chhindwara, and Betul. One of the most instructive papers on this subject is that written by Captain Brooke—then Deputy Commissioner of Seoní. The reader will find Captain Brooke's account reproduced in full in Geoghegan's *Silk in India*, and epitomised in Wardle's *Wild Silks of India*, as also in the *Indian Museum Notes*. Since one or other of these publications is likely to be in the hands of the reader specially interested in Silk, it does not seem necessary to again republish Captain Brooke's remarks, even although they perhaps deal with the subject in greater detail than is the case in any other single article that can be discovered. Good accounts, however, occur in several of the Settlement Reports, such as those for Bilaspur, Chanda, Sambulpur, etc., and these, in an abbreviated form, are given in the Gazetteer of the Central Provinces. Since, however, the Settlement Reports are not very accessible,

REARING in Central Provinces.

the following statements regarding Bilaspur and Chanda may be here republished :—

Bilaspur. 1973 Conf. with p. 102.

BILASPUR.—"One interesting item of forest resource remains to be referred to—the *tassah* cocoons, which supply the useful silk so esteemed by the community. The Bhumias and other hill-men collect these during the monsoon and are marvellously active and shrewd in finding them in the jungles. They are found chiefly on the *saj* tree (**Terminalia tomentosa**). In the month of August, the primitive huts of these wild races are invaded by rearers of the *tassah* worm, from the more open portions of the district. These men come to purchase, and a party usually consists of seven or eight persons. A sufficient stock having been obtained these bearers return to their selected locality, which is a tract of stunted *saj* trees covering 8 or 10 acres near a village skirting the forest. Here in September they tie the cocoons to a series of strings, each string stretching from a branch of one tree to a different branch in another, the cocoons thus suspended looking from a distance like a great row of eggs. By degrees the moths cut through the cocoons, during which process they are closely watched, and after they have paired, the females are placed in earthen vessels (*ghurras*) in which they lay their eggs and die. The males fly away. The eggs are kept in the huts of the people generally in cloth and incubated by heat. They are little round dots about the size of mustard seed. In eight or ten days the worm is formed, and as each female moth placed in the vessel deposits about a hundred eggs, a great outturn is obtained. The worms thus incubated are taken out and placed on *saj* trees, on the leaves of which they feed. They are small tiny insects at first, but they grow in size till they attain the thickness of a man's finger and are perhaps two and a half inches long. At this stage they are very prettily marked, but in three months they have attained their full size and then commence their cocoons which are finished in two days. It is quite an interesting spectacle to see these insects busily employed, throwing one thread round their bodies and then another until they are completely encased in their silken home. A period of some four months elapses, *viz*., from September to December, from the time the moth breaks out of the old cocoon to the formation by the freshly generated worm of the new one, through the processes of incubation, development, etc. The new cocoons are sold to the silk-weavers who steep them in hot water, mixed with tamarind pods or leaves, in order to communicate to the thread additional strength and elasticity, when the thread is carefully wound off and manufactured into the light textured *tassah* silk. One piece requires on an average some 800 cocoons, and as the writer estimates the probable amount of silk weaved at 10,000 pieces, the annual supply, to admit of this, must be something like 8,000,000 cocoons, the outturn probably of some 80,000 moths. It is strange that the '*khewuts*' who rear the worms, instead of depending annually on the Bhumia supply from the wilds, do not themselves maintain a permanent stock to breed from. They urge that experience has not proved that process profitable; but the writer fancies the fact is, it would entail too much system to satisfy their tastes. As it is, while employed on rearing, they remain away from their homes, confine their diet to rice and salt, and depend on the prayers of the Bhumia '*Bygas*' for success. The absence of this last element has, in every instance, it is alleged, been followed by failure" (*Settl. Rept., pp.* 77-78).

Chanda. 1974

CHANDA.—"*Tasar* cocoons are numerous in the forest, and are gathered from the middle of June to the middle of February. They are found chiefly on the *en* tree and are egg-shaped, their length varying from 1¼ to 2 inches, and the diameter of their greatest circumference from ¾ to 1¼ inches, while each yields about ⅜ *mashas* of silk thread. The worm is also extensively bred in captivity in most of the pergunnahs north-east of Chanda, the culture being exclusively carried on by men of the Dheemur caste. Upon the moths emerging from the cocoons they are paired, male and female, the sex being known by the colouring of the body and wings; and each pair is placed in a box formed of *chor* leaves about a hand-breadth square, but no provision is made for feeding them. In about ten days the female lays her eggs varying in number from 150 to 200, and shortly afterwards dies as does her male companion. In another ten days the eggs are hatched with the exception of about 25 per cent., which generally fail. A space covered with *en* trees, cut down to a few feet from the ground and throwing out bushy branches, is now selected in the vicinity of the silkworm-grower's village, and on these the young worms are placed, and the bushes are, whenever practicable, protected by fine nets and watched day and night; the great enemies of the worm being birds, heavy rain, and unusual cold. In about two months the worms begin to form cocoons, which are completed in from eight to ten days, of a colour usually dirty white, sometimes, though rarely, a dull yellow. Three sets are produced during the year, in each of which the quality of the silk remains constant, but the quantity varies,

SILK: Tasar.

Rearing of the Tasar Silkworm

REARING in Central Provinces.

increasing with the inreased life of the worm before it passes into the chrysalis stage. The average crop of a grower may be taken at 5,000 cocoons, and of these he retains 50 or 100 for future operations and sells the remainder, the present price being 250 for the rupee. They are bought by the *koshkatees* and are thus manipulated: a handful of the ashes of the Aghara plant or of castor-oil seed, is tied in a rag and placed in an earthen vessel. Above the ashes are arranged 200 or 300 cocoons, and the vessel is filled with water, in the proportion of a quarter *pailee* of water to 100 cocoons. A fire is then lit round the vessel, and the mass very slowly boiled, until all the water has evaporated, upon which the cocoons are taken out and arranged singly on a bamboo mat, where they are sprinkled with cold water. They are then placed on cowdung ashes and kept there perfectly dry, when they are ready for reeling off. The quantity of silk thread thus produced annually is very considerable, and the yield may be indefinitely increased" (*Settl. Rept.*, *110-111*).

The Proceedings of the Government of India do not possess any very important additional information regarding the *tasar* silk of these Provinces than has already been conveyed. In 1873 **Mr. J. F. Muir**, Personal Assistant to the Chief Commissioner, expressed the hope that the improved Italian method of reeling the cocoons (referred to by **Mr. Massa** in a communication to the Government of India), would soon be made public, since, if it could be cheaply performed by the ordinary reelers, a great improvement in the trade might be anticipated by the production of a superior fibre to that turned out by the crude system presently known to the people. In a previous communication (see **Geoghegan**, Section VII, p. 129), an attempt was made at estimating the annual production of *tasar* cocoons by the amount of the known manufactures at the chief centres. In a letter dated May 28th, 1873, **Mr. Lindsay Neill** regretted the inability of the Chief Commissioner to furnish more definite particulars of the annual production. In 1876 the Government of India required a few maunds of *tasar* cocoons to be collected in the Central Provinces, and in ordering these opportunity was taken to call once more for any available information on the annual production. The following may be given as the substance of the replies:

Chanda. 1975

Chanda.—Outturn of *tasar* cocoons in the Khalsa is estimated at 375 maunds and the price at R32 per maund.

Betul. 1976

Betul.—Cost, inclusive of the transport charges to the Itarsi Junction, R35 per maund.

Seoni. 1977

Seoní.—The estimated average outturn is 80 maunds per year: this outturn could be increased up to 150 maunds if the cultivation of *tasar* in the reserved forests were permitted. Price per maund R40

Raipur 1978

Raipur.—The increased attention paid to lac has caused *tasar* cultivation to recede. Price per maund R40. The industry in this district is not likely to expand even if the demand for cocoons were brisk.

Sambalpur. 1979

Sambalpur.—Annual outturn of cocoons is not known, but the Deputy Commissioner thinks he could collect 1,000 maunds of cocoons. Price R60 per maund of unpierced cocoons of the small variety, and R40 of the large variety at Sambalpur.

In a further communication, in reply to a letter from the Government of India in which information was desired on the subject of the lowest price at which the cocoons could be landed at the railway stations, the Chief Commissioner reported that this could not be done at R30 a maund, the lowest estimate having been R37. At the same time it is explained that in transit the cocoons often sustain a serious loss in weight from drying. A consignment, for example, reached Chanda about the middle of March, and up to 6th May, the dryage had been equal to half their weight, or from 7 maunds and 19 seers to 3 maunds and 35 seers for the total number of the cocoons of 70,850. The loss in number of cocoons from Mul on transit to Chanda was only 170 (a loss admissible from defective package), so that it is believed the great loss in weight mentioned was ertirely owing

Loss in weight through drying. 1980

Conf. with pp. 112, 113, 126.

REARING in Central Provinces.

to dryage and not to theft or falsification of the purchase. Accepting that explanation as correct, the inference to be drawn is of no small importance. A maund of cocoons in March (presumably shortly after collection) was equal to 9,443 cocoons, whereas a maund from the same consignment weighed in May equalled 19,543 cocoons. In connection with Bengal the opinion has been quoted that a maund of *tasar* cocoons was equal to 4,000. That there must be some mistake in these figures is perhaps the most convenient explanation, but there is just the possibility of the weighments at Mul and Chanda being approximately correct, and that the loss in weight is either due to criminal damping, so as to increase the weight or to the natural drying of the cocoon, consequent on the death of the chrysalis. That 4,000 fairly dry ordinary-sized *tasar* cocoons would, however, weigh 82lb or one maund seems highly improbable from the fact determined at Chanda by the weighment made on the 6th May in the presence of the Deputy Commissioner. But, on the other hand, **Mr. Coldstream's** estimate (*p. 125*) of 106 cocoons to the seer would be 4,250 to the maund. Whatever be the cause of the dryage it is clear that a maund of fresh cocoons weighed at the place of collection would not by any means weigh a maund at its destination, so that a more trustworthy method of dealing in such a commodity would be a valuation by number not weight.

Cocoons to Maund. 1981 *Conf. with pp. 112, 113, 126.*

Valuation by Number Desirable. 1982

III.—NORTH-WEST PROVINCES & OUDH.

IN N.-W. P. & OUDH. 1983

In its reply to the Resolution of the Government of India (alluded to above), the Government of the North-West Provinces and Oudh replied, in November 1877, that the worm occurs to such a small extent in these Provinces that it would be needless to make any proposals or suggestions regarding the development of the trade. Enquiries were made from the Commissioners of Meerut, Bareilly, Allahabad, Benares, Jhansi, and Kumaon, being the divisions in which there are large areas under forest, and also from the officers of the Forest Department. It appears from their reports that *tasar* silk is either not found, or is not known, in any district except Mirzapur, Lalitpur, and Garhwal, the quantity found in the last two being very insignificant. The only place in the North-West Provinces, the report continues, where any appreciable amount of *tasar* cocoons is gathered, is the wild tracts lying to the south of the Mirzapur district, which belong more properly to the region of Central India than of the North-West Provinces. Cocoons are there collected by Kols, Chamars, and other jungly castes in the month of October, and are sold at the local bazárs to *putwas* and other traders, who take them for sale to the Patna and Gya districts, and also to Ahraura Bazár in Mirzapur. From the last place the greater part of the cocoons are exported to Bengal and the Central Provinces, and a small part reeled there. The silk thus reeled is sent in its raw state to Benares, Patna, and other districts. No weaving of cloth worth the name is carried on, there being only one family in Ahraura engaged in the manufacture. The selling price of cocoons in Mirzapur is from R5 to R5-8 per 1,000, but the trade is by no means extensive. Speaking of the lower Himálayan tracts, **Capt. Hutton** wrote: "In the Dehra Dún and extending up the hill-side to about 4,500 feet, or perhaps more, we have two species of *tusseh*, one of which is also found in Central India; what the other is I am not yet prepared to say. Here, however, we have no artificial crossing, so that our species may be regarded as types."

Decline of Azimgarh Tasar Trade. 1984 *Conf. with pp. 59, 148, 157, 158. Conf. also with pp. 113, 114-117, 126.*

The most interesting information as yet published regarding these Provinces will be found in **Geoghegan's** *Silk in India*. Thus at page 103 he writes:—

"The *tusser* appears at one time to have been largely manufactured in the

SILK: Tasar.

Rearing of the Tasar Silkworm

REARING in N.-W. P. & OUDH.

south-east corner of the tract now embraced in the North-Western Provinces. In an abstract of the results of the survey and settlement of Azimgarh in 1837, **Mr.** (now **Sir R.**) **Montgomery** estimates the quantity of *tusser* annually manufactured in that district at 318, 772 pieces (of a size not stated), and large quantities, both of cloth of pure *tusser* and of a cloth called *soosee* ('being a cotton *dosotee* cloth with a stripe of coloured silk through it'), used to be made at Ahrourah in the Mirzapore district. English piece goods, however, seem to have nearly driven these fabrics out of the market, though the price has fallen 50 per cent. The worm is still bred in the jungle tracts of Mirzapore, and the yield of the season 1870-71 is said to have been 1,500,000 cocoons. The silk is valued at R4 per seer, but the price has fallen, for it used to be as much as R7 a seer. Both cocoons and raw silk are exported; silk to Azimgurh and Benares, and both cocoons and silk to the Central Provinces. The Sanitary Commissioner of the North-Western Provinces, in the report of his last tour, mentions a town in Azimgurh as the seat of a mixed cotton and tusser manufacture, the fabrics being exported and getting even as far as Bombay."

It would thus seem that *tasar* silk was formerly of far greater importance in these Provinces, than it is at the present day. **Dr. Bonavia,** in a report on sericulture in Oudh, published 1863, says of the *tasar* worm that it was reported to feed on the *ber* tree (**Zizyphus Jujuba**) and the *sál* (**Shorea robusta**). He adds, "hardly any experiments have been made with it. In former times the cocoons used to be cut spirally into a thin strip, and used for tying the barrels of matchlocks to their stocks."

Conf. with p. 102.

IN PANJAB. 1985

IV.—THE PANJÁB.

Mr. Baden Powell, in his *Panjáb Products* (a work which was issued originally as the Catalogue of the Lahore Exhibition of 1864), remarks that the samples of silk and cocoons shown were all the produce of the mulberry-feeding insect. Though subsequent writers have alluded to *tasar* (or a special form of that insect formerly recognised by certain entomologists as **Antheræa sivalika**) as plentiful throughout the lower Himálaya and other mountainous tracts of the Panjáb, it would appear that the Natives, if they were aware of the existence of the insect, attached to it no economic value. To **Mr. W. Coldstream** (while Deputy Commissioner of Hoshiarpur) is due the merit of having prominently drawn attention to the fact that the insect was not only plentiful, but might be even more successfully semi-domesticated than had been found to be the case in the experiments conducted at Poona.

Conf. with p. 237.

The Financial Commissioner of the Panjáb recently published (March 1884) a revision of **Mr. Coldstream's** views and results, which that gentleman had drawn up for a monograph on Panjáb silks. The paper in question deals so fully and completely with the subject that it is felt an abridgment of it might mar its utility and therefore, at the risk of repeating statements regarding the *tasar* insect which appear in other sections of this article, **Mr. Coldstream's** essay on the subject may be here given *in extenso*:—

Distribution. 1986

"GEOGRAPHICAL DISTRIBUTION IN PANJÁB.—The Tasar moth is common in the sub-montane districts of the Panjáb. As far as my observations extend, it is commonest in the lowest or outermost range of the Himálayan chain, *viz.*, in the Siwaliks. But it is found pretty abundantly in the lowest and most southern valleys of the main range, and it also extends to the plains.

"It may probably be found in the hills at as great an elevation as 3,000 or 4,000 feet. How far it extends into the Punjáb plains, I do not know.

Species. 1987

SPECIES AND NAME.—"The species of *tasar* found in those regions, is, I believe, that which is now appropriately recognised at the British Museum as **Antheræa sivalika**: under **Captain Hutton**'s classification it used to be **Antheræa paphia.** I do not think the natives have any name for it, except the general one of *bhambiri*, a word applied promiscuously, I believe, to moths and butterflies of some size.* The cocoon they call *tutti*, *kaunta*, and *kaintr* in the Simla hills.

* It is also called *Joadri*, I find, in the Simla hills.

REARING in Panjab.

Colour. 1988

COLOUR NOT A SPECIFIC DIFFERENCE.—"I shall not here attempt a description of the insect in its various stages, but merely note that the female *imago* or moth is most commonly of a bright yellow colour, while the male, which is much smaller, is of a much duller tint inclining to brick red. But a pinkish shade, and an ashen grey tint, are not at all uncommon among the females. Nor do I think that this variation in colour, considerable though it is, is accompanied by any difference which should be regarded as specific.

Metamorphoses. 1989

METAMORPHOSES.—"This species is bivoltine, that is, it goes through the cycle of its various transformations twice in the year, and I believe, no oftener. The moth bursts from its cocoon in the beginning of the rains, in June or very early in July, and the eggs then laid produce a worm which forms a cocoon, which bursts in August or September. In October the second generation is laid to sleep in its cocoons for the long winter and spring months, and does not wake till next season's rains begin.

Silk-producer. 1990

THE INSECT NOT KNOWN AS SILK-PRODUCER.—"The ignorance which prevails in the country-side on the subject of the insect is astonishing. The villager knows the '*kaunta*' or '*tutti*' (cocoon), which he will describe as a kind of leathern covering (*chamra*) of an insect (*kira*), useful in days gone by for cutting into thin strips and binding the barrel to the stock of the matchlock. But that a beautiful moth emerges from this, not one in a thousand knows. The reason of this curious ignorance, I believe, is simply that the moth invariably emerges in the evening or later on in the night, either when it is dark, or is beginning to darken, and the process or its emergence is unlikely, in the natural state, ever to be witnessed.

Enquiry. 1991

HISTORY OF ENQUIRY REGARDING TASAR IN THE PANJAB.—"The existence of the *tasar* cocoon in the Province, and its possible economic importance was well known twenty-six years ago. I find in the office of the Simla District, a letter addressed to the Deputy Commissioner by Sir Donald McLeod, then Financial Commissioner for the Panjáb, of 27th January 1858, in which he informs that officer that 'a method having been lately discovered by which the fibre of the *tasar* or wild silkworm may be detached from the cocoon and adopted for the loom without unwinding it, which last process is so difficult as to be almost impracticable, certain parties in Calcutta have taken up the matter, and are anxious to obtain the cocoons of the above worm in any quantities that may be obtainable.' Mr. Cope, then established as a merchant at Harriki, was, it was said, prepared to purchase *tasar* cocoons. The price he offered was R4-8-0 per maund, at Kalka; the records do not show whether any cocoons were then collected. But Sir Donald then Mr. McLeod, evidently continued to interest himself in the matter, for I find that, in the preface to a small volume entitled *Miscellaneous Papers* on silk, printed at the *Lahore Chronicle* Press in 1859, he recommended the plan of rearing the worm on trees in the open air, the only plan which even at this day appears to promise any hope of success.

"For seventeen years nothing more was heard of *tasar* in the Panjáb. In 1875 or 1876 some enquiries were made about it, and, I believe, so complete was the ignorance on the subject, that it was reported to the Government of India that *tasar* did not exist in this province. Soon after, however, this was shown, I believe, by the late Mr. F. Halsey, to be a mistake.

Experiments. 1992

ORIGIN OF EXPERIMENTS BY WRITER.—"It was that gentleman who in 1876 first showed me a *tasar* cocoon, and suggested the plan of open air cultivation. He had tried it himself, I think, only to a very limited extent. He was busy with mulberry silk at his filature near Pathánkot, but he did not hope much from *tasar*, and recommended me to leave it to his relative Major Coussmaker at Bombay. I had, however, collected some cocoons, and I determined to try what could be done; and in that year (1876), and down to 1881, with the exception of a season (1879), when I was on furlough, I carried on experiments either at Hoshiárpur or Lahore.

Collection. 1993

COLLECTION OF COCOONS.—"Cocoons I collected in considerable quantities through zaildárs and lambardárs, who got villagers to collect them from the *beri* trees, and sent them to me. I had in one season 5,000 cocoons collected, of which 1,000 had the chrysalis alive in them. On one occasion, of which I took a note, two men searched for two days and got 66 cocoons. This will show that in the Hoshiarpur district the insect is plentiful. The cocoons are most easily collected in the cold weather, when the trees are leafless or comparatively so. In the summer and rains the cocoons are very difficult to distinguish among the mass of foliage.

Keeping. 1994

KEEPING OF COCOONS.—"The cocoons, when kept for breeding purposes, can be kept in large cages of bamboo wicker work, say 3′ × 2′ × 2′, or (which I found

Rearing of the Tasar Silkworm

REARING in Panjab.

the most convenient way) can be laid on shelves in a verandah enclosed with fine twine netting (such as is used for fish nets). The object is that the moths should not escape when they emerge from the cocoons, but pair as soon as possible. From a large batch of cocoons a few will come out in May and early part of June, but as soon as the first shower of rain falls, numbers of cocoons will burst

Bursting. **1995**

BURSTING OF COCOONS.—"The moths always leave the cocoon between 7 and 9 o'clock in the evening. Late in the afternoon the upper end of the cocoons which are to burst that evening, though they have been eight months hard and dry, will be seen to become moist and soft. The cocoons, if laid on a flat surface, will occasionally roll about, while gradually the soft end becomes more and more protuberant, till the moth pushes his head through, disengages himself from his prison, and emerges like a huge grub with his undeveloped wings, soft and shrivelled looking. He immediately seeks a surface (the under side of its own cocoon, if fixed and firm in preference to any other) on which it can fasten its feet firmly and hang with its back downwards. In this position gravity helps to unfold and smooth out the wings, which in the course of an hour or two are quite developed. In a verandah enclosed by netting, the females would often, after emerging, cling to the netting. Wild males were thus attracted, I believe, occasionally. This happened at Bharwain (4,000 feet above sea level) but not, I think, at Hoshiárpur.

Emerging. **1996**

TENDENCY OF MALES TO EMERGE FIRST.—"I have noticed,at least once, a tendency to a large number of males emerging first, whilst subsequently females appeared in large numbers, without a corresponding number of males. I have not discovered the reason for this. The moths, male and female, only live a few days. They eat nothing during this period. The female dies almost immediately after laying her eggs. The male insect also dies very soon.

Collection. **1997**

"COLLECTION OF EGGS.—The female begins to lay her eggs about the second day after she has left the cocoon, and it is, therefore, of importance that she should pair as soon after her appearance as possible. She will lay most of her eggs on the second or third day, whether she has paired or not. She lays from 150 to 250 eggs. An average of the eggs of 24 moths gave 185 eggs per moth, another average of 84 moths gave 151 eggs per moth. In an experiment in 1876 with 1,000 cocoons, all the worms were hatched out by 6th August, the first eggs having been laid about the 8th or 9th of July. I found it the best way, in order to collect the eggs easily, to put the female, after she had paired, into a small basket by herself, or with one or two other females. In from two to four days, she had laid all her eggs, and was dead or dying.

"I then collected the eggs, which were sticking in little agglutinated masses of from three to fifty eggs, to the bottom and sides of the basket, and put them aside in batches according to date of their being laid.

Hatching. **1998**

HATCHING OUT OF WORMS.—"The worms hatch out of the egg the eighth or ninth day. The cycle would be somewhat as follows:—

First cycle.

Moth emerges, say, 1st July.
Pairs and lays „ 2nd „
Worms hatch out 11th „
Cocoons formed (33 days after) 13th August.

Second cycle.

Moth emerges . . 25th August.

Treatment. **1999**

TREATMENT OF EGGS.—"I tried two plans of disposing of the eggs to be hatched one was putting 50 or 100 eggs into a leaf basket or little dish (*duni*), such as natives use for carrying curds and other eatables, and hanging this on the tree which was to be the home of the worm, and on which it was eventually to form its cocoon; several such leaf baskets, each containing 50 or 100 eggs, could be hung on a *beri* tree 20 or 30 feet high.

"The other plan was to keep the eggs in flat open baskets (*chabri*), and as soon as they begin to hatch out, to put some fresh *beri* leaves in the basket, feed them thus for a day or two, and then put them on a tree.

"Both plans answered well. But the young worms should be put on the trees before they are many days old.

Domestication. **2000** *Conf. with pp. 106, 108, 128-29.*

TASAR MOTHS WILL NOT FULLY DOMESTICATE.—"The worms do not flourish under cover or kept in trays and fed on *beri* leaves cut from the tree. I have fully tried the experiment of rearing the worms by bringing food to them under cover. The worms reared in the open grew twice as fast and were far more healthy. The hand-reared worms grew slowly, did not attain the same size as those fed in the open, and never attained the bright translucent shade of green which characterises the healthy worm. Their skin seemed to remain opaque. The house-fed worms gradually died away every morning, a few were found dead in their baskets; and very few of them spun cocoons. They seem to require the open air and the dew of heaven. They will not domesticate. Rain seems to revive them and do them good.

REARING in Panjab.

REARING OF WORM.—"The small brown caterpillar, only about one-sixth of an inch long, when hatched, soon grows under favourable circumstances, quite out of recognition. Its brown changes to a delicate light green, and curious bright spots of a metallic lustre develope at the spiracles; when full grown, it is from 4 to 5 inches in length.

The Worm. 2001

GROWTH OF WORM.—"Several hundred worms on a tree soon strip it of its leaves, and it is a curious sight to see a tree suddenly cropped of its foliage, and nothing but branches and leaf-stalks remaining, while, the fat worms move anxiously about the tree in quest of more leaves, and at last descend towards its trunk; instinct prompting them no doubt to seek another tree, even if necessary by travelling over the ground.

Food. 2002

BERI TREE (**Zizyphus Jujuba**), THE FOOD OF THE WORM.—"The great suitableness of the *beri* tree as a source of food supply for the worm, is that, after being denuded of its leaves in the rainy season, it will, in a few days, send forth an abundant flush of fresh leaves, which will soon be ready in their turn to be browsed down.

"**Lagerstrœmia indica.**—Besides the *beri* tree, the only tree or bush on which I found the worm to thrive was the **Lagerstrœmia indica,** called in the Panjáb by gardeners *Sáwani* from its flowering in *Sáwan* (July—August). On this shrub it did very well, but not so well as on the *beri* (**Zizyphus Jujuba**). I do not think the **Lagerstrœmia** recovers itself so quickly as the **Zizyphus,** though on this point I am not sure.

OTHER TREES TRIED.—"**Bombax malabaricum** and, I think, **Carissa Carandas,** I also tried, but I did not find that the worms took to them readily. The *beri* is evidently the Panjáb tree for the *tasar*. Mr. **Stewart,** Collector of North Arcot, in his No. 2798 of 4th September 1883, mentions that there the *tasar* thrives on the *jáman*. I do not think it is found on the *jáman* in the Panjáb.

Conf. with p. 101.

BERI NURSERY FOR TASAR.—"I planted, close to my house at Hoshiárpur, a little grove of *beri* trees only a few yards apart. They established themselves easily, and in three or four years were ready as feeding grounds for the worms; some of them reaching a height in that time of 10 to 12 feet, and becoming quite well developed little trees.

BERI GARDENS SUITABLE FOR TASAR REARING.—"In the Panjáb, groves of grafted or Cabuli *bers* are very common. The trees spread thick and low with long pendulous branches arching over to the ground and thickly covered with leaves. Such groves could, I believe, most easily be converted into *tasar* plantations. The trees are not high, so that the watchers' task would be easy, and the supply of succulent leaf abundant, while, as the *beri* crop is over early in the year, the trees becoming in autumn a pasture ground for *tasar* worms would not, I believe, impair their fruit-producing powers.

"I had also a small plantation of **Lagerstrœmia** bushes. I had not a full opportunity of judging of their recuperative powers, as I left the district before a fair trial could be given to them; but I see that in Bombay it is found that the **Lagerstrœmia** flushes well.

Conf. with p. 102.

NUMBER OF COCOONS PER TREE.—"I cannot say how many cocoons I have taken off one tree, but the number is upwards of 200. In the Administration Report of Chota Nagpore for 1882-83, the average yield per tree at Singhbhoom, stated as 30 cocoons per tree, seems to me very small. One hundred per tree is there deemed a bumper crop. On the other hand, I have never taken 500 cocoons off one tree, as Mr. **Stewart** has done in North Arcot. But my later experiments were all made on small trees planted for the purpose, and the number of cocoons per tree will mostly depend on the size of the tree.

Yield. 2003

MATURITY OF WORM.—"The worms arrive at full growth at from 25 to 40 days, and forthwith spin their cocoons.

Maturity. 2004

ENEMIES OF TASAR WORM.—"During the feeding period they are liable to the attacks of enemies; sparrows pick them off the tree, ants climb up and torment or injure them and at the higher elevations (for instance at the Bharwain, in Hoshiárpur district) the worms were mercilessly attacked and killed by a species of Ichneumon fly or hornet of a brown colour. Birds can be kept off by watchers, ants can be checked by a small ring of water being kept at the bottom of the tree;

Enemies. 2005

SILK: Tasar.

Rearing of the Tasar Silkworm

REARING in Panjab.

the wasps, however, are extremely difficult to circumvent. My experience in the matter of rearing the worms has been much the same as that of **Mr. Stewart** of North Arcot, but his report does not mention the hornet. I found that a number of young worms died from excessive heat, a number also died from being spiked with thorns of the tree. Their being so impaled looked like a species of suicide. Very young worms sometimes begin to die in quantities in September from heat. A thatch was put up for them.

WATCHER EMPLOYED.—"A boy on R3 or R4 per mensem I found could watch a good many trees, and was useful in conveying he worms on little bits of branches from a stripped tree to a fresh one.

"During one season's experiments, I had a number of young *beri* trees actually covered and enclosed over with a huge netting supported on poles. I do not now consider this to be at all necessary. The watch boy is sufficient protection.

Outturn. 2006

TOTAL OUTTURN.—"As regards the outturn of cocoons thus reared, compared with the original stock, I have not been at all successful. I have even finished with fewer cocoons than I started with; indeed I have hardly in a single instance, if ever, increased the original stock at all. But some experience has been gained as to the lines on which future experiments should proceed, when conducted by those who can devote more leisure to the task.

Reeling. 2007

DIFFICULTIES OF REELING.—"After the cocoons are collected or gathered, the reeling process must engage attention. A great deal has been said about the difficulties of reeling off the *tasar* cocoon, and this is especially alluded to in the circular of the Secretary to the Government of India, Department of Agriculture and Commerce, No. 36 F. S., dated 25th May 1883. I did not find much difficulty in having it done, and I think it was well done too. I got for the work one of the silk-rearers from the neighbouring district of Gurdaspur. He had, I think, seen *tasar* cocoons before; at least he knew how to manage them. He reeled off a clean lustrous thread of four strands, keeping four cocoons bobbing in a dish of boiling or very hot water in front of him. I do not believe he used *sajji* or anything else to dissolve the natural cement, the hot water seemed to act as a sufficient solvent.

DIFFICULTIES OVERCOME.—*Rate of Reeling.*—"The rate at which he worked was about 50 cocoons per diem, yielding 2½ tolas of silk. This would be a seer (valued, I think, at R10) in 32 days. This is not a very remunerative rate, five annas per diem: but no doubt men accustomed to the work would work faster, and even four annas a day would be good wages for women and children.

"Nothing has been allowed in this for firewood, which in the hill villages would cost very little.

Floss. 2008

TASAR FLOSS OR WOOL.—"This man also turned out beautiful *tasar* wool and floss; the cocoons being teased out instead of being reeled out. This carded or teased *tasar* was clean and soft and elastic like cotton wool.

Cocoons to Maund. 2009

Conf. with pp. 112, 113, 120.

WEIGHT OF COCOONS.—"As regards the weight of coc ons, I found that 106 full cocoons weighed 1 *seer*, 245 empty cocoons 1 *seer*.*

Outturn. 2010

OUTTURN OF SILK.—"As regards produce, I found that twenty-one cocoons gave one *tola* of wound silk; 800 cocoons would give one pound of silk. But by another calculation I made, 700 would make up one pound. In Chota Nagpur 436 go to the pound. The cocoons there are, I believe, larger.

REPORT MADE OF THESE EXPERIMENTS.—"A report on my experiments in rearing *tasar* was submitted to the Commissioner of the Jullundur Division, with my No. 447 of 26th August 1876.

"This report, and further notes on the subject, was published in Vol. VI., Part 4, of the Journal of the Agricultural and Horticultural Society of India.

Specimens. 2011

SPECIMENS EXHIBITED AT LAHORE AND CALCUTTA EXHIBITIONS.—"Specimens of cocoons, moths, drawings of the worm, silk reeled, silk carded, and silk spun, with cloth woven from Tasar of Hoshiarpur district were all exhibited by me in a case at the Lahore Exhibition, 1881, and many of these specimens were also taken to the Calcutta Exhibition now open.

EXPERIMENT AT LAHORE IN 1880.—"**Mr. Spooner**, of the Agri.-Horticultural Gardens at Lahore, conducted an experiment one year at Lahore under my directions. He understands the system. The experiment was not a particularly successful one, but several small *beri* trees in the Agri.-Horticultural Gardens were covered with cocoons.

Prospects. 2012

Conf. with pp. 74-5, 105, 114-117, 129, 132.

PROSPECTS OF A TASAR INDUSTRY IN THE PUNJAB.—"As regards the general prospects of the industry, I am inclined to agree with those who think *tasar*

*This would be, say, 4,200 of the former and 9,800 of the latter per maund.—*Ed., Dict. Econ. Prod.*

REARING in Panjab.

would not flourish alongside of mulberry silk; but mulberry silk is not indigenous, and disease and other difficulties attend its cultivation, and the collection of fresh leaves daily is expensive. Although we do not know fully the difficulties which may eventually attend the domestic cultivation of *tasar* on the lines proposed, *i.e.*, rearing it on trees in the open, we do know that the worm is indigenous, that it is possible to rear it in this way to at least some extent, that this method of rearing costs absolutely nothing, except the labour of a boy watcher, and that the value of *tasar* seems to be rising in the European market.

Nurseries. 2013 *Conf. with pp. 101-2.*

WATERING FOR NURSERIES NOT REQUIRED IN PUNJÁB.—"I observe that the Collector of North Arcot, in his No. 2798, dated 4th September 1883, proposes to plant and water plantations of **Eugenia Jambolana** (*jáman*) and **Zizyphus Jujuba** (*beri*). **Major Coussmaker** also proposed to plant hedges of **Lagerstrœmia indica.** In the Punjab, at least in the submontane districts, no watering of the *beri* is necessary. If a sapling is planted in the rains in fair soil, it will establish itself and be a vigorous young tree in three years. Here we have evidently an advantage over Madras.

PROFITS REALISABLE IN BENGAL.—"A writer in the *Indian Agriculturist* of 15th January 1884, who evidently has much experience of *tasar* in Bengal, writes that the industry is a most profitable one, but is strangled by royalty charged (by the Government forest officers apparently), as well as by zamíndars on *tasar* rearing.

WHETHER TASAR CAN BE PROFITABLY REARED IN PUNJAB REMAINS TO BE PROVED.—"Of course it entirely remains to be proved whether the silk can be produced, or the cocoons reared so as to be commercially profitable. But I do not think the establishment of *tasar* sericulture, as an industry for the submontane districts of North India, is to be despaired of. Considering the very inexpensive process of rearing the cocoons, that it could be done by women and children, that the worm is indigenous to the province, and that the attention of European manufacturers appears to have been directed to the commodity, it seems certainly possible that an industry may be developed. I think experiments should be encouraged. Natives should, if possible, be induced to take an interest in it, but this will not be probably till European skill has shown the way to a profit. The conditions of successful rearing have been approximately, but not fully, gauged. As above stated, my experiments as regards outturn in proportion to original stock were not at all successful. Still a large number of cocoons were reared, and there is no apparent reason why, when nature has been further interrogated, the secret of preserving most of the worms should not be discovered and success achieved.

Major Coussmaker's CONCLUSIONS NOT APPLICABLE TO PUNJÁB.—In my No. 1391 of 28th August last, to the address of the Junior Secretary to the Financial Commissioner of the Punjáb, I gave reasons why I thought **Major Coussmaker's** advice, given in his final report on *tasar* experiments in the Bombay Presidency, to discontinue the attempts to extend this kind of sericulture should not be held applicable to Upper India. **Major Coussmaker** (see Circular No. 36 T. S. of the Secretary to Government of India, Revenue and Agricultural Department, dated 25th May 1883) gave as one reason for deprecating the continuance of the experiment that the cocoons of Bombay were smaller and contained less silk than those of other parts of India, and he had been unsuccessful in improving them by importing the larger varieties from other parts. This would obviously be no reason for discontinuing the experiment in the Punjab.

"Another reason given by **Major Coussmaker** was that the expense of collecting the cocoons or rearing them in a state of semi-domestication was great. This I have not found to be the case in the Punjab. A first batch of seed cocoons would cost a little no doubt, but the cost of rearing the insect in a state of semi-domestication, on the plan I pursued at Hoshiarpur, is certainly very small indeed.

CONCLUSION.—"On the whole, though I cannot say I have absolutely ascertained the conditions of success, I have seen so much in the course of my experiments as to make me believe it possible that a kind of cottage industry of rearing *tasar* requiring absolutely no capital, and capable of being conducted by women and children, may some day arise, if pains are taken, by experiment, and the offer of rewards, to ascertain these conditions, and to introduce the industry to the notice of the natives.

"The wild tribes of Central India rear the cocoons; why should not the cottagers in the Punjáb hills?"

V.—BOMBAY & SIND.

IN BOMBAY & SIND. 2014

Colonel Sykes was, perhaps, the first officer who specially described the *tasar* silk and cocoons of Bombay. He says it is known as *kolísura*

SILK: Tasar.

Rearing of the Tasar Silkworm

REARING in BOMBAY and SIND.

silk-worm of the Deccan. **Sir George Birdwood**, in the Report of the Victoria Museum, Bombay, for 1859-60, details how his attention was first drawn to *tasar* silk by **Mr. Heycock**. From his remarks it would appear the cocoons were never used in Bombay until an officer who had served in Bengal prior to his removal to Bombay had drawn attention to them (see **Wardle's** *Wild Silks of India, p. 71*). The facts published on the subject, up to the date of **Major Coussmaker's** experiments (already incidentally alluded to), are, however, scarce worth reviewing. **Major Coussmaker**, under instructions from the Bombay Government, and in consequence of recommendations from the Government of India, performed at Poona an extensive series of experiments which were undertaken with the primary object of ascertaining whether, under a system of domestication or semi-domestication, it would be possible to produce the cocoons at a more profitable rate than the present system of collection from wild sources. It may briefly be stated that while **Major Coussmaker** made many important discoveries, and solved most of the obscure problems of the life-history of the insect, his results demonstrated conclusively that in Poona at least the insect could not, under any degree of domestication, be reared profitably. The effects of the nature of the food given to the worm on the character of the cocoons spun, constituted one of the chief features of **Major Coussmaker's** enquiry. He found that the colour of the cocoon was capable of modification and that by regulating the food a marked influence was exercised on the nature of the gum (or cement as it has been called) which is discharged by the worm during the formation of its cocoon.

Feeding. 2015

Feeding the Worms.—The worms fed on the *ber* tree (**Zizyphus**) were in some respects superior to all the others, and came to resemble very closely the soft pale-coloured cocoons of the Panjáb, the insects of which naturally feed on the *ber*. **Major Coussmaker** says of one experiment with this tree that the worms spun "larger than any that I had gathered from off the other trees." Perhaps the best plant, however, when facility of cultivation and yield of leaf is taken in conjunction with merit as food material was found to be **Lagerstrœmia indica**. Speaking of that shrub, **Major Coussmaker** remarks: "**Mr. Lyle** by accident found that the worms throve very well on **Lagerstrœmia indica**, a very leafy, ornamental, flowering shrub found in most gardens; both he and I put some caterpillars on these trees, and found that they grew enormously and spun very large cocoons." In a subsequent report he again wrote: "The **Lagerstrœmia** bush proves an excellent food; it flourishes so quickly that a plant 2 feet high, after being fed off quite bare, cut back, and repotted, was again in thick leaf in a fortnight, and the same batch of worms stripped it again. In changing the plants, and in daily examining the cages, a few accidents occurred; but 100 worms yielded 71 cocoons."

It would thus appear that in regions where the **Lagerstrœmia** can be easily grown, the most likely way to domesticate the worm, should that result be ever again contemplated, would be by the formation of a large plantation of these bushes. Such a plantation might be produced on very nearly the same lines as a tea garden, since the plants need never exceed 3 or 4 feet in height, and be, therefore, far more manageable than a jungle of trees, such as the species of **Terminalia** on which the Natives of Chutia Nagpur and the Central Provinces mostly rear their worms. But **Major Coussmaker** in his early experiments made certain practical observations which may be here briefly dealt with.

Breeding. 2016

Rearing and Breeding.—It was found that the worms could not be reared in houses unless these were open all round, so as to allow of the most perfect circulation of air. "Although I failed in rearing," he writes, "I succeeded in breeding and in procuring fertile eggs. During the hot weather

REARING in Bombay.

there were no wild males flying about, so I found that it was not much use tying the females out, but during the rains I was quite successful. I therefore, from February to May, turned all the moths, as they came out, into a bedstead covered with mosquito curtains, and a fair proportion paired there; but after May I adopted the following system, which I shall always follow in future. I rigged up swing trays and in the neat trays resting on them I placed the cocoons, covering the whole with bamboo chicks fastened up like a pent-house about 3 feet high. The moths, after they came out of the cocoons, crawled up the chicks and there hung while their wings were expanding. I found that several pairs took place in these cages; each morning I looked at them, and found that some pairs had taken place; these were left undisturbed, and all the unpaired females tethered to a small frame of trellis-work. At dark this frame was hung up to a tree, and all the unpaired free males liberated near it. I generally found in the morning that the majority of the tethered females were paired; the frame was then brought in-doors and hung up out of the way. Great care was always taken that no pair should be separated by force; they were always allowed to free themselves, and after they had done so, the females were put under inverted baskets to lay their eggs, while the males were put into a basket to be liberated at sunset. By systematically carrying out this method, the majority of females paired and their eggs proved fertile, but the average was less than I had formerly noticed, only 106 per moth. I did not have the eggs laid counted, only the number of worms hatched."

"My head silk-worm tender, continues **Major Coussmaker**, was a Mahrátta widow. She tried every contrivance in the way of closing windows and doors, hanging up wet cloths, putting *kaskas tatties* to the doors, sprinkling the twigs, dipping them in water; but it was of no avail. We had no disease, no epidemic, no bad smell, but simply starvation. We kept accounts of deaths in some instances, and compared the proportions of deaths among the different sizes, and found that two-thirds of those which died were under a week in age; but of 170,634 hatched between 1st April and 10th September, only 2,623 grew up and spun cocoons, and this we laid entirely to the difficulty of getting suitable food."

So far **Major Coussmaker's** experiments might not unjustly be characterised as failure to meet the position created, than failure of the experiments. A greater number of worms had been hatched than could be nourished by the food provided. The worms were fed in variously constructed houses, cages, baskets, or nets, but for the most part the food had to be collected from wild sources and carried great distances. The food was thus probably procured at as great a cost, as the collection of wild cocoons could have been. When brought to the worms the leaves, or twigs with attached leaves, rapidly dried up and were unpalatable. Some of them, such as the leaves of **Ficus Benjamina** and **F. Tsiela**, on drying, often curled up and injured the caterpillars, especially so during the moulting time. The practical observation made about this time by **Major Coussmaker**—that although the worms could be fed on a large series of leaves, when once they had started to eat a particular leaf, they could not be induced to change to another sort—increased the difficulty, since it was not always found possible to procure a continuous supply of fresh leaves of the particular kind and at the exact stage of growth which the worm relished best. As exemplifying these difficulties, **Major Coussmaker** tells us that at times he had to send as far as 20 miles for his leaves.

Conf. with p. 102.

Failure to Domesticate.

2017

Conf. with pp. 106, 108, 128-29, 132.

Some of **Major Coussmaker's** early observations seem to point also to the hopelessness of striving after complete domestication of the *tasar* worm. The simple handling of the cocoons was found sufficient to cause

SILK: Tasar.

Rearing of the Tasar Silkworm

REARING in Bombay.

the insect to prematurely effect its escape. This fact would, therefore, favour the notion that if success is to be attained, breeding must necessarily form part of the scheme, so as to secure the escape at the season best suited. While in native opinion fresh stock, brought from the jungles, is a periodic necessity, still with the hill tribes of Chutia Nagpur and the Central Provinces, breeding forms an integral part of the procedure.

The reader has doubtless formed the opinion that so far **Major Coussmaker's** experiments have by no means demonstrated that breeding in connection with a plantation of, say, 300 acres of **Lagerstrœmia** or other suitable bushes, might not be made profitable. While that is so, **Major Coussmaker's** more recent experiments, which will now be dealt with, proved that in Poona at least that also was a mistaken opinion. The insect he found was fairly plentiful from Tanna along the districts bordering on the sea to Akola, a distance of 300 miles. Poona was thus at least within one of the wild areas of the insect, and had a fair chance to prove climatically a hopeful region for successful experiment. In the *Poona Gazetteer* a detailed account is given of **Major Coussmaker's** endeavours, and a sketch is furnished of the thorough way he went about the business together with his ultimate results and final conclusions. We are there told that "Descriptive circulars were sent in English, Marathi, Gujaráti, and Kanarese, offering to buy seed cocoons at 1*s*. (8 as.) and burst cocoons at 6*d*. (4 as.) the hundred. Fortnightly reports were called for on facts which came to the notice of the Native officials who were specially entrusted with the enquiry. By these means a general interest in the collection of *tasar* cocoons was aroused, and at a cost of £16-8*s*. (R164) **Major Coussmaker** received 62,216 cocoons by rail, post, cart, and head-load. Most of these cocoons came from the Konkan forests. Their collection from the trees and moving them, shutting them up in baskets and bags, and generally distributing them had the effect of repeatedly bringing out the moths during the months of February and March. Upwards of 100 moths were out every night, and whenever a fresh batch of seed-cocoons arrived, whatever the temperature or the time of the year, moths came out in large quantities. The details of the eight months ending September 1876 show that on an average 529 females paired and 21,329 worms were hatched every month. The total results of these eight months' breeding were 4,097 females, 4,231 males, 2,045 paired, and 170,634 worms hatched. Of that large number of worms only 2,623 grew up and spun cocoons. On a range of hills a few miles out of Poona, **Major Coussmaker** found a grassy tract with many bushes and saplings of **Terminalia, Lagerstrœmia,** and **Carissa**. Here he turned out some thousand worms, and set men to watch them during the day. For some five months they did well. Then a very hot fortnight set in, the saplings and small bushes lost their leaves, and almost all the worms died. **Major Coussmaker** thought the failure was entirely due to the unprecedented drought. In that view he was doubtless correct. But the worm, judging from the millions of cocoons annually collected from the wild state, must not always be subject to such calamities. The out-door experiment alluded to should probably be viewed in the light of demonstrating two facts—(*a*) that the climatic peculiarities of certain parts of India, such as Poona, may be unsuited for the semi-domestication of the worm; and (*b*) that the system of rearing would have to be so regulated that the worms were hatched at the season of the year that gave them, for the locality of experiment, the best chance of the required climatic conditions. All past experience would seem to show that they suffer far more from drought than from excessive humidity.

It must not, however, be thought that **Major Coussmaker** failed to

REARING in Bombay.

realise the necessity of ascertaining whether better results could be obtained by the formation of a regular plantation of the food necessary for the worms. The following review of his experiments with that object in view may be extracted from the *Poona Gazetteer*:—

Nurseries. 2018

"In 1879, **Major Coussmaker** resumed his experiments. He set aside fifty cocoons of the year 1878 crop for breeding. He also got from others a good supply of moths, many of which he allowed to escape, as he had not food for many caterpillars. He kept some 10,000 eggs, hoping to find food for them in Poona. But he failed to get more than 500 good cocoons from them, of which he had kept only a hundred. As before, the great difficulty was to secure an unfailing supply of suitable food. To improve his supply, with the first promise of rain in June, **Major Coussmaker** set aside about one-sixth of an acre in his garden with a southerly aspect. This he cleared of trees and bushes and laid it out in ridges 4 feet wide with side gutters. On these ridges he planted 340 feet of *dháyti*, **Lagerstrœmia indica**; 270 feet of *ber*, **Zizyphus Jujuba**; 90 feet of *karvand*, **Carissa Carandas**; 107 feet of *ain*, **Terminalia tomentosa**; 15 feet of *arjun* or *sádada*, **Terminalia Arjuna**; and 46 feet of *nándrúk*, **Ficus Benjamina**. He found *dháyti* the most suitable plant. With liberal water it constantly threw out shoots covered with leaves which the worms ate greedily. The plant could be easily grown from the root. The *bor* was liked by the worms, but the leaves were small and thinly scattered and were soon eaten. The *karvand* was leafier, but a slow grower. The *ain* and *arjun* had larger leaves, but were slow growers. The *nándrúk* was a failure; it did not thrive and was not eaten. A *dháyti* plantation with *bor* and *karvand* hedges would yield plenty of food after the beginning of its third rains. **Major Coussmaker** kept all his seed cocoons hung on a wall out of reach of rats. So long as they were left undisturbed the moths came out only during the regular season. Large numbers died when cold October east wind set in. But the chief causes of death were preventible—shortness of food and attacks of insects, birds, mice, and other enemies.

Conf. with pp. 101-2.

"In 1880-81, **Major Coussmaker's** crop of cocoons failed. He thought this failure was the fault of the cages. These were tarred screens of split bamboo. They kept out rats, mice, birds, squirrels, and lizards, but they were too dark; the plants did not thrive, and the worms were always trying to escape. He made the cages longer and put netting at the top, and everything throve till some wasps and other insects punctured and killed most of the silkworms. He had about 30,000 clean perforated cocoons weighing about 60 pounds. He thought it best to go on collecting until he got about a hundred weight. In 1881, though the results were better, **Major Coussmaker** did not succeed in gathering a full season's crop of cocoons of his own rearing. His food-supply was perfect, and the cages kept out all the larger enemies of the worm; still there was much sickness and many deaths. Only 1,000 cocoons were gathered. His first batch of worms hatched on the 2nd of May and the first cocoon was spun on the 6th of June. The last batch of worms hatched in the middle of November, but they gradually dwindled and came to nothing; the last worm died on the 8th of December. The whole season's collection amounted to 60,000 cocoons, double of the 1880 collection. It was chiefly received from the Forest Department who sent 58,000 cocoons. **Major Coussmaker** had all these cocoons cleaned of extraneous matter. The outturn for the two years, 200lb of clean cocoons, was sent to **Mr. Thomas Wardle** of Leek in England. This was sold to **Messrs. Clayton Marsdens & Co.** of Halifax at 1*s.* 3*d.* the pound. The spinners reported that the fibre was somewhat coarser than most *tasar* waste, and the cocoons had been opened, but this was not a serious drawback to its spinning qualities. At this time, in **Major Coussmaker's** opinion, the prospects of the *tasar* silk industry were promising, every year showing an improvement. **Major Coussmaker** laid out a sixth of an acre as a *dháyti* or *gulmendhi* plantation. The land was laid out in ridges 7 feet wide with a gutter of one foot between. The *dháytis* were put into a trench of good soil mixed with manure in the middle of each one foot apart. Where the ground was not filled with the cages, on each side of the *dhaytis* on the ridges vegetables were grown. Care was taken to lay out the ground in the way best suited for watering. The cages were tarred rectangular pieces of split bamboo screen work, a cheap light material neither liable to be hurt by the weather not to be gnawed by rats. In making the cages he tied the screens together, making the sides 3 feet high and the ends 6 feet wide. The cage could be put up over the whole length of the hedge and was divided into 12 feet sections.

SILK : Tasar.

Rearing of the Tasar Silkworm

REARING in Bombay.

From side to side, arched over the top of the hedge, pieces of rattan had their ends fastened to the screens and the middle to a light ridge pole which rested on triangular screens. Over these hoops coarse open cotton was spread. By this arrangement nothing touched the shrubs which were uniformly cut to a height of 4 feet, and nothing tempted the worms to leave their food. There are three screens under the triangles. The middle screen was fixed and the two smaller screens were fitted with string hinges, allowing boys to go in and clean on both sides of the hedges without injuring the shrubs. When hatching, the worms were put on the plants near the door, and they ate away steadily crawling to the next when the first twig was stripped. As fast as they were eaten the bare twigs were cut off and fresh ones grew. After a few weeks the hedge was as thickly covered with leaves as when the caterpillars were put in, and this process went on as long as the rearing of the worms was continued. When the twigs in any section of the screen were stripped, the screen was taken down and shifted along the hedge or to some new place. As a rule, little water was required. In July 1882, Government held that the experiments conducted by **Major Coussmaker** proved that *tasar* silk could be grown with success in the Deccan. They proposed to continue the experiments, and hoped they would lead to the considerable growing of *tasar* silk. In 1882 **Major Coussmaker** increased his **Lagerstrœmia** plantation to 1,500 feet and his **Zizyphus** hedge to 300 feet. In February 1883, before retiring from the service, **Major Coussmaker** in the final report expressed his opinion that *tasar* silk-growing would not pay. Large imports from China had lowered the price of *tasar* waste in England, the Bombay cocoons were small and yielded little silk, and the gathering of wild cocoons or the rearing of worms were both costly, 6*d*. (4 annas) a hundred was the cheapest rate at which forest cocoons could be gathered, and this was too high to admit of profit. The people did not find it pay them to leave their regular work and gather cocoons. It was only by the personal exertions of the Forest officers that so much had been gathered. **Major Coussmaker** had nearly every year tried to increase the size of the cocoons by bringing large cocoons from Sambalpur, Yauntara, Maúbhum, and other places, but with no success. The moths had paired readily with the small Deccan variety; the worms had hatched, but there was no difference in the cocoons. **Major Coussmaker** believed that the smallness of the Deccan cocoon was due to the climate and perhaps in a less degree to the food. As far as outturn went, the result of rearing the *tasar* silkworm was satisfactory. Within six weeks **Major Coussmaker** had been able to gather three cocoons from each foot of hedge. In 1882, the first worm hatched on the 9th of May and the first cocoon was gathered thirty-two days later. The worms of this batch numbered 380, and 349 of them spun cocoons, beginning on the 7th and ending on the 24th of June. They consumed 110 feet of **Lagerstrœmia.** Of 1,800 feet of **Lagerstrœmia,** one-half was sufficiently grown to yield a steady supply of food. From these 900 feet between May and October **Major Coussmaker** gathered 5,678 cocoons. Of these only about half, which were almost all gathered before the end of July, were sound and perfectly formed. Later in the season, without any apparent cause, he lost many hundreds of worms in all stages, some being the progeny of moths of the preceding year. Still many cocoons were spun, some of which were very fine, but the majority were weak and thin. These facts, his own former experience, and the information received in letters and printed reports, showed that no reliance could be placed on any but the first crop of the season, the progeny of the moths which rest in their cocoons during the cold and hot seasons, and which emerge early in the monsoon when the first showers of rain fall. Throughout the whole monsoon and often at other times, when disturbed, moths continue to appear, but with an unsatisfactory result and much loss of life. Enough cocoons were spun to ensure a supply of seed-cocoons, but not enough to call a crop. **Major Coussmaker's** arrangements had succeeded in guarding the worms and ensuring a steady supply of food. The labour bill was reduced to a minimum; one woman and one boy could easily look after at least an acre of hedge and keep the enclosures in repair. At the same time if the south-west rains did not break early and heavily, the hedges would have to be watered and the expense of enclosing would be very great. So long as *tasar* continued cheap, this system could not pay. Crows, sparrows, squirrels, and rats gather near dwellings and must be kept out. **Major Coussmaker** succeeded in keeping the worms safe from their enemies, but the process was costly. **Major Coussmaker** having wound up his series of experiments, handed his plantation of **Lagerstrœmia** and **Zizyphus** bushes, together with the bamboo screens and iron rods which he used for his enclosures, to the Superintendent of the Central Jail at

Domestication of Tasar will not pay.
2019
Conf. with pp. 106, 126, 129.

REARING in Bombay.

Yaravada. There is land attached to the jail, and the head jailor took an interest in silk experiments."

In concluding this notice of the *tasar* silk of Bombay, it may be said **Major Coussmaker's** final report was submitted to the Government on the 28th of April 1883, in which he deprecated any continuance of the experiment. He attributed his failure to three causes:—

"(1) A very large supply of wild silk is now exported from China, and the price of 'tasar waste' has in consequence fallen very low in England.

"(2) The *tasar* cocoons of the Bombay Presidency are smaller, and contain less silk than those found in other parts of India; and all efforts to improve them by importing some of the larger variety from other parts of India have failed, the climate being the insurmountable obstacle, and the difference of food being perhaps also to account.

"(3) The expense of collecting the cocoons, or of rearing the worms in a state of semi-domestication, as I did, is great."

VI.—BERAR & BARODA.

IN BERAR & BARODA. 2020

Some useful information was published about 1841, both by **Dr. Walker** and **Dr. Smith**, regarding the *tasar* insect as met with in the Nizam's dominions. These papers were, however, originally published by the Agri.-Horticultural Society of India, and have been reviewed by **Geoghegan**, so that it does not seem necessary to deal with them in this work. It is almost enough, in fact, to affirm that the insect exists in Berar, but the reader, if he so desires, will find further particulars in the annual report of the Bombay Museum for 1859-60, in which the author of that report, **Sir George Birdwood**, discusses the *tasar* silk of the Deccan. More recently the subject has been carefully gone into by **Captain Catania**. A detailed account of the explorations conducted by that officer, was recently published in the *Deccan Times*, from which it would appear he has given the subject much careful study and made extensive collections.

Of the *tasar* of Baroda a brief notice will be found in the Gazetteer. It is said to occur fairly plentifully in the forests. About the end of May or the beginning of June the moth appears and shortly after lays its eggs on the leaves of the trees on which the worm feeds. It is further stated that there are two crops of the cocoon during the year. The moths appear from the cocoons in May, June, and July, the worms of which eat for fifty days, form cocoons, and the moths again appear in August, September, and October, the cocoons being again spun up to December.

VII.—MADRAS.

IN MADRAS. 2021

Dr. Geddes was apparently the first officer who drew attention to *tasar* silk in Madras, but the most detailed article the writer has come across, is one written by the late **Dr. Shortt**, regarding the insect as found in the jungles to the west of Orissa. This will be found by the reader in **Geoghegan's** most valuable compilation of the literature of this subject, pp. 144-146. **Dr. Shortt** also contributed an instructive paper on the *tasar* silk of Madras to the *Madras Monthly Journal of Medical Science* in 1871. As that paper is, however, more of an entomological than practical interest, and adds little or nothing to the facts given above, it need not be here reproduced. In the Madras Exhibition Jury Reports for 1855 a brief notice of *tasar* silk occurs: "It does not appear," says the report, "that silk in any quantity has been obtained from this source in the Madras Presidency. Considerable quantities of the small silk cloth worn by the Brahmins at their meals are imported into the Northern Circars from Cuttack. The only use to which the cocoons appear to be turned is that of a ligature for matchlocks. They are cut spirally into long narrow bands, with which the barrels are tied to the stock."

SILK: Tasar.

Rearing of the Tasar Silkworm

REARING in Madras.

The Board of Revenue on January 26th, 1877, reported to the Government, in reply to the Resolution of the Government of India in which information was called for regarding *tasar* cocoons and silk "that it appears *tasar* silk exists in most of the districts of the Presidency, but nowhere in abundance. In Ganjam it is found in the Maliah tracts, . . . the cocoons do not appear to be collected for the manufacture of fabrics, as the raw material is imported from the Central Provinces and Bengal into Berhampore for the weaving which is carried on there.

"Mr. Goodrich reports that the *tasar* worm is found in the extreme north west of Jeypore and in the adjacent districts of Bustar; that the cocoons are collected to a certain extent, but are sent northwards to be reeled. He mentions having seen the cocoon in great abundance in Kalahandy, whence the cocoons are sent to Sonepore on the Mahanadi. The price of the best *tasar* silk cloth at Sonepore is stated to be a rupee a yard, the width being a yard and a half. The only portion of the Kistna district in which the tasar silkworm is found is the Vissanapett division, but the use and value of the silk is not understood. The Deputy Tahsildar states, however, that in the Nizam's dominions, adjoining the Kistna district, considerable attention is paid to the breeding of the worm, and the process of reeling is reported to be understood. The worm is also found to a limited extent in all the other districts, except the Godavery, Bellary, Chingleput, Tanjore, Tinnevelly, Trichinopoly, Nilgiri, and Salem districts, but it appears to be confined to a few scattered villages and to be generally scarce. No mention is made of the cocoons being utilised except in the Kadiri taluq of Cuddapah, and in the Head Assistant Collector's division of North Arcot. The manufacture of silk from the domesticated worm at Kadiri is stated to have been discontinued of recent years, owing to the mulberry trees having died; but in former days, when the industry was in a more thriving condition, it appears that cocoons of undomesticated species were collected and attention paid to breeding from them. In North Arcot it is stated that no attempt is made to reel off the silk, but the filaments are torn to pieces and then subjected to a process, which the Collector terms 'flipping,' like cotton, *i.e.*, beaten with a bow, it is presumed, after which the silk is spun into thread and used in the manufacture of coarse cloth. The Collector of Salem reports that silk-reeling was formerly carried on in the Oossoor *taluq*, but has been discontinued, owing to the want of success experienced. The Collector of Coimbatore considers that the difficulties in the way of collecting the cocoons of the wild species at the exact period, when they are fit for use, and in protecting the worms in a semi-wild state from destruction, are insuperable. Mr. Thomas draws attention to the utilisation of the *tasar* worm in the manufacture of catgut for fishing, and suggests the advisability of encouraging the manufacture of this article as a means of employing women and children and the aged and infirm. The practical question is, as he observes, whether there is any probability of a sufficient market being found for such a product, and in the absence of statistics no conclusion can be arrived at on this point.

Silkworms of South India 2022 *Conf. with pp. 69, 76, 82, 90.*

"From Dr. Bidie's report it appears that there are only three important species of undomesticated silk-producing moths in Southern India. Dr. Bidie states that the tasar moth is very generally distributed over the Presidency, but is nowhere very abundant, and the cocoons are not sufficiently abundant to repay the cost of collecting. On a microscopic examination of the filaments taken direct from the tasar cocoon, however, he has failed to detect any such quality in structure as is mentioned in paragraph 15 of the Resolution of the Government of India. The second species mentioned, *viz.*, Actias selene, is less common than the *tasar*; and the third* is not found in this Presidency, but is common in Coorg. Dr. Bidie understands that the silk of the former has been reeled at Pondicherry, but considers that that of the latter could never be turned to any practical purpose. The specimens of cocoons received from the districts were sent to Dr. Bidie for examination. In reply, he states that the cocoons received from Cuddapah, Kistna, Nellore, and Madura are those of the common *tasar* of fair quality, but rather small, those from North Arcot and Kurnool rather coarse, those from Ganjam and Vizagapatam large, dark in colour, with much glutinous matter; the single cocoon sent from Malabar is dark and imperfect. He adds that the cocoons forwarded from Ganjam and Vizagapatam, which differ considerably from those sent from other districts, may be the product of the common *tasar*, the extra size and dark colour being due to food and climatic peculiarities, or they may be the product of other species. Specimens of the moth are required to decide the point; the Collectors of Ganjam, Vizagapatam, and Malabar will be requested to procure them for further investigation.

* Cricula trifenestrata ?—*Ed., Dict. Econ. Prod.*

REARING in Madras.

"From the information now furnished, it is clear that a traffic in the cocoons of raw silk of the undomesticated worm is never likely to be established in this Presidency."

In a later communication (1886), the Director of Revenue Settlement and Agriculture again raised the question of the Madras participation in the *tasar* silk trade, by the submission of specimens of Chingleput *tasar* to Mr. J. Cleghorn for favour of report on their merit. Mr. Cleghorn is a well-known expert on the subject and the patentee of an improved method of reeling *tasar*. Mr. Cleghorn's report was so favourable that he added that if these cocoons could be had "in sufficient abundance, he was confident Messrs. Jardine, Skinner & Co., the managing agents of Messrs. R. Watson & Co., would take all that the district could produce or would, if it could be proved that a minimum supply of 15,000 *kahons* of cocoons were available annually, and provided that the labour available could also be utilised for manufacturing purposes, build a factory on the spot." This favourable result led to an enquiry which occasioned the submission of a report to the Madras Government in which the Director Agriculture says :—

"In 1885, a large quantity of *tasar* cocoons was collected from several districts for the London Exhibition, and out of this collection some 500 cocoons were forwarded to Mr. Cleghorn, who had kindly promised to reel silk from them and to report upon their marketable value. The cocoons, however, had not been collected with care in the first instance, and they were forwarded several months after their collection. The result of the reeling was therefore not satisfactory. Subsequently, in 1886, a supply of fresh and unpierced cocoons was obtained from the Vizagapatam and Chingleput districts and sent to Mr. Cleghorn, who pronounced them to be very good, particularly those collected from the Kambakam jungles in the latter district. Mr. Cleghorn reported that the silk reeled from these cocoons was superior to any *tasar* that he had seen in Bengal, and added that Messrs. R. Watson & Co. would be glad to purchase all that the district could supply at R7 or R8 a *kahon*, and that they would even, if a minimum supply of 15,000 *kahons* (a *kahon* = 1,280 cocoons) could be ensured, be prepared to build a factory on the spot, if the requisite labour could be secured. Mr. Cleghorn's report and the samples of silk furnished by him are submitted to Government.

"On receipt of Mr. Cleghorn's report, the Collector of Chingleput was requested to state whether he could make arrangements, either through the Forest Department or otherwise, for the systematic collection and regular supply of cocoons, but he replied that further inquiry had led him to doubt the possibility of his being able to supply cocoons in sufficient quantities to be of practical utility, as the headmen of the jungle tribes had assured him that they could not collect more than 2,000 cocoons a year.

"The Conservator of Forests, Southern Circle, was then asked if his department would take up the business, which promised to be fairly remunerative. It was pointed out to the Conservator that the trees, on the leaves of which the *tasar* silkworms feed, grow in abundance naturally and could be planted in any number in the jungles, and that it was worth the while of the Forest Department not only to collect the cocoons, but even to rear the worms with a view to supplying cocoons.

"Colonel Campbell Walker reports that large numbers of cocoons could be collected in most of the coast districts, but he thinks it impossible to guarantee anything like a minimum supply of two millions or so, though that number might probably be worked up to in two or three years, if the matter received the attention it deserved. The best plan, in his opinion, would be to employ an intelligent special officer of the rank of Forest Ranger, who would institute inquiries, collect reliable information, and make arrangements with the hillmen or jungle tribes for breeding the worms, etc., and collecting the cocoons in good order. He is not able, however, to spare any man from his staff, as forest work is very pressing in all districts. He therefore suggests that this department should pay for a man of the stamp required, or allow him the services of one who, after a little training under the District Forest Officer of Chingleput, may be employed on this work. Colonel Campbell Walker also suggests that it should be ascertained whether, until the Forest Department is able to collect cocoons in sufficient quantity, Messrs. Watson & Co. or the Mylitta Silk Mills Company, Madras, will be prepared to purchase the cocoons which it may be possible to collect.

SILK: Tasar.

The Tasar Silkworm

REARING in Madras.

"Accordingly samples of the kind of cocoons examined by Mr. Cleghorn, as well as of the larger kinds met with in villages along the coast, which had not been sent to him, were furnished to the Mylitta Silk Mills Company, and they were asked to reel silk from them and to state whether they would be prepared to buy cocoons of similar description and at what prices. They have furnished samples of the silk reeled by them, which are herewith submitted for inspection, and have informed the Board that they will take the cocoons, if offered in lots of 250 or 300 kahons, at R4 per kahon for the inland cocoons and R5 for the coast ones.

"The Board consider that the time has arrived for making some systematic arrangements for the collection of cocoons and for rearing silkworms through the aid of the Forest Department, and as a first step towards the creation of this important industry, which bids fair to give employment to a large number of people, at no distant date, they recommend the appointment of a special officer on R80 a month, to be eventually raised to R100, as proposed by Colonel Campbell Walker. Of course more than one officer will be required when the industry has been established in Chingleput and extended to other districts, but one will be enough to begin with. The Board also recommend the purchase of two suffocating machines of the description mentioned by Mr. Cleghorn (the cost of each being about R60).

"A small sum of money will be required for buying, suffocating, packing, and carrying the cocoons. The exact sum cannot be stated. The Collector of Chingleput paid 4 annas per 100 cocoons for collecting, but this was a large amount paid on a special occasion, and with a superintending officer and the assistance of the Forest Department, it is probable that the cost could be greatly reduced. Even if it were not, the loss to Government in the first two years or so could hardly be more than R1,100 * a year, and it would disappear when he supply of cocoon is worked up to two millions after two or three years as expected by Colonel Campbell Walker.

	R
* Special officer	960
Contingencies	140
TOTAL	1,100

	R
Cost of collection of two million cocoons, at 4 annas per 100	5,000
Cost of special officer	960
TOTAL	5,960
Price of two million cocoons, at R4 a kahon	6,250
	+290

"The small balance of R290 will more than cover the cost of suffocating, drying, and packing and carriage to Madras. But there is no reason why, if the matter receive proper attention, the outturn of cocoons should not be very much larger than two millions. Moreover, the rate of R4 offered by the Mylitta Silk Mills Company is decidedly low and will probably be raised by them when they see that they can count upon being supplied with certainty.

"If Government approve of the above proposals, the Board would meet the cost of the experiment from the provision of R3,000 for the improvement of agriculture and trade."

An extremely instructive paper on *The Wild Silks of Malabar* was published by Mr. R. Morgan (District Forest Officer) in 1883. In his remarks on the *tasar* worms, Mr. Morgan goes into great detail on the subject of the enemies of the worm, and offers a suggestion on the subject of how these dangers may be guarded against by a system of domestication. Mr. Morgan may be allowed, however, to state his case in his own words:—

"The *tusser* silkworm is fairly abundant throughout the district; but the cocoons are exceedingly difficult to find except in February, when the deciduous forests are bare of leaves. The cocoons are quite equal in size to any I have ever collected in the Kurnool forests, or received from Bengal, if not a trifle larger. For an account of some of the experiments I have carried out with this silkmoth, *vide* G. O. No. 946 of the 9th August 1880.

"There are millions of trees in the forests of Malabar that furnish the favourite food of the tusser, and there is no doubt that if a method could be discovered o protecting the larva from its innumerable enemies, at a moderate cost, the cult

REARING in Madras.

vation of this and some other wild silks would pay fairly well, and a thriving industry be established.

"To protect the larvæ placed on trees in the open air from their enemies, which are legion, I am positively convinced will not only never pay, but is nearly impossible. Let us consider first what these enemies are. Here is a list of them:—

Insects.	Animals.	Birds.	Reptiles.
Ichneumon flies of 5 species.	Rats of 4 or 5 species.	All species of insectivorous or partially insectivorous birds.	Lizards of several species.
Hornets of 12 species.	Bats of all species.		
Wasps of 16 species.			
Ants of 4 species	Tree cats.		
Mantis of 6 species.	(*Musanga.*)		

"The most deadly of all these enemies are the ichneumon flies, and next the hornets, wasps, and ants. Eternal war is waged by these creatures against all insect creation, and if it was not for their services, man would have but a poor chance against his insect foes, notwithstanding the great and efficient help that he is afforded by the feathered creation.

"The ichneumon fly, of which there are several species in Malabar, hunts chiefly by day. Watch it, and you will see it examining every leaf and bud on a tree. When it espies a caterpillar, it alights near it, and walks round deliberately examining it, to see if it has been previously punctured by any other ichneumon. If it has,—and this is denoted by a series of small black spots along the side of the larvæ*—the ichneumon leaves it at once and searches for another. If not, the fly darts its sharp-pointed oviposite into the side of its writhing victim, and lays an egg between the skin and the stomach of the larvæ. It repeats this operation till it has laid a number of eggs proportionate to the size of the larvæ, and then leaves it for another.

"The ichneumon larvæ are hatched in a few days, and prey on the juices of the caterpillar which goes on eating, apparently well, till the larvæ of the ichneumon, having reached their full size, eat their way out of the caterpillar, which dies. The larvæ of the fly burrow into the ground, and there undergo their transformations. Occasionally, the fly larvæ do not leave the caterpillar till it has spun its cocoon, in which they find themselves imprisoned, and after reaching the imago stages, die in it.

Conf. with p. 102.

"Of the thousands of worms I put out in my compound in Wynaad, on trees of **Lagerstrœmia Flos-Reginæ** not *one* lived to spin its cocoon. Day after day I used to inspect them carefully, and note the destructions going on. Hornets, wasps, and ichneumons kept buzzing about the trees all day, carrying away the larvæ, whilst armies of ants marched up and down the trunks, conveying the helpless little creatures to their underground stores.

"Now, how are we to contend successfully with these infinite hordes of enemies? Well, this is what I propose:—

"First, a wooden building must be put up. It should be 100 feet long by 30 feet broad. It should be moveable, and be merely a framework. The roof should be of tarpaulins running on rings, so that the building may be made watertight during very heavy rains. The floor should be of Portland cement, and have ant-traps, *i.e.*, little hollows 6 inches deep by 2 feet in length at intervals of every 10 feet round it, which can be baited with fish or meat.

"The Portland cement floor should extend for a distance of 10 feet all round *outside* the building, which must be surrounded with a Portland cement channel full of water, 6 inches deep by 2 feet wide. The object of this trench of water is to cut off all communications from outside by which noxious creeping insects might gain access to, and injure, the silkworms in the building.

"Breeding cages made of galvanised wire-netting of very fine mesh, each 15 feet by 4 feet, and raised 2 feet off the ground, should be arranged in the building, allowing sufficient space between the cages for the (silkworm) attendants to pass freely, and distribute leaves to the worms.

* This would seem to be a mistake, as these are the stigmata for admitting air to the tracheal system.—*Ed., Dict. Econ. Prod.*

SILK: Tasar.

Rearing of the Tasar Silkworm

REARING in Madras.

"The upper edge of each breeding cage should be lined with glass to prevent the larvæ from crawling on to the lid above, and this glass should be greased to prevent the worms getting any foothold.

"Every breeding cage should be a framework of teak, with the galvanised wire gauze nailed on its inner side. At the bottom there should be a moveable drawer, which should be drawn out and the excretæ removed every morning.

"Adjoining this building there should be a nursery for the young worms.

"This nursery should be an oblong frame, 30 feet long by 10 feet high by 15 feet wide, covered with wire gauze and with wooden stands in it, on which should be arranged small trees of **Lagerstrœmia microcarpa** grown in tubs. The floor should be of Portland cement and the building should have a moat full of water round it.

"The eggs to be tied (in leaf cups) to the branches of these trees and the larvæ allowed to denude them of leaves, when fresh trees in tubs can be substituted, and the worms transferred by placing green leafy boughs on the denuded shrubs on to which they will at once crawl, and can thus be bodily transferred to the fresh trees without there being any necessity for handling them.

"When the worms in the nursery are one inch in length, they should be transferred to the large breeding-room, and placed in the cages, where they should be supplied twice daily with a liberal allowance of leafy twigs, at 6 A.M. and 4 P.M. The twigs should be placed on a barred frame at the bottom of the breeding cage, the frame being one foot above the drawer at the bottom of the cage. This will allow the excrement to fall through, and at the same time keep the worms clear of it, as it is poisonous to them.

"In very hot, sunny weather, and in very wet weather, the tarpaulins should be drawn over the roof; at other times it should not, for the worms benefit greatly by the admission of sunlight, which, if long withheld, kills them, and by light showers of rain. Although the tusser worms do not object to hot sun, yet, if the building were not covered in such weather, say, from 8 A.M. to 4 P.M., the leaves given to the worms would wither, and be no longer fit food for them.

"I would establish such a breeding-house in a sheltered position, and in a forest where ample food for the worms for a radius of 5 miles around could be easily secured.

"Of course, capital is required to start such a scheme, and it could not be well done for less than R10,000 to commence with, for galvanised wire gauze is very expensive. I wrote to a manufacturer near Easton Road, London, and I find the very lowest wholesale rates, with discount, are considerable.

"If an experimental trial, on a small scale, were made, I am confident that, so far as success in breeding the worms is concerned, it would be achieved. It remains to be proved, however, what it will cost to turn out the raw produce, and what comparison the cost will bear to the selling price in the market."

IN MYSORE & COORG. 2023

VIII.—MYSORE & COORG.

A fairly extensive official correspondence exists on the subject of the *tasar* of these Provinces. **Dr. Shortt** specially described the insect found on the coast tracts, and the subject appears to still excite a certain amount of interest. One of the most recent reports is that published by **Mr. J. Cameron** in 1887-88. It seems probable that nothing of a very specially local nature can, however, be said of Mysore *tasar* than will be found in the sections of this article devoted to Madras and to Bombay.

IN ASSAM. 2024

IX.—ASSAM.

For many years it has been known that the *tasar* silk-worm exists in Assam, and that the *muga* and *eri* are, by the people of that Province, preferred to it, so that the *tasar* may be said to be a neglected wild product. The late **Mr. Stack's** remarks on this subject constitute at once the most complete and most recent account of the product, so that it may suffice to reproduce here the passages of his valuable note on the wild silks of Assam which deal with *tasar* :—

"(a) *Kutkuri.*—The wild silk-worm called *kutkuri* is believed to be the same as the common *tasar* of Bengal. Its food is principally the *kutkuri* (**Vangueria spinosa**) from which it takes its name, or else the plant called

REARING IN ASSAM.

(erroneously) the wild rhododendron (**Melastoma malabathricum**), the Assamese name of which is *phutuka*. It has been cultivated in the palmy days of the Assam silk industry, but it is now almost entirely neglected, as being inferior to *muga*, and also, perhaps, because it yields only three broods in the year. Its habits are now known to a few old people only in Jorhat. Mr. Buckingham, to whom I am indebted for most of my information about this worm, says that the *kutkuri* is common in the wild state in the neighbourhood of Jorhat. It is also common in Cachar, but there no use is made of it. Mr. Buckingham notes the following experiment with this silk-worm :—

June 16th	Obtained the moth.
" 17th	Eggs laid (378).
" 23rd	" hatched.
" 28th	First moulting.
July 3rd	Second "
" 7th	Third "
" 13th	Fourth "
" 18th	Spinning began.

"The worms were fed on the *phutuka*. Worms put outside while very young were speeidly devoured by ants; but if kept in-doors till the second moulting, they were then found to do very well on the bushes. Mr. Buckingham adds :—

"I reared ten worms in this way, and all, except one, made their cocoons between the leaves of the shrub, one solitary worm descending and making its cocoon in the grass. The natives had previously informed me that this wild species of worm was less liable to the attacks of crows, bats, &c., than tame species were, and it was curious to watch how the worm, at the slightest show of danger, let go the leaf or stem with all its front legs, hanging on by its holders behind, and in this position, with its head slightly curled round and its front legs well tucked up, it took an experienced eye to detect the difference between the leaf of the tree and the worm."

"An experiment made by Krishna Kanta Ghugua with worms got from the jungle in September gave results as follow :—

October 7th	Spinning began.
April 10th	Moth emerged.
" 12th to 15th	Laid eggs.
" 23rd	Worms hatched.
" 29th	First moulting.
May 4th	Second "
" 10th	Third "
" 15th	Fourth "
" 21st	Spinning began.

"According to this experiment the chrysalis state of the moths lasts six months.

"The only point in which the *kutkuri* cocoons seemed to Mr. Buckingham to differ from those of the Bengal *tasar*, was that the *tasar* cocoon was rather closer spun and more compact, and less pointed at the ends than the *kutkuri*; but the colour was as nearly as possible the same.

The silk is ranked below *muga* in value, being coarse though glossy, and so strong that the natives compare it to rhea thread. Tne *phutuka* is one of the commonest wild shrubs in Assam, and the worm could probably be cultivated at very little cost, but the silk could not compete with the cheaper and better tasar supplied by Bengal.

Deo-Muga. 2025 *Conf. with p. 5.*

(*b*) The *Deomuga.*—Another worm which appears to be simply a variety of the *tasar*, feeding on the *phutuka* like the worm just described, is counted by the Assamese as a distinct species, and known by the

SILK: Tasar.

Peculiarities of the Tasar Insect

REARING in Assam.

name of *deomuga*. It must not be confounded with the genuine *deomuga* described further on. An experiment made with cocoons of this (so-called) *deomuga* by Krishna Kanta Ghugua furnished the following record:—

August 14th	Cocoons obtained from the jungle.
„ 17th	Moths emerged.
„ 19th, 20th, and 21st	Laid eggs.
„ 27th	Worms hatched.
„ 31st	First moulting.
September 5th	Second „
„ 10th	Third „
„ 15th	Fourth „
„ 22nd	Spinning began.
March 13th	Moths emerged.

"Here also the period of the chrysalis was about six months.

"Some of the cocoons were boiled in potash water for two hours, and a fine thread, resembling that of the *muga*, was reeled off them.

"(c) The *Sálthi*.—The wild silkworm called *sálthi* is also a species of tasar. It is called *deomuga* by the Kacharis, but must not be confounded with the *deomuga* proper, which is described below, and which is a **Bombyx**. The *sálthi* worm feeds on the *kamranga* (**Barringtonia racemosa**) and the *hidál*. The worm itself is very rarely met with, but herd-boys and wood-cutters occasionally bring home the cocoons, and the silk obtained from them can be used for mixing with *eri*. To extract it, the cocoon has first to be boiled in a strong alkaline solution, and afterwards bruised in a mortar. The hollow cocoon is often converted into a tobacco-box, or is used to keep lime in for eating with the betel-nut, or as a cup for dipping oil out of a jar. The habitat of the worm is the jungle at the foot of the Bhután Himálayas. The chrysalis of this species, as of all the wild silkworms, is eaten with much relish by the Kacharis."

From the above passage it will be seen that Mr. Stack regarded the *tasar* silks of Assam as comprised under three distinct forms of the insect. The fact of the third kind feeding on a plant not hitherto recorded as that on which the *tasar* is found, would justify the suspicion that it may be specifically distinct.

PECULIARITIES. Crops. 2026

PECULIARITIES OF THE TASAR INSECT & OF ITS SILK.

Crops of the Tasar.—The *tasar* silk-moth has, generally speaking two crops a year, but instead of being bivoltine in its wild state, it is most probably quadrivoltine. The cocoons are purchased by the rearers in May and June from persons who collect them in the jungles. The larger ones are, generally speaking, females, and as much as 8 to 10 cowries are paid a piece for these, while the smaller or male cocoons only fetch 4 to 5 cowries.

The crops* may be traced out as follows:—

2027

1st Crop.—From the *Dhaba* or seed-cocoons in Bhagulpore, the *Ariya* or *Ranwat* in Seoní, the insects emerge in June, eggs are produced, then worms, and by July these pass again into the chrysalis, coming out as

* Since writing the above, the author learns from the Rev. A. Campbell that, after studying the worm for a great many years, he is of opinion that there are only three main crops yearly, but that individuals are to be found at all seasons of the year. He starts the rotation from the second crop above, thus: 1st crop spins cocoons in September; 2nd (a small proportion of 1st crop) emerges in October and spins cocoons about January; and 3rd crop, moths from 1st and 2nd crops coming in about June which brings us to the 1st crop again. A proportion of the September cocoons only emerges in August.

PECULIARITIES.

perfect insects in three weeks, that is, in August. This is the first or *Bhadeli* crop, from *Bhadon*, August. The *Bhadeli* cocoons are not sold except to rearers. They are preserved, and from them a fresh supply of insects is obtained, the perforated cocoons being then sold at a low rate.

2028

2nd Crop.—The *Bhadeli* insects lay their eggs, and in due course these hatch and worms are obtained which pass into chrysalis in September, the cocoon being mature in October, or, in some districts, not until November. This is the second crop known as the *Kartic* or *Katkahi*, because it appears in the month of *Kartic* (October and November).

Captain Brooke, in his interesting account of the *tasar* industry of Seoní in the Central Provinces, published by **Geoghegan**, describes another crop:—

2029

3rd Crop.—In Nagpur seed-cocoons from the *Kartic* crop are reserved, and in due course these produce eggs, worms, and a crop of cocoons which mature in January. This crop is accordingly known as the *Magh* or *Maghur*.. [**Mr. Campbell** says this is also the case in Chutia Nagpur.—*Ed., Dict. Econ. Prod.*]

2030

4th Crop.—**Captain Brooke** infers, and apparently correctly, that in its wild state the *tasar* insect is quadrivoltine, the *Dhaba* or May seed-cocoons being obtained from the *Magh*, so that the *Dhaba* is really the fourth crop.

Entomologists seem to regard the insect as *bivoltine*, but the reports from different parts of India are most conflicting. It is remarkable that so much confusion should exist regarding the life-history of so very important an economic insect. As practised by the Natives of India, the rearing of the *tasar* cocoon crop occupies about five months a year, commencing from the bursting of *Dhaba* cocoons in June to the sale of the *Kartic* crop in the end of September or beginning of October. **Dr. Buchanan-Hamilton** enters into a discussion regarding the different modes of obtaining seed, the important facts of which agree with all other accounts. Wild cocoons are sometimes collected and sold to the reeler, but as a rule they are sold to the rearer. These are called *Dhaba*. The silk produced from this would, accordingly, be *Dhaba* silk, but that which finds its way into the hands of the weaver is chiefly the *Sarihan* silk, or that produced from the first and second crops above discussed. Should seed-cocoons be preserved from the *Kartic* crop over till next May in place of fresh *Dhaba* seed, the silk produced from this source is known as *Langga*. So much has the insect deteriorated, by this temporary domestication however, that this class of silk is regarded as very inferior. The success of the *tasar* silk industry is dependent on the fact that, unlike the mulberry and other domesticated worms, it is never likely to be visited by a serious plague. Fresh silk seed is always procurable, and from our interminable forests this is ever likely to remain the case.

Collection of Eggs.

2031

COLLECTION OF EGGS.—In Seoní **Captain Brooke** (*see Geoghegan, p. 146*) reports that the insects are in a state of partial domestication, being tended in all their stages, the rearers depending upon the wild supply for their seed-cocoons. The seed-cocoons are placed in baskets, which are generally, for this purpose, large and flat. The insects escape from the cocoons during night, and in some districts the males are allowed to fly away, in others all are confined together in a room.* Whichever

* **Rev. A. Campbell** informs the author that the Santals have a system by which they cause the *tasar* insects to escape simultaneously from all the cocoons of a batch. It is, of course, advantageous to the rearer to have all his worms spinning about the same time. The Santal method of effecting this result is to smoke the cocoons by burning underneath them a quantity of *sál* resin (**Shorea robusta**).

SILK: Tasar.

Peculiarities of the Tasar Insect

PECULIARITIES.

course be followed the males soon discover the females and perform their mission. In 15 or 20 hours after their escape from the cocoons, the females are picked up and placed in closed baskets ending in long, narrow mouths, carefully lined with fresh leaves. Sometimes earthen pots lined with leaves are preferred. In the course of a day the females commence to deposit their eggs, they lay from 50 to 200 during the first three or four days of their brief existence, and perish in eight or ten days more.

Superstition. 2032 *Conf. with pp. 151, 157.*

Superstition regarding the Worms.—Throughout India a strict and severe superstitious observance is preserved, from the period of hatching until the cocoons are collected. The men engaged in this trade lead lives of the strictest abstinence during this period, and so distasteful is the necessity for this observance, that, as compared with other industries, silk labours under a considerable disadvantage. It is remarkable that this religious observance is not confined to *one* race of men, nor to any particular religious community, nor, indeed, is it restricted to a particular species or class of insects. It is almost universal. In the month of April, while in Manipur (during the Burma-Manipur Expedition), the writer expressed a wish to see the process of domestication adopted in that State. He was near a silk-cultivating village (Susikamai) at the time and was shown the worm and cocoon on condition that he would not approach the house. A seat was assumed upon a wall near by, and the worms and cocoons were brought out for inspection. A woman also came and showed the process of reeling. News of this fact spread to the Maharajah, and his Durbar, in great alarm, asked that on no condition was even the most accidental enquiry to be made regarding the worms, in case, as had happened on a former occasion, they should take revenge on the intrusion and die off, to the ruin of a large population of cultivators.* An edict preceded the writer's every movement, prohibiting, with severe penalty, any person from showing the worms or the cocoons or answering even questions addressed to them on the subject of silk. **Dr. Buchanan** accounts for the origin of these observances, as instituted, to preserve a monopoly in the hands of a certain community, who took pains to make every one else believe that they and they alone could successfully rear the insect. **Mr. Baden Powell** describes, in his *Panjáb Products*, the successful introduction near Amritsar of the mulberry silkworm by **Jafir Ali**, a Kashmiri. That gentleman, to preserve a monopoly, adopted at once the practices of the professional silkworm-rearers. He would allow no one, not even his sons, to approach the worms in case of the evil eye proving fatal to his crop. In most parts of India women are supposed by the silk cultivators to be unclean, and are accordingly not allowed to see the worms, and the men who tend on these will not approach a woman in case of being defiled. In Manipur, however, the writer found women busily reeling the cocoons and tending the worms at the same time, so that in that little State the restrictions against women do not prevail.

In Manipur. 2033 *Conf. with pp. 15, 35, 55.*

PRODUCTION.

Production of Silk.

Eggs. 2034

THE EGGS.—The *Eggs* are small, white, flattened, oval bodies, deposited in masses which often adhere together. They are biconvex, and nine, if arranged in a row, will measure one inch. On the ninth day the eggs are hatched within the baskets above described.

Worm. 2035

THE WORM OR LARVA.—At first, when the worm escapes from the egg, it is so small that it can hardly be seen. It at once commences to eat the leaves lining the baskets, but, as the baskets are at this stage placed on the trees, it soon attacks the fresh leaves thus supplied, and rapidly

* The collection of cocoons, caterpillars, etc., made, had accordingly to be returned.—*Ed.*

PRODUCTION.

increases in size. It moults five times, at intervals of from 5 to 8 days, and commences to construct its cocoon in about 36 to 40 days, after the date of hatching. When full grown it is about 4 inches in length, is of a pale green colour, has 12 joints marked with reddish spots, and a reddish yellow band which runs along either side. It is so heavy, when mature, that it is compelled to walk along the delicate twigs, suspended from below by its feet. Birds and ants are its greatest enemies.

Cocoon. 2036

THE TASAR COCOON.—During the long period the insect remains in the lethargic condition, it is absolutely necessary that the cocoons should be strongly and firmly attached. Were they loosely fixed to leaves, as with many other species of silkworm, in the course of a few months, the leaves being caducous, the cocoons would be precipitated to the ground, where, of necessity, the creature would perish. But this is entirely prevented, for the *tasar* worm not only spins a closely-woven and firmly-cemented cocoon of the appearance and consistence of the shell of an egg, but the cocoon is suspended by an elegant and ingenious cord from the twigs around which a strong loop is formed. This suspensor is generally about 3 inches in length, the loop being flattened on the top of the twig to a considerable extent, so as to make the suspended cocoon less likely to be dashed backward and forward. In fact, it soon becomes so firm that the cocoons remain suspended rigidly from the often leafless twigs like so many fruits.

Cement. 2037

Conf. with p. 144, 150.

Cement of Tasar Cocoons.—The cocoon itself is almost perfectly oval, smooth, of a grey colour, with darker veins reticulating across its outer surface. The largest are about 2 inches long and 1¼ broad, the average size about 1½ inches long. The inner layer of fibre is quite loose and forms a soft cushion for the insect within. For a long time this layer was all that could be utilised as a silk fibre, but recent modes of separating or decomposing the cement have rendered it possible to utilise almost the entire shell. **Major Coussmaker** has paid special attention to the subject of the *cement by which the tasar worm consolidates its cocoon*, and he has been able to arrive at an interesting conclusion, namely, that this substance consists of the excrements from the alimentary canal, and that its nature and colouring or injurious power, greatly depends on the food upon which the worm has been fed. He is of opinion, consequently, that judicious feeding will greatly lessen this difficulty, by altering the nature of the cement. The chemical analyses performed under **Major Coussmaker's** instructions revealed the fact of the cement containing the acid urate of ammonia. The silk, of which the pedicel or suspensor is composed, as also the outer shell, is of a reddish colour, and is built up of short broken fibres firmly cemented together. The inner layer is much finer, and entire.

Spinning. 2038

Process of Spinning the Cocoon.—"Each species of silk-worker has two stores of silk, one on each side of the alimentary canal, and below its mouth it has two so-called spinorates or orifices, through which the silk issues simultaneously in fine parallel filaments. As the silk is drawn out of these stores, the worm coats it with a varnish technically called "gum," which contains a brownish-yellow colouring matter.

"The *tasar* worm, in spinning its cocoon, takes short sweeps of its head from side to side, depositing the silk very closely in parallel fibres, which take a zigzag course round the cocoon as he does so. It has been thought that the worm twists or spins the silk as it exudes it, but this is not the case. Besides the gum which coats the silk, the worm secretes at intervals a cementing fluid, which it kneads by an expanding motion of its body through the whole cocoon to consolidate and harden it. This cement gives to the cocoon its drab colour" (*Wardle*).

Period. 2039

Period required.—When about to spin its cocoon the worm, as if to screen itself, first binds together a few leaves within which it commences

Peculiarities of the Tasar Insect and of its Silk.

PRODUCTION.

its operations.* The cord or suspensor is next prepared. The cocoon is then proceeded with, and at first it is so transparent that the entire movements of the creature may be carefully studied. By-and-bye it becomes quite opaque† through the coatings of cement with which it binds the threads together, and in the course of a few days it is perfectly hard. It requires in all 15 days to construct its cocoon.

Cocoon. 2040 Utilisation, 2041

Utilisation of the Cocoon in India.—The hard outer layer or cocoon shells of this moth are now largely carded and spun into *tasar* silk, but from almost time immemorial, they have been used for the formation of strong bands or strings, by being carefully clipped off round and round. These straps the Natives regard so strong as to resist both fire and water; they were formerly, and are even still, used in the Deccan to fasten the barrels of matchlocks to their stocks.

Escape of Insect. 2042

NATURAL SOFTENING OF THE COCOON TO ALLOW ESCAPE OF PERFECT INSECT.—**Mr. E. C. Cotes** writes: "With regard to the natural solvent of this cement, the observations of the writer seem to show that the solvent fluid, which is stored in a large bladder like dilatation in the lower portion of the digestive tract of the future moth, can be freely poured out through the anus of the moth into the chrysalis case; but the chrysalis case itself prevents its passing into the lower portion of the cocoon. Now, the moth emerges through a longitudinal dorsal slit in the thoracic segments of the chrysalis, and, in its struggles to extricate itself, it forces this fluid between its body and the chrysalis case through the slit, on to the cemented wall of the cocoon, precisely in the spot where the moisture appears, and the softening of the cement takes place prior to the moth's working its way through. It would at first sight appear likely that this excreted fluid, being milk-like in consistency, would, in bathing the abdomen of the moth inside the chrysalis case, stain the delicate scales and hairs with which the moth is covered: that this is not the case, however, is at least indicated by the fact that much of the fluid has been found inside chrysalis cases from which unstained moths have just emerged."

Cement. 2043 *Conf. with p. 143.*

The perfect Insect. 2044

THE PERFECT INSECT.—The *escape of the perfect insect* from the cocoon is caused through its secreting the fluid (discussed above) which softens the cement on a spot on the apex of the cocoon. It is quite a mistake to think that it eats its way through. It has no mouth, properly so speaking, and certainly nothing by which it could cut the cocoon. When softened the insect simply forces its escape by displacing the fibres. In this process it is supposed to be aided, however, by its wing spines.

On escaping from the cocoon it discharges its duty, the perpetuation of the species, during a brief existence of 10 or 12 days. It neither requires food nor is it provided by any process by which it could eat or digest food, hence, having accomplished its mission, it perishes.

THE FIBRE. 2045

The Tasar Silk Fibre.

Mr. Thomas Wardle, in his interesting *Hand-book on the Wild Silks of India*, states that "there is a striking peculiarity about the fibre of *Tusur* silk. I have carefully and thoroughly examined it many times

* **Rev. A. Campbell** is of opinion that without this temporary enclosure the worm could not turn round to spin on all sides.

† **Rev. A. Campbell** informs the author that the cement appears to be applied to the cocoon after it has been spun. About the second day the cement from the inside begins to permeate to the outside. When the Santals wish to say that a worm has finished its cocoon, they speak of it as having applied "the lime." In fresh cocoons the cement looks not that unlike lime.

TASAR FIBRE.

under the microscope, and find undoubtedly that it is almost flat and not round, as is the case with the silk produced by the mulberry-fed worm. There is no doubt that it is to this property that *Tusser* silk owes its glossy or vitreous look, reflecting a little glare of light from the angle of incidence on its flat surface, whilst the mulberry-silk fibre, being round, reflects the light in all directions. By some this property is considered a drawback, but by the time the fibre has become modified and the flatness diffused in the loom, I think the lustre of the cloth is enhanced by it. This tape-like appearance gives the fibre this disadvantage, that it is less homogeneous than the round fibre of the mulberry-silk, and I find an undoubted tendency in it to split up into smaller fibrets, of which the fibre is evidently composed, causing the silk to swell out when subjected to severe dyeing processes, particularly the bleaching one of recent date, thus giving a substantial and important reason why its coloured cements should be removed." It has already been said that to Captain Mitchel apparently belongs the merit of having first pointed (1867) that the fibre of the *tasar* cocoon was flat instead of round.

Tasar Silk Flat. **2046** *Conf. with p. 100.*

English and Italian improvements in the *reeling* and *spinning* of the Tasar cocoon have produced, within the past few years, a complete revolution in the European demand for this common Indian insect. While experiments to improve the rearing and, if possible, to domesticate the worm, have, in India, failed financially, there seems every probability that a reaction may one day take place. The demand for *tasar* cocoons and *tasar* silk seems likely to become each year more urgent, but it may be doubted, how far the so-called *tasar* of Europe is, and must in the future be, the produce of **Antheræa paphia.** A careful perusal of Mr. Wardle's interesting *Hand-book* forces upon one the conviction that since recent discoveries render it possible to spin even the waste particles rejected from reeling, every fibre of the cocoon being now utilised, a yearly increasing demand will be made for this and all other wild silks. But it is absolutely necessary to impress upon the people of India the distinction between reeled *tasar* silk and spun *tasar* silk. By the former the cocoons, after being boiled in an alkaline solution, have the original thread drawn out from the interior, a process which could only be carried to a certain extent, all experiments having failed to soften the cement so as to allow of the entire cocoon being reeled. Indeed, even were it possible economically to soften the cement entirely, a large proportion of the outer shell could not be reeled owing to its being composed of broken or short threads. By the Indian, and indeed by the old European, mode of reeling the *tasar* cocoon, about 1℔ of reeled silk was all that could be prepared from 11℔ of the cocoons; the remaining 10℔ were technically known as *tasar*-silk waste. Mr. Wardle says that a few years ago this *tasar*-waste was valueless and lay about our English ports, for some time, quite unsaleable. It now is greedily bought up for 2*s.* a ℔, as also the waste from all other kinds of silk. This is due to the fact that these waste cocoons can be carded and spun into thread like other fibres, instead of being reeled; thus not only utilising the waste but opening up a complete new silk industry, of spun and carded goods. From these spun silks many new fabrics, which are rapidly gaining public favour, have come into existence. It will be seen below, however, that from time immemorial the people of Assam have carded and spun their *eri* silk, so that Europe is only now following their example.

Spun Silk. **2047** *Conf. with p. 163.*

Mr. Wardle gives the dimensions of a single fibre as $\frac{1}{820}$ inch if taken from the outside of the cocoon, $\frac{1}{580}$ inch from the inner part; the tension or limit of stretch before breaking, of a single fibre, one foot long, 1·9 inch, the fibre being taken from the outside of the cocoon, 2·7 inch when

Measurements of the Fibre. **2048**

TASAR SILK FIBRE.

from the inner part, the strength of a single fibre taken from the outside of the cocoon 6⅝ drams avoirdupois and 7¾ drams if taken from the inner part of the cocoon; and the dimensions of the cocoon 1½×⅞ inches. Captain **Mitchel**, so early as 1867, had measured the *tasar* silk fibre, and he records the fact that the finest specimen seen by him was only $\frac{1}{35,000}$th (*sic*) of an inch in thickness.

Conf. with p. 100.

REELING.

Native mode. 2049

Native Mode of Reeling the Silk.

It has already been stated that, according to **Rumphius**, the Natives of India understood, 200 years ago, how to reel *tasar* silk. They were able to draw out of the cocoons a continuous thread 20 to 30 yards long. The Native method of reeling, as prastised at the present day, is probably very nearly the same as that witnessed by **Rumphius**. In Bengal about 400 cocoons are placed in an iron pot along with 7½ seers of water, in which a small piece of potash has been dissolved. The bottom of the pot is protected by a small piece of mat, to save the cocoons from being burned. The cocoons are boiled for one hour. The alkaline water is then poured off and the cocoons transferred to a clean pot, where they are left standing over for three days, exposed to the sun, a thin cloth being tied over the mouth of the pot to prevent them being soiled by dust or birds and insects. On the fourth day they are again boiled with 2½ seers of water for about an hour, and thereafter poured into a basket where they are allowed to cool. They are then washed with cold water and spread out upon a floor of cow-dung ashes to dry, a cloth being stretched across to keep them clean. In six hours they are ready to be reeled, but should experience show that some are still not ready, these are carefully picked out and exposed for a longer period to the action of the sun.

Each cocoon is now carefully picked by the hand, so as to remove the waste outer shell known as *jhurí*. This substance is sold at a small rate to potters to make the brushes with which they apply the pigments to their wares. The outer continuous fibre of each cocoon is then sought with the hand, and those from 5 to 10 cocoons (according to thickness of required thread) are twisted together by being rubbed across the left thigh. The thread thus formed is wound upon a crude spindle, which is twirled in one hand while the fibres are twisted by being rubbed upon the thigh with the other. While being reeled the *tasar* cocoons are not placed in hot water, but are left quite dry dancing about in a basket. The first or finest thread removed in this way is called *lak*. After the removal of the *lak* there remains a coarser thread which is next reeled. This is known by the same name as the waste, namely, *jhurí*. This coarser thread is sold to men who prepare silk strings. The perforated cocoons are also reeled, but they bring a much lower price, because the fibre has to be so often joined that the thread is very inferior.

Working in this way a woman will boil, dry, and reel about one rupee's worth or 400 cocoons in 10 days, or 1,200 a month. These will yield about 2·247℔ of fine thread (*lak*) worth R5-6, and 1½ annas of *jhurí*. The cost of pots and firewood leave a profit of R1-8 to R1-12 per mensem (*Dr. Buchanan*). In the account of *tasar*-reeling in the *Gazetteer of the Santal Parganas* it is stated that an average *kahan* of 1,280 cocoons yields from 1½ to 2 seers of *tasar* silk. This would be a considerably higher return than that given by **Dr. Buchanan-Hamilton**.

One of the greatest difficulties in reeling *tasar*, after the cocoons have once been softened, is to make the separate strands cohere in the reeled thread or "single." This difficulty does not exist in the case of mulberry silk, where, unlike the *tasar*, the cement is only softened during the process

REELING

Native Mode.

of reeling. By some of the new methods of reeling *tasar*, this difficulty, it is understood, is now, however, overcome by passing the "single" through a special gummy preparation, which not only causes adhesion but imparts a gloss to the fibre that greatly enhances its value. Report has it that the gum of a **Sterculia** (*katila*) has been found peculiarly suitable for this purpose.

Italian mode. 2050

Italian, French, and other Improved modes of Reeling.

In principle this is identical with that described as practised by the Natives of India, namely, the extraction or uncoiling of the natural fibre from the cocoon (each of which, as prepared by the insect, is composed of two filaments), the fibres from a required number of cocoons being wound together and slightly twisted into a thread known as a "single." In practice, it is very different however. The fibre is cleansed of all its impurities, which, by the Native process, are left adhering to it. A fixed and definite number of fibres are wound together into the "single," which is of uniform thickness, and much finer than can be produced by the Natives of India. In other words, the one produces a careless or accidental thread, and the other an accurate and definite one. A skein of 1,000 yards in length of the ordinary native-reeled "single" weighs from 9 to 15 drams, technically known as 152 to 255 "deniers." From *tasar* cocoons reeled by the Italian process Mr. Wardle obtained a size of 51 deniers or 3 drams per 1,000 yards. The *tasar* fibre is about $\frac{1}{710}$ part of an inch, or three times as thick as ordinary silk, so that 51 deniers would, for such a fibre, be regarded as a good practicable result. The denier is equal to about 0·825 grains.

The cocoons are boiled for a considerable time in an alkaline solution, to which some glycerine may be added. After being boiled they are conveyed to a basin over which a semi-rotating brush is so adjusted as to brush off the outer waste shell and ultimately pick out the continuous threads. When these have been found the cocoons are transferred to the reeler. A number of cocoons, with the ends of their found fibres twisted together, are placed in the hot-water basin of the reeling machine; four or five of these are passed through the agate centre guides and the *croiseur*, and are thus cleansed, and to a required extent twisted, before being conveyed to the reel. The reel is driven by a handle or windlass, and the connection between the fly wheel and the reel is such that the reeler may stop the action at any moment, having a lever near by which throws the reel out of gear, should any necessity arise for stopping the machine. The moment a thread breaks, or whenever a cocoon is reeled out, the end of a fresh one is quickly presented and the action continued.

Prepared by this mode the thread is cleaner and devoid of smell; it takes colour more rapidly, and without requiring to be bleached, the lighter shades of colours may be given to the silk.

The Italian-reeled fibre, the primary thread, or "single," produced by reeling, has now to go through the process technically known as "throwing." Two or more "singles" are "thrown" together, and spun or twisted into a yarn. For many years English spinners could only produce the "tram" or weft required by the silk-weavers. The finer and more delicate "organzine" or warp had to be imported from Italy and France. John Lombe of Derby managed to become possessed of the secret however, and from that date it spread rapidly over the world. The tram or weft yarns are composed of two or more singles, only slightly twisted together, being left loose and open so as to cover more freely the warp. Warps are rarely composed of more than two singles, and for fine warps a "single" alone is used. It is much more difficult, therefore, to pro-

SILK: Tasar.

Italian, French, and other Improved

REELING.

Italian mode.

duce the warp which has to go through six processes, *viz*., winding, cleaning, spinning, doubling, spinning, and reeling; the warp has eight turns in the inch, weft only four. Reeling is as a rule performed by quite a distinct person from the spinner, and the singles reach the latter firmly twisted into "knots" and tied up in batches known as "books." The Italian-reeled *tasar* is as pure as ordinary silk, and only loses two ounces a ℔ on being dyed, while Native-reeled *tasar* loses as much as seven and never less than five ounces a ℔. That is to say, the European-reeled yarn loses 12½ as compared to 37½ per cent. The books of thrown silk, as they reach the weaver, are known as "hard yarn." For most fabrics they have to be softened by being boiled, a process which brings out the brilliancy of the fibre as well as softens the yarn. By the process of softening Native yarns lose seriously in weight, and thus not only are Native-reeled singles and thrown yarns unsuited for the majority of European purposes, but on being purified they lose so seriously as of necessity to cause their commercial value to be considerably below that of European-reeled and thrown silks. So great are these disadvantages that the future of *tasar* silk depends more upon the efforts put forth to improve the reeling than upon improvements in the breed of the insect.

Expectations of a great Future.

2051

Conf. with pp. 106, 193, etc.

Since the above abstract of the main ideas brought to light on the subject of reeling of *tasar*, appeared in the Calcutta International Exhibition, the subject has been greatly advanced and several patents taken out, many of which (such as that discovered by **Mr. Cleghorn**), produce very nearly all that could be desired. Indeed, it may be said that, within the past few years the subject has, both in India and Europe, made rapid strides, but most of the discoveries bearing on this subject are at the present day confidential, and cannot, therefore be discussed in this place. **Mr. Wardle** alludes, in the highest terms, to the improvements made in India in reeling *tasar* silk, as manifested by the sample shown at the Jubilee Exhibition, Manchester. Thus, for example, he says: "Tusser silk, too, as adapted for export, has been, during the last ten years, very slow in taking root in India, and large supplies have had to be obtained from China to meet the gradually growing European demand. At the Paris Exhibition of 1878, **Sir Philip Cunliffe-Owen** determined, with my assistance, to give this silk an opportunity of asserting itself, and afterwards, in the Indian Museum, at South Kensington, he took care that its capabilities and uses should be conspicuously displayed. Not a little of the industrial growth of this useful though wild silk is due to his encouragement, and now in this Exhibition can be seen the fruits of all the care which has been bestowed upon it in various ways, and I am more than pleased to state that India now has an enormous and yearly increasing demand." So sanguine a view of the Indian trade would scarcely seem justified by the facts brought together in this paper, though improvements have doubtless taken place. But **Mr. Wardle** continues in another pasage, "A number of gentlemen in India are viëing with each other to improve the methods of reeling and with singular success. In the Exhibits of Tusser raw-silk are shown results which, a few years ago, would have been thought to be impossible. Already this silk is capable of far more extended use than ever before, and although it cannot be expected, on account of its structure and properties, to take the place of the more beautiful silk of the **Bombycidæ**, it has its uses, and those in a much higher degree than it was ever thought susceptible of." So, again, **Mr. Wardle** says of India, "Her tusser silk is now an established and well-rooted industry, a few years ago in export non-existing." **Mr. Wardle** catalogues a fairly extensive series of *tasar*, cocoons and silk, shewn at the Manchester Exhibition, as, for example, *cocoons* from Singbhum, Ranchi, Manbhum, Birbhum, Burdwan, Darjil-

Decline in Production.

2052

Conf. with pp. 59, 75, 106, 120-1, 157-158, also 112, 114-117, 125, etc.

PRESENT POSITION of INDIA TASAR.

ing and Gaya in Bengal; from Ganjam, North Arcot, Dudhi, Yerak, Caddapah, and Madras in South India; from Phillaur and the Beas in the Panjáb. *Tasar* Silk (dyed and undyed) and silk waste and thread from the 24-Pergunnahs, Birbhum, Fatwa, Burdwan, Singbhum, Manbhum, Ranchi, Gaya, Murshidabad, Midnapur, Orissa, Bankura, and Shahabad in Bengal. Commenting on some of the exhibits he says of **Mr. T. F. Peppe's** Shahabad samples that they were reeled by Natives who used castor-oil and the alkaline earth *sajji mati*—to aid in the process. A *tasar* raw-silk of improved reeling was shown by **Messrs. Louis Payen & Co.**, from their Filature at Berhampur in Bengal. A case of samples from **Messrs. Robert Watson & Co.**, Surdah in Rajshahi; from **Mr. T. F. Peppe's** Arrah (a case of cocoons, silk, waste, and fabrics of *tasar*). Also a large assortment of *tasar* cocoons, silk (bleached white) dyed various shades and of piece goods by **Joshua Wardle & Sons.** Concluding this section of his Catalogue, **Mr. Wardle** deals with certain exhibits designated 'European Utilisation of Tusser Silk.' This embraces certain stuffs after the manner of Madras Muslins, the woven patterns of which are *tasar* silk variously coloured. This was lent by **Messrs. Alexander Jamieson & Co.**, of Glasgow. A *tasar* silk rug, designed by **Mr. William Morris**, and also a large *tasar* silk rug dyed in permanent Indian dyes; these were exhibited by **Mr. Wardle** himself. But in this section of the Exhibition were also shown Fichus or Shawls of Leicester manufacture made of *tasar*; Chenille *tasar* silk shawls of German manufacture; Lyons manufactures of *tasar*; and seal cloth of Yorkshire make. This fabric, **Mr. Wardle** adds, has a deep plush pile very much resembling seal skin, but much healthier to wear."

Price of Tasar must be below Mulberry Silk.
2053
Conf. with pp. 75, 106, 115, 121, 126, 151, 158-59.

M. Natulis Rondot, in letter (*see Proceedings Agri-Horticultural Society of India, 1886 p. 62*) on the subject of *tasar* manufactures, alludes to certain important features of the trade which seem pertinent, regarding which it need scarcely be added that he accepts the term "*tasar*" as applicable to all silks which in the present article have been designated "wild silks." He remarks: "The demand for Raw Silks of wild worms continues very active, and is likely to remain so, provided, of course, the prices are not very high, as the silk of the wild worms of India is inferior to the silk of the domesticated mulberry worms. As soon as the price of *tusser* silk approaches the prices of China or Japan silks, preference is given to the latter, and this preference is justified.

Tasar mixed with other Silks.
2054

"The *tusser* silks have, at the present time, the benefit of a move in fashion, but sufficient quantities are not received to meet the wants of manufacturers, and cloths of *tusser* silk are made mixed with silks other than *tusser*.

"This will seem strange to you, but three things must be considered—

"*1st*—The great influence which fashion has on manufacture.

"*2nd*—The large quantity of silks used by the manufactories of Lyons.

"*3rd*—The smallness, up to date, of the production of silks known as *tusser*.

"It is because I know that the main object of the Society is to help the development and prosperity of Indian industries, with a view to the public interest, that I have urged that your Society should advise its members and correspondents not to venture, except with a knowledge, in the silk trade. The nature and quality of the cocoons are of great importance, and it should at the outset be known what value the materials procurable have in an industrial point of view.

"We have as much interest in France to be well provided with cocoons

HIGH EXPECTATIONS DEPRECATED.

and silks, as the growers and merchants of India have in finding a large market for their cocoons and silks. But there are cocoons and silks for which we do not care, because they do not suit our requirements, and it is as well that the fact should be known in your quarters."

EUROPEAN MANUFACTURES. 2055

The European * Manufactures of Tasar Silk.

In addition to the remarks already made on this subject, the following passage may be reprinted from the author's Calcutta International Exhibition Catalogue which appeared in 1884:—

Imitation Seal-skin. 2056

1st—Imitation seal-skin cloth.—The use of this fabric for cloaks and mantles for winter wear has already commanded a regular and established place in the market. The Tasar-spun thread for this purpose has a much closer resemblance to the true seal-skin than could be produced from any known species of reeled silk, and it is, moreover, much more durable.

Utrecht Velvet. 2057

2nd—Tasar spun yarn also bids fair to become an important substitute in the manufacture of Utrecht velvet.

Carpets. 2058

3rd—It promises to become extensively used in Carpet manufacture, excelling all other silks in possessing rigidity, a quality indispensible in a carpet fibre. The brilliancy with which the silk-coloured threads enliven carpets and other mixed fabrics seems certain to give birth to a totally new and unlimited industry.

Mr. Wardle, from whom the above information regarding the manufactures of the *tasar* has been derived, urges the absolute necessity of pressing upon the people of India this new discovery, with the view of encouraging them to preserve the vast quantities of cocoon waste, the supply of which in Europe will be the only impediment to the development of this new industry. China is already alive to this position, and at present the waste and perforated cocoons used in the spinning trade are chiefly imported into Europe from that country.

EUROPEAN Grenadine. 2059

Reeled *tasar* silk has also undergone immense improvements, and is largely made into silk fringes and into the woollen cloths known as grenadine or mandarin. It has been contended that this new impulse to the *tasar*

* Since the above was passed to the Press, the author has received a copy of Mr. T. Wardle's lecture read before the Society of Arts (*London, May 14th, 1891*), in which he enumerates many new purposes for which *tasar* silk has found a ready market. These may be mentioned by name in the order dealt with by Mr. Wardle:—Tusser Pile Fabrics; Tusser Embroidery; Trimming Materials; Handkerchiefs; Tusser Carpets and Rugs; Opera Shawls; Lace; Elastic Webs; Printed Tusser silk; Embossing with Tusser silk, etc. In reply to Lady Egerton's enquiry as to the suitability of *tasar* for dress Mr. Wardle said: "A fabric, such as that required for the best purposes of dress, would never equal those of the best silk; but there were a great many purposes for which it could be used, chiefly for furniture silks, and for cloaks and shawls, and even for dress secondary only to the best silk; but one of its chief uses was for trimmings and chenille, in which very successful effects were produced, both on the Continent and in England."

While Mr. Wardle in the opening sentences of his lecture gave the distinction, which he has not before made sufficiently clear (*see* his Handbook, pp. 14, 39, etc.,) into Chinese and Indian *tasar*, he unfortunately did not preserve that distinction when dealing with the manufactures. This is the more to be regretted since, while India might attempt the cultivation of the Chinese worm in the temperate tracts of the country, her interest in her indigenous tropical worm are of an altogether independent nature. In other words, if suitable for the same purposes, these two forms of silk (from a sericultural point of view) are more dissimilar than the *tasar* is from the mulberry-feeding insects of India. Both may and are reared in this same localities, but the Chinese and the Indian *tasar* never could be produced side by side.

EUROPEAN MANUFACTURES.

trade largely took its birth from the Paris Exhibition, where these facts were first prominently published. Mr. Wardle (*Handbook, p. 39*) gave the average London consumption for the four years ending 1877 as 238 bales, for 1878 (the year of the Paris Exhibition) it became 736 bales, while in 1879 the consumption was increased to 1,142 bales.

But amongst the siks of European commerce Indian *tasar* occupies perhaps the least important position. This is doubtless due chiefly to three causes, namely, *first*, that one maund of cocoons of *tasar* yields about 4 seers of silk-reeled fibre or $\frac{1}{10}$th of the weight, while mulberry cocoons give almost half weight of fibre. This difficulty has now, however, been all but removed. While the entire weight cannot be reeled, every particle of the cocoon can be utilised. The *second* great difficulty to the development of the Indian *tasar* silk trade is the imperfect and faulty system of Indian reeling. This fact is at once established by the published figures of the sales of *tasar*-reeled fibre, the Italian or improved fibre obtaining three or four times the price of the ordinary Native-reeled silk. What seems wanted therefore is to introduce the Italian *or other improved* processes of reeling the cocoon (so far as that may be found possible), and to instruct the Natives to carefully preserve the waste or outer shell, when the wild *tasar* industry might become one of some importance. It might afford remunerative employment for a certain percentage of the population of our lower hilly undulations, who, by nature, are opposed to agricultural labour, but who are driven out of the silk market through their low-land neighbours having taken to rearing the domesticated mulberry silk worm as an auxiliary to their other employments. The *third* great difficulty, and the one which is perhaps of chief importance, lies in the fact that the insect is a wild one and has hitherto not proved amenable to *complete* domestication, but the social customs and habits of the people of India are inimical to changes and hence oppose progression.

Conf. with p. 149.

Dyeing and Bleaching of Tasar Silk.

DYEING & BLEACHING. 2060

It was for long thought that an utterly insurmountable obstacle existed to the development of the *tasar* silk trade in the difficulty experienced in causing the fibre to take the lighter shades of colour. That is to say, it was regarded as impossible to bleach the fibre so as to fit it to take the lighter shades. Within the past few years, however, this has been so far overcome that time may be stated to be all that is required to secure complete success, in other words to allow of the development of scientific principles which have already been recognised as having vanquished the difficulty of bleaching. There are two widely different modes by which this most desirable object has been accomplished. Allusion has already been made to Major Coussmaker's discovery, regarding the matter voided by the insect, during the construction of its cocoon. This was stated to constitute the cement used by the worm in consolidating its cocoon, and to be the substance which imparted the objectionable colour to the fibre. Systems of improved feeding were stated to greatly alter the nature of the excrements, and so completely was Major Coussmaker able to carry out this idea that he produced cocoons perfectly free from the objectionable colouring matter, the worm having been taught to void the injurious materials before constructing the cocoon. If it were possible to completely domesticate the Indian insect and to produce its cocoons profitably, this discovery would doubtless prove a convenient and practicable solution of the difficulty of bleaching, but unfortunately the cost of production renders domestication unprofitable.

Cement of Cocoons. 2061 *Conf. with pp. 143, 144.*

Cost of Production. 2062 *Conf. with pp. 132, 149.*

The natural cocoon has now, however, been found to be capable of bleaching or of parting with its colour whenever the fibre is brought into contact with nauscent oxygen. This was first discovered by M. Tessie du Motay,

SILK:
Tasar.

Dyeing and Bleaching

DYEING & BLEACHING.

who used permanganate of potash for this purpose,—one of the most powerful oxidising bodies, upon organic matter. Unfortunately, however, this substance injures the fibre, but the re-action establishes a principle which seems likely to be applied successfully with some other re-agent. Binoxide of barium, by simple contact with the fibre, accomplishes the same purpose, but it is expensive. **Motay**, however, obtained a gold medal at the Paris Exhibition of 1878, so that he may be accepted as the pioneer in the effort to bleach *tasar* silk. The advances in this branch of sericulture have been so rapid that it may, in fact, be said that *tasar* cocoons can be now economically and cheaply bleached.

Major Coussmaker, in his final report of the Bombay *tasar* silk experiments, conducted by him, furnishes, for example, the following particulars regarding the bleaching of the silk:—

"I take this opportunity of putting on record some information which **Mr. Wardle** has kindly placed at my disposal regarding the bleaching of *tasar* silk. It is well known that the great objection to this silk was its natural colour and reputed inability to take dye satisfactorily. **Mr. Wardle** has made a persistent series of experiments to overcome these defects, and informs me that he has now perfectly succeeded, and there will always be a demand for clean, evenly reeled *tasar* silk thus bleached. His method is as follows: Mix in water well impregnated with soft soap, carbonate of soda crystals equal in weight to one-quarter of the silk to be bleached, and immersing the silk, boil it for a little more than half an hour. Then wash it thoroughly first in hot and afterwards in cold water, until it is perfectly clean. Next put it into a narrow deep vessel, and having added a few drops of ammonia to the liquid peroxide of hydrogen, so as to make it alkaline, pour it over the silk, until it completely covers it. The silk while undergoing this treatment must be kept perfectly in the dark, and if the peroxide does not remain alkaline, a little more ammonia must be added. Keep the silk in this state for 24 hours, then heat up the mixture, till its temperature reaches 120°-130° Fahr., and let it remain at this heat for 12 hours, when it will be found that the silk has become bleached.

"He also gives the following instructions for preparing peroxide of hydrogen: Mix one part of hydrochloric acid with six parts of water. Add to this, *very gradually, finely powdered* barium dioxide (commercial), until the acid is *almost* neutralised. Stir the liquid all the time, taking care not to add the barium quickly for fear of raising the temperature. Next add baryta water, a concentrated solution of Barium monoxide, to the mixture. At first a dirty-looking brown precipitate will be formed, showing that the oxides of iron or other metals which exist as impurities in the commercial barium dioxide are being got rid of. In a short time, carefully watching through the sides of the glass vessel in which you are making the preparation, you will see that a white precipitate is beginning to form; directly this appears, pour off the liquid into a clean glass vessel, filtering it at the time and continue to add the baryta water *very gradually*, as long as any precipitate forms; you can tell when this stops by taking a drop out, and adding baryta water to it, if it has no effect you will see that no more precipitate will fall in the bulk.

"Let the white crystalline precipitate stand for a short time, and then pouring off the liquid add a little pure water, and pour off that also. Repeat this operation several times, so as to be sure that the precipitate is thoroughly cleansed, adding less and less water each time, till the pure precipitate alone remains.

"Take 7 or 8 parts of pure water and one of the purest sulphuric acid available, mix them and let them get *perfectly cool*.

"When quite cool, add the white precipitate very gradually until the mixture is neutralised, great care being taken to let the precipitate fall in slowly and uniformly to settle at the bottom of the vessel. If the liquid be at all discoloured, filter it again. This makes the pure solution of peroxide of hydrogen, which must be decanted into stone jars and put away till wanted in a cool dark place."

Although particulars of the various processes now in use for bleaching this silk have not as yet been published, it may be said that both in France and in England the object in view has been completely attained. *Tasar* may now be bleached pure white and thus rendered capable of being dyed into the lightest shades. The commonly accepted view that the Natives of India could not dye *tasar* silk, prior to the modern advancements which

DYEING & BLEACHING.

have been accomplished in Europe, is, however, quite incorrect. They have from time immemorial thoroughly understood numerous and complicated methods of applying most if not all of their permanent tinctorial reagents to this silk. Indeed, it may safely be said that the progress made by mulberry silk, in displacing tasar from popular demands, operated prejudicially on the *tasar*-silk dyer's industry. In some of the great *tasar* centres of trade, a century ago, the dyer's art was in a flourishing state. The earlier records of such districts as Bhágalpur speak not only of thousands of *tasar* looms daily at work, where there are now not hundreds, but they deal with the marvellous results attained by the skilled *tasar*-silk dyers—a trade practically extinct. From one end of India to the other similar records may be found, but in the reports of dyeing it is rare that the exact processes adopted with *tasar*, as distinct from other silks, are given by writers on the subject. An extensive series of reports might, however, be quoted in which it is said the dyers of this and that town are expert in the art of imparting all shades of colour to *tasar*. Thus, for example, **Dr. G. Smith** in 1859 (*Jour. Agri.-Hort. Soc., Ind., XI., 426*) wrote of Hyderabad that "the dyers dye *tasar* silk all colours except green." He gives the process to be pursued in producing various shades of red, orange, yellow, and black. Space cannot be afforded to republish **Dr. Smith's** paper, but from this reference to it, the reader who may be specially interested in the subject should have no difficulty in seeing the original. Suffice it to say that lime (*chunam*) would appear to have been an essential element in the Hyderabad methods of dyeing *tasar*. In the case of the production of shades of red and yellow, alum was also used along with the lime, but for black, an alkali (prepared from the leaves of the *palas*, **Butea frondosa**) was substituted. It may in passing be mentioned that the Hyderabad black *tasar* dyes, used in **Dr. Smith's** time, were derived from indigo, but apparently no fermentation took place. The seeds of **Morinda tinctoria** were regarded as an indispensible element in the preparation. These, we are told, afforded a "glutinous fluid," the function of which it is difficult to discover. The red dyes described by **Dr. Smith** were obtained from lac in combination with an infusion of *old* tamarinds. New tamarinds, we are informed, were injurious. Throughout India, wherever *tasar* silk dyeing is practised, the use of tamarinds is highly spoken of in all light shades. Whether these have in any way a bleaching property or merely clear or brighten the dye, it would be difficult to say without actual experiment.

Bleaching.
2063
Conf. with pp. 25, 57, 156.

A fairly extensive series of articles similar to that by **Dr. Smith** might be quoted on the subject of *tasar* dyeing. In many of these the remark occurs that the silk is first bleached* before being dyed for the lighter shades. The authors of these reports do not appear, however, to have observed that the process of bleaching was by a long way the most interesting feature. The writer has, therefore, failed to find any particulars regarding the bleaching of *tasar*, but it would appear highly likely that a fresh investigation conducted throughout India might even now suffice to elicit from the veterans of the tinctorial craft, some particulars of the decayed if not forgotten art—forgotten through the lesser demand for *tasar* goods. For this purpose, perhaps, no more hopeful province exists, for special enquiry, than the Central Provinces. Some few years ago the Government of India instituted an investigation on the subject of the dyes and dye processes of the country. As the result, **Sir E. C. Buck** brought out his report on the *Dyes & Tans of the North-West Provinces;* **Mr. L. Liotard** published his *Memorandum* on the *Dyes of Indian Growth &*

* **Liotard,** Memo. on *Dyes of Indian Growth, p. 121.*

SILK: Tasar.

Red, Yellow, and Blue Dyes

DYEING & BLEACHING.

Dyers sent to Bengal. 2064 Conf. with p. 185.

Production, and **Dr. H. McCann** wrote his *Dyes & Tans* of Bengal. Any person taking the trouble to turn over the pages of these publications will be satisfied that while to **Mr. Wardle** is largely due the honour of having first successfully used Indian permanent dyes on *tasar* silk in Europe, to the Native dyers of India is no less due the merit of having inherited and improved, from their ancestors, it may be blindfoldly, all and perhaps more than has been as yet discovered by their more fortunate brethren of the West, through scientific methods. If by way of illustrating the full force of this contention, the pages of **Mr. Liotard's** *Memorandum* (published in 1881) be turned over, and special attention be paid to the remarks regarding the processes of dyeing known and practised in the Central Provinces, it will be seen that in one of the great tasar-producing Provinces nearly as good results are, and have for centuries been, obtained in the dyeing of that silk as prevail with the more easily-dyed mulberry silk. Thus:—

Red Dyes. 2065

Red Dyes on Tasar Silk.

In Seoní (*Liotard, p. 63*) lac, alum, turmeric, and *lodh* (**Symplocos** bark) used to produce shades of red on *tasar*, though the colour is not durable. But if after being so dyed it be boiled in tamarind "until the dissolution of the red colour is stopped, and white or unclean water becomes visible," "the colour thus given to the silk will be durable." It is called the *lukhia* shade by dyers.

In Sambalpore the process is slightly different. The ingredients are lac, myrobalans ash, and *lodh* bark. From these materials a dye-stuff is prepared which will keep good for a month. This is manufactured by the *turkari* and sold to the *koshta*. When about to be used by the *koshta* the hanks of *tasar* are dipped in the dye-stuff for six hours, washed out, and boiled with tamarinds (the green fruit preferentially). Any desired shade can be produced.

Yellow. 2066

Yellow Dyes on Tasar.

Throughout the Central Provinces, and, indeed, with the Indian dyers generally, the flowers of **Butea frondosa** are viewed as of great value in *tasar* silk-dyeing. By some of the processes described by **Mr. Liotard** (*p. 90*) the colour is permanent, by others not. The mordant employed is alum, but in some cases Sesamum oil is also used. The *kámela* dye (the powder from the fruits of **Mallotus philippinensis**) is also one of the yellow dye-stuffs for *tasar*. The dye powder is for this purpose mixed with the ash of the wood of the myrobalan (**Terminalia Arjuna**), the two powders being thrown in water and allowed to stand by till the sediment subsides. The water is then decanted into another vessel in which is placed finely powdered *lodh* bark. The *tasar* silk is next soaked for six hours in the preparation; it is taken out and dried, put back again in the fluid, dried, again submerged until the liquid is used up or the fabric obtains the desired shade. It is then fixed and is of a brilliant yellow.

Blue. 2067

Blue Dyes on Tasar.

The process of dying *tasar* with indigo differs in no essential feature from that with cotton or mulberry silk. The dyers of the Central Provinces, however, use very largely the seeds of **Cassia Tora** along with the indigo.

Red & Yellow-Orange 2068

Mixtures of Red & Yellow-Orange on Tasar.

In the Central Provinces the dyers produce any shade of orange on *tasar* silk by adding to the flowers of **Butea frondosa** a certain amount of lime.

DYES.

The process pursued is otherwise identical with that followed in the production of bright yellow tints. If a deep orange colour be desired, the *kamela* powder is combined with the **Butea** flowers (*Liotard, p. 125*).

Black and Mixtures of other Colours.

Black. 2069

In the Central Provinces *tasar* is often dyed a yellowish-brown colour. For this purpose the barks of *rohan* (**Soymida febrifuga**) and *lodh* (**Symplocos racemosa**) are used. About two seers of each of these barks are boiled in 10 seers water, and when half the quantity of the water has evaporated it is ready for use. The colour produced is fixed with catechu and lime, but alum may be employed in place of these mordants. The silk is first saturated in the mordant and dried a little, then placed in the dye solution (*Liotard, p. 127*). The process of obtaining black has already been discussed in the remarks regarding Hyderabad. In Mysore gallnuts are used for that purpose, the silk being first steeped in rice water in which iron has been kept for eight or ten days.

Mixtures of Blue & Yellow-Greens.

Blue & Yellow-green. 2070

On this subject Mr. Liotard writes:—"Thus in some of the districts of the Central Provinces *bleached tasar silk*, after having been coloured with indigo, is dyed as follows: twenty *tolas* of turmeric and 5 *tolas* of alum are pounded and dissolved in 3 seers of water; then ½ seer weight of tasar, coloured at first with indigo, is dipped in the yellow dye three or four times being dried before each dipping; and lastly, the *tasar* is boiled with tamarind juice, and receives a permanent green colour." In the Nizam's Territory "a similar process obtains, with this difference that lemon juice is used with the turmeric instead of alum, and that alum takes the place of tamarind juice. Sometimes, instead of indigo, the flowers of the 'Marking-nut-tree,' called in the vernacular *kakarsingi*, or **Semicarpus Anacardium**, are used." The author does not recollect of having heard before of a blue dye being obtained from the flowers of **Semicarpus**, and its vernacular name is not *kakarsingi* in any part of India. That name is almost universally given to the galls on **Pistacia integerrima**, a Himálayan tree. These galls are exported all over India and used as a mordant in dyeing. They are especially mentioned as employed by the silk-dyers of Bombay in the production of the shades of green from indigo and some yellow tinctorial substance. Though not specially mentioned as employed with *tasar* silk, Mr. Liotard further informs us that the silk-dyers of the Panjáb, in the production of shades of green, use *akalbír* (**Datisca cannabina**), turmeric, alum, and mica in the production of the famed pistachio green of that Province. He makes no mention of indigo, but presumably the fabric is first dyed blue. In Sialkot the root of rhubarb is said to be used as the yellow dye for silk intended to receive a green shade.

But the reader is, perhaps, satisfied, from the examples given, that the Natives of India are by no means ignorant of methods of dyeing *tasar* silk. Nearly every tinctorial substance can be fixed, more or less permanently, on that fibre. Indeed, some dyes which are fleeting on cotton are fixed or almost so on *tasar* silk. It is a common statement by European writers, however, that *tasar* silk shows a stronger affinity for fleeting (*petit teint*) dyes than for those of a more permanent nature. Hence the ease with which it lends itself to aniline colours. One feature already commented on in all the Indian reports on methods of dyeing *tasar* silk, is the almost universal use of lime or of tamarind, separately or together.

At the Colonial and Indian Exhibition there were shown in the Silk Court many samples of fabrics of varying structure, dyed in every shade

SILK: Tasar.

Dyeing of Tasar Silk.

DYES.

of colour. Many of these were produced by **Mr. Wardle**, a gentleman who, of English dyers, has certainly taken a prominent place in advancing the knowledge of this subject. **Mr. Wardle** would himself, however, be the last person to allow the labours of the great French pioneers to be forgotten.* Some sixty years ago **M. Loiseleur-Deslongchamps** directed the attention of the French manufacturers to the subject of *tasar* and other wild silks.

Bleaching. *Conf. with pp. 25, 57, 153.* **2071**

In 1849 **M. Guinon**, a distinguished dyer of Lyons, applied himself to the study of the gum and colouring matter of *tasar* silk. He was successful, and published several processes for extracting the gum, and bleaching and dyeing the silk. It remained, however, for **M. Tessie Du Motay** to place the bleaching of *tasar* silk on a sound scientific basis and, therefore, to render it possible to dye it in any shade and with any tinctorial material applicable to silk generally. The enlightened action taken by **Mr. Wardle** and his associates at the Paris Exhibition and on all subsequent occasions prevented the entire European *tasar* interest from becoming confined to France. British manufacturers have fully held their own in the new trade, and the reaction of the interests aroused has been distinctly felt in India during the past eight or ten years. It is thus, through no ignorance of the popular revival that has given vitality to the so-called *tasar* trade, that the author of this review has been constrained to assume a tone of deprecation. He is fully aware of the value of the new trade, but is unable to see in it anything as yet that could be construed into justifying high expectations for India. In every aspect of the subject this country has unmistakably shown the strongest proofs of decline. The collection of cocoons is no longer profitable, a result due perhaps far more to the prosperity of India and consequent higher value of labour, than to depreciation of the value of these wild products. *Tasar* silk-weaving has undoubtedly greatly contracted within the past half century, and everything points to a further decline rather than a revival of the village hand loom with its crude though durable silk productions. With such results, is it to be wondered at that the dyer's art has also felt the effects of changes so radical in their influences as to have converted the expert weavers in silk, wool, and cotton into common agriculturalists? It is the struggle between steam-power and hand labour. If the price paid for cocoons will not repay domestication and production within a limited area, the labour and expense of collection over extensive forests will only be remunerative at certain specially favourable points along the lines of communication and export. The fact that a totally distinct trade in a domesticated insect pays in China (the so-called Chinese *tasar*) is no argument that Indian *tasar* should be equally remunerative. It would not be abandoned in favour of agricultural pursuits if the latter were not more profitable and more congenial to the people. Chinese *tasar* is obtained from a domesticated or semi-domesticated worm, reared on the village trees, hence each artizan or cultivator can add to his earnings by the production of a few cocoons without having to look to these as his sole source of livelihood. So far as has yet transpired, such a state of affairs is not attainable in India, with the *tasar* worm. Agriculture is opposed to the existence of the trees on which it feeds. The extension of agricultural influence means, therefore, the expulsion of *tasar* production further and further into the depths of the forests. The col-

Conf. with pp. 59, 75, 132, 149, 151.

* Since passing to Press the above remarks on the dyeing of *tasar*, the writer has had the pleasure to receive a copy of **Mr. T. Wardle's** lecture, read before the Society of Arts at its meeting of the 14th May 1891, in which **Mr. Wardle** offers certain remarks on the progress made in dyeing *tasar* silk. He alludes to the large and expensive collection of Indian dye stuffs that was sent to him for the purpose of investigation. Commenting on his report on thesehe says: "In parenthesis, I

DYES.

lector of *tasar* cocoons is himself actuated by religious restrictions that alone are likely to prove fatal to progression for many years to come, perhaps for centuries. Thus, for example, of Bhágalpur, (one of the great *tasar* silk-producing districts), Sir W. W. Hunter says: "Women, who would seem to be the best fitted for the work of rearing and superintending *tasar* worms, are entirely excluded, and even the wives of those engaged are not permitted to approach the workers. The low castes are excluded, as their appetites are defiled by gross impurity of animal food. The workers eat sparingly, once a day, rice cleaned without boiling (*alwá dhán*), and seasoned only with vegetables. They are not permitted to employ the washerman or the barber." Is it to be wondered at, therefore, that an occupation so repulsive should not be chosen if other modes of living can be and are accessible? (*Conf. with p. 142.*)

TRADE IN TASAR & OTHER WILD SILKS.

TRADE IN WILD SILKS. 2072

Conf. with pp. 70, 72, 73, 98-99.

It has already been stated that the earliest direct reference to the *tasar* silk of Bengal is that given by Rumphius. He carefully studied the insect in April 1691, and stated that the dirty yellow garments made of it were technically known as *gingangs* (*ginghams*). At the same time Rumphius alludes to the Tonkin trade in the same form of silk, so that it seems safe to assume that *tasar* silk was contemporaneously known in India and Tonkin. The first official record of *tasar* is stated by Roxburgh to be that of Mr. M. Atkinson, who in 1796 wrote: "I send you herewith for Dr. Roxburgh a specimen of *bughy tusseh* silk . . . . . There are none of the Palma Christi species of *tusseh** to be had here. I have heard that there is another variation of the *tusseh* silk-worm in the hills near Banglipur." In *Milburn's Oriental Commerce* (*Ed. 1813*) under the article "Calcutta," occurs a list of the Government Customs dues on imports and exports in which *tasar* silk appears. This is called "*Tusha*," and is classed into "*Tusha*" and "*Chassam*." These articles were to pay 7½ per cent. duty on the fixed valuation of 5 annas per seer for the former and 3 annas per seer for the latter. In that work also occurs much useful information regarding the rise and progress of the Indian and English silk trade. Regarding *tasar* it is remarked that it is "found in such abundance over many parts of Bengal, and the adjoining provinces, as to have afforded to the Natives, from time immemorial, a considerable supply of a most durable, coarse, dark-coloured silk,

feel bound to mention that a work of this laborious and valuable nature has, as far as my experience goes, never had such scant treatment at the hands of the Government of any country. My book, as far as any usefulness it possesses either to India or to Europe, is a dead-letter, and it had been more economical never to have published it. It, so far, has been love's labour lost.

"The method of publishing it was of the most parsimonious kind, entirely counteracting its usefulness. In vain I urged that it ought to be published in England, and to have the methods by which I succeeded in obtaining the beautiful colours you see here inserted in the book."

This was clearly an unfortunate mistake, for to the people of India the expenditure in making the collection (sent to Mr. Wardle) was spent in vain when all they were told, for example, was that *tasar*, dyed with "fir bark" obtained from Peshawar, took a yellowish-drab colour by process "GX." But in the preface to the Report Mr. Wardle informs the public that it was "not within the limits of ordinary prudence to publish the means by which I obtained the best results, because to have done so would have laid open the business methods to the public which are private property and in which the successful nature of such a business depends." The writer alludes to these facts by way of apology for his—not only in connection with *tasar*-silk, but throughout this work,—having been debarred from freely quoting Mr. Wardle's methods of using the Indian dye-stuffs.

* Eri silk, see below at p. 162.

SILKS: Wild.

Trade in Tasar

TRADE.

commonly called *tusseh* silk, which is woven into a kind of cloth called *tusseh duties*, much worn by Brahmins and other sorts of Hindus. This substance would, no doubt, be highly useful to the inhabitants of many parts of America, and the south of Europe, where a cheap, light, cool, durable dress, such as this silk makes, is much wanted."

Decline and its Causation. **2073** *Conf. with pp. 59, 115-18, 121, 145, 149, etc.*

The study of most articles of Indian trade reveals the fact that there are two markets of independent and often co-equal importance—the foreign and the internal. To an Empire of such magnitude as India, the latter is frequently the more important of the two, though it is a common error of popular writers to judge the Indian trade by its foreign imports and exports. In no instance, perhaps, would this one-sided aspect prove more disastrous than in that of *tasar* silk. While within recent years there has been created a foreign market for Indian *tasar* raw silk and cocoons—a trade which, moreover, has begun to show signs of again declining—there has coincidently occurred an alarming falling off in the internal consumption of these cocoons and a corresponding decrease in local manufactures. So much so is this the case that from every corner of the Empire comes the statement that at present market rates it will not pay to collect the cocoons, and that the manufacturer's price is now no longer obtainable for his goods, so that reelers and weavers alike have had to abandon their ancestral trades and take to other more profitable forms of earning a living. Neglect of such a state of affairs, and the bald presentation of the paltry returns of a new foreign export trade in cocoons, would naturally convey a false impression of the Indian *tasar* trade. But this is what has alone been represented to the public by most of the recent writers. That the internal *tasar* trade of India is steadily declining will have been abundantly shown by the provincial chapters of this article, more especially by the series of notes extracted from the annual reports of the internal trade of Bengal. Were it necessary to produce further proof of this statement, as much again could be written as has already been given. One or two examples need only be added. The first, from an unofficial source and bearing a date prior to the rise of the foreign export trade of which so much has been said. In the *Saturday Review* for October 14th, 1876, the following occurred: "The work of the *tasar* silk-weavers has so fallen off that the Calcutta merchants no longer do business with them." But an even stronger example may be given in the extinction of the Azimghur manufactures. This has already been briefly mentioned in connection with the remarks on the *tasar* silk of the North-West Provinces. **Royle's** comments on this subject may be here given since they exhibit at the same time the then recorded exports of *tasar* piece goods:—

Decline of Azimghur Tasar Trade. **2074** *Conf. with pp. 121-22.*

"The Tusseh silk is still better known, having also afforded the Natives in Bengal, &c., a coarse durable silk, which is much esteemed in India, both for ladies' and children's dresses; for the latter especially, on account of its cheapness and durability. This silk will probably become an extensive article of commerce, as some of it having been sent on speculation to Paris in its unbleached state was there employed as a covering for parasols, and was found to answer so well, that an instantaneous demand for it sprung up. The price advanced, and the quantity imported into Europe, has greatly increased; in 1835, only 158 pieces, in 1836, 850, in 1837, 2,647, and in 1838 no less than 4,249 pieces were imported. Other uses will probably be found for it when it is better known, and from the extensive tracts over which the Tusseh silkworm is distributed, we know that this commerce is susceptible of great extension. As an instance of the quantities in which the Tusseh silk is produced in India, we may adduce the evidence of **Mr. R. Montgomery**, of the Bengal Civil Service, for one district only. In an abstract statement in 1837, of the results of

TRADE.

the Survey and Settlement of the district of Azimghur, giving accurate returns of the total area cultivated, culturable, and of waste land, with the revenue of the district, **Mr. Montgomery** has also given an estimate of the quantity of cloth, silk, and Tusseh manufactured, which **Mr. Tucker** and himself obtained from looms at work. In this statement 318,772 pieces of Tusseh silk are given as the quantity produced annually."

Royle is probably in error in regarding all *tasar* cloth as necessarily made of pure *tasar* silk, but **Montgomery** tells us there were 3,121 looms at work on silk and *tasar* goods, and that the total piece goods manufactured (cotton, silk, and tasar) was 999,436 pieces valued at R22,72,308. What is the state of affairs at Azimghur now, fifty years subsequent to the date of the first survey of that district? The Government reports, of the North-West Provinces as a whole (not of one district of them), affirm that "No weaving of cloth (that is to say, silk or *tasar,—Ed.*) worth the name is carried on, there being only one family in Ahraura engaged in the manufacture." On the other hand, we are assured by writers who apparently think only of the prosperity of the *tasar* industries of Europe that the Indian *tasar* silk trade is now "established" and "well rooted." **Mr. Wardle**, for example (*Jubilee Exhibition Catalogue, pp. 25 and 26*), says that at "the Paris Exhibition of 1878 **Sir Philip Cunliffe-Owen** determined, with my assistance, to give this silk an opportunity of asserting itself." "Not a little of the industrial growth of this useful though wild silk is due to his encouragement." "I am more than pleased to state that India now has an enormous and yearly increasing demand." Perhaps the reader may be disposed to regard **Mr. Wardle's** opinion as that of an enthusiast. His "enormous and yearly increasing demand" would appear scarcely borne out by the actual facts of the trade. It is true of China, but not of India. The Chinese *tasar* is the produce of **Antheræa pernyi** mainly, not of **A. paphia**, so that although vulgarly called *tasar*, it is a perfectly distinct fibre from the true *tasar*. *Tasar* is, in fact, exclusively derived from India, and the foreign trade in the fibre is very insignificant indeed. Not only so but the story of Azimghur could be told of any and every district in India, for the Indian trade in *tasar* silk has seriously declined within the past 20 or 30 years.

Expectations of a Future for Tasar.
2075
Conf with pp. 106, 148, 156, 161, 193.

So far, therefore, for the evidence of the present position of the Indian internal trade in *tasar*. The facts regarding the foreign trade may be briefly reviewed. Through the kindness of **Mr. J. E. O'Conor** the writer has fortunately been furnished with actual figures, so that this part of the trade is not liable to the charge of being local opinion. Deeming it likely that the necessity might shortly arise for the publication of returns to show the share taken in each class of silk in the grand totals which have alone hitherto been published, **Mr. O'Conor** directed the Customs' authorities to distinguish wild from domesticated silks in their records of exports. On pages 198, 199 200, will be found tabular statements of the total export trade, in which the shares taken by each class of silk in the grand total are exhibited. The main facts of the wild silk trade may, however, be here republished by itself:—

SILKS: Wild.

TRADE.

Trade in Tasar

	Exports of Wild Silks to Foreign Countries.									
	1886-87.		1887-88.		1888-89.		1889-90.		1890-91 (eleven months).	
	lb	R	lb	R	lb	R	lb	R	lb	R
Raw Silk. (Tasar, Cricula, Muga, Eri, etc.)	38,875	*1,95,704*	91,699	*4,50,343*	51,595	*2,38,941*	91,124	*4,12,803*	85,273	*3,63,482*
Chasam. (Above kinds) . . .	349,271	*3,55,875*	252,342	*2,65,560*	233,090	*3,12,027*	341,558	*3,49,045*	150,459	*1,50,611*
Cocoons. (Above kinds) . . .	54,994	*64,676*	135,207	*1,65,749*	334,038	*5,09,388*	223,620	*3,55,365*	110,844	*1,60,106*
Total .	443,140	*6,16,255*	479,248	*8,81,652*	618,723	*10,60,356*	656,302	*11,17,213*	346,576	*6,74,199*
Percentage of the above to the total exports of all forms of silk	27·98	12·72	29·49	18·22	29·16	20·44	31·41	17·46	22·03	10·54

TRADE.

Comment on these figures seems almost unnecessary. The internal trade in *tasar* goods has for many years been steadily declining. But even if that were not so, the export trade in raw *tasar*, regarding which so much has been written, is not very important. Taking the averages of the past five years' returns, the *tasar* trade is about one-fourth the value of the saltpetre or of the lac exported annually, and only half the value of ginger, or of the bones. And, though even thus in total value relatively quite unimportant, it has as yet shown no tendency to, or even capability of, expansion. So early as 1810 the Honourable the East India Company began to record their exports to London of Tasar silk. In the year 1814, or, say, 76 years ago, these amounted to 7,872℔ of *reeled tasar*, so that the increase has been about 40 per cent. on the quantity of *all* reeled wild silks exported in 1889-90. It should be carefully observed that the chief item in the above table is *chasam* or waste materials of wild silks, not necessarily all *tasar*. The expansion of the trade has thus been remarkably slow, considerably below that of any other article of Indian commerce. Its expansion within the last twenty years has, of course, been more rapid than prior to that date, but infinitely less than can be shown with almost any other article, the present value of which, like that of the wild silks, is under £70,000. But instead of showing a continued though slow increase during the past five years, it has manifested a distinct decline. Success in the European *tasar* manufactures has, in fact, been attended by an inverse ratio of failure in the Indian. The weavers of this silk, who, once upon a time, were rich and influential members of the Indian city communities, have had to close their looms and seek other and more lucrative employments—a truly disappointing picture to contrast with the triumphant statements of the prosperity which we are told has been effected through the enlightened action of the helping hand of European enterprise.

Expectations of a Future for Tasar.
2076
Conf. with pp. 106, 148, 159, etc.

It may, in fact, be reiterated that "the alarming expansion" of the European demands for *tasar*, which we have been recently told is spreading dismay among the European rearers of mulberry silk, if such exists, is due entirely to the increased exports of Chinese wild silks (vulgarly called *tasar*) which are mainly derived from an insect as different from Indian *tasar* as it is from the mulberry silk-worm.

Since the above was passed to Press, the author has had the pleasure to receive **Mr. T. Wardle's** lecture on *Tasar* Silk (delivered before the Society of Arts on May 14th, 1891). The remark seems justified that it is somewhat unfortunate **Mr. Wardle** did not make a greater effort to refer the statistical information he publishes to the two all-important sections—Indian and Chinese *tasar*. He devotes, it is true, separate sections of his lecture to what were meant to represent these two forms of silk, but, unfortunately, he leaves the reader in a state of considerable indecision as to the share in the trade taken by each of the countries named. In the one section, for example, he quotes a statement of the "shipments of raw silk, waste silk, and pierced cocoons, from the Provinces of Manchuria, Chili, and Shantung for the ten years ending 31st December 1888," which shows an expansion from 169,496℔ to 2,874,766℔. That "enormous increase, it is said, in this business is mainly due to the application of tussur spun silk to the seal and plush trades." In another part of his lecture **Mr. Wardle** discusses "the quantities used in Lyons" of *tasar* silk. "The Conditioning House, he tells us, registered for 1890 a total quantity of 306,152 kilogrammes or 673,534℔ of tussur silk, thus divided: Organzine 13,347 kilos, or 29,363℔; Tram 69,028 kilos. or 151,864℔; Raw tussur 223,776 kilos or 492,307℔. This is an average of 92 bales of 140℔ each per week, and it is confirmed by the Chamber of Commerce's published report of last year. Compare this with the whole year of 1879, when only 53 bales or 7,420℔ were imported, and they certainly were not all used. For the week ending April 11th last, Lyons capped its record. It received and conditioned 136 bales of tussur silk, or 39,040℔, against 178 bales or 39,160℔ of silk of French growth, and 38 bales or 8,360℔ of Italian silk for the same week." Now were these quantities obtained from India? The table given at page 160 of the total exports of all forms of wild silks from India will be found to show con-

ERI SILK. 2077

THE ERI SILK-WORM.

(For Tasar see pp. 96–161; for Muga pp. 174–183.)

Vern.—*Eri*, Ass.; *Arindi*, Beng. (**Attacus ricini**).

The name *eri*, given to this form of silk, is derived from the fact of its feeding chiefly on the castor-oil plant, which in Assam bears that name, and in certain other parts of India it is also known by words traceable to the same root.

Habitat.—It may be said that commercially this silk is obtained in Assam, but it is also found in various districts of Bengal, such as Rungpur, Dinájpur, Purnea, Bogra, Jalpigori, Darjiling, Gya, Shahabad, Chittagong, and Puri. **Dr. Bennett** and other writers have spoken of it as found in Dinapore, but **Hutton** regards this as a mistake. It may, however, be said to extend sparsely from the region above indicated throughout the lower Himálaya and mountains of India wherever the castor-oil plant is cultivated. Nepál and Kumáon are specially mentioned as places where *eri* silk is procurable, though it is probable **A. cynthia** is the insect seen in these localities. (*Conf.* with remarks above under that species, pp. 85-86, also 87.)

CULTIVATION in Assam. 2078

Extent of Cultivation in Assam.—"The *eri* worm is cultivated to a greater or less extent in every district of the province. Being regarded as of doubtful purity, it is left principally to Rábhas, Meches, Kacháris, Mikirs, Kukis and other non-Hindu tribes. In the submontane country inhabited by the Kacháris and their cognates, along the north of the districts of Goálpára, Kámrúp, Darrang, and Lakhimpur, almost every house has its patch of castor-oil plant, on which *eri* worms are fed. In some parts of this region the Marwari traders make advances to the cultivators in October, when the revenue is falling due, and take repayment afterwards in thread or cloth, and both these products are commonly exposed for sale in the petty markets, in the same manner as other articles of village merchandise. A good deal of *eri* is also produced in the district of Sibságar, and in Upper Assam generally the *rayat* may be seen swathed in a warm sheet of coarse *eri* cloth in the winter mornings and evenings. Throughout the whole range of the southern hills, from the Mikir country to the Gáro, *eri* thread is in great request for weaving those striped cloths in which the mountaineers delight An estimate of 183 cwt. (250 maunds) has been furnished for the outturn of the North Cachar section of these hills, and a similar amount for the Khási Hills district. The Mikirs, Kukis, and Gáros cultivate the worm for themselves, but the handsome and durable cloths worn by the Khasis and Santengs are woven of thread procured from Mikir

siderably lower figures than given by **Mr. Wardle** as imported by Lyons alone. Indeed, in a further sentence, **Mr. Wardle** says: "The Lyons statistics show a rapidly increasing consumption of 150,000lb of Indian tussur for last year for that city." If this new statement be meant as in round figures the totals of the organzine and tram, then there would appear to be no doubt that **Mr. Wardle** accepts as correct the report that Lyons received in 1890, 673,534lb of *tasar* from India, whereas the total Indian exports of all forms of wild silks, for the whole world, were considerably less than that amount. But there is still another obscure point in the figures quoted by **Mr. Wardle**, namely, the share of the Chinese exports of 2,874,766lb actually taken of the so-called "Chinese tasar." If this includes, as seems likely, a large amount of mulberry waste and pierced cocoons, may not a proportion of the so-called imports by Lyons of Indian *tasar* embrace ordinary waste or *chasam?* The enormous expansion of the waste or spun silk trade of India suggests at least the caution against forming too hard-and-fast opinions as to its destination and utilisation. The term *tasar* or *tussur* in European commerce means practically any silk that is intended to be spun instead of reeled. Is this the meaning to be put on **Mr. Wardle's** statistical information regarding the *Tasar* Trade?

and Kuki breeders inhabiting the lower hills on the northern and southern faces of the range. All these people eat the chrysalis with avidity, considering it especially delicious in the form of curry. *Eri* is but little cultivated in the plains of Sylhet and Cachar" (*Stack*).

CULTIVATION in Assam.

HISTORY. 2079

Historic Records.—The earliest mention of this form of silk in European commerce occurred in the year 1679 when the Fort St. George Agent wrote that large quantities were produced in Gooraghat. In December of that year he ordered 600 pieces of *arundí* and four bales of *arundí* yarn to be provided by the Maldah factory and to be sent to Europe. Interest in the subject, however, died out till **Sir W. Jones** in 1791 and **Roxburgh**, 1804, wrote of it. **Dr. Buchanan-Hamilton** suggested that it might be mixed with wool, and accordingly recommended that a few hundred weights might be sent home to test its usefulness for that purpose. **Mr. Benthall**, writing of 1837, says, the people rarely sell the silk or cocoons, as it is all required for local use. It is referred to in the Board of Trade Proceedings of 1819, and **Mr. Glass** early in the century sent to Europe a small consignment of *eri* silk. **Mr. Hugon** of Assam devoted, perhaps, more attention to the subject of this silk than did any of the other early writers. He estimates the production in Nowgong-Assam to have been during his time 1,000 maunds. **Mr. G. Eveleigh** (in a paper published in 1843 in the *Jour. Agri.-Hort. Soc., Ind.*) states that he was enabled to wind the *eri* silk by feeding the worms uniformly on moist leaves, and giving them occasionally mulberry leaves. Lukewarm water was used to soften the cocoons. Shortly before the date of **Mr. Eveleigh's** paper **Captain Jenkins** in the Journals of the same Society offered a reward of R600 for an effectual and cheap solvent for the cocoons. The Society also promised a gold medal for the same discovery. Neither prizes appear to have been contested for successfully. **Mr. Brownlow** alludes to *eri* silk in Cachar, and says the cocoons are softened before being carded in a solution of cow-dung and water. The *eri* worm seems to have been introduced into Malta, France, Italy, and to have been reeled in Malta. A French writer in 1860 stated that failure to reel the cocoon was due to its being open. It thus fills with water instead of floating like the mulberry cocoon. The above brief historic notices have been taken mainly from **Geoghegan's** *Silk in India* (pp. 25 & 155), a work which should be consulted for further particulars. **Mr. Geoghegan's** report appeared in 1872, and it is the publication from which the bulk of the facts since published and republished time after time, have been drawn.

Spun Silk. 2080 *Conf. with p. 145.*

But to continue the historic record of publications on *eri* silk. **Ainslie** (*Materia Medica, I., 254*) briefly alludes to this insect in his account of **Ricinus communis**, so that it would appear that very nearly as much was known at the beginning of the present century as has since been brought to light. The fact that it was spun and not reeled was at all events made freely known, so that the claim advanced by certain manufacturers in Europe of having discovered this new mode of utilising wild silk is not well founded. Thus, for example, in **Milburn's** *Oriental Commerce* (Ed. 1813) it is said of *eri* silk that "Feeding these caterpillars with the leaves of **Ricinus** or *palma Christi* plant, will, therefore, make it doubly valuable where they know how *to spin and manufacture the silk*. Their cocoons are remarkably soft, and white or yellowish, and the filaments so exceedingly delicate, as to render it impracticable to wind off the silk; it is, therefore, spun like cotton. The yarn, thus manufactured, is woven into a coarse kind of white cloth, of a seemingly loose texture, but of incredible durability. Its uses are for clothing for both men and women; and it will wear constantly ten, fifteen, or twenty years. The merchants also use it for packing cloths, silks, or shawls. It must, however, be always washed in cold water;

Food of the Eri Silkworm.

HISTORY.

if put into boiling water, it makes it tear like old rotten cloth." **Hutton** seems to have arrived at the opinion that although it had been found possible to reel both the *eri* and the atlas cocoons, it was probable the old system of carding and spinning the fibre would be found the preferential mode of utilising these silks (*Jour. Agri.-Hort. Soc., Ind., XIII., 56*). One of the most valuable papers on the silk of Assam (and the one from which many subsequent writers appear to have drawn largely) is an official letter to the Government of India by **Mr. S. O. B. Ridsdale**, dated 1st July 1879. That communication submits a Resolution of the Assam Government to which are appended two notes—one answering certain question regarding the wild and domesticated insects, and the other "on a few of the more important facts known on the subject of practical sericulture in the province." Very little has since been made known regarding the two chief wild (or so-called wild) silks of Assam, *viz.*, *eri* and *muga,* than will be found in **Geoghegan's** *Silk in India* and the official publication just mentioned. While repeating, though amplifying and confirming the main facts regarding the life-history of each of these insects, the more recent official correspondence gives statistical information and other details which render their publication in this work preferable to the accounts furnished by the older writers. It, therefore, seems sufficient to have referred the reader to the leading publications on this subject without specialising the share which each writer took in developing the existing knowledge.

About the same time that **Mr. Ridsdale** submitted to the Government of India his report, **Mr. T. Wardle** read a paper (2nd May 1879) before the Society of Arts, London, on the *Wild Silks of India,* which ultimately matured into his little book on that subject. In the discussion which followed the reading of the paper **Mr. Francis Cobb** said of *eri* that **Attacus ricini** "was a worm which was a good colonist; it might be taken to almost any of the colonies and it would thrive." "**Mr. Wardle** had shown how valuable that particular silk was, and it might be grown to a very large extent in the colonies." Whether the cultivation of *eri* is likely to be extended to the British colonies is a point on which no evidence at present exists, but it may fitly be said that from a chronological point of view the record of published information regarding this silk closes with the appearance of **Mr. Wardle's** *Wild Silks of India.* The more recent Government publications amplify the details here and there, but are more valuable, it might almost be said, as exhibiting the failures which have attended the efforts hitherto made to extend *eri* cultivation, than as affording any new facts.

Food. 2081

Food of the Worm.

The following may be given as the list of plants upon which, writers on this subject say, the eri silk-worm feeds:—

1. **Ailanthus excelsa,** *Roxb.*
2. **A. glandulosa,** *Desf.*
3. **Coriaria nepalensis,** *Wall.*
4. **Gmelina arborea,** *Roxb.*
5. **Heteropanax fragrans,** *Seem.*
6. **Jatropha Curcas,** *Linn.*
7. **Ricinus communis,** *Linn.*
8. **Zanthoxylum alatum,** *Roxb.*
9. **Zizyphus Jujuba,** *Lamk.*

Of these plants by far the most important are **Ricinus communis,** the Castor-oil, and **Heteropanax fragrans,** the *keseru,* of Assam writers. There are doubtless many others on which the wild insect feeds in the jungles, but these are of less importance. In connection with this subject **Mr. Lepper** says that the worm may be changed from one to the other as circumstances require, an opinion in which the Native rearers do not concur. It is, however, important to note that **Mr. Lepper** was disposed to think that the *keseru* would prove the more profitable plant should *eri*-culture assume commercial importance. The castor-oil plants, he says,

FOOD.

would be a great nuisance and an unhealthy jungle to cultivate. The *keseru* can be had in great abundance; the seedlings are very hardy and can be easily transplanted from the jungles; the plants in their second year can be plucked for leaf, and in their third year do not suffer even from hard plucking; and the plants constantly throw out fresh leaf, so that they are very favourable to rearing of worms in different ages. These are certainly important considerations when it is added that the plant is a perennial by nature, instead of, as in the case of the castor-oil, a perennial chiefly through cultivation. Though one of the forms of castor-oil lasts for a few years, and many writers speak of the wild stock as normally a perennial, still it is more an annual or biennial, and is consequently less able to withstand the extremes of climate—heat, cold, humidity, and drought—than a perennial plant would be if indigenous to such environment. Before these considerations are urged too strongly, however, it seems necessary to throw out the caution that it has as yet been by no means demonstrated that the insects thrive as well on *keseru*, and it would seem fairly established that in the wild state the insect shows a decided preference for the castor-oil. The native rearers also would appear to regard that as the most satisfactory plant on which to feed it.

The subsequent experiments made after **Mr. Lepper** (the Indian Agent in Assam of **Mr. Lister** of Bradford) had left the country and abandoned the enterprise may be said to have been characterised by failure on every point. Seed of Patna castor-oil was imported and cultivated, but the high rainfall was found to destroy the plants, the Assam and Cachar so-called indigenous stock having been found less affected by the extreme humidity that prevails for months. **Mr. Lister** does not appear, however, to have abandoned the hope of seeing Assam come forward as a country from which a large supply of silk could be depended upon by the British manufacturers. Though he discontinued direct efforts at cultivation, he offered to purchase all the *eri* cocoons that could be procured. The encouragement thus given seems to have stimulated the Government to make an effort to ascertain whether Assam planters might not be induced to enter on the undertaking. The small sum of R1,500 was set apart for this purpose and on certain terms it was paid over to **Mr. F. F. Mackenzie** of Cachar, a gentleman who had given much careful study to the habits of the insect and with a small stock had in few years greatly improved the breed. Several of his observations are of the greatest value. He noted, for example, that in warm, moist climates the size of the worms and of the cocoons were increased, but that in colder climates a finer and strong fibre was produced. His experiments were conducted with the castor-oil plant chiefly, and in the case of the larger and final trial, disaster and complete loss was caused through the appearance of a pest that destroyed the stock of plants and left the caterpillars nothing to eat. Millions of caterpillars (since identified as those of **Achæa melicerte**, *Drury*, emerged, he says, from the bamboo jungle around, and attacked the castor-oil plants eating every leaf, bud, soft stalk, and even parts of the bark. They arrived during the night, and, when discovered next day, had overspread some 3 or 4 acres. All available hands were immediately put on to pick them off the plants and kill them, and by evening, though many thousands had been destroyed, the numbers left appeared scarcely diminished. Next morning it was found that they (the caterpillars) had increased in number during the night, and, in spite of the most strenuous exertions, by the morning of the third day there was literally *not a leaf left* throughout the whole area. Commenting on this calamity **Mr. Darrah** (the Director of Land Records and Agriculture) suggests that in future experiments the sites selected for experiment should be chosen at a greater

A Pest fatal to Eri Silk.
2082

SILK: Eri.

Rearing of the Eri

FOOD.

distance from the jungles, and should be surrounded by a belt of ploughed up land kept free of weeds, so that the approach of the pest could be seen and intercepted. But a careful perusal of all the experiments hitherto made in Assam do not, in the writer's opinion, justify the adverse conclusion arrived at by **Mr. Mackenzie** when he says, "The matter is now set at rest and the object of the experiment gained, for I hold it satisfactorily proved that it is much too risky an investment for any capitalist to take up. I myself shall certainly never again attempt it." The same pest often attacks the castor-oil crops in other parts of India, but a loss occasionally sustained has not interfered with the annual cultivation of the plant, and, moreover, there has transpired nothing that would justify the conclusion that the danger is not remedial, even supposing castor-oil be demonstrated as best suited for the *eri* silkworm. A far greater danger lies in the tendency to disease, but this danger is by no means greater in the case of *eri* than any other silk-worm. The amount paid for *eri* silk has been more than doubled within the past few years, and it would seem that the subject should still present many attractions that are well worthy of more energetic investigation than has as yet been bestowed on them. Few industries of commercial importance have attained to that position without having had to contend against disadvantages and dangers far more serious than those hitherto brought to light in the experiments made to remove *eri* cultivation from the peasant's to the planter's hands.

For information regarding the methods pursued in India in the cultivation of the castor-oil plant, the reader is referred to the article **Ricinus communis** (*Vol. VI., Pt. I., pp. 506-557*), and for the vernacular names and other facts about the other plants mentioned above, as eaten by the *eri* silk-worm, to their alphabetical positions in this work.

Cost. 2083

Estimated Cost of Production.

Mr. Mackenzie furnishes the following estimate:—

	R	a.	p.
"(1) One acre of land cleared and planted with castor-oil plants	20	0	0
(2) Cultivation during one year	9	0	0
(3) 20,000 worms, 10 cycles, feeding and manipulation during one year	90	0	0
(4) Buildings, trays, original supply of eggs	70	0	0
(5) Superintendence, &c.	15	0	0
(6) Boxes, packing freight to London	12	0	0
TOTAL .	216	0	0

200,000 cocoons (20,000 × 10 cycles) = 200℔, @ R140 per maund = R350.

Therefore the profit from one acre of castor-oil plants in one year would be R134. This expenditure is rather over than under-estimated. The buildings would last for three years, although the whole of their cost is debited against the first year. The castor-oil plants put out in the first year would also suffice for two years, while no account has been taken of the profit from sale of the castor-oil seed produced from these plants. I have had pierced cocoons valued in Calcutta and also at the late Manchester Exhibition, the average being 2*s.* 6*d.* per ℔."

Rearing. 2084

Rearing of the Worm.

In 1884, **Mr. Stack**, then Director of Land Records and Agriculture, Assam, published a carefully prepared note on the subject of the silk in

REARING.

that Province. A recent pamphlet issued by **Mr Darrah**, the present Director, republishes the chief paragraphs on *eri*, remarking that with the exception of the question of price, the facts given by **Mr. Stack** are applicable to the present day. It may serve the purpose of the present article to reproduce in this place the main ideas conveyed by **Mr. Stack** regarding this silk-worm :—

"The *eri* worm is a multivoltine, and is reared entirely in-doors. The castor-oil plant grows abundantly in the *rayat's* garden, springing up from dropped seed in every little patch of unoccupied land around his house. The tending of the worms devolves principally upon the women of the family, and goes on all the year round. As many as eight broods can be obtained in twelve months, but the number actually reared never exceeds five or six, and depends a good deal upon the quantity of food which chance has provided for the worms, since no care is taken to ensure a supply by planting out trees. It is the autumn, winter, and spring broods, spinning their cocoons in November, February, and May respectively, which are chiefly destined for use, and of these the spring cocoons are the most numerous, and yield the most silk. The broods of the rainy months—June to September—are reared for the purpose of perpetuating the stock. But both breeding and spinning, to a greater or less extent, go on all the year round.

Treatment. 2085

"TREATMENT OF COCOONS FOR BREEDING.—Cocoons reserved for breeding are placed in a round basket woven of bamboo, with a narrow mouth, and are hung up in the house out of the way of rats and insects. After about 15 days in the hot season, and 20 to 30 days in the colder months, the moths emerge, and are allowed to move about in the basket for four-and-twenty hours. The females, distinguished by their larger body and broader and flatter abdomen, are then tied to pieces of reed or *ulu* grass by a ligature passing under the shoulder-joint of a pair of wings on one side of the body only, leaving the pair of wings on the other side free. Ten moths will thus be tied to a piece of reed 2 feet long. The males, though left at liberty, do not attempt to fly away, but remain with the females to which they have attached themselves, until the latter have laid their eggs, when the males depart. If some of the females, as may easily happen for want of any criterion of sex in the cocoon, are unprovided with males, they are exposed on the eaves of the house in the evening, and are visited by any stray males that may be in the vicinity. The female lays about 200 eggs in three days, and the life of the moth lasts a day or two longer.

Hatching. 2086

"HATCHING AND NURTURE.—The eggs are picked off the straws, wrapped in a piece of cloth, and hung up in the house. The period of hatching varies with the season : in the month of May, with an average temperature of 83° Fahr., it has been found not to exceed a week, but in the winter it is about fifteen days, and in the months of medium temperature nine or ten days is the usual term. When the eggs begin to hatch, the cloth is opened, and tender leaves of the castor-plant, previously crushed between the fingers to render them still softer, are supplied to the young worms for food, and subsequently they are transferred to a bamboo tray suspended in a place of safety. As the worms grow stronger, older leaves are given to them. Their supply of food is occasionally intercepted by swarms of caterpillars appearing on the castor-oil plant about the month of June. These must be carefully removed from the leaves that are given to the silkworms and the leaves themselves washed in water. It is at seasons like this that the leaves of a variety of trees are used as substitutes for the favourite food of the worm.

Diseases. 2087

"DISEASES AND ENEMIES.—Large numbers of the worms are lost by

SILK: Eri.

Rearing of the Eri

REARING. Diseases.

disease, of which neither the nature nor the remedy is known, but which probably has its origin in uncleanliness.* No care is taken to remove the excreta, nor are the dead worms regularly rejected. The native account of the disease is simply that the worm ceases to eat and withers away. Some good effects are said occasionally to follow from sprinkling water in which *tulsi* leaves have been steeped over the worms among which this disease has made its appearance. The ichneumon fly is a deadly enemy. Its bite, which leaves a black mark, usually proves fatal to the worm at the next moulting; and if the wound has been inflicted after the last moulting, the worm spins a smaller cocoon, and dies before it is completed, leaving the eggs of the fly to hatch inside the cocoon. Rats are still more destructive, sometimes sweeping off an entire brood in a single night. The cultivator is careful to abstain from praising his crop of worms, lest any of these calamities should overtake them.

The worm. 2088

"LIFE OF THE WORM.—The number of moultings is four, known locally as *háludia, duirkáta, tinirkáta,* and *chárikáta;* the first term denotes the yellow colour of the worm, the three others merely mark the order of the moultings. **Mr. Thomas Hugon**, who held the office of Sub-Assistant (corresponding to the present office of Assistant Commissioner) in the Nowgong district, contributed a very carefully-written paper upon the silkworms of Assam to the Proceedings of the Asiatic Society of Bengal for 1837, whence the following description of the *eri* worm is taken: 'The caterpillar is first about a quarter of an inch in length, and appears nearly black.' (The colour is, perhaps, more exactly described as a blackish-yellow.) 'As it increases in size, it becomes of an orange colour, with six black spots on each of the twelve rings which form its body. The head, claws, and holders are black; after the second moulting, they change to an orange colour, that of the body gradually becomes lighter, in some approaching to white, in others to green, and the black spots gradually become the colour of the body. After the fourth or last moulting, the colour is a dirty white or a dark green. On attaining its full size, the worm is about 3½ inches long.' According to one series of observations, it would appear that in the hot months the first change of skin occurs three days after hatching, and the rest follow at intervals of three days, while the worm begins to spin on the fourth day after the final change, or the fifteenth day after hatching. In the cooler months, the period before each moulting is four or five days, making twenty to twenty-five days between hatching and beginning to spin; and in the winter season the worm lives a whole month, or even longer.

"After the final moulting, the worms are transferred from the tray to forked twigs of the castor-oil plant, with the leaves on, suspended across a piece of reed. As the worms attain maturity, they cease to feed, and crawl to the top of the fork; and if held up to the ear and gently rolled between the fingers, their bodies emit a crackling or rustling sound. They are now placed on the *jáli*, which consists of a bundle of dried plantain leaves, or of branches of trees with the withered leaves attached, and this also, like the feeding-tray, is suspended from the roof within doors. Here they begin to spin, usually on the same day, and not unfrequently two worms will select the same leaves as their covert, and join their cocoons together. The time occupied in spinning is three to six days.

Cycle. 2089

"DURATION OF CYCLE.—It will be gathered from the foregoing that a complete cycle of the insect may be as long as twelve weeks in winter

* See the remarks regarding flacherie at the conclusion of this section on the opposite page.—*Ed.*

REARING : Cycle.

or as short as six weeks in summer, while in the intermediate months it varies between these extremes. The maximum and minimum periods are shown in the subjoined table :—

	Minimum days.	Maximum days.
"Hatching	7	15
As a worm	15	32
Spinning cocoon	3	6
In the cocoon	15	30
As a moth (up to laying of eggs)	3	3
TOTAL .	43	86"

Experiments performed by **Mr. F. F. Mackenzie** of Cachar during 1887-88 resulted in the following being given as the time required for the life cycle of the insect :—Chrysalis, 18 to 19 days; Imago, 3 to 4; Egg, 7 to 8; Worm, 21 to 24; Total Period, 49 to 55 days. In **Mr. Mackenzie's** experience the warmer and moister the climate in which the worms are bred, the shorter will be the cycle (*i.e.*, the period of life from egg to moth), and also the larger the worms and cocoons produced.

Flacherie. 2090

In the chapter on **Pests (Insect)** (*Vol. VI., Pt II., 151*), the reader will find much information of interest, but in passing it may be said that **Mr. Cotes** of the Indian Museum has ascertained the nature of the disease that killed off so many of the *eri* worms reared by **Mr. Mackenzie** in Cachar. **Mr. Cotes** says it was undoubtedly *flacherie.* He was able to make out the chain ferment of **Streptococcu s bombyces** characteristic of that disease. He adds that **Mr. Mackenzie's** description of the symptoms, even if the ferment had not actually been made out, would leave no room for doubt. **Mr. Cotes** adds : "I am particularly interested in finding this undoubted case of *eri* worms affected by *flacherie*, a disease which is so intimately connected with the fermentation of the mulberry leaf, that it might have been supposed that it would not affect the *eri* worm, which feeds on a totally distinct plant."

In amplification of the facts given above by **Mr. Stack** regarding breeding, the following passage may be given from **Dr. Buchanan-Hamilton's** account of the process pursued in Eastern Bengal :— "The cocoons preserved for breeding having produced moths, which are very beautiful, the impregnated females cling to a small twig that is hung up near them, deposit their eggs round it in spiral rings, and then die clinging to the stick. These twigs are often sold at markets, and, with the dead moths hanging round, make a very curious appearance. A breeder having procured one of these twigs scrapes the eggs into a piece of cloth, which he lays on a wide-mouthed basket which is supported at some distance from the floor in one end of his hut. The eggs are soon hatched, and the worms are daily supplied with fresh leaves, and kept clean. The worm grows rapidly, and when ready to spin, some twigs are put into the basket to assist its operations. The cocoons that are to be spun are thrown into boiling water, and the threads of from five to six are wound into one by means of the common silk-reel of Bengal. This forms a coarse, rough thread of a dirty white colour, and totally destitute of the silky lustre. A seer of 96 sicca weight ($2\frac{484}{1000}$lb) of this thread is worth from annas 12 to R1, but it is very seldom sold, and the people who keep the insect in general rear no more than is just sufficient to make cloths for their own family. The cloth last very long, owing to which quality it is probable that some use might be found for this material in our manufactures at home."

SILK: Eri.

Rearing of the Eri

PRODUCTION.

Production of Silk.

This may be discussed under the following headings taken from Mr. Stack's note:—

Cocoon. 2091

"THE COCOONS.—The dimensions of a full-sized cocoon are about 1⅕ inch in length by ⅘ inch in diameter. The cocoon without the chrysalis weighs five grains. It is destitute of floss. Its proper colour is white, but a large proportion of the cocoons are of a dark brick-red colour, for which it is difficult to account. Mr. Hugon, after noting that the colour of the mature worm is either dirty white or dark green, adds—'The white caterpillars invariably spin red silk, the green ones white.' However this may be, it is at least certain that worms of the same brood, fed on the same leaves, will produce dark and light cocoons indifferently. The dark colour can be purged away by boiling the cocoon in alkali water. It is said that in some places where cocoons are sold the white cocoons are sorted out, and command a higher price. There seems to be reason to believe that, with proper care in providing the worms with suitable shelter for spinning, the proportion of white cocoons could be increased, and the quality also of the silk could be improved. Mr. C. H. Lepper, who in 1872 attempted the experimental cultivation of the *eri* worm in the Lakhimpur district on a considerable scale, found that darkness in the place of spinning was a favourable condition. Some cocoons spun in a wine-case nearly filled with loose shreds of newspaper, and with the lid closed, proved to be perfectly white and exceptionally good." In subsequent experiments performed by Mr. Mackenzie it was found that by careful selection and breeding, the size of worms and cocoons could be greatly increased, and the tendency to the objectionable red colour eliminated.

Spinning. 2092

"MODE OF CARDING & SPINNING.—In preparing the cocoons for use, the first step is to destroy the life in the chrysalis. For this purpose exposure to the sun during one or two days is usually sufficient, and this is the method preferred by the cultivators, as enabling them to keep the cocoons longer, and avoiding the discoloration which is caused by fire. When fire has to be employed, it is applied under bamboo trays upon which the cocoons are placed. Cocoons intended for immediate use are boiled for two or three hours in an alkaline solution of the ashes of the plantain-stem in water, which serves the double purpose of killing the chrysalis and softening the cocoon. Usually, however, the cultivator keeps his cocoons until he has a stock sufficiently large to make it worth his while to begin to spin. He then boils them in the solution described above; or the ashes used may be those of grass, rice-straw, or the stems and leaves of the castor-oil tree, or of various other plants. In this way cocoons several years old, if they have been kept uninjured, can be softened and rendered capable of spinning. After this process, the cocoons are opened, and the chrysalis is extracted; they are next washed white, slightly kneaded in the hand, dried in the sun, and are then ready for use.

Reeling. 2093

"REELING.—The *eri* cocoon has been successfully *reeled* in Italy, and experiments have shown that it can be reeled in India, but the only method employed by the cultivator is that of spinning off the silk by hand. At the time of spinning, the empty cocoons are placed in an earthen bowl containing water, with which a little cow-dung is sometimes mixed. Each cocoon is taken up separately, and the silk is drawn off in a coarse thread, nearly as thick as twine. Uniformity of thickness is roughly preserved by rubbing the thread between the finger and thumb, and in this way also new cocoons are joined on. It is said that six spinners can spin about 4 chitaks (8 ozs.) of thread in a day, consuming

PRODUCTION.

Reeling.

thereby some 1,200 to 1,500 cocoons. A seer (2lb) of empty cocoons will yield about three-quarters of a seer of thread."

In a letter to the Agri.-Horticultural Society (February 1843), **Mr. George Eveleigh** stated that he had succeeded to reel this silk by simply placing them in lukewarm water. He attributes his success to his having fed the worms occasionally on mulberry leaves and given them leaves at all times in as moist a state as possible. "This," he says, "appears much to increase the growth of the worms, ... and the silk appears greatly improved both in quantity and quality, and there appears to be less deposit of the resinous matter in the silk, which is therefore more easily wound."

Much doubt may be admitted as justifiable as to the desirability of attempting to reel this cocoon. Modern uses of the silk have rendered its utilisation quite possible and advantageous by being carded and spun. A far greater difficulty lies in inducing a sufficient cultivation.

Value of Cocoon. 2094

VALUE OF THE COCOON, THREAD & CLOTH.—It may be instructive to give **Mr. Stack's** original statement on this subject, and to follow with **Mr. Darrah's** more recent remarks.

"Cocoons prepared in the manner above described are sold at R2½ to R3 per seer of about 3,600 cocoons, but the waste cocoons out of which the moth has been allowed to make its way (*khola* cocoons) can be had at about one-fourth of this rate. Pierced *eri* cocoons sell in Calcutta at R60 to R70 the maund (82lb). Cocoons containing the desiccated chrysalis sell at the rate of 1,200 to 1,500 the rupee, or about 9 annas per seer of 700 cocoons. These prices, however, are liable to great fluctuations, and it must not be supposed that there is anything like a fixed rate for cocoons. They are nowhere offered for sale in open bazár; and whether they can be procured in the villages or not depends very much upon the character of the season. If the brood has been a plentiful one, the superfluous cocoons are for disposal; if not, the cultivator will not part with those which he has reserved for his own use. The value of the thread varies from R4 to R7 per seer, and the most important fabrics woven from it are waistcloths (*dhoti*) and sheets (*bor kapor*). The latter are large pieces of cloth, about 6 to 7 yards long, by 4 to 4½ feet in width, and their price varies from R7 to R20, according to quality. The cloth is often extremely coarse, and of a dark colour and open texture, but it is always very durable, and the texture grows closer by wearing, as the nap or floss rubbed off the thread serves to fill the interstices. A superior piece of *eri* cloth, on the other hand, is nearly as white as linen, and fine enough to make a travelling dress for a lady. One excellent quality of these fabrics is their exceeding durability. An ordinary *bor kapor* is reckoned to last thirty years."

Commenting on the above **Mr. Darrah** writes that while in 1884 empty or pierced *eri* cocoons sold in Assam at R50 to R60 a maund, the price ruling at the present date is rarely under R100, the better and whiter cocoons often fetching R130 a maund. "The reason of this rise in price is the greater demand in England, and the cause of the greater demand is the discovery of machinery adopted for utilising the cocoon. It must be remembered that the cocoon practically cannot be reeled, it must be spun, and it is only within recent years that spinning machinery has been improved sufficiently to enable a proper use to be made of the intractible *eri* cocoon. The sort of material required is thus described by a merchant engaged in the manufacture of spun silk: 'The class of silk called spun silk is made by a combing and carding process out of the refuse of thrown silk and out of cocoons that are damaged and not windable, also out of pierced cocoons, as we name those from which the

PRODUCTION. Value of Cocoon.

moth in the order of nature has escaped. It is in this latter condition that I think wild silks should be found somewhere in India, and this is what I principally want I do not wish to wind such silk, but to spin it into fine thread. It is no matter how broken and rough it may look or how much it is knocked about, torn, or crushed. I only want it as free as possible from the dead bodies of the worms and of such foreign matters as sand or branches. It will not look like silk at all till the gum and dirt are boiled and worked out of it. You observe I ask nothing from India that requires skilled labour or machinery. *Eri* I like best for its whiteness; I believe it breeds frequently, but I do not believe any amount of cultivation could get thrown silk out of it,—I mean of course to be of any commercial value.' "

Fibre. 2095

THE FIBRE.—Its thickness is $\frac{1}{1500}$ of an inch when taken on the outside of the cocoon, and $\frac{1}{1450}$ in the inner part. The average value of the fibre may be stated to be from 12 annas to one rupee per 2½lb. **Mr. T. Wardle** obtained, in 1879, through the Government of India, 70lb of *eri* cocoons: he had them carded or dressed and spun, and reported that he knew of no silk better adapted for spinning. "The staple obtained from the first draft's operation is glossy, long, and very fine. Its fineness is owing to that of the ultimate fibre. It is about one-half finer than *tasar* silk, although not more than two-thirds as fine as the Bengal mulberry-fed silk of **Bombyx mori** or Silk of Commerce. The after or shorter drafts are also of much importance as showing the economising of the shorter fibres after the longer ones have been removed. These are used for less important manufactures than the long staple. Nothing is wasted in the modern mode of spinning. The yarns made of these fibres are of great regularity and fineness, proving this silk capable of uses for spinning and weaving purposes to an unlimited extent."

Outturn. 2096

Outturn of Silk.

"In the absence of any large markets, and indeed of any regular trade in either the thread or the cloth, it would be quite useless to attempt to conjecture the probable outturn of *eri* silk in Assam. An estimate of 25½ cwt. (35 maunds) has been furnished for the produce of Kámrúp, 177 cwt. (242 maunds) for Darrang, and 205 cwt. (280 maunds) for Nowgong, but the latter district probably produces less *eri* than either of the other two, and the estimates may be regarded as mere guess-work. In no district does the produce do much more than supply local wants. A trade in cocoons, to the extent of 400 or 500 cwt. yearly, has sprung up between Goálpára and Calcutta, whence the cocoons are shipped for England. They are said to come chiefly from Upper Assam. The cloth which finds its way to the shops of the Marwari traders is by them exported to Bengal. The mountaineers of Bhután who visit the plains in the winter carry away with them a considerable quantity both of cloth and yarn. The quantity of cloth is estimated at 2,000 pieces, while the yarn is dyed by the Bhutias and woven into gaily-coloured coats and striped cloths, some of which find their way back to the bazárs of Assam. The value of the silk thus exported from the three Bhutia fairs in the Darrang district last year was returned as R43,000, and probably we may allow as much more for the Bhutia trade in Kámrúp. As regards its use in the province, however, the general opinion is that the native *eri* is being supplanted by cotton goods from England. It is alleged that the cloth is procurable with more difficulty now than formerly, and it is certain that the price has risen greatly within the last thirty years. If we go back so far as fifty years, we find the yarn selling for R2 a seer in 1834. There is, however, reason to doubt whether *eri*

OUTTURN of Silk.

was more easily procurable then than it is now, and perhaps the rise of price is chiefly to be explained by the influx of money which has accompanied the development of tea cultivation. It is impossible to say whether the actual outturn is less or greater now than at any former period. There is no natural obstacle to an increase of production to any imaginable limit."

In the article which appeared in the Catalogue of the objects shown at the Calcutta International Exhibition (1883-84), the writer gave the following facts on the subject of the annual outturn of *eri* silk. These passages are here republished in amplification of the above passage from Mr. Stack's note, since no more recent information has been brought to light:—

2097

In Assam, about 54,000lb of *eri* silk, in the raw state, unreeled, can be obtained annually from the districts of Kámrúp, Darrang, Nowgong, and Lakhimpur, and about 30,000lb from the Jaintia hills. In Goálpára and Sibsagar, the production of silk is carried on to a very limited extent, chiefly for home consumption. In Cachar the silk is worked up, for their own use, by the hill tribes in almost every village of the hills in the northern parts of that district. From Sylhet no information has been furnished, but here, as well as throughout Assam, the necessary food of the *eri* worm grows in abundance.

2098

In Bengal, the Dinagepur district can supply about 13 maunds of *eri* cocoons annually in the winter. In Rangpur about 30 to 35 maunds are produced, but it is difficult at present to obtain any supply, as the Natives are unwilling to sell the cocoons; they prepare therefrom cloths for their own use. In the Bogra and Julpiguri districts the silk is worked up for home consumption only, the quantity produced being about 18 maunds of cocoons in Bogra and 40 to 50 maunds of thread in Julpiguri. In the Darjíling Terai about 10 to 12 maunds of cocoons could be annually obtained. In the Chittagong district a small quantity of the silk is produced, and the thread is made into twine for fishing purposes and sold to the extent of R500 annually in the local bazárs. In Purneah the worm is reared to a very small extent for the silk, which is used in home consumption only. In Gya the silk is worked in certain wild tracts. In Shahabad the quantity produced amounts annually to about 9,000lb. In Pooree the worm, though entirely neglected, is common, especially in the Khurda estate; and in this latter place if a demand arises a new useful industry could easily be opened to the Natives.

(The above facts regarding Bengal appeared originally in an official communication from the Bengal Government in April 1880.)

DYEING. 2099

Dyeing of Eri Silk.—Under this head it is necessary to have recourse to Mr. Wardle's little book on *The Wild Silks of India*, and to relate in his own words the result of his careful, comprehensive experiments:—

"The dyeing of *eri* silk much resembles the dyeing of *Tusser*. Whether owing to the flatness of its fibre, or to the nature of its sericine, it is far behind mulberry silk in its natural affinity for dye-stuffs. Heat and the media of mineral salts, however, are the principal agents in bringing the fibre into a dye-receiving subjection. * * * The dyeing baths have to be much stronger in tinctorial matter than those for mulberry silks. It follows, therefore, that there is an unavoidable increase in the cost of dyeing *eri* silk, as is also the case in *Tusser* silk, and to about the same extent. Probably I shall not be far from accuracy in stating that *eri* silk requires twice as much dye-stuff as mulberry silk, thereby causing the dyeing to cost considerably more. The *eri* cocoons being of two kinds, some of them rust colour and others white, cannot be dyed into pale colours without bleaching, which again adds to the cost of

DYEING of Eri Silk.

dyeing. It bleaches very well with the bioxide of barium process, and takes excellent colours in pale tints afterwards. For dark shades bleaching is not necessary, nor would it be necessary for paler shades in silk spun from the white cocoons if they could be kept separate from the brown ones. * * * I have succeeded in imparting a variety of colours to this silk which leave little or nothing to be desired. As far as I can learn, I believe this is the first time in Europe that *eri* silk has been white."

2100

THE MUGA SILK-WORM.

(For Tasar see pp. 96-161; for Eri, pp. 162-174.)

The Múga or Múnga Silk-Worm (Antheræa Assam).

Vern.—*Múnga, múga,* Ass.

Habitat.—This insect is met with chiefly in Assam, but it extends east to the Naga hills and the mountains of North Burma, including Sylhet and Cachar, and south to Tipperah. In Cachar and the Naga hills it is not known whether or not the cocoons are even collected, although, in some districts, they are quite plentiful. In his *Hand-book of the Indian Wild Silks,* **Mr. Wardle** gives a map for *múga* silk, in which he shows it to be found, in addition to the above localities, far away to the west in Dehra Dún, and across the Peninsula at Dhurrumpur in the Bombay Presidency. This is apparently founded upon a remark made by **Captain Hutton** that it is met sparingly at Dehra Dún; but in the text of his work **Mr. Wardle** makes no mention of these localities. **Hampson**, in the Appendix below (*p. 237*), gives the distribution of this species as from Kangra to Assam.

Cultivation. 2101

Extent of Cultivation.—The *múga* silk-worm is, to a certain extent, domesticated in Assam, being reared in houses, but it is found to produce better and more productive cocoons when allowed to shift for itself on the trees around the cultivator's house. It is stated to have five broods a year. The breeders of Upper Assam annually import their seed cocoons from Kámrúp, all attempts to successfully perpetuate the species in domestication having failed. Breeding cocoons cost R2 per thousand.

Mr. Stack says the name *múga* is derived from the amber colour of the silk, and is frequently used to denote silk in general, so that *eri muga* means *eri* silk, *kutkuri muga, tasar* silk, and so on. The genuine *múga* is distinguished by the title *sumpatia múga* or silk yielded by the worm that feeds on the *sum*-leaf. **Mr. Stack** adds: "It is a multivoltine worm, and is commonly said to be semi-domesticated, because it is reared upon trees in the open air; but in fact it is as much domesticated as any other species, being hatched in-doors, and spinning its cocoon in-doors, while during its life on the tree it is entirely dependent on the cultivator for protection from its numerous enemies." The Jorhat district of Assam may be said to be the most important region in the production of *múga* silk.

HISTORY. 2102

Historic Records.—It is probable that the earliest account of this silk is that dating from 1662 in connection with Mir Jumla. **Geoghegan** says that the earliest definite notice of it is that given by **Dr. Buchanan-Hamilton.** Then there followed **Captain** (the late **General**) **Jenkins'** more detailed account (*Jour. Agri.-Hort. Soc., Ind., Feb. 1833*), and still later that by **Hugon & Helfer** (*Jour. Asiatic Soc., Beng., VI., 43 (1837)*. **Mr. C. Brownlow** (*Jour. Agri.-Hort. Soc., Ind, XIII., 392-415*) gives some useful information on *múga*, especially on the subject of the identity of the wild insect of Cachar with the domesticated condition in

HISTORY.

Assam. The valuable statement which will be found quoted below, from the pen of the late Mr. **Stack** (formerly Director of Land Records and Agriculture in Assam), may be said to be an amplified and in some respects greatly improved version of **Hugon & Helfer's** paper which appeared originally in the *Journal of the Asiatic Society of Bengal.*

Food of the Worm.

FOOD. 2103

The worm is described by **Sir D. Brandis** (*Indian Forester, V., 35*) and by **Mr. Hugon** and other authors as feeding on the following trees:—

1st—**Cinnamomum obtusifolium,** *Nees.*

Ram-tezpat, BENG.; *Patichanda,* ASS.

The *múga* silk-worm sometimes feeds upon the leaves of this tree

2nd—**Cylicopodaphne nitida,** *Meissn.*

Kotoloah, ASS.

This is most probably the *kontuloa* referred to by **Hugon.**

3rd—**Michelia Champaca,** *Linn.*

Champa, or *champaca,* BENG.; *Titasappa,* ASS.; *Oulia champ,* NEP.

Captain Jenkins says the *múga* silk-worm feeds upon this tree, but there is very probably some mistake regarding this statement, the species found upon the *champa* being most probably quite distinct from the ordinary *múnga.*

4th—**Machilus odoratissima,** *Nees.*

Súm, ASS.; *Kawala,* HIND.; *Dingpingwait,* KHASIA.

This is the chief plant upon which the *múga* silk-worm feeds. It grows gregariously, forming forests, and is often cultivated around villages to feed the worm.

5th—**Symplocos grandiflora,** *Wall.*

Bumroti, ASS.; *Moat soom,* PHEKIAL.

A handsome tree or large bush which **Mr. Mann** says is sometimes used to feed the *múnga* (*múga*) silk-worm. Two other members of this genus are employed as food for the small yellow silk-worms (*eri*), *viz.*, **S. cratægoides,** *Ham.*, and **S. ramosissima,** *Wall.* Could it be possible that **Mr. Mann** mistook large *eri* worms for the *múnga,* for it would seem unlikely that the *múnga* feeds upon anything but laurels.

6th—**Litsœa citrata,** *Blume.*

Adakuri, edenkuri, mezenkuri, ASS.; *Siltimber,* NEPAL; *Terhilsok,* LEPCHA.

In Assam the leaves are largely used to feed the *múga* silk-worm; in fact, this tree is next in importance to the *sum* for this purpose. With regard to the species of **Litsœa** (=**Tetranthera**), it is not quite clear if the vernacular names have been given correctly. If *sualu* is synonymous with *adakura,* they refer to **L. polyantha,** *Juss.*

Sir D. Brandis makes no mention of the *champa* tree being used in Assam to feed the *múga* worm, while **Captain** (the late **General**) **Jenkins** says: "The silk produced from the worm feeding upon this plant gives the finest and whitest silk, used only by the Rajah and great people, and is called *champa-pattea múnga.* The thread is sold at from R11 to R12 a seer. With the exception of this plant and the species of **Symplocos** referred to above, the *múga* silk-worm seems to feed entirely upon species of laurel. This is a somewhat remarkable fact, which of itself circumscribes the home of the *múga* worm, and removes that insect in a marked degree from all the other silk-worm moths. One could hardly imagine a creature which displays so decided a preference for dry, evergreen, aromatic leaves, ever taking to any other kind of food, and there would, from that fact alone, seem some doubt regarding the *champa* tree as a source of food for the *múga* worm. There is, in fact, every probability that the *champa*-feeding worm (if such exists) is a perfectly distinct species, and since it is

Conf. with pp. 71, 176, 183.

SILK: Muga.

Rearing of the

FOOD.

reported to yield the finest *múnga* silk, it seems highly desirable that special attention should be given to this subject. It is the more probable that this may be found a correct conjecture, since up to within a few years the **Antheræa mezankuri** was supposed to be the same species as the common *múnga*. Mr. Hugon places the *champa*-reared *múnga* on a par with the *mezankuri*, and regards both as 50 per cent. finer than the ordinary *múnga*. Hampson, it will be seen by the Appendix, accepts **A. mezankuri** as a synonym for **A. assama.**

7th—Litsæa polyantha, *Juss.*

Sualu, Ass; *Haura*, CACHAR; *Bolbek*, GARO.; *Meda, gwa, singraf, marda, kerauli, patoia*, HIND.; *Mendah, kari, leja*, GONDI.

Upon the aromatic leaves of this plant the *múga* silk-worm is stated to feed in Assam.

8th—Litsœa salicifolia, *Roxb.*

Diglotti, ASS.; *Sempat*, NEPAL; *Diglilati*, MECHI.

An evergreen tree of the Eastern Himálaya and East Bengal, upon the leaves of which the *múga* silk-worms are sometimes fed.

Since the above remarks on the food of the múga silk-worm appeared in the Catalogue of the Calcutta International Exhibition, Mr. Stack has furnished the following facts regarding the subject:—

Conf. with pp. 71, 183.

The *sum*-tree (**Machilus odoratissima**) constitutes its favourite food; but in Lower Assam it is extensively bred on the *suálu* (**Tetranthera monopetala**). The leaves of certain other forest trees—the *dighlati* (**Tetranthera glauca**), the *pátichanda* (**Cinnamomum obtusifolium**), and the *bamroti* (**Symplocos grandiflora**)—can be eaten by the worm in its maturer stages if the supply of its staple food begins to fail; but the *sum* and the *suálu* are the only trees upon which the worm yielding the ordinary *múga* silk (as distinguished from *champa* and *mezankuri*, which will be mentioned hereafter) can be permanently reared. The *sum*-fed worm is considered to yield the more delicate silk, and *suálu* trees on the edges of *sum* plantations are generally left untouched, though small plantations of *suálu* only may occasionally be met with. It will thus be seen from the paranthesis above, that Mr. Stack has, as the writer anticipated, isolated the *champa* and *mezankuri* insect from the ordinary *múga*, as special forms assumed by the insect when fed on these trees. His account of this subject will be found in the concluding paragraph of the present article.

REARING. 2104

Rearing of the Worm.

The following account from Mr. Stack's *Silk in Assam* gives the main facts on this subject that have up to date been published:—

Broods. 2105

"NUMBER OF BROODS IN A YEAR.—Five successive broods are distinguished by vernacular names roughly denoting the months in which the worms are bred and spin their cocoons. These are the *kátia* brood, in October—November; the *járua* in the coldest months (December—February); the *jethua* in the spring; the *aharua* in June—July; and the *bhádia* in August—September. But it is only in a few parts of the Assam Valley that this regular succession of broods is maintained. The *aharua* and *bhádia* broods are reared chiefly in the district of Kámrúp, whence cocoons are exported for the *kátia* brood in Upper Assam. In Darrang and Sibságar the only broods for use are the *kátia, járua,* and *jethua;* while in Lakhimpur only the *járua* and *jethua* are generally in fashion. The worm is said to degenerate if bred all the year round in Upper Assam; and another reason for the discontinuance of breeding in the summer is that the *sum* forests are flooded by the rains, the watching of the worms becomes more troublesome, and losses increase. Hence the breeders of Upper Assam generally go down to Kámrúp or Nowgong to

Múga (or Múnga) Silkworm. (*G. Watt.*) SILK: Muga.

REARING of Muga Silk-worm.

buy cocoons at the beginning of the cold season. Occasionally, a *bhádia* brood of inferior quality is reared in Sibsagar on a high-lying patch of *sum* land. Even in Jorhát, the centre of the cultivation of the *muga* silk-worm, one-fourth of the breeding cocoons, it is estimated, are imported from Kámrúp. The price of cocoons thus purchased varies from two to four rupees the thousand, according to the supply. Sometimes the worms themselves are sold, at the rate of 100 to 150 per rupee.

Treatment. 2106

"TREATMENT OF COCOONS FOR BREEDING.—The cocoons intended for breeding are placed in trays of woven bamboo, and hung up safely within the house. The period of the chrysalis lasts about a fortnight in the warm months, and three weeks or a few days longer in the cold season, when the room in which the cocoons are kept has to be warmed by a fire, and they are sometimes suspended near the hearth.

"If the cocoons are kept in a covered basket, the moths are allowed to move about inside it till the day following their emergence; but where open trays are employed, the female moths, recognisable at once by their bulkier body, are immediately tied by a thread passing round the thorax behind the wings to single pieces of straw, which are hooked on a line stretched across the room; or several moths may be fastened in this way to a bunch of straws 18 inches long by one in diameter. Straws black with smoke are usually selected, from a notion that the colour helps to reconcile the moth to captivity. The male moths are left free, and some of them make their escape into the open air, but the majority remain attached to the females. Any deficiency in the number of males is supplied by placing the females outside the house in the evening, when unattached males will discover and consort with them. A song chanted by the cultivator is supposed to attract the males on such occasions. Each female produces about 250 eggs in three days, and the life of the moth lasts one or two days longer, but eggs laid after the first three days are rejected, as likely to give birth to feeble worms.

Growth. 2107

"GROWTH OF THE WORMS.—The pieces of straw, with the eggs deposited on them, are carefully taken down and placed in basket covered with a piece of cloth. The room in which they are kept is heated by a fire in winter, or the eggs may be laid in a place warmed by the sun, but not directly exposed to his rays, the heat of which would prove destructive. They ought to be kept in the dark as much as possible. The period of hatching lasts from seven to ten days, according to the time of year. In the summer months, it is not necessary to keep the eggs in-doors at all, and they can be placed on the tree at once, with due precautions, however, against sun, rain, and dew; and even in the winter a small proportion of the eggs may be placed out unhatched, together with the young worms. Generally, however, the worms are hatched in-doors. 'On being hatched' (says Mr. Hugon) 'the worm is about a quarter of an inch long; it appears composed of alternate black and yellow rings. As it increases in size, the former are distinguished as six black moles, in regular lines, on each of the twelve rings which form its body. The colours gradually alter as it progresses, that of the body becoming lighter, the moles sky blue, then red, with a bright gold-coloured ring round each.' The worm passes through four moultings, known respectively as *chaiura, duikáta, tinikáta,* and *maiki-chál-káta*. The full-grown worm, when extended in the act of progression, measures about 5 inches long, and is nearly as thick as the forefinger. Its colour is green, the under part being of a darker shade, while the back is light green, with a curious opaline or transparent tinge. Excluding the head and tail, the body is composed of ten rings, each having four hairy red moles, with bright gold bases, symmetrically disposed round its edge; a brown and yellow

SILK: Muga.

Rearing of the

REARING of Muga Silk-worm.

Growth.

stripe extends midway down each side from the tail to within two rings of the head, and below it the breathing-holes are marked by a series of seven black points; the head and claws are light brown, the holders dark green, with black prickles, the tail pair widening above into green circles enclosing a large black spot. Two sizes of the full-grown worm are distinguished. The *borbhogia* is 5 inches long, the *horubhogia* somewhat shorter; and a similar difference is observed in the size of the cocoons. It is not necessary that the worms should complete their growth on a single tree. If the leaves be exhausted, they descend the trunk till they are stopped by a coil of straw rope, or by a band of plantain-leaves, which serves to arrest them till they can be gathered and transferred to another tree. This may be done either by simply placing them on the trunk and leaving them to crawl up (and if so treated it is said they will refuse to ascend a tree which has already been stripped of its leaves), or by means of a triangular tray, which is pushed up at the end of a long bamboo, and hooked on to one of the upper branches. The latter is also the method employed in putting the young worms on the tree for the first time. Young trees are preferred to begin with, and generally trees from three to twelve years old are considered the best. Old trees are avoided, as they harbour ants, and the moss on their branches impedes the movements of the worm. The worms feed from about eight o'clock in the morning till near noon, and again from three till sunset. During the intervening hours they descend the trunk to bask in the sun and at night they take shelter under the leaves. A dropping sound like that of light hail is heard under the tree at feeding time, and is caused by the pea-like excrement (*lád*) of the worms, which is constantly falling to the ground.

Diseases. 2108

"ENEMIES AND DISEASES.—During their life in the open air the worms are exposed to the attacks of various enemies, among whom the crow and kite are the most persistent and destructive, but the *sáksákia*, or wandering pie, by day, the aziola, or 'little downy owl' (*pesa*), and the large frugivorous bat (*bandali*) by night, are also to be dreaded. The insects which do most damage are the wasp, the ichneumon fly, and a red ant called *ámruli*, but the latter is dangerous to the worm only in its earlier stages. The result of a bite is a blackness extending from the injured part over the whole body, which gradually withers away. The cultivators wage war against the ants with fire and hot water, or skewer bits of fish on the trunk to attract them and prevent them from ascending the tree; the pellet-bow is used against birds by day, and a tall clapper of split bamboo, pulled by a string from within the watcher's hut, serves to frighten away nightly marauders; but with all these precautions the losses by theft are considerable. This constant watching becomes very troublesome, especially in the months of inclement weather, and is usually left to the children and old people, where there are any in the family. Continued heavy rain is apt to wash the worms off the trees, but they can shelter themselves under the leaves against passing showers, and, in fact, light rain in October and November is considered favourable to the growth of the winter brood. A hailstorm is the greatest calamity of all, for it not only kills numbers outright, but so weakens others that they die before maturity, or spin imperfect cocoons, and the weakness is even said to be transmitted to the moth, should any emerge. The worms, finally, are subject to a disease called 'the swelling' (*phula-róg*), for which no remedy is known. In Upper Assam this epidemic occasionally destroys the worms on acres of *sum* forest together, and even where the mortality is less wholesale, the silk-producing power of the survivors is found to be impaired. The worms often die off in large numbers without any swelling

REARING of Muga Silk-worm.

or other external symptoms, merely ceasing to feed, and perishing apparently of inanition, and in this case also the yield of silk from the surviving portion of the brood is poor. Apart from these causes, a difference is said to be noted in the productive powers of worms of the same breed. It is alleged that some worms can be distinguished as destined to die immature: these are called *hahoya* and *bisa;* others, called *phutuka,* spin cocoons yielding an imperfect quantity of silk.

Spinning. 2109

"THE SPINNING OF THE COCOONS.—The period from hatching to maturity varies from 26 days in summer to 40 days in winter. The moultings are completed about a week or ten days before the end of this term. There is no difficulty in discerning when the worm is ready to begin its cocoon, because it invariably descends the tree to the edge of the plantain-leaf band, and there remains motionless, grasping the bark with its holders only, while the forepart of the body is raised and thrown slightly back. Another sign is said to be a peculiar sound yielded by the body when lightly tapped. Worms which show these symptoms are removed at nightfall, or if left over night they begin to make their preparations for spinning in a roll of grass tied round the tree for that purpose. Being carried to the house, the worms are there placed on a bundle of branches with the dry leaves attached, or in a basket with a bundle of leaves suspended over it, into which the worms crawl. From four to seven days are spent in spinning the cocoon.

"A complete cycle of the insect lasts about 54 days in the warm months and 81 days in the cold season. The maximum and minimum periods are shown in the subjoined table :—

	Minimum days.	Maximum days.
Hatching	7	10
As a worm	26	40
Spinning cocoon	4	7
In the cocoon	14	21
As a moth (up to laying eggs)	3	3
TOTAL	54	81"

Transition. 2110

TRANSITIONS IN THE FORM & SIZE OF THE WORM, COCOON, & EGGS.—"On being hatched this caterpillar is composed of alternate black and yellow rings, but as it grows older the black bands are reduced to black spots or moles in regular lines, on each of the twelve rings which form the body. As it matures, the colours change still further, the main colour becoming light greenish-yellow, with brilliant red moles, each having a gold edge around it and four sharp prickles and a few black hairs. When full grown it is about 4 inches long. The eggs are hatched in 10 days, the moths remaining within the cocoon for 16 to 20 days.

"The cocoon is fawn-coloured, large and thin, devoid of the curious suspensor so characteristic of the *tasar* cocoon. The short period of lethargy does not necessitate so much care in the construction and protection of the chrysalis as is displayed by the instinct of the *tasar* worm in the formation of its cocoon"

Production of Silk.

PRODUCTION.

Treatment of the cocoons, &c. 2111

On this subject Mr. Stack furnished the following particulars :—

"THE COCOONS.—The *múga* cocoon is in size about $1\frac{3}{4}$ inches long by one inch in diameter. In colour it is a golden yellow; but there are usually a number of dark cocoons in every brood, for which no satisfactory reason can be assigned. The difference does not seem to be due to any of the con-

PRODUCTION.

Treatment.

ditions of food or breeding. A large proportion of the dark cocoons which come into the market, however, are no doubt to be accounted for by discoloration in the process of firing. Boiling in alkali water is the method employed to restore dark cocoons to their proper colour. With the living chrysalis inside, the cocoon weighs about 66 grains, with the dead and dried chrysalis 27½ grains, and the empty cocoon from which the moth has made its escape weighs 6 grains only. The ordinary selling-rate for cocoons with the desiccated chrysalis is R2 the thousand, but they can often be bought in the villages for 700 to 800 the rupee. The waste or perforated cocoons from which the moth has escaped can be had for about R2 the seer, containing nearly 3,000 cocoons. There is, however, no regular market for cocoons, and persons wishing to procure a stock must visit the villages where the worms are bred, and make their own bargain with the cultivators, and waste *múga* cocoons do not seem to be easily procurable by any artifice.

Reeling. 2112

"REELING.—The silk of the *múga* cocoon is reeled. The life in the chrysalis having been destroyed by exposure to the sun, or by fire, the cocoons are boiled in an alkaline solution. When required for use, their floss is plucked off, and they are placed in a pot of warm or cold water. Two persons are employed, one to take the silk from the cocoons, the other to reel it. The former brings together the filaments of silk from a number of cocoons, varying from 7 to 20, and hands them to the reeler, who rubs them into a thread by rolling them on his thigh with the palm of his right hand and the under part of the forearm (which usually suffers more or less from the operation), while with his left hand he turns the fly-wheel of the primitive reeling apparatus that stands beside him, an axle turning in the notches of two uprights, with the aforesaid wheel at one end, or often merely a cross-stick in the middle to serve the purpose of a fly-wheel. In this way, the whole of the cocoon can be unwound, except the innermost layer next to the chrysalis. The thread is reeled off on the axle, in skeins of about half a seer at a time. The quantity of silk yielded by the cocoons varies according to the brood. The cold-weather brood gives the least, and is usually reserved for breeding, only the inferior cocoons being spun. The *kátia* and *jethua* broods yield the most silk. A thousand cocoons of the *járua* brood will yield about 2 chitaks of thread, and of the *kátia* or *jethua* brood 3 to 4 chitaks.

"Opinions differ as to whether old cocoons can or cannot be reeled. The cultivator does not usually keep his cocoons so long as a year, unless he is accumulating a stock very slowly. But it would seem that reeling is practicable up to two years at least, and that, if carefully kept, cocoons of even four or five years old can be reeled, and will give silk in no respect inferior to that yielded by fresh cocoons. The experiment, however, is one which is not often made.

"No part of the *múga* cocoon is rejected as useless: the floss plucked off before reeling, the silk of the shell immediately surrounding the chrysalis, and the cocoons kept for breeding, after the moth has forced its way through them, though unfit for reeling, are spun by the hand into a coarser kind of thread, called 'waste' or *era*, which is used for mixing with *eri* thread, or is woven by itself into rough but warm and durable fabrics.

"The price of *múga* thread varies, according to quality, from R8 to R12 per seer. The latter is the ordinary rate in Sibságar bazár at present. In 1876 I find the price quoted as R7. Waste *múga* thread can be bought for R4 the seer.

"The cloth woven from *múga* yarn has a bright yellow colour and a pretty gloss. It stands washing much better than other silks, keeping

PRODUCTION.

Reeling.

gloss and colour to the last. It is usually sold in pieces about 5 yards long by 4 feet broad, and the price varies from R1-8 to R2 per square yard.

"A curious tradition is preserved of the *khesa*, or 'raw' *múga* silk, which used to be manufactured in the days of the Ahom kings. The worms intended to yield this product were kept alive for three days after completion of their growth, without being allowed to spin their cocoons. The result is said to have been the accumulation of the silk fibre in the body of the worm. Their heads were then plucked off, and the bodies thrown into a vessel of warm water, and the ends of the fibres being extracted, the silk was reeled off in the ordinary way. This kind of silk was reserved for the exclusive use of royalty."

Fibre. 2113

THE FIBRE.—The soft loose fibre from the inner part of the *múga* cocoon is "thrown," in Assam, into a simple kind of yarn, and in this condition it is largely exported. The fabrics made from it are worn by the middle class, the *eri* silk fabrics by the poor. The outer fibre is about $\frac{1}{1480}$ inch in diameter, and bears a strain of $2\frac{1}{8}$ drams, while the inner fibre is $\frac{1}{1080}$, and will support 3 drams. The tension of the outer fibres is about one inch to the foot, while the inner fibre is about $1\frac{3}{4}$ inches. The fibre is not only much finer than the *tasar* silk, but it is round, like that obtained from the mulberry-fed worms. It will show the difference between the *munga* and *tasar* silks to give here the measurements of the latter so as to allow of comparison:—

From edge to edge of the *tasar* fibre, Mr. **Wardle** says, the diameter is $\frac{1}{770}$th part of an inch taken from the outer fibres, and from the inner fibres $\frac{1}{710}$; the former bear a strain of 7 drams, while the fine and uniform fibres from the inner layer of the cocoon bear as much as 8 drams. The thickness of the *tasar* fibre is about $\frac{1}{1900}$ part of an inch.

The mulberry silk-worm of Bengal produces a fibre $\frac{1}{2000}$ part of an inch in diameter for its outer fibres and $\frac{1}{2200}$ for its inner.

Mr. T. Wardle in a lecture read before the Silk Conference held in connection with the Colonial and Indian Exhibition of 1886, at which **Sir E. C. Buck** presided, said that embroidery was carried on in India on a great scale. The Natives ladies of Assam decorate in this way presentation garments, and the turbans of Dacca which find their way to Arabia are embroidered, he said, with a silk that "I feel sure, will, one day, be in considerable demand in Europe—I mean the silk produced by the worm of the **Antheræa assama**, the *múga* silk-worm. It was not known to the English until recently that any of this silk was exported."

Markets and Extent of Production.

Markets. 2114

There is no large market where either the cocoons, the thread, or the cloth can be purchased wholesale, but cocoons and thread are to be procured in small quantities at most of the petty village fairs in the *múga*-producing districts, and some stock of cloth can usually be found at the head-quarters stations, particularly in Sibságar and its two sub-divisional stations of Golághát and Jorhát, the latter of which, in the days of native rule, was the grand centre of the silk industry of Assam. The Marwari traders pick up silk in the villages, and the cultivators also come in occasionally, as their needs impel them, and sell to the traders in their shops; it is altogether a casual kind of commerce, fluctuating greatly from year to year, but never attaining any regular flow or considerable demensions. The export of *múga*, unlike that of *eri*, is principally in the form of thread, which goes to Calcutta for local consumption, or for export to the Persian Gulf; it is too dear for the English market, though the Calcutta prices are quoted as low as R6 to R11 per seer.

SILK: Muga.

Rearing of the Múga (or Múnga) Silkworm.

PRODUCTION.

Markets.

"*Múga* is less widely spread than *eri*, and the annual outturn is probably less; but there are no means of estimating its amount. The outturn of the Sibságar district is supposed to be 205 cwt. (280 maunds) of silk (only half the leased *sum*-lands being assumed as actually under *múga* cultivation); that of Darrang is shown as 8 cwt. (11 maunds), and of Kámrúp as 15 cwt. (20 maunds), though Darrang produces more than Kámrúp; but these estimates are quite untrustworthy. Sibságar is the great *múga*-growing district of the Assam Valley; next to that, the south-western portion of the Mangaldai sub-division, and the western part of Kámrúp to the south of the Brahmaputra, where the Rani mauza especially is celebrated for supplying breeding cocoons to Upper Assam. *Múga*-breeding is also carried on to a considerable extent in the closely populated tract in the centre of the Kámrúp district north of the Brahmaputra. There is a good deal of *múga* cultivation in Lakhimpur, where the *sum*-tree grows wild in great profusion; and the more closely-peopled mauzas of Nowgong also contribute largely to the stock of *múga* in the province. The worm seems to be unknown in Sylhet and Cachar, while the hill districts do not produce the trees on which it feeds."

Hugon wrote in 1834 that "*Múga* forms one of the principal exports from Assam. The average quantity passed at Goálpára during the last two years that duties were levied was 257 maunds, valued at R56,054. It leaves the country principally in the shape of thread, most of it going to Berhampore" "The total quantity that leaves the province may, I think, be estimated at upwards of 300 maunds." "The quantity produced in the province may be reckoned at 600 to 700 maunds."

Outturn. 2115

"OUTTURN PER ACRE OF *sum*-TREES.—Various estimates have been made of the quantity of silk yielded by the worms fed on an acre of trees. Mr. Hugon reckoned 50,000 cocoons yearly per acre, yielding more than 12 seers of silk, in value R60, and as there were then 2,000 acres of *sum*-trees in Nowgong, the total produce of that district in 1837 would appear to have been 600 maunds, or double that of Sibságar at the present day. The Sibságar estimate, however, rests on the hypothesis of 18,000 cocoons per acre, giving 3 seers of thread. The great discrepancy between these two calculations would appear at first sight to point to a remarkable falling off in the productivity of the worm, but it is probably to be explained by other causes. The number of trees assumed per acre, for instance, makes a substantial difference in the result. The Sibságar mauzadárs return the average number as 80, whereas in the plantations of Lower Assam a quarter of an acre sometimes contains a larger number than this. Mr. Hugon, indeed, estimates 4,000 cocoons to a seer of silk, against the modern Sibságar estimate of 6,000; but it is more probable that both guesses are inaccurate than that the *muga* cocoon has decreased in size by one-third during the last half-century. As regards the total outturn of the Assam Valley, the earliest information we have is that recorded by the chronicler of Mir Jumla's invasion in 1662, who remarks that the silks are good, but that the people produce little more than they require for use. Mr. Hugon says that the price of *múga* thread rose from R3-8 to R5 per seer in the three years ending with 1837, and since then the price has doubled, but we cannot infer a diminished production from this fact alone. The trade was no brisker in Mr. Hugon's time than it is now. Merchants requiring silk were obliged to make advances to the cultivators. Coming down to the present day, it is impossible to say with certainty whether the cultivation of *múga* has declined or not, but if the general opinion of the country be accepted, it must be believed that a large falling-off has actually taken place. The area of *sum*-bearing lands rented from Government in the Sibságar district was 15,907 acres in 1876,

PRODUCTION.

Outturn.

against 12,393 acres in 1881, and in Lower Assam also patches of *sum* plantation, evidently intended for the breeding of silkworms, will often be found lying unused. On the other hand, there is ample room in the Assam Valley for the extension of *múga* cultivation. The virgin *sum* forests of Lakhimpur are capable by themselves of sustaining millions of worms, if the industry should ever receive that quickening impulse of which it stands in need."

It would seem probable that the facts given above by **Mr. Stack** regarding the production of this silk, review not only the statements made by **Hugon** but those made also by **Hopkinson**. The latter writer, while Chief Commissioner of Assam, stated that the *sum* forests of that province covered some 34,000 acres, of which 18,000 were assessed and yielded a revenue of R28,000. By far the greater portion of the assessed area, he informs us, was then in Sibságar district, and gave employment to 48,000 person, who, however, were not employed wholly on *múga* silk rearing. The outturn, he adds, was estimated at 100,000lb.

Champa & Mezankuri Muga.

CHAMPA MUGA. 2116

Conf. with pp. 71, 175, and 176.

Mr. Stack writes as follows on the forms of *múga* silk:—"An account of *múga* silk would not be complete without a few words on the two varieties assumed by it when the worm is fed on the *champa* (or more properly *chapa*) and the *mezankuri* or *adakuri* (**Tetranthera polyantha**). *Champa* silk seems to be quite forgotten now. It is described as a very fine white silk, which used to be worn only by the Ahom kings and their nobles. *Mezankuri* silk is still to be procured, but with great difficulty. In 1881 there does not seem to have been a single piece obtainable in Jorhát. One of the reasons alleged for this falling-off is that the new rules restricting clearances in the forests are unfavourable to the growth of the *mezankuri* tree. This tree springs up spontaneously in abandoned clearances, and it is in this early shrub-like stage that it is fit for the worms to feed on. In its second year, the worms fed on it give coarser silk; in the third year, the silk is hardly distinguishable from the common *múga*. Thus, the mature tree is quite out of the question, and as the *mezankuri* is never cultivated, forest clearances were the only places where the breeders could look for young trees. When fed on the *mezankuri*, the *múga* worm spins a fine silk of almost pure white, about thrice as valuable as the common *múga*, in fact the most costly of all the silks of Assam. The thread was selling at R24 the seer in Jorhat in 1883. This silk is altogether an article of luxury." **Captain Hutton** upholds the view, *viz.*, that **Moore** was in error when he assigned the *mezankuri* cocoons to a separate species from the ordinary *múga*; and, as already remarked, this opinion is also held by **Hampson** and most modern entomologists.

TRADE IN SILK & SILK-GOODS.

TRADE. 2117

In a Report on the Proceedings of the East India Company regarding the Trade, Culture and Manufacture of Raw-silk, published in 1836, we are told the transactions of the Company in raw-silk were "inconsiderable in extent before the middle of last century." Until the establishment of regular places for its preparation, under the management of their own servants, the East India Company, in common with other Europeans who then had factories in India, or resorted there for the purpose of trade, provided their investments by purchases in the market, or by contracts with Native dealers and others. The chief places then producing silk were Cossimbazar, Commercolly, and Rungpur. **Tavernier** wrote of Bengal silk in the seventeenth century that a village in the kingdom of Bengal (**Kásimbázar**) can furnish 20,000 bales annually, each bale weighing 100

SILK.

Trade in Silk

TRADE.

Historic Records.

2118

livers of 16 ounces. The Dutch generally took, he says, either for Japan or Holland, 6,000 to 7,000 bales of it, and they would have liked more, but the merchants of Tartary and of the whole of the Mongol Empire opposed their doing so. It is difficult to see how Tavernier was able to get the figure of total production, but his remarks show the importance of the Bengal mulberry silk trade during his time and his notice of it is perhaps the earliest direct reference by a European writer. The description of silk exported from Bengal by the English and Dutch Companies was of the kind now known by the technical term of "*Country wound,*" being drawn from the cocoons and reeled into skeins after the crude manner immemorially practised by the Natives of India. That kind of silk was suited for but few of the articles then manufactured in England, the chief consumption being for sewing-silk, buttons, and other small articles of haberdashery, etc. Its inequality induced the Company to endeavour to improve the methods of reeling, and, accordingly, in 1757 they sent out to Bengal Mr. Richard Wilder on a mission "to examine into the causes of the defective quality of Bengal raw-silk; for which purpose he was directed to proceed to Cossimbazar (at that time the Company's chief factory in Bengal) in order that he might be on the spot where the raw-silk was produced." In another part of the present article the chief historic facts connected with the effort to improve the reeling and to produce a better quality of silk will be found briefly reviewed. It does not seem necessary, therefore, to do more here than to add that from 1757 the modern export trade in silk may be accepted as dating. In 1768 so greatly had the traffic advanced that the Court of Directors informed their Indian representatives that they looked "to the increase in raw-silk chiefly, for the means of bringing home their surplus revenue, the importation being a national benefit and the consumption being far less limited than that of manufactured goods."

Country wound.

2119

Early attempt to improve reeling. *A.D. 1757.* **2120**

Conf. with pp. 51, 62, 185, 186.

In Milburn's *Oriental Commerce* (published 1813) there occurs a most instructive article on *The Rise & Progress of the Silk Trade.* It deals with the silk of all the world; but more especially that of India, China, and Japan, and therefore it would perhaps be superfluous to reproduce it in its entirety. It is a review of all the proceedings of the East India Company and of the various political and social influences in Europe that fostered or retarded the growth of the silk trade. From that article most modern writers appear to have drawn largely, and as the work is by no means readily accessible, it may be as well to give here an abstract from it (by way of introduction to the Indian trade), of the various passages, in chronological order, which deal with the silk of this country:—

A.D. 1621. **Turkey Trade** **2121**

1621.—"Previous to the commencement of the trade between England and the East Indies, England was dependent on Turkey for the silk consumed in her manufactures. Raw-silk was, therefore, considered as the article of most consequence, and great exertions were made, and expense incurred, in forming establishments in Persia, with a view of securing a certain and regular supply from thence for the use of the manufactures in England. Mr. Munn states the importation to be 107,140lb, which cost in India 7*s.* per lb, and that the selling price in England was 20*s.* He also states that many hundreds of people were constantly employed in winding, twisting, and weaving silk in London."

A.D. 1680. **Commencement of East India Company's Trade.** **2122**

1680.—"The Turkey or Levant Company began to complain about this period of the East India Company, on account of the great quantities of silk they imported by way of the Cape of Good Hope, which had formerly been imported solely from Turkey. They made a formal complaint to the King on the subject, in which they stated their exports to amount to £500,000, consisting principally of woollen manufactures and other English wares; in return for which they imported raw-silk, galls, drugs, cotton, &c., all of which being manufactured in the kingdom, afforded bread and employment to the poor. On the other hand, they stated that the East India Company exported much bullion, with an inconsiderable quantity of cloth; in return for which they imported a deceitful kind of raw silk, which was an inevitable destruction to the Turkey trade. It was also stated at this period, that the

Company had sent to India throwsters, weavers, and dyers, and had set up a manufacture of silk which, by instructing the Indians in these manufactures, and by importing them so made, tended to impoverish the working people in England.

TRADE:

Historic Records.

Efforts to improve Indian Silk. 2123

"The East India Company's reply was to the following effect: 'That with respect to the Turkey Company's objections to the importation of raw silk it is essential to the good of the kingdom that since their importation of it, the silk manufactures have increased from one to four. With respect to the quality of the India raw-silk, it is the same as with all other commodities, some good, some bad, and some indifferent. With respect to the sending to India throwsters, weavers, and dyers, the whole is a mistake, excepting only as to one or two dyers usually sent to Bengal, and to no other part of India; and this for the nation's, as well as the Company's, advantage, especially as to plain black silks, generally exported again.' This defence was deemed satisfactory, and the complaints of the Turkey Company were dismissed."

Dyers sent to Bengal. 2124

Conf. with p. 154.

A.D. 1697.

1697.—"The trade with India being at this time carried on by two rival Companies, there was such an excessive importation of the various kinds of East India and China wrought silks, and the prices were so reduced, as to occasion great loss to the importers; and it interfered so much with the home manufactures of silks as to occasion great discontent among the manufacturers throughout England, more particularly in London, where they became very outrageous, and carried their violence so far as to attempt to seize the treasure at the East India House, and had almost succeeded in it, but were in the end reduced to order.

Cheapness of Indian Silk. 2125

Herba. 2126

Conf. with Vol. IV., 223; also Vol. VI., Pt. II., 460.

"The great cheapness of Indian silks occasioned the wear of them to become almost universal throughout England; to remedy which, an Act was passed for encouraging the home manufactures, which enacted, 'that from Michaelmas, 1701, all wrought silks, Bengals, and stuffs mixed with silk or herba, of the manufacture of Persia, China, or the East Indies, should be locked up in warehouses till re-exported; so that none of the said goods should be worn or used, in either apparel or furniture, in England, on forfeiture thereof, and also of £200 penalty on the person having or selling any of them.' 11 & 12 Will. iii, Chap. 10."

A.D. 1749.

Reduction of Duty on Chinese Silk. 2127

1749.—"The duties on raw-silk imported from China were now reduced to the same duty as Indian raw-silk, and with a view to encourage the growth of silk in America, a law was passed admitting it to be imported, duty free, if properly certified to be the growth and produce of the British Colonies.

"The following accounts of the quantities of raw silk imported, and from what parts, with the remarks thereon, are principally extracted from the reports made by the East India Company at various periods on the subject of the silk trade.

"The silk imported into Great Britain in this year, and from whence was as follows:—

Total Imports into Great Britain. 2128

	℔		℔
Flanders	1,407	Italy	36,301
Spain and Portugal .	2,564	East India (including China) . . .	43,876
Straits	14,897	Turkey	132,894

forming a total of 231,939℔, by which it appears that the imports from Turkey were more than one-half from the East Indies and China less than one-fifth, and from Italy less than one-sixth."

A.D. 1765.

Exports from Bengal 80,340lb. 2129

1765.—"From 1751 to 1765, the silk imported from Bengal rose, on an average, to about 80,340 small pounds of 16 ounces each per annum; and no doubt but the quantities from other parts increased in an equal, if not a superior, degree of proportion.

"When the Company obtained possession of the Bengal Provinces, it was judged expedient to extend their commercial concerns, with a view of realising the surplus revenues of India. The article of raw-silk appeared the most eligible for the interests of that country; first, as affording the means for extending cultivation; and, secondly, by creating additional employment for the natives. In 1766, there was an import of 195,637 small ℔; and, on an average of the five succeeding years, from 1767 to 1771, it increased 327,630 small ℔ per annum. But as the quality of the article was of the kind at present known by the technical term of "Bengal wound," or silk reeled in the rude and artless manner immemorially practised by the natives of that country, it was suited but to a few articles of home manufacture, the principal consumption being in sewing silk, buttons, twist, and other articles of haberdashery, of comparatively limited demand. The market by these increased imports became so overcharged as to cause a reduction in the price; the silk which in 1765 sold at 27*s*. per

Price to be paid 27s. per great pound.

SILK.

Trade in Silk

TRADE:

Historic Records.

Introduction of Italian Reeling.

2130

Conf. with p. 184.

great ℔ sold in 1771-72 at 18*s.* 6*d.* per great ℔, notwithstanding the sales in these periods fell short of the imports by 171,807℔.

"Experience having thus shewn that the article had been pushed to its utmost extremity without effecting the desired end, a plan was next suggested for introducing into Bengal the mode of winding practised in the filatures, or winding-houses of Italy and other parts of the continent; which, if carried into execution, might create an opening for a still further consumption, by its becoming a substitute for some of the silks of Italy, Turkey, and Spain, to divers of the manufacturing branches of which, in its then state, it was not applicable. This plan on being maturely considered was deemed sufficiently eligible, and the needful measures were taken for carrying it into effect.

First consignment of Filature Silk made in *A. D. 1772.*

2131

"Although the first consignment of filature-wound silk reached England in 1772, yet it was not till 1775 that the new mode could be considered as in full operation. In the intermediate period much time was unavoidably taken up in erecting buildings, fitting up furnaces, reels, &c., and in instructing the natives, who are most scrupulously averse to innovations of every kind; to which, above all, may be added, that the country was recovering but slowly from the calamitous effects of a most dreadful famine, which swept off millions of the lower class of inhabitants, and occasioned a considerable defalcation in every species of its productions. From these causes the imports of silk from Bengal, from 1772 to 1775, the filature included, as will appear from the tables of silk imported, were so circumscribed as not to exceed, on an average, 187,494 small ℔, by which the price came round to about 24*s.* per great ℔, which in general may be considered as about its fair level.

"The new mode of winding being now sufficiently established, and the country recovered from its enfeebled state, the meditated competition took place, and was pursued with such energy and effect, as politically to answer for the most sanguine expectations, although the issue commercially, from a variety of concurring circumstances, was unfortunate in the extreme. From 1776 to 1785 the imports from Bengal appear to have been, on an average, 560,283 small ℔, while those from Italy, Turkey, etc., did not exceed 282,304℔. During this period the raw-silk was provided by contract, and by which the Company sustained a loss of £884,744.

Average Imports by Britain of Bengal Silk 560,283lb, or twice as large as those from Italy, Turkey, &c., &c.

2132

"The result of this successful effort was quickly seen in the declension of the trade from Aleppo, Valencia, Naples, Calabria, and other parts; from many of which that formerly furnished very considerable quantities, not a single bale was imported for many years; so that, generally speaking, the silk manufactured in England is now furnished from the northern provinces of Italy, Bengal, and China.

"The following is a statement of the quantities of raw-silk imported into England from Bengal; the quantities sold; the prime cost thereof, including duties, freight and charges; the sale amount thereof, the discount deducted; and the loss sustained by the East India Company for ten years previous to the abolitions of the system of supplying it by contract:—

Losses caused through the Contract System of purchasing Silk.

2133

(Conf. with profit, p. 190.) *A.D. 1785.*

SEASON.	IMPORTED.	SOLD.	Total of Prime Cost, Duties, Freight and Charges.	Sale amount, Discount deducted.	LOSS.
	Small ℔ of 16 ounces.				
	℔	℔	£	£	£
1776 . .	515,913	311,551	409,851	365,653	44,198
1777 . .	563,121	547,045	440,877	323,031	117,846
1778 . .	602,964	589,245	472,114	325,505	146,609
1779 . .	737,560	596,343	421,899	299,053	122,846
1780 . .	235,216	574,065	288,933	217,599	71,334
1781 . .	785,673	553,863	629,438	481,584	147,854
1782 . .	77,610	292,141	64,160	56,752	7,408
1783 . .	611,071	592,831	480,515	388,233	92,282
1784 . .	1,149,394	486,336	874,097	779,626	94,471
1785 . .	324,307	576,175	252,617	212,721	39,896
TOTAL .	5,602,829	5,119,595	4,334,501	3,449,757	884,744

and Silk Goods. (*G. Watt.*) SILK.

TRADE: Historic Records.

"Previous to the year 1786, the Company's investments of raw-silk were chiefly provided by contract. Under this system, particularly in the earlier part of its establishment, the most flagrant abuses prevailed: the Company's interests were sacrificed; the manufacturers were oppressed; and, as a natural result, the goods were furnished of debased qualities, and at extravagant rates of cost.

A.D. 1782.

"In 1782, during Mr. Hastings' administration, the evils before alluded to received a partial check; but it was reserved for the wisdom of Lord Cornwallis' Council, seconded by the active superintendence of the Board of Trade, appointed by His Lordship under a new establishment, to effect that radical reform which has since happily been introduced into every branch of the Company's commercial concerns. In short, by abolishing the mode of provision by contract and substituting in its room that of agency corruption has ceased; errors have been corrected; the manufacturers have been relieved; the fabrics have been restored; and goods of the choicest description have been supplied at genuine and reduced rates of cost. The benefits which have resulted to the Company from the success of His Lordship's measures, as far as they regard the article under consideration, are conspicuously exemplified by the statement, which follows hereafter, of the cost, amount sales, &c., in the years 1786 to 1803 inclusive.

Lord Cornwallis abolishes Contracts in Silk. 2134

"The era of the commencement of the cotton fabrics upon an enlarged scale may be dated about the year 1787; and although, since that time, it will be seen that the imports of silk from Bengal have fallen considerably short of what they were in former periods, it will also be found that they have been more than equal to what the market has required. The issue of the sales is an infallible test for ascertaining the quantum of demand. It appears that in the five years commencing 1783 and ending 1787 the Company sold 2,437,384 small ℔. And in the subsequent five years, *viz.*, from 1788 to 1792, only 1,693,784 small ℔ a quantity less by one-third; and the trade was at so low an ebb, that though the average quantity of Bengal silk sold was 338,757 small ℔, it was stated that no reduction of price would add a single ounce to the consumption. With respect to the imports from China, it will be seen that the quantities sold in the last five years were, on an average, 220,526 small ℔."

Decline in Demand for Bengal Silk. 2135

1793.—"The French Revolution gave a severe check to commerce. Its influence was felt not only in England, but in every market upon the Continent. A general alarm prevailed; mercantile transactions were in a great degree suspended, and manufactures in general were nearly at a stand. The silk trade participated largely in this scene of distress, and experienced a more than ordinary depression. Great numbers of weavers were out of employ; the buyers were loaded with a heavy stock of silk upon hand; the East India Company had a large quantity in their warehouses unsold; and the imports in the approaching season were expected to be considerable. Under these circumstances a memorial was presented to the Company by the trade, requesting that the September sale of 1793 might be altogether dispensed with. The Company's exigencies were too pressing to admit of this request being complied with; but, with a view of affording the buyers every relief within their power, it was agreed to postpone the sale for four months. At the end of that period no favourable change having taken place, the Company, from their increasing difficulties, found themselves under the unavoidable necessity of forcing the market. A sale was made, and the silk actually disposed of; but at such reduced rates, that the loss upon the quantity sold on February 25, 1794, was £47,746.

A.D. 1793. Effects of the French Revolution. 2136

"With a view of guarding against future losses, and of enabling Bengal to avail herself of the advantages she was found in a capacity to derive from the increased produce of raw silk, the Company proposed the measure of causing the surplus quantity of silk *beyond what the markets could take in its raw state*, to be thrown into *organzine* in England, for the purpose of its being brought into use as a substitute for part of the thrown silk imported from Italy; and upon consulting some intelligent persons in the silk line, there seemed good reason to conclude that it would be found sufficiently adapted to the warp of ribbands. An experiment was therefore made; and although the issue was in every respect encouraging and satisfactory, and the legality of the Company's proceedings were strongly combated by those interested in the imports from Italy, yet there were many persons in the trade decidedly hostile to the undertaking, and who confidently pronounced it was impossible it could ever be brought to answer. Among the various objections that were urged, it was asserted, in a memorial from the silk merchants addressed to the Lords of Trade, 'that as Bengal raw-silk had, in the general opinion, attained its utmost possible state of perfection, it could only, when worked into organzine, be used, in a few articles of the silk manufacture; that in most others, from its irremediable deficiency of staple, it could not be substituted for Italian organzine and that the attempt to introduce it into a more general consumption would produce the

Introduction of Trade in Thrown Indian Silk. 2137

Memorial against the new Departure. 2138

Further Improvement Doubted. 2139

Conf. with pp. 57, 59, 62, 184, etc.

SILK. | Trade in Silk

TRADE: Historic Records.

greatest discontent and tumult among the journeymen weavers, particularly of Spitalfield, who universally reprobated Bengal organzine.' The object was, however, too important, and the prospect too flattering, to be hastily abandoned. Further trials were made, and in proportion as the article became more known, and the views of the Company were better understood, much of the prejudice that had been excited against the measure, subsided.

"The quantities of thrown silk, imported into England in the under-mentioned periods, appear to have been as follow, of which at least nine-tenths have been from Italy:—

Indian Thrown Silk used for Warp of Ribbands. 2140

			Small ℔.
In the 10 years 1776 to 1785 inclusive on an average per annum			. 392,918
In the 5 years 1786 to 1790	ditto	ditto	. 391,746
In the 2 years 1791 to 1792	ditto	ditto	. 453,535

"The silk, when thrown, is used for the warp in the manufacture of ribbands and broad goods, in each branch of which the consumption is considered nearly equal.

"In 1796 the reputation of the article was so far established that a great number of the most eminent houses in the various branches of silk manufacture presented to the Court of Directors a memorial, dated London, February 5, 1795, of which the following is a copy:—

Memorial in favour of Bengal Silk, thrown in England. 2141

"'We, the undersigned silk manufacturers understanding from the reports published by the East India Company, that the Bengal Provinces are capable of furnishing a more abundant supply of raw-silk than hitherto, are of opinion, if due attention is paid, in the first instance to reel the same of proper sizes, that, after making a due provision for singles, trams, and sewing silks, the surplus, by being thrown into organzine in this country, can be successfully brought into use in our respective manufactories to a very considerable extent in lieu of part of the thrown silk presently supplied from Italy. Considering, therefore, the measure now carrying on by the East India Company as highly laudable, and meriting every degree of support, we trust they will persevere in the same with firmness, being well convinced that it cannot fail of proving highly beneficial to the national interests.

"'*First*, by giving to a country which makes part of the British dominions, the advantages desirable from the production of a commodity which forms the basis of one of the most important of the national manufactures.

"'*Secondly*, by creating employment at home for a numerous class of our poor, particularly women and children in the throwing of it into organzine.

"'*Lastly*, by affording a large and more certain supply to the manufacturers in general, it may have a tendency to lower the prices of the raw material, and in future to shelter the silk market from the alarming fluctuations that have repeatedly taken place, and probably increase greatly the general consumption of the silk-manufacturers.'

Bengal Silk greatly improved. 2142

Conf. with pp. 57, 62, 184, 185, 186, 187.

"Thus pointedly called upon by the principal consumers, the weight of whose testimony was sufficient to silence all doubts with regard to the propriety of the measure, the Company sent directions to the Bengal Government to extend their consignments to 4,000 bales per annum. Instructions were also forwarded, requiring them to pay the most unremitting attention to the quality; means were also suggested for remedying existing defects, and samples were transmitted for their guidance in regard to sizes: in consequence of which, the quality has in general been in a progressive state of improvement, and in some instances has arrived at such a degree of excellence, as to rival the most perfect productions of Italy.

Organzine of Bengal Silk largely used. 2143

"In consequence of this improvement, the use of Bengal organzine has not been confined merely to the warp of ribbands but it has been introduced with equal success in Sarcenets, Florentines, modes, handkerchiefs, velvets, &c.; and Bengal raw and thrown silks in their present improved state are fully competent to most of the material and extensive purposes to which the raw and thrown silks of Italy have hitherto been exclusively applied; and if they could be furnished in sufficient quantities, they would supplant at least three-fourths of the silks at present drawn from Italy."

A. D. 1803.

1803.—"It appears that from the period the measure of throwing Bengal silk into *organzine* was resorted to, in 1794 to 1803, a period of ten years, there were thrown 1,453 bales, or rather more than 140 bales per annum; and that this quantity

TRADE:

Historic Records.

sold at the sales for £268,395, the whole of which sum would have gone to the aggrandizement of Italy. This sum may be thus divided:—

Gain to Britain and Loss to Italy in the Organzine Trade.
2144

	£
Bengal benefited in the prime cost, or, in other words, in the manufacture and culture	124,711
And the remainder, which is thus appropriated, is added to the riches of the country, *viz.*, the charge of throwing which gave employment to the industrious poor	78,167
In freight and duties which benefit navigation and assist the revenue	25,066
And after deducting from the sale amount 5 per cent. for the charges of merchandise which are principally labour . .	13,420
It yielded the Company a profit in the last five years of £28,688, from which deducting a loss in the former five years of £1,637, leaves a net gain of	27,031
TOTAL . .	268,395

"The following is an account of the quantities of raw-silk imported into Great Britain from Bengal, China, Turkey, and all other places, in the thirty years, 1773 to 1802 inclusive, likewise the quantity of thrown silk imported during the same period:—

Imports of Great Britain.
2145
Conf. with table on *p. 195.*
Exports from Bengal, *Column III.*
2146

YEAR.	RAW SILK IMPORTED BY GREAT BRITAIN.					Thrown silk imported.	Total of raw and thrown silk imported.
	Bengal. I.	China. II.	Italy and Turkey. III.	Other Parts. IV.	Total. V.		
	Small lb	Small lb	Small lb	Small lb	Small lb	Small lb	Small lb
1773 . .	145,777	203,401	187,099	6,190	542,467	234,906	777,373
1774 . .	213,549	276,781	220,933	2,610	713,873	428,978	1,142,851
1775 . .	208,881	167,229	272,782	13,380	662,272	411,895	1,074,167
1776 . .	515,913	244,839	515,235	22,048	1,298,035	454,414	1,752,449
1777 . .	563,121	221,902	350,640	42,451	1,178,114	396,543	1,574,657
1778 . .	602,964	266,678	133,636	12,558	1,012,836	186.512	1,199,348
1779 . .	737,560	234,906	850	130,503	1,103,819	383,042	1,486,861
1780 . .	235,216	301,300	844	209,557	445,617	487,678	933,295
1781 . .	785,673	301,301	23,878	288,906	1,701,058	443,384	2,144,442
1782 . .	77,610	79,725	37,894	178,084	373,313	331,685	704,998
1783 . .	611,071	241,107	140,866	129,758	1,122,802	495,203	1,618,005
1784 . .	149,394	100,602	262,419	74,688	1,587,103	406,468	1,993,571
1785 . .	324,307	98,920	245,230	25,996	694,453	344,251	1,038,704
1786 . .	252,985	59,551	222,175	35,101	569,812	361,448	931,260
1787 . .	178,180	366,878	185,983	21,583	752,624	389,381	1,142,005
1788 . .	305,965	312,182	148,922	23,207	790,276	306,640	1,096,916
1789 . .	427,263	257,022	148,582	23,881	856,748	393,258	1,250,006
1790 . .	320,826	216,005	194,974	25,953	757,758	508,005	1,265,763
1791 . .	373,503	203,539	294,103	38,288	909,433	470,195	1,379,628
1792 . .	380,107	104,830	358,500	45,881	889,318	436,875	1,326,193
1793 . .	736,081	165,435	110,276	8,216	1,020,008	241,955	1,261,963
1794 . .	521,460	99,356	44,911	17,501	683,328	330,978	1,014,306
1795 . .	380,352	154,590	80,579	110,995	726,416	336,995	1,063,411
1796 . .	347,936	12,968	19,045	107,682	487,631	398,948	886,579
1797 . .	92,204	78,520	4,058	91,494	266,276	401,662	667,938
1798 . .	353,394	136,196	...	241,295	730,885	402,917	1,133,802
1799 . .	644,819	63,604	11,455	520,594	1,240,832	467,349	1,708,181
1800 . .	583,086	92,385	40,239	117,862	833,572	333,717	1,167,289
1801 . .	444,862	131,335	62,264	193,503	831,964	275,149	1,107,113
1802 . .	242,809	75,588	179,009	193,395	692,801	396,210	1,089,011

A.D. 1802.

"The greater part of which is consumed in the manufactures, as will appear from the following accounts of the quantities of the different kinds of raw and thrown silk

SILK.

Trade in Silk

TRADE: Historic Records.

exported in the years 1790 to 1796 inclusive, being three years previous to the year in which the war commenced, and three years after:—

A.D. 1700.

	Bengal.	China.	Italy and Turkey.	Thrown.	Total.
	lb	lb	lb	lb	lb
1790	43,500	10,758	15,285	20,064	89,607
1791	36,456	8,209	21,847	22,428	88,940
1792	13,406	5,310	15,798	10,579	45,093
1793	19,397	3,572	5,590	2,607	31,166
1794	61,989	7,502	13,643	24,385	107,519
1795	39,547	3,622	11,640	27,425	82,234
1796	70,113	7,279	11,289	38,927	127,608

"The exports in 1796 were to the following places: Russia 24,964lb, Holland 299lb, Germany 10,577lb, Italy 1,007lb, Gibraltar 3,084lb, Turkey 944lb, and the remaining 86,733lb to Ireland."

Profit of Silk Trade since introduction of Agency System.
2147
(Conf. with previous loss, p. 186.)

"From the establishment of the agency system in Bengal, which took place in 1786, till 1803, the Company's investments of raw-silk in general have been productive, as will appear from the following account of the prime cost, including freight and charges in each year; the sale amount in England during the same period, and the profit or loss arising in each of the above years:—

Season.	Prime cost including freight and charges.	Sale amount.	Profit.	Loss.
	£	£	£	£
1786	192,898	198,507	5,609	...
1787	133,795	145,712	11,917	...
1788	212,357	221,888	9,531	...
1789	276,732	289,271	12,539	...
1790	268,790	302,993	34,203	...
1791	290,159	320,395	30,236	...
1792	262,902	276,317	13,415	...
1793	274,553	221,329	...	53,224
1794	290,419	309,743	19,324	...
1795	378,512	381,385	2,873	...
1796	335,315	327,427	...	7,888
1797	262,917	258,644	...	4,273
1798	277,990	322,873	44,883	...
1799	324,460	390,149	65,689	...
1800	208,969	297,645	88,676	...
1801	262,428	395,410	132,982	...
1802	156,502	269,249	112,747	...
1803	195,117	292,659	97,542	...
TOTAL	4,604,815	5,221,596	682,166	65,385

A.D. 1803.

Improvement of Indian silk.
2148
Conf. with pp. 57, 62, 184, 185, 186, 187, 188.

"leaving a net profit in 18 years of £616,781, which on an average is £34,266 per annum, or about 13 per cent.

"During the seven years, 1786 to 1793, the provisions were made under many unpropitious circumstances, such as storms, inundations, droughts. &c., the silks which were produced during that period yielding a net profit of £117,450. In 1793, 1796, and 1797 there were losses; during this period the silks were gradually approaching to that degree of excellence to which they have ultimately arrived. In the six years, 1798 to 1803 inclusive, it appears by the above statement, that the silk, which amounted in prime cost and charges to £1,425,466, produced at the Company's sales £1,967,985, leaving a net profit of £542,510, on an average £90,419-16*s*. 4*d*. per annum."

Net Profit £90,419 per annum.
2149

TRADE: Historic Records.

Exports other than the Company's. 2150

"The following is a statement of the value of raw silk exported from Bengal exclusive of the East India Company's, in the years 1795-6 to 1805-6 inclusive:—

	Sicca rupees.		Sicca rupees.
1795-6	5,81,183	1801-2	13,65,882
1796-7	3,40,975	1802-3	16,38,467
1797-8	6,12,253	1803-4	19,10,398
1798-9	6,67,300	1804-5	33,82,000
1799-1800	14,33,751	1805-6	30,86,491
1800-1	10,51,957		

"Forming a total in 11 years of 1,60,70,657, of which only 40,13,177 sicca rupees was exported to London, the remainder to the coasts of Malabar and Coromandel, the Gulfs of Arabia and Persia, and a small portion to Pulo Pinang and places to the eastward."

A.D. 1810.

1810.—"Previous to the year 1801, the private imports of raw-silk from Bengal were very trifling, nor were they much extended in the two succeeding years; but the whole is brought into one point of view by the following account of the raw-silk imported from Bengal on account of the East India Company, and that imported by individuals in private trade and privilege and sold at the Company's sales, in the years 1801 to 1810 inclusive; together with the sale amount of each, and the average price per pound at each of the Company's sales during the same period:—

Total of all sales in Bengal Silk. 2151

Years.	Company's Silk. Quantity.	Company's Silk. Sale amount.	Company's Silk. Average.	Privilege Trade. Quantity.	Privilege Trade. Sale amount.	Privilege Trade. Average.	Total. Quantity sold.	Total. Sale amount.
	Small lb	£	£ s. d.	Small lb	£	£ s. d.	Small lb	£
M. S. 1801 .	193,569	162,332	0 19 0	39,234	30,294	0 15 5	444,862	382,638
S. S. — .	210,467	188,688	0 17 11	1,592	1,324	0 16 7		
M. S. 1802 .	142,083	147,544	1 0 9	24,120	26,626	1 2 1	244,809	265,382
S. S. — .	66,932	80,250	1 4 0	11,674	10,962	0 18 9		
M. S. 1803 .	118,661	112,834	0 19 1	20,571	17,484	0 17 0	324,764	311,791
S. S. — .	137,199	142,830	1 0 10	48,333	38,643	0 16 0		
M. S. 1804 .	164,705	163,533	0 19 10	129,824	71,899	0 11 2	565,494	473,057
S. S. — .	195,066	184,073	0 18 10	75,999	53,552	0 14 1		
M. S. 1805 .	42,255	45,534	1 1 8	70,648	44,228	0 12 6	267,900	227,409
S. S. — .	154,977	137,647	0 17 10	...	...	...		
M. S. 1806 .	54,271	47,149	0 17 4	167,894	125,398	0 14 11	504,281	372,126
S. S. — .	87,454	72,907	0 16 8	194,662	126,672	0 13 0		
M. S. 1807 .	156,516	99,695	0 12 8	113,729	75,359	0 13 3	666,904	463,609
— .	170,235	110,884	0 13 0	226,424	177,671	0 15 8		
M. S. 1808 .	146,012	181,584	1 4 10	57,701	91,397	1 11 8	423,668	564,561
S. S. — .	184,138	244,791	1 6 7	35,817	46,789	1 6 1		
M. S. 1809 .	172,734	250,208	1 8 11	47,398	67,272	1 8 3	303,855	444,010
S. S. — .	58,632	90,589	1 10 11	25,091	35,941	1 8 7		
M. S. 1810 .	......	......	......	52,987	82,929	1 11 4	306,535	444,939
S. S. — .	162,885	237,746	1 9 2	90,663	124,264	1 7 5		

SILK. **Trade in Silk**

TRADE: Historic Records.

	£	s.	d.
"From the foregoing statements some idea may be formed of the profit which was attached to the trade in raw-silk, as carried on by individuals from Bengal; taking the exports from Bengal of six years, 1800-1 to 1805-6 inclusive, as the prime cost, and which amounted to Sicca R35,59,269, which at 2s. 6d. per rupee is	444,908	12	6
The sales of privilege and private trade silk in the six years, 1801 to 18c6, was	547,082	0	0
Leaving an apparent profit, in the period of six years, of	102,173	7	6

Silk exported by Private Agencies inferior to that of the Company. 2152

"From which are to be deducted the freight, amounting to about 3 per cent., the premiums of insurance, commission, fees of office, &c. It must, however, be observed that the above period includes the years in which the silk trade was very much depressed, and when much of the silk imported by individuals turned out very inferior to the Company's, being badly worked, foul, and gouty, and partaking largely of those defects for which Bengal silk was formerly so much reprobated."

"From the foregoing accounts it appears that in the thirty-five years, 1776 to 1810 inclusive, the sale amount of the raw-silk imported from Bengal into Great Britain, on account of the East India Company, in the private trade of the Commanders and officers of their ships, and in the privilege tonnage allowed to individuals by the Act of 1793, was as follows:—

	£
In 10 years, 1776 to 1785 inclusive	3,449,757
In 18 ,, 1786 to 1803 ,, ...	5,221,596
In 7 ,, 1804 to 1810 ,, ...	3,115,044
Forming a total in 35 years of .	11,786,397

A.D. 1810.

Silk Trade of Great Britain exclusive of India & China. 2153

exclusive of prize and neutral property, which has been, to a very trifling extent, in the years 1793 to 1810 amounting to only £6,455."

"The following is an account of the value of the raw and thrown silk imported into Great Britain, exclusive of East India and China silk, in the years ending 5th January, 1807-8-9, likewise of the value of the manufactured silk-goods, and raw and thrown silk exported during the same period, taken from the papers annually laid before the House of Commons:—

	IMPORTED.	EXPORTED.		
	Raw and thrown silk.	Silk manufacture.	Raw and thrown silk.	TOTAL.
	£	£	£	£
1807	1,222,022	833,035	99,062	932,097
18c8	711,242	804,178	118,891	923,069
1809	343,901	473,078	67,053	540,131

"It is impossible to ascertain, with any degree of correctness, the extent of the manufactured goods consumed in Great Britain; but from the general use of silk in every class of society, from the throne to the cottage, the quantity must be immense."

Attained a degree of perfection supposed to have been impossible. 2154

Conf. with pp. 57, 62, 184, 185, 186, 187, 188, 190.

"Owing to the unremitting care and attention that have been f r years past given to raw-silk, both at home and abroad, it has been progressively improving, and continues to improve in its quality. It has already attained a degree of perfection which it was formerly pronounced to be altogether incapable of reaching, and is perhaps susceptible of still further improvements under the vigilant and active superintendence of the Company."

"The principal market which the English manufacturers have hitherto looked up to for a very large proportion of this important raw material, has been Italy In the present state of continental affairs, it is impossible to calculate upon events. Circumstances may arise, and recourse may be had to measures, the direct operation of which might tend to check or impede the supplies which have hitherto been drawn from that quarter. In this view of things, Bengal raw-silk has a claim not only to commercial, but to great political, consideration. *The deep-rooted prejudices that*

TRADE : Historic Records.

formerly prevailed against it, are daily vanishing, and the article is proportionably rising in the public esteem; but it is evident that its future success will altogether depend upon the degree of attention that shall continue to be paid to its quality. If there shall be the least relaxation on this important point, the character to which it has arrived by slow gradations, will at once be lost, and the flattering hopes which the Company have been looking to, of retrieving the heavy sums which have been sunk in bringing it to its present state of perfection, will be annihilated."

The proficiency attained had cost the Company £1,000,000.

2155

Conf. with pp. 13, 184-6, 190.

The passages, which the writer has italicised above, show how thoroughly the silk trade of India was understood at the beginning of the century. The East India Company had begun to realise that the preservation of a high quality was an essential in the competition with the silks of Europe. The statistical information given will show how rapid a progression had been made by Italy, and later on by France. Turkey had practically ceased to be of moment, but new competitors sprung into existence, as the demands for silk widened into numerous and distinct markets. So far the Company had conquered deep-rooted prejudices, as the result of an enlightened policy. The quality of the silk had been improved by the education of the Natives in the European methods of reeling, and by the introduction of superior breeds of silk-worms. All this was, however, recognised as contrary to the wishes and inclinations of the Indian people, and degeneration was foreseen as inevitable, if at any future period the degree of attention, which had hitherto been bestowed on it, should be relaxed. The reader, who may have followed the facts placed before him in the various chapters of this article, will doubtless concur with the author in the opinion that while spasmodically and more as the result of private than official enterprise, effort has been made to revive or preserve the trade of India in reeled silk, it has nevertheless seriously declined in quantity. But influences have for years been in operation which were destined to change the silk trade of India. With the advance of manufacturing skill in Europe, new uses were discovered for the various kinds of Indian silk. Special properties were also seen to be possessed by the superior silks of Italy and France with which India might relinquish all hope of successful competition. Hence it came about that while for forty years after the date of Milburn's review, the total value of the Indian exports manifested an apparent expansion, that result was due to the growth of a traffic in waste and wild silks, and not to the improvement of the mulberry worm. India drifted into a new trade, and did not possess the means of preventing the revolution that was taking place. Indeed, it may be said that coincident with this change in the nature of the exports, India gradually became a large importing country, its own mulberry plantations being no longer able to meet even its local demands. That this change was brought about by a countless series of influences, there can be no manner of doubt.

A change in the character of India's Silk Trade.

2156

Conf with pp. 32, 62, 159, 197, 201.

During the period alluded to, European production was also greatly expanded, and changes in the constitution of the East India Company threw open to free trade the vast resources of China and Japan. These countries did not, however, come prominently forward until the outbreak of disease was feared to imperil the European supply. The alarm thus aroused temporarily revived the Indian trade, and it placed China and Japan upon a permanent footing. In India the point appears to have been gradually reached when it became no more a profitable undertaking to contend against climatic disadvantages, and the apathy and indifference of the Indian growers. A market had, however, been discovered for even the most inferior qualities and waste materials of silk, and the export in these was found to be more profitable than the troublesome and expensive production of mulberry silk under direct European supervision. It thus transpired that the necessity never occurred to return in real earnest to the endea-

Competition with China & Japan.

2157

TRADE: Historic Records.

vours towards improvement, which, at the beginning of the century, appear to have been crowned with the success aimed at. The Company had established Indian silk as a distinct factor in the then markets. But its policy, had it been confronted with the seriously altered state of affairs, which has transpired, would very probably have been similar to that pursued by the private enterprise, which has ruled the Indian silk trade during the period of direct British Government in the destinies of India. The majority of the parties most interested in silk are now loud in their protestations, that what they want is not so much superior quality as the removal of the disease and fly-pest which retard a greater production. The Indian silk trade, both foreign and internal, seems thus to have changed in every aspect, and it may be accepted as highly problematic if it is ever again likely to return to the position aimed at by the East India Company, when the Indian exports overthrew the Turkish monopoly, and contested the supply to the British manufacturers with Europe and China. The present position of the trade may, indeed, be accepted as one of those often inexplicable adjustments between supply and demand which oppose an infinitely more formidable front than any supposed inherent incapacity to change or to improvement. The interests of the Indian people, through the vast advances in agricultural commerce, have gradually been revealed as more intimately associated with other articles of production than with silk. Their time and resources have been proved to them, as more profitably spent in other enterprises until it has actually become no longer a paying undertaking to collect even wild silks. Fashion and the idiosyncrasies of demand generally, have procured to India what is practically a new silk trade, and one which, if ever again lost, will leave this country with its comparatively unimportant traffic in reeled mulberry silk, in a position considerably lower than before the date of the earliest endeavours to place it among the important silk-producing countries of the world. The fancy cotton textiles of Europe have largely displaced from India, as a whole, the once extensive market for silken garments and have ruined the home manufactures. The demand for silk fabrics may be viewed as having survived in Burma alone, and the future enhancement of the production in reeled mulberry silk may be said therefore to have to contend against the loss of the Indian demand, the superior and more extensive production of Europe, China, and Japan, and the climatic and other disadvantages which tend constantly to lower the quality and hence to necessitate expensive and troublesome renewal of stock. Indian mulberry silk has, however, at present a distinct place in the manufacturing industries, due largely to its peculiar glossiness—a property which recommends it for certain purposes for which other silks are not so suitable. But the major portion of the export trade in Indian silk is comprised under the heading "Waste Silk," and although the demand for the articles embraced under that designation has continued to increase for some years past, a change in fashion might ruin this feature of the Indian trade within a few years' time. If such a calamity ever occurs, silk might be assigned a place among the articles of Indian export which are not of sufficient importance to call for separate recognition. The date when waste and wild silks first began to be exported is probably now impossible to determine definitely. In one of the official returns of the East India Company, the exports are given of *tasar* from 1810 to 1816; the first official recognition of waste silk having become an important item in the Indian export trade occurs about forty years ago; the export of what was recognised as "Country Wound Silk" disappeared from the returns of trade in 1825, when filature, or as it was then called *Novi* silk, took its place. It is thus probably safe to assume that the Indian export

Present demands 2158

The effect of European Cotton goods on Indian Silk market. 2159

Glossiness of Indian Silk. 2160

Waste Silk Trade. 2161

TRADE: Historic Records.

trade in Raw and Waste Silk gradually assumed its present character from about the time India passed out of the hands of the East India Company. On account of the difficulty which thus exists in separating the returns into the items comprised under the heading "Raw Silk," it may be as well to give here the figures collectively down to 1835 in continuation of the table at page 189, which exhibits the imports into Great Britain from 1773 to 1802:—

Statement of the Quantities of Raw Silk imported into London from Bengal from 1803 to 1835.

Exports from Bengal. 2162 *Conf. with table on p. 189. (Column I.)*

Years.	I. Company's Bengal Raw Silk imported.	II. Private Bengal Raw Silk imported, warehoused by the Company.	III. Total Company's Import and Private Import, warehoused by the Company.
	lb	lb	lb
1803	336,189	68,904	405,093
1804	415,917	205,793	621,710
1805	460,303	375,601	835,904
1806	235,215	173,308	408,523
1807	225,984	267,601	493,585
1808	325,243	53,225	378,498
1809	116,124	46,623	162,747
1810	373,598	211,120	584,718
1811	258,953	145,803	404,756
1812	558,862	423,565	982,427
1813	831,891	252,459	1,084,350
1814	722,727	114,239	836,966
1815	522,810	279,476	802,286
1816	381,215	398,549	779,764
1817	373,459	128,876	502 335
1818	758,116	402,860	1,160,976
1819	553,105	197,922	751,027
1820	811,875	259,572	1,071,447
1821	817,625	172,838	990,463
1822	845,382	197,235	1,042,617
1823	850,668	310,518	1,161,186
1824	660,012	271,637	931,649
1825	699,230	220,206	919,436
1826	898,388	338,635	1,237,023
1827	926,678	99,361	1,026,039
1828	1,039,623	96,686	1,136,309
1829	1,129,710	258,044	1,387,754
1830	1,096,071	90,092	1,186,163
1831	1,030,280	64,597	1,094,877
1832	750,828	205,625	956,453
1833	698,851	52,129	750,980
1834	757,517	53,124	810,641
1835	721,509	6,026	727,535

A.D. 1835.

As at the present day, so during all the past history of the Indian silk trade, the exports from Bengal are practically the total exports from India. It would, however, be now incorrect to give the imports of Bengal silk into London as anything like approximately the total exports from India. During the time of the East India Company Indian goods were entirely consigned to London, and were distributed to other parts of the world from the Company's warehouses.

A.D. 1855.

The returns of trade from 1835 to 1855 are believed to manifest no special feature that calls for comment, and it is assumed that a fairly ac-

SILK. **Trade in Silk**

TRADE: Historic Records.

curate conception has already been conveyed of the rise and progress of the foreign trade in Indian silk. In the table below, the exports of "Raw Silk" are carried from 1855 down to 1891. The figures for the first nineteen years of that period are published in pounds sterling, and to keep up the comparison, those of the succeeding years have been expressed according to the conventional formula of R10 to £1. It will, however, be recognised that while down to 1874 that ratio of the two currencies was very nearly correct, subsequent to 1874 a reduction, based on an average rate of exchange of 1*s*. and 6*d*. to the rupee, would more nearly express the actual money value of the Indian exports. It must also be explained that an error is admitted by the writer in the figures given from 1855 to 1869, since he has been unable to separate from the totals the quantities, in each of these years, of foreign raw-silk re-exported from India. The subsequent years are the actual exports of Indian raw-silk. The average annual re-exports of foreign raw-silk for the past twenty years has been about 130,000lb, so that the quantities shown from 1855 to 1869 should be accepted as considerably above the actual exports of Indian produce, How far this error applies to the figures given from 1773 to 1869 the writer has been unable to discover, but it may be safe to assume that, since the opening of the Suez Canal, India has year by year been a less and less important emporium for the Asiatic export traffic.

Re-exports, Foreign silk. 2163 *Conf. with* *pp. 28, 202.*

Quantity and value of the Exports of "Raw-Silk" from India from 1855 to 1891.

Year.	Quantity.	Value.	Year.	Quantity.	Value.
	lb	£		lb	£
1855-56	1,148,841	707,706	1873-74	2,223,917	1,143,744
1856-57	1,756,778	782,140	1874-75	1,656,015	766,461
1857-58	1,580,463	766,673	1875-76	1,310,569	415,961
1858-59	1,217,438	725,655	1876-77	1,417,893	776,903
1859-60	1,670,698	817,853	1877-78	1,512,819	703,549
1860-61	1,955,656	1,036,728	1878-79	1,329,599	570,229
1861-62	1,101,844	686,083	1879-80	1,401,506	526,157
1862-63	1,228,684	822,892	1880-81	1,302,576	548,201
1863-64	1,369,556	954,649	1881-82	1,117,026	388,263
1864-65	1,582,341	1,165,901	1882-83	1,359,433	544,143
1865-66	1,445,153	745,352	1883-84	1,602,812	627,611
1866-67	2,145,354	811,798	1884-85	1,564,901	463,561
1867-68	2,226,201	1,553,229	1885-86	1,438,767	332,251
1868-69	2,463,937	1,362,381	1886-87	1,583,924	484,339
1869-70	2,368,452	1,422,076	1887-88	1,625,177	480,810
1870-71	2,131,399	1,258,527	1888-89	2,121,914	518,750
1871-72	1,893,322	1,081,097	1889-90	2,089,762	639,817
1872-73	2,231,578	1,256,356	1890-91	1,573,214	521,069

A.D. 1873.

A comparison of the figures of the Indian *Raw Silk Trade*, from 1773 to 1891, will reveal the fact that, during the first half of that period, the exports gradually increased, until in 1829 they reached the by no means inconsiderable amount of 1,387,754lb. But what is of even greater importance there is reason to believe that practically the whole of these exports were in reeled mulberry silk. The exports of wild silk were inconsiderable, and prior to 1857 "Waste Silk" was in Europe a useless bye-product, and was not likely therefore to have been exported from India. The entire returns of raw silk, sent from India during the administration of the East India Company may, therefore, be accepted as having been reeled silk.

TRADE in RAW SILK.

But to convey a more precise conception of the modern Indian trade in silk, it is necessary to treat the subject under two sections—(I) Raw Silk and (II) Silk Manufactures.

RAW SILK.

I.—Foreign Exports by Sea.

FOREIGN Exports. 2164

An effort has been made in the remarks above to sketch out the chief historic features of the growth of the Indian export trade in Raw Silk. It has been shown to have greatly expanded under the fostering care of the East India Company, but that, with the discovery of a means of utilising waste silk, it changed its character rapidly. It may, in fact, be said that at first the English market was supplied by Turkey, and that the Hon'ble the East India Company successfully competed with that source of supply. Gradually the Turkey imports declined; but about the same time the renewed effort to produce silk in Italy and France was crowned with complete success. At one time, through the improvements effected in reelling silk in India, this country was able to hold its own against the Continental silks, but the superior quality of Italy and France soon began to tell, hence India steadily gave indications of being no longer able to hold the high position it had attained, when it had to compete mainly with Turkey. It had thus lost its hold even before the discovery of the value of waste and wild silks, but from 1858 down to the present day a continuous series of changes have characterised the trade, until many writers have spoken of Indian silk as completely ruined. The effort has been made in this article to show that what has actually taken place has been a complete change in the nature and location of the traffic more than a positive decline. This has doubtless resulted in the ruin of many persons who were formerly flourishing traders, but it would be incorrect to speak of the silk interests of India as totally destroyed. The waste silk of the Indian village reeling basins and of the filatures of this country have found a ready sale. A demand has been created for wild silks, and a trade has come into existence in cocoons of all classes. The reeled silk of India has been at the same time recognised as possessing special features of its own for which there is at the pesent moment a distinct demand, even although in many respects it is inferior to the silks which are being poured into England from Italy, France, China, Japan, etc. The market for Indian reeled silk has been thus circumscribed, and the exports of that class have declined in almost the same ratio that the exports of waste and wild silk and cocoons have expanded. At the same time a great improvement has been effected in the quality of Indian reeld silk, both Mulburry and Tasar.

A change in the character of Indian silk trade. 2165 *Conf. with pp. 32, 59, 62, 72, 106, 118, 121, 148, 156, 187, 193, 201.*

The following table gives the *Raw Silk* (*Indian Produce*) *exported* from India during the years 1879 to 1891:—

Year.	Silk.	Chasam.	Cocoons.	TOTAL.
	lb	lb	lb	lb
1879-80	563,210	788,481	49,815	1,401,506
1880-81	550,665	733,464	18,447	1,302,576
1881-82	339,322	749,121	28,583	1,117,026
1882-83	501,576	834,405	23,452	1,359,433
1883-84	672,710	886,045	44,059	1,602,814
1884-85	531,205	950,983	82,713	1,564,901
1885-86	358,071	1,023,807	56,889	1,438,767
1886-87	449,515	1,020,595	113,814	1,583,924
1887-88	453,568	998,235	173,374	1,625,177
1888-89	433,473	1,313,874	374,567	2,121,914
1889-90	593,425	1,233,494	262,843	2,089,762
1890-91	454,280	983,039	135,895	1,573,214

A.D. 1800.

FOREIGN TRADE in RAW SILK.

Exports.

The following table analyses the returns of the Raw Silk Export Trade of the years 1885 to 1890 :—

Analysis of the Exports from India of Raw Silk for the past five years, designed to show the Countries to which consigned and the Provinces from which exported.

Countries to which exported.		I. 1885-86.	II. 1886-87.	III. 1887-88.	IV. 1888-89.	V. 1889-90.
United Kingdom.	Quantity lb	63,453	74,497	122,139	135,039	167,365
	Value . R	437,273	685,943	10,21,523	10,62,427	12,86,623
France .	Quantity lb	159,807	167,367	208,437	244,245	299,272
	Value . R	10,62,296	12,39,992	14,51,624	17,23,889	22,59,883
Italy .	Quantity lb	124,470	203,552	110,421	44,187	112,129
	Value . R	8,85,349	16,47,721	9,51,282	3,39,777	10,06,664
Turkey in Asia.	Quantity lb	5,753	3,548	10,316	6,642	7,429
	Value . R	24,213	16,537	52,277	33,542	33,693
Aden .	Quantity lb	802	...	1,275	977	1,119
	Value . R	3,050	...	4,957	4,120	4,190
United States.	Quantity lb	...	...	...	1,631	5,354
	Value . R	...	...	...	11,162	38,872
Others .	Quantity lb	3,786	551	980	2,383	6,111
	Value . R	16,771	2,703	2,378	3,554	2,534
XI.—Total of Reeled silk.	Quantity lb	358,071	449,515	453,568	433,473	593,425
	Value . R	24,27,444	35,92,821	34,85,642	31,77,955	46,32,975
XII.—Total of Chasam and Cocoons.	Quantity lb	1,080,696	1,135,409	1,171,609	1,688,441	1,496,337
	Value . R	8,95,068	12,50,571	13,22,466	20,07,553	17,65,203
XIII.—Grand Total of Silk of all kinds.	Quantity lb	1,438,767	1,583,924	1,625,177	2,121,914	2,089,762
	Value . R	33,22,512	48,43,392	48,08,108	51,87,508	63,98,178

Provinces from which exported.		VI. 1885-86.	VII. 1886-87.	VIII. 1887-88.	IX. 1888-89.	X. 1889-90.
Bengal .	Quantity lb	351,155	447,354	445,831	426,133	581,864
	Value . R	24,06,581	35,88,259	34,67,320	31,66,336	45,66,132
Bombay .	Quantity lb	6,476	2,161	7,737	7,322	1,866
	Value . R	19,530	4,562	18,322	11,544	7,177
Sind .	Quantity lb	440	...	...	...	...
	Value . R	1,333	...	...	...	...
Madras .	Quantity lb	...	...	...	...	9,695
	Value . R	...	...	...	...	59,666
Burma .	Quantity lb	...	...	...	18	...
	Value . R	...	...	...	75	...
Total of Reeled Silk.	Quantity lb	358,071	449,515	453,568	433,473	593,425
	Value . R	24,27,444	35,92,821	34,85,642	31,77,955	46,32,975
Chasam and Cocoons.	Quantity lb	1,080,696	1,135,409	1,171,609	1,688,441	1,496,337
	Value . R	8,95,068	12,50,571	13,22,466	20,07,553	17,65,203
Grand Total of Silk of all kinds.	Quantity lb	1,438,767	1,583,924	1,625,177	2,121,914	2,089,762
	Value . R	33,22,512	48,43,392	48,08,108	51,87,508	63,98,178

S. 2165

FOREIGN TRADE in RAW SILK.

Exports.

Through the kindness of Mr. J. E. O'Conor the writer has been furnished with the following more detailed analysis of he exports from India, in which is exhibited the relative amounts of wild and domesticated silks which are included under he headings Raw Silk, Silk Waste (or Chasam), and Cocoons:—

Exports of Raw Silk (Indian Produce) from British India to Foreign Countries in 1886-87 to 1890-91.

COUNTRIES.	1886-87.		1887-88.		1888-89.		1889-90.		1890-91. (11 months.)	
	Quantity.	Value.	Quantity.	Value.	Quantity	Value.	Quantity.	Value.	Quantity.	Value.
	℔	R	℔	R	℔	R	℔	R	℔	R
Silk, Raw (excluding Tasar and other Wild Silks)—										
To United Kingdom	65,483	6,27,355	114,546	9,89,537	119,988	9,84,160	125,658	10,82,148	112,676	9,71,914
,, France	140,915	11,19,913	134,143	10,85,842	214,406	15,96,855	256,953	20,83,601	137,647	10,92,106
,, Italy	203,403	16,46,721	109,824	9,46,882	44,187	3,39,777	112,129	10,06,664	114,052	9,99,430
,, Other Countries	839	3,128	3,356	13,038	3,297	18,222	7,561	47,759	4,632	29,850
TOTAL	410,640	33,97,117	361,869	30,35,299	381,878	29,39,014	502,301	42,20,172	369,007	30,93,300
Silk, Raw (Tasar, Cricula, Muga, Eri, &c.)—										
To United Kingdom	9,014	58,588	7,593	31,986	15,051	78,267	41,707	2,04,475	112	672
,, France	26,452	1,20,079	74,294	3,65,782	29,839	1,27,034	42,319	1,76,282	80,329	3,40,832
,, Turkey in Asia	3,220	15,737	9,066	47,375	6,376	32,242	6,386	29,000	4,832	21,978
,, Other Countries	189	1,300	746	5,200	329	1,398	712	3,046	...	...
TOTAL	38,875	1,95,704	91,699	4,50,343	51,595	2,38,941	91,124	4,12,803	85,273	3,63,482
Silk, Chasam (excluding Tasar, &c.)—										
To United Kingdom	112,301	1,03,856	164,075	1,53,356	227,434	2,17,267	104,474	1,04,910	237,418	2,43,297
,, France	558,804	6,34,822	580,138	6,65,828	851,138	9,08,242	782,401	8,96,365	593,712	6,63,943
,, Other Countries	219	151	1,680	1,000	2,212	3,715	5,061	3,268	1,450	726
TOTAL	671,324	7,38,829	745,893	8,20,184	1,080,784	11,29,224	891,936	10,04,543	832,580	9,07,965

SILK. Trade in Silk

FOREIGN TRADE in RAW SILK Exports.

Exports of Raw Silk, etc.—Contd.

Countries.	1886-87.		1887-88.		1888-89.		1889-90.		1890-91. (11 months.)	
	Quantity.	Value.	Quantity.	Value.	Quantity.	Value.	Quantity.	Value.	Quantity.	Value.
	lb	R	lb	R	lb	R	lb	R	lb	R
Silk, Chasam (Tasar, &c.)—										
To United Kingdom	127,419	1,02,751	150,024	1,54,144	164,803	2,15,510	270,631	2,76,335	109,330	99,669
,, France	221,852	2,53,124	102,318	1,11,416	68,287	96,517	70,927	72,710	40,776	49,665
,, Other Countries	...	...	...	...	...	...	...	...	353	1,277
Total	349,271	3,55,875	252,342	2,65,560	233,090	3,12,027	341,558	3,49,045	150,459	1,50,611
Silk, Cocoons (excluding Tasar, &c.)—										
To United Kingdom	43,815	55,135	23,404	33,395	10,433	17,305	37,923	54,950	15,075	21,416
,, France	14,829	34,596	14,763	37,578	30,094	41,559	1,300	1,300	9,976	11,377
,, Other Countries	176	1,460	...	...	2	50	...	...	...	...
Total	58,820	91,191	38,167	70,973	40,529	58,914	39,223	56,250	25,051	32,793
Silk, Cocoons (Tasar, &c.)—										
To United Kingdom	32,689	37,899	82,298	98,073	272,326	4,26,335	221,108	3,52,725	107,841	1,57,077
,, France	21,035	25,737	52,609	67,326	61,712	83,053	2,512	2,640	3,003	3,029
,, Other Countries	1,270	1,040	300	350	...	...	...	...	...	...
Total	54,994	64,676	135,207	1,65,749	334,038	5,09,388	223,620	3,55,365	110,844	1,60,106
GRAND TOTAL	1,583,924	48,43,392	1,625,177	48,38,108	2,121,914	51,87,518	2,089,762	63,98,178	1,573,214	47,08,257

FOREIGN TRADE in RAW SILK.

Exports.

In order to show the relative importance of the wild to the domesticated silks, the following statement may be furnished of the chief items of export shown in the above table :—

	1886-87.	1887-88.	1888-89.	1889-90.	1890-91 (11 months).
	Quantity.	Quantity.	Quantity.	Quantity.	Quantity.
MULBERRY SILKS—	℔	℔	℔	℔	℔
Raw	410,640	361,869	381,878	502,301	369,007
Chasam . . .	671,324	745,893	1,080,784	891,936	832,580
Cocoons . . .	58,820	38,167	40,529	39,223	25,051
TOTAL .	1,140,784	1,145,929	1,503,191	1,433,460	1,226,638
WILD SILKS—					
Raw	38,875	91,699	51,595	91,124	58,273
Chasam . . .	349,271	252,342	233,090	341,558	150,459
Cocoons . . .	54,994	135,207	334,038	223,620	110,844
TOTAL .	443,140	479,248	618,723	656,302	346,576
GRAND TOTAL .	1,583,924	1,625,177	2,121,914	2,089,762	1,573,214
Percentage of wild silks on the grand total . .	27·98	29·49	29·16	31·41	22·03

Altered nature of India's silk trade. 2166 Conf. with pp. 32, 59, 62, 158, 193, 197, etc.

The reader will now appreciate the full force of the remark that the Indian export trade in raw silk has, during at least the past twenty or thirty years, changed its character. The exports of what appears to have been reeled mulberry silk began to exceed one million pounds in weight about 1820, and in 1857 (the year when the utilisation of waste and wild silk was discovered) the Indian exports stood at one million seven hundred thousand pounds. Twenty years later the declarations of mulberry reeled silks had fallen to about one quarter that amount, and during the past four years they have averaged four hundred thousand pounds. It has thus transpired that while the grand totals of the Indian exports in raw silk have remained practically stationary, from some years past, this has been due to the growth of the new trade in waste and wild silk and cocoons. At the same time the exports in reeled *tasar* have slowly improved: they were 7,872℔ in 1814, and during the past five years they have averaged 73,000℔, or an increase of less than 1,000℔ per annum. The error of disregarding this altered nature of India's trade in raw silk might be exemplified by an extensive list of quotations from writers in Europe. One may suffice. **Mr. T. Wardle**, in his *Descriptive Catalogue of the Silk Section of the Jubilee Exhibition, Manchester*, says: "India sends to Europe but very little raw silk now. It was only 457,600℔ in 1885; in 1874 it was 2½ million ℔, and in 1870, 2¼ million ℔, against an annual export from China to Europe in 1883 of 7,000,000℔, and from Japan of 3,000,000℔." **Mr Wardle**, thus, inadvertantly, while proving the decline of the Indian reeled mulberry silk trade, compares apparently the exports of reeled silk in 1885 with the total of all raw silk in 1874. The correct figures for these years will be found in the table given above, p. 196.

Re-exports. 2167

RE-EXPORTS BY SEA.

But India imports a considerable amount of raw silk from foreign countries, and re-exports again to foreign countries a certain proportion of these imports. The following are the amounts of these re-exports since 1871-72 :—

SILK. | **Trade in Silk**

FOREIGN TRADE in RAW SILK.

Re-exports.

1871-72	℔ 94,545
1872-73	„ 142,361
1873-74	„ 168,313
1874-75	„ 74,754
1875-76	„ 106,744
1876-77	„ 150,597
1877-78	„ 145,186
1878-79	„ 205,116
1879-80	„ 271,556
1880-81	„ 207,030
1881-82	„ 157,485
1882-83	„ 163,912
1883-84	„ 130,373
1884-85	„ 142,184
1885-86	„ 84,457
1886-87	„ 124,605
1887-88	„ 109,209
1888-89	„ 111,832
1889-90	„ 116,261

These re-exports are silk imported mostly from China, the Straits, Persia, Egypt, and Japan, and exported again from India to the United Kingdom, Egypt, Aden, Arabia, and France. These transactions to and from India are made in the order of importance in which the countries have been mentioned above.

Conf. with pp. 28, 196.

It has already been explained that a serious error is often made by writers on Indian Silk through their giving as the exports from India, the totals of the Indian and Foreign Raw Silk. They thus raise incorrectly the importance of India as a source of the commodity. Below will be seen the total import traffic, and it will be observed that these re-exports leave a balance of foreign silk in the country quite as large, if not larger than the total exports of Indian grown silk. (*Conf.* with the illustration in the concluding paragraph of the Foreign Trade on page 204).

Land Routes. 2168

FOREIGN EXPORTS BY LAND ROUTES.

The traffic in raw silk from India across the frontier is not very extensive: the average total exports during the past five years were a little over 6,000℔. The chief countries to which raw silk is consigned are Bhután, Kashmír, Sewestan, and Tibet. The exports in 1889-90 amounted to 3,584℔, valued at R7,47,065.

If now it be desired to ascertain the total exports of all kinds of raw silk from India, it would only be necessary to add together, for any one year, the exports of Indian produce given in the table at page 195 to those of foreign silk re-exported shown above with those carried across the frontier by land routes. Thus the total exports in 1889-90 were 2,209,607℔.

Imports by Sea. 2169

II.—Foreign Imports of Raw Silk by Sea.

The following were the imports during the past twenty-one years, by far the major portion of which came from China to Bombay :—

1869-70	℔ 2,019,974
1870-71	„ 2,328,854
1871-72	„ 1,799,591
1872-73	„ 1,930,910
1873-74	„ 2,282,758
1874-75	„ 2,467,255
1875-76	„ 2,457,244
1876-77	„ 1,461,069
1877-78	„ 2,102,930
1878-79	„ 1,813,993
1879-80	„ 2,005,070
1880-81	„ 2,511,802
1881-82	„ 1,760,595
1882-83	„ 2,386,150
1883-84	„ 2,210,893
1884-85	„ 1,831,702
1885-86	„ 1,732,559
1886-87	„ 1,737,891
1887-88	„ 2,598,597
1888-89	„ 2,045,569
1889-90	„ 2,360,467

By carrying the statistics of the trade further back it is seen to have increased more rapidly, during the previous period of twenty years, than during that for which the figures have been given. In 1850-51, for example, it stood at 1,259,974℔.

FOREIGN TRADE in RAW SILK.

Imports by Sea.

The following table may be given as an analysis of the returns during the past five years :—

Analysis of the Imports of Raw Silk for the past five years designed to show the Countries from which obtained as well as the receiving Provinces.

Countries from whence imported.		I. 1885-86.	II. 1886-87.	III. 1887-88.	IV. 1888-89.	V. 1889-90.
CHINA — Hong-Kong	℔	1,223,921	1,220,556	1,722,987	1,187,551	1,420,724
	R	46,54,889	54,31,236	71,33,185	47,67,403	58,36,951
CHINA — Treaty Ports	℔	202,242	205,804	320,596	303,231	3[illegible]3,776
	R	10,64,110	10,37,979	15,76,944	15,21,367	17,43,615
Straits Settlements	℔	257,589	244,675	484,314	459,859	540,168
	R	12,89,527	11,13,920	26,93,819	24,57,370	27,56,804
Persia	℔	15,725	41,064	38,259	27,119	19,319
	R	78,588	2,07,595	1,91,708	1,37,491	91,746
Egypt	℔	11,545	10,634	10,423	4,145	18,563
	R	64,651	64,212	61,680	34,432	1,60,278
United Kingdom	℔	10,292	5,166	2,433	1,097	3,079
	R	50,415	33,696	10,783	5,368	10,221
France	℔	1,000	Nil	250	404	829
	R	5,000	Nil	750	1,443	5,599
All other Countries	℔	2,163	2,101	16,373	55,123	13,600
	R	8,436	8,382	59,681	1,20,437	63,939
XI.—Total of Reeled Silk	℔	1,724,477	1,730,000	2,595,635	2,038,529	2,360,058
	R	72,15,616	78,97,020	1,17,88,547	90,45,311	1,06,69,153
XII.—Total of Chasam and Cocoons	℔	8,082	7,891	2,962	7,042	409
	R	3,550	36,743	14,663	14,075	1,300
XIII.—GRAND TOTAL OF RAW SILKS OF ALL KINDS	℔	1,732,559	1,737,891	2,598,597	2,045,569	2,360,467
	R	72,19,166	79,33,763	1,17,43,210	90,59,386	1,06,70,453

Provinces to which imported.		VI. 1885-86.	VII. 1886-87.	VIII. 1887-88.	IX. 1888-89.	X. 1889-90.
Bengal	℔	106,134	36,765	34,428	13,359	12,783
	R	1,97,563	81,386	89,363	30,070	28,372
Bombay	℔	1,440,514	1,555,135	2,198,246	1,678,932	1,956,846
	R	58,86,001	69,13,580	91,81,250	67,82,260	81,80,274
Sind	℔	3,457	1,443	1,118	945	1,120
	R	18,220	6,300	5,875	5,820	7,320
Burma	℔	174,372	136,657	361,843	315,293	389,309
	R	11,13,832	89,5,754	24,52,058	22,27,161	24,53,187
XI.—Total of Reeled Silk	℔	1,724,477	1,730,000	2,595,635	2,038,529	2,360,058
......	R	72,15,616	78,97,020	1,17,28,547	90,45,311	1,06,69,153
XII.—Total of Chasam and Cocoons	℔	8,082	7,891	2,942	7,042	409
......	R	3,550	36,743	14,663	14,075	1,300
XIII.—GRAND TOTAL OF RAW SILKS OF ALL KINDS	℔	1,732,559	1,737,891	2,598,597	2,045,569	2,360,467
......	R	72,19,166	79,33,763	1,17,43,210	90,59,386	1,06,70,453

NOTE.—The difference between the figures in XI. from XIII. for each of the years is the amounts of chasam and cocoons. The figures in columns I. to X. are the returns of reeled silk.

SILK. **Trade in Silk**

FOREIGN TRADE in RAW SILK.

Imports by land. 2170

FOREIGN IMPORTS BY LAND ROUTES.

The imports of raw silk by land routes are much more important than the exports. The major portion consists of two classes of silk, *viz.*, that known in the bazárs of Upper India as "Bokhara silk" which comes to India *viá* Kabul or Ladakh, and "Kashmir silk." In Upper India, Amritsar is the chief mart to which these silks are consigned. During the past five years the imports (treating Upper Burma as no more a foreign country) have been—

YEARS.	Quantity in lb.	Value in rupees.
1885-86	165,760	6,50,248
1886-87	114,352	5,50,780
1887-88	61,376	3,04,706
1888-89	14,324	60,820
1889-90	45,024	2,45,605

It would thus appear that the imports by land into India are manifesting a serious decline. A small traffic into Burma takes place from China and Siam. The average imports from Siam for the past five years has been 3,136lb.

Net balance in favuor of Foreign Imports. 2171

From the facts given in this section it will be learned that the total imports of all forms of raw silk were last year 2,405,491lb, which, by deducting the total exports 2,209,607lb, left a net import of 195,884lb. It will thus be seen that India should be regarded as a recipient rather than participator in the world's supply of raw silk. (*Conf. with pp. 228, 229*)

INTERNAL TRADE. 2172

III.—Internal Trade in Raw Silk.

Having now shown the foreign transactions, it may be as well to endeavour to exhibit some of the chief features of the inter-provincial exchanges. The returns available are, however, very incomplete, as they are the records of the traffic on the chief routes only, such as by ship along the coast or up the rivers and by rail. The road trade of India is nowhere so accurately recorded as to allow of even an approximate statement of the exports from the producing districts. Local consumption cannot, therefore, be determined, since the multitude of small transactions carried from village to village entirely escape registration. The returns of coastwise and of rail and river trade, so far as they go, may, however, be accepted as accurate, and the facts there manifested convey a fairly complete conception of the relative values of the producing areas, and they point also to the chief consuming centres. In using the figures, with the object of discovering these all-important aspects of the internal trade, it is necessary to bear in mind that the imports by the port towns (more especially Calcutta) are intended to meet the foreign demands.

The trade may be viewed according to two sections—Coastwise and Rail & River.

Coastwise. 2173

BY COASTWISE.

Indian Produce.—Madras (after Bengal itself) is the chief consuming province of Bengal silk, though lesser quantities of it go to Bombay and Burma. The Bombay exports are comparatively unimportant, and they are consigned mainly to Sind, Cutch, and Kathiawar. In using the coastwise trade returns it is practically sufficient to deal with the exports, since the consignments from one province may be said to be the imports by another. Of course that view of the coastwise traffic admits one or two insignificant

INTERNAL TRADE in RAW SILK.

Coastwise.

errors, due to so many ships being always at sea, and the destination of a consignment being often altered by telegraphic instructions. In drawing up the tables given below of coastwise transactions the trade between one port and another within each presidency has been excluded from consideration, so as to show the actual exchange between the provinces. For the past five years the gross exports in Indian raw silk have been as follows :—

YEARS.	Quantity in lb.	Value in rupees.
1885-86	246,867	11,04,108
1886-87	285,631	14,69,523
1887-88	340,400	17,14,772
1888-89	262,714	13,41,496
1889-90	250,691	12,31,148

BOMBAY SUPPLY.

2174

Conf. with pp. 30, 52, 209, etc.

This trade may be said to have seriously declined, since the exports ten years ago were much larger than at present. Thus the provincial exports were in 1877-78, 692,000lb. A noticeable feature of this decline is the disappearance practically of the coastwise exports from Bengal to Bombay. Thus, not only is the great mulberry silk-producing area (Bengal) unable to meet the demands of the chief manufacturing centre (Bombay), but it has apparently lost what hold it once enjoyed of that market. The exports from Bengal to Bombay from 1878 to 1890 were as follows :—

	lb.		lb.
1878-79	31,938	1884-85	32,968
1879-80	30,399	1885-86	27,768
1880-81	3,633	1886-87	8,286
1881-82	25,626	1887-88	*Nil*
1882-83	1,398	1888-89	765
1883-84	82	1889-90	*Nil*

Reason of China Traffic.

2175

The fact that China retains so firm a hold in the Bombay market is largely due to the freight being less from that country than the railway charges from Bengal. The remarks regarding the supply of Bombay with sugar are (it will be found) instructive, because due to similar causes to that of the silk supply. (*Conf. with Vol, VI., pt. II, 37, 213, 326, 328, 329 etc.*) The steamers that trade with China from Bombay have often to return almost in ballast, and their return freights are, therefore, exceptionally low.

The following may be given as analysis of the chief inter-provincial transactions in Indian raw silk coastwise during 1889-90 :—

Importing Provinces.	Exported from Bengal.		Exported from Bombay.		Exported from Sind.		Exported from Burma.		Total Imports.	
	lb	R	lb	R	lb	R	lb	R	lb	R
Bengal	...	...	...	...	...	...	91	320	91	320
Bombay	...	...	...	...	588	500	...	...	588	500
Madras	237,639	11,66,177	450	3,500	...	...	...	...	238,089	11,69,677
Burma	7,599	38,828	...	...	...	...	...	...	7,599	38,828
Sind	...	...	...	...	...	...	...	...	...	...
Cutch	...	...	2,363	12,270	...	...	...	...	2,363	12,270
Kattywar	...	...	1,961	9,553	...	...	...	...	1,961	9,553
TOTAL	245,238	12,05,005	4,774	25,323	588	500	91	320	250,691	12,31,148

SILK.

Trade in Silk

INTERNAL TRADE in RAW SILK.

Coastwise.

Foreign Produce.
2176

Foreign Produce.—As might be expected, Bombay port town is chiefly concerned in this trade. Excluding the inter-provincial exchanges the following were the amounts imported by the provinces of India during the past five years:—

	lb	R
1885-86	103,869	3,56,390
1886-87	145,482	5,41,537
1887-88	137,327	5,08,832
1888-89	118,162	4,04,461
1889-90	113,185	3,92,088

These imports were drawn almost entirely from Bombay port town and very probably, therefore, consisted largely of Chinese silk. The following analysis may be given of last year's trade:—

Imported by	Exported from Bombay.		Exported from Burma.		Total.	
	Quantity.	Value.	Quantity.	Value.	Quantity.	Value.
	lb	R	lb	R	lb	R
Bengal	...	...	196	1,414	196	1,414
Bombay	...	...	985	3,240	985	3,240
Sind	87,338	2,43,797	...	...	87,338	2,43,797
Madras	309	861	91	250	400	1,111
Diu	18	40	...	...	18	40
Goa	327	1,013	...	...	327	1,013
Cutch	14,103	82,382	...	...	14,103	82,382
Kathiawar	9,818	59,091	...	...	9,818	59,091
Total	111,913	3,87,184	1,272	4,904	113,185	3,92,088

The small amounts from Burma, Calcutta, Madras, and Bombay are somewhat interesting, since the chief share of the Bombay and of the Calcutta manufactures is taken by the Burmese. The following exhibits the distribution of the imports into the Provinces and Native States during the past five years:—

Years.	Bengal.	Bombay.	Sind.	Madras.	Goa.	Cutch.
1885-86	5,019	4,645	70,005	1,011	459	14,240
1886-87	7,419	5,364	97,689	3,966	116	18,445
1887-88	3,042	2,330	98,888	2,620	715	19,787
1888-89	*Nil*	684	95,330	459	994	11,552
1889-90	196	985	87,338	400	327	14,103

Years.	Kathiawar.	Burma.	Diu.
1885-86	8,490	*Nil*	*Nil*
1886-87	11,153	1,330	*Nil*
1887-88	10,370	350	25
1888-89	9,143	*Nil*	*Nil*
1889-90	9,818	*Nil*	18

INTERNAL TRADE in RAW SILK.

Sind, Cutch, and Kathiawar are by far the most important consuming areas of the foreign silk exported coastwise from Bombay.

Road, River & Rail. 2177

BY ROAD, RIVER & RAIL.

The trade in raw silk carried along these channels of exchange may be said to be referable to two sections, *viz.*, the transactions in Indian Raw Silk and the distribution of the imported Foreign Raw Silk. The latter is very much less important than the former, since the foreign imports are mainly into Bombay town, and are intended to feed the local mills. These imports, accordingly, disappear very largely from the registration of internal traffic. The Madras seaports and the port town of Bombay show a surplus of exports over imports (in Indian Raw Silk) by internal routes, but that is the result of the coastwise trade which deposits in these towns large quantities which are carried to the interior by road, rail, river, and canal. With the exception, therefore, of these port towns, the provinces which show an excess of exports over imports may be viewed as the producing areas of India. These are Bengal, the Central Provinces, and Assam. On the other hand, the value of the provinces of India as consuming areas (with the exception of Calcutta, which is the chief emporium in the foreign export trade) may be accepted as denoted by the extent of their excess of imports over exports. These are mainly Bombay, Madras, the Panjáb, and the North-West Provinces. In order to demonstrate these features of the Indian internal trade in raw silk, the following returns for the year 1888-89 may be here given :—

PROVINCES WHICH MANIFEST A NET EXPORT—THE PRODUCING AREAS.			PROVINCES WHICH MANIFEST A NET IMPORT—THE CONSUMING AREAS.		
Provinces and Towns.	Export and import in maunds of 82½ ℔.	Net export in ℔.	Provinces and Towns.	Import and export in maunds of 82⅓ ℔.	Net import in ℔.
INDIAN PRODUCE.					
Bengal . . . {	32,656 5,862		Calcutta . . {	31,108 1,588	
Net export .	26,794	2,200,505	Net import .	29,520	2,435,400
Madras Sea-ports . {	2,233 1,963		Madras Presidency {	2,298 255	
Net export . .	270	22,275	Net import .	2,043	168,547½
Bombay Port Town {	3,516 702		Sind . . . . {	42 2	
Net export. .	2,814	232,155	Net import .	40	3,300
Karachi . .	1	82½	North-West Provinces and Oudh. {	1,422 1,021	
Central Provinces . {	295 122		Net import .	401	33,082½
Net export. .	173	14,272			

SILK.

Trade in Silk and Silk Goods.

INTERNAL TRADE in RAW SILK.

Road, River & Rail.

Provinces which manifest a net export—the producing areas.			Provinces which manifest a net import—the consuming areas.		
Provinces and Towns.	Export and import in maunds of 82½ lb.	Net export in lb.	Provinces and Towns.	Import and export in maunds of 82½ lb.	Net import in lb.
INDIAN PRODUCE—*contd.*			Panjáb . .	1,347 180	
Assam . . .	4,302 11		Net import .	1,167	96,277½
Net export. .	4,291	354 007½	Berar—Net import	71	5,857½
Nizam's Territory .	60 49		Bombay Presidency	3,224 404	
Net export. .	11	907½	Net import .	2,820	232,650
			Rájputána and Central India.	132 16	
			Net import .	116	9,570
FOREIGN PRODUCE.					
Bombay Port .	6,992 17		Bombay Presidency	2,732 7	
Net export. .	6,975	575,437½	Net import .	2,725	224,812½
Karachi—Net export.	1,114	91,905	Sindh . . .	1,071 2	
Calcutta—Net export	247	20,377½	Net import .	1,069	88,192½
Bengal . . .	32 9		North-West Provinces and Oudh.	45 11	
Net export. .	23	1,897½	Net import . .	34	2,805
Central Provinces—Net export.	71		Panjáb . . .	4,331 8	
Berar—Net export.	20	1,650	Net import. .	4,323	356,347½
Rájputána and Central India.	5 2				
Net export. .	3	247½			
Nizam's Territory—Net export.	97	8,002½			

The above table, following the course pursued in connection with coastwise trade of the provinces, has separated the inter-provincial exchanges

S. 2177

INTERNAL TRADE in RAW SILK. Road, River & Rail.

(which are carried along rail and river routes) into two sections, *viz.*, transactions in raw silk of Indian and of foreign produce. Bengal is the chief exporting province in Indian raw silk, and Bombay Town the chief distributing centre of foreign raw silk. Leaving Calcutta out of consideration for the present (since it is the emporium in the foreign export trade), it may be said that the Panjáb and Bombay Presidency (as distinct from its port town) are about equally important consuming provinces of the silk carried by internal routes. During the year under notice the Panjáb received a net import of 356,347½℔ foreign and 96,277½℔ Indian raw silk; Bombay Presidency during the same period took 224,812½℔ foreign and 232,650℔ Indian. It may thus be safely assumed that silk manufacture is more extensive in these than in any of the other non-silk-producing provinces.

In fact, Bombay Presidency and the Panjáb between them use up, as a rule, the entire foreign raw silk which is exported from the port town of Bombay by rail. The imports of foreign silk by Calcutta are not very extensive. In the year under review (1888-89), it obtained 13,359℔ direct from foreign countries, but it drains annually by coastwise and transfrontier routes certain supplies which go to make up the exports from Calcutta to Bengal. These exports in the year here dealt with amounted to 20,377℔, which amount, except about 2,000℔, was utilised in Bengal. The exports from Bengal in foreign silk mostly find their way to the Panjáb.

Since detailed information is not available regarding the local manufactures of the various provinces of India, we are driven to accept the indications obtainable from the movements of raw silk as denoting local consumption. It would be, however, a false conclusion to suppose that since the exports of raw silk from Bengal, Assam, and the Central Provinces far exceed the imports, these provinces possess no local manufactures. An error would also be involved by the acceptance of the rail-borne returns of provinces having a sea-board as of co-equal importance with those of interior provinces, such as the Panjáb and the North-West. To arrive at some idea of the value of the Indian provinces as consuming areas for raw silk, it becomes necessary, in fact, to draw up a balance sheet of all their transactions. The admitted defects, due to the absence of returns of road traffic and our ignorance of the extent of local production, render it, however, undesirable to place implicit confidence on any such analytical system in the determination of provincial consumption of raw silk. The main facts given in the previous pages regarding three of the provinces may, however, be tabulated, in order to exemplify the method that might be pursued with all the others.

PANJAB. 2178

THE PANJAB TRADE IN RAW SILK.

Balance Sheet of the recorded Transactions with the Panjáb in 1888-89.

Imports.	Quantity in ℔.	Exports.	Quantity in ℔.
By Transfrontier Routes .	10,192	By Transfrontier Routes—	
By Rail and River, etc.—		(*a*) Indian Produce .	336
(*a*) Indian Produce .	111,127	(*b*) Foreign ,, .	...
(*b*) Foreign ,, .	357,307	By Rail and River—	
		(*a*) Indian Produce .	14,850
	478,626	(*b*) Foreign ,,	660
Deduct—Exports .	15,846		
			15,846
Net Imports .	462,780		

INTERNAL TRADE in RAW SILK.

Panjab.

The Panjáb having no sea-board, the above are the only routes of trade which are registered. It may thus be inferred that the silk goods of the Panjáb are largely made of mulberry silk, since the external supplies of the province are mainly drawn from foreign countries in which mulberry is chiefly produced. Amritsar is the great silk-manufacturing town of the province. The bulk of the foreign silk imported by Bombay is drawn from China, and that being so, it may safely be affirmed that the silk exported from Bombay to the Panjáb is Chinese silk, but whether mulberry or some of the other Chinese silks cannot be determined * A striking feature of this Panjáb trade is the fact that year by year the imports from Kashmír, Afghánistán, and Bokhara have declined, as those from Bombay have expanded. Great efforts were some few years ago made to develop the local industry of growing mulberry silk, but it may be said these failed, and that the amount of silk produced in the province is comparatively small, so that the net import of 462,780℔ may be accepted as fairly representing the Panjáb manufactures.

BOMBAY.

2179

Conf. with pp. 30, 55, 209.

BOMBAY TRADE IN RAW SILK.

A similar table may be given for the Bombay Presidency and for the Port Town of Bombay. In this case, however, the chief transactions are maritime to and from foreign countries and to and from other presidencies by coastwise routes A certain percentage of the latter transactions are to and from Bombay Presidency coast towns (other than the so-called port town), and these should be debited, or credited, as the case may be, to the presidency, but the difficulty in separating these small amounts from the greater traffic with the port town has recommended it as the safer course to treat the entire maritime trade of Western India as with its port town.

Balance Sheet of the recorded Transactions with the Bombay Presidency in 1888-89.

Imports.	Quantity in ℔.	Exports.	Quantity in ℔.
By Transfrontier Routes—		By Transfrontier Routes—	
(*a*) Indian . .	*Nil*	(*a*) Indian . .	*Nil*
(*b*) Foreign . .	*Nil*	(*b*) Foreign . .	*Nil*
By Rail and River—		By Rail and River—	
(*a*) Indian . .	265,980	(*a*) Indian . .	33,330
(*b*) Foreign . .	225,390	(*b*) Foreign . .	577
	491,370		
Deduct—Exports .	33,907		33,907
Net Imports .	457,463		

It would thus appear that the Presidency of Bombay obtained in 1888-89 a net import, 457,463℔. An examination of the incomplete statistics that exist regarding more recent years manifests the same fact, namely, that the Bombay Presidency manufactures use up about half a million pounds of raw silk, of which fully one-half that amount is Indian silk obtained through the port town. The returns of the coastwise trade show the imports of Indian raw silk to have seriously declined, but in no single year have they anything like equalled the amount returned as having been conveyed by road and rail from the port town to the presidency. The source from which the port town draws its supplies of Indian raw silk is thus inexplicable. The foreign raw silk used up in the presidency is also drawn from the extensive foreign supplies which Bombay town has

Indian Silk used in Bombay.

2180

* The waste silk spun in Bombay appears to be obtained from **Attacus cynthia** and **Antheræa pernyi**—see pp. 86, 213.—*Ed., Dict. Econ. Prod.*

INTERNAL TRADE in RAW SILK.

been exhibited as obtaining mainly from China. In the *Annual Report of the Rail and Road-borne Trade of Bombay*, it is shown that the EXPORTS from the Port Town to the Presidency were consigned to the following districts:—

Indian. 2181

Indian Raw Silk—To Gujarát and Kathiáwár, 142,560lb; to the country south of Narbada and below the Gháts, 97,927½lb; to Khándesh, Násik, and Ahmednagar, 6,930lb; and Poona and Sholápúr, 18,315lb. The presidency also imported 247lb from Madras, making the total gross receipts of 265,980lb. The IMPORTS by the port town from the presidency amounted to only 30,607½lb, which quantity was derived from the districts already named.

Foreign. 2182

Foreign Raw Silk.—The EXPORTS from the port town in this class of silk went to Gujarát and Kathiáwár, 21,945lb; to the country below the Gháts, 9,487½lb; to Khandesh, Násik, and Ahmednagar, 80,107½lb; and to Poona and Sholápur, 113,850lb. The IMPORTS by the port town from its presidency by rail and road are unimportant, 247½lb, having in the year in question been obtained from Gujarát and Kathiáwár. The districts named may, therefore, be accepted as the great manufacturing centres of the Bombay Presidency, after its port town. The following balance sheet may be given as representing the transactions of the Bombay Port Town:—

Balance Sheet of the Transactions with the Port Town of Bombay in Raw Silk during 1888-89.

Imports.	Quantity in lb.	Exports.	Quantity in lb.
BY SEA FROM FOREIGN COUNTRIES.		BY SEA TO FOREIGN COUNTRIES.	
Foreign Produce—		Re-exports, Foreign Produce—	
Silk	1,678,932	Silk	109,182
Waste . . .	*Nil**		
Cocoons . . .	7,030	BY SEA TO FOREIGN COUNTRIES.	
BY COASTWISE FROM INDIAN PROVINCES.		Indian Produce—	
Indian Produce—		Silk	7,322
Silk	3,102	Waste . . .	2,881
		Cocoons . . .	*Nil*
BY COASTWISE FROM INDIAN PROVINCES.		BY COAST TO INDIAN PROVINCES.	
Foreign Produce—		Indian Produce—	
Silk	5,257	Silk	11,600
BY RAIL AND RIVER, ETC.			
Indian Produce . .	57,915	BY COAST BY INDIAN PROVINCES.	
Foreign ,, . .	1,402		
	1,753,638	Foreign Produce—	
Deduct the Exports .	1,119,804	Silk	121,909
Net Import, being amount available for local demands	633,834	BY RAIL AND RIVER, ETC.	
		Indian Produce . .	290,070
		Foreign ,, . .	576,840
			1,119,804

*There must be some mistake in the official returns, since a large proportion of the silk imported into Bombay from China is Waste, not Reeled Silk. See foot-note to previous page.—*Ed.*

SILK.

Bombay Trade in Raw Silk.

TRADE in RAW SILK.

Foreign.

In the above table (as, indeed, in all the remarks hitherto offered regarding internal trade), the figures for the year 1888-89 have been given, in preference to those of more recent years, since for that year alone do we possess an imperial review of the rail, road, and river traffic. The reader will, however, discover that wherever possible more recent returns have been exhibited, and that these have been shown in such a manner as to preserve the relation of the selected year of detailed review (1888-89) to all the past and current periods in the historic sketch here offered of India's trade in silk. By the balance sheet given, Bombay port town is shown to have retained a net import of 633,834℔ of silk, an amount which, it is presumed, meets the demands of its local mills. Of the IMPORTS OF FOREIGN SILK it has already been indicated by the table at p. 200, that by far the major portion comes from China, but a small amount of foreign silk is recorded as imported coastwise. The following items of this trade may be noted: 4,418℔ exchanged between British ports within the presidency, not necessarily, therefore, entirely imported by the port town of Bombay; 602℔ obtained from Sind; and 237℔ from Cutch. By rail and road Bombay port town obtained 247⅓℔ from Gujarát and Kathiáwár, 660℔ from the Panjáb (Kashmír and Bokhara silk), and 495℔ from Calcutta.

Of INDIAN SILK, Bombay port town is shown to have obtained 3,102℔ coastwise and 57,915℔ by rail and road. The largest share of the former was obtained from British ports within the Presidency, and may, to a considerable extent, be locally produced and probably wild silks. It was, however, an intra-provincial exchange, and should not necessarily be accepted as entirely imported by the port town. Of the latter by far the most important item was an import by the port town of 30,607⅓℔ from the presidency. This may be largely locally produced silk, and perhaps mainly waste and wild silks. The next largest portion of these imports came from Calcutta, 15,840℔. It has already been shown that the coastwise exports from Calcutta to Bombay port town have declined since 1878-79 from 31,938℔ to be at most nominal. The returns of the rail-borne trade in raw silk from Calcutta to Bombay only began to appear separately in 1888-89, so that it is not possible to say whether it is expanding or contracting. In 1889-90 it stood at 18,975℔. It is thus, so far as can be learned, stationary, and at most only nominal in extent. Bengal, therefore, may with perfect fairness be said to take little or no share in the supply of Bombay with raw silk.

The chief producing area of India has thus failed to compete with China and other foreign countries in the supply of silk to the chief manufacturing centres of India. The balance sheet of Bombay port town traffic exhibits, however, certain important items of export trade which may be here discussed. One of the most striking features is the fact that for foreign silk Bombay is to a considerably larger extent an emporium of distribution than a local market of consumption.

If the imports and exports of foreign and Indian silk be balanced separately, Bombay port town is seen to make a net import of 884,690℔ foreign silk out of a total import of 1,692,621℔, but the ultimate balance between both kinds of silk results in the net import to the Bombay town of only 633,834℔. A certain amount of Indian silk is thus manufactured in Bombay, which releases a corresponding quantity of foreign silk. Of the countries and provinces to which Bombay exports foreign silk, it has already been shown, in connection with rail-borne traffic, that Bombay Presidency and the Panjáb consumed in 1888-89 about equal quantities, making a total of a little over half a million pounds. The foot-note on the last

TRADE in RAW SILK.

Foreign.

page suggests that there exists very probably some mistake in the distribution of the Bombay imports of raw silk, since the official returns show no waste silk as received. This must doubtless be incorrect, since one of the most striking differences between the Calcutta and the Bombay power-looms is the fact that the latter manufactures its silken goods for the Burmese market very largely of Chinese waste silk. The writer had recently the pleasure to inspect the Sasson Mills, and he was much struck with the extent to which these mills employed Chinese waste, and, to a large extent, the waste of what would in Europe be designated *tasar* silk.

Conf. with pp. 86, 210.

The re-exports to other foreign countries took place mainly with the United Kingdom, Arabia, Egypt, and France in the order of importance in which named. Of the exports coastwise the major portion in the year in question went to Sind, 95,330℔; to Cutch, 11,552℔; and Kathiáwár, 9,143℔. Of the exports of Indian silk, the largest consignments were made to Kathiáwár (4,535℔), to Cutch (3,240℔), and to Sind (2,611℔). The exports by rail, as already explained, were consigned to the Bombay Presidency.

CALCUTTA TRADE IN RAW SILK.

CALCUTTA. 2183

An effort may now be made to convey some idea of the Calcutta trade. From what has already been stated it will have been discovered that Calcutta (indeed Bengal) takes but a very small amount of the imported foreign silk. In this respect it assumes a very distinct position from Bombay. The one city is the seat of India's export traffic to foreign countries in locally produced silk, the other is the emporium of foreign grown silk. At the present day imported silk to perhaps a larger extent than Indian feeds the power and hand looms of this country. Though not strictly speaking correct, yet for all practical purposes it may, therefore, be said that the returns of Calcutta trade in raw silk consist of two classes of the fibre—the Bengal grown mulberry and the wild silks of Assam, Bengal, and the Central Provinces. It is, perhaps, needless to remind the reader, however, that of the returns of exports to foreign countries less than one-third the total weight consists of reeled mulberry, the remainder is waste silk, wild silk (reeled or otherwise), and cocoons. (*Conf. with the tables at pp. 198-100*). It may almost, in fact, be said that by far the most important item of India's exports of silk to foreign countries is waste or *chasam*. Thus, for example, of the exports for the four years ending March 31st, 1890, the following were the shares taken by the *chasam* (*e.g.*, both mulberry and wild waste):—1886-87, total exports 1,583,924℔, *chasam* 1,020,595℔; 1887-88, total 1,625,177℔, *chasam* 998,235℔; 1888-89, total 2,121,914℔, *chasam* 1,313,874℔; and 1889-90, total 2,089,762℔, *chasam* 1,233,494℔. The export traffic in *tasar* and other wild silk *chasam* may be said to have been stationary or practically so during the period named: it averaged about 300,000℔, or a little more than one-third the weight of the mulberry *chasam*. During the same period the exports in reeled mulberry silk have averaged a little over 400,000℔, while the corresponding trade in reeled *tasar* and other wild silks has amounted to close on 70,000℔. The total trade in wild silks has, therefore, during the years 1886-87 to 1889-90, amounted to an average of 549,353℔. But the averages here discussed represent the whole of India, not Calcutta; they are given to allow of easy comparison with the actual return of the Calcutta trade.

The following may be given as a balance sheet of the Calcutta trade during the past six years:—

SILK.

Calcutta Trade in Raw Silk.

TRADE in RAW SILK. Calcutta.

Analysis of the Calcutta Traffic in Raw Silk for the years 1884 to 1890.

	IMPORTS.						EXPORTS.					
	1884-85.	1885-86.	1886-87.	1887-88.	1888-89.	1889-90.	1884-85.	1885-86.	1886-87.	1887-88.	1888-89.	1889-90.
	Mds.	Mds.	Mds.	Mds.	Mds.	Mds.	Mds.	Mds.	Mds.	Mds.	Mds.	Mds.
By East Indian Railway . .	13,640	11,263	13,975	12,215	13,569	13,704	3,764	2,763	2,896	1,763	1,288	1,554
,, Eastern Bengal State Railway .	7,596	5,068	7,131	7,988	12,936	8,848	113	95	230	651	246	135
,, Boat . .	4,911	4,841	4,718	4,534	2,751	2,639	48	86	165	69	26	151
,, Inland Steamer .	711	1,530	1,536	2,438	1,852	3,397	*Nil*	*Nil*	35	29	28	46
,, Road . .	19	45	...	...	...	...	301	...	...	...	...	...
,, Sea (Coastwise & Foreign) .	2,812	1,400	874	631	367	276	22,253	19,542	21,858	22,365	26,139	25,540
TOTAL { Mds.	29,689	24,147	28,234	27,806	31,475	28,864	26,479	22,486	25,184	24,877	27,727	27,426
TOTAL { lb .	2,375,120	1,931,760	2,258,720	2,222,480	2,518,000	2,309,120	2,118,320	1,798,880	2,014,720	1,990,160	2,218,160	2.194,080
Net Import { Mds.	3,210	1,661	3,050	2,929	3,748	1,438	...	...	...	...	...	...
Net Import { lb .	256,800	132,880	244,000	232,320	299,840	115,040	...	...	...	...	...	...

NOTE.—To reduce maunds to pounds multiply by 82⅙.

S. 2183

TRADE in RAW SILK. Calcutta.

In order to prevent any misconceptions regarding the facts shown in the above table with those of the earlier tables, it may be as well to explain that, under the headings Imports and Exports by Sea, are included only such as take place with Calcutta. Coastwise imports or exports between Bengal ports and foreign countries, or along the coast from other provinces, have been excluded. These transactions would naturally, however, appear in any statement of the Bengal trade as a whole, so that the Calcutta figures are slightly lower than those given for Bengal. But the smallness of difference between the tables, manifests the important fact that so completely do the external transactions of Bengal take place to and from Calcutta, that the above table might, without involving any very serious error, be accepted as expressing the entire external transactions of the province. But it would convey an absolutely misleading impression of the Bengal manufactures, to accept the net import (shown above as retained in Calcutta) as the entire available surplus for Bengal manufactures. There is only one power loom silk factory in Calcutta, so that the balance shown may be described as the amount required to supply that mill and to make up the reserve stock. On a further page will be found certain details regarding the Bengal filatures and silk mills, so that it is, perhaps, unnecessary to say anything further in this place than to add that Bengal silken goods are manufactured very largely of Bengal filature silk, with the exception of course of those made of waste and wild silks. There still exists also a considerable trade in the coarse textiles of native-reeled silk, both pure and mixed, but the returns from the various districts of Bengal complain of the decline of Native manufactures. That the Bengal filature silks are gaining ground both in India and abroad is undeniable, yet **Mr. Wardle** tells us (*Jubilee Exhibition Catalogue, p. 18*) that from "Deccan to Calcutta and from Calcutta to Benares and on to Peshawar, I found either China or Bokhara silk; and so down Rájputána to Ahmedabad, Baroda, Surat, Yeola, and Thana, everywhere the Native silk avoided, and everywhere the same reason given, its want of thread regularity." There would appear to be but two possible explanations of such an opinion: that the entire Indian statistics and literature of silk are incorrect, or that **Mr. Wardle** from a deep-seated preconceived notion allowed himself to be misinformed (*see p. 228.*) The above balance sheet of the trade of Calcutta shows that the imports of foreign silk into Bengal have, during the past six years, declined from, say, 260,000 to 20,000fb, a fact which might be accepted as proving that Bengal silk is holding its own in Bengal.

Bengal Silk holds its own in Bengal. 2184

Space cannot be afforded to deal with the other provinces of India on the same scale as has been devoted to the silk of the Panjáb, Bombay, and Bengal. But the tables furnished will, it is hoped, enable the reader to elaborate whatever information he may require. For example, he need have little trouble in discovering the sources from which Madras or the North-West Provinces draw their supplies. The former almost exclusively from Bengal, for there is little or no local production, except in *tasar* silk. The latter has a limited local production of mulberry silk, but, like the Panjáb (regarding which so much has of late been written as a future field of mulberry silk production), its plantations and silk farms cannot be regarded as having advanced very far beyond the experimental stage. It is a matter of supreme congratulation for India, however, that among the names of the pioneers of the development of her silk industry, it is possible to include that of **Mr. Cunliffe Lister**. In a letter addressed to **Mr. T. Wardle** (*January 2nd, 1889*), in which he deprecates the idea that the surest way towards progression is to improve the systems of peasant cultivation, **Mr. Lister** adds: "I have paid for my learning, as for several years I joined the Government in giving prizes; but I soon saw that it was a perfect waste of time and money. Then it was that I determined to try what could be done by having everything carried out in a proper, business-like manner, and I am now, as I think, on the point of having a great

Madras supplies of Silk. 2185

Conf. with pp. 12, 17-18, 35-36, 61-62.

TRADE in RAW SILK. N.-W. Provinces. 2186

success, after years of trouble and expense." The hopeful tone in which Mr. Lister speaks of his present efforts in Dehra Dún encourages the expectation that it may, in the near future, be possible to regard the North-West Provinces as an important producing area. That this result will be mainly due to Mr. Lister's dauntless courage and limitless resources, may be seen from one fact. At a meeting of the Silk Association of Great Britain (June 1888), Mr. Wardle (the President of the Association) announced that Mr. Lister informed him that his experiments at Dehra Dún had cost him £50,000, and that, though he regarded the money as practically wasted, he was now at last about to have a splendid year. That statement was hailed by the members of the Association as a prognostication of a future vastly extended trade with India. It is, therefore, all the more disappointing to have to add that within the past few days (June 1891) the Indian newspapers (upon what authority the writer does not know) have announced that the experiments of 1890 were so far from being the great success anticipated, that Mr. Lister had resolved to give silk culture in India but one year's more chance. Failure will certainly not mean in his case want of enterprise or of enlightened enthusiasm and determination of purpose to accomplish whatever object he has set his mind upon attaining. The announcement, which has only just appeared, that Her Majesty the Queen-Empress has been graciously pleased to raise Mr. Lister to the peerage, will, therefore, be viewed by all concerned in the silk industries of the British Empire as a well-merited recognition of the services of a gentleman who has done more for the silk interest, not of India but of the world, than has ever before been achieved by a single individual.

SILK MANUFACTURES. 2187

SILK MANUFACTURES.

In pursuance of the system established in connection with the TRADE IN RAW SILK, India's interest in SILK MANUFACTURES may be referred to two sections—INDIAN and FOREIGN. It has been admitted that the available information regarding local silk manufactures is too imperfect to admit of an attempt being made to furnish a statistical statement that would manifest either its present position or past history. Numerous writers deal with the silk manufactures of isolated districts or towns, but a concentrated and concomitant enquiry, throughout the country, has never been made. While we have certain statistics of one province for one year, and similar facts regarding another for a different period, we do not possess particulars of any one feature of the silk manufactures of all India for one and the same time. In their Annual Administration Reports, as also in the statements of Internal Trade, the Local Governments continue to give, year after year, the figures of certain features of the silk trade (mostly of raw silk), while they do not touch on other equally important aspects of the subject. Each annual report is framed on the pattern of its predecessors; it consistently gives certain facts and omits others. Indeed, it might be said of all such reports that they furnish the most meagre generalizations only, regarding *local silk manufactures*. The best sources of information are the special District Manuals and Gazetteers, but these unfortunately have appeared at different times over perhaps the past twenty or thirty years, so that, as already remarked, it is impossible to give a statement of the Indian manufactures for any desired period. But it may fairly be said that one feature stands out more prominently than all others as characteristic of the multitudinous writings that have appeared, namely, a consensus of opinion that the local silk manufactures of India have been completely ruined by the growth of the British imports of cotton and silk goods.

Indian 2188

INDIAN MANUFACTURES.

FILATURES & MILLS.—It has been accepted that one of the most satisfactory methods of forming an opinion as to the present position of the Indian manufactures of silk is through the trade returns of the movements

TRADE in MANUFACTURES.

Indian.

of raw silk. It has been shown by the table at page 204 that in 1888-89 Calcutta had a net import by land routes of 2,435,400℔, and Bengal Province a net export of 2,200,505℔. That amount at least was, therefore, produced in Bengal. That there must have been more produced, however, it need only be necessary to remind the reader of the fact that local manufactures must to a large extent draw their supplies from local production, the silk being carried along road and river without being registered. But if we accept the net export (which has been registered as carried by rail and river) as a large portion of the production, it becomes possible to form some idea of Bengal silk manufactures. During the calendar year of 1889 there were in Bengal 82 filatures. These gave employment to 21,654 persons, of whom 7,413 were permanent hands; the balance employed during the working season. They produced 610,653℔ of silk, valued at R47,81,570. Of these filatures twenty-one constituted three Joint Stock Companies with a capital of R7,45,000; the remainder were private concerns. It should be carefully noted that the returns of the filatures are for the calendar year, and the figures of trade for the official year which ends 31st March. Any comparison of these returns, therefore, involves an error which could only be corrected by taking the average for a number of years. But accepting that error as unavoidable, the following were the exports from Calcutta to foreign countries: 381,878℔ reeled mulberry silk and 51,595 reeled *tasar* and other wild silks. If now we assume that these were entirely drawn from the filature production, there would have remained in India, say, 250,000℔ of reeled Bengal silk. It is not known how much of the net export from Bengal (2,200,505℔) consisted of reeled silk, but it included the filature production, so that the balance of all forms of raw silk would have been the amount available against the exports to foreign countries by sea of waste and wild silk, the exports to foreign countries by land routes of all forms, the exports coastwise of all forms, and the exports by rail and river of all forms. These exports amounted to a little over two million pounds gross, or a little over a million and a half net, so that it seems safe to infer that the local consumption in Bengal was about 300,000℔ of Bengal silk. If to that figure be added the local consumption not registered as carried by trade routes, it is probably admissible to conclude that the consumption in Bengal of Bengal-grown mulberry silk exceeds rather than falls short of that figure. Its silk manufactures, therefore, are a little more than half as valuable as those of Bombay town, and about equally valuable with those of the Bombay Presidency and of the Panjáb. The importance of allowing a large margin for peasant reeled silk and for village hand looms may at once be exemplified by the facts already shown, *viz.*, that while Bombay Presidency and the Panjáb neither possess filatures nor silk mills, their consumption of raw silk, as registered by rail, amounted in 1888-89 to 457,463℔ and 462,780℔ respectively. A similar extensive consumption must, therefore, be conceded as at least possible in Bengal, and hence (as the great producing province which possesses facilities for local trade that escape all registration) an annual consumption of 300,000℔ would appear a safe though low estimate. Some further notion of the extent of hand loom manufactures in Bengal may be learned by the fact that (if the coastwise trade to Burma be removed from consideration) Bengal exports more silk goods to Europe than Bombay. (*Conf. with the remarks below, pp. 227-228.*)

It seems unnecessary to go over the facts already reviewed regarding Bombay and the Panjáb. Suffice it to say that these provinces consume large quantities of silk, and that perhaps next to Burma the Panjáb might be given as the province of India where relatively the highest amount of silk is used. The Bombay mills and one Calcutta mill are almost entirely concerned in the Burmese supply. Within recent years, however, a marked decline has taken place in the Panjáb manufactures. The Bombay mills are apparently proving year by year more disastrous to the village and town industry of the Panjáb, much after the same fashion as Chinese

SILK. — Silk Manufactures

TRADE in MANUFACTURES. Indian.

silk appears to be checking the imports into the Panjáb of Kashmír and Bokhárá silks. These silks are brought to Bombay as return cargoes by ships that might have to return almost in ballast. They take opium and cotton to China, but have difficulty in getting goods for their homeward journey. The freight from China to Bombay is, accordingly, less than the railway charges from Bengal, and to this fact largely must be attributed the fact of Bengal silk not finding a market in Bombay.

In spite of all the changes of fashion, in which highly coloured cotton goods are admittedly displacing the locally produced silken garments, Bengal and the Panjáb are still important consuming provinces of silk. As already pointed out, from the consideration of other features of the trade, it is difficult to understand **Mr. Wardle's** remark, in its bearing at least upon Bengal, that "everywhere," he found, "the Native silk avoided, and everywhere the same reason given, its want of thread regularity." The defect complained of could certainly not apply to Bengal filature silk, but applicable, though doubtless it be, to Native reeled silk, it is impossible to believe what appears to be **Mr. Wardle's** final conclusion that it is nowhere used in India. The coarse appearance of Native woven goods, from Native reeled silk, it might be said, is almost, by religious prejudices, preferred to the more carefully produced cloth imported from foreign countries or made by power loom mills. At all events the consumption of Indian silk must still be very great, as may be seen by the percentage of persons dressed in such silken garments as they frequent the bathing ghâts along the rivers of Bengal. Opinions of trade based on personal observation of that kind could not, however, be advanced in any estimate framed on statistical returns, and it is only mentioned as an indication of the extent of local manufactures which, it is believed, would be very generally concurred in by residents in the Lower Provinces. To a far less extent would the observation be applicable to the North-West Provinces, the Central Provinces, and Central India, where the middle classes of the community prefer or are contented with coloured cotton garments.

While statistics are not available regarding the number of silk hand looms in India more than of the peasant's reeling basins certain details are forthcoming regarding the mills. In the Statistical Tables of British India, six mills are shown to have been working. The following facts regarding these may be here given :—

Province.	Name of Mills.	Capital (if a Joint Stock Company).	Average number of persons employed daily.	YEARLY OUTTURN.	
				Quantity. Yards.	Value. R
		R			
Island of Bombay	Sasson & Alliance Silk Mills Company, Ld.	10,00,000	718	567,607	7,83,488
BENGAL. Burdwan	Silk Clothing Manufactories.	...	891	289,535	1,67,333
BENGAL. 24-Pergunnahs, Calcutta.	Silk Factory in Ultadanga.	...	350	84,395	1,02,852
BENGAL. Manbhum	Tasar Cloth Factory.	...	100	18,000	13,500
BENGAL. Beerbhum	Tasar Manufactories.	...	2,100 (a)2,520	12,600	9,450
BENGAL. Hooghly	Tasar and Silk Cloth Manufactories.	...	850	206,000	1,53,600
TOTALS	6 mills.	...	7,529	1,178,137	12,30,223

(a) Employed during the working season.

TRADE in MANUFACTURES.

Indian.

The first and the third of the above mills are worked by steam and with the most recent European machinery; they might not incorrectly be described as concerned almost entirely in the manufacture of cloth required for the Burma market, while the second and the sixth might, perhaps, be said to work mainly for the local Bengal market and the foreign exports from Bengal. They are, in fact, Native factories owned and worked by Natives and by Native appliances. They are largely concerned in the *tasar* silk manufactures. It would be beside the scope and character of this work to deal with the silk manufactures of India from an art point of view, but, as denoting some of the chief centres of manufacture, the following brief notice of the Indian silks shown at the Jubilee Exhibition may be here given :—

"The collection in itself may be considered a typical one in design and colouring, and it gives an accurate idea of almost all kind of fabrics of which silk forms the whole or part. It comprises the *corah* silks of Bengal, rudely produced by looms that would raise the smile and wonder of Europeans; the coarse *tusser* fabrics woven in the same and other districts; the magnificent *kinkhabs* of Benares, Ahmedabad, and Surat, in which gold and silver form such important decorative features; the plainer silks of Delhi; the delicate and beautiful silks of Thana (a very ancient Christian settlement); the rich fabrics of Yeola, incomparable for living beauty and Arabian grace of design; the ruder though not less interesting silks of Peshawar and the surrounding country; the satins of Azimgarh, Ahmedabad, Surat, Dhranghra, and Kathiáwár; the wonderfully constructed patterns of the *patolo* weaving with 'tie and dye,' warp and woof; the silks of Berhampur, Cambay, Cutch, Indore, Kathiáwár; and Bombay,—all testify not only to the skill acheived by Indian dyers and weavers during many ages, but also for the fascinations which have held these people spell-bound in the production of their fabrics of mystery and beauty.

"The printed silks of India, too, are by a long way not the least of the interesting decorative work. It is a great pity that anything should have superseded the permanent and striking prints of the old-fashioned pocket handkerchiefs. I have seen them printed on the squat tables of the Calcutta printers, with indescribable interest, who use their prettily-sculptured little blocks with a dexterity and exactness marvellous to see, requiring no pin points to guide them in their repeats of patterns."

To attempt to review, however briefly, the numerous district accounts of the silk manufactures of India would occupy more than the already undue space which has been devoted to silk. The reader must, therefore, consult the Gazetteers and District Manuals, etc., of which the following may be specially mentioned :—

Madras :—Manuals, Coimbatore District, by **F. A. Nicholson**; North Arcot, by **A. F. Cox**, 339.

Panjáb Gazetteers :—Provincial Vol., 1888-89, 149; Jalandhar District, 46; Shahpur District, 76; Hazara District, 144; Lahore District, 97; Gurdaspur District, 72; Mooltan District, 108; Amritsar District, 41; Ludhiana District, 158; Delhi District, 136; Kohat District, 114.

Bombay Gazetteers :—Ahmedabad, iv., 135; Cutch, v., 126; Kolaba, xi., 131; Khandesh, xii., 180; Thana, xiii., Pt. i., 378-385; Nasik, xvi., 155; Dharwar, xxii., 375; Bijapur, xxiii., 371.

Sir W. W. Hunter's Statistical Account of Bengal :—Birbhum, iv., 338, 374; Maldah, vii., 94; Rangpur, vii., 304; Murshedabad, ix., 148; Pabna, ix., 332; Santal Parganas, xiv., 338; Hazaribagh, xvi., 168; Lohardagga District, xvi., 346.

Mysore & Coorg, i., 436; ii., 37, 473.

TRADE in MANUFACTURES.

Indian.

North-West Provinces & Oudh Gazetteers:—Agra, vii, 555; Azamgarh, xiii., 125.

Milburn's Oriental Commerce, 1825, 300, 301-303, 304, 496-497.

Journals of Indian Art, January 1885, 33; September 1886, 115; July 1888, 58, 61; October 1888, 68; July 1889, 10, 14.

uropean.
2189

EUROPEAN MANUFACTURES.

There may be said to be two natural divisions of this subject—(*a*) European uses for Indian silk, and (*b*) the silk goods imported into India. The latter will be taken up, however, under the heading TRADE IN SILK GOODS (p. 224), so that the former only need be here dealt with. The information available regarding the European uses for Bengal silk is so uniformly alike in all technical works on the subject, that they need scarcely be taken up in detail. When the East India Company went with spirit into the subject of improving their silks, they were, in the first instance, mainly concerned in contesting the Turkey supply to England. French and Italian silks had not then attained the vast importance that they have since done—at least in England. In 1749, for example, the imports by Great Britain from Italy were only $\frac{1}{6}$th the value of those from India. The taste for silken goods had not been created as an attainable luxury of the community at large. The purposes for which silk was mainly in demand were quite different then from what they are to-day. Indeed, it seems fairly clear that, had the present market existed, a little over a century ago, when the East India Company drove Turkey silks out of England, neither India nor Turkey could have enjoyed the positions of importance they respectively held. While not desiring to dispute entirely the oft repeated assertion that Bengal silk has degenerated, it seems fairly admissible to affirm that to a total change in the silk market is largely due the fact that Bengal silk has drifted into a peculiar and limited market for which it possesses special merits. If this idea be accepted, it would be possible to account for a retrogression in the trade without contending that, through the ignorance or indifference of the capitalist, presently interested in the silk trade, the quality of the article had so declined that it had become a drug on the market. Even in India it is freely admitted that Kashmír and Bokhárá silks are far superior to those of Bengal. These, like the silks of France and Italy, are obtained from an altogether different worm from that of Bengal. It has been satisfactorily shown that even in the Panjáb it is hopeless to expect to be able to rear **Bombyx mori**, while no one has seriously thought such a result attainable in Bengal. Nor is there anything to show that the silk of Bengal during any period of its past history has been otherwise than the sub-tropical and mostly polyvoltine insects at present reared. The silk produced by these worms can at once be recognised by the expert wherever they are obtained, so that the question of the degeneration or not of Bengal silk can alone be determined when the annual (**B. mori**) silk is excluded from the comparison. It would seem absurd to argue that, although polyvoltine silkworms, such as those of Bengal, produce a fibre which has certain definite and well-recognised properties, with greater care, they might be made to compete successfully with a silk of an altogether different nature, and one which may be accepted as having given birth to many of the modern phases of the trade. That there is a distinct market for Bengal silk there can be no manner of doubt. It is significant, for example, to note that, in connection with the proceedings of a Committee of owners and agents of Bengal filatures, on the subject of the disease of the worm, a unanimous opinion should have been given that what was wanted was not a new and improved race of worms, but prevention from disease, or, in other words, increased facility for the production of a quality of silk for

TRADE in MANUFACTURES.

European.

which there was a good demand. By the advocates of the theory of decline in quality two opposite opinions are advanced. It is held by the one school that the depreciation is due to defective manipulation and carelessness, by the other, and with apparent reason, to be a natural consequence of climatic conditions which operate injuriously upon an exotic insect. That filature silk is superior to and has driven Native reeled silk practically out of the foreign market seems admitted on all hands. But that the capitalists, interested in filatures, could be so blind to their own interests as to refuse to adopt, from time to time, as necessity arose, modern improvements; in other words, that the spirit of competition which characterises all branches of British enterprise should be latent in the silk trade, seems absurd. The effects of climate on the insect, therefore, call for more serious consideration, but in even this aspect of the question the peculiar silk which it is desired to produce must not be lost sight of. Would the periodic importation or careful breeding of stock result in the production of a silk of a sufficiently improved quality to repay the cost entailed? When the Honourable the East India Company tried to improve the breed of the worms, they may be viewed as having attempted to compete year after year with the then growing recognition of the superior silks of Europe—silks that were destined to and which have since circumscribed the Indian produce to a definite market. It seems more than likely that the chief feature of all such experiments would lie in the success which might be attained in combating disease, than in any advantage from a slight improvement in quality. After a fairly careful perusal of the very extensive literature that exists on this subject, the author feels constrained to hold that Bengal silk of its present quality had better be accepted as a factor in the trade. **Hutton** stoutly upheld the view that if India was to maintain its own against the superior quality of silk being produced by France and Italy, the production of the fibre would have to migrate from Bengal to the warm temperate tracts of the Himálaya. But here, again, the contention was that **Bombyx mori** would have to be substituted for **B. fortunatus** and **B. crœsi**. Subsequent results have so far proved the accuracy of **Hutton's** opinions, for, as he foretold, the Panjáb has proved unsuited to the worm. Experiments are now being vigorously prosecuted in the region recommended by **Hutton**, *viz.*, Dehra Dún, but from the brief notices which have as yet appeared, it may be said that only doubtfully good results have been even there attained. Indeed, it is highly likely that if India is ever to compete with Europe, the attempt will have to be made in Kashmír and other such portions of the Himálaya that possess a climate and soil more nearly approximate to that of Europe than exists in the Dún. But is it necessary that India should accept the position that competition with Europe is an essential feature of its silk enterprise? Such degeneration as may be admitted to have taken place might be held to be due to the fall in the price of silk throughout the world. The improvement in quality, and the cheapening of prices which became possible in Europe, have rendered it problematic how far it would be remunerative to undertake now the similar expensive measures to guard against degeneration which the East India Company could afford to combat. Much higher prices were obtained in Europe by the Company than can be got now, and the Bengal silk was at the same time procurable in India at a much lower figure than it is at the present day. But the eyes of all interested parties are naturally turned on **Mr. Lister's** undertaking in Dehra Dún. If he succeeds, India may then have two distinct, but by no means antagonistic, silk centres of enterprise. If such a state of affairs becomes an established commercial fact, the character of the exports may be materially changed, and Indian silk would then very probably appear on the tables of European brokers who at present take no cognisance of it. At the present day it may be said that the bulk of the exports are in waste and wild silks. The reeled silk of Bengal finds a ready sale for a few restricted

SILK. **European Uses**

TRADE in MANUFACTURES.

European.

uses, and it thus seems probable that while the market for that class of fibre may never expand very greatly, it will always remain open, so that whether India can or cannot produce superior qualities of the silk of **Bombyx mori**, its polivoltine insects will continue to hold their present position. **Mr. Wardle**, in his *Jubilee Exhibition Catalogue,* gives the following instructive information regarding Bengal Silk :—

"I have had many opportunities of observing silk-reeling in both France and Italy for many years past, and I felt that if the same care and appliances were used in India as in these countries, silk of proper quality could be obtained.

"In that I perfectly succeeded, and in no case in India was it told me that an improved thread was not the result—in every case it was admitted.

"I found by using exactly the same appliances as those of Italy that there was no difficulty in unwinding or reeling the cocoon with almost perfect regularity, and in order that this possibility may be exemplified to those who may have before doubted it, the Silk Section Committee have decided not only to have this system in daily practical operation but also two of the best French methods. A large quantity of Bengal cocoons have been sent from the Bengal filatures of **Messrs. Robert, Watson & Co.**, of Rajshaye, and a French *fileuse* is reeling them into raw silk daily during the time the Exhibition is open. At the other end of the reeling-stand is the Italian method in operation, and cocoons from Cyprus, Adrianople and elsewhere are also being reeled. *Tussur* and *Muga* cocoons are also occasionally being reeled, and on short notice being given any person can see the reeling of *Tussur* or *Muga* cocoons in operation.

"One thing has been proved, and I have been permitted to bring it home to the minds of impartial and unprejudiced manufacturers in England and on the Continent, that the Bengal cocoon has not the inherent imperfections which it was thought pertained to it, and that there is a prospect of a greatly enlarged output of silk from Bengal for several important purposes in the silk trade of Europe and America.

"*First, Sewing Silk.*—Several of the best manufacturers of silks for sewing purposes in Leek have assured me, after full examination and trial, that this silk is peculiarly applicable to their trade.

"**Mr. S. Gibson**, a Leek manufacturer, writes:—

"'I am very pleased with the five bales of Bengal silk I have just bought; they work very freely, almost running from beginning to end of the skein without breaking down, which means winding without loss. The strand is of nice even size, suitable for the Leek trade, free from rough or slubby places, so much so as to render one important process in the manufacturing unnecessary, *viz.*, cleaning. I am working these new Bengals in both the bright and washed state, and they are coming out at about one-half the cost of original Bengals. If this improved method is maintained, it must have a serious influence on the China and Canton raws.'

"**Mr. S. Goodwin**, another silk manufacturer, and President of the Leek Silk Association, writes:—

"'I have worked the sample skein of Bengal raw-silk, and am pleased to say that it is simply perfection. As to reeling, I may say that it wound almost without a break from end to end.'

"This skein was reeled by me in Bengal with the Tavelette Keller, also erroneously called Consono.

"The silk of the Bengal worm, by its greater elasticity, is much better adapted for sewing silk than any other. I have estimated, in experiments conducted during the last few days, the tension of the bare, or double fibre, deposited by the silkworm, of the Bengal Madrassee or hot-weather cocoons, the Bengal Desi or November bund cocoons, and of Italian cocoons. The results are shown in the following table, each figure being the average of numerous determinations, and representing the number of centemetres which three decimetres of the bave is capable of stretchin before it breaks:—

Cocoons.	Tension at the end of the cocoon bare which is at the surface of the cocoon immediately beneath the superficial loose fibres or waste.	Tension at the middle of the cocoon bare.	Tension at the end of the cocoon bare which is nearest the telette or inner envelope.
Madrassee Cocoon . .	5·5	9·0	5·0
Desi Cocoon . . .	7·0	7·2	5·9
Italian Cocoon . .	4·5	6·1	4·4

TRADE in MANUFACTURES.

European.

"*Second, Organzine and Tram for Weaving.*—Mr. Nicholson, silk manufacturer, Macclesfield, in a letter to me, writes the following:—

"'In answer to your inquiry, I consider that good Surdah raw, when well reeled with plenty of spin upon it, will work well. It will then be a good substitute for Italian, its cheapness being the reason for its use.'

"In addition to this, I may say that Mr. Nicholson is speaking of a Bengal silk that was not reeled by the Italian method. I contend that there would be no greater evenness of thread in the Italian silk over that of Bengal if the Italian method of reeling were used.

"Messrs. G. Davenport & Co. of Leek, to whom I sent a portion of the 10 to 12 deniers, which I saw reeled by the Tavelette Keller in Bengal, have thrown it into organzine and tram, and sent me the following report: 'The slip winds beautifully. Enclosed are samples of two threads tram and a 500 yards skein of organize. The silk is very clean. We consider it equal to ordinary Italian. It was running for an hour and only broke down once.'

"Now it is necessary to say that, even with every improvement, Bengal silk may not be expected to rival or supersede the *finest* qualities of silk in the market. A reason or two from me may suffice. It will never be as white as China Silk, because one is from a yellow cocoon and the other from a white one. It will not 'boil-off' or condition as well as the silks of Italy, China, or Japan, because it contains more gum or *grés* than these, and this brings me to an entomological point, namely, it is probably not of the same species; but of this further on.

"I must guard myself against being thought slack in acknowledging the claims which at least three Bengal firms have on the consideration of European manufacturers for careful reeling. The excellence of the silk produced at Surdah and its allied factories in the district of Rajshahi and other parts of Bengal, that of the Bengal Silk Company's factories, the chief of which is at Berhampur, and that of the well-known firm of Messrs. Louis Payen & Co., are too well known to need mention. I acknowledge with much pleasure the kindness I received from these firms, and they are well deserving of the confidence of all interested in silk.

"But apart from the efficiency of these well-known firms, there remains the much larger native industry, the reeling that is carried on in the numerous villages under the shade of banyan, palm, and mango groves. I visited many of these, and found the appliances very rough and rude, the reeling by them varying from 10 to 20 cocoons in almost as many seconds.

"In the Rajshahi district alone, out of ninety-seven filatures, sixty-three are native and the remaining thirty-four European, 11,000 to 12,000 natives being employed in silk-reeling in this district alone, 150 square miles of which exist under mulberry cultivation.

"If in these villages native filatures can be induced to improve their reeling, a largely extended industry lies waiting for them in their own country; for it goes without saying that the resources of China and Bokhara would not be drawn upon if Bengal silks were of the required quality. Many native manufacturers assured me they would much prefer to buy Indian silk if only the quality were good enough. The consumption of silk for native uses alone is enormous. All Hindus wear it at meals and worship. The Muhammadans wear *mashru*, or cloth of cotton-warp and silk weft, the wearing of pure silk fabrics being forbidden by the Koran.

"Since I was in India another silk-reeling machine has been brought to my notice by the Under-Secretary of State for India. Its inventor, Mr. Serrell of New York, claims for it that unskilled labour can be used with its aid in reeling cocoons. This machine is automatic, and by means of a feeble electric current which controls the feed, takes up another cocoon thread when one breaks, so that a new cocoon is added whenever required to keep up the size of the thread. It is stated to do two and a half times as much work as the present system, at a saving of 2*s.* 9½*d.* in wages, and one reeler can attend to six bassines.

"I am sorry it has not been possible to have this machine at work in the Exhibition side by side with the other three, but I have lately seen it working in London, and have been very much struck with its automatic action. It was reeling Bengal cocoons without any difficulty, and it seems to me to be an appliance particularly suited to silk-producing countries where labour is dear, such as in our Colonies.

"By this apparatus the cocoons are softened in a few seconds, and the reelable ends found without crushing the cocoons. They are then transferred to the machine, and placed singly in slots on a revolving disc. The thread is taken automatically from each cocoon by a self-acting hook immediately it is needed at the tavelette, and this want is indicated by tension drums, in electrical connection with the hook, whenever the combined thread of raw silk becomes less than its normal strength.

"Explanatory notes and printed particulars are published, and may be had, as

SILK. | **Trade in Silk Goods.**

TRADE in MANUFACTURES.

well as all information respecting this interesting invention, from Mr. F. B. Forbes, 5, St. James' Place, London, S. W.

"It appears to be able to do what Mr. Serrell claims for it, and it promises fair to be a valuable addition to the Bengal silk industry, and I commend it to those interested for a thorough investigation of its capabilities."

SILK GOODS. 2190

TRADE IN SILK GOODS.

In dealing with this subject, it may, perhaps, be the most instructive course to refer the available information to two sections, *viz.*, FOREIGN and INTERNAL. Each of these sections would, accordingly, embrace the imports and exports between India and foreign countries, or between the various provinces of India. The other method of treatment would be to exhibit in one place all the available facts regarding the trade in silk piece goods (both pure and mixed), also thread and the smaller articles designated in the statistical returns as "Other Sorts."

FOREIGN. Imports. 2191

I.—Foreign Trade in Silk Manufactures.

IMPORTS.—The following table may be given of the Imports of Silk Manufactures during the past five years:—

YEARS.	Thread for sowing.	Piece goods.	Goods of silk mixed with other materials.	Other sorts.
		Quantities.		
	lb	Yds.	Yds.	lb
1885-86	5,954	8,999,359	2,174,429	2,788
1886-87	10,318	10,541,862	2,626,011	3,616
1887-88	13,301	11,760,401	3,970,372	1,603
1888-89	8,163	10,952,732	4,223,332	4,497
1889-90	10,382	11,426,168	3,978,949	7,420
		Values.		
	R	R	R	R
1885-86	56,136	94,42,414	15,73,325	18,555
1886-87	94,327	1,16,20,741	20,90,162	32,120
1887-88	1,26,647	1,44,77,510	28,16,701	17,321
1888-89	72,353	1,34,96,746	32,11,321	41,241
1889-90	1,11,929	1,42,30,272	33,57,605	81,335

Accepting the returns of the last year in the above series as fairly characteristic of the trade, its present position may be briefly reviewed.

Silk Thread. 2192

Silk Thread.—The major portion came from the United Kingdom, the next important source being from the Straits Settlements, then Hong-Kong, Austria, and last of all France. Fully two-thirds were taken by Bombay, the balance in almost equal proportions going to Burma and Bengal.

Pure Silk Piece Goods. 2193

Pure Silken Piece Goods.—Out of the total of close on 11½ million yards, imported by India last year, a little over 5 million yards came from the United Kingdom, the next most important European country being France with 616,356 yards. The French share in this trade seems to be declining. In 1885-86 it stood at over a million yards, and in the succeeding year it was just under two million yards; while Great Britain furnished in the former year 3,987,356 and in the latter 4,337,199 yards. Of the other European countries, which supply India with silk piece goods, Italy stands next in importance. In 1889-90 India obtained from thence 241,234 yards, being fully a hundred thousand more than in 1885-86. The Italian trade thus shows signs of improvement. After Italy comes Egypt, that country having furnished India with 80,672 yards in 1889-90. The Egyptian trade is, however, very uncertain. It stood at 17,267 yards in 1885-86, 135,040 yards in 1887-88, and 153,166 in 1888-89. The other countries that supply India with silk piece goods are Austria, Germany, Persia, Japan, Ceylon, Zanzibar, and Belgium,—these countries contributing in the order named from 25 to 2,000 yards.

TRADE in MANUFACTURES.

Foreign Imports.

Pure Silk Piece Goods.

But the countries mentioned make up about 6½ million yards only of the total imports in 1889-90; the balance come from Hong-Kong, the Treaty Ports of China, and from the Straits Settlements. Not only therefore does China furnish India with, say, one-half the amount of raw silk that is actually manufactured in this country, but with a large quantity of piece goods. What is more significant, the imports from that country have been steadily improving, as will be seen from the following figures of the trade from 1885 to 1890:—

Quantities and values of the Pure Silk Piece Goods imported by India from China, the Straits Settlements, the United Kingdom, and France during the past five years.

YEARS.	CHINA.				Straits Settlements.		United Kingdom.		France.	
	Hong-Kong.		Treaty Ports.							
	Yds.	R	Yds.	R	Yds.	R	Yds.	R	Yds.	R
1885-86	3,059,021	33,19,838	73,553	49,432	622,030	7,00,314	3,987,356	38,98,687	1,057,594	12,22,887
1886-87	3,585,110	39,83,684	155,046	81,266	213,290	2,75,707	4,337,199	50,14,674	1,957,928	19,26,853
1887-88	3,897,735	45,95,852	206,328	1,23,173	607,233	7,66,576	5,684,085	71,79,518	979,485	13,23,309
1888-89	3,163,652	36,32,537	240,560	1,48,062	424,731	6,77,251	5,432,230	70,21,814	1,222,007	14,68,548
1889-90	4,636,450	50,25,645	241,393	1,39,621	420,648	6,36,165	5,108,307	71,18,268	616,356	8,35,549

SILK. **Trade in Silk Goods.**

TRADE in MANUFACTURES.

Foreign Imports.

Pure Silk Piece Goods.

It will thus be learned that China and the Straits furnished India, during the years exhibited in the above table, with very near as much silk piece goods as did the United Kingdom and France. How much of the exports from Great Britain were in reality French goods re-exported it is impossible to tell; but as the figures stand, England has more to fear from China than from France, in competition for the Indian trade.

The distribution of the very large quantity of pure silk goods imported annually into India is a matter of considerable interest. It will be seen from the table which follows that Burma takes the largest share in the trade, but the actual amount of foreign silk goods consigned to Burma is larger even than shown by a little under half a million yards. This is explained by the fact that of the amounts shown as consigned to Bombay and Bengal, a large re-export exists coastwise, as also a considerably larger quantity of Indian made piece goods (*Conf.* with tables at pages 227 and 228) :—

Analysis of the Imports into India of Foreign Pure Silk Piece Goods so as to show the shares taken by the Provinces.

Years.	Bengal.	Bombay.	Sind.	Madras.	Burma.
		Quantities.			
	Yds.	Yds.	Yds.	Yds.	Yds.
1885-86	1,014,949	3,565,243	18,102	50,519	4,350,546
1886-87	2,088,369	4,198,785	14,247	39,503	4,200,958
1887-88	885,685	4,771,369	47,244	35,083	6,021,020
1888-89	1,185,553	4,211,295	26,841	23,156	5,505,887
1889-90	658,599	5,497,872	16,713	21,176	5,231,808
		Values.			
	R	R	R	R	R
1885-86	11,65,403	39,63,374	25,833	59,798	42,22,006
1886-87	19,16,762	47,50,506	26,497	50,921	48,76,055
1887-88	10,80,110	57,05,798	75,805	42,277	75,73,920
1888-89	13,45,277	48,82,338	36,214	33,285	71,99,637
1889-90	8,03,234	60,60,487	27,054	31,662	73,07,835

Mixed Silk. 2194

Silk mixed with other materials.— As will be seen from the table at page 221, the trade in the goods included under this heading is only about one-third as extensive and valuable as that of pure silk piece goods. It has, however, one somewhat remarkable feature. During the past five years it has averaged 2,400,000 yards, of which France furnished 2,200,000 yards. The French trade has also steadily improved from 1,239,045 to 2,931,926 yards, while the imports from the United Kingdom have correspondingly declined from 665,940 to 283,309 yards. The next most important country in the supply of the goods of this class is Italy, the imports from which have improved from 70,028 to 382,279 yards, but in 1888-89 the Italian imports were 582,958. A similar improvement has taken place in the imports from Austria and Germany,—from the former country from 123,040 to 213,618 yards, and from the latter country from 4,805 to 92,620 yards. It may be added, in concluding this brief notice of India's import trade in mixed silken piece goods, that the supply drawn from China and the Straits Settlement is unimportant.

The distribution of these imports in India may now be noticed. Of the total for 1889-90 Bombay took 3,404,869 yards and Bengal 555,485, the balance in about equal proportions went to Madras, Sind, and Burma.

Other Silk Manufactures. 2195

Other Sorts of Silk Manufactures.—From the table at page 221,

TRADE in MANUFACTURES. Foreign Imports. Other kinds.

will be seen the articles registered in the trade returns under this designation are not very important. They are almost entirely drawn from the United Kingdom, and are imported mainly by Bombay. In the year 1889-90 Bombay took R67,227 worth of these goods, as against R12,859 consigned to Bengal, R987 worth to Burma and R263 worth to Sind.

Having now placed before the reader some of the more salient features of the imports of foreign silk munufactures, it may be desired to contrast these with the exports from India to foreign countries by sea of Indian manufactured silk.

Exports. 2196

EXPORTS OF INDIAN PRODUCE.—The following table exhibits the trade under each of the four classes of silk manufactures :—

Exports from India during the past five years in Indian manufactured silk.

YEARS.	Thread for sewing.	Piece goods.	Goods of silk mixed with other materials.	Other sorts.
		Quantities.		
	℔	Yds.	Yds.	℔
1885-86	698	3,728,213	146,726	17,217
1886-87	28	3,161,179	189,310	36,896
1887-88	113	3,522,528	210,782	20,411
1888-89	1,033	2,807,203	169,035	15,225
1889-90	Not shown.	2,330,360	184,500	Not shown.
		Values.		
	R	R	R	R
1885-86	2,625	31,12,554	1,64,103	16,554
1886-87	118	29,22,861	2,24,000	34,057
1887-88	531	35,21,757	2,50,737	19,947
1888-89	6,531	26,84,652	1,81,602	12,087
1889-90	Not shown.	23,98,929	2,08,284	Not shown.

It seems scarcely necessary to comment on each of these separate headings. The trade in piece goods is the only important item. Of the pure silk piece goods, out of an average during the past five years of, say, 3,100,000 yards, the United Kingdom took 2,160,000 yards and France an average of about 700,000 yards. The other countries, in the order of importance as consuming centres of Indian silk piece goods, were Arabia, Turkey in Asia, Persia, Mauritius, and Egypt. Judging from the figures given, the export trade in this class of Indian silk manufactures seems if anything to be declining.

In order to manifest the share taken by the provinces of India, it may be said that Bengal exported out of the total an annual average for the past five years of 2,920,000 yards, Bombay of 154,000 yards, Madras of 25,300 yards, Burma of 1,131 yards, and Sind of 666 yards. The actual exports from Calcutta in 1889-90 were, however, 2,117,218 yards, and from Bombay 201,020 yards. In the table at page 217, which exhibits the number of silk mills in India and their estimated outturn, it will be seen that the outturn (presumably mainly of silk piece goods) in 1889 was estimated at 1,178,137 yards, of which 567,607 yards were turned out by the Sassoon Mills in Bombay. It has been affirmed that that mill runs almost exclusively for the Burma market, as well as the Ultadanga Mills near Calcutta. The last-mentioned mill had an outturn of 84,395 yards. That this view is correct of these two mills will be found confirmed by the magnitude of the

TRADE in MANUFACTURES.

Foreign Exports.

coastwise exports from Bombay and Bengal to Burma. It thus seems a safe suggestion to offer that it may be accepted that of the Calcutta exports of 2,117,218 yards, given above, and of the Bombay exports of 201,020 yards, something like 80 per cent. were produced by village manufacturers. In the passages which follow regarding the coastwise and internal trade, the distribution of the total Indian production of silk piece goods will be more fully exemplified, but the facts here dealt with may be accepted as affording an additional confirmation of the statement made that the Bengal manufactures from Bengal grown silk are far more extensive than might be supposed from the deprecatory remarks made on this subject by **Mr. T. Wardle** (*see p. 215*). It may, in fact, be said that, though there is a marked decline in the demand for silk goods in India as a whole, the power looms that exist in the country have not seriously interfered with the import trade in foreign silk goods, nor materially disturbed the proportion of the present trade that falls to the lot of the village weavers of the country.

Re-exports. 2197

RE-EXPORTS.—The following table gives the re-exports from India of foreign silk manufactures. The returns of this nature manifest the extent to which India stands as an emporium in the distribution *to* foreign countries of the goods she imports *from* Foreign countries:—

YEARS.	Piece goods.	Goods of silk mixed with other articles.	Thread for sewing.
	Quantities.		
	Yds.	Yds.	℔
1885-86	335,654	61,416	1
1886-87	338,536	61,850	767
1887-88	415,423	69,984	373
1888-89	585,021	83,059	131
1889-90	564,152	53,032	Not shown.
	Values.		
	R	R	R
1885-86	3,23,077	42,069	42
1886-87	3,21,115	47,785	6 996
1887-88	3,91,384	70,615	3,282
1888-89	5,54,876	89,015	630
1889-90	51,9,555	58,023	Not shown.

Comment on the trade shown by the above table is, perhaps, unnecessary. European, Central Asiatic, and Chinese piece goods and thread are consigned by India to the following countries, the figures given being those for 1889-90: Ceylon (145,902 yards), Egypt (73,635 yards), Persia (61,963 yards), Arabia (44,007 yards), Malta (38,539 yards), Aden (30,986 yards), Zanzibar (30,643 yards), the United Kingdom (26,174 yards), Turkey in Asia (16,949 yards), Mauritius (14,684 yards), Hong-Kong (12,573 yards), Straits Settlements (12,106 yards), France (6,935 yards), &c., &c.

If now it be desired to ascertain how far the remark made regarding raw silk is confirmed by the trade returns of silk manufactures, *viz.*, that India is a consuming rather than a producing country, it is only necessary to add together the exports of Indian produce to the exports of

TRADE in MANUFACTURES.

Trans-frontier. 2198

foreign produce, and see how far the total exceeds or falls short of the imports:—

Trade by sea. Year 1889-90.	Thread for sewing.	Piece goods, pure silk.	Piece goods, mixed silk.	Other sorts.
	lb	Yds.	Yds.	lb
Exports, Indian Produce	*Nil*	2,330,360	184,500	*Nil*
Exports, Foreign Produce	*Nil*	564,152	53,032	*Nil*
TOTAL	*Nil*	2,894,512	237,532	...
Imports, Foreign Produce	10,382	11,426,168	3,978,949	7,420
Net Import	10,382	8,531,656	3,741,417	7,420

In the concluding paragraph on the foreign trade in raw silk, page 204, it is shown that the imports from all sources were in 1889-90 2,405,491lb, which, by deducting the exports to foreign countries, left 195,884lb to the credit of India. A correction would have to be made to the above table for the foreign transactions across the frontier by land routes. This trade is not very important, as will be seen from the table below which shows a net export of piece goods for 1889-90 to the weight of 31,248lb, a surplus which was valued at R4,74,633.

FOREIGN TRADE IN SILK PIECE GOODS ACROSS THE LAND FRONTIER.—A difficulty arises in connection with this as also the traffic carried along the railways and rivers, in that the returns are expressed in weight instead of yards or pieces of cloth.

Foreign Imports and Exports by Land Routes.

	1887-88.		1888-89.		1889-90.	
	Quantity.	Value.	Quantity.	Value.	Quantity.	Value.
	Cwt.	R	Cwt	R	Cwt.	R
Imports	64	71,286	17	22,112	75	1,12,335
Exports	427	5,01,332	459	5,40,911	354	5,86,968
Net Exports	363	4,30,046	442	5,18,799	279	4,74,633
Ditto in lb	40,656		49,504		31,248	

II.—Internal Trade in Silk Manufactures.

INTERNAL. 2199

In discussing the Indian internal trade in silk manufactures, it is necessary to follow the same course as has been adopted for raw silk, namely, to separate the transactions recorded by sea coastwise, or by rail, road, and river, into two sections, *viz.*, Indian and Foreign produce.

Coastwise. 2200

COASTWISE TRANSACTIONS.—It has already been explained that the imports into the provinces may be accepted as equivalent to the exports. That is to say, that the returns of imports, for example, into Bengal from Bombay are the recorded exports from Bombay to Bengal. That is not, strictly speaking, correct, since the destination of goods may be altered, though recorded as an export destined for a certain port. Int he returns of imports coastwise, the trade in silk goods are grouped into two sections, traffic in India and in foreign manufactures. Each of these is again broken into two sub-sections—pure silk piece goods and mixed silk goods. In drawing up the following tables the imports from one province to another

Trade in Silk Goods.

TRADE in MANUFACTURES.
Inter-provincial Exchanges
2201

have alone been taken into account, the exchanges between ports within a province being excluded. Thus, for example, the trade from Chittagong to Calcutta has not been taken into account, since the goods carried between these two ports are accepted as still in Bengal. The object in thus treating the returns is to exhibit the degree of indebtedness of one province to another in the supply of silk piece goods :—

Inter-provincial Exchanges in Silk Goods, being the Imports into one Province from another, coastwise.

	Provinces into which imported.	Indian Produce.									
		1885-86.		1886-87.		1887-88.		1888-89.		1889-90.	
		Yds.	R	Yds.	R	Yds.	R	Yds.	R	Yds.	R
Pure Silk Piece Goods.	Bengal	1,620	2,052	8,849	11,773	960	3,960	4	7	265	1,110
	Bombay	782	8,200	15,543	12,222	2,508	3,379	2,129	3,321	1,001	1,338
	Sind	182,993	1,33,903	25,737	1,78,241	177,017	1,48,682	114,751	91,032	77,021	66,215
	Madras	99,389	66,864	40,337	43,851	30,178	29,233	47,965	42,511	14,838	19,842
	Burma	697,258	7,86,923	891,174	9,73,079	691,235	8,06,880	767,802	10,15,281	717,480	9,16,323
	Total	989,089	9,97,942	1,186,640	12,19,166	901,895	9,92,134	932,651	11,52,153	810,605	10,04,828
Mixed Silk Piece Goods.	Bengal	1,600	5,000	*Nil*	*Nil*	*Nil*	*Nil*	*Nil*	*Nil*	105	75
	Bombay	756	907	888	963	7,086	5,415	2,704	2,298	23,061	10,784
	Sind	3,696	5,112	4,995	7,116	5,600	5,200	3,994	5,027	3,548	3,286
	Madras	133	266	252	90	484	1,050	508	373	162	180
	Burma	53,373	40,137	44,439	42,554	19,054	17,637	16,788	19,350	28,980	25,302
	Total	73,958	51,422	50,574	50,723	32,224	29,302	23,994	27,048	55,856	39,627

TRADE in MANUFACTURES.

Inter-provincial Exchanges.

Inter-provincial Exchanges in Silk Goods, being the Imports into one Province from another, Coastwise—contd.

Provinces into which Imported.		Foreign Produce.									
		1885-86.		1886-87.		1887-88.		1888-89.		1889-90.	
		Yds.	R	Yds.	R	Yds.	R	Yds.	R	Yds.	R
Pure Silk Piece Goods.	Bengal	4,568	*7,396*	17,164	*18,278*	75,797	*46,643*	11,942	*19,995*	6,080	*5,942*
	Bombay	1,364	*1,540*	273	*552*	2,628	*3,278*	4,767	*6,111*	26,856	*17,209*
	Sind	110,279	*1,02,890*	141,658	*1,41,658*	287,986	*2,72,214*	250,633	*2,19,011*	285,452	*2,22,424*
	Madras	35,936	*37,043*	35,112	*41,404*	38,073	*37,690*	39,112	*44,557*	39,870	*37,778*
	Burma	400,831	*4,06,440*	207,694	*2,31,178*	600,413	*6,86,443*	348,009	*3,64,114*	248,759	*2,50,709*
	Total .	552,978	*5,55,309*	401,901	*4,32,970*	1,004,897	*10,46,268*	654,103	*6,53,788*	607,017	*5,34,062*
Mixed Silk Piece Goods.	Bengal	*Nil*	*Nil*	3,418	*2,480*	14,052	*4,926*	7,619	*6,601*	*Nil*	*Nil*
	Bombay	838	*792*	2,706	*1,272*	4,959	*5,290*	19,932	*14,272*	4,185	*3,936*
	Sind	5,515	*7,444*	15,499	*16,585*	14,554	*15,409*	15,521	*15,267*	17,366	*22,592*
	Madras	*Nil*	*Nil*	*Nil*	*Nil*	60	*120*	220	*165*	*Nil*	*Nil*
	Burma	17,452	*14,356*	26,989	*17,818*	28,986	*22,093*	27,823	*17,306*	19,148	*11,650*
	Total .	23,805	*22,592*	48,612	*38,155*	62,611	*47,838*	67,115	*63,611*	40,699	*38,178*

SILK.

Trade in Silk Goods.

TRADE in MANUFACTURES.

Inter-provincial Exchanges.

It will be observed that by far the most important feature of the coastwise trade is the imports of Indian and foreign piece goods into Burma from the provinces of India. This subject has already been briefly alluded to as supplementing the direct imports by Burma of foreign piece goods. If, therefore, it be desired to ascertain the total Burmese consumption of pure silk piece goods, it would be necessary to add together the direct imports from foreign countries to the coastwise imports from Indian provinces of foreign and Indian goods.

YEARS.	Foreign Imports direct.	Imports coastwise, foreign goods.	Imports coastwise, Indian goods.	Total Imports into Burma.
	Yds.	Yds.	Yds.	Yds.
1885-86 . . .	4,350,546	552,978	989,089	5,892,613
1886-87 . . .	4,200,958	401,901	1,186,640	5,789,499
1887-88 . . .	6,021,020	1,004,897	901,895	7,927,812
1888-89 . . .	5,505,887	654,103	932,651	7,092,641
1889-90 . . .	5,231,808	607,017	810,605	6,649,430
Averages for five years	5,062,043	644,179	963,976	6,670,399

From these returns of silk piece goods it will be seen that since the average annual import is 6,670,399 yards, that quantity may, roughly speaking, be accepted as the average annual consumption. The exports by land routes to foreign countries do not seriously disturb that estimate, since it is believed these are more than compensated for by local manufactures.

Burma supply. 2202

The following facts exhibit the shares taken by the provinces of India in the supply of pure silk piece goods to Burma:—

INDIAN PIECE GOODS.

	1885-86.	1886-87.	1887-88.	1888-89.	1889-90.
	Yds.	Yds.	Yds.	Yds.	Yds.
Bengal	484,514	698,006	535,711	415,208	393,855
Bombay	96,475	93,810	56,504	223,414	212,257
Madras	116,269	99,286	78,948	129,180	111,374

FOREIGN PIECE GOODS.

	1885-86.	1886-87.	1887-88.	1888-89.	1889-90.
Bengal	308,554	130,827	221,643	189,119	67,636
Bombay	88,056	76,867	378,770	158,530	180,163
Madras	4,221	...	...	400	960

It will thus be seen that during the past five years the trade in both classes has been expanding from Bombay and declining from Calcutta. This same observation may be made regarding the imports (shown on page 226) from foreign countries into Bengal and Bombay. The former are manifesting a distinct indication of declining in almost an inverse ratio with the improvement in the returns for Bombay. Whether these changes in the trade be accepted as of a permanent and serious nature or not, one observation remains that cannot be doubted, *viz.*, that Burma is by far

TRADE in MANUFACTURES.

By Rail & River.
2203

the most important Indian market for silk piece goods. The traffic in mixed silk piece goods is too unimportant to call for any special remarks.

RAIL & RIVER TRANSACTIONS.—Following the course already pursued with raw silk (p. 207), the following balance sheet may be given of the recorded traffic by these routes. It need only be necessary to remind the reader that these returns are given by weight, not by measurement. It is accordingly difficult to arrive at a comparative figure that would reduce these returns, so as to allow of an estimate of the consumption of Indian and foreign piece goods by the interior provinces.—

PROVINCES WHICH MANIFEST A NET EXPORT—THE PRODUCING AREAS.			PROVINCES WHICH MANIFEST A NET IMPORT—THE CONSUMING AREAS.		
Provinces and Towns.	Export and import in maunds of 82½℔	Net export in ℔.	Provinces and Towns.	Import and export in maunds of 82½℔.	Net Import in ℔.
INDIAN PRODUCE—					
Bengal . . . {	4,643 438		Madras . . {	495 16	
Net Export .	4,205	343,912	Net Import .	479	39,517½
Assam . . . {	281 7		Bombay . . {	444 209	
Net Export .	274	22,605	Net Import .	235	19,388
Madras Sea-ports . {	1,585 24		Sind . . {	39 2	
Net Export .	1,561	128,782½	Net Import .	37	3,052½
Bombay Town . {	1,051 187		N. W. Provinces and Oudh. {	1,191 212	
Net Export .	864	71,280½	Net Import .	979	80,767½
Karachi . . {	16 2		Panjáb . . {	770 264	
Net Export .	14	1,155	Net Import .	506	41,745
			Central Provinces {	26 3	
			Net Import .	23	1,897½
			Berar—Net Import	17	1,402½
			Rájputána and Central India—Net Import.	83	6,847½
			Nizam's Territory—Net Import.	13	1,072½

SILK.

Trade in Silk Goods.

TRADE in MANUFACTURES.

By Rail & River.

PROVINCES WHICH MANIFEST A NET EXPORT—THE PRODUCING AREAS.

Provinces and Towns.	Export and import in maunds of 82½ lb.	Net export in lb.
FOREIGN PRODUCE—		
Madras Seaports Net Export.	36	2,970
Bombay town . {	5,080 6	
Net Export .	5,074	418,605
Karachi—Net Export	361	29,782½
Calcutta . . {	324 43	
Net Export .	281	23,182½
Central Provinces {	2 2	
Net Export .	...	...

PROVINCES WHICH MANIFEST A NET IMPORT—THE CONSUMING AREAS.

Provinces and Towns.	Import and export in maunds of 82½ lb.	Net import in lb.
Mysore . . {	1,125 3	
Net Import .	1,122	92,565
Calcutta . . {	3,877 453	
Net Import .	3,424	282,480
Madras Presidency—Net Import.	15	1,237½
Bombay . . {	4,816 1	
Net Import .	4,815	397,238½
Sind . . .	341	28,132½
Bengal . . {	16 10	
Net Import .	6	495
N.-W. Provinces and Oudh.	109 3	
Net Import .	106	8,745
Panjáb . . {	435 6	
Net Import .	429	35,392½
Berar—Net Import	5	412½
Rájputána and Central India—Net Import.	9	742½
Nizam's Territory—Net Import.	3	247½
Mysore—Net Import	23	1,897½

TRADE in MANUFACTURES. By Rail & River.

It will be seen that Bengal and Assam are the chief exporting provinces. The Madras sea-port towns have to be left out of consideration, as these merely distribute the foreign and coastwise imports to the presidency. The Bombay town exports are, however, of more importance, since these (both Indian manufacture and foreign) are mainly consigned to the Bombay Presidency and the Panjáb. Of the foreign piece goods exported from Bombay town, 4,816 maunds went to the presidency and the balance mainly to the Panjáb. Of the exports of Indian goods, 526 maunds went to the Panjáb, and 434 maunds to the Bombay Presidency. It is currently reported that every year shows a brisker trade from Bombay to the Panjáb, so that the competition of power-looms seems to be telling against the local hand-loom productions. But the Panjáb also draws largely from Calcutta (123 maunds Indian and 223 maunds foreign), and considerably smaller quantities from Karachi and Bengal. As already pointed out, therefore, the Panjáb, after Burma, may be accepted as the next most important consuming province for silk in India.

APPENDIX.

2204

Complete List of the Indian Bombycidæ and Saturniidæ by G. F. Hampson, Esq.

BOMBYCIDÆ.

1. **BOMBYX MORI,** *L.; Syst. Nat., II., p. 817.* (*See pp. 2, 15.*)
   ,, **crœsi,** *Hutt.; Trans. Ent. Soc., 1864, p. 312.* (*See pp. 2, 10.*)
   ,, **fortunatus,** *Hutt.; Trans. Ent. Soc., 1864, p. 312.* (*See pp. 2, 12.*)
   ,, **sinensis,** *Hutt.; Trans. Ent. Soc., 1864, p. 313.* (*See pp. 2, 18.*)
   ,, **textor,** *Hutt.; Trans. Ent. Soc., 1864, p. 313.* (*See pp. 2, 19.*)
   ,, **arracanensis,** *Hutt.; Trans. Ent. Soc., 1864, p. 313.* (*See pp. 2, 7.*)
   Habitat.—China and, in a domesticated state, Europe, India, etc. *See pp. 2, 5, 22—67.*

2. **ECTROCTRA DIAPHANA,** *Hampson: Moths of India, ined.*
   Habitat.—Momeit, Burma.

3. **OCINARA SIGNIFERA,** *Wlk.: Journ. Linn. Soc., vi., p. 130.*
   ,, **lactea,** *Hutt.: Trans. Ent. Soc., 1864, p. 328.*
   ,, **diaphana,** *Moore; Lep. Atk., p. 83.*
   Habitat.—North-West Himálayas; Khasias; Nilgiris; Andamans; Sumatra; Borneo. *See p. 3.*

4. **OCINARA APICALIS,** *Wlk.; Journ. Linn. Soc., vi., p. 130.*
   ,, **signata,** *Wlk.; Journ. Linn. Soc., vi., p. 131.*
   ,, **moorei,** *Hutt.; Trans. Ent. Soc., 1864, p. 326.*
   ,, **lida,** *Moore; Cat. E. I. C., p. 381.*
   Bombyx plana, *Wlk.; Cat. xxxii., p. 575.*
   Habitat.—Masuri; Hongkong; Borneo; Java. *See p. 3.*

5. **OCINARA VARIANS,** *Wlk.; Cat., v., p. 1153.*
   Naprepa albicollis, *Wlk.; Cat., vi., p. 171.*
   ,, **cervina,** *Wlk.; Cat., xxxii., p. 489.*
   Chazena velata, *Wlk.; Char. Undescr. Het., p. 21.*
   Habitat.—Philippines; China; Formosa; throughout India and Ceylon; Borneo. (*See p. 5.*)

6. **GUNDA JAVANICA,** *Moore; P. Z. S., 1872, p. 576.*
   Habitat.—Sikkim; Java.

List of Indian Bombycidæ

LIST.

7. **GUNDA APICALIS**, *Hampson*; *Ill. Het., ix., ined.*
Habitat.—Ceylon. *See p. 5.*

8. **GUNDA SIKKIMA**, *Moore*; *P. Z. S., 1879, p. 406.*
Norasuma var egata, *Hampson*; *Ill. Het., ix., ined.*
Aristhalla thwaitesii, *Moore*; *Lep. Ceyl., ii., p. 136.*
Habitat.—Sikkim; Ceylon. *See p. 5.*

9. **THEOPHILA HUTTONI**, *Westw.*; *Cab. Or. Ent., p. 26.*
„ **sherwillei**, *Hutt.*; *Trans. Ent. Soc., 1864, p. 324.*
„ **affinis**, *Hutt.*; *Geoghegan, Silk in India, App. A, p. 3.*
„ **bengalensis**, *Hutt.*; *Trans. Ent. Soc., 1864, p. 322.*
Bombyx religiosa, *Helf.*; *Journ. A. S. B., 1837, 41, 3.*
Habitat—North-Western Himálayas; Sikkim; Assam. *See pp. 3-4.*

10. **THEOPHILA LUGUBRIS**, *Drury*; *Ins. Exot., iii., p. 28.*
Habitat.—Madras.

11. **MUSTILIA FALCIPENNIS**, *Wlk.*; *Cat. xxxii., p. 581.*
„ **castanea**, *Moore*; *Lep. Atk., p. 82.*
Habitat.—Sikkim; Bhotan.

12. **MUSTILIA SPHINGIFORMIS**, *Moore*; *P. Z. S., 1879, p. 407.*
Habitat.—Masuri.

13. **MUSTILIA HEPATICA**, *Moore*; *Lep. Atk., p. 82.*
„ **columbaris**, *Butl.*; *P. Z. S., 1886, p. 387.*
Habitat.—Murree, Sikkim.

14. **ANDRACA BIPUNCTATA**, *Wlk.*; *Cat. xxxii., p. 582.*
„ **trilochoides**, *Moore*; *P. Z. S., 1865, p. 820.*
Habitat.—Sikkim; Assam.

15. **ANDRACA TRISTIS**, *Feld.*; *Reis. Nov. pl. 95, f. 4.*
Habitat.—Ceylon (or ! S. Africa).

SATURNIIDÆ.

1. **ACTIAS SELENE**, *Hubn.*; *Samml. Schmett. I., pl. 172, f. 3.*
Plectropteron dianæ, *Hutt.*; *Ann. N. H., 1846, p. 60.*
Habitat.—China; throughout India, Ceylon, and Burma. *See pp. 68-69.*

2. **ACTIAS MÆNAS**, *Doubl.*; *Ann. N. H., 1847, p. 95.*
„ **leto**, *Doubl.*; *Proc. Ent. Soc., 1847, p. li.*
Habitat.—Sikkim; Butan; Khasias; Sibsagar. *See p. 67.*

3. **ACTIAS IGNESCENS**, *Moore*; *P. Z. S., 1877, p. 602.*
Habitat.—Andamans. *See p. 67.*

4. **ATTACUS ATLAS**, *Linn.*; *Mus. Lud. Ulr., p. 366.*
„ **silhetica**, *Helf.*; *Journ. A S. B., vi., p. 41.*
„ **taprobanæ**, *Moore*; *Lep. Ceyl. ii, p. 124.*
Habitat.—Throughout India, Ceylon, and Burma. *See pp. 80-85.*

5. **ATTACUS EDWARDSII**, *White*; *P. Z. S., 1859, p. 115.*
Habitat.—Sikkim; Shillong. *See p. 86.*

6. **ATTACUS CYNTHIA**, *Drury*; *Exot. Ins., pl. 6, f. 2.*
„ **vesta**, *Wlk.*; *Cat. xxxii., p. 525.*
„ **canningi**, *Wlk.*; *Cat. xxxii., p. 525.* *See p. 84.*
„ **walkeri**, *Feld.*; *Wein. Ent. Mon. vi., p. 34.*
„ **pryeri**, *Butl.*; *Ill. Het. iii., p. 11.*
Habitat.—Japan; China; Himálayas; Assam; Cachar; Java; Nilgiris. *See pp. 85-86, 162.*

and Saturniidæ. (*G. F. Hampson.*) SILK.

LIST.

7. **ATTACUS RISINI,** *Boisd.; Ann. Soc., Ent. Fr. 1854, p. 755.*
   „ **guerini,** *Moore; P. Z. S., 1859, p. 269. See p. 87.*
   „ **lunula,** *Wlk.; Cat. v., p. 221.*
   „ **obscurus,** *Butl.; Trans. Ent. Soc., 1879, p. 5.*
   „ **iole,** *Westw.; P. Z. S., 1884, p. 144.*
   Habitat.—China; Sikkim; Bengal; ? Ceylon. *See pp. 87, 162-174.*

8. **ANTHERÆA ROYLEI,** *Moore; P. Z. S., 1859, p. 256.*
   Habitat.—Masuri; Kangra; Sikkim; Khasias. *See pp. 76-78.*

9. **ANTHERÆA ASSAMA,** *Westw.; Cab. Or. Ent., p. 42.*
   „ **mesankooria,** *Moore; Wardle's Wild Silks, p. 5. See p. 71.*
   ? „ **perotteti,** *Guér.-Mag.; Zool., 1843, pl. 123. See pp. 75, 76.*
   Habitat.—Himálayas from Kangra to Assam? Pondicherry. *See pp. 70, 71, 75, and 174-183.*

10. **ANTHERÆA ANDAMANA,** *Moore; P. Z. S., 1877, p. 602.*
   Habitat.—Andamans. *See p. 70.*

11. **ANTHERÆA KNYVETTI,** *N. Sp.*

♂ and ♀ very much resemble A. **roylei** from which the species differs in being tinged with pink and yellow, and in the antimedial line of the hindwing being further from the base and touching the ocellus. From **paphia** it differs in having the ground colour greenish grey; the costal grey band of the forewing only extends half way from the base; the occelli of both wings are small, each hyaline spot having a dark lemule on its inner edge; the marginal line dark; the sub-marginal line of hindwing placed far from the margin as in **roylei.**

The larva differs from that of **roylei** in having sepia streaks on the head; silver spots on the lateral band of 4th and 5th somites only, instead of the 4th, 5th, 6th, and 7th; the apical tubercles on 2nd and 3rd somites blue instead of green. It feeds on the Wild Cherry "not the Mohwa," Oak and Birch, and spins a small hard *pedunculate cocoon* instead of a double walled non-pedunculate cocoon.

Habitat.—Sikkim. *Exsp. 156, M. M. See pp. 70, 71, 76-7.*

12. **ANTHERÆA HELFERI,** *Moore; P. Z. S., 1859, p. 257.*
   Habitat.—Sikkim. *See p. 71.*

13. **ANTHERÆA PAPHIA,** *Linn.; Syst. Phal., 4.*
   „ **mylitta,** *Drury; Exot. Ins., ii., pl. 5, 61.*
   „ **cingalesa,** *Moore; Lep. Ceyl. II., p. 122.*
   „ **sivalica,** *Moore; Wardle's Wild Silks, p. 5. (See pp. 45, 79, 122.)*
   „ **nebulosa,** *Hutt.; Moore. Wardle's Wild Silks, p. 5. See p. 71.*
   „ **fraterna,** *Moore; P. Z. S., 1888, p. 402.*
   Habitat.—Throughout the greater part of tropical India. *See 72 and 96-161.*

14. **ANTHERÆA FRITHI,** *Moore; P. Z. S., 1859, p. 256.*
   Habitat.—Sikkim; Bhutan. *See p. 70.*

15. *Hope; Trans. Linn. Soc., 1845, p. 132.*
   Habitat.—Sikkim. *See p. 94.*

16. **SATURNIA THIBETA,** *Westw; P. Z. S., 1853, p. 166.*
   Rinaca extensa, *Butl.; Ill. Het. v., p. 61.*
   Habitat.—Thibet; Sikkim; Assam. *See p. 88.*

17. **SATURNIA ANNA,** *Moore; P. Z. S., 1865, p. 818.*
   Habitat.—Sikkim. *See p. 95.*

18. **SATURNIA GROTEI,** *Moore; P. Z. S., 1859, p. 265.*
   Habitat.—Himalayas from Simla to Sikkim. *See p. 95.*

S. 2204

LIST.

19. **SATURNIA LINDIA,** *Moore; Trans Ent. Soc., 3rd Ser., ii., p. 424.*
   „ **hockingii,** *Moore; P. Z. S., 1888, p. 402.*
   Habitat.—Kulu. *See p. 95.*

20. **SATURNIA PYRETORUM,** *Westw.; Cab. Or. Ent., p. 49.*
   „ **cidosa,** *Moore; Trans. Ent. Soc. (3), II., p. 423.*
   Habitat.—China; Sikkim. *See pp. 95, 96.*

21. **SATUR IA SIMLA,** *Westw.; Cab. Or. Ent., p. 41.*
   Habitat.—North-Western Himálayas. *See p. 87.*

22. **SATURNIA HUTTONI,** *Moore; Trans. Ent. Soc. (3), i., p. 321.*
   Habitat.—North-Western Himálayas. *See p. 94.*

23. **SATURNIA CACHARA,** *Moore; P. Z. S., 1872, p. 578.*
   Habitat.—Cachar. *See p. 88.*

24. **SATURNIA STOLICZKANA,** *Feld.; Reis. Nov. pl. 87, f. 3.*
   Neoris shadulla, *Moore; P. Z. S., 1872, p. 577.*
   Habitat.—North-Western Himálayas; Yarkand; Ladak. *See p. 94.*

25. **LOEPA KATINKA,** *Westw.; Cab. Or. Ent., p. 25.*
   „ **miranda,** *Moore; Trans. Ent. Soc. (3), ii., p. 424.*
   „ **sikkima,** *Moore; P. Z. S., 1865, p. 818.*
   Habitat.—Sikkim; Assam; Yunan; Nilgiris; Java. *See p. 94.*

26. **LOEPA ? SIMPLICIA,** *Mass. & Weym.; Beittrs Zur. Schmelt pt., ii., p. 20.*
   Habitat.—East Indies.

27. **LOEPA NEWARA,** *Moore; P. Z. S., 1872, p. 578.*
   Habitat.—Nepal; Sikkim. *See p. 95.*

28. **SALASSA LOLA,** *Westw.; Cab. Or. Ent., p. 25.*
   Habitat.—Sikkim; Silhet. *See p. 95.*

29. **SALASSA ROYI,** *Elwes.; P. Z. S., 1887, p. 447.*
   Habitat.—Sikkim. *See p. 95.*

30. **CRICULA TRIFENESTRATA,** *Helf.; Journ. A. S. B., vi., p. 45.*
   „ **burmana,** *Swinh.; Trans. Ent. Soc., 1890, p. 198.*
   Habitat.—Throughout India, Ceylon, and Burma; Andamans; Java. *See pp. 88-93.*

31. **CRICULA DREPANOIDES,** *Moore; P. Z. S., 1865, p. 817.*
   Habitat.—Sikkim. *See p. 88.*

BRAHMÆIDÆ.*

1. **BRAHMÆA WALLICHII,** *Gray; Zool. Misc., p. 39.*
   „ **conchifera,** *Butl.; Ill. Het., v., p. 95.*
   „ **rufescens,** *Butl.; Ann. N. H., 1880, p. 62.*
   Bombyx spectabilis, *Hope; Trans. Linn. Soc., 1841, p. 443.*
   Habitat.—Nepal; Sikkim; Assam. *See p. 96.*

2. **BRAHMÆA HEARSEYI,** *White; Proc. Ent. Soc., 1862, p. 26.*
   „ **whitei,** *Butl.; Ill. Het., v., p. 119.*
   Habitat.—Masuri; Sikkim; Burma. *See p. 96.*

(*W. R. Clark.*)

SILVER.

2205

Silver, *Ball in Man. Geol. Ind., III., 231.*

Vern.—*Chandi, rupa,* HIND.; *Tara, chandi,* BENG.; *Rupa,* DEC.; *Chandi,* GUZ.; *Velli,* TAM.; *Venni,* TEL.; *Fazzeh, faddah,* ARAB.; *Sim, nokra,* PERS.; *Rupya, sveta, raiata,* SANS.; *Peddi,* SINGH.

References.—*Mason, Burma and Its People, 564, 729; Ainslie, Mat. Ind., II., 562; U. C. Dutt, Mat. Med. Hind., 61; Atkinson, Him. Dist.*

* The *Brahmæidæ* spin no cocoon and are, therefore, not silk worms, though by some authors they have been quoted as such.

(X., N.-W. P. Gaz.), 279; Forbes Watson, Ind. Survey, 406; Ayeen Akbary, Gladwin's Trans., I., 15; II., 42, 183; Linschoten, Voyage, to East Indies (Ed. Burnell, Tiele & Yule), I., 21, 80, 109, 128, 147, 153; II., 294, 295; Tavernier, Travels to E. Ind., I., 8; II., 25, 162, 270, 281; Marco Polo, Travels (Yule's Ed.,) II., 325, 327; Man. Madras Adm., II., 35; Gribble, Man. Cuddapah, 45; Bomb. Admin. Rep. (1871-72), 373, 384; Gazetteers:—Bombay, IV., 127; Panjáb, Lahore, 99; Delhi, 127; Hazara, 143, Peshawar, 152; Mysore & Coorg, I., 432; Balfour, Cyclop. Ind., III., 641.

HISTORY. 2206

History.—Silver is one of the most anciently known of metals. In the Bible it is frequently mentioned; from the time of Abraham onwards it has been a common object of barter and traffic, and Solomon is said to have hired fleets of ships from Tarshish for the purpose of obtaining from Ophir the silver used in decorating his palace and temple. The latter place has been identified by some writers as a port or district on the Malabar coast; but at least it is undoubted that **Pliny**, who wrote A.D. 77, referred to India as a country whence silver was obtained for the use of the Romans. He says, "The Dardanians inhabit a country the richest of all India in gold mines, and the Setæ have the most abundant mines of silver;" and in another passage, "In the country of the Nareæ, on the other side of the mountain Capitalia (*i.e.*, Mount Abu), there are a very large number of mines of gold and silver in which the Indians work very extensively." **Mr. Calvert** (*Kulu and its Beauties*) has suggested that the country of the Setæ was probably Kulu, the *Wazir-i-rupi* or silver country of the Wazirs, and undoubtedly argentiferous galenas do occur there, but **General Cunningham** was inclined to think that the Setæ referred to by **Pliny** were the native bankers or Seths who have for ages held in their own hands the wealth of India. The silver mines of the Nareæ have been identified with some of the ancient mines in Southern India, in which argentiferous galena was found, and some of which are in the vicinity of the country of the Nairs or Malabar.

There seems good reason to believe that the Chinese, at any rate, at one time, obtained silver from Malabar, but **Marco Polo**, in his *Travels*, states that ships coming there from the East to trade brought gold and silver to exchange for the products of the country, so that in his time Malabar had probably ceased to be an exporting centre. In the *Ain-i-Akbari*, written about the end of the sixteenth century, silver mines are mentioned near Agra and in Kashmír, but the bulk of the silver used at this period in India seems to have been imported.

Tavernier, in his *Travels*, says in one passage that Japan is the only Asiatic country where silver occurs, but afterwards cites Bhután and Assam as possessing silver mines.

OCCURRENCE 2207

OCCURRENCE.—Silver in India occurs most commonly in ores of lead and copper, but there are several recorded cases where native silver has been said to have occurred, and to them reference will first be made. The account of the occurrence of silver is an abstract of that given in *Ball's Economic Geology of India (Vol. III., Man. Geol. Ind).*

Native Silver Madras. 2208

A.—NATIVE SILVER.—*Madras: Mysore.*—According to **Sir W. Ainslie**, **Captain Arthur** discovered thin plates of native silver adhering to some cubical crystals of gold, and also silver in the condition of chlorides associated with iron pyrites. **Dr. Scott's** assays of the Karnúl galena indicate an extraordinarily rich ore (374 oz., 175 oz. 3 dwts.; 165·76 oz. of silver in the ton of lead), but **Dr. Balfour's** original examination showed no traces of silver. The Karnúl galena silver mines deserve their title if these amounts were really found.

On the authority of a letter from **Dr. Roxburgh**, **Dr. Heyne** states that on the analysis by the Assay Master of Bengal, some Cuddapah

SILVER.

Occurrence of Silver

OCCURRENCE. Native Silver.

galena from an abandoned mine, formerly worked by Tippu Sultan, contained 11 per cent. of silver, or more probably 11 oz. to the ton, since 11 per cent. gives the incredible amount of 246·4lb to the ton. Silver alloyed with Wynaad gold gives an average of about 15 per cent., but it is probably less.

Rajputana. 2209

Rajputana: Ajmere District.—Silver is said to have been found in Lakhan Koti, a sandy well in the western portion of the city, but the amount realised was insignificant.

Bombay. 2210

Bombay: Dharwar District.—In the streams from the Dambal or Kappatgode hills, Captain **Newbold** found small quantities of silver associated with copper in gold dust which was being washed. A fragment of grey ore and some white metallic particles, which were thought might be silver, were also met with. **Mr. Foote**, under similar circumstances, subsequently found particles of native silver. A rumour that silver occurs in the volcanic island of Adjar, off Kattywar, needs confirmation.

Argentiferous Galena. Upper Burma. 2211

B.—ARGENTIFEROUS GALENA.—*Upper Burma: Shan States.*—In these States there are some smelting works for the argentiferous galena which occurs in the limestones near a village called Kyouktat. The precise position of the mines and process of extracting silver from galena was ascertained by **Mr. Fedden**. Having put the ore with charcoal and a proportion of broken slag into a clay cupola or blast furnace, 3 feet high and from 14 to 16 inches in diameter, two women, standing on a staging, work wooden cylinders with pistons, which produce a blast. As the galena is reduced by the sulphur being driven off, the metal accumulates at the bottom of the furnace, and is then run out and cast in pigs, which, on setting, are removed to a refining shed and placed in a reverbratory furnace in which lumps of charcoal are kept on fire-clay supports just above the molten metal. Litharge forms on the surface and is removed by an iron roller, and the process is repeated till all the lead has been extracted without the aid of cupellation, and pure silver remains of such condition that unless alloyed by silversmiths it is not used for currency and jewellery. The litharge subsequently reduced is converted back to metallic lead. The smelter buys up all argentiferous and cupriferous refuse from the silversmiths' shops and separates the metals in his furnace.

The following table showing the amount of silver per ton of (α) lead, (β) copper, (γ) percentage of silver alloyed with Indian gold, will suffice to convey to the reader the localities where such alloys have up to the present been principally found. They are abstracts of similar tables found in **Ball's** valuable work :—

TABLE (α.)

District.	Name of Mine.	Oz, Dwt. Grs.	Analyst.	Date.	Reference.
		MADRAS.			
Kadapah . .	Jungumraz-pilly	10 14 0	P. W. Wall .	1858 . .	P. W. Wall.
Do. . .	Do. .	13 13 0			
Do. . .	Do. .	22 7 0	F. R. Mallet	May 1879 .	W. King.
Do. . .	Baswapur .	35 7 0	P. W. Wall.	1858 . .	Madras Journ. Lit. and Sci., XX, 284.
Karnul . .	? .	374 0 0	Dr. Scott .	...	Balfour, Cyc. Ind., Art. Galena.
Do. . .	? .	175 3 0	Do. .	...	
Do. . .	? .	165 15 0	Do. .	Jan. 1859 .	Rep. to Off. in ch. Govt. Central Mus.

SILVER.

OCCURRENCE

Argentiferous Galena. Upper Burma.

in India. (*W. R. Clark.*)

District.	Name of Mine.	Oz. Dwt. Grs.	Analyst.	Date.	Reference.
		BENGAL.			
Santal Pergunnahs.	Bairuki*.	29 8 0	Johnson and Mathey.	Oct. 1859	J. Barratt.
Do.	Laksimpur near Nia Dumka.	50 6 3	A. Tween	Dec. 1860	...
Bhagalpur	? Gangue	52 8 14	F. R. Mallet	May 1871	Commissioner, Bhagalpur.
Do.	Phaga	103 2 12	Chem. Ex. to Govt.	...	(? 55 chittacks to the ton.)
Manbhum	Dhadka	119 4 16	A. Tween	1869	V. Ball.
Do.	Do.	99 0 0	F. R. Mallet	April 1881	...
		CENTRAL PROVINCES.			
Sambalpur	Jhuman	12 5 0	A. Tween	Feb. 1875	V. Ball.
Hosangabad	Joga	21 3 0	F. R. Mallet	June 1879	G. J. Nichols.
Do.	Sleemanabad	19 12 0	A. Tween	June 1870	T.W.H.Hughes.
		REWAH AND BUNDELKHAND.			
Rewah	Bengowak	{ 8 6 14 5 4 12 }	A. Tween	June 1870	Dr. Stratton.
Jhansi	...	19 12 0	Do.	May 1873	B. W. Colvin.
		RAJPUTANA.			
Meywar	...	10 12 9	A. Tween	Oct. 1873	Pol. Ag., Meywar.
		PANJAB.			
Hazara	{ Kakal Gashi Habibulla }	7 8 11	F. R. Mallet	Sept. 1877	Financial Comr, Panjáb.
Sirmur	..	{ 13 10 0 26 15 0 24 10 0 }	A. Tween	Dec. 1865.	
Simla	Subathu	10 12 12	Do.	1869.	
Kulu	Parbatti river Do.	22 17 8 89 16 16	Do.	1869	F. Schiller.
	Do. Do.	13 14 9 17 19 8	Do.	March 1870.	
	Do. Do.	81 16 14 38 19 6	Do.	May 1870	F. Calvert.
	Ballarag	65 6 16	Do.	Oct. 1873.	
	Koman Kotkundi.	25 6 8	Do.	Nov. 1873.	
	?	46 1 4	Do.	May 1874.	
	?	50 12 16	Do.	June 1874.	
	?	18 1 14	Do.	Do.	
	? No. 8 lode	20 8 8	Do.	Do.	
Lahoul	Shigri	89 0 0	T. W. H. Hughes.	Sept. 1874	J. Calvert.
		NORTH-WEST PROVINCES.			
Masuri	?	16 6 16	A. Tween	Feb. 1873	Col. Need.
		NEPAL.			
Nepal	?	14 14 0	A. Tween	Jan. 1863	Col. Lal Singh.

* Per ton of ore.

SILVER.

OCCURRENCE

Argentiferous Galena.
Upper Burma.

Occurrence of Silver in India.

District.	Name of Mine.	Oz. Dwt. Grs.	Analyst.	Date.	Reference.
		BURMA.			
Tonghu . .	...	20 8 7	A. Tween .	May 1871 .	Capt. Cooke.
Do. . .	Between Phagat and the Yongaten.	5 14 0	D. Waldie .	1854 . .	Letter from E. O'Riley, Esq., to Chief Commissioner.
	Do. .	9 0 0			
	Do. .	Trace.			
	Do. .	5 8 0			
Amherst . .	Teetameelay Hill.	8 3 8	Do. .	March 1873.	
Do. . .	Teetalay Hill .	Trace.			
Do. . .	Meezine Hill .	14 14 0	Do. .	Dec. 1872.	
Moulmain .	...	19 5 14	A. Tween .	1864.	
Tavoy . .	...	16 7 9	Do. .	Sept. 1853 .	Asstt. Secy., Govt. of India.
Do. . .	...	Trace.			
Tenasserim .	? .	4 14 7	Do. .	July 1875 .	Col. Duncan, Secy. to Chief Commissioner.
	...	31 0 16	Do. .		
? . .	? .	12 17 8	Do. .	Oct. 1875.	
	King Island .	13 1 8	Do. .	March 1872.	
	Do. .	11 8 16			
	Do. .	11 0 0			
	? .	12 5 0	Do. .	Do.	
Bhamo . .	...	58 14 8	Do. .	June 1863 .	Dr. Williams.
Yunan . .	Kyet You 3 days' march to north of Momein.	104 10 6	Do. .	March 1870 .	Dr. J. Anderson.
Do. . .	Ponsee silver mines, Khakhyee Hills.	73 10 0			

TABLE (β).
BENGAL.

District.	Name of Mine.	Oz. Dwt. Grs.	Analyst.	Date.	Reference.
Santal Pergunnahs.	Bairuki . .	52 5 0	Johnson and Mathey.	Oct. 1856.	
Do. .	Do. (Pacos)	85 0 0	H. Piddington.	...	Journ. As. Soc. Bengal, Vol. XXI., 74.
Singhbum .	...	2 5 17 1 2 20 0 19 14	Messrs. Phillips and Darlington.		
		BURMA.			
...	Yoonzalin River	31 6 12	D. Waldie .	1870 . .	Procdgs. As. Soc. Bengal, 1870 279 (O'Riley, etc.).

TABLE (γ).
MADRAS.

District.	Locality.	Gold.	Silver.	Analyst.	Date.	Reference.
Wynaad .	Alpha reef . .	67·07	32·93	A. Tween .	Dec. 1874 .	W. King.
	Devala reef .	93·	7·			
	Do. .	84·7	15·33	Do. .	Sept. 1870 .	J. W. Minchin.
	Monarch . .	82·69	11·32	Do. .	Jan. 1875 .	W. King.
	Do. . .	87·96	12·04			
	Devala surface .	90 9	8·67			
	Devala reef .	86·86	10·96			
	Average .	84·74	14·03			
Kolarin, Mysore.	...	91·66	15·3	Do. .	Sept. 1875 .	W. King.

District.	Locality.	Gold.	Silver.	Analyst.	Date.	Reference.
			BENGAL.			
Jashpur .	...	94·64	5·15	A. Tween .	Dec. 1874 .	Col. Dalton.

MEDICINE. 2212

Medicine.—According to Sanskrit writers silver, before administration as a medicine, should be purified by being converted into a black oxide, and resublimed, the process being repeated fourteen times. The properties of silver thus prepared are said to be similar to those of gold, but inferior. It is generally recommended for use, in combination with other metals, such as gold, iron, etc. The dose given is from one to two grains, and it is said to be useful in various nervous diseases. In India the nitrate of silver is occasionally prepared by Muhammadan physicians, and used by them in the treatment of nervous diseases. In European medicine, the nitrate and oxide of silver are the only salts used. The nitrate is applied externally in substance, as a caustic and escharotic; internally it is given in small doses as a tonic, antispasmodic, and sedative, while the oxide is analogous to the nitrate in action, but milder in its operation. Both are apt, but more particularly the nitrate, when administered over a long period of time, to produce a blue discoloration of the skin.

DOMESTIC. 2213

Domestic and Sacred.—Silver is largely employed not only in the coinage of India, but also for purposes of personal adornment in the massive ornaments worn by the upper and middle classes. It is unnecessary in a work like the present to enter into a detailed account of the art-industries, such as the silver filagree work of Cuttack, Dacca, and Delhi, the jewellery of Southern India and the enamelled silver-work of Hyderabad and Multan, for accounts of which works, the reader is referred to the Journal of Indian Art.

TRADE. 2214

Trade.—During the official year 1889-90, the imports of silver into India, from all sources, amounted to 117,175,091 tolas, valued at R12,38,84,740; of these 117,173,091 tolas, valued at R12,38,82,740 were privately imported, while 2,000 tolas, valued at R2,000, were imported by Government. The chief importing countries were the United Kingdom (with an import valued at R9,76,91,280), Hongkong (R61,38,640), and Aden (R40,86,093). During the same period, there was an export of 14,125,028 tolas, valued at R1,41,25,028, of which R1,38,61,963 worth was exported by Government, and R6,44,020 by private individuals. Nearly half this amount R60,22,210 went to Ceylon, and smaller quantities from 10 to 13 lakhs to the Mauritius, Arabia, Persia, and the Straits Settlements.

Sinapis nigra, *Linn.;* see **Brassica nigra,** *Koch.;* CRUCIFERÆ; Vol. I., 530.

S. ramosa, *Roxb.;* see **B. juncea,** *H. f. & T.;* Vol. I., 528.

SIPHONODON, *Griff.; Gen. Pl., I., 370, 998.*

[CELASTRINEÆ.

2215

Siphonodon celastrineus, *Griff.; Fl. Br. Ind., I., 629;*

Syn.—ASTROGYNE CORIACEA, *Wall.*

Vern.—*Myouk-opshit,* BURM.

References.—*Kurz, For. Fl. Br. Burm., I., 254; Gamble, Man. Timb., 83.*

Habitat.—An evergreen tree, frequent in the tropical forests of the eastern slopes of the Pegu Yomah and Martaban, and distributed to Java.

TIMBER. 2216

Structure of the Wood.—Pale yellowish, heavy, of a coarse unequal fibre, hard, and rather brittle.

SISAL HEMP.

2217

Sisal Hemp.—In continuation of the article on this subject, (Vol. I., 143)—**Agave sisalana**—it may be said that experiments in the cultivation of this valuable fibre plant were begun in India in 1890, when a case of living plants, from the Director of the Royal Botanic Gardens, Kew, was forwarded to the Superintendent of the Botanic Gardens, Calcutta, for the purpose of introducing the Sisal hemp industry into India. No details, however, of this experiment have as yet been published. A report on the Sisal Hemp industry of the Bahamas (written by **Mr. George Preston**) was presented to both Houses of Parliament in March 1890, No. 85.

FIBRE. 2218

Sissoo, see **Dalbergia Sissoo,** *Roxb.;* LEGUMINOSÆ; Vol. III., 13.

SISYMBRIUM, *Linn.; Gen. Pl., I., 77.*

2219

Sisymbrium Irio, *Linn.; Fl. Br. Ind., I., 150;* CRUCIFERÆ.

Syn.—S. IRIOIDES, *Boiss.*

Vern.—(Seeds=) *khúbkalán,* HIND.; *Naktrúsa, jangli sarson,* (seeds=) *khúb kalán, khâksi,* PB.; *Parjan,* MERWARA; *Jungli surson,* SIND; (Seeds=) *khákshí,* BOMB.; (Seeds=) *Rán-tikhí,* MAR.; (Seeds=) *khákshí,* PERS.

References.—*Boiss., Fl. Orient., I., 218; Stewart, Pb. Pl., 15; Dymock, Mat. Med. W. Ind., 2nd Ed., 57; S. Arjun, Bomb. Drugs, 12; Murray, Pl. and Drugs, Sind, 50; Dymock, Warden & Hooper, Pharmacog. Ind., I., 121; Year-Book, Pharm., 1878, 289; Gaz., N.-W. Prov., IV., lxvii.; Ind. Forester, XII., 5.*

Habitat.—A tall, annual or biennial herb, found in Northern India from Rájputana to the Panjáb, and distributed westward from Afghánistán to the Canary Islands.

MEDICINE. Seeds. 2220

Medicine.—The SEEDS are officinal in the Arabic and Muhammadan Pharmacopæias. In the older Hindu Materia Medicas, there is no mention of this drug, which appears to have been introduced by the Muhammadans as a substitute for **S. officinale.** It is described by them as having expectorant, stimulant, and restorative properties; it is also used externally as a stimulant poultice (*Dymock*).

SPECIAL OPINION.—§ "A decoction of the seeds is used as a febrifuge" (*Surgeon-Major C. W. Calthrop, M.D., Morar*).

2221

S. Sophia, *Linn.; Fl. Br. Ind., I., 150.*

References.—*Stewart, Pb. Pl., 15; Aitch., Bot. Afgh. Del. Com., 34; Gaz., N.-W. P., X., 305; Journ. Agri.-Horti. Soc. Ind., XIV., 11.*

Habitat.—An annual herb, found on the Temperate Himálaya from Kumáon to Kashmír, at altitudes between 5,000 and 7,000 feet, and in Western Tibet at altitudes between 9,000 and 14,000 feet. It occurs also in the Panjáb on the salt range and near Peshawar, and is widely distributed in Europe, North Africa, and North and South America.

MEDICINE. Seeds. 2222

Medicine.—Its SEEDS are used medicinally as a substitute or adulterant for those of the preceding (*Stewart*).

SKIMMIA, *Thunb.; Gen. Pl., I., 302.*

2223

Skimmia Laureola, *Hook. f.; Fl. Br. Ind., I., 499;* RUTACEÆ.

Syn.—LIMONIA LAUREOLA, *Wall.;* AN-QUE-TILIA LAUREOLA *Dcne.;* LAUREOLA FRAGRANS, *Roem.*

Vern.—*Chumlani,* NEPAL; *Timburnyok,* LEPCHA; *Nehar, gurl pata* KUMAON; *Ner, barrú, shalangli* (?,) PB.

References.—*DC., Prod., 1, 536; Brandis, For. Fl., 50; Gamble, Man. Timb., 61; Darjiling List, 14; Stewart, Pb. Pl., 29; Agri.-Horti. Soc. Ind. Jour., XIV., 13; Gaz., N.-W. P., X., 307.*

Habitat. A glabrous shrub, found throughout the Temperate Himálaya from Marri to Mishmi, at altitudes between 6,000 and 10,000 feet, and on the Khásia mountains between 5,000 and 6,000 feet. It is distributed to Afghanistán.

MEDICINE. Leaves. 2224

Medicine.—"The LEAVES, when crushed, have an orange-like smell, and are burned near small-pox patients with a view to curative effects. As they are dotted, this may have arisen from the 'doctrine of signatures'" (*Stewart*).

FOOD. Leaves. 2225

Food.—The hill tribes often eat the LEAVES in curries.

TIMBER. 2226

Structure of the Wood.—White, soft with distinct, white concentric rings, when fresh, has an aromatic scent. Weight 42lb per cubic foot.

DOMESTIC. Timbe . 2227 Leaves. 2228

Domestic and Sacred.—The TIMBER is used to make hoe and axe handles (*Gamble*). The LEAVES are burnt as incense (*J. F. Duthie*).

(*G. Watt.*)

SKINS.

2229

In commerce the term "Skins" is generally applied to the hides of Sheep and Goats, though doubtless those of Antelopes, Deer, Tigers, Leopards, etc., would be so designated. Ornamental skins may, however, be more naturally classed with "Furs," and it is probable that these, as also calves skins, do not materially upset calculations based on the assumption that the skins recorded in Indian trade statistics are those of "Sheep and Goats," more especially the latter.

The reader may be able to link together the details of this trade by consulting the numerous articles which have already appeared in this work on kindred subjects, more especially the following:—**Camel** (*Vol II., 63*); **Crocodile** (*Vol. II., 592*); **Deer** (*Vol. III., 55*); **Fur** (*Vol. III., 458*); **Hides** (*Vol. IV., 248*); **Leather** (*Vol. IV., 605*); **Oxen** (*Vol. V., 659*); **Rhinoceros** (*Vol. VI., Pt. I., 489-90*); and **Sheep & Goats** (*Vol. VI., Pt. II., 549-672*). **Tigers, Cats & Civets**, *Vol. VI., Pt. III., 9-16*.

2230

Skins.—To arrive at an approximate conception of the magnitude of the traffic in skins, it becomes necessary to endeavour to obtain some idea of the number of sheep and goats in India. In the Agricultural Returns of the provinces of British India. Bengal and Assam have not hitherto furnished estimates of these animals, and the whole of the Native States have been left out of consideration. It may, however, be said that, of the provinces of British India, for which returns of sheep and goats have been published, there were last year in round figures 28,000.000. Madras was estimated to have had 11,000,000, the North-West Provinces and Oudh 6,000,000, and the Panjáb 6,400,000. A provision of 10,000,000 for Bengal and of 2,000,000 for Assam would not, therefore, seem excessive, so that British India may be put at 40,000,000 of sheep and goats. If now to this be added 10,000,000 for the Native States, the total for all India would be 50 millions. The returns of trade in Skins show that a provision of that number would by no means be a high one. Thus, for example, there left India last year by marine traffic alone 19,244,116 dressed skins and 3,906,711 raw skins. This would, therefore, represent very nearly 50 per cent. of the estimated stock, but the local consumption of skins in India itself must be very great. A reproduction of 3 in two years is given by most writers as a good result with sheep, but a deduction has to be made for the male animals, so that a reproduction of 50 to 60 per cent. would, very probably, be the utmost that could, with safety, be accepted. It has, for example, to be borne in mind that a ewe and she-goat do not come into bearing until they are two years old, so that one year at least is lost. On the other hand, India draws supplies of skins from foreign countries, both coastwise and from across the land frontier, so that the home and foreign consumption is not alone dependent upon internal production. But when a provision has been made for imports, it is felt that the recorded internal and external transactions are so great that either the estimate of 50 million stock must be very considerably under the mark, or that reproduction must be far higher than is generally

SKINS.

Indian Transactions in Skins.

TRAFFIC.

believed. Some difficulty is experienced, however, in following out the transactions in skins, so that errors are probably unavoidable. In the foreign trade skins are recorded by number and also in cwt., and in the rail returns by mds. (presumably of 82℔). After testing a considerable number of the returns so as to arrive at a uniform standard, the writer has been led to believe that a cwt. of tanned skins may be accepted as representing 116, a md. 163; a cwt. of raw skins would probably contain 103 and a md. 144. If these figures be accepted as fairly correct, it may be inferred that the total number of "skins" carried by road, rail, and river last year came to 114,754,525. But of course that figure by itself is of little value, since duplication of records is involved. It is the total transactions (imports and exports) to and from all provinces and ports. It, therefore, very largely includes the amount from which the foreign exports were drawn, as well as that used up in India itself, with this qualification that the returns published by the railways, etc., by no means represent the total transactions. The writer can produce no authoritative balance sheet of the Indian trade in skins, but after the most careful consideration of all the sources of information at his disposal, he believes that future and more exhaustive enquiry will reveal the fact that the Indian consumption of skins cannot be far short of 20,000,000 annually. If that figure be admitted as worthy of consideration, alongside of the actual record of foreign exports, then India might be said to have, last year, used up and exported collectively close on 45,000,000 skins.

Of the provinces of India, Madras is by far the most important in the art of tanning skins. Indeed, the coastwise traffic, as also the rail returns, show that it draws largely on Bengal, the Central Provinces, and Bombay in its supply of raw skins, so that its local flock of 11 million sheep and goats is by no means sufficient to meet the demands of its tanners. It has been pointed out under "Hides" that the foreign exports in these articles have recently shown, if anything a decline rather than an expansion. The traffic in tanned skins, on the other hand, has year by year manifested a steady tendency to expand. This is the more satisfactory when it is recollected how persistently the foreign exports from India are in raw rather than manufactured articles. India possesses a rich store of valuable tanning materials. The development of a foreign traffic in tans labours, however, under many disadvantages. They are essentially bulky materials of an inflammable nature. The immense distances they have to be carried operate prejudicially when placed in competition with the fairly extensive supply of cheap tans that are already at the disposal of the European tanners. To foster, therefore, the art of skin-tanning is a direct gain to India. Not only is a lucrative industry thereby created, but a home market is afforded for indigenous tans. The reader will find, under the article **Leather** the leading particulars regarding Indian methods and materials of tanning, but he might also consult the chapter "**Tanning Materials**" in the writer's volume of "*Selections from the Records of the Government of India*" for the year 1888-89, pp. 83-111. Useful information will also be found under **Acacia Catechu** and the other species of **Acacia** (*Vol. I., 17-61*); **Cæsalpinia coriaria** (*Divi-divi, Vol. II., 6-9*); **Cassia,** various species, but more especially **C. auriculata** (*Vol. II., 210-226*); **Diospyros Embryopteris** (*Vol. III., 141-145*); **Mangrove** (*Vol. V., 157*); **Rhus** (*Vol. VI., Pt. I., 495-502*); **Shorea robusta** (*Vol. VI., Pt. II., 672-679*); **Terminalia** (*Vol. VI., Pt.III., 16-41*); etc., etc. The tanning material that seems to find most favour in Madras, in the preparation of the skins that are exported from that Presidency, appears to be **Cassia auriculata.**

Very little more can be said on the subject of Indian skins, but the reader may be able to derive useful particulars by the study of the following tables of the trade:—

TRADE.

Imports.
2231

Tabular Statement of the Indian Transactions in Skins with Foreign Countries by Sea.

	IMPORTS.				EXPORTS.			
	Raw.		Manufactured.		Raw.		Manufactured.	
	No.	R	No.	R	No.	R	No.	R
	I.	II.	III.	IV.	V.	VI.	VII.	VIII.
1885-86 . .	204,987	1,35,675	126,618	2,40,405	4,947,579	35,40,831	17,491,513	1,81,04,645
1886-87 . .	329,037	2,21,478	161,146	2,97,436	4,077,259	32,03,441	18,219,437	1,89,71,043
1887-88 . .	384,342	2,57,028	203,187	3,63,837	3,817,124	30,36,756	18,760,809	1,92,08,570
1888-89 . .	328,949	2,59,429	124,902	2,57,433	4,356,561	38,19,563	19,946,919	2,04,83,429
1889-90 . .	451,839	3,13,320	160,899	3,31,205	3,906,711	33,53,985	19,244,116	1,99,00,995

Exports.
2232

The exports of foreign skins (raw and tanned) are very unimportant, so that the net exports may be accepted as the difference between the columns above; thus the figure in column I. for 1889-90 deducted from that in V., for the same year, would leave the net export of raw skins. By adding together for any one year the figures in columns II., IV., VI., and VIII., the gross value of all the Indian maritime foreign transactions in skins would be obtained. Thus, for example, the traffic for 1889-90 will be seen to have been R2,38,99,505, or at the nominal value of R10 to £1 sterling, it may be said to have been £2,389,950.

The countries from whence and to which the Indian skins are obtained and consigned may now be exhibited, as also the provinces to which and from whence exported. For this purpose it is unnecessary to do more than analyse the returns for the last year in the above series, *viz.*, 1889-90, and to do so by reference to value only.

Analysis of the Imports and Exports of Raw and Manufactured Skins by Sea for the year 1889-90.

Countries from whence imported or to which exported.	IMPORTS.		EXPORTS.	
	Raw.	Manufactured.	Raw	Manufactured.
	R	R	R	R
United Kingdom	...	2,98,494	6,08,341	1,48,28,347
Austria	...	...	3,394	2,70,429
Belgium	...	...	...	3,95,683
France	...	10,874	1,47,915	2,09,031
Germany	...	10,844	...	...
United States	7,067	...	25,74,388	35,74,062
Ceylon	59,439	2,124	2,656	9,687
Persia	88,951	5,031	...	...
Arabia	37,223	...	...	...
Turkey in Asia	71,007	3,172	...	...
Straits Settlements . . .	30,376	...	...	5,93,403
Other Countries	19,257	666	17,291	20.353
TOTAL .	3,13,320	3,31,205	33,53,985	1,99,00,995

It will thus be seen that Great Britain takes the chief part in the imports of manufactured skins as in the exports of both raw and manufactured skins. Of the countries that supply India with raw skins,

SKINS.

with Foreign Countries.

TRAFFIC.

Persia, Turkey in Asia, Ceylon, Arabia, and the Straits are the chief, but the traffic with the United States (shown above) may be regarded as very significant. It may now be useful to exhibit the degree of participation by the various provinces of India in these foreign transactions.

Analysis of the Imports and Exports of Raw and Manufactured Skins by Sea for the year 1889-90.

Provinces into which the Imports are made and from which the Exports take place.	IMPORTS.		EXPORTS.	
	Raw.	Manufactured.	Raw.	Manufactured.
	R	R	R	R
Bengal	1,060	1,09,418	32,74,235	1,67,864
Bombay	1,84,833	1,75,933	4,250	39,92,348
Sind	30,545	881	21,415	550
Madras	96,882	17,231	54,085	1,57,40,233
Burma	...	27,742	...	...
TOTAL	3,13,320	3,31,205	33,53,985	1,99,00,995

Coastwise.
2233

Of the imports of raw skins it may be said that the supplies drawn by India from Persia, Arabia, Turkey in Asia, and the Straits Settlements go mainly to Bombay, and that Ceylon and the other countries mentioned in the table furnish Madras with supplies. The Panjáb, Sind, and the North-West Provinces, it will be seen in a further paragraph, procure foreign skins from across the land frontier. Of the manufactured skins by far the major portion comes from the United Kingdom, and is received by Bombay and Bengal in approximately equal proportions. Of the exports it will be seen that Bengal has the largest traffic in raw skins and Madras in manufactured skins.

Turning now to the coastwise transactions (or the interchange from province to province along the seaboard of India), it may be said that the magnitude of the local demand for skins in India itself may be inferred by the very extensive traffic not only from province to province, but from port to port, within each province. Thus, for example, the raw skins exported from the provinces of India in 1889-90 were valued at R14,26,615 and from port to port, within the provinces, at the additional sum of R2,43,846, or, say, 3,274,700 skins. In tanned skins the traffic was much smaller, the exports from the provinces having been valued at R2,60,815, within the provinces at R1,97,155, or, say, 495,600 skins. But of the former Sind headed the list, having last year exported R9,44,293 worth of raw skins, mostly to Bombay; then Bengal with an export of R4,36,381 worth, mostly to Madras. Of the latter (tanned skins) Madras headed the list, R4,13,872 worth, of which R1,80,582 worth were consigned to Bengal and R1,93,171 worth to ports within the presidency. It is perhaps unnecessary to exhibit the returns of other years, the features shown having been more or less constant for some time past, with this modification that the demand by Madras for raw skins has for some years manifested a decided tendency to expand. Bengal, on the other hand, has lost ground steadily in the production of tanned skins and become more and more dependent on Madras for its supplies.

But a conception of the Indian traffic in skins would be incomplete without particulars of the trans-frontier (land) transactions. The following table, therefore, shows the imports and exports during the three years ending 31st March 1890:—

TRAFFIC.

Trans-frontier.

2231

Statement of the Trans-frontier (Land) Imports and Exports of Skins (Sheep, Goats, and small animals) during the past three years.

Countries from whence imported or to which exported.	IMPORTS.						EXPORTS.					
	1887-88.		1888-89.		1889-90.		1887-88.		1888-89.		1889-90.	
	No.	R	No.	R	No.	R	No.	R	No.	R	No.	R
	I.	II.	III.	IV.	V.	VI.	VII.	VIII.	IX.	X.	XI.	XII.
Lus Bela	15,258	9,546	20,093	13,908	12,821	8,707	...	...	...	...	...	...
Sewestan	7,900	2,160	10,800	3,186	10,800	4,576	...	...	...	...	...	...
Tirah	1,500	616	300	61	6,800	1,831	...	...	...	...	...	...
Kabul	31,900	9,916	31,400	9,249	62,600	19,956	25,500	8,853	19,300	6,336	17,000	5,592
Bajaur	400	132	...	...	32,600	10,656	...	...	...	...	...	...
Kashmir	244,600	85,319	316,800	1,16,994	518,600	1,91,048	2,500	855	400	150	...	...
Ladakh	...	...	700	405	1,100	675	10,900	7,365	15,500	10,520	5,900	3,600
Tibet	477	382	321	170	720	609	1,063	740	2,258	1,157	1,368	817
Nepal	88,289	1,41,163	32,895	62,896	65,037	1,14,352	62	31	500	1,055	30	15
Sikkim	1	2	...	...	10	19	4	7	...	...	38	69
Trans-frontier by Sind-Pishin Railway	199,000	27,060	180,200	24,510	152,700	20,790	4,700	640	4,500	610	6,400	890
TOTAL	559,325	2,76,296	593,509	2,31,377	8,63,788	3,73,219	44,720	18,491	42,458	19,828	30,736	10,983

SKINS. **with Foreign Countries.**

TRAFFIC.

Trans-frontier.

In the published returns of the above traffic the number of skins is sometimes given, at others their weight in cwt. It is also not stated whether any of the skins were tanned, so that in expressing the returns to number of skins errors are unavoidable. In the opening paragraph of this article it has been stated that 103 raw skins very probably go to the cwt. To simplify the calculations, as also to be on the safe side of under, rather than over, stating the case, the writer has allowed 100 skins only to the cwt. It may safely be assumed, therefore, that the difference between the totals of columns V. and XI. very fairly represent the net import of skins in 1889-90, *viz.*, 833,052. The gross import shown in column V. was distributed as follows:—

	No. of Skins.
Into the Panjáb	632,500
,, Sind	165,521
,, Bengal	61,360
,, the North-West Provinces	4,407
TOTAL	863,788

A very large share of the Panjáb imports doubtless re-appears in the rail-borne transactions as an export from that province. But, before proceeding to deal with the internal trade, it may be added that. leaving out of all consideration transactions that bring to British India (from Native and Foreign States within India) certain supplies, it has been shown that the imports by sea from foreign countries (mostly down the Persian Gulf) amount to fully half a million skins, and that there is carried to India across the land frontier about one million more, so that these two quantities may be taken as the direct additions to local production.

A review of the rail, road, and river-borne transactions in skins is not only very difficult, but open to grave doubt as to its accuracy. The figures which may be here dealt with are those published for the year 1888-89. It may be convenient to discuss the traffic under the two sections—"Tanned Skins" and "Raw Skins." The total imports of *tanned skins* by all Provinces, Native States, and port towns came to 3,34,879 maunds (or, say, 54,585,277, skins), valued at R2,22,62,391 (say £2,226,239). The imports, on the other hand, of *raw skins* as recorded by rail, road and river traffic came to 4,17,842 maunds (say 60,169,248 skins), valued at R1,00,51,814. The total value of the transactions recorded as having been carried by rail, road and river came, therefore, to R3,23,14,205 (or, say, £3,231,420). If now from that record of the rail-borne traffic the imports by the port towns be separated, some conception of local consumption may be obtained, since from these port town imports were mainly drawn the foreign exports. In the year 1888-89 the imports by the port towns were valued at R2,37, 99,258 (or, say, £2,379,925), and the foreign exports of that year came to R2,37,86,130 (or, say £2,378,613). But it must be recollected that the port towns drew in addition very considerable supplies by coastwise transactions, so that these two sources of supply furnished, after the foreign exports were withdrawn, the amounts necessary to meet the local demands of the port towns. There thus remains only one other consideration in order to exemplify the assertion that the total Indian demand for skins may be put at 20,000,000. The transactions with the provinces are mainly in raw skins, the foreign transaction chiefly in tanned skins. The force of this observation will at once be apparent, if the weight (or estimated number) and the value of tanned skins (given above) be compared with the weight and value of raw skins. The former will be seen to be twice as valuable as the latter, hence half the value would represent approximately

TRAFFIC.

the same number of skins. But it must be added that the numbers accepted above as equivalent to a cwt. or maund have been arrived at by the study of the returns of foreign trade. It would appear that the statistics of rail-borne trade, when expressed in number of skins, by the aid of the accepted equivalent to the maund, become unnecessarily high, although, as already remarked, these returns involve an extensive duplication. This source of error is, however, mitigated by abstracting the returns of the port towns so as to show the net imports from the provinces. In point of value these were found to be, say, £2,300,000, or, say, two-thirds of the total declared value of the rail-borne transactions in skins. The foreign exports were valued at, say, £2,300,000, and, as already shown, that valuation was the price of 23,000,000 skins—the number registered by the custom-house authorities. It, therefore, follows that since raw skins—the skins chiefly used up in the provinces of India—are only half the price of tanned skins, that the provincial transactions (registered by rail, road, and river), which came to over £1,000,000 (nominal) valuation, must be accepted as at least equivalent to 20,000,000 skins. But there must be a very extensive local consumption of skins that are produced, tanned and used up, without being registered as passing along any channel of transport, so that it would be quite safe to assume a very much more extensive consumption than the figure mentioned. Thus, for example, it might be urged that in India, at least 100,000,000 persons possess or purchase one pair of shoes a year. The Hindus would have no objection to pig skin being employed in the manufacture of their shoes, but Muhammadans could not wear shoes made of that skin. The Hindus, on the other hand, could not, by the injunctions of their religion, wear shoes made of cows' leather, so that some idea of the number of skins likely to be used up in India, in the manufacture of shoes alone, may be arrived at. But even such an estimate would leave out of all consideration the numerous other uses of skins, such as saddlery, irrigation water-bags, grain and meal bags, water-carrier's bags (*mashak*), etc., etc. With the exception of the horse and cattle hides and pig skins used in saddlery, the other purposes of leather are almost entirely met in India by skins. It follows that while the Indian consumption might easily be 40 million, it certainly cannot be less than 20 million annually. This much may safely be affirmed that the number of skins exported and used up in India annually could not be afforded by a smaller stock of sheep and goats than the figure estimated, namely, 50,000,000.

(*W. R. Clark.*)

SLATE.

2235

Slate, *Ball in Man. Geol. Ind., 551.*

ROOFING SLATE, SLATE, CLAY-SLATE, *Eng.*, ARDOISES *Fr.;* SCHIEFER, *Germ.;* LAVAGNA, *Ital.*

Vern.—*Chapar* (= roofing), PB. HIM.; *Sil*, HIND.; *Kalpalagi*, TAM.; *Rati palaka*, TEL.

References.—*Baden Powell, Pb. Prod., 38, 56; Forbes Watson, Indust. Survey, 410; Settlement Reps.:—Panjáb, Hazara Dist., 9; Central Prov., Dumoh Dist., 88; Gazetteers, Panjáb, Gurgaon, 14; Administration Reports, Bombay (1872 73), 365; C. P., 80; Gribble, Man. Cuddapah Dist., 28; Balfour, Cyclop. Ind., III., 672.*

Occurrence. 2236

Occurrence.—The following note upon the occurrence of Slate in India has been kindly furnished by **Dr. Medlicott,** late Director of the Geological Survey:—

"The greater part of the rocks used in India under the name of slate hardly come within the meaning of the term, being generally flags of

SLATE.

Occurrence of Slate in India.

OCCURRENCE

greater or less thickness, which are obtained by splitting the rocks along planes of stratification or lamination. Slates, on the other hand, are obtained always by splitting the rocks along the planes of cleavage, which, in general, are transverse to the stratification planes, and in many cases so well developed that the rock may be split up to a paper-like thinness. A good distinction between flags and slates is that the latter have in a wap to be torn asunder as the splitting chisel is driven down in the rock masses, whereas a blow of a hammer, or a single stroke of the chisel, will, generally, disassociate the flags at once.

"In this way, there are really very few tracts where true slates are to be found, and even then, their texture is in all cases inferior to that of the average slates of the British market. The Himálayan region is the best of these tracts, and the favoured localities are in the Simla and Gurdaspur districts, as about Dharmsala and in the Kangra Valley. The latter region offers such favourable opportunities for export, and the slates are so fairly good, that systematic quarrying and dressing are carried on here by the Kangra Valley Slate Company. The Kangra Valley slates appear to be often cleaved with the lamination, so that they would, in this case, be called flaggy or bedded slates, or slaty flags according as they are split along lamination or cleavage planes.

"In nearly all other regions the rocks are more properly clay-slates, and then their fissility depends on the extent to which they are cleaved; though this is seldom sufficient to allow of any but small and not very thin slates, which are necessarily somewhat soft.

"In the North-West Provinces some slate was raised at Chitali (Kumáon) for roofing buildings at Ranikhet, giving slabs about a foot square and a quarter of an inch thick. Slates are also stated to be obtainable at Dhari in the Bel Patti of Gangoli, in Borarao Patti, and in Naini Tal. A thin dark blue slate, inferior to that from the Chitali quarry, occurs at Jobha in Garhwál. Slates were formerly quarried near Simla, but have been superseded by iron for roofing purposes.

"In the Kharakpur Hills of Bengal, there is a band of slates from 6 to 12 feet in thickness, which is traceable for many miles along their northern margin. With care these rocks can be split up in thicknesses of ⅛ of an inch; but the surfaces are scaly and rough.

"In the districts of Chutia Nagpur, Manbhum. and Singbhúm, flaggy, laminated, slaty rocks occur in many places among the sub-metamorphic rocks. With a little polishing, slates suitable for drawing on can be made from the rough material occuring near Chaibassa.

"Still further west, in the Central Provinces, imperfectly cleaved slates are to be found both in the Chanda and Chindwara districts.

"In the Ulwar State of Rájputana, slates are said to occur at Bílashpur in Ramgarh, and at Mandan in the north-west corner of the State.

"In the Bombay Presidency, fairly promising slates occur among the transition rocks between Surajpur and Jambughora, north-east of Baroda. There are also some slates which might answer for roofing purposes near Bagh. Again, in the South Mahratta country, at Belgaum, slates of the Kaladgi series of rocks are reported as having been formerly worked for roofing.

"One of the largest areas of cleaved or slated rocks occupies considerable portions of the Cuddapah, Kurnool, and Kistna districts of the Madras Presidency; but there are few if any true slates. The prevalent rock is coarse clay-slate badly cleaved, but much jointed, and thus offering little chance of good specimens of this economic product. Quarries might with careful searching be opened up in the clay-slate basin of Wontamitta, south-west of Cuddapah on the line of railway, and at many places in the Nullamallay hills. Much confusion of terms has arisen both in Kurnool and Cuddapah through the vulgar notion that the fine limestone flags of both places are slates."

(*For further information, see the Manual of the Geology of India, and publications of the Geological Survey of India.*)

DOMESTIC. 2237

Domestic Uses.—In India, slates are in many localities used for roofing and flagging, but the trade in them is almost purely local. The coarse slates of the Kharakpur Hills in Bengal are to some extent used for curry platters and plates. Slates that with a little polishing would be suitable for drawing upon have been found in Chutia Nagpur.

SLEEPERS.

Sleepers, Railway. Timbers used for—

2238

Acacia arabica, *Willd.*
A. Catechu, *Willd.*
Anogeissus latifolia, *Wall.*
Barringtonia racemosa, *Blume.*
Bassia latifolia, *Roxb.*
Boswellia serrata, *Roxb.*
Calophyllum inophyllum, *Linn.*
Careya arborea, *Roxb.*
Cedrus Deodara, *Loudon.*
Cinnamomum glanduliferum, *Meissn.*
Dalbergia latifolia, *Roxb.*
D. Sissoo, *Roxb.*
Eucalyptus Globulus, *Labill.*
Eugenia Jambolana, *Lamk.*
Hardwickia binata, *Roxb.*
Hopea parviflora, *Beddome.*
Lagerstrœmia parviflora, *Roxb.*
Melanorrhœa usitata, *Wall.*
Mesua ferrea, *Linn.*
Mimusops littoralis, *Kurz.*
Odina Wodier, *Roxb.*
Pterocarpus Marsupium, *Roxb.*
Shorea robusta, *Gærtn.*
Tectona grandis, *Linn.*
Terminalia Chebula, *Retz.*
T. tomentosa, *Bedd.*
Xylia dolabriformis, *Benth.*

SMILAX, *Linn.; Gen. Pl., III.. 763.*

2239

The Editor having been favoured with a proof copy of the forthcoming volume VI. of the *Flora of British India*, the synonymy of the following useful species of this genus have been taken from that work.

Smilax China, *Linn.; DC., Monograph. Phanerogam., I., 46;*
CHINA ROOT. [SMILACEÆ.

2240

Syn.—S. JAPONICA, *A. Gray*; COPROSMANTHUS JAPONICUS, *Kunth.*

Vern.—*Chob-chíni*, HIND.; *Chob-chíni, shúk-china*, BENG.; *Chob chíni*, PB.; *Chob-chíni*, BOMB.; *Paringay*, TAM.; *Pirangi chekka, gáli chekka*, TEL.; *China-alla*, SING.; *Chobachini*, SANS.

References.—*Roxb., Fl. Ind., Ed. C.B.C., 725; Sir W. Elliot, Fl. Andhr., 57, 154; Pharm. Ind., 227; Flück. & Hanb., Pharmacog., 712; Fleming, Med. Pl. & Drugs (Asiatic Reser., XI.), 186; Ainslie, Mat. Ind., I., 70, 592; O'Shaughnessy, Beng. Dispens., 645; Irvine, Mat. Med. Patna, 21; Medical Topog., Ajmir, 131; U. C. Dutt, Mat. Med. Hind., 264, 295; S. Arjun, Cat. Bomb. Drugs, 135; Murray, Pl. & Drugs, Sind., 25; Dymock, Mat. Med. W. Ind., 2nd Ed., 838, 839; Birdwood, Bomb. Prod., 85; Baden Powell, Pb. Pr., 379; Linschoten, Voyage to East Indies (Ed. Burnell, Tiele and Yule), II., 107-112.*

Habitat.—A climbing, thorny shrub, frequent in Japan, and found also in China and Cochin China. **Fluckiger & Hanbury** describe this species as occurring also in Eastern India, but more probably **Roxburgh's** species **S. glabra** and **S. lanceæfolia** are alluded to, or perhaps the allied species **S. ferox,** *Wall*, since **DeCandolle** mentions no Indian habitat for the plant; nor, in fact, does **Sir J. D. Hooker** (in the forthcoming volume of the *Flora of British India*) give an Indian habitat for the China Root.'

OIL. 2241

Oil.—**Baden Powell** mentions this plant among his medicinal oils.

MEDICINE. Roots. 2242

Medicine.—The tuberous ROOTS at one time were officinal in all the European Pharmacopæias, and were considered a most valuable remedy

MEDICINE.

in venereal and rheumatic disorders. The drug has, however, now fallen into disuse in Europe, although in Chinese and Indian medicine it still enjoys a very high reputation and is considered a valuable remedy not only in syphilis and rheumatism, but also an excellent demulcent and an aphrodisiac. The TUBERS are to be met with in all the Indian drug bazárs. They are usually peeled and trimmed and are consequently of irregular form, resembling a piece of heavy pinkish white wood. No active principle has been separated from them.

Tubers. 2243

SPECIAL OPINIONS.—§ "It is highly prized by Natives as an antisyphilitic, and I have seen some good cures effected by it" (*Honorary Surgeon E. A. Morris, Tranquibar*). "Generally used by Natives in leprosy and chronic skin affections with success. Is a valuable alterative; has a depressant power on the heart, and so care should be taken not to administer the drug to persons of weak constitutions" (*E. W. Savinge*). "The hard tuberous root of **S. China** or the *Chob-Chíní* of the bazár is a cheap and useful drug. It is an alterative and nutrient tonic, and as such a pretty good substitute for Cod-liver oil, Jamaica Sarsaparilla, and Iodine. I have used it with satisfaction in some cases of secondary and tertiary syphilis, rheumatism, scrofula, and consumption. It may be used in powder or decoction, the latter being prepared as follows:—Four ounces of the root in powder boiled on a slow fire with thirty-two ounces of water, till the liquid is reduced to twelve ounces and strained when cool. Dose of the powder from one to three drachms; and of the decoction, from one to three fluid ounces; three or four times a day" (*Honorary Surgeon Moodeen Sheriff, Khan Bahadur, G.M., M.C., Triplicane, Madras*).

TRADE. 2244

Trade.—The drug is brought to India by coasting steamers from the ports of China and the Straits Settlements. The value of the unpeeled tubers is from R3 to R4 per maund of 37½℔; the peeled are worth about R9. In 1881-82 Bombay imported 945 cwts., valued at R12,692 (*Dymock*).

2245

Smilax glabra, *Roxb.; Fl. Br. Ind., VI., 302.*

Vern.—*Bari-chobchini*, HIND.; *Hariná-shúk-china*, BENG.; *Hazina*, GARO.

References.—*Roxb., Fl. Ind., Ed. C.B.C., 725; DC., Monog. Phan., I., 60; Voigt, Hort. Sub. Cal., 649; Pharm. Ind., 227; Flück. & Hanb., Pharmacog., 712; Year-Book Pharm., 1874, 42; Drury, U. Pl., 396; Agri-Horti. Soc. Ind. Jour., X., Sel., 6 (New Series), I., 178; III., Pro., 59.*

Habitat.—A scandent shrub with no dose roots; found in Eastern Bengal, Sylhet, and the Garo and Khásia hills, Tenasserim, etc., and distributed to China.

MEDICINE. Tubers. 2246

Medicine.—It has large TUBERS, not to be distinguished by the naked eye from those of the preceding species; indeed, some of the imported China root is probably derived from this and not from **S. China,** since the learned authors of the *Pharmacographia* state that **Dr. Hance** of Whampoa received a living specimen of China root which proved to be that of **S. glabra.** A decoction of the fresh ROOT is used by the hill tribes of Assam for the cure of sores and venereal complaints.

Root. 2247

2248

S. lanceæfolia, *Roxb.; Fl. Br. Ind., VI., 308.*

Vern.—*Hindi-chobchini*, HIND.; *Gutea-shuk-china*, BENG.

References.—*Roxb., Fl. Ind., Ed. C.B.C., 725; DC., Monog. Phan., I., 57; Voigt, Hort. Sub. Cal., 649; Drury, U. Pl., 396; Agri.-Horti. Soc. Ind. Jour. (New Series), III., Pro., 59.*

Habitat.—A climbing shrub found in the Eastern Himálaya, from Sikkim to Bhután, at altitudes between 3,000 and 7,000 feet, and in the Khásia hills between 4,000 and 6,000 feet. Also frequent in the Naga

Hills, Manipur, and Burma. It is distributed to the Island of Hong-Kong and China.

Medicine.—This species also has tuberous ROOTS, like those of **Smilax China**, the JUICE from which is taken inwardly for the cure of rheumatic pains, and the REFUSE, after extracting the juice, is laid over the parts most pained (*Roxburgh*).

MEDICINE. Roots. 2249 Juice. 2250 Refuse. 2251

Smilax macrophylla, *Roxb.; Fl. Br. Ind., VI., 310.*

2252

Syn.—S. OVALIFOLIA, *Roxb.;* S. ROXBURGHU, *Kunth.;* S. RETUSA, *Roxb.;* S. PROLIFERA, *Wall.;* S. PROLIFERA & S. OVALIFOLIA, *Herb. Ind. Or., Hook. f. & T. T.;* S. GRANDIS, *Wall.;* S. GRANDIFOLIA, *Voigt.*

Vern.—*Jangli-aushbah, chobchini,* HIND.; *Kúmarika,* BENG.; *Atkir,* SANTAL; *Chobchini,* NEPAL; *Gúti, gútwel, gholyel,* MAR.; *Malait-támara,* TAM.; *Konda dantena, konda támara, konda gurava tige, sitapu chettu, kistapa tamara, kummara baddu,* TEL.; *Kal-támara,* MALAY; *Kú-ku,* BURM.; *Maha-kabarosa,* SING.

References.—*Roxb., Fl. Ind., Ed. C.B.C., 726; DC., Monog. Phan., I., 199; Dalz. & Gibs., Bomb. Fl., 246; Sir W. Elliot, Fl. Andhr., 91, 94, 95, 97, 102, 169; Mason, Burma & Its People, 501, 813; Dymock, Mat. Med. W. Ind., 684, 839; Lisboa, U. Pl. Bomb., 179; Agri.-Horti. Soc. Jour., X., 347; (New Series). III., 59; Gazetteers:—Mysore & Coorg, I., 70; N.-W. P. (Him. Dist.), X., 319; Ind. Forester, VII., 94.*

Habitat.—A very common climber, with the stout petiole narrowly sheathing;—a native of Eastern Bengal and Assam, also Chutia Nagpur, to the Western Peninsula, Madras, Burma, and the Malayan Peninsula. It also occurs very plentifully on the lower Himálaya from Kumaon eastward, ascending to about 6,000 feet.

Medicine.—In some parts of India the ROOTS of this species are used as a substitute for Sarsaparilla in the treatment of venereal disease. Among the Santals they are applied, for rheumatism and pains, in the lower extremities. The inhabitants of Nepál give them, in doses of three *mashas,* for the treatment of gonorrhœa and other discharges from mucous membranes.

MEDICINE. Roots. 2253

Food.—At Poona, the LEAVES and ROOT of this plant were eaten during the famine of 1877-78. The black BERRIES are also commonly eaten by Natives.

FOOD. Leaves. 2254 Root. 2255 Berries. 2256

Domestic.—The long STEMS are used about Bombay for tying bundles and the loads upon carts (*Dymock*).

DOMESTIC. Stems. 2257

S. prolifera, *Roxb.; Fl. Br. Ind., VI., 312.*

2258

Syn.—S. LAURIFOLIA, *Roxb.;* S. LAURINA, *Kunth.;* S. OVALIFOLIA, *A. DC., Monog. Phan., I., 199;* S. UMBELLATA, *Herb. Heyne.*

Habitat.—An extensive climber, with large auriculate sheaths at the base of the petioles. Frequently confused with **S. macrophylla** under the name **S. ovalifolia**. It is, by **Sir J. D. Hooker**, said to occur over practically the same area as **S. macrophylla**. *viz.,* Tropical Western Himálaya, Sylhet, Bengal, Behar, Burma, the Deccan peninsula, and Ceylon. It seems likely, however, that this form has not been separately recognised by the Natives of India, and that its vernacular names and uses are likely to be identical with those already recorded.

SMITHIA, *Ait.; Gen. Pl., I., 516.*

Smithia sensitiva, *Ait.; Fl. Br. Ind., II., 148;* LEGUMINOSÆ.

2259

Syn.—S. ABYSSINICA, *Hochst.*

Vern.—*Oda-brini,* HIND.; *Nulla kashina,* BENG.

References.—*Roxb., Fl. Ind., Ed. C.B.C. 573; Mason, Burma & Its People, 478; Lisboa, U. Pl. Bomb., 291; Gazetteers:—Bombay, XV., 432; N.-W. P., X., 308; Agri.-Horti. Soc. Ind. Journ., IX., 415.*

Habitat.—An annual herb, found all over India from the Himálaya where it ascends to 3,000 feet to Ceylon and Burma. It is distributed to Madagascar, Abyssinia, Java, and China.

FOOD & FODDER Leaves. 2260

Food and Fodder.—The LEAVES are eaten as a pot-herb. Cattle are fond of it, and it makes excellent hay (*Roxb.*).

mut, see **Fungi** and **Fungoid Pests,** Vol. III., 457.

Snakes, see **Reptiles,** Vol. VI., 428-435.

Soap, see **Oils,** Vol. V., 461.

Soapstone, see **Seatite** pp. 347-358.

SOLA.

2261

Sola, see **Æschynomene aspera,** *Linn.*; LEGUMINOSÆ; Vol. I., 125.

DOMESTIC. Pith. 2262

Domestic Uses.—[Besides being used for the purposes referred to in a previous volume, *sola* PITH is largely employed by entomologists for lining the drawers in which their specimens are placed. It has of late come into use in Europe as a substitute for sponges and tangles in the manufacture of tents for dilating the os uteri, also in the preparation of corn-plasters.—*Ed.*]

2263

SOLANUM, *Linn.; Gen. Pl. II., 888.*

A genus of shrubs, herbs, or small trees, comprising about 700 species. They are found chiefly in the hotter parts of the globe, and are most numerous in America. From an economic point of view, several members of the genus, such as the potato and brinjal, are most important, since they are the vegetables which are, perhaps, more generally used than any others by a large section of the inhabitants of India. The brinjal is, however, of greater value than the potato, though it seems possibl, that this may not much longer be the case, for potatoes are rapidly gaining a distinct place in popular estimation.

2264

Solanum coagulans, *Forsk.; Fl. Br. Ind., IV., 236;* SOLANACEÆ.

Syn.—S. SANCTUM, *Linn.*

Vern.—*Maraghúne, bari mauhari, mahori, tingi,* PB.

References. *Boiss., Fl. Orient., IV., 286; Stewart, Pb. Pl., 160; Gazetteer, N.-W. P., X., 314; Agri.-Horti. Soc. Ind. Journ. (New Series), I., 96; Ind. Forester, XII., App., 18; XIV., 390.*

Habitat.—A herbaceous plant, nearly allied to **S. Melongena,** but more rigid and tomentose, found in Western India, the Panjáb, and Sind; distributed to South-West Asia, Arabia, and Egypt.

FOOD. Fruit. 2265

Food.—The FRUIT is, in some places, eaten by the Natives either fresh or in pickles.

2266

S. crassipetalum, *Wall.; Fl. Br. Ind., IV., 232.*

Syn.—BASSOVIA (?) WALLICHII, *Dunal.*

Habitat.—A shrub 2 to 9 feet high, very common at altitudes between 3,000 and 8,000 feet, on the Temperate Himálaya from Nepál to Bhután.

FOOD. Leaves. 2267

Food.—Sir J. D. Hooker says that the people of Sikkim cook and eat the LEAVES of this plant.

2268

S. dulcamara, *Linn.; Fl. Br. Ind., IV., 229; Bent. & Trimen, Med.* [*Pl., III., 190.*

DULCAMARA OR BITTER SWEET.

Syn.—SOLANUM LYRATUM, *Thunb.;* S. PERSICUM, *Willd.;* S. LAXUM, *Royle, Ill., 279.*

Vern.—*Rúba barik* (=the leaves), PB.

References.—*DC., Prod., XIII., Pt. I., 78; Boiss., Fl. Orient., IV., 285; Stewart, Pb. Pl., 159; Pharm. Ind., 179; Flück. & Hanb., Pharmacog., 405; O'Shaughnessy, Beng. Dispens., 462; Dymock, Mat. Med. W. Ind. 638; Year-Book Pharm., 1874, 627; Smith, Ec. Dict.,*

Habitat.—A climbing shrub, found in the Temperate Western Himálaya at altitudes between 4,000 and 8,000 feet. It is frequent in the regions between Kashmír and Garhwál, and has been observed by **Sir J. D. Hooker** as far east as Sikkim. In distribution it extends to Europe, Western and Central Asia, China, and Japan.

MEDICINE. Dulcamara. 2269

Medicine.—Among the Natives of India, DULCAMARA is considered alterative, diuretic, and diaphoretic; it is regarded as useful in constitutional syphilitic affections, chronic rheumatism, and especially so in psoriasis, lepra, and other obstinate skin diseases. Dulcamara is usually administered by Natives in the form of a decoction, but **Dr. Dymock** says that the dried FRUIT, *anab-us-salab* (foxes' grapes) are imported from Persia into Bombay. In Europe, dulcamara was a favourite drug in the middle ages. It was one of the ingredients of a celebrated nostrum used by the Welsh physicians for the bite of a mad dog, but in modern practice it is but seldom used, and its action is quite unknown. Although it is reputed to operate as a diuretic and diaphoretic, and in excessive doses as an acronarcotic poison, yet it is but little esteemed, and several physicians have given large doses both of the decoction and of the fresh BERRIES without any ill effects.

Fruit. 2270

Berries. 2271

CHEMISTRY. 2272

CHEMISTRY.—The bitter sweet taste of dulcamara appears to be due to a bitter principle which yields, by decomposition, sugar and *solanine*, an alkaloid which has been found in several species of **Solanum**. Later chemists, however, are of opinion that solanine is itself a peculiar conjugated compound of sugar and an alkaloid called by them *solanidine*, and in 1852 still a third alkaloid called *dulcamarine* was detected in the stems of the 'bitter sweet' which differs in its reactions from the other two and was thought by its discoverer to be the source of the peculiar taste to which this plant owes its name (*Pharmacographia*).

Solanum ferox, *Linn.; Fl. Br. Ind., IV., 233; Wight, Ic., t. 1399.*

2273

Syn.—S. LASIOCARPUM, *Blume*; S. HIRSUTUM, *Roxb.*; S. STRAMONIFOLIUM, *Dunal.*

Vern.—*Ram-begun*, BENG.; *Sin-ka-de*, BURM.; *Malla-cattú*, SING.

References.—*DC., Prod., XIII., Pt. I., 252; Kurz, For. Fl. Burm., II., 226; Ind. Forester, II., 26; Agri.-Horti. Soc. Ind. Trans., IV., 148; Journ., VI., 50; X., 26; Thwaites, En. Pl. Zey., 216; Rheede, Horti. Mal., II., 35.*

Habitat.—A stout herbaceous plant, frequently found in the tropical zones of Eastern and Southern India, from Assam to Ceylon and to Tenasserim. It is distributed to Hong Kong.

MEDICINE. Berries. 2274

Medicine.—The BERRIES are used medicinally by the Natives, but are not considered of much value.

FOOD. 2275

Food.—They are also said to be sometimes eaten (*Agri.-Horti. Soc. Journ.*).

S. gracilipes, *Dcne.; Fl. Br. Ind., IV., 237.*

2276

Syn.—S. JACQUEMONTI, *Dunal.*

Mr. C. B. Clarke in the *Flora of British India* remarks, "This is believed to be **S cordatum**, *Forsk.*, an Arabian species, of which no authentic example can be seen.

Vern.—*Howa, marghí pal, kaurí búti, kandiári, pilak, valúr, patrawála damá, hálún, gágra,* (Bazár leaves=) *gákra.*

References.—*Boiss., Fl. Orient., IV., 286; Stewart, Pb. Pl., 159; Baden Powell, Pb. Pr., 363; Agri.-Horti. Soc. Ind. Journ. (New Series), I., 96.*

Habitat.—An under-shrub of Western India, found in the Panjáb and Sind, and distributed to Baluchistán.

MEDICINE. Fruit. 2277 Leaves. 2278 FOOD. 2279

Medicine.—The FRUIT is said to be collected by the *hakims* for application in otitis. The LEAVES are used as a drug by the Native doctors in the Panjáb (*Stewart*).

Food.—It is eaten by the Natives in some parts of the Panjáb.

2280

Solanum indicum, *Linn.; Fl. Br. Ind., IV., 234; Wight, Ic., t. 346.*

Syn.—S. VIOLACEUM, *Jacq.*; S. CUNEATUM, *Moench.*; S. CANESCENS, *Blume*; S. HEYNII, *Roem. & Sch.*; S. PINNATIFIDUM & AGRESTE, *Roth.*; S. ANGUIVI, *Bojer.*; S. HIMALENSE, *Dunal.*; S. JUNGHUHNII, *Miq.*

Vern.—*Barhantá, birhatta,* HIND.; *Byákura, byakur, gurkámái,* BENG.; *Tid bhagnri,* ASSAM; *Cheru-chunta,* MAL. (S P.); *Ringli,* C. P.; *Katang-kári,* N.-W. P.; *Kandyári,* PB.; *Ringani, dorli, motiringi,* BOMB.; *Mulli, pappara-mulli,* TAM.; *Tella mulaka, kákamuchi, tellanéla mulaka,* TEL.; *Tibbatu,* SING.; *Vrihati, bhantáki,* SANS.

References.—*DC., Prod., XIII., Pt. I., 309; Roxb., Fl. Ind., Ed. C.B.C., 191; Kurz, For. Fl. Burm., II., 226, 227; Thwaites, En. Ceyl. Pl., 217; Dalz. & Gibs., Bomb. Fl., 174; Sir W. Elliot, Fl. Andh., 76, 131, 132, 177, 178; Rumphius, Amb., V., t. lxxxvi.; Pharm Ind., 181; Ainslie, Mat. Ind., II., 207; U. C. Dutt, Mat. Med. Hind., 210, 324; Dymock, Mat. Med. W. Ind., 2nd Ed., 637, 638; Year-Book Pharm., 1880, 250; Baden Powell, Pb. Pr., 362; Drury, U. Pl. Ind., 396, 408; Atkinson, Him. Dist. (Vol. X., N.-W. P. Gaz.), 750; Boswell, Man. Nellore, 138; Gazetteers:—Bombay, XV., 439; Panjáb, Gujrat, 12; N.-W. P., I., 83; IV., lxxv.; X., 314; Mysore & Coorg, I., 63; Agri.-Horti. Soc. Ind. Jour. X., 26.*

Habitat.—A very common under-shrub, found throughout tropical India, ascending to 5,000 feet. It is distributed to Malaya, China, and to the Philippines.

MEDICINE. Root. 2281

Medicine.—In Hindu medicine the ROOT is largely used in combination with more important drugs. It is one of the ingredients of the preparation known as *dasamula,* which is so frequently alluded to in Sanskrit works on medicine. Although not usually prescribed alone it is regarded as expectorant and useful in cough and catarrhal affections (*U. C. Dutt*). According to the author of the *Makhzan* it is cardiacal, aphrodisiacal, astringent, and resolvent, useful in asthma, cough, chronic febrile affections, colic, flatulence, and worms. Among the Natives of the Central Provinces the smoke of the burning FRUIT is used as a remedy for toothache. In the practice of European practitioners it does not appear to be used, although **Ainslie** mentions it as prescribed by Indian doctors in the form of decoction to the quantity of half a tea-cupful twice daily, in cases of dysuria and ischuria. According to the same writer, **Horsfield**, in his account of Java medicinal plants, says that the root, taken internally, possesses strongly exciting qualities. **Rumphius** states that it was used among the Amboyans in cases of difficult parturition.

Fruit. 2282

FOOD. Fruit 2283

Food.—The FRUIT is in many parts of India used as a vegetable.

2284

S. Melongena, *Linn.; Fl. Br. Ind., IV., 235; Wight, Ill., t. 166.*

EGG-PLANT, BRINJAL; ALBERENGENA, BERENGENA, *Sp.*; BERINGELA, *Port.*; MELANGOLUS, MERANGOLUS, *Low Latin*; MELANGOLA, MELANZANA, MELA INSANA, *It.*; AUBERGINE, MELONGÈNE, MERANGÈNE, BELINGÈNE, ALBERGAINE, *Fr.* The botanical name is not Latin, but a factitious rendering of *melanzana*. The Italian *mela insana* is an attempt to give a meaning to the foreign name by connecting the vegetable with its reputation of being indigestible.

Syn.—S. INCANUM, *Linn.*; S. INSANUM, *Linn.*, also *Roxb.*; S. ESCULENTUM, *Dunal*; S. LONGUM, *Roxb.*; S. TRONGUM, *Lamk.*; S. FEROX, *var.* β, *Kurz*; S. TORVUM, *var.* INERME, *Dalz. & Gibs.*

Vern.—*Baigun, bangan, bhantá, brinjal, baingan, badanjan,* HIND.; *Begún, kúli-begún, bong, bartakú, mahoti hinpoli.* BENG.; *Jati bengani,* ASSAM; *Valúthala,* MAL (S.P.); *Báigun,* URIYA; *Brinjál, bhanta,* N.-W. P.; *Baigan, bhutta,* KUMAON; *Bádanján bostani, tukm-i-bádanján rumi,* KASHMIR; *Bengan,* PB.; *Wangan,* SIND; *Wangi,* DECCAN; *Baigana, vánge, waingi, bengan,* BOMB.; *Vángi,* MAR.; *Vengni, ringni, vanták, rigana,* GUZ.; *Kuthirekai, valúthalay, vankaya,* TAM.; *Chiri vanga, vanga chettu, niru vanga, metta vanke, wang-kai, vanga-chiri-vangu,* TEL.; *Badane kayi, badnekái, dodda badane,* KAN.; *Trong,* MALAY.; *Kha-yan,* BURM.; *Wambatu,* SING.; *Vártáku* (according to **Dutt**) [or *bhantáki* (according to **Yule & Burnell**)] *hindira, vangana, vardákú, bhandáki, jukutam, hingoli* (according to **Sir M. Williams**), SANS.; *Badangan, badinjan, kahkam, kahkab,* ARAB.; *Bádanján, bádingán,* PERS.

References.—*DC. Prod., XIII. Pt. I. 368; Roxb., Fl. Ind., Ed. C.B.C., 190, 191; Kurz, For. Fl. Burm., II, 226; Dalz. & Gibs., Bomb. Fl., 61; Stewart, Pb. Pl., 160; DC., Orig. Cult. Pl., 287; Mason, Burma & Its People, 471, 798; Sir W. Elliot, Fl. Andhr., 43, 115, 136, 189; U. C. Dutt, Mat. Med. Hind., 323; Year-book Pharm., 1876, 232; Birdwood, Bomb. Prod, 170, 171; Baden Powell, Pb. Pr., 265, 363; Drury, U. Pl. Ind., 398, 399; Atkinson, Him. Dist. (Vol. X., N.-W. P. Gaz.), 703, 750; Useful Pl. Bomb. (Vol. XXV., Bomb. Gaz.), 167; Econ. Prod., N.-W. Prov., Pt. V. (Vegetables, Spices, and Fruits), 13, 20; Manual and Guide, Saidapet Farm, Madras, 67; Bomb. Man. Rev. Accts., 102; Kumaon, Official Rept., 279; Bengal Admin. Rept., 1882-83, 13; Settlement Reports:—Panjáb, Kangra, 25; N.-W. P., Kumaon, App. 33; Central Provinces, Chanda, 82; Reports of Agricultural Dept. (Bengal), 1886, 64; (Madras), 33; Reports of Experimental Farms (Madras), (1876), 44; (1879), 111; Gazetteers:—Bombay, VIII., 183; N.-W. P., I., 83; IV., lxxv.; X., 703; Mysore & Coorg, I., 55, 63; II., 11; Hunter, Orissa, II., 179; App. VI.; Agri.-Horti. Soc: Ind., Trans., I., 41; III, 10, 61, 69, 194; Jour. IV., 227; IX., Sel., 58; X., 26; XIII., Sel., 54; XIV., 15, 59; (New Series), IV., 24, 25; V., 32, 38; Ind. Forester:—III., 162; IX, 162.*

Habitat.—Not known wild in India, but cultivated throughout both Peninsulas, and in all the warmer regions of the globe. According to **DeCandolle**, wild forms of **S. insanum**, *Roxb.*, which the former botanist regards as belonging to the same species as **S. Melongena**, have been found in the Madras Presidency and in Burma, and he considers that from this fact its indigenous habitat may be Asia. Other botanists fix its origin in Arabia; but **Mr. C. B. Clarke**, in the *Flora of British India*, concludes his remarks on the subject by saying that its original habitat appears very uncertain. As an escape from cultivation it often becomes intensely prickly and the peduncle carries one to five fruits. [**Sir Walter Elliot** does not give this plant any Sanskrit names—a fact which may be regarded as denoting his want of faith in the names commonly quoted as referring to it. The majority of the vernacular names in use in India are derivable from the Arabic, but **Yule & Burnell** think the Arabic came from the Sanskrit *bhantáki*. Most of the European names are referable to the Arabic through the Persian.—*Ed., Dict.*]

CULTIVATION 2285

CULTIVATION.—In ordinary garden cultivation the seeds are sown at the beginning of the rains, and the plants are put out at the distance of a foot and a half apart. Though like all other vegetables benefited by a rich soil, the Brinjal succeeds well in common garden earth. The vegetable comes into season in August and remains in season from that time till the end of the cold weather. It is very valuable for the table for two or three months, as during that time few other vegetables are obtainable.

In districts of Bengal, Madras, and Bombay, chiefly those in the vicinity of large cities, the Brinjal is cultivated as a field crop, and the fol-

CULTIVATION

lowing account of its experimental cultivation as such is taken from the Report of the Agricultural Department of Bengal (1886) :—

"*Begun* must be regarded as one of the most important of indigenous vegetables. It is more or less cultivated throughout the division, but it is only in the Burdwan and the Hooghly districts that it is grown as a field crop. The best *begun* is that produced on both sides of the Damudar.

Soil. 2286

Soil.—The best *begun* soil is a high well-drained sandy loam. In clay soil the plant as well as the fruit are of stunted growth, and in soils containing excessive organic matter the plants grow too luxuriantly to bear much fruit.

Varieties. 2287

Varieties.—The different varieties of *begun* are *muktakesi, elokesi, chhatere*, and *makra*, of which the first and last are regarded as the best. There is also a variety of *begun* known as *douko* or *kuli begun*, of which the fruits are small and spindle-shaped, and which grows in bunches.

Seeds. 2288

Seeds and Seedlings.—Seeds are generally preserved by the rayats for themselves from the previous crop. For this purpose fruits are allowed to ripen on the trees till they become golden yellow; when they are fully ripe they are removed, and allowed partially to rot. This is effected by cutting the fruits right through the middle and keeping them in a heap for a day or two. The seeds are then removed, washed in water, and dried in the sun. The seeds are first sown in a nursery, for which a cool and shady place is chosen. The soil must be sandy loam which is repeatedly ploughed or hoed by the *kodali*. The clods are carefully broken, and the ground is enriched with ashes, well-rotten dung, and the earth taken from the floor of the cowshed. It is absolutely necessary that the soil should be well ærated and sun-burnt. After the soil has been well prepared for the nursery the plot of ground is divided into a number of small beds about 5 feet square each. The time of sowing the seeds extends from the middle of March to the middle of May, more seeds being sown by the beginning of May than before or later. If moisture sufficient for the proper germination of seeds does not already exist in the soil, as is generally the case at this time of the year, the ground is watered, and after the excess of moisture has evaporated, then the seeds are carefully sown at evening. It is necessary to distribute the seeds as evenly as possible and to cover them lightly. *Begun* bears pretty thick seeding. For a few days it will be necessary to water the ground every evening, but an excess of moisture must be avoided. If the site of the nursery be much exposed to the sun, the ground is covered with date leaves or straw. The plants will come out in three or four days. Watering is continued as often as the plants require it. There may be a heavy fall of rain, then look after the drainage. Weeding may be necessary if the ground had not been properly prepared. The seedlings are sometimes much injured by various insects.

Tillage. 2289

Tillage.—The best way to prepare the land intended for *begun* is to hoe it once by the *kodali* in the middle of December. It is necessary that this should be done as early as possible, as the fresh earth thrown up by the *kodali* requires to be well weathered. Hoeing by a *kodali* is an expensive operation, requiring 8 to 16 men per bigha, but experienced cultivators consider this operation as indispensable, for the plough does not go deep enough. But hoeing is seldom given, and the preparation of the *begun* field does not generally begin till pea and other *rabi* crops are off the field. The land is then ploughed at intervals of a week or so, and after the second ploughing the ladder is passed over the field each time it is ploughed. The land ought to be ready for planting by the middle of May, if not earlier.

Transplanting 2290

Transplanting.—Drains are made all round the field, and if the length

CULTIVATION

of the field be great, a few water-channels are also made across it. Now, by means of the *kodali* it is necessary to make a number of deep furrows called *juli* or *nol*, at intervals of some 36 inches. The *begun* seedlings are planted in holes made at the bottom of these furrows at intervals of nearly 36 inches. The transplanting, if possible, should be done after a good shower of rain. In some places after the field has been well prepared and levelled, the seedlings are planted on the flat at intervals of 36 inches either way. The water-channels are made afterwards.

Manuring. 2291

Manuring.—When cow-dung is applied to *begun* fields, it should be given only in a well-rotten state. *Begun* fields are sometimes top-dressed with finely powdered old mustard cake at the rate of 2 maunds per *bigha*. The cake is applied to each plant separately, and not by spreading it over the whole field. Castor cake is not used in *begun* fields. This manure as well as cow-dung when fresh tends to encourage the growth of the vegetative part of the plant at the expense of its reproductive powers.

After-Treatment. 2292

After-treatment.—When the plants have taken root the space between the rows is hoed by the *kodali*, and after about a fortnight the field is hoed a second time, and the plants are earthed up converting the furrows into ridges. Irrigation is not generally necessary before the month of Kartik (October-November), but if there be no rain for some time irrigate once, and then hoe and earth up the plant once more. In Kartik and thence, forward irrigate once a month. The plants will begin to bear fruits in the middle of September, and will continue to give fruit till the middle of March. The late planted trees will not dry up till the middle of April.

Begun fruits may be obtained almost all the year round by planting the trees at different times.

Varieties.	Time of sowing seed.	Time of planting.	Give fruit.
Ordinary (early) .	February-March	April-May	August to January.
Ditto (late) .	April-May	September-October	October to March.
Kuli . .	September-October	October-November	February to May.

To obtain fruit from May to September trees that have ceased to give fruit early, say at the middle of February, should be pruned, manured, and irrigated Fresh shoots will be thrown out which will flower and bear fruit, but of somewhat inferior quality.

Diseases. 2293

Diseases, &c.—The seedlings, while still in the nurseries, are sometimes much injured by a kind of green-coloured grub resembling much in appearance those feeding on cabbages. The young plants are sometimes totally destroyed by red ants. There is another insect which destroys both the plant and the fruits. Stagnant water is very injurious to *begun* plants. *Begun* plants are sometimes attacked by a disease known as *tulshimara* among the rayats, because the leaves of plants so attacked resemble in appearance those of *tulshi* (**Ocimum villosum**). Such plants will yield no fruit, and as soon as the disease makes its appearance, are uprooted and thrown away.

Cost. 2294 Outturn. 2295

Cost of Cultivation, Outturn.—The following figures giving an account of the approximate cost of cultivation are taken from the Report of the Bengal Agricultural Department, quoted above:—

	R	a.	p.
Ten ploughings at 6 annas each	3	12	0
Planting, 6 men	1	8	0
Hoeing 4 times, 4 men each time	4	0	0
Oil-cake, 2 maunds	2	6	0
To irrigate 3 times, 2 men each time	1	8	0
Plucking	1	0	0
Rent	2	0	0
	16	2	0

SOLANUM Melongena.

Cultivation of the Brinjal in India.

CULTIVATION

More complete returns of the cost of cultivation and the outturn are obtainable from the Madras Experimental Farm Report for 1876. Two plots of brinjal were grown during the year 1875, and the following is the cost of cultivation and value of outturn:—

PLOT I.

Area of plot . . . . 1 acre.
Date of planting . . . . 11th March 1875.
„ harvesting . . . . May 15th to October 10th.

Cultivation expenses.

	R	a.	p.
Seedlings, value of	1	8	0
Manure, 600lb of oil-cake at 100lb per R1 and 6 loads of farm manure at 12 annas per load	10	8	0
Cattle labour, in ploughing and working the land: 6½ pairs of cattle with drivers, at 12 annas per day	4	14	0
Watering, 14 waterings at R1 each	14	0	0
Manual labour, planting seedlings in the field, spreading manure, &c., 11 men, at 3 annas per day	2	1	0
Hand-hoeing and digging during growth of the crop, 43 men at 3 annas, 5 women at 2 annas, and 2 boys at 1½ annas per day	8	14	0
	41	13	0

As the effect of the manure would operate over, at the least, three years, two-thirds of its cost must be deducted, thus reducing the cultivation expenses to R34-13-0. The plant began to bear in May, and continued to bear until the end of October; during this period the total weight of fruit gathered was 3,168lb, which was sold in the field at an average of 7½ pies per viss, realizing R41-4-0.

PLOT II.

Area of plot . . . . 2,068 square yards.
Date of planting . . . . 20th April 1875.
„ harvesting . . . . June 7th to November 9th.

Cultivation expenses.

	R	a.	p.
Ploughing	3	0	0
Manuring and planting	2	14	0
Watering	11	0	0
Hoeing	1	12	0
Manure, 10 loads of farm-yard manure	5	0	0
	23	10	0
The crop was sold in the field for	45	1	6

MEDICINE. Leaves. 2296 Seeds. 2297 FOOD. 2298

Medicine.—The LEAVES are said to have narcotic (*Baden Powell*) and the SEEDS stimulant properties (*Atkinson*).

Food.—Brinjals are much eaten by the Natives, whenever procurable, and by the Europeans in India during the summer months when other vegetables are not to be had. The Natives use them (*a*) in curries, (*b*) roasted in hot ashes and mashed with salt, onions, chillies, and lime-juice, or mustard oil, (*c*) cut into slices and fried in oil, (*d*) pickled while young and tender with mustard oil, chillies, salt, etc. (*L. Liotard*). By Europeans they are usually prepared by being half-boiled, the interior scooped out and mashed with pepper, salt, and butter, then replaced and baked.

Solanum nigrum, *Linn.; Fl. Br. Ind., IV., 229; Wight, Ic., t. 344.* 2299

Syn.—S. RUBRUM, *Miller*; S. TRIANGULARE, *Lamk.*; S. VILLOSUM, *Lamk.*; S. INCERTUM, *Dunal.*; S. NODIFLORUM, *Jacq.*; S. ROXBURGHII, *Dunal.*

Vern.—*Makoi,* HIND.; *Gurkámái, kákmáchi, mako, tulidun,* BENG.; *Pich kati,* ASSAM; *Kámbei, kwan saf safei, káchmách, riaungi,* (the fruit=) *mako,* PB.; *Kározgi,* PUSHTU; *Kan perún,* SIND; *Kámuni, gháti, mako,* BOMB.; *Munna-takali-pullum, manattak-kali,* TAM.; *Kanchi-pundu, káchi, kámanchi, kasaka, kamanchi chettu, kakamachi, gáju chettu,* TEL.; *Kalú-kung-waireya, tibbatu,* SING.; *Kákamáchi,* SANS.; *'Anb-us-sá'lap,* ARAB.

References.—*DC., Prod., III., 150; Boiss., Fl. Orient., IV., 284; Roxb., Fl. Ind., Ed. C.B.C., 190; Kurz, For. Fl. Burm., II., 224; Stewart, Pb. Pl., 160; Aitchison, Rept. Pl. Coll. Afgh. Del. Com., 91; Sir W. Elliot, Fl. Andhr., 52, 56, 75, 80, 81, 125; Rheede, Hort. Mal., X., t. 73; Rumphius, Amb., VI., t. 26, f. 2; Lace, Flora of Quetta, MSS.; Pharm. Ind., 181; O'Shaughnessy, Beng. Dispens., 462; Irvine, Mat. Med. Patna, 65; Medical Topog., Ajmir, 145; U. C. Dutt, Mat. Med. Hind., 210, 302; S. Arjun, Cat. Bomb. Drugs, 98; Dymock, Mat. Med. W. Ind., 2nd Ed., 638; Baden Powell, Pb. Pr., 362; Useful Pl. Bomb. (XXV., Bomb. Gaz.), 167, 202; Boswell, Man., Nellore, 125; App. to Note on Condition of People of Assam (vide Agri. File 6 of 1888); Gazetteers:—N.-W. P., I., 83; IV., lxxv.; X., 314; Mysore & Coorg, I., 83; Agri.-Horti. Soc. Ind., Jour. (New Series), I., 96; Ind. Forester, III., 238; XII., App. 18.*

Habitat.—A herbaceous or suffrutescent plant, found throughout India and Ceylon up to altitudes of 7,000 feet. It is distributed to all temperate and tropical regions of the world.

MEDICINE. Berries. 2300

Medicine.—In Sanskrit works of medicine, the BERRIES of this plant are described as having tonic, alterative, and diuretic properties, and are said to be useful in anasarca and heart-disease. They are used for these purposes by Native practitioners of the present day, and are given as an ordinary domestic remedy by many of the Natives, who regard them as cool and moist, useful in fevers, diarrhœa, and ulcers. They are employed both externally and internally in disorders of the eye-sight and in hydrophobia (*Baden Powell*). The JUICE of the plant is also considered medicinal, and is given in doses of from 6 to 8 ounces in chronic enlargements of the liver. After expression it is warmed in an earthen vessel until it loses its green colour and becomes reddish brown; when cooled, it is strained and administered in the morning. It acts as a hydragogue, cathartic, and diuretic. In smaller doses it is used as an alterative in chronic skin diseases such as psoriasis (*Dymock*). **Rumphius**, who describes this species, also says that in his time it was used by the Amboyans to bring out the eruption in the small-pox of children, and that the Chinese employed the juice of the leaves to alleviate the pain in inflammation of the kidneys and bladder and in virulent gonorrhœa, diseases which he declares were much more prevalent and severe in that country than in Europe.

Juice. 2301

CHEMICAL COMPOSITION.—It was from this species that the alkaloid solanine was first isolated in 1820 by **Defosses** of Bisançon (*v.* **Solanum dulcamara**).

Chemistry. 2302

SPECIAL OPINIONS.—§ "The whole PLANT is much used as an article of diet for dropsical patients and those suffering from chronic inflammation of the liver, &c." (*Assistant Surgeon Mokund Lall, Agra*). "Diuretic, absorbent, and laxative. Useful in dropsy" (*Assistant Surgeon Nihal Sing, Saharunpore*). "The decoction of its ROOT mixed with a little *goor* is given to produce sleep" (*Surgeon H. W. Hill, Manbhum*). "An extract made of the LEAVES, FRUIT, and tender portions of the STEM, has been employed here and found to be a very useful and effectual laxative and diuretic,

Plant 2303 Root. 2304 Leaves. 2305 Fruit 2306 Stem 2307

MEDICINE.

especially in dropsy, connected with heart disease. The extract, however, will not keep long, soon becoming mildewed. To prevent this, some country spirit was invariably added to it, when the extract is about to become of a proper consistence. Dose of the extract: From ½ to 2 drachms once or twice a day" (*Apothecary J. G. Ashworth, Kumbakonam, Madras*). "I have used with success a decoction in jaundice, chronic enlargement of the liver, in combination with acid nitro-muriatic. It has also been found useful in hepatic dropsy" (*Assistant Surgeon Bhagwan Das, Rawalpindi, Panjáb*). "The juice is extensively used as an external application in dropsy" (*Assistant Surgeon T. N. Ghose, Meerut*). "The leaf is chewed in ulcerated states of the tongue, in stomatitis, and dyspepsia" (*Native Surgeon T. R. Moodelliar, Chingleput, Madras Presidency*). "Diuretic, cathartic. Fr sh juice dose 2 to 8 ounces. Extract 1 to 2 drachms" (*Apothecary T. Ward, Madanapalle, Cuddapah*).

FOOD. 2308 Leaves. 2309 Shoots. 2310 Berries. 2311

Food.—The LEAVES and tender SHOOTS are boiled like spinach, and thus eaten in many parts of India. When ripe the BERRIES are often eaten by children, without any bad effects.

Solanum spirale, *Roxb.; Fl. Br. Ind., IV., 230.*

Vern.—*Bagua*, SILHET; (The root=) *mungas kajur*, PATNA.

References.—*Roxb., Fl. Ind, Ed. C.B C., 230; Kurz, For. Fl. Br. Burm., II., 227; Irvine, Mat. Med. Patna, 68.*

Habitat.—A herb, or under-shrub, of Assam, the Khásia mountains, East Bengal, and Upper Burma; found at altitudes up to 3,500 feet.

MEDICINE. Root. 2312

Medicine.—The ROOT is said by Irvine to be given in Patna as a narcotic and diuretic in doses of from gr. ¼ to grs. viii.

2313

S. torvum, *Swartz.; Fl. Br. Ind., IV., 234; Wight, Ic., t. 345.*

Syn.—S. STRAMONIFOLIUM, *Roxb., Fl. Ind., Ed. C.B.C., 192;* S. FERRUGINEUM, *Jacq.;* S. PSEUDOSAPONACEUM, *Blume;* S. WIGHTII, *Miq.*

Vern.—*Gota begún*, BENG.; *Súndai kai*, TAM.; *Chúndai kai*, TEL.

References.—*DC., Prod., XIII., Pt. I., 260; Roxb., Fl. Ind., Ed. C.B.C., 192; Kurz, For. Fl. Br. Burm., II., 225; Dalz. & Gibs., Bomb., Fl., 175; Lisboa, U. Pl. Bomb., 202; Alpin, Rep. on Shan States (1887-88); Gazetteers, Mysore & Coorg, I., 63; Ind. Forester, III., 238.*

Habitat.—A shrub, 8 to 12 feet high, or flowering as a herb; found throughout the tropical regions of India, except the western desert tract. In Bengal it is very common. It is distributed to Malaya, China, the Philippines, and Tropical America.

FOOD. Fruit. 2314

Food.—The FRUIT attains a considerable size and is used as a vegetable, especially in times of scarcity.

2315

S. trilobatum, *Linn.; Fl. Br. Ind., IV., 236; Wight, Ic., t. 854.*

Syn.—S. ACETOSÆFOLIUM, *Lamk.;* S. CANARANUM, *Miq.*

Vern.—*Nabhi-ánkuri*, URIYA; *Uchchinta, uste, tella uste, mulla-muste;* TEL.; *Tudavullay, thuthuvelai*, TAM.; *Alarka*, SANS.

References.—*DC., Prod., XIII., Pt. I., 287; Roxb., Fl. Ind., Ed. C.B.C., 192; Kurz, For. Fl. Br. Burm., II., 224; Sir W. Elliot, Fl. Andhr., 118, 180, 185, 187; Bidie, Cat. Raw Prod., S. India, 39; Ainslie, Mat. Ind., II, 427; W. W. Hunter, Orissa, II., 181, App. II.; Nellore, Manual, 129; Gazetteer, Mysore & Coorg, I., 55, 63.*

Habitat.—A scandent under-shrub, 6 to 12 feet high; frequently found in the Western Deccan Peninsula from the Konkan southward. It is met with also in tidal swamps in the Malay Peninsula from Arracan to Malacca, and in Ceylon.

MEDICINE. Roots. 2316 Leaves. 2317 Shoots. 2318

Medicine.—The ROOTS, LEAVES, and tender SHOOTS of this creeper are all used in medicine by the Tamuls; the two first, which are bitter, are occasionally given in consumptive cases in the form of electuary, decoction, or powder (*Ainslie*).

Food.—Among the Teligus, the LEAVES are much used as a vegetable, and their consumption is believed to improve the intellect.

FOOD. Leaves. 2319

2320

Solanum tuberosum, *Linn.; Fl. Br. Ind., IV., 229.*

POTATO, *Eng.;* POME DE TERRE, *Fr.;* POME DE TERA, *It.;* KARTOFFEL, *Germ.*

Vern.—*Alú,* HIND.; *Alú,* BENG.; *Álú,* N.-W. P.; *Álú,* KUMAON; *Álú,* PB.; *Batata,* BOMB.; *Batate,* MAR.; *Batata,* GUZ.; *Wallarai kilangu,* TAM.; *Álú guddalu, útalay gudda,* TEL.; *Batáte, álú,* KAN.; *Rata innala,* SING.

References.—*Stewart, Pb. Pl., 160; Aitchison, Flora of Lahoul, in Journ. Linn. Soc., X., 74; DC., Orig. Cult. Pl., 45; Mason, Burma & Its People, 466, 798; Sir W. Elliot, Fl. Andhr., 14; Birdwood, Bomb. Prod., 171; Baden Powell, Pb. Pr., 258, 362; Atkinson, Him. Dist. (X., N-W. P. Gaz.), 314, 703, 750; Useful Pl. Bomb. (XXV., Bomb. Gaz.), 167; Econ. Prod., N.-W. Prov., Pt. V. (Vegetables, Spices, and Fruits), 19; Royle, Prod. Res., 68, 69, 228, 360; J. G. Baker in Journ. Linn. Soc., XX., 489; Grigg, Manual of Nilghiri Dist., 134; Report, Govt. Garden and Park, in the Nilgiris (1881-82), No. 6; Man. Bomb. Rev. Accts., 101; Settlement Reports:—Panjáb, Kangra, 25, 28, 80; Lahore, 9, 12; Hazara, 88-89; N.-W. P., Kumaon, App., 33; Central Provinces, Chhindwara, 28; Jubbulpore, 86; Port Blair, 1870-71, 25; Kumaon, Off. Rep., 279; Gazetteers:—Bombay, VIII., 183; XVI., 103; Panjáb, Simla, 59; Lahore, 87; N.-W. P., IV., lxxv.; Mysore & Coorg, I., 63; Rep. Agri. Dept. Beng., 1886, 26, (app.) lix., lxxxvii., xc., xciv.; 1886-87, 9, 14; Assam (1884-85), 3, 21; (1885-86), 19; (1887-88), 21; Bombay, (1881-82), 4; (1883-84), 12; (1884-85), 20; (1888-89), 27; Agri.-Horti. Soc. Ind., Trans.:—VIII., 307-313; Jour., II., Pro. 89, 293, 417-425; Sel., 417-426; III., Pro. 61, 230; IV., Sel. 28; V., Sel., 37; VI., 10-13; Pro., 31; VIII., Sel., 157, 158; IX., Pro., 26, 102, 123, 129, 130; X., 63-65; 83, Sel., I.; XIII., 380; New Series, III., Sel., 44-48; IV., 34, 35, Pro., 10, 11, 47, 48; V., 35, 43, xlix., l. Pro., xiii., xiv.; VI, Pro., xxxvi.; VII., 61-66, Pro. 147, 159; VIII., Pro., 15, 16, 58; Tropical Agriculturist, March, 1889, 624, 637; Ind. Forester, XIII., 190-192; Indian Agri., May 15th, 1886; July 20th, 1887; June 2nd, 1888; The "Englishman," August 19th, 1887; Jour. Royal Hort. Soc., XI., 4 & 5; Smith, Ec. Dict., 337.*

Habitat.—With reference to the indigenous habitat of the potato, **DeCandolle** states that the only locality in which it is found really wild, in a form which is still seen in our cultivated plants, is Chili, and that it is very doubtful whether its natural home extends to Peru and New Grenada. According to **Mr. J. G. Baker**, however, in the Journal of the Linnæan Society undoubted forms of **Solanum tuberosum,** have been found wild in Lima and New Grenada also, but the plant is everywhere, one occurring at a comparatively high altitude and in a dry climate, and is met with nowhere in the near neighbourhood of the coast. Specimens, referred by **M. DeCandolle** to **S. tuberosum** which had been obtained near the coast of Valparaiso by **Mr. Alexander Colcleugh,** Secretary to the Legation at Rio Janeiro and by **Mr. Darwin** on the sea-coast at the Chonos Archipelago, have been referred by **Mr. J. G. Baker** to another species—**S. Maglia,** *Schecht,* thereby verifying **Mr. Darwin's** sagacious remark in the last sentence of his description of the Chonos plant:—"It is *remarkable* that the same plant should be found on the sterile mountains of Central Chili, where a drop of rain does not fall for more than six months and within the damp forests of these western islands."

HISTORY. 2321

HISTORY.—It is proved without a doubt that at the time of the discovery of America, the cultivation of the potato was practised with every appearance of ancient usage in the temperate regions extending from Chili to New Grenada at altitudes varying with the latitude. The true potato was not, however, apparently known to, or cultivated by, the aborigines of the Eastern temperate region of South America, nor by the

Cultivation of the Potato.

HISTORY.

inhabitants of Mexico, before the advent of Europeans by whom the cultivation was extended to these regions. In Europe the potato was first introduced, at some period between 1535 and 1585, into Spain whence its cultivation spread into Portugal, Italy, France, Belgium, and thence into Germany. The introduction into Ireland by **Sir Walter Raleigh's** companions, in 1585 or 1586, was independent, but considerably after the era of its first cultivation in Spain. It was on this occasion obtained apparently from Virginia or Carolina, whither it had most probably been spread from its original home by some Spanish or other travellers, since there seems to be no evidence of its cultivation in North America by the aboriginal races.

The date of the introduction of the potato into India seems unknown, and very few facts can be gathered to give grounds for even an approximate date. It must, at any rate, have been widely cultivated in India before the beginning of the eighteenth century, since **Roxburgh**, who wrote at the end of that period, says that it was in his time cultivated largely during the cold weather and produced abundant tubers, and that this cultivation must have been going on for some considerable time, since the Hindus make it an article of their diet, a thing which they will not do with any vegetable that has not been introduced at least fifty years. The probability is that the cultivation of the potato was introduced into India from Spain, whether directly or indirectly it is impossible to say, some time between the end of the sixteenth and the beginning of the eighteenth centuries.

CULTIVATION in Bengal. 2322

CULTIVATION IN INDIA.

At the present day it is cultivated more or less in all parts of India, both in the plains and in the hills, up to an altitude of 9,000 feet. The following extracts from Agricultural Reports, District Manuals, and other sources, will convey to the reader some idea of the importance and extent to which the potato is now cultivated in India.

Bengal.—The account here given will be found to be a reprint of what occurs in the Agricultural Report of Bengal (1886):—

"This valuable crop is largely cultivated in the Hooghly and Burdwan districts, and the cultivation of it is yearly increasing. The spots that have generally been selected for its cultivation are the old beds of rivers which have either changed their course or have altogether dried up. Most of these places may be connected by a line which, commencing close to Baidyabati station on the East Indian Railway, passes along the Tarakeswar Railway line almost throughout its length; between Haripal and Tarakeswar it changes its direction, and coming close to the important village of Chakdighi in the Burdwan district, passes along the Chakdighi-Memary road; at Memary it crosses the East Indian Railway line and comes to Satgachi; it then changes its direction again, and passing along the Burdwan-Culna road *viâ* Bohar, comes to within a few miles of the sub-divisional town of Culna on the Hooghly. The places most famous for potato culture are Nalikul, Haripal, and Singhur in the Hooghly, and Memary, Satgachia, and Bohar in the Burdwan districts.

"The best potato soil is a sandy loam having as fine a texture as possible. The soil must not be saline, nor must it contain too much iron. Soils containing *kankar* or nodules of carbonate of lime are also considered by the ryots unsuited for potatoes. As artificial irrigation is indispensable in the cultivation of potatoes—canals, tanks, jhils, or some other form of reservoir for water must be close by. The ryots mostly make use of tanks or small rivers in irrigating potato fields. Irrigation from wells is not resorted to, but the practice of potato-growers in the Patna and the Shahabad districts shows that where labour is cheap and the water is within

CULTIVATION in Bengal.

a reasonable depth, this mode of irrigation can be profitably adopted. All the necessary requirements for the successful cultivation of potatoes are met with near small rivers or their old beds. Sandy soils near large rivers are coarse-grained. The south-eastern part of the division along the Hooghly below Howrah, and along the lower courses of the Damudar or Rupnarayan, being subjected to strong tidal currents, is not suited to growing good potatoes on account of the soil containing too much common salt. Laud along the upper courses of these two rivers, as well as that along the Burrakar, Ajai, and the More, is too coarse-grained and contains too much of iron. Here and there, however, detatched areas are to be met with all over the division where all the conditions requisite for profitable potato culture exist.

"The different varieties of potatoes grown here may be classed under two heads. The *deshi* or country varieties including *usha, jhati, benajhara, goro*, and *thikre* and the imported varieties known as the Bombay. Of the country varieties the *benajhara* is the best. The Bombay variety is a middle-sized, thick-skinned, and mealy potato, and is grown to a very limited extent, and its cultivation being almost entirely confined to the villages of Singhur, Haripal, and Nalikul. Generally speaking, the seed used by the ryots is very bad, no care being taken either to select good or to introduce new seed.

"Generally the smallest potatoes are used for seeds, but the middle-sized tubers are admitted to be better. Three different kinds of seed are used. If a potato plant be raised, it is found that there are small potatoes close to the surface; then there are the large and middle-sized potatoes and below them again small potatoes. For seed purposes are used the small potatoes close to the surface as well as those found below the large potatoes. Of these two, the former, when partially exposed to light and turned somewhat green, are considered very good for seeds, while those found below the large potatoes are regarded as the worst for this purpose. But the best potato seeds are obtained in the way described below. While the plants are still quite strong and healthy the earth is removed from around the large tubers, which, though somewhat immature, are raised and sold in the early market at a high price. The parts of the plants immediately above the ground are then covered with earth and the fields are watered, so that a second crop of middle-sized potatoes is thus obtained and kept for seed.

"Much care is taken by the ryots in the preservation, though little or none in the selection of seed. After potatoes have been sorted those kept for seed are dried in the sun and spread over a bamboo structure (*machan*) in a cool part of the house. The potatoes must not be kept in heaps, but spread as thinly as possible. If sufficient space be available, they should be so kept that one may not touch another. At the setting in of the rains potatoes begin to rot, and then the rotten and diseased ones should be picked off as frequently as possible, for one rotten potato will spoil all with which it comes in contact. Potatoes kept for seeds are sometimes much injured by a disease known as *ladhara*, the lac disease. The first symptom of this disease is the appearance of a white powdery substance like wheat-flour around the buds or the eyes, which when pressed with the fingers become red. The potato so attacked rots, becoming quite watery within. The disease makes its appearance in the months of Ashar, Sraban, and Bhadra, and is very infectious: when it once makes its appearance, all the potatoes in the village, and sometimes even in the neighbouring villages, are sure to be attacked by it. Generally out of 10 maunds of potatoes kept for seed purposes, 7 maunds are ordinarily available for planting at the sowing time, but when the lac

SOLANUM tuberosum.

Cultivation of the Potato

CULTIVATION in Bengal.

disease makes its appearance, it often happens not even a single potato remains.

Rotation. 2323

"*Rotation.*—Potatoes are often in this division grown on the same field year after year. A newly broken field does not give a good outturn. Potatoes are generally grown after *aus* paddy as a second crop, but a field which grows only potatoes gives a better crop, and on such fields potatoes may also be grown much earlier.

Outturn, 2324

Tillage. 2325

"*Tillage.*—Early planted potatoes are likely to bring greater profit to the cultivator, but early sowing is attended with some risks. For the early market potatoes must be put in immediately the rains have ceased, for should there be heavy showers after the potatoes have been planted, whole fields may be totally injured, thus requiring replanting or growing with some other crop. Of the two varieties, the *deshi* and the hill potatoes, it is necessary that the latter be put in earlier, because while by proper irrigation the plants of *deshi* potatoes may be kept up till the end of Falgoon (middle of March, the hill potatoes will die off by the end of Magh (middle of February); and if put in late, there will not be sufficient time for the proper growth of the tubers. Potatoes are sometimes put in as early as the end of Bhadra (middle of September), but the most general time for planting is the end of Aswin (middle of October). When potatoes follow *aus* paddy, as is generally the case, as soon as the latter has been harvested, the land should be ploughed repeatedly. The object to be kept in view in preparing lands for potatoes is to pulverize the soil thoroughly, and to cultivate as deeply as possible. It is also necessary completely to eradicate the weeds and the roots of the paddy. All this is effected by the repeated use of the plough and the ladder. Clubs are sometimes used in breaking the clods, especially if the land be heavy. The furrows should be made light, but as close as possible, and after every ploughing the ladder should be passed. It is also necessary that after the paddy has been harvested the soil should be at least once thoroughly dried by exposure to sun and air. This is done by not passing the ladder over the field after the first two ploughings, but laying it in open furrows. Sometimes when the paddy is harvested, the land is left too dry to work with the plough: it will then be necessary to water the field before commencing with the ploughing. Where fields are set apart for the cultivation of potatoes after potatoes, without the intervention of any other crop, there is much time to work the land. This is done at Nalikul, the greatest potato-growing village in Bengal. Here potato fields are sometimes enriched by green manuring. At the beginning of Jaistya (middle of May), indigo seeds are thickly broadcasted in the fields intended for potatoes. At the end of Ashar (middle of July), the plants are ploughed in and mixed with the soil.

Planting. 2326

"*Planting.*—When the land has been well prepared, it is levelled as perfectly as possible by passing the ladder several times over the field, which is then divided into a number of plots by water channels. A number of main channels are first drawn from the headland to the bottom of the field, dividing it into a number of long strips about 40 feet broad. Smaller water channels are then made across these strips at a distance of 15 feet from one another. Now by means of a piece of rope and two sticks the spaces between the smaller water channels are marked by a number of lines parallel to the main water channels, and at intervals of 22 inches; lines are also made parallel to the smaller water channels equidistant from two contiguous ones. Finely powdered oil-cake is now spread along these lines, and the soil turned up by the *kodali*. The *kodali* must be made to go as deep as possible. Potatoes are then planted by hand along these lines. The seeds must be put in deep enough to save them from the scorching sun. Small entire potatoes, of which the shoots

CULTIVATION in Bengal.

have developed to 2 or 3 inches, are generally planted in case of the country varieties. Not to break the shoots the earth over them is placed very carefully, but it is at the same time necessary that this should be well pressed to save the tender shoots from the sun. These two conditions are fulfilled by making the earth over each of the seeds into a hemispherical ball. The Bombay potatoes are not planted entire, but are cut into pieces containing one or two buds. In case of the country variety the *rayats* prefer middle-sized potatoes with only one strong bud. A potato with a number of eyes gives weak plants. The potatoes are planted at a distance of about 7 inches in the rows.

After-treatment. 2327

"*After-treatment.*—In case of the *deshi* potatoes the plants will come over ground in a week; the Bombay variety will take a little longer. If the plants do not come out in due time, it will be necessary to apply water, which is done in the following manner. The main and the smaller channels are filled with water, and then by means of a shallow earthen vessel called *shara*, water is thrown into the space between the minor channels, one-half of which is watered by the channel on one side, and the other half by the channel on the other side. If a crust be formed by this watering, it should be broken. The plants now generally come out, but if there be failure in any place, the process should be repeated. Plants, however, which come out without watering, shooting through the dust as it is called by the ryot, are much stronger. After a few days again apply water in the way described above, hoe the space between the rows of potato plants by a *kodali*, and stir up the soil along the lines of plants. When the plants have grown to a height of about 6 or 7 inches, the operation of earthing up should begin. The field is watered as before, the spaces between the rows are hoed, and the plants slightly earthed up. This first earthing up, called *kani mati*, is done by placing a little earth with a corner of the *kodali*. After this operation has been done, the field appears as if laid in ridges, and as if the plants had been planted on these ridges

"The next operation is called *sara mati*, or finishing the earthing up. The field is once more irrigated, but henceforward by letting water pass through the furrows between the ridges, the plants are top-dressed with powdered oil-cake, and finally earthed up. This finishes all the important operations in potato cultivation; the only thing that now remains is to irrigate the field once in every seven to ten days.

Manures. 2328

"*Manures.*—The manures applied to potatoes are well-rotten dung, mustard and castor cakes, and ashes, more fields being manured with castor cake than with any other manure. Dung is used only in a well-rotten state, and that a long time before potatoes are put in, so that it may be well mixed with the soil. Green manuring with indigo seeds is, so far as I have observed, resorted to in one or two of the best potato-growing villages. The quantity of manures used depends not on the requirements of the fields or the crop, but on the means of the cultivators. Generally about 20 maunds of dung and 8 to 10 maunds of oil-cake are used per bigha. Half of the oil-cake is used at the time of planting, and the other half as a top-dressing when the plants are a second time earthed up.

Harvesting. 2329

"*Harvesting.*—Except when the potatoes are intended for the early market, the ryots do not lift them till the plants die and dry up. The general harvest time is the end of Falgun. After the potatoes have been raised they are sorted, the largest and the middle-sized ones are sold, and the smallest are kept for seed and household use.

Diseases. 2330

"*Diseases*—The most dreadful potato disease known to the ryot is *dhasa* or rot. This disease first makes its appearance on the root just below the

CULTIVATION in Bengal.

ground, which rots, and the upper portion of the plant dries up. Hundreds of potato fields are sometimes totally destroyed by this disease, and the loss is immense. It does not seem that the disease is induced or even influenced by any peculiarities of the soil or climate. It is very probably propagated through the tubers, and in the tubers if not induced, is at least much aggravated, by continuous heavy showers and high temperature in Sravan and Vhadra. Another disease from which potatoes in Bengal sometimes suffer is curl, by which the leaves are curled up, and the plants at first become stunted in growth, and afterwards dry up altogether. Of the insect pests, a kind of red ant sometimes does great injury, making holes through the tubers at first, and then causing the death of the plants. These insects are found more in coarse-grained red sand than in any other soil. Fresh dung applied to potato fields is supposed to discolour the tubers, and gives rise to various kinds of insects.

Cost. 2331

Approximate cost of potato cultivation.

	R	a.	p.
6 ploughings at 6 annas	2	4	0
40 baskets of dung	2	8	0
To apply the same, 2 men at 4 annas each	0	8	0
Oilcake, 8 maunds	9	0	0
To plant, 8 men	2	0	0
1½ maunds of seeds at R5 per maund	7	8	0
To irrigate, 2 men	0	8	0
Ditto	0	8	0
Hoeing, 5 men	1	4	0
To irrigate, 2 men	0	8	0
1st earthing, 4 men	1	0	0
To irrigate, 2 men	0	8	0
2nd earthing, 4 men	1	0	0
To irrigate, 8 times more	4	0	0
To lift, 8 men	2	0	0
Rent	2	0	0
	37	0	0

"*N.B.*—The cost is very variable. Sometimes each time the field is irrigated the cultivator has to pay more than a rupee, and the seeds often cost as much as R20 to R25. The price of the oilcake in some years rises to R1-8 a maund. If a field is to be planted with potatoes for the first time, instead of 6, some 20 ploughings will be needed. The cost therefore may amount to nearly R60.

"The yield again is even more variable. In some of the places now and then it exceeds even 80 maunds per *bigha*. If the plants are attacked with *dhasa*, it may come down to below 10 maunds. I think I shall not be much out of the mark if I put down on the average the cost of cultivation at R35, and the outturn at 50 maunds per *bigha*, worth as many rupees."

Assam. 2332

Assam.—Potato cultivation is largely carried on in Assam, chiefly in the Khásia hills, although it grows well, too, on the *chars* of the Brahmaputra. In the Khásia hills, the methods of cultivation are very rude. The crop is generally grown from small seed, all the best potatoes being kept for sale, and is produced on land which is utilised for the same purpose from year to year without intermission. Further, two crops are constantly taken off the same land in the year, and it not infrequently happens that to allow of the ground being prepared in time for the autumn crop, it is necessary to take up the early crop before the potatoes have properly ripened and the haulms show signs of decay. The natural result of this is a potato that will not keep properly, but which rots after even a few days' storage (*Reports of Director, Land Records and Agriculture, Assam*). Potato disease has been present in Assam to a limited extent since

1885, but it was not till 1887-88, when the crop turned out an almost utter failure, that attention was directed to its ravages and attempts made to prevent them in future. Since that time, old fields have been largely given up and the importation of fresh seed from other parts of India has been beneficial, and in the Report of the Director, Land Records and Agriculture (1889-90) it is reported that "the disease which has annually attacked the potato crop in the Khásia hills for the last three years appears to be disappearing."

CULTIVATION in Assam.

North-West Provinces.—In his account of the vegetables of the North-West Provinces **Mr. Atkinson** writes:—"The potato flourishes well in the hills at Naini Tal, Almora, Paori Lohughat and beyond Masúri, and also in the plains. The potatoes grown in the hills undoubtedly attain a larger size and have a better flavour, but owing to the frequent sowing of the same seed in the same places, have of late become waxy and watery. To remedy this a large quantity of new seed has been imported from England to renew the stock. The average price of hill potatoes in the season at Masúri is 4*d.* to 5*d.* a stone of 14lb, of the plains variety, in the plains 2*d.* to 3*d.* Drill cultivation is that usually adopted, and in favourable years it is a profitable crop. In the Gazetteer of the Himálayan districts, North-West Provinces, **Mr. Atkinson** remarks that the potato, which now forms an important article of export from Kumáon, was introduced in 1843 by **Major Welchman**, and that the seed is from time to time renewed by fresh importations."

N.-W. Provinces. 2333

Panjab.—Both on the plains and in the hills potatoes are largely cultivated in the Panjáb, especially in the country around the larger cities Thus in the Gazetteer of the Lahore District we find it stated:—"This esculent has become quite an article of commerce, and Natives are beginning to consume it largely; they are grown in some quantities around the city of Lahore, and are procurable all the year round; but from August till December they are imported from the hills, as during the hot weather those grown in the plains become watery and bad." In the Panjáb hills potato cultivation is carried on to a considerable extent. Thus the account given in the Settlement Report of the Hazara District is as follows:—"The cultivation of potatoes in the hill tracts adjoining Murree was introduced shortly after annexation. The crop was at first cultivated by Hindus from the Panjáb, who leased fields from the resident villagers at cash rents; but the villagers are now beginning to cultivate it themselves. Its success is very variable, so also is the price for which it sells. It also requires much more tending than the maize crops and exhausts the land to a greater degree. But despite these drawbacks it is probable that its cultivation will extend considerably, more especially as its consumption by Natives is on the increase."

Panjab. 2334

Central Provinces.—Potato cultivation is carried on to a considerable extent in several districts of the Central Provinces, but of late years owing to the constant use of seed grown in the same ground, the tuber has deteriorated much both in size and quality. It is generally sown in October and taken up in February or March. It is planted in rows for convenience of irrigation, but after being sown the rows are not usually further earthed up as in England.

Central Provinces. 2335

Bombay.—The Annual Report of the Director, Land Records and Agriculture, Bombay (1888-89), gives the following account of potato cultivation in that Presidency:—"The crop was raised over 11,700 acres against 12,200 acres in last year, chiefly in the following districts: Poona (9,800 acres), Ahmadnagar (550 acres), Satara (530 acres), Ahmedabad (200 acres) Kaira (190 acres). Apart from its use as a food on ordinary days, it is in extensive use all over the country among Hindus on days of fast, when

Bombay. 2336

SOLANUM tuberosum.

The Potato as an Article of Food.

CULTIVATION in Bombay.

grains are forbidden. The Mahábaleshwar potato is well known. In Ahmedabad it is largely cultivated, especially on the dry beds of rivers which are peculiarly well adapted for its growth. At the Bhadgaon Farm in Khandesh it was grown so successfully some years ago, and the *rayats* took up its cultivation so readily, that Páchora has now become a market for potatoes. The potato is extensively cultivated in the Poona District, chiefly in Khed, Sirur, Junnar, and Purandhar *tálukas*. The produce finds its way to the Poona and Bombay markets. In the Khed *táluka*, where the area under potato (7,000 acres) is the largest, it is one of the best paying agricultural products. Two varieties are grown—one the mealy, which has a thin, smooth, light brown peel, the other having a thick, rough, dark peel and becoming waxy when cooked. For the better kind, the country is indebted to the late **Dr. Gibson** who, in 1838, imported the Nilgiri and, again, in 1845, an excellent stock of the Irish potato. Potato is produced in the deep red soils which prevail in the north-east of the *táluka*. Land yielding potato is usually cropped twice during the year. As a dry crop it is planted in July and gathered about the end of September, and as a garden crop it is planted in December and gathered in February. Owing to the ready sale it meets in the Poona and Bombay markets, and the large profits it yields, the cultivation of potatoes has very considerably increased, having in some instances entirely superseded that of sugarcane. The produce of the Khed *táluka* is now sent by the cultivators direct from the Telegaon station to Bombay. The potato merchants of Bombay write their orders by post direct to the principal cultivators. From experiments conducted by the Assistant Settlement Officer, **Mr. Whitcombe**, the yield of potatoes per acre from dry crop land was found to be 12,000lb and from irrigated land 18,000lb. The prices ruling since 1848-49 have fluctuated from R2 to R7 per *palla* of 240lb."

Madras. 2337

Madras.—In this Presidency it is chiefly in the Nilgiri Hills that potatoes are cultivated, although attempts at their introduction have been made elsewhere with a varying amount of success. Even from the Nilgiri hills it is reported that this crop, although it has long been cultivated, "has, of late years, been more or less of a failure owing to *potato disease* and from constantly growing the tubers in the same piece of ground by which the soil gets exhausted" (*Manual of Nilgiri District*).

Burma. 2338

Burma.—"The introduction of potato cultivation in British Burma has engaged the attention of the Provincial Agricultural Department, and of the Agri.-Horticultural Society, since 1882. Seed potatoes were procured from England and Scotland, but the trial did not give good results at first. In 1883 an outturn of nineteen-fold in the Karen hills was the highest; the size of the potato dug was generally small, although there were many gathered, which went five, six, and eight to the pound. The people of several villages in the Karen hills sowed the tubers and obtained yields with more or less success. The Agri.-Horticultural Society of the Province, in its Report for 1883, wrote:—"It is a source of satisfaction to know that the initiatory measures taken by this Society in introducing this excellent tuber have at last resulted in one large Military cantonment being regularly supplied."

MEDICINE. Tubers. 2339

Medicine.—The TUBERS when dried are often employed as a substitute or adulterant for salep.

FOOD. 2340

Food.—The potato is now used as an article of food by almost all classes in India when attainable, but are said by Natives to have a tendency to cause indigestion and flatulence. As already stated it is eaten extensively, especially by Hindus on fast days when grains are forbidden.

Solanum verbascifolium, *Linn.; Fl. Br. Ind., IV., 230; Wight, Ic., t. 1398.*

2341

Syn.—S. PUBESCENS, *Roxb.*; S. ERIANTHUM, *Don.*

Vern.—*Dursul,* NEPAL; *Sivor,* LEPCHA; *Asheta,* N.-W. P.; *Chichora,* C. P.; *Kala mewa, tiári, olá, kharawine,* PB.; *Rasagadi mánu,* TEL.

References.—*DC., Prod., XIII., Pt. I., 114; Kurz, For. Fl. Br. Burm., II., 225; Gamble, Man. Timb., 273; Dalz. & Gibs., Bomb. Fl., 175; Stewart, Pb. Pl., 160; Sir W. Elliot, Fl. Andhr., 88, 163; Atkinson, Econ. Prod. N.-W. P., Pt. V., 91, 96; Gazetteers:—N.-W. P., I., 83; IV., lxxv.; X., 314; Mysore & Coorg, I., 56; Agri.-Horti. Soc. Ind. Journ., X., 26; XIII., 317; XIV., 68; Ind. Forester, XI., 4.*

Habitat.—A shrub or small tree, frequently met with throughout India in the Tropical and Sub-tropical zone. It is distributed to South-East Asia, Malaya, North Australia, and Tropical America.

MEDICINE. 2342

Medicine.—It is used medicinally by the Natives, but its properties are unimportant.

FOOD. Fruit. 2343

Food.—In Southern India it is cultivated for its FRUIT, which is eaten in curries.

TIMBER. 2344

Structure of the Wood.—Light yellow in colour, soft.

2345

S. xanthocarpum, *Schrad. & Wendl.; Fl. Br. Ind., IV., 236; Wight, Ic., t. 1401.*

Syn.—S. JACQUINII, *Willd.*; S. DIFFUSUM, *Roxb.*; S. VIRGINIANUM, *Jacq.*; S. ARMATUM, *Br.*

Vern.—*Kateli, katai, ringni,* HIND.; *Kantakari,* BENG.; *Rengnie, khat-khatayа,* PATNA; *Rangaini janum,* SANTAL; *Bheji begun, ankránti,* URIYA; *Ringni,* C. P.; *Warúmba, mahori, kharián maraghúne, mamoli, choti mauhari, harnauli, kandiári, pilak,* (the fruit=) *katela, batkateya,* PB.; (The root=) *aderay-ja-denay,* SIND.; *Ringni,* DECCAN; *Bhúringni, ringni,* BOMB.; *Bhuiringani,* MAR.; *Bhoyaringani,* GUZ.; *Cundung katric, kandan-kattiri,* TAM.; *Pinna mulaka, nela mulaka, jiddu us te, vankuda, nella molunga,* TEL.; *Ella-battu, kattu-wel-battu,* SING.; *Kantakari, nidigdhika,* SANS.

References.—*DC., Prod., XIII., Pt. I., 302; Roxb., Fl. Ind., Ed. C.B.C., 191; Kurz, For. Fl. Burm., II., 224; Thwaites, En. Ceyl. Pl., 217; Dalz. & Gibs., Bomb. Fl., 175; Stewart, Pb. Pl., 161; Sir W. Elliot, Fl. Andhr., 74, 131, 153, 179, 188; Pharm. Ind., 181; Ainslie, Mat. Ind., II., 90; U. C. Dutt, Mat. Med. Hind., 208-210, 303, 311; S. Arjun, Cat. Bomb. Drugs, 99; Murray, Pl. & Drugs, Sind, 158; Dymock, Mat. Med. W. Ind., 2nd Ed., 636; Year-Book Pharm., 1880, 250; Trans. Med. & Phys. Soc. Bomb., New Series, No. 4, 154; Baden Powell, Pb. Pr., 363; Drury, U. Pl. Ind., 397; Useful Pl. Bomb. (XXV., Bomb. Gaz.), 202; Boswell, Man., Nellore, 140; Baroda Durbar, Calcutta International Exhibition, No. 168; Settlement Reports:—Central Provinces, Chanda, App. VI; Gazetteers:—Bengal, XV., 439; Bombay, V., 27; XVIII., 48; Panjáb, Karnal, 16; N.-W. P., I., 83; IV., lxxv.; X., 314; Mysore & Coorg, I., 63; Hunter, Orissa, II., 158; App. IV., 180, App. VI; Agri.-Horti. Soc., Ind.:—VI., 50; X., 26; New Series, I., 96; Ind. Forester, III., 238; IV., 232, 234; VI., 239; XII., App., 18.*

Habitat.—A very prickly diffuse herb, commonly met with throughout India, from the Panjáb and Assam to Ceylon and Malacca, and distributed through South-East Asia, Malaya, Tropical Australia, and Polynesia.

MEDICINE. Root. 2346

Fruit. 2347

Medicine.—The ROOT of this species is, like that of **S. indicum,** much used in Hindu medicine as one of the ingredients of the highly esteemed *dasamula* (see **Desmodium gangeticum**). It is believed to have expectorant qualities, and is largely given in cough, asthma, catarrh, and pain in the chest. The FRUIT is thought to possess properties similar to those of the root. Both root and fruit are given in medicine in various forms, such as decoction, electuary, *ghrita,* and are usually prescribed in combination with other expectorants and demulcents (*U. C. Dutt*).

In the works of Muhammadan Medical writers, three kinds of **Solanum**

Medicinal uses of the Kantakari.

MEDICINE.

are described, all of which have somewhat similar medicinal properties. Their *Hejazi*, according to Dymock, is **Solanum xanthocarpum**, and is recommended by them for use in asthma, cough, dysuria, catarrhal fever, leprosy, costiveness, and stone in the bladder (*Mat. Med. W. Ind.*).

In Southern India Ainslie notices the use of the root as an expectorant, and says it is much given in the form of decoction, electuary or pill in coughs and consumptive complaints, and in humoral asthma.

Stems. 2348 Flowers. 2349 Seeds. 2350

A description of the uses of **S. xanthocarpum** is given in the supplementary list of the Indian Pharmacopœia, where it is stated that, in addition to having expectorant and diuretic properties assigned to it by the people of India, the STEMS, FLOWERS, and fruit have, in the opinion of some European practitioners, bitter and carminative qualities and are useful in those forms of burning of the feet (*ignipeditis*) which are attended with a vesicular watery eruption. It is also remarked that fumigations with the vapour of the burning SEEDS are in high repute among the Natives as a cure for toothache, probably because they act as a sialogogue. No chemical examination of the plant appears to have been made.

SPECIAL OPINIONS.—§ "The decoction of *kantakari* is much used as a diuretic in both active and passive dropsies. It is advantageously given in combination with alcohol and other mineral diuretics, and during its use, milk diet should be prescribed" (*Surgeon E. S. Brander, M.B., F.R.C.S.Edin., I.M.D., Rungpore*). "A strong diuretic in cases of dropsy. Useful in chronic diarrhœa" (*Civil Medical Officer Forsyth, F.R.C.S. Edin., Dinajpore*). "Decoction of the plant is used as a febrifuge" (*Surgeon A. C. Mukerji, Noakhally*). "Decoction of *kakhkari* commonly used by native *baids* in fever and cough" (*Assistant Surgeon S. C. Bhattacharji, Chanda, Central Provinces*). "It is said to possess stomachic and antibilious properties, when prepared and administered in the following manner :—The pods are freed from the seeds, boiled with butter-milk and salt, then dried in the sun by day and soaked in butter-milk by night. It is so treated for four or five days, and then fried in *ghí* and eaten" (*Surgeon Major D. R. Thomson, M.D., C.I.E., Madras*). "The seeds are rubbed with water, and applied over local inflammations and swellings" (*Surgeon Major Robb, Civil Surgeon, Ahmedabad*). "A decoction of *kantikari* is given with benefit in cases of cough and cold attended with liver" (*Assistant Surgeon R. C. Gupta, Bankipore*). "*Kanticary, khatpapra*, and *goluncho* : I have heard that native physicians employ these in the form of decoction with good effect, in chronic malarious fevers. *Kanticary* promotes the secretion of urine, and, therefore, may be useful in dropsy cases, gonorrhœa, &c." (*Surgeon J. Ffrench Muller, M.D., I. M. S., Saidpore*).

FOOD. Seeds 2351

Food.—In some places the SEEDS are eaten.

SOLENANTHUS, *Ledeb.; Gen. Pl., II., 848.*

2352

Solenanthus sp., BORAGINEÆ.

Vern.—*Lendi, lenwa*, (root=) *mulin*, PB. ; *Shomá, dimmuk*, LAD.

Reference.—*Stewart, Pb. Pl., 155.*

Habitat.—A perennial herb, found in parts of the basins of the Chenab and Jhelum, at 4,000 to 10,000 feet, and in Ladak up to 16,000.

MEDICINE. Root. 2353

Medicine.—"The pounded ROOT is applied to abscesses, and is exported to the plains" (*Stewart*).

Solenostigma Wightii, *Bl.*; see **Celtis Wightii**, *Planch.*; [Vol. II., 244; URTICACEÆ.

Soma, see **Ephedra**, *Linn.*; GNETACEÆ; Vol. III., 246.

SONCHUS, *Linn.; Gen. Pl., II., 528.*

[*1142*, COMPOSITÆ.

Sonchus arvensis, *Linn.; Fl. Br. Ind., III., 414; Wight, Ic., t.* 2354

Syn.—S. ORIXENSIS, *Roxb.;* S. LONGIFOLIUS, *Wall.;* S. WIGHTIANUS, *DC.*

Vern.—*Sahadevi bari,* HIND.; *Ban-pálang,* BENG.; *Birbarangon,* SANTAL; *Nalla tapeta,* (?), PB.; *Bhangra, kala bhangr, a jángli tamáku,* TEL.

References.—*Roxb., Fl. Ind., Ed. C.B.C., 593; Voigt, Hort., Sub. Cal., 432; Sir W. Elliot, Fl. Andhr., 127; Baden Powell, Pb. Pr., 355; Gazetteer, N.-W. P. IV., lxxii.; Jour. Agri.-Horti. Soc. Ind., XIV., 6.*

Habitat.—A milky herb, with creeping root-stock, met with throughout India, both wild and in cultivated places, scarce in the plains, common in the Himálaya and the Khásia mountains up to 8,000 feet.

Medicine.—Among the Santals the ROOT is given in jaundice (*Rev. A. Campbell*). In the Panjáb this plant is credited with similar medicinal properties to those of **Lactuca scariola** (see Vol. IV, 578).

MEDICINE. Root. 2355

Fodder.—Cattle are fond of every part of the plant, which, on being wounded, discharges a MILKY JUICE. This thickens into a substance resembling opium (*Roxburgh*).

FODDER. Milky Juice. 2356

S. oleraceus, *Linn.; Fl. Br. Ind., III., 414; Wight, Ic., t. 1141.* 2357

MILK THISTLE.

Syn.—S. CILIATUS, *Lamk.;* S. ROYLEANUS, *Wall.*

Vern.—*Titaliya,* PATNA; *Dodak,* PB.; *Ratrinta,* TEL.

References.—*DC., Prod., VII., 184; Voigt, Hort. Sub. Cal., 432; Stewart, Pb. Pl., 130; Year-Book Pharm., 1876, 230; Irvine, Mat. Med. Patna, 117; Gazetteer, N.-W. P., IV., lxxiii.; Agri.-Horti. Soc. Ind. Journ., XIV., 6.*

Habitat.—An annual weed, found in fields and cultivated places throughout India, ascending to 8,000 feet in the Himálaya. It is found wild or introduced in all temperate and many tropical countries.

Medicine.—The ROOT and LEAVES are used by the Natives of Bengal in infusion as a tonic and febrifuge (*Irvine*). The ancients and European herbalists of the middle ages valued this PLANT highly as a galactagogue and in diseases of the liver.

MEDICINE. Root. 2358 Leaves. 2359 Plant. 2360

Food and Fodder.—The Kashmíris are said to use it as a vegetable (*Stewart*). Cattle are very fond of it.

FOOD & FODDER. 2361

SONNERATIA, *Linn. f.; Gen. Pl., I., 784.*

[*t. 340.;* LYTHRACEÆ.

Sonneratia acida, *Linn. f.; Fl. Br. Ind., II., 579; Wight, Ic.,* 2362

Syn.—RHIZOPHORA CASEOLARIS, *Linn.;* AUBLETIA CASEOLARIS, *Gærtn.*

Vern.—*Orcha, archaká,* BENG.; *Sundarí gnuá,* URIYA; *Ta-bu, ta-mu,* BURM.; *Gedde-killala,* SING.

References.—*Roxb., Fl. Ind., Ed. C.B.C., 405; Brandis For. Fl., 242; Kurz, For. Fl. Burm, I., 526; Gamble, Man. Timb., 205; Dalz. & Gibs., Bomb. Fl., 98; Mason, Burma & Its People, 534, 746; Murray, Pl. & Drugs of Sind, 193; Drury, U. Pl., 399; Lisboa, U. Pl. Bomb., 80, 81; Gazetteers:—Bombay, XV., 434; W. W. Hunter, Orissa, II., 176 (app. VI.); Ind. Forester, VI., 124; VIII., 401; Agri.-Horti. Soc. Ind., Trans., VII., 60; Rheede, Hort. Mal., III., t. 40; Rumph., Amb., III., t. 74.*

Habitat.—A small evergreen tree, met with in the tidal creeks and littoral forests of India, Burma, and the Andamans. It is distributed to Java and Siam.

Medicine.—The FRUIT is used as a poultice in sprains and swellings. The fermented JUICE of the fruit is said to be useful in arresting hœmorrhage.

MEDICINE. Fruit. 2363 Juice. 2364

FOOD. Fruit. 2365 TIMBER. 2366 DOMESTIC. Leaves. 2367 Wood. 2368

Food.—In the Sundarbuns the acid-bitter FRUIT is eaten. It is also used as a condiment by the Malays.

Structure of the Wood.—Grey, soft, even-grained. Weight 31lb per cubic foot.

Domestic.—A kind of silkworm feeds upon the LEAVES. The WOOD is said by **Beddome** to be used for models. It is largely employed as fuel, and in Ceylon and Sind is used as a substitute for coal on coasting and river steam boats, a purpose for which it is said to answer better than any other kind of wood.

2369

Sonneratia apetala, *Ham.; Fl. Br. Ind., II., 579.*

Vern.—*Keowra*, BENG.; *Khírwá*, URIYA; *Kam-ba-la*, BURM.

References.—*Roxb., Fl. Ind., Ed. C.B.C., 405; Kurz, For. Fl. Burm., I., 527; Gamble, Man. Timb., 205; Mason, Burma & Its People, 420, 746; Lisboa, U. Pl. Bomb., 80; W. W. Hunter, Orissa, 175 (app. VI.); Ind. Forester I., 8; Agri.-Horti. Soc. Ind. Trans., VII., 60; Jour., IX., (Sd) t. 2, 57.*

Habitat.—A moderately-sized tree, met with in the tidal creeks of Bengal, in the Deccan Peninsula as far as the Konkan, and in Burma.

FOOD. Fruit. 2370 TIMBER. 2371 DOMESTIC. 2372

Food.—The sub-acid green FRUIT is used in Burma in curries.

Structure of the Wood.—Sapwood grey, heartwood reddish-brown, moderately hard. Weight 44lb per cubic foot.

Domestic.—The wood is used for house-building and as planking for packing cases. It is not much valued.

2373

SOPHORA, *Linn.; Gen. Pl., I., 555.*

A genus of trees or shrubs belonging to the Natural Order LEGUMINOSÆ, and comprising above twenty species, widely distributed in warm countries. Ten species are indigenous to India, none of which are of much economic value, but it seems possible that **Sophora japonica**, a valuable tree the timber of which is used in Japan for many domestic uses and whose flower-buds form the principal source of the Chinese yellow dye, may be acclimatised in some parts of India.

2374

Sophora alopecuroides, *Linn.; Fl. Br. Ind., II., 250;* LEGUMINOSÆ.

Vern.—*Guázárákha*, PUSHTU. [*MSS.*

References.—*DC., Prodr., II., 96; Mr. J. H. Lace, Fl. of Quetta Dist.,*

Habitat.—An undershrub of the temperate regions of Western Tibet, Afghánistan, and Baluchistán.

FODDER Pods. 2375

Fodder.—The PODS are browsed by sheep and goats (*Lace*).

2376

S. mollis, *Graham; Fl. Br. Ind., II., 251.*

THE HIMALAYAN LABURNUM. [*Wall.*

Syn.—EDWARDSIA MOLLIS, *Royle, Ill., t. 32, f. 2;* SOPHORA HOUGHIANA,

Vern.—*Kún, kohen, málan, búna, bankeintí, tilún, tarní, brisarí, kathí,* PB.; *Arghawán*, AFGH.

References.—*Stewart, Pb. Pl., 68; Gamble, Man. Timb., 116; Brandis, For. Fl., 132; Aitchison, Bot. Afgh. Del. Com., ; Gazetteers:—N.-W. P., IV., lxxi.; X., 309; Panjáb, Bannu, 23; Dera Ismail Khan, 19; Ind. Forester, V., 179, 184, 186.*

Habitat.—A low spineless shrub of the plains and low hills of North-Western India from Hazára and the Salt Range to Kumáon and Nepál, ascending to 4,000 feet in altitude.

FODDER. 2377

Fodder.—Goats browse on it, but it is said to be fatal to other animals

Sophora robusta, *Roxb.;* see **Ormosia robusta,** *Wight:* Vol. V., 494.

SOPUBIA, *Hamilt.; Gen. Pl., II., 970.*

Sopubia delphinifolia, *G. Don.; Fl. Br. Ind., IV., 302;* SCROPHULARINÆ.

2378

Syn.—GERARDIA DELPHINIFOLIA, *Linn.*; G. HEYNEANA, *Benth.*; EUPHRASIA COROMANDELIANA, *Roth.*

Vern.—*Dudhâli,* BOMB.

References.—*Roxb., Fl. Ind., Ed. C.B.C., 491; Dalz. & Gibs., Bomb. Fl., 182; Dymock, Mat. Med., W. Ind., 581; Gazetteers:—N.-W. P., I., 83; IV., lxxv.; X., 314; Bombay, XV., 439.*

Habitat.—A handsome, erect, annual herb, found in Banda, Behar, the Deccan Peninsula, from the Konkan southwards, and Ceylon.

MEDICINE. Juice. 2379

Medicine.—The JUICE of the plant is applied by field labourers in the Deccan to their feet to heal sores caused by exposure to moisture. It is astringent and stains the skin at first yellow and afterwards a black colour (*Dymock*).

(*G. Watt.*)

SORGHUM, *Pers.; Gen. Pl., III., 1135.*

2380

A genus of grasses which belongs to the Tribe ANDROPOGONEÆ. It embraces one or two of the most important millets, as, for example, the *Juár* of Indian agriculture—a cereal which, after rice, is perhaps the most valuable single article of food in this country. If Bengal (the great rice-eating province) be left out of consideration, *Juár* takes the first place as the staple of Indian diet, and ranks a long way before wheat, barley, or Indian corn. The millions of the population of India live far more on millets and pulses than on rice, wheat, and barley. The chief millets are *Juár* and *Bájra,* but of the former there are numerous varieties and races, and even (according to some botanists) several species. Some of these are grown because of the sugar contained in their stems, others because of their grain, while a third class may be said to be cultivated exclusively as green fodder for cattle.

VARIETIES. *Conf. with pp. 282, 291, 293, 296, 305, 308, 309, 310, 311, 312.* 2381

Dr. Roxburgh, at the beginning of this century, referred the Indian forms to ANDROPOGON and described five species under the section now isolated as the genus SORGHUM, namely, A. BICOLOR, *Roxb.*, A. SORGHUM, *Roxb.*, A. CERNUUS, *Roxb.*, A. SACCHARATUS, *Roxb.*, and A. LAXUS, *Linn.* Hackel, the most recent author who has written on this subject, has returned these and many other forms to the genus ANDROPOGON and treated them as cultivated states of one common species under the oldest name for the series, *viz.*, ANDROPOGON ARUNDINACEUS, *Scop.* (= A. HALEPENSIS, *Sibth.*, HOLCUS HALEPENSIS, *Linn.*, SORGHUM HALEPENSE, *Pers.*).

It would be beyond the scope of this work to deal with purely botanical problems, and therefore the advisability or otherwise of Hackel's reduction of the genus SORGHUM need not be gone into further. But the probability of all the cultivated forms being but varieties or races of one species,—and that a plentiful wild plant in India,—has a distinctly practical significance. The following may be given as the chief varieties recognised by Hackel:—

2382

Var. I., saccharatus, *Kœrnick.*

Syn.—HOLCUS SACCHARATUS, *Linn.*; H. DOCHNA, *Forsk.*; SORGHUM SACCHARATUS *Pers., non Roxb.*; ANDROPOGON SACCHARATUS, *Pers., non Roxb.*

This form is said to be extensively grown in Africa and America, the plant of the former country being the *Imphee* (SYN.—S. CAFFRORUM, *Kunth.*), and of the latter the *Sorgho,* which is mainly cultivated on account of sugar, but the grain and straw are also of great value.

Var. II., technichus, *Kœrnick.*

2383

This is spoken of as a form with lax inflorescence, having long branches. It is extensively cultivated in the Mediterranean regions and in North America. It is the plant chiefly used for brooms. (*Conf. with p. 295.*)

Varieties of

2384

Var. III., vulgaris, *Pers.*

Syn.—HOLCUS SORGHUM, *Linn.;* SORGHUM VULGARE, *Pers.;* ANDROPOGON SORGHUM, *Roxb.*

This is said to have a more or less crowded obovate panicle of generally pale-coloured spikelets. Under this **Hackel** mentions three sub-varieties:—TYPICUS, with yellowish spikelets; ÆTHIOPS, with blackish or reddish spikelets; and JAPONICUS, with whitish spikelets, but larger and broader than in TYPICUS. The figure 15 in **Professor Church's** *Food Grains of India* may be taken to represent the type form of this plant—an erect, elongated panicle built up of compact racemose spikes, which, in their ultimate sections, appear almost sessile.

Var. IV., niger, *Kunth.*

2385

This is said to differ chiefly from VULGARIS in the darker colour of the inflorescence and grain.

Var. V., cernuus, *Kœrnick.*

2386

Syn.—HOLCUS CERNUUS, *Ard.;* S. CERNUUM, *Host., also Reich. It., t. 80, f. 466;* A. CERNUUS, *Roxb.;* A. CERNUUS, *Kunth.;* S. BICOLOR, *Willd.*

This is said to be very much like VULGARIS, except that the stalk of the panicle is thick and rigidly reflexed, so that the inflorescence becomes nodding. It is reported to be cultivated in Africa, Egypt, Portugal, Greece, Asia Minor, Persia, India, etc. In one part of his work **Hackel** speaks of this as grown in the higher mountainous parts of India. That remark is probably in allusion to **Roxburgh's** Manipur habitat, but, in another page, **Hackel** quotes **Wallich's** No. 8777 *K.*, as an example of CERNUUS—a specimen which was collected on the plains of the North-West Provinces, namely, at Allahabad. **Duthie & Fuller's** Plate VI. (*Field and Garden Crops*) as near as possible represents the plant which it is here accepted as that denoted by **Hackel.**

Var. VI., Durrha, *Forsk.*

2387

Syn.—HOLCUS DURRHA, *Forsk.*

This is an African, Egyptian, Abyssinian, and Persian form, which has a very crowded panicle. It is, says **Hackel,** one of the most important articles of food in Africa.

Var. VII., Roxburghii, *Hackel.*

2388

Syn.—A. SACCHARATUS, *Roxb., non Pers.* (? A. BICOLOR, *Roxb.*).

This is a remarkable form, with long, pointed, very hairy spikelets, and a wide-spreading inflorescence, each division of which is borne on a long naked peduncle or pedicel. The reader will find that a white and ared-grained form of the plant, recognised as referable to this place, are described below—the equivalents of A. SACCHARATUS, *Roxb.* and A. BICOLOR, *Roxb.* **Hackel** refers to certain African (Niger) sheets of this plant, which the writer has had the pleasure of consulting at the Royal Herbarium, Kew. These have an undoubted affinity to S. HALEPENSE, so that, if **Hackel's** reduction of all the cultivated forms to that species, be not considered desirable (by future investigator), it seems likely that in this one instance, at all events, his views may be upheld. But, many years ago, **Sir William Hooker** arrived at the same opinion (see *Flora Nigirtiana, p. 574*). "This is certainly," says **Sir William,** "the species so widely diffused in Africa, which goes by the name of S. SACCHARATUM, but I have much doubt whether it be more than a large variety of S. HALEPENSE. It grows to the height of 6 or 8 feet, with very ample spreading panicles; . . . The Nun specimens, from the inundated banks of the river, are still more luxuriant, and their large spikelets, about 4 lines long, clothed with red-brown hairs, give them a very rich aspect. Several varieties of the S. VULGARE, *Linn.*, are generally cultivated in Guinea, as well as the S. SACCHARATUM." The peculiarity of the hairs on the long pointed spikelets (which form a lax inflorescence), are perhaps the most distinctive characteristics of this plant. It would seem likely that Figure 16 in **Professor Church's** *Food Grains of India* is intended to represent one of the numerous forms of ROXBURGHII, but for the fact that the spikelets are too rounded and apparently glabrous. In some respects the description given by **Roxburgh** of his A. BICOLOR agrees more nearly with S. HALEPENSE (*var.* ROXBURGHII) than with S. VULGARE. The writer has experienced some difficulty in finally deciding this point. A. BICOLOR, *Roxb.*, has, from the description, the lax verticelled inflorescence of this form. **Roxburgh** lays

Conf. with pp. 279, 280.

stress on the value of the number of valves (or glumes) in the hermaphrodite and neuter flowers, in the distinction of the various species of this genus. A. SORGHUM, *Roxb.* (*e.g.*, SORGHUM VULGARE), has three valves to the neuter flowers, while those of A. BICOLOR and A. SACCHARATUS like the wild species S. HALEPENSIS) have only one or two. The inflorescence of these two forms is also very similar, so that it is probable A. BICOLOR should be regarded as the red or dark-coloured form, while ROXBURGHII (*e.g.*, A. SACCHARATUS) is the white state of the cultivated series here dealt with.

The reader will find further particulars below regarding A. SACCHARATUS, *Roxb.*, as met with in India. The chief object of the present chapter is to exhibit the views held by **Hackel**, on the varieties of this important group of millets. In conclusion, therefore, it need only be added that **Hackel** mentions three other forms even more obscurely defined than those given above.

Conf. with chapter on Roxburghii, p. 281, also under Bombay p. 312.

These are WIGHTII—met with in South India; GLOBOSUS—a Serampore (Calcutta) plant, described by **Voigt** as having almost orbicular blunt spikelets; and MILIIFORMIS—a Bengal plant with small spikelets, white and very hairy, specially around the margins of the glumes. The sheets of what may be this last plant (in the Kew Herbarium), have, however, exceptionally large spikelets, so that not only regarding these three plants has the writer experienced the greatest possible difficulty in recog nising **Hackel's** forms, but he is by no means certain that he can distinguish more than three, if not rather only two, *viz.*, VULGARIS, SACCHARATUS, and ROXBURGHII, the last two being very probably only different cultivations from S. HALEPENSE. It is quite likely, however, that all the forms may, as **Hackel** thinks, have to be regarded as varieties and races of one species, and *that* a plentiful wild plant in India and Africa, the grains of which in both countries are, in times of scarcity, regularly collected as an article of food. The late **Mr. Bentham** in his review of **Targioni-Toz-Zelti's** *Historical Notes on Cultivated Plants*, advanced the opinion that all the cultivated forms of SORGHUM would have to be regarded as derived from S. HALEPENSE. The form which approaches nearest to S. HALEPENSE is the Indian cultivated plant which **Roxburgh** doubtfully assigned to HOLOCUS SACCHARATUS, *Linn.* In part support of there being possibly two species, however, the fact may be mentioned that agriculturally there are in India two crops of this millet, and many forms which the Natives readily recognise as belonging to the one or the other crop. It would thus seem that India has as strong a claim as Africa to being regarded as the home of certain of the cultivated forms of this millet. **DeCandolle** assigns the whole series to Africa.

Conf. with Sir Hooker's opinion pp. 278, 280.

Two crops. *Conf. with pp. 282, 283, 291, 293, 296, 308, 309, 310, 311, 312, 313, 314.*

Sorghum bicolor, *Willd.*; GRAMINEÆ.

2389

2390

Syn.—HOLCUS BICOLOR, *Linn.?*; ANDROPOGON BICOLOR, *Kunth.*, *Non Roxb.*; (according to **Hackel**, in *DC.*, *Monog. Phan. VI.*, *519*); A. CERNUUS, *Roxb.*; A. LAXUS, *Linn.* (according to **Sir W. Elliot**).

Vern.—*Kalá múcha?*, *killo-debdháor*, *ded ún*, BENG.; *Sundia* (*Broach*, according to **Dalzell & Gibson**), BOMB.; *Gaddi janumu* (according to **Elliot**), TEL.

References.—*Roxb.*, *Fl. Ind.*, *Ed. C.B.C.*, *90-91*; *Birdwood*, *Bomb. Pr.*, *112*, *113*; *Sir W. Elliot*, *Fl. Andhr.*, *56*; *also in Trans. Bot. Soc.*, *VII.*, *282-285*; *Bomb. Gaz.*, *II.*, *404*; *Agri.-Hort. Soc. Ind.*:—*Trans.* (*Royle*), *VIII.*, *96*.

Habitat.—According to **Roxburgh Andropogon cernuus** (a variety of **Sorghum**) is "cultivated by the inhabitants of Manipur, Kúnkó, and other mountainous districts immediately east of Bengal." **Hackel**, who places it in his variety **cernuus**, mentions one of **Wallich's** plants as representing this form, the specimen alluded to having been collected at Allahabad. It is thus probable that the form here indicated is in reality cultivated throughout India. The only character mentioned by **Hackel** that would seem to isolate it from the other states of **vulgaris** is the fact that the inflorescence is reflexed. **Roxburgh**, on the other hand, says that the grain is milky white; that the stems branch freely from the root; that the lower joints throw out copiously verticils of roots (if the plant be suffered to remain for more than one year on the soil); that the panicles are large, oval on more slender branches than in **vulgaris**, "but in such as terminate the primary stems, the form is obscure, from the drooping habit of their branches;" that the awn is so small and short as to be hid within the glume; and that the neuter flowers are very minute, and consist of only one or two slender villous calycine glumes.

2391

This seems to be nothing more than a white-grained state of **S. vulgare** with a rather dense and heavy head of fruits which, by its weight more than anything else, frequently tends to droop. If this be so, it is fairly frequent throughout India. The name **bicolor**, if justified by being the oldest, is probably the least appropriate. **Roxburgh's bicolor** is very generally recognised as a dark two-coloured grain, quite distinct from the variety here dealt with. (*Conf.* with the remarks under **S. halepense** and the section on Bombay **S. vulgare**).

Hooker & also Bentham held same opinion, *Conf. with pp. 278, 279.*

2392

Sir Walter Elliot, in a very able paper (*Trans. Bot. Soc., VII., p. 284*), discusses the seven chief forms of **Sorghum**, recognised by **Kunth**, *viz.*, **Andropogon Sorghum,** *Brot.*, **A. niger,** *Kunth.*, **A. cernuus,** *Roxb.*, **A. bicolor,** *Roxb.*, **A. rubens,** *Kunth.*, **A. saccharatus,** *Roxb.*, and **A. caffrorum,** *Kunth.*, as forming a series of cultivated races which, if they might not be viewed as referable to one species, at most but constitute two species represented by **S. vulgare,** *Pers*, and **S. saccharatum,** *Pers.* He thus held very nearly the same view as **Hackel** has recently advanced. In his *Flora Andrica*, however, **Sir Walter** places **A. laxus,** *Roxb.*, as a synonym for **A. cernuus,** *Roxb.*, but modern writers (and apparently correctly) give **A. laxus** as a synonym for **S. halepense**. **Royle**, in a paper on "*Corn and Pasture Grasses of India,*" says, "**A. cernuus** is a distinct species." **Dalzell & Gibson** describe, under the name **Holcus cernuus,** a Bombay cultivated **Sorghum,** which the writer is disposed to transfer to **A. saccharatus,** *Roxb.* (**A. vulgare** *var.* **Roxburghii,** *Hackel*). The above is briefly the chief notices that exist, in Indian works, of the plant under consideration. As already stated it possesses very little claim to being viewed as anything more than a race of **S. vulgare**, but is apparently a very prolific plant, especially if fodder be the object aimed at.

FOOD & FODDER. *Conf. with pp. 287, 294, 304, 305, 306. 309, 311.* **Grain.** **2393**

Food and Fodder.—The GRAIN appears to be the chief form of *juár* grown by the hill tribes on the eastern frontier of India. Cattle are said to be very fond of the fodder, and, from its manifesting a power to survive for more than one year and to branch freely from the ground, **cernuus** is perhaps, better suited than any other form for the purpose of fodder.

2394

Sorghum halepense, *Pers.; Duthie, Fodder Grasses of N. Ind., 40.*

In the United States it is known as JOHNSON GRASS, CUBA GRASS, MEAN'S GRASS.

Syn.—HOLCUS HALEPENSIS, *Linn.;* ANDROPOGON HALEPENSIS, *Sibth.* A. LAXUS, *Roxb.;* A. SORGHUM, *Biot.*, var HALEPENSIS (according to Hackel); SORGHUM GIGANTEUM, *Edgew.*

Vern.—*Barú*, HIND.; *Kálá-múcha*, BENG.; *Galla jári, padda jalla gudi* (CHÁNDA), C. P.; *Bájra, barru, bara* (BANDA), BUNDEL.; *Buru, rikhon, da*, BHABAR.; *Bikhonda*, KUMAON; *Brahám*, KASHMIR; *Barú, barwa, brahám*, PB.; *Barúa*, PUSHTU; *Bowári*, MERWARA; *Kartál*, (BERAR) DECCAN; *Gaddi janú*, TEL

References.—*Roxb., Fl. Ind., Ed. C.B.C., 91; Stewart, Pb. Pl., 262; Ainslie, Mat. Ind., II., 112; Indian Forester, XII., App., 25; Gazetteers:—N. W. P., (Bundelkhand), I., 85; (Agra), IV., lxxx.; (Himálayan Districts), X., 321, 692; Settl. Repts., C. P., (Belaspore), 77; Pánjab (Karnal), 19; (Hoshiarpur), 14.*

Habitat.—A tall, perennial grass, with creeping rhizomes and numerous suckers. Common all over India in cultivated and uncultivated ground. **Roxburgh** says, it (**A. laxus**) grows in hedges, on banks of water-courses, and on land that has lately been cultivated. **Atkinson** alludes to its occurring wild in Bhabar. **Ainslie** says that in Rajmahal (according to **Hamilton**) a kind of bread, which is very palatable, is made of three seeds *junerah*, *butah*, and *búra*. He affirms that the first of these is **Sorghum saccharatum,** the second **Zea Mays,** and the third **Fennisetum typhoideum.** As having a possible bearing on the origin of the chief cultivated millets, it may be pointed out that the name *búra* is more generally restricted to this wild grass, than assigned to the spiked millet (*bájra*). This might be viewed as giving countenance to the idea of **S. halepense** having been displaced from general use among the aboriginal tribes of India on the introduction of better millets from foreign countries.

The wild form of this plant seems to be characterised by having flat leaves with a prominent mid-rib below. The panicle is large and pyramidal, with long naked branches arising more or less in whorls and bearing secondary panicles, the ultimate divisions being composed of four to six pedicelled racemose spikes. Spikelets more or less elongated and pointed: hermaphrodite ones sessile, more ovate oblong than the lateral pedunculate male spikelets, three-valved and awned: male spikelets prominently veined especially on the apex, two-valved; glumes (or valves) clothed with long-spreading brownish hairs, especially at the nodes (or joints) of the panicle. Grain oblong.

FOOD & FODDER. Seed. 2395

Food and Fodder.—**Tod**, in his 'Rajasthan,' mentions that the SEED of this grass is collected, mixed with *bájra* flour, and eaten by the poorer classes in Bikanir. It is considered a good fodder grass both for grazing and for hay, but is held to have injurious effects if eaten when too young or when the plants are stunted by drought. Thus, **Stewart** remarks "it is at times browsed by cattle, but I was told in Hazara that after eating it they sometimes had fatal head affections." It is largely used both as a fodder grass and for hay in the United States and Australia, and is much sought after by cattle. In these countries no mention seems to be made of the injurious properties, which in India it is supposed to possess.

DOMESTIC. 2396

Domestic.—Native pens are made from the stems.

2397

Var. Roxburghii, *Roxb.; Flora Indica, Ed. C.B.C., 91.*

Vern.—*Deo-dhan*, HIND.; *Tilya*, PB.; *Shállú, sundia*, BOMB.; *Tella jonna, dévatá dhányamu*, TEL.

This has already been indicated as the form described by **Roxburgh** under the name of **A. saccharatus.** The characters given for the plant by **Roxburgh** are briefly—Panicles verticelled with ramifications drooping: calyx hairy: corolla awnless: hermaphrodite flowers three-valved, neuter one-valved. Margin of the leaves hispid, the rest smooth with a white nerve on the upper surface. Sheath as long as the panicle when young, afterwards the whole is completely bent down from the weight of the seed. The author, it will be seen from the remarks below, has provisionally placed **A. bicolor,** *Roxb.*, as one of the sugar-yielding forms of this variety. The definition given by **Roxburgh** for that form differs in no essential feature from what has been recorded above as the characters of **A. saccharatus,** *Roxb.* (**S. helepense,** var. **Roxburghu**).

Roxburgh says the present plant is extensively cultivated over various parts of India, during the rainy and cold seasons, upon land which is too high for rice. It is, he adds, the *deo-dhan* (which may, perhaps, be rendered God's-grain), a name which probably points to an ancient cultivation. The writer, while admitting below that **DeCandole** is probably correct in regard-

The Deo-dhan or God's grain.

ing certain qualities of *juár* as introductions to India, is strongly disposed (as already stated) to accept this at least as a purely indigenous plant, derived by the cultivation **S. halepense.** It would seem absurd, in fact, to reject the possibility of that plant (which, even in its wild state, is regularly resorted to as an article of human food) having been cultivated by the Natives of India long anterior to any possible importation of African millets. And the more so since many of the forms of the cultivated plants, here grouped under the name suggested for them by **Hackel**, are clearly very closely allied botanically to the wild (indigenous) species. This idea obtains still further support from the fact that the millets of this section are, by the Indian agriculturists, not only recognised as forming a series distinct from the group (which, for convenience of expression, may be designated the African Sorghums), but they are grown upon a different class of soils and during a different season of the year. They are mostly sown in Autumn and reaped in early Spring, being thus sub-temperate, like wheat and barley (crops grown during the same months), while the African Sorghums are sown in Spring and reaped in early Autumn. Thus, through their continuing on the field during the hot season, the African Sorghums may not incorrectly be characterised as tropical plants. **Dalzell & Gibson** identified a cultivated grain (met with in Bombay) as **Holcus cernuus,** *Roxb.*; but, as already remarked, this was more probably one of the forms of **Roxburghii.** These authors tell us that in Western India it is known as *shállú* (Eastern Deccan and Gujarat) and they add, "The stalk, as a forage for cattle, is unsurpassed. It contains a great deal of saccharine matter and is thus very nutritive. In the Broach Collectorate, the *shállú* is grown in the Dejbarra and other districts, under the name *sundia.*" In the Gazetteers and other modern works *shallú* is spoken of as "the late or *rabi* variety." In the section of this article below devoted to the Bombay *juár* cultivation, the reader will discover that in many districts of Western India the *rabi* crop of *juár* is not only far more important than the *kharif*, but that it is the staple article of food with the people. In this respect, therefore, Bombay differs essentially from a great part of the rest of India. **Elliot**, writing of Southern India, speaks of "**S. vulgare** as the early *jowarí* (*cholam*) and **S. saccharatus** as the late." In the *Flora Andhrica*, he says (*pp. 46, 177*), that the Mahratta name *sálú* is in South India often used to denote "wild rice."

Rabi crop of Juar. *Conf. with pp. 279, 283, 291, 293, 296, 305, 308, 310, 311, 312, 313-314.* **2398**

It is, perhaps, unnecessary to quote other passages in this place, in support of the contention that it is here desired to make, namely, that there are in India two widely different groups of *juárí* which correspond to the types **S. halepense (A. saccharatus,** *Roxb.*), and **S. vulgare,** *Pers.* Unfortunately, however, this distinction has not been observed by popular writers, so that the information on methods of cultivation and seasons of sowing and reaping, etc., etc., given below (under **vulgare**), include the crop that should have been here dealt with, as well as that of **vulgare** proper. This being so it is unnecessary, perhaps, to discuss the special properties of this series.

Varieties. *Conf. with pp. 277-9, 291, 305, 308, 309, 310, 311, 312.* **2399**

Food and Fodder.—To preserve the uniformity of this article it need, therefore, be only said in this place that the GRAIN is of course eaten, but that the plant is, in many parts of India, even more highly valued as a FODDER. Like the species or variety next to be dealt with (in alphabetical sequence) it branches far more freely than does **S. vulgare.** It therefore yields a larger amount of fodder, and the stems, being charged with sugar, they are more greedily eaten by cattle than are those of the coarser-stemmed *juári*. **Hackel** remarks that **S. saccharatum,** *Pers.*, is by no means the form best suited for cultivation as a sugar-producer. Other forms of

FOOD Grain. 2400
FODDER. Plant. 2401

vulgare, he says, are quite as rich if not richer in SUGAR. It is thus possible that were it desired to pursue in India the line of research presently being prosecuted in many parts of America, *viz.*, to produce sugar from Sorghum, some of the indigenous forms of this plant would be more likely to prove successful than the imported stock presently being experimented with. As a matter of fact, it would appear sugar has regularly been made in India from certain forms of Sorghum long before that idea was thought of in America or elsewhere. There are two chief forms of what may be called the indigenous sugar-yielding Sorghum :—a white-seeded (**A. saccharatus**, *Roxb.*), and a red-seeded (**A. bicolor**, *Roxb.*). Several writers, for example, mention the much-famed sugar-candy of Bikanír as being, and as always having been, made from these two forms of **Sorghum.** Thus **Thakur Jainarian Singh**, in a letter which appeared in the *Indian Agriculturist* (*July 13th, 1889*), says:—

Sugar. 2402 *Conf. with Vol. VI., Pt. II., 14, 122, 126, 224, 232. Conf also with pp. 295, 311.*

"As far as I have been able to ascertain, there are several varieties of Sorghum that are sown in the parts of the country called Hariana, on the border of the Bikanir State, and is styled 'Alapur Joar.' It is of two kinds—one yielding white, and the other red, seed. This is largely cultivated, and is the sole saccharine produce of those parts. You are, perhaps, aware that Bikanír produces the famous crystal-white sugar-candy which has no parallel in India. Some years ago I myself cultivated a field of Sorghum. It yields two crops in one season, and is sown in rows in May, in well-manured ground, which requires irrigation till the commencement of the rains. It becomes ripe for pressing in November. After being cut the shoots come up, and the second crop is ready by the beginning of February. As far as I have been able to learn, the cultivation of this Alapur Joar is limited, and serves only for purposes of local consumption, for people do not cultivate it largely owing to the scanty means of irrigation they possess. So far as we are aware this is the first time that this particular Sorghum has been brought to public notice, and it appears to be desirable that the several Agricultural Departments should experiment with the plant."

Bikanir Sorghum Sugar. *Conf. with p. 296.* **2403** **Two crops by Ratooning.** *Conf. with pp. 279, 282, 291, 293, 296, 308, 310, 311.*

The suggestion offered above that the red-seeded sugar-yielding Sorghum of India is **A. bicolor**, *Roxb.*, is based purely on the description given by Roxburgh of that form agreeing more nearly with **A. saccharatus**, *Roxb.*, than with any other plant. The writer is confined, in the preparation of the present article, to reviewing the literature that exists, supplemented with the study of herbarium specimens. One writer (*Agri.-Hort. Soc. Ind., XI., Proc., 1859-60*) affirms of the imported Imphee that "it appears to be the **S. bicolor**, known by the Natives under the name of *Kalo-debdhan*." The more obscure features of this subject will never, however, be cleared up until some officer travels over the length and breadth of India in order to study and compare the living plants. In conclusion, it may be recommended that the Bikanír sugar-yielding Sorghums seem more worthy of experimental cultivation at the Government farms than do the foreign races presently being tried. In Bombay, occasional mention is made of the *shállú* form of *juár* as being used in sugar-making. The whole of the *shállú* Sorghums (except perhaps the Imphee and Sorgho, which have come to be regarded as *shállú* canes) seem to belong to this series, and any one of these, by careful cultivation selection, etc., might be developed into a rich sugar-yielding plant (*Conf. with the remarks in the Bombay section of* **Sorghum vulgare** *below, pp. 312-5*).

Plants most deserving of experimental cultivation, 2404

Sorghum saccharatum, *Pers.*

2405

BROOM CORN OF CHINESE SUGARCANE; IMPHEE and SORGHO; the PLANTER'S FRIEND, ETC.

Syn.—HOLCUS SACCHARATUS, *Linn.*; ANDROPOGON SORGHUM, *Brot.*, *var.* SACCHARATUS, in (*Hackel Monogr.*); A. SACCHARATUS, *Kunth, non Roxb.*; S. KAFFRARIUM, *Kunth.*

NOTE.—The reader will observe that although to the Agriculturist the Imphee and Sorgho are different things, they must be viewed as races of one plant. They do not

SORGHUM saccharatum.

History of the Foreign Shallu

differ from each other to so great an extent as is observable in the Indian forms of this plant. Indeed, it is highly probable that **A. saccharatus,** *Roxb.*, and **Holcus saccharatus,** *Linn.*, may have to be treated as one and the same, the former being the Indian, and the latter the African, cultivated races of the wild species **S. halepense.** The Natives of India do not regard the imported plants as anything more than forms of their *rabi* crop of *juár*. Accordingly, the vernacular names mentioned below are in reality those of **Sorghum halepense,** *var.* **Roxburghii,** of the classification here adopted. Indeed, the writer has followed the usual course of separating Imphee and Sorgho, from the corresponding Indian forms, more with the view of isolating the exotic stock than from any belief that they are botanically distinct.

Vern.—*Deo-dhan,* HIND.; *Shálú,* DECCAN; *Deo-dhán,* BOMB.; *Salú,* MAR.; *Tella jonna, dévatádhányamu,* TEL.; *Pyoung,* BURM.; *Kao-liang,* CHINESE.

References.—*DC., Orig. Cult. Pl., 382; Mason, Burma & Its People, 476, 817; U. C. Dutt, Mat. Med. Hindus, 296; Sorgho and Imphee, H. S. Olcott, 1857; Birdwood, Bomb. Prod., 109; Baden-Powell, Pb. Pr., 236, 237; Manual and Guide, Saidapet Farm, Madras, 41; Short, Man. Ind. Agri., 312; Settlement Report: Central Provinces, Nagpúr, 273; Experimental Farm Reports, Madras (1882-83), 37, 38; (1883-84), 70; (1884 85), 7, 19, 28, 29, 48; (1885-86), 32; Bombay, (1884-85), 17; (1885-86), 22, 40; (1886-87), 10, 16; N.-W. P. (1882), 50; (1883), 22; (1884), 11, 13; (1885), 25; (1886), 16, 20; Bengal (1886), 28, App. ii, xxvii., xcv.; (1887), 12; Assam (1883-84), 15; (1884-85), 18; (1885-86), 4, 19; Nagpúr, (1883-84), 5; (1885-86), 4; Cawnpúr (1882), 7, 10; (1883), App., iii.; Khandesh (1885-86), 10; Agri.-Horti. Soc. Ind. Jour., XI., 202, 204, 297, 306; Pro. (1859), 35, 56, 57, 68; (1860), 42, 44, 110; XII., Pro. 3, 45, 46; New Series, III., Sel. 27 32, Pro., 45, 46; V., Pro. (1876), 22; VI., 129 131; VII., 360; Indian Agriculturist, 12th April, 1890; Indian Forester, I., 314; III., 51; Smith, Ec. Dict., 63, 65; Note on sorgho (Dept. Rev. and Agri.), F. G. Wigley; also an extensive official correspondence from 1871 to 1891.*

Habitat.—Indigenous probably to the east coast of Africa, but cultivated nearly over the whole world, having a distribution very similar to wheat. Certain forms are grown for fodder, others for grain; but the chief interest in the plant may be said to centre in the American experiments, to develope from this stock a semi-temperate sugar-yielding cane. Both Sorgho and Imphee are being experimentally grown in India with indifferent results.

RACES. 2406

Races of Sorgho and Imphee.—For the purpose of this work it is almost sufficient to say that under the two forms indicated above an extensive series of races have been developed, each having properties claimed to be more valuable than those possessed by all the others. In few subjects of agricultural enterprise, in fact, has so much attention been given with less results than to the present. At one time it was thought that the sugar-cane planters of the West Indies would find themselves beaten by American skill and ingenuity. But already indications are not wanting that the hopes of success with **Sorghum** as a sugar plant were sanguine in the extreme. **Mr. Wray,** in a paper on Imphee, describes sixteen different forms, but in the various reports that have appeared, especially in connection with the Department of Agriculture in America, from about the year 1879, descriptions and figures of a far larger number have been published. For information regarding these the reader should consult the library of technical works that exist.

HISTORY. 2407

History.—The cultivation of Sorgho in India dates back only to the year 1858, when seed was obtained by the Madras Government from America. America had obtained its supply from France, which, in its turn, had got seed from China in the year 1851, hence the name of Chinese Sugar-cane. The Madras Government also procured, about the same time, a supply of **Imphee** seed from the Cape of Good Hope. The seed thus secured

HISTORY.

was distributed all over India for experiment, so as to determine the value of Sorgho. In the *Journal of the Agri. and Hort. Soc.*(*XI., pp. 997, 1861*) will be found the results of the efforts put forth and the various opinions held regarding this crop. Experiments were tried at Moorshedabad, Backergunge, Champarun, Midnapur, Rajshaye, Beerbhoom, Rungpore, Baraset, Dacca, Sarun, Nuddea, Behar, Bancoorah, Mymensingh, Tirhoot, Patna, Shahabad, Purneah, Jessore, Bograh, Maldah, Burdwan, etc., etc., which, from various causes, were attended by only partial success or complete failure. For some time after this the cereal fell into complete disuse, with the exception of experiments made by Government, the general opinion being that, as far as the production of sugar was concerned, it was not equal to the indigenous sugar-canes, although a use might be found for it as a forage plant. The next that is heard of Chinese Sorgho was in 1870, when the results of experiments with it, at the Sydapet Experimental Farms, Madras, were made known. The following year **Colonel Boddam**, who had been sent on a special mission to Mysore, drew public attention to Sorgho as a forage crop, enumerating amongst its advantages "that the crop is not cut all at once, but as the plant sends out successively a series of six to eight canes, these should be cut successively when the flower expands, so that you can obtain a quantity of green food spread over several weeks; and it is, when cut in this manner, that the enormous weight of fodder is realised; whereas with *jawari*, the stalk is cut down once and there is an end of it." Since then very favourable accounts have from time to time been given by individual cultivators, and various attempts have been made to introduce Sorgho as a staple product throughout India. Still it may be said that, with the exception of experiments conducted at the various Government farms, little advance in the extended cultivation of Sorgho for the last thirty years at least, has taken place. As will be shown under CULTIVATION it occupies a mid position between that of a saccharin producer and a fodder grass, not coming up in either condition to equal other plants.

OIL. Leaves. 2408

Oil.—In several of the American reports mention is made of a vegetable wax which may be collected from the LEAVES of this plant. It does not, however, appear that that substance is of any practical value.

DYE. Seed. 2409

Dye.—The SEED contains a colouring matter, which has been used as a dye. **Dr. Henri Erni**, Chemist (Dept. of Agri., U. S) in 1864, isolated the dye and gave particulars of its tinctorial value. He mentions that the seeds, boiled in vinegar with a little sulphuric acid, yield a deep orange colour. He employed dilute chloride of tin as the mordant applied after the colour. Cotton and silk took a red colour and wool a beautiful purple. The shades of colour might, he adds, be modified to any extent by the use of other mordants. Cotton showed the least affinity and wool the strongest for **Sorghum** dye. From the STALKS can also be extracted the same dye in the following manner:—"The pressed canes are left to ferment in heaps until the colour changes to a red or reddish brown. They are then cut up, dried, and washed. The colour is extracted by a weak dye of caustic potash. By neutralising the alkaline solution by a weak solution of oil of vitriol, the colour falls in the form of red flakes, which are easily soluble in alcohol, alkalies, and diluted acids (*Report of the Commissioner of Agri. for the year 1862, p. 535*).

Stalks. 2410

FIBRE. 2411

Fibre.—In Illinois, United States, 1862, two mills had been established for the sole purpose of manufacturing paper from these canes. Crushed canes have also been used for fuel and manure (*F. G. Wigley*).

FOOD & FODDER. 2412

Food and Fodder.—Of all the uses to which this plant may be put, that of growing it as a forage crop will probably be found the most important. Almost all parts of the plant are consumed by cattle. Either in

SORGHUM saccharatum. **Cultivation of the Chinese Sugar-cane**

FOOD & FODDER. Leaves. 2413 Stalks. 2414 Grain. 2415 Seed. 2416

the dry or green state the LEAVES and STALKS are readily eaten; while the GRAIN can be given to poultry, pigs, horses, etc. Mr. F. G. Wigely (*Note on Sorgho*) says that the SEED furnishes a "very good flour, which makes excellent bread. This flour, however, has a violet tint running through it, which comes from the hull of the seed, and this is the principal reason why Sorgho bread, although said to be very palatable, is not popular. Bread and cakes, made from this flour in the United States, have been pronounced to be second only to those from wheat flour. The seeds of Sorgho contain a starch of good quality. This starch, however, unless its extraction from the seed is very carefully performed, is always more or less tinged with the colouring matter which abounds in the seed." The flesh and even the bones of animals fed on this grain are said to be stained by the colouring matter which occurs everywhere—in the grain, stems, leaves, etc. It does not seem likely, therefore, that a plant with such an objection to it is likely to readily displace the numerous allied India forms, which, while being equally nutritious, do not, as a rule, possess this objectionable property. Some idea of the extent of the cultivation, however, of Sorgho and Imphee in America may be learned from the fact that in 1885 600,000lb of sugar were mauufactured from the sugar-yielding forms of this plant. The difficulty in crystallizing the sugar was felt to be very serious for some time, but this was duly overcome. But nearly every report that has as yet appeared, alludes to the acid nature of the saccharine juice and to the low percentage of crystallizable sugar that it contains. In fact it may be said that the almost universal opinion seems to be that it possesses no advantage over sugar-cane.

Conf. with para. p. 292, regarding Dye.

CULTIVATION 2417

CULTIVATION.

Sorgho and Imphee have been introduced into India, and are now to some extent cultivated in every province. The information available regarding the success or otherwise of this new crop chiefly exists in the Annual Reports of the Government Experimental Farms and in official correspondence. The leading facts which had been brought to light up to 1877 were thrown into the form of a *Note on Sorgho* drawn up by Mr. F. G. Wigley, and published by the Revenue and Agricultural Department of the Government of India. It does not seem necessary to re-publish the facts there set forth, even although the material which has since appeared is of a more fragmentary or spasmodic character. Sorgho and Imphee cultivation have attained importance alone in America—if China and Africa be for the moment left out of consideration. In India interest in these crops cannot be said to have passed beyond the experimental stage. In America (especially in the United States) the most advanced chemical and agricultural methods and appliances have been brought to bear on the endeavour to mature and perfect the industry to take the place, if possible, of sugar-cane. The conclusions arrived at may be said briefly to be that, if grown primarily for the juice, the yield is inferior in quality to that of cane. The syrup is often distinctly acid, while the difficulty to prepare crystalline sugar from it is greater than with the time-honoured cane.

As a fodder plant it may be said that it is never likely to successfully contest the field with the fodder varieties of *juar*. The liability to disease and pests is very much greater than with the indigenous stock of India, and, when grown for both grain and fodder, no form of Sorgho or Imphee can be compared with the crops already in India. It is commonly held that Sorgho or Imphee are stronger and more prolific than any of the Indian plants, and may be ratooned freely, that is, made to yield several flushes of

CULTIVATION

Conf. with pp. 280, 294, 304, 305, 306, 309, 311.

fodder off the same root. Through cultivation of *juar*, primarily as a grain crop (fodder being but a by-product so to speak), the Indian forms have mostly lost the power of branching freely. But many of the Indian races are quite as prolific as Sorgho, and they are even regularly ratooned. All that seems necessary to produce a form that would meet the wishes of those who have called out the loudest for Sorgho, would be a more careful study of the Indian plants, with selection and adaptation to the purpose desired. The further argument that Sorgho comes into season when *juár* is not available, can be urged only in such tracts of India as possess at present but one crop of *juar*. From the remarks, which will be found below, some provinces, such as Bombay, have regularly two crops of *juár* or even a third, so that in India during no season need the cultivators be without fresh *juár* fodder.

Two crops. 2418

Conf. with pp. 282, 283, 291, 293, 296, 308, 399, 310, 311, 313-314.

The following selection from the papers consulted by the writer is intended to exhibit the chief ideas brought out, but is not a comprehensive review of all the reports that have recently appeared on Sorgho and Imphee :—

N.-W. Provinces & Oudh. 2419

North-West Provinces and Oudh.—The reports of the Department of Agriculture, which have appeared within the past decade, have contained numerous passages on the experiments performed at the Cawnpore Experimental Farm and elsewhere to acclimatize the better qualities of Imphee and Sorgho. At first these were framed in the language of high expectancy, but gradually less and less interest was manifested. The following passage from the report of 1882 gives briefly the chief arguments that have been advanced in favour of this crop :—

"Perhaps the most important of the experiments with products were those with different kinds of Sorgo. Even in America the exact value of this plant as a sugar-producer has not been finally determined, but its great agricultural importance is widely recognized. It has three great advantages over the cane—in yielding a grain fit for human consumption; excellent fodder for cattle; and in taking up the land for four months only instead of a whole year. In addition to this, it requires hardly any manure and no irrigation. Of three varieties tried, the amber and the red, while they yielded a rather less proportion of juice and gur to the whole plant than sugarcane, were not inferior to it in the proportion to the cleaned canes. The gur, though not yielding so large a proportion of crystal as ordinary sugar and possessing a peculiar acidity, was well-flavoured and commanded a higher price in the market. For eating and various manufacturing purposes it appears to have a good future."

It is, perhaps, unnecessary to furnish other passages. In none of the more recent reports has anything been announced that would warrant the opinion that the subject has been advanced beyond the experimental stage.

Central Provinces. 2420

Central Provinces.—The reports from the Experimental Farm, Nagpore, for the years 1883-84 and 1885-86, are as follows :—

"The cultivation of Sorgho for sugar, I am sorry to say, resulted in failure. Some excellent American Sorgho seed was received from the Government of India, which gave a really splendid crop of stalks, but the proper time for crushing it was allowed to pass, and when the preparations for sugar-making were ready, only the smaller after-shoots were sufficiently juicy to be worth pressing, and a very large proportion of these were diseased. In consequence no syrup could be manufactured which could compare in any way with that I saw later at the Cawnpore Farm, although boiling was carried on with the Cook's evaporator, which proved so successful at Cawnpore, and under the direction of a trained Cawnpore man whose services were lent me. A fresh trial will be made during the current year, for, if Sorgho can be readily made to yield an eatable sugar, it promises to be of real importance in the agriculture of these provinces, where irrigation is, as a rule, so difficult to provide, and the sugarcane is so sparsely grown in consequence (*1883-84, p. 5*)." "Seed was obtained from the Cawnpore Farm and sown on a small area. The plants came up well. After a growth of $3\frac{3}{4}$ months when the seed had matured, the crop was cut and stalks being stripped and topped were crushed for sugar in an iron roller sugar-mill. The juice was boiled in the American (Cook's)

CULTIVATION

evaporator, and syrup of a very fair quality was obtained. But it was found impossible to make it crystallize."

Bombay and Sind. 2421

Bombay and Sind.—One of the earliest notices of Sorgho and Imphee (in this Presidency) is that given in the Satara Gazetteer (*Vol. XIX., 168*), where mention is made of the attempt in 1860 to introduce Sorgho and Imphee. Interest in the subject appears, however, to have died out until aroused once more by the experiments conducted at the experimental farms and gardens under the supervision of the Director of Land Records and Agriculture. The results attained as yet do not, however, appear to have been greater than in other parts of India. It certainly cannot be said that the Natives of Bombay have found it to their interests to cultivate either of these crops in place of those which, for many centuries, have been grown and perfected to local requirements and conditions.

Madras. 2422

Madras.—The following are some of the opinions formed regarding Imphee and Sorgho resulting from the cultivation of these cereals on the Experimental Farm, Sydapet:—

"'Planter's Friend' was first tried in 1875, and the results were satisfactory; it has been found to yield from 4,000 to 12,000lb of straw, besides a crop of grain per acre; it is, however, especially a fodder-producer, and to obtain the best results should be cut when in flower. The straw of this variety is very rich in saccharine matter. The results which have been obtained show abundantly that both "Sorghum" and "Planter's Friend" are most valuable fodder-producers; the former gives a heavier yield of grain than the latter, but it may be remarked here that the grain of both varieties is at present unsaleable at Saidapet for use as food for man; that of the Sorghum, because of the difficulty in removing the husk; that of the Planter's Friend, because of a bitter taste which it possesses. For the production of sugar neither crop has been investigated on this farm, but there is every reason to believe that it would be a most valuable addition to the crops of the country in this respect; occupying the ground about one hundred days, it will, without irrigation, produce in that time about 8 or 10 tons of green Sorghum fit for crushing. From an experiment made, with very imperfect means of extraction, it was found that the crude juice showed specific gravity of 1·90 or 12° density on the Beaume scale, which is higher than the average commonly obtained in America. A supply of the seed of some of the best sugar-yielding varieties has been indented for from the United States, but unless some adequate means of crushing the cane and purifying and evaporating the juice be also supplied, it will not be possible to do much in this direction."

More recent experiments have tended to modify the good opinion formed of Sorghum as a sugar-producer. In the report of 1885 the following account is given: "**Sorghum saccharatum** and *amber cane* both grow very well in this Presidency. The difficulty has been to extract a marketable sugar from them. The purchase of a set of the Blymyer machinery has been sanctioned by Government, and when this is received, the question of the value of these products, as sugar-producers, will be finally settled."

"It may here be mentioned that spirit manufactured from **Sorghum saccharatum** was found, on an analysis by the Chemical Examiner, to be of good quality. It was very clear and tasted much like rum, but, after being opened for a short time, it deposited a gelatinous-looking substance. What this was and what caused it has not yet been ascertained. **Messrs. Minchin Brothers,** who manufactured it, propose to make further experiments and have promised to communicate the results to me."

Assam. 2423

Assam.—Report of the Agricultural Department, Assam, 1884, p. 18:—

"In February 1884, 2 maunds of Sorgho were received from the North-Western Provinces and distributed to the Deputy Commissioners, Lakhimpur, Sibságar, Nowgong, and Darrang, half-a-maund to each. In Lakhimpur the seed was sown in April, and three months after the crop was gathered. No attempt was made to extract sugar, as the rain did serious injury. **Mr. Grant,** of Cumatoliah, was partially successful. He used the stalks as fodder, and found the grain was appreciated by his coolies. In

in Assam. (G. Watt.)

SORGHUM vulgare.

CULTIVATION in Assam.

Sibságar, for some unaccountable reason, scarcely any of the seed germinated, and the experiment was a dead failure. In Darrang **Dr. Campbell** found that the yield of grain amounted to a little over 6 maunds per acre. Owing to a misunderstanding, no sugar was made from the stalks. In Nowgong some of the grain was distributed to mauzadars, and some was tried in April in the jail garden, where **Babu Nandeswar Phukan** made the most successful experiment of the year with this crop. A small patch (1,696 square feet) only could be sown, but this yielded grain at the rate of 15 maunds 16 seers (1,262lb), unstripped canes at the rate of 346 maunds (254 cwt.) stripped canes at the rate of 205 maunds (151 cwt.), and *gúr* (the usual semi-liquid kind used in the Province) at the rate of 738lb per acre. The average length of the canes was 5 feet 7 inches. The actual cost of the experiment was R3-10, and the value of the grain and *gúr* produced R4; but, of course, on a larger scale the proportional cost would be materially reduced, while the proportional value of the outturn would remain the same.

"Sorgho possesses some advantages which may render it a popular crop. (1) It is grown with ease on poor soil, careful ploughing and manuring not being necessary. (2) It occupies the ground for only 3 to 3½ months. (3) It yields an edible grain as well as a supply of sugar. (4) The latter comes into the market much before ordinary cane sugar. Another advantage which is sometimes claimed for it, *viz.*, that cattle will eat the megass, was not found to be the case at Nowgong. This may be, however, because the bullocks were not used to it. All cattle will not take to cotton-seed or oil-cake when these are first offered, though most will eat these commodities greedily when once the taste for them has been acquired. A further supply of seed was distributed early in 1885, and the result will be embodied in next year's report.

"In continuation of the arrangements made in the previous year, a supply of Sorgho seed was sent to the Deputy Commissioners of Kámrup, Darrang, Sibságar, and Lakhimpur. The Deputy Commissioner of Nowgong had saved a quantity from his former crop, and did not, therefore, require a fresh supply.

"In Kámrúp the experiments failed. The sub-divisional officer of Barpeta reports that the seed refused to germinate. At Palasbari and Choigaon the seed sprouted but, for some unaccountable reason, the young plants never came to maturity. At the Gauháti Jail the crop grew well, but no statistics were collected, and no attempt at the manufacture of sugar was made.

"In Darrang, a portion of the seed sent was put down, but it is reported not to have germinated.

"In Lakhimpur half the seed was tried in the jail, and by mauzadars; none of it appears to have come up. The other half was distributed by the Assistant Political Officer at Sadiya to the Miris, who put it down as directed, but, naturally enough, kept no details of the result. They obtained a crop, however, but did not appear to care for the produce.

"In Sibságar the seed was distributed to tea-planters and others, but in only one instance did it germinate and the plants come to maturity. As a rule either the plants did not appear above the ground, or, having done so, rotted off. In the one successful instance, although clear instructions were sent with the seed, no attempt was made to extract sugar.

"At Nowgong, the seed preserved from the previous year was distributed to mauzadars, and also tried within the precincts of the jail. The mauzadars' results were unsatisfactory and somewhat unreliable, but in the jail 3 kottas 2 lessas (8,928 square feet = ·205 of an acre) yielded 170lb of grain, 4,704lb of stripped stalks, and 219lb of *gúr*. This yield of *gúr* was 4·65 per 100lb of stalk. Last year the yield was 4·37. The outturn of *gúr* per acre was 1,024lb during the year under report. In 1884 it was only 738. It is worth comparing these figures with those for cane. Considering the great difference in cost of cultivation, Sorgho ought to be a popular crop; but prejudice is very difficult to overcome, and the ryot regards every new thing with a certain amount of suspicion, not perhaps altogether unreasonable."

2424

Sorghum vulgare, *Pers.; Duthie, Fodder Grasses of N. Ind., 41.*

THE INDIAN OR GREAT MILLET; GUINEA CORN.

Syn.—HOLCUS SORGHUM, *Linn.;* ANDROPOGON SORGHUM, *Brot.;* A. SORGHUM, *Roxb.;* A. BICOLOR, *Roxb.*

SORGHUM vulgare.

Varieties and Races

Vern.—*Juár* (?=a blade of barley), *joar, jowári, jondla, janera, jundri, juvari jondhla,* HIND.; *Juár, kurbi, chari* (stalks), *kasa-jonar,* BENG.; *Chavela,* MAL (S.P.); *Jowár, phag, thuthera,* C. P.; *Juár, junri, choti juár, bajra-jhopanwa,* N. W. P. & OUDH; *Júnali* (BHÁBAR), *jowár,* KUMAON; *Joár, junri, choti juár, choti junri, bájra-jkupanwa, chari, ka,* PB.; *Jowár, jáwars, jaori-turkimani, jaor, kios-a-gi* (the last name means "*bent-grass*" according to Aitchison), AFGH.; *Jowári, jondla,* DECCAN; *Jowári, joar, jondla, kangra, jaundri, karbi* (=stalks), BOMB.; *Juári, jondhala, shálu, kadval,* MAR.; *Jowár, sundia,* GUZ.; *Cholam,* TAM.; *Talla, jonna, bonda-janu* (the plant), *tella-janular* (the grain), *konda* (white), *tella, janu* (plant), *jonnalu,* TEL.; *Jolah, shálu, kenjol, yengara, nirgol,* KAN.; *Pyoung,* BURM.; *Zúrna* (= Sanskritized form of the Arabic name *dhura*), *yavanála, rakta khurnah,* SANS.; *Dhúra,* (*zúra*), *taam, jawars* (=smaller millet), *dhurat,* ARAB.; *Kaydi, durra,* EGYPT; *Kao-liang* (=tall millet, according to Bretschneider), CHINESE.

References.—*Roxb., Fl. Ind., Ed. C.B.C., 90, 557, 701; Stewart, Pb. Pl., 262; also Foods of the Bijnor District, 467; Aitchison, Rept. Pl. Coll. Afgh. Del. Com., 123; DC., Orig. Cult. Pl., 380; Mason, Burma and Its People, 476; Sir W. Elliot, Fl. Andh., 59, 75, 95, 139, 163; also Trans. Bot. Soc., VII., 282-287; Medical Topog., Oude, 80; Stewart, Sorghum and its Products; Collier, Sorghum; Birdwood, Bomb. Prod., 113, 128-130; Baden-Powell, Pb. Pr., 383; Atkinson, Him. Dist. (Vol. X., N.-W. P. Gaz.), 692; Duthie & Fuller, Field and Garden Crops, I., 25; Useful Pl. Bomb. (Vol. XXV., Bomb. Gaz.), 186, 208, 276; Royle, Prod. Res., 214; Manual and Guide, Saidapet Farm, Madras, 40; Church, Food-Grains, Ind., 80; Buchanan-Hamilton, Kingdom of Nepal, 227; Kumaon, Official Report, 279; Man. Madras Adm., Vol. I., 288; Nicholson, Man. Coimbatore, 216-218; Morris, Account Godavery, 68; Moore, Man., Trichinopoly, 71; Bombay, Man. of Rev. Accts., 101, 102; Settlement Report:—Panjáb, Lahore, 9; Jhang, 85, 92; Dera Ghazi Khan, 9; Dera Ismail Khan, 344; Kangra, 25; Montgomery, 107; N.-W. P., Azamgarh, 116; Allahabad, 24, 25; Bareilly, 82; Central Provinces, Baitool, 63; Chanda, 80, 97, 98; Chhindwara, 22; Hoshungabad, 282, 285, 288; Nagpúr, 272; Upper Godavery, 35; Nimar, 192; Wurdha, 63, 67; Madras, West Delta Taluqs, 137; South Arcot Dist., 108; Gazetteers:—Bombay, II., 63, 65, 269, 271, 273, 277, 280, 284, 287, 291, 295, 390, 536, 538, 541, 544, 547; IV., 54; VIII., 182; XII., 149; XIII., 289; XVI., 91, 99, 323; XVIII., 262; Panjáb, Karnal, 172; Hoshiarpur, 94; N.-W. P., I., 57, 85, 198; III., 225; IV., lxxx; X., 321, 692; Mysore and Coorg, I., 451; II., 11; Crop Experiments, Bombay Presy. (1883-84), 30, 33, 37, 64-66; (1884-85), 15; (1886-87), 8; Agri.-Horti. Soc. Ind.:—Trans., III., 66, 147, 189; VI., Pro. 189; VIII., 22, 93, 96; Jour., IV., 52-55, Sel., 151, 152; VII., 8; X., 357, Pro. 90; XII.. Pro., 11; XIII., 387, Sel., 50, 51; New Series, II., Sel., 57, 58, 65-70; III., Sel., 29-31; VI., 1-5; VII., 236, Pro., 7, 8; Ind. Agri. Gazette, (1887), 652; Indian Forester, I., 314; XII., App., xxv.; Experimental Farm Reports—Bengal (1885-86), 28, App. ii, xxvii, xcv; (1886-87), 12; Bombay (1883-84), 8; (1884-85), 17, 20, 21; (1885-86), 21; (1886-87), 10, 24, App. vi; Madras (1871), 11, 18, 21; (1873), 32; (1875), 28; (1877), 96; (1879), 40, 104, 110; (1882), 43; (1884), 26, 48; N.-W. P., (1877), 25, 71; (1878), 2; (1881), 15, 21; (1882), 31, 34, 47; Assam (1886), 23; Khandesh (1885-86), 5, 11, 34; (1887), 3; C. P., (1885), 6, 9, 12; Statements, D. E. F.; Nagpúr (1885-86), 3; Berars (1872), 63; Hyderabad (in Sind) (1885-86), 34; (1886-87), 3, 13; (1887-88), 9; Smith, Ec. Dict., 272, 273.*

Habitat.—A tall, handsome grass, extensively cultivated throughout India, and chiefly for its grain—the Great Millet. Whether originally a native of India seems very doubtful. No Indian writer records having found it either wild or naturalized. The resemblance in general characteristics between the fruiting-spike of this millet and that of Indian-corn has led to these grasses being spoken of by comparative names. To this fact is due the existence of vernacular names for Indian-corn—a grass of undoubted modern introduction into India. But, as indicated above, there would appear abundant evidence that a species or form of **Sorghum** was cultivated in India long anterior to the introduction of the presently grown forms of *juár* (see the remarks under **S. halepense**).

Varieties and Races.—So very imperfect is the available material on this subject that the writer feels he has peshaps ventured as far as he is justified, without having very specially investigated the living plants. **Hackel's** views have already been briefly exhibited and certain departures from them, which the writer has felt called upon to advance, have been exemplified. These may be briefly recapitulated. There are two chief forms of *juár*, the one a *kharif* and the other a *rabi* crop. The former—the *kharif*—includes an extensive series of forms of the present species, one of which has already been separately dealt with, *viz*, **S. bicolor**, *Willd.* (**A. cernuum**, *Roxb.*). The latter—the *rabi*,—the forms of **S. halepense**, represented by the types **Roxburghii, saccharum, caffrorum,** etc. Speaking generally, the forms of **S. vulgare** may be characterised by the form and colour of the grain and the shape of the spike. The grain may be pure white, straw-coloured, orange, brick-red, black or parti-coloured. The white grained forms are considered the best, and some of these seem to belong to the **S. Roxburghii** series. **Duthie & Fuller** (*Field and Garden Crops*) speak of three chief forms as met with in the North-West Provinces. These are (1) the double-seeded, which has two grains within the husk; (2) the dwarf, grown in the Allahabad district, in which the stalks do not grow to a greater height than 3 to 4 feet instead of 7 or 8; and (3) the variety, known as *cháhcha*, met with in the Cawnpore district in which the grain is completely covered by the husk, and which is said to suffer less from the depredations of birds than the ordinary kinds.

VARIETIES. *Conf. with pp. 277-279, 282, 283 291, 293, 296, 305, 308, 309, 310, 311, 312.* **2425**

2426

2427

2428

It will be seen from the review of **Hackel's** Monograph given above (*pp. 277-279*) that that botanist recognises three chief sub-varieties of **var. III. vulgaris**. If these could even be defined, they would probably only be recognisable when a limited number of dried botanical specimens were being examined. They possess, moreover, little or nothing of a practical or agricultural character, and may therefore, so far as India is concerned, be rejected. The three forms mentioned by **Messrs. Duthie & Fuller** appear to be but three out of many that might be given from the wider area of the entire Empire. Most agricultural crops or races of grain-yielding plants are scarcely recognisable in the herbarium. They do not admit of definitions by which they could be recognised by any person but the cultivator who has acquired with *juár*, as with rice and many others, the intuitive power of selecting the peculiar form suitable to his individual fields and capabilities.

Roxburgh attempted to group the cultivated **Sorghums** under four species, two of which (**A. Sorghum** and **A cernuus**) the writer regards as forms of **S. vulgare**, and the other two (**A. bicolor** and **A. saccharatus**) as forms of **S. halepense.**

History.—From the remarks already made, the reader may have inferred that the contention has been advanced that since at best one group of the forms of *juár* appears botanically to have been derived from the abundant wild species **Sorghum halepense**, it would be but natural to suppose that some of the cultivated **Sorghums** had been developed in India. And, in support of the opinion that certain forms are purely indigenous, while others are imported, there are many considerations of much value. **Sir Walter Elliot** pointed out that the most general Sanskrit name for the crop, *Yavana*, denotes in other connections a Greek, Muhammadan or stranger, while its Persian name *juár-i-hindi* shows that it reached Persia, at least, from India. **DeCandolle** lays a certain amount of stress on "the absence of a Sanskrit name as rendering the Indian origin doubtful." He

HISTORY. **2429**

Yavana. **2430** *Conf. with Zea mays.*

SORGHUM vulgare.

History of the Juar Millet.

HISTORY.

also cites the fact that **Roxburgh** does not admit his having seen it wild in India. But if the reduction of **Roxburgh's** names (here given) be accepted, then his **Andropogon laxus** (**Sorghum halepense** as it is now called) would be the wild state of at least two of his cultivated forms. Other authors besides Elliot give the above Sanskrit name to the crop, but even if that be rejected, there would seem certain features of the India side of the subject of sufficient weight to corroborate the botanical. An ancient cultivation may be inferred from the very extensive series of forms, recognised by distinct names and adapted to the climates and soils of the particular tracts of country in which they occur. There is little similarity also in the generic names given to the crop in the chief languages of India. Such names as *Juár, Cholum, Talla, Jonna, Shálú, Phag, Thathera, Chavela, Kenjol, Nirgol,* etc., show little indications of derivation from a common external source or of their being modern. It has been contended that the Hindi word *juár* is derived from two roots that mean ' a blade of barley." This seems to the writer a highly improbable derivation for one of the most general of all Indian names for this plant. But even, if correct, the millet is thereby compared with perhaps the oldest cultivated cereal. The vernacular names for this millet are mostly meaningless words, which denote the peculiar grain in question and nothing else. The traditions of many of the aboriginal tribes of India point to **Sorghum halepense,** the *Barú, Kála-múcha, Gaddi-janú, Galla-jári, Rikhonda, Braham, Kartal,* etc., as having afforded food to man before a cultivated **Sorghum** came to be recognised. The Indian form of all cultivated **Sorghums** that bears the closest resemblance botanically to the wild species (or *galla-jári*) is very generally designated in India *Deo-dhan* or *Shállú.* Wild rice in South India is also called *Sálú,* and in the Panjáb that name is applied to **Setaria italica**, but, if *juár* had come to India from Africa, it may with perfect fairness be pointed out that it is significant no form of the Arabic name *Dhúra* exists for it in any of the languages or dialects of India. Some writers, it is true, have given *Zúrna* or *Zúra* as the Sanskrit for this grain, but, if that be the case, neither *Dhúra* nor *Zúra* has given origin to any of the Indian names *Zúra* or *Zúrna* is, moreover, clearly derived from the Arabic *Dhúra.* The Arabic word has, on the other hand, passed into the Egyptian and perhaps also the Hebrew, so that it seems almost justifiable to say that the aboriginal people of India knew of, and perhaps cultivated, their indigenous **Sorghum** long anterior to the Aryan invasion. Indeed, it may be assumed that the Sanskrit people first learned of this grain in India, but gave themselves very little concern regarding it. But, indeed, the absence of any allusion to it in the classic literature of the Sanskrit people, can hardly be advanced as proof positive that it was unknown to them. The religious associations of the grain, the observances of cultivation, and the multiplicity of forms of the crop, all point to an antiquity quite as great as can be shown for most other articles of the humbler phases of life. The absence of any historic indications of an ancient importation and the presence in India of an abundant wild species that affords a large, conspicuous and edible grain seem, when taken in conjunction with the arguments already advanced, conclusive evidence in support of the opinion that many of the forms of this millet are beyond doubt natives of India.

2431

DYE & TAN. 2432 *Conf. with p. 286.*

Dye and Tan.—In the *Mysore & Coorg Gazetteer*, Vol. I., p. 451, Mr. Lewis Rice states that at *Harihar*, a set of people called *Muchikar* prepare a very pretty kind of red morocco from goat-skin, employing the meal of this plant in the process. The leather prepared from the skin "is laid in the shade, and when dry is rolled up and kept in a house for two or three days, in a place secure from smoke and from insects; it is then soaked for eight hours in pure water, and scraped with a piece

DYE & TAN.

of earthen-ware till it becomes quite white. Before the leather is dyed, it is soaked for one night in a *pakha* seer of water which has been mixed with a handful of *cholam* meal and warmed on the fire; in the morning it is taken out and dried with a piece of cloth: when well dried, it is soaked again for half an hour in water with which one seer of tamarinds has been mixed; it is then spread on a mat and the colour applied." The colour is, however, obtained apparently from lac. On a sheet of African **Sorghum** (in the Kew Herbarium), the remark has been recorded that it affords a red dye. It is thus just probable that the Mysore use of **Sorghum** flour may exercise some tinctorial influence. (*Conf. with p. 286.*)

FIBRE. Stems. 2433

Fibre.—The STEMS have been recommended as possible of value as a paper-making material. It seems probable, however, that even were they much richer than they are in fibrous material, their value as fodder would preclude their being available for the paper-maker.

MEDICINE. Grain. 2434

Medicine.—"The GRAIN is reckoned heating" (*Stewart, Pb. Pl. p. 262*).

FOOD & FODDER. 2435

Food and Fodder.—The Great Millet, *juár*, is so important an article of food with the labouring classes of India that it need scarcely be dealt with in this paragraph. The whole article, here given on **Sorghum**, is an exemplification of the value of this food-stuff. For Museum purposes the grain should be grouped into the two great sections which have already been established, namely, the *kharif* crop, or that which ripens in autumn, and the *rabí* crop, or that which ripens in spring. Generally speaking the former embraces the numerous forms of **S. vulgare**, or the millet with compact heads, and the latter mostly the forms of **S. halepense** *var.* **Roxburghii**, or the varieties with a lax, much branched inflorescence. A place has already been assigned for the **Roxburghii** series, but, as explained elsewhere, this distinction, though fully recognised by the Indian cultivators, has not been adhered to by the writers whose publications afford the material from which the present review has been compiled. In consequence, the chapter on Cultivation below, doubtless embraces both crops, and, when offered for sale, the two grains are rarely distinguished; and, indeed, the forms of both classes are so much alike that they can with difficulty be recognised when presented in their prepared state, ready for consumption. Speaking popularly, the grain of the **vulgare** series is more rounded, and, when seen with the adhering floral envelopes, is less hairy than in the **Roxburghii** series. In both sections the greatest possible variation occurs in the colour of the floral envelopes and in the colour of the cleaned or husked grain.

Two Crops. *Conf. with pp. 279, 282, 283, 286, 291, 304, 310, 311, 312-314.* 2436

Sorghum holds in India a position very much like that of oats in Scotland. Indeed, by the Natives of India, it is regarded as more wholesome than wheat or rice, because more easily digestible. It is ground into meal and eaten as cakes or porridge. The parched grain with salt, *gúr*, chillies, etc (as flavouring ingredients), is also made into many special dishes. **Professor Church** gives the nutrient-ratio of *juár* as 1 : 8¼ and the nutrient-value as 86. It contains, he tells us, ·86 per cent. of phosphoric acid and ·21 per cent. of potash. The following analysis, **Professor Church** furnishes, as that of this grain:—

CHEMISSTRY. 2437

	In 100 parts.	In 1 lb.
Water	12·5	2 oz. 0 gr.
Albuminoids	9·3	1 ,, 214 ,,
Starch	72·3	11 ,, 248 ,,
Oil	2·0	0 ,, 140 ,,
Fibre	2·2	0 ,, 154 ,,
Ash	1·7	0 ,, 119 ,,

SORGHUM vulgare.

Chemistry of the Juar.

CHEMISTRY.

Professor Church, in his corresponding analyses of Indian oats, wheat, and rice, shows the following amounts of albuminoids, starch, and oil:—

	Albuminoids.	Starch.	Oil.
Oats	10·1	56·0	2·3
Wheat	13·5	68·4	1·2
Rice	7·3	78·3	·6

The Professor explains, however, that the sample of oats analysed compared very unfavourably with the oats of Europe, where the average percentages are 12 albuminoids, 6 oil, 11 fibre, and 3 ash. The Professor further adds that sometimes the percentage of albuminoids rises to 15½ and that of oil to 7. The nutrient-value of Indian wheat expresses as 84·6 and of rice at 86⅓. It will thus be seen that, although certain writers have spoken of the poor Natives of India as *living upon inferior grain,* (through the greed of the money offered them for their wheat and rice) the chemist's investigations establish (one of the great staples of Indian food, as quite as wholesome as the more expensive food-stuffs exported from India. Indeed, it may safely be said that to the agricultural community of India, *juár* is a more useful crop than wheat, since its straw constitutes the chief staple fodder. It is, therefore, food not only for man but for his indispensably necessary bullocks which, for a large portion of the year, fail to obtain sufficient grass and leaves from the natural herbage of the country. This will at once be understood when it is recollected how completely everything is burned up for months together. Were the Natives of India therefore to take to eating wheat more generally than they do, they would have to grow special food for their cattle—a state of advanced agriculture beyond the means of the Indian peasant. *Juár,* when exported to Europe, appears to be used mainly for feeding poultry. Some of the more succulent forms of **Roxburghii** are grown for sugar or are eaten like cane in the fresh state as sweetmeats.

Wheat versus *Juár.* 2438 *Conf. with p. 296.*

SPIRIT. 2439

Spirit.—The Karens of Burma are said to distil a kind of whisky from the grain.

FODDER. *Conf. with pp. 287, 304, 305, 309, 311.*

Fodder.—As a fodder *juár* may be given green or after being dried and preserved for months. Some varieties are specially grown for fodder, and the kinds preferred for this purpose are the sugar-yielding forms, which, like their presumed ancestral state (**S. halepense**), manifest a strong tendency to branch at the nodes or joints, and to root at the kneeing angles of the lower branches. Many of these special fodder forms may be propagated by ratooning and, if left in the soil, will survive for more than one season and yield several crops of fodder. The dry STALKS and LEAVES are chopped to form the ordinary cattle fodder of the country, for some months of the season, being known by the name of *karbi.* Occasionally, in parts of the North-West Provinces and Oudh and the Panjáb, *juár* is grown solely for cattle fodder, in which case the stalks are cut while green before the seed matures (*Duthie & Fuller*). In this case it is usually sown in the hot weather before the rains, it requires irrigation, and is cut early enough to be succeeded by one of the cold weather crops. The outturn per acre is, on irrigated land, 300 maunds of green fodder (known as *chari*), equal to 100 maunds of dry fodder; and on unirrigated land, 280 maunds, equal to 90 maunds of dry fodder. In Montgomery district, 40

Stalks. 2440

Leaves. 2441

FODDER. 2442

seers of seed are sown per acre for *chari*. Sown July-August, it is cut May-June, and given to cattle mixed with *turi*. The stalks are called *tánda*, and when green are sometimes eaten as sugar cane. They are the best fodder obtainable, and are worth R12 to R16 per acre. A bullock will eat twice the weight of *juár* stalks when dry, as it will of broken wheat straw (*turi*), say, 30 seers daily (*Gaz., 110*). Of Dera Ghazi Khan it is said that in the Sind circle *juár* is grown for fodder and is not a good grain-bearing crop. Of the Central Provinces (Chhindwara) it is state that the stems known by the name of *kurbí* constitute the chief food of the cattle during the months when pasture is not to be had. Of Nimar it is reported that the stalks (*karbí*) form a valuable cattle fodder, fetching when sold, which is rarely the case, about R4 or R5 an acre. In connection with the Bombay Presidency the subject of *juár* fodder is more frequently mentioned. The following further selection of opinions on the subject of *juár* fodder may be offered :—

2443

The stalks (*kadbi*) of a variety called *nilva* in Khandesh district are soft and easily chewed, and form a favourite food of cattle (*Gaz., 149*). The stalks (*kadbi*) are a valuable fodder in Nasik (*Gaz., 99*). The fodder is prized for milch cattle. *Jári* is the only cereal whose straw, or *kadba*, is used as fodder in the natural state. The broken straw of other cereals mixed with chaff is called *bhuska* or *bhuskat*. *Jvári* stalks are stacked and thatched in the rainy west; in the drier east they are stowed in long grave-like ridges and covered with clods of black soil (*Gaz., Poona, II., 40*). *Shálu*, or the late variety of Indian millet, is grown specially for cattle fodder in garden lands in Kolhápore district (*Gaz., 167*).

2444

The ryots of Gudalur (Madras Presidency), whose working cattle are better than those generally used in the Presidency, grow small plots of an unirrigated variety of *cholam*, called *adai-pyir*, as fodder. In the *Saidapet Experimental Farm Manual and Guide*, it is said that, if intended for fodder, *cholam* should be cut when two-thirds grown, and will, when under irrigation, yield as far as 50,000 to 70,000℔ of fodder per acre in a year. If unirrigated during the seven or eight months, it will produce four cuttings, yielding about one-third of that amount. It may be added that some of the above passages would seem to refer to the variety **Roxburghii** and not to *juár* proper.

DOMESTIC. Flowering Spikes. 2445

Domestic and Sacred.—After the removal of the grains, the FLOWERING SPIKES become hard and rigid. They are used for making brooms, and are imported into England from the various countries in which the millet is grown for the purpose of being made into carpet brushes. (*Conf. with var.* **technicus**, *p. 277.*) The harder CULMS are sometimes, like those of the wild species (S **halepense**), employé as pens. **Smith** (*History of Bible Plants, p. 214*) has endeavoured to show that the stalks of this millet were very probably the reed of **St. Mathew**, and that the spikelets on its top were very likely the hyssop of **St. John**, mentioned at the crucifixion. The hyssop (*Esob* of the Hebrews) of **Moses** was a word used to denote any common article in the form of a broom or a material suitable for that purpose. If this view be accepted, the cultivation of **Sorghum** in Palestine may be regarded as very ancient.

Culms. 2446

2447

The religious rites, or superstitious observances, which in India are so marked a feature in the carrying out of almost every agricultural proceeding, especially the sowing or harvesting of the time-honoured crops, are fully recognised in the cultivation of *juár*, in areas widely remote from each other. The following may be cited :—**Mr. Wright** (*Memo. Agri., Cawnpore*), quoted in **Duthie & Fuller**, *Field and Garden Crops*, thus describes the threshing of the *juár* in that district : "The *juár* was heaped by the cultivator in the shape of the figure 8, one end towards the Ganges,

DOMESTIC.

and a sickle and a branch of *madár*, in honour of *Shaikh Madár* (a local saint), stuck up in it. All round the heap a line of cow-dung was traced, and the smoke of a sacrificial fire made to blow upon the heap to keep off evil spirits (*jins*). A double handful of grain was given in honour of *Shaikh Madár*, one to the village minstrel (*bhát*), one to the Brahmin, one to the family priest (*parohit*), and half a seer to the village carpenter, blacksmith, barber, and water-carrier."

CULTIVATION Grain. 2448

CULTIVATION.

Judri, under a great variety of names, is cultivated extensively throughout nearly the whole of India. Although a widely grown and much valued cereal, the uses to which it is put vary very greatly in the different districts, depending largely on the relative wealth of the Native population. In some districts, it is grown primarily on account of the value of the GRAIN (consumed by the poorer Natives) and for the straw which is given as fodder to the cattle. In other districts, the crop is cut green and made into FODDER, the grain not being valued as an article of food. In India as a whole indigenous forms of **Sorghum** are not very extensively grown on account of their SUGAR, but this use of the crop was known and practised in India long before it was thought of in America. (*Conf. with pp. 283, 312-3*). Apart from the highy nutritious nature of the grain, which on this point compares very favourably with the more expensive staples of rice or wheat, *juár* has many good qualities which have led to its extensive cultivation by an important section of the Native population. It grows on most soils, even very poor soils, although black cotton soil gives the best returns. The more trouble the cultivator takes to plough deeply and break up the soil, the better an outturn is he likely to receive; probably in consequence of the firmer roots sent out by the plant. Although quickly killed by over-soaking its roots, it amply repays the trouble and expense of judicious irrigation. It is generally sown, mixed with other grain, as a *kharif* or hot weather crop, but occasionally as a *rabí* or cold weather crop, that is to say, sown in autumn and reaped in spring.

Fodder. 2449

Sugar. 2450

Both a Kharif & a Rabi Crop. 2451 *Conf. with pp. 279, 282, 283, 287, 291, 293, 305, 308, 309, 310, 311-312, 313-14.*

AREA. 2452

AREA, OUTTURN, AND CONSUMPTION.

Although it seems desirable that this section should be established (in conformity with the treatment pursued with other important articles of food), it has to be admitted that the most meagre information only exists regarding the subjects that should be here treated of. In the published *Agricultural Statistics of India*, *Juár* and *Bájra* are classed under the heading "Oth r food-grains, including pulses," and in only one or two provinces has it been found possible to procure separate returns of these all-important millts. To the vast majority of the people of India (if Bengal be excluded from consideration) *juár* and *bájra* are of greater importance than rice, wheat, or barley. **Sir Walter Elliot** speaks of *juár* "as the staple dry grain of India, and, indeed, of all tropical countries of Asia and Africa." It would, therefore, seem desirable that it should in future be separately returned. Thus, for example, while in Bombay and Sind during 1889-90 wheat occupied 2,311,558 acres, there were 8,282,876 acres of *juár*; in Madras during 1889-90 there were 6,458,668 acres of rice and 18,902 acres of wheat, with 4,276,509 acres of *juár* (*chólam*). It is needless to multiply such examples in justification of the importance of the *juár* crop. It need be necessary only, therefore, to add that the pecuniary value of the wheat crop s, of course, much greater than the *juár*, but wheat in India as a whole is grown far more as a source of revenue than as an article of diet to the actual

CULTIVATION

Area.

cultivator. In India the ratio of wheat to rice, as articles of diet, is as 1 to 4¼ (6 million tons of wheat to 25⅜ million tons of rice). The annual consumption of all grains and pulses is probably over 55 million tons, so that of the food of the people of India the millets and pulses have to at least provide 26 tons.

2453

The following table exhibits the total *surveyed* area devoted in India, in the year 1889-90, to "Other food-grains, including pulses," as also, so far as has been found possible, the shares of that area taken by *juár* and *bájra* :—

Area devoted in 1889-90 to "Other Food-Grains and Pulses," distinguishing, when possible, Juár and Bájra.

	Juár.	Bájra.	"Other Food-Grains and Pulses," including Juár and Bájra.
	Acres.	Acres.	Acres.
1. Bengal	No returns available.		
2. Bombay and Sind	8,282,876	5,727,306	18,649,125
3. Madras	4,276,509	2,596,947	14,102,265
4. Panjáb	2,369,400	2,273,302	10,790,973
5. North-West Provinces	3,147,523*	1,965,471*	17,112,490
6. Oudh	550,000*	No returns	6,659,654
7. Central Provinces	No returns.		4,330,418
8. Central India and Rájputana	No returns available except Nos. 9 and 16.		
9. Ajmír	No returns.		177,288
10. Berar	No returns.		2,921,446
11. Nizam's Territory	No returns.		
12. Mysore	No returns.		
13. Coorg	No returns.		1,638
14. Assam	No returns.		57,024
15. Burma	No returns.		1,032,699
16. Pargana Manpur	No returns.		3,070
TOTAL	...	...	75,841,694

2454

The reader might, perhaps, consult in this connection the discussion given under Rice (**Oryza sativa**, *Vol. V., pp. 519-527, also 537-539*), where the areas under wheat and rice have been shown and the value of millets dealt with. It will there be seen that the area annually made to yield rice in Bengal has been given at 42,000,000 acres, so that, with the 27,866,447 acres under that crop in the other provinces of India, rice assumes by far the most important position of any single article of food. The second place should, however, be accorded to *juár*, since of the provinces in the above table, for which returns are available, there were over 19,000,000 acres devoted to this crop alone. It seems thus likely that were returns available for the other provinces, the total area under it would exceed rather than fall short of 25,000,000 acres. In Bengal generally *juár* is, by no means, very important, except among the hill tribes. In the Central Provinces, in much of Central India and Rájputana, in the Nizam's Territory and Mysore, it is at least quite as important as in the North-West Provinces and the Panjáb. This may be seen in the

* Figures taken from Messrs. Duthie & Fuller's *Field & Garden Crops*, published in 1882.

CULTIVATION

Area.

case of the Central Provinces, from the fact that, while rice and wheat together have an area of 7,900,000 acres, the other food-grains have more than half that area. Still more striking are the figures for Berar, 850,000 acres devoled to wheat and rice with 2,921,446 acres to other food-grains. An allowance of close on 10,000,000 acres for the provinces not shown in the above table would probably not far exceed the amount annually devoted to this food-grain, in the provinces for which returns are unavailable. A little more than half that acreage has, however, been allowed for in the assumption of a total area of 25,000,000 acres, devoted to this crop annually in India.

2455

But it has to be pointed out that *juár* is very largely grown as a mixed crop in association with one or other of the extensive series of pulses available for the cultivator's selection. *Arhar* (**Cajanus indicus**) is in the North-West Provinces generally associated with it, the yield per acre of the *juár* being decreased thereby 25 per cent., though the mixed crop is generally regarded as more profitable, or, at all events, safer. The acreage yield is also very much higher in irrigated than unirrigated land. Thus **Messrs. Duthie & Fuller** give the yield as 10 maunds of grain and 60 maunds dry fodder from irrigated land, and 8 maunds grain and 45 maunds fodder from unirrigated land. Less than half the acreage of the North-West Provinces is a pure crop, the other half *juár-arhar*, and the balance cultivated with *chari* (*juár* fodder that is cut before the grain matures). But of the pure and mixed crops, by far the major portion is usually unirrirgated, so that a yield of 6 maunds an acre might be assumed as a fair average for these provinces especially when it is recollected that the area in the four permanently-settled districts has not been taken into consideration. A yield of 6 maunds all over would give an outturn of 22,188,000 maunds, or 792,428 tons, for these provinces alone. If the same factor (6 maunds an acre) be accepted for the rest of India, the total outturn for the whole empire, *viz.*, 25,000,000 acres of *juár*, would be 150,000,000 maunds, or 5,357,142 tons. From the passages quoted below, on the cultivation in the various provinces of India, the reader will be able to judge whether the allowance of 6 maunds be over or below the average, and the correction can accordingly be made on the estimate here framed. It would seem fairly safe to say that the production of this grain alone, in India, amounts to at least 5,000,000 tons a year, and since very little is exported that amount may be assumed as consumed. From this estimate, however, the entire population of Bengal should be eliminated, since little *juár* is grown or consumed in the Lower Provinces. The people of the Panjáb also eat a very considerable amount of wheat, and the well-to-do everywhere use rice, wheat, and other food materials, eating very little, if any, millets. The total area under millets may be put at 40,000,000 acres and under pulses at 37,000,000 acres, so that it may safely be said these vast areas afford in normal years quite sufficient food for the community that may be viewed as dependent on millets and pulses as their staple articles of diet. But the *juár* crop is not alone of value as food for man. Its stems constitute the chief fodder for the cattle of a large portion of India. Loss of the *juár* crop first assumes the manifestations of famine by the starvation and death of the cattle.

Yield. 2456

Total Outturn. 2457

Diseases. 2458

DISEASES AND PESTS OF THE JUAR CROP.

It will be found that the district and provincial accounts (quoted below) of the cultivation of this plant very frequently allude to the diseases to which it is liable. These accordingly afford local particulars, and it, therefore, remains to discuss in this place the leading scientific facts that have been brought to light on the subject. The diseases and pests of the Sorghum crop may be said to be four-fold, namely, due to (I.) Fungoid growths;

DISEASES.

Fungoid. **2459**

Rust, **2460**

(II.) A parasitic flowering plant; (III.) Pests in the form of insects, birds, squirrels, etc.; and (IV.) Climatic disturbances.

(I.) **Fungoid.**—Under this section three diseases have to be dealt with, *viz.*, Rust, Smut, and Bunt.

RUST—**Puccinia Penniseti.**—Through the indefatigable researches of the late **Surgeon-Major Barclay** we now possess scientific details regarding this disease. **Dr. Barclay** says: "This affection, known locally as *Kani*, is possibly **Puccinia Sorghi,** *Schweinitz*, and I have named it so provisionally, but it is quite possibly a new species. I am the more inclined to think it is a new species, because I have never seen nor received specimens on **Zea Mays** in India; and as the latter is very extensively cultivated, this is unexpected on the assumption that the fungus is **P. Sorghi.** Still, as I have not had good opportunities for obtaining information about the existence of any Rust on **Zea,** it is quite possible that it exists. Assuming the fungus on **Sorghum** to be **P. Sorghi,** the Indian species differs especially in having considerably larger *uredo* and *teleutospores*; in the teleutospores not being thickened at the free ends; and in the spores being associated with *paraphyses.*"

It is significant that **Dr. Barclay** should have (shortly after the appearance of the paper from which the above has been abstracted) obtained abundant evidence in support of the suspicion he entertained that the rust on **Sorghum vulgare** was not **Puccinia Sorghi,** but a new and hitherto undescribed species of fungus. Near Simla he found a specimen of **Zea Mays** (the Indian Corn) attacked by **Puccinia, Sorghi,** *Schw.*, or at all events, by a Uredine that much more closely resembled **Puccinia Sorghi** than does the fungus on **Sorghum vulgare** itself. During **Dr. Barclay's** last excursion throughout India (in connection with the Leprosy Commission) he was able to collect much information on the diseases of the chief crops of India, but which has been largely lost through his premature death, shortly after his return to Simla. While at Erode (in Madras), for example, he collected a Uredine on **Pennisetum typhoideum** (*Bájra*) which proved identical with that which he had previously found on Poona specimens of the *shallú* crop of **Sorghum** (?) **vulgare.** He was thus enabled to arrive at the definite opinion that the Indian disease on *Juár* and *Bájra* is not the Uredine found in America and other countries, but a distinct species for which he suggested the name **Puccinia Penniseti.** While, on the other hand, he became convinced that the true **Puccinia Sorghi** was, in India, the rust that attacked Indian-corn. The writer had the pleasure to enjoy the late **Dr. Barclay's** confidence and friendship. During many botanical excursions he discussed with him the value that might be placed on the fungoid diseases of plants as supporting other arguments that might be advanced in tracing out the nativity of crops. The problem that distressed **Dr. Barclay** most was the fact that while the berberry Æcidium-bearing bushes of the Himálaya were yearly attacked, the rust of the Himálayan wheat-fields was in no way connected with the berberry. During our last botanical excursion, however, the writer had the good fortune to find in a field close to Muttiyana (Simla) a crop of wheat attacked with both **Puccinia Rubigo-vera** and **P. graminis,** and underneath the crop a **Bromus** and **Polygonum aviculare,** also attacked in a remarkable manner. Without the aid of the microscope **Dr. Barclay** was unable to do more than speculate as to whether it was possible the **Polygonum** (which he had never before found attacked by a Uredine) could be the cause of the rust on the wheat of the plains of India. Shortly before his death **Dr. Barclay** told the writer, however, that he had been able to prove that the **Polygonum** very probably had nothing whatsoever to do with **Puccinia graminis,** but was a well-known species found on the same

SORGHUM vulgare.

Diseases and Pests

DISEASES.

Fungoid.

Rust.

Polygonum in Europe. There remained, however, he added, the fact that the hosts of **P. graminis** as well as of **P. Rubigo-vera** were both very likely to be found at Muttiyana. **Dr. Barclay** was hopeful that he would thus soon solve the problem of the wheat rust of India. Unfortunately, his death within a few days after the conversation detailed above, closed a brilliant career, all too short for the obligations laid upon him by the necessities of India in a field of research in which **Dr. Barclay** had no equal and no successor. But to conclude this brief reference to the rust of **Sorghum, Pennisetum,** and **Zea,** it may be added that **Dr. Barclay** was disposed to regard the peculiarities indicated, as favouring the opinion that all three plants were exotics in **India,** but it may be added that it is significant that **P. Penniseti** should have been found on the peculiar crop of **Sorghum** which the writer is disposed to regard as a native of India and as quite distinct from **S. vulgare.**

The importance of the study of rust may be inferred by the following passage, which, it is believed, is by no means an account of an exceptionally bad manifestation :—" *Rust.*—This disease appeared first on an early sown crop of **Sorghum,** which was grown from imported seed, from which it is presumable that the fungus germs were derived, and, finding in the plants a prepared ground whereon to work, they some time before the seed was fit for harvest, so completely enveloped the plant with a red covering, which was increasing and destroying its value as fodder, as to necessitate its being cut down; not, however, before the disease had spread further, and in all our **Sorghum** it appeared, as well as in the 'Planter's Friend,' but not to so great an extent in the latter as the former. This rust is probably closely allied to corn rust, which is common in damp districts in England, but it is of a darker colour, being more purple than red" (*Rept. Agri. Dept., Madras, 1878-79*)

It is, perhaps, unnecessary to have to say that the fungus here dealt with belongs to a widely different family from that of the diseases described below under their English names Smut and Bunt. It is a Uredine, and, like most of the members of that great family, exists in two (in some forms more than two) stages—the active disease and the resting condition—as they may be called. In many Uredines the latter stage migrates from the crop to a weed, or it may subsist on the decaying portions of the crop. Thr chief rust of wheat leaves the crop and assumesits second stage on the berberry. To accomplish a cure, therefore, the life-history of the particular form of rust has to be worked out. To pickle the seeds (a procedure perfectly rational in smut and bunt) would, with a uredinous disease, be to pay the penalty of unpardonable ignorance, since the resting stage of the disease might be all the while maturing on the weeds around the field or on the decayed matter of the last crop, and be thus awaiting the germination of the expensively purified seeds to renew the activity of the pest. Extermination of the second host, or, if (as in the case of **Sorghum** rust) the fungus be autœcious, thorough freedom from the field of the decaying matter of the previous crop, are the only rational cures.

The explanation that the rust that did so much harm in Madras was brought with certain seed is highly improbable unless the seed was sown mixed with dried fragments of the leaves. It seems more likely that rust is always more or less present in the *juár* plant of the plains of India (perhaps to less extent than in wheat), but that it only developes into an epidemic extent under certain climatic conditions. The imported crop may also have been a more favourable host for the disease than the Indian forms.

Smut. 2461

SMUT—**Ustilago carbo** (*Conf. with Vol. III., 457*).—This fungus may be said to appear when the **Sorghum** sets its flowers. The whole inflorescence, and even neighbouring leaves and twigs, become covered with a black powdery substance. This partially or completely destroys

DISEASES.

Fungoid.

Smut.

the particular head so attacked, but the injury thus effected is not all. The spores are blown from the affected heads by the wind, and being minute, they get between the husk and the grain in other spikes. There they rest for the greater part of a year and undergo no further change. On the seed being sown next year the spore, if not destroyed, will commence its new life with the germinating embryo to which it attaches itself. Growing up through the tissue of the stem it finally reappears in the flowering spike, to disseminate the fresh showers of black spores that carry on the process. Year after year, if not checked, the proportion of diseased stems may increase until a crop may be seen that is practically worthless. One peculiarity of Smut may be here further added, namely, that while no structural difference has as yet been recognised in the smuts met with on wheat, oats, barley, sorghum, etc., the smut that has appeared on one of these crops does not seem to possess the power to infect another. Each crop may, therefore, be assumed to possess its own peculiar form of **Ustilago carbo** or Smut.

With reference to the appearance of this disease in India, numerous reports might be quoted and with advantage could space be afforded. The following may perhaps suffice: **Mr. W. R. Robertson, M.R.A.C.,** Superintendent, Government Farms Madras (*Agri. Report, Madras, 1878*), wrote:—

"I have the honour to forward, for the Board's inspection, a few heads of *cholam* attacked by a blight called "*Smut*," a fungoid disease which appears to be very common this season in this district. It will be observed that the grain is very much changed from its ordinary shape, and that many of the grains are filled with a fine black granular powder (spores). The blight is of common occurrence, but I do not remember any year in which I have seen so much of it as this. In most countries this blight is well known to agriculturists, but it occurs to a serious extent chiefly only in those countries in which agriculture is yet in a backward state. In some parts of England its occurrence is certainly far from unfrequent, but this is only where an antiquated system of agriculture is still pursued, or where injudicious attempts are made to grow cereal crops unsuited to the climatical conditions. However, in England, there has been a very marked disappearance of the disease during the past twenty or thirty years, which must be attributed to the progressive development of a superior agricultural practice. Straw and grain infested by the spores of 'Smut' are not thereby, it is believed, rendered at all injurious when used as the food of live-stock, neither is the grain rendered unfit for human food as is that infested by the spores of the fungoid disease called 'Ergot.' The loss that results from "Smut" is chiefly in the diminution of the yield of grain; and when, as is sometimes the case, from 20 to 30 per cent. of the heads of grain in a field are rendered abortive, the loss is a serious one. As regards the prevention of this blight, it appears to be generally admitted that superior culture with the protection of the seed by the use of chemical dressings are the most efficient means. Of course, nothing can effectually protect a field of grain if the crop in a neighbouring field or in the locality has been attacked by the blight; hence the attempt to root out the disease in a locality must be general on the part of all the farmers. It was not usual in England until lately to dress seed in view to protect the crop from "Smut," but the practice is now general. It is true that the means taken are not so efficient as could be desired, for the sporules are so liable to become dispersed before harvest and during the harvesting operations. However, it has been shown that, when seed infested by the spores of the blight have been sown, the blight has been reproduced in the crop and that, when infested seed has been properly dressed with a chemical solution, the blight has not appeared. There are many different descriptions of seed preparations used. Those in which sulphate of copper forms the chief or only ingredient are the most to be relied on. The following process might, I think, be adopted with every hope of success in the treatment of seed infested by the spores of Smut:—'For 50lb of seed take 3 ounces of sulphate of copper which disolve in one quart of hot water; when the solution is quite cold pour it over the seed, with which mix it thoroughly; when quite dry the seed is fit for sowing'" (*Conf. with Rept., Agri. Dept., Madras, 1878-79*).

As alluding to Smut and, perhaps, also to Rust and Bunt, the following passage may now be quoted from **Mr. Nickelson's** *Manual of Coimbatore,*

SORGHUM vulgare.

Diseases and Pests

DISEASES.

Fungoid.

Smut.

Madras, as it shows the diseases which are known to attack the *juár* crop in that district :—

" *Sembei* is rust and is common in cold misty weather. *Kariputtei* is mildew ; it is not known whether this is the developed form of *Sembei.* Instead of grain, the ears are filled with a blackish brown powder like smut in wheat. *Navei-puchi* is a light green insect beginning as a tiny worm, *Kuanthi-Paluva,* which feeds and grows on the tender plant ; it developes at the time of flowering if the weather is misty. A similar cause is alleged for *Asugani* (*Pen* or *Porigan*), which is a minute black insect."

Very little is known for certain regarding the insect pests of this crop, but the reader should consult the special paragraphs below on that subject.

Bunt. 2462

BUNT—**Tilletia caries.**—The spores of this fungus are slightly larger than those of Smut, and their surfaces are covered with a reticulation. They ripen and are dispersed about the time of the harvest. They attach themselves to the outside of the grain and remain until it is sown next year. With the germinating seedling the spore renews its activity, attacks the embryo, and grows up within the tissue of the plant, but in doing so it undergoes a succession of changes and thereby effects a very serious injury to the host, upon whose tissue it literally feeds. The spore first sends out a little tube, from the top of which arise from four to eight cells, and these form the spores of the second generation. They unite by cross tubes into H-shaped figures, then fall off and give rise to spores of the third generation, which in their turn produce spores of the fourth generation. From the last mentioned spores the *mycelium* arises—a network of exceedingly delicate tubes which, like that of Smut, permeates the entire substance of the host and pushes forward as if its aim had been the destruction of the grain, for, on the fruits forming, the fungus attacks their substance and produces within each grain a baneful crop of fresh shores which, by the wind, is sown on the unaffected spikes or ears of corn. The grain attacked by *bunt* looks, however, externally as if perfectly healthy. It is, if any thing, plumper and rounder. On pressing it, however, it bursts and emits an offensive smell and a slimy or greasy dark coloured powder. Little wonder therefore that, the farmer fears Bunt more than Smut, for, the latter is seen while the former all unsuspected, may be reaped, and on the thrashing floor the baneful spores sown on the entire crop of grain. Bunt has been recorded on more than one occasion as doing serious damage to the **Sorghum** crop of India (*Conf. Field & Garden Crops*), but it does not seem necessary to republish the local reports. From the brief account here given of Smut and Bunt, the reader will be able to appreciate the value of the pickling as it has been called (in official correspondence) of the seed of **Sorghum** and other crops. The difficulty in India rests in the fact that while one cultivator may be induced to adopt such scientific measures, his neighbour by refusing to do so undoes all the good effects of the pickling. It has, however, been pointed out that the germs of both Smut and Bunt are so sensitive to hot water and so easily removed mechanically by washing, that steeping or washing the grain for a short time, in slightly warm water, would do much good where chemical agents to destroy the adhering fungoid germs were beyond the means of the cultivators In some parts of India (as in the Panjáb, p. 308) to wash the seed before sowing is a common practice. For the results of experiments and informatton regarding the materials used in India to pickle the seed, the reader should consult the Experimental Farm Reports for Bombay, 1885-86, 1886-87, 1887-88, etc. Carbolic acid, sulphate of copper, common salt, etc., are the substances that have been chiefly experimented with.

Parasites. Striga. 2463

II. **A Parasitic Flowering Plant—Striga.**—The reader who may be interested in this subject will find full details under **Saccharum** (sugar-

DISEASES. Striga.

cane) (*Conf. with Vol. VI., Pt. II., 126*), where the same parasite, by attacking the roots of either or both crops, does such damage as to often effect total destruction. Many popular writers affirm that **Striga** acts by depriving the crop of the natural nourishment of the soil, but on the other hand, the species of **Striga** are known botanically to be root parasites and therefore to suck the sap from the foster-plants to which they attach themselves. Although no Indian writer has especially noted the direct attachment of the **Striga** weed to the roots of **Sorghum** or **Saccharum**, yet there is little doubt that it is in that way that the weed effects its destructive mission. The observation that the parasite is destroyed if cotton be rotated with **Sorghum** would be accountable for by the affirmation that it is unable to subsist on the roots of cotton and, being an annual, the seeds germinate and die, from want of a suitable host on which to feed and produce their fruits and seeds. The Director of Land Records and Agriculture, Bombay (*Report, 1886-87, p. 10*), refers to this disease:—

Conf. with p. 310.

"One-third of the crop of the general farm was devastated. The experiments to study the habits of the parasite have been continued. Plots sown with maize followed by gram and with *jowári* were attacked, though in the case of maize the parasite did not show till late. It was ploughed in before the gram was sown, and did not reappear. In the *jowári* plot it held sway in spite of all efforts to keep it down. In another plot, last year sown with *jowári*, when the parasite appeared. Cotton is supposed to keep it in check. Sub-soiling was tried as a preventive in another plot. Gram was the crop taken. The result was very favourable, though not conclusive, because the parasite does not feed on gram. *Jowári* should be tried after sub-soiling. Horse-dung is locally supposed to favour the parasite. The result did not in any way support the supposition, for, though *jowári* was sown, it was uninjured."

Insects, Birds, &c. 2464

III. **Pests in the Form of Insects, Birds, Squirrels, etc.**—Most writers say that birds and squirrels are perhaps the worst enemies to **Sorghum.** Their depredations necessitate constant watching for at least 25 days before the crop is cut, a circumstance that enhances greatly the cost of cultivation. But this danger and trouble is perhaps so evident that it needs no further explanation. It is otherwise with the insect pests. Some of these are fully understood, others are so obscure that much difference of opinion prevails as to whether the poisonous property (spoken of as possessed at times by the stems, when used as fodder) is due or not to an insect. In the special article on **Pests** (*Vol. VI., Pt. I., 147*) it is stated that the larvæ of a moth known in the North-West Provinces as *bhaunri* (not as yet identified) attacks the *juár* stalks in much the same fashion as the sugar-cane is tunnelled by the "sugar-cane borer." These larvæ, in fact, bear so strong a resemblance to those found in cane that **Mr. Cotes** suggests, that they may also set up decompositions sufficient to cause the poisonous properties regarding which so much has been written. It may be added that the prevalent belief among the Natives is that the poison is the result of an insect, and it is worthy of note that it occurs at the same peried and under similar circumstances as in cane, *viz.*, during an exceptionally dry season. The following two passages may be accepted as representing the somewhat exhaustive controversy that exists on this subject:—

"The most peculiar of the diseases to which *juár* is liable is that which makes the young stalks poisonous to cattle, if eaten by them when semi-parched from want of rain. Of the fact there can be no doubt; in the scarcity of 1877 large numbers of cattle were known to perish from this cause, their bodies becoming inflated after a meal of the young *juár* plants, and death ensuing shortly afterwards, apparently in severe pain. A good explanation is not, however, forthcoming. The opinion universally accepted by Natives is that young *juár* when suffering from deficiency of rain becomes infested with an insect called *bhaunri*, to which its poisonous effect on cattle is due. Immediately rain falls the insect is said to perish, and unless the ears

SORGHUM vulgare.

Diseases and Pests

DISEASES.

Insects, Birds, &c.

have appeared before the rain failed, the crop often recovers itself, and yields a good outturn of grain" (*Duthie & Fuller*).

A totally different explanation of the great mortality amongst cattle, in the year 1877, is given in a paper by **Veterinary Surgeon J. Anderson** *Agri. and Hort. Soc., Journ.; VI., Part I, p. 1, New Series*). The article is too long to give *in extenso*, but the following extracts will show the opinions arrived at by that writer :—

"From my recent investigations and experiments, I have come to the conclusion that *jowar* is not poisonous. Some stocks here and there contain insects, others a fungus, both of which are supposed by many to be the medium of poison. Even if they were poisonous, they are not found in sufficient quantities to prove injurious, and account for such wholesale mortality. The prevailing idea is that the *jowar* has become poisonous not only from want of the usual rains, but also from the scorching effects of the unusual return of the hot winds; or that a poisonous gas engendered in the stalk by the action of the heat, etc., etc. The insects are of the HEMIPTERA family commonly met with on plants, and the fungus, when submitted to a microscopical examination, consisted almost entirely of round spores. **Dr. Franklin**, who examined it, cannot say to what species it belongs." "I look upon *jowar* as a destructive substance or thing, and not as a poison. It destroys life by acting mechanically on the system, just as a sponge dwells in the stomach and kills on being retained there. When *jowar* has been eaten, it generally produces 'hoven' distension of the first stomach as known by the generation of a large quantity of confined air, a product of fermentation arresting the natural function of rumenation and digestion which causes the animal to swell even to a state of suffocation or rupture of some part of the stomach or intestines to death." "'Hoven,' there fore, is a very common and exceedingly fatal form of disease, or rather accident. The stomach becomes surcharged with flatus, becomes paralysed as it were, and is thus rendered incapable of expelling its contents. Consequently rumenation and digestion cannot be carried on properly, but fermentation goes on instead, by which a large quantity of gas is immediately generated which, from want of ventilation, causes the animal to swell to a state of suffocation and sudden death is the consequence.

Conf. with pp. 287, 294, 306, 309, 311.

"The symptoms are very alarming and rapid in their course; and may come on sometimes, even if only a very small quantity of *jowar* has been eaten. The left flank swell up accompanied by distressed and labored breathing, panting and great depression, disinclination to move, with prominent and blood shot eye; the animal staggers, falls down, and dies of suffocation." "Hoven is the result of irregular and bad feeding, or from cattle eating food which they are unaccustomed to; or from eating partially decayed or unripe fodder of any description (stunted *jowar*), and particularly, succulent shoots which spring up after the first showers of rain. Cattle are sure to overgorge themselves, more especially half-starved ones, or eat so greedily that it is not sufficiently masticated, and before it can be properly prepared by rumenation, fermentation takes place, and carbonic acid gas is eliminated. Hoven is the result, appearing like an epizootic, affecting many animals at the same time and place. As this complaint is preventable, the assamis should hinder their cattle from eating *jowar* by herding them, or by fencing in the *jowar khets*. This has been done in many places with marked success . . . . . The first thing to be done is to arrest the process of fermentation and restore healthy secretion and action of the stomach. To do this the air must be displaced by eructation or through the intestines, or by opening the stomach with a trocar and canula the latter should be left in the wound for some time, so that the gases generated may escape as far as formed." "I have given an experimental bullock as much as six seers at one time, of what was supposed to be poisonous *jowar*, and it had no effect whatever. I attribute this simply to the fact that the animal ate it surely, masticated it properly, and ruminated quietly. The same animal has had it dried in the form of *bhoosa* and in decoction. The experimental sheep will not eat more than a stalk at time, and after being starved for 24 hours. A strong decoction had no effect on them either. The great loss through neglect and ignorance ought to rouse cattle owners from their apathy, and show them the necessity of storing fodder in plentiful seasons, preserving pastures and fencing in crops."

Climate. 2465

IV. Climatic Disturbances, such as want of rain, excess of humidity or damp, cloudy weather, and extreme or unnaturally high temperature, etc.—It will be observed from the remarks already offered that whether due to the presence of an insect or to some physiological change in the growth of the plant, owing to climatic disturbances, the *juár* stems are

DISEASES.

Climatic.

not always liable to cause injury to cattle. The occurrence of the poisonous property (as it has been called) is simultaneous over a large tract of country, appearing and disappearing within certain fixed limits of time. The inference is, therefore, unavoidable that the stems have been affected or altered in some way. Both the theories (advanced above) are, therefore admissible, namely, that certain climatic conditions pre-dispose the stems to an epidemic development of the insect pest or alter the nature of the stem or of its sap, so as to favour the fatal fermentation to which **Veterinary Surgeon Anderson** attributes the disease. The only observation that seems consistent to both theories is that the disease follows unusually high temperatures. The *juár* is mostly grown on high lands as a *kharif* crop and is only very occasionally irrigated. It is dependent on the rains for its moisture, and if these fail famine both to cattle and men must supervene over large tracts of India. A delay of the rains or an unusually high temperature must materially alter both the yield and the quality of the crop. It would thus seem, that the effect of climatic disturbances, both as favouring one or other of the diseases dealt with above, and as modifying the quantity and quality of the crop, has not received that degree of consideration which it demands. Need it therefore be added that the study of the races of **Sorghum**, in relation to climate and soil, is of the very first importance.

Conf. with pp. 277, 282, 291, 308, 309, 310, 311.

CULTIVATION in BENGAL. **2466**

Cultivation in Bengal.—Although grown by the hill tribes of Chutia Nagpur, Rajmahal, and the Tarai—to a limited extent—*juár* cannot be regarded as an important article of food in Bengal. It is more generally cultivated in the portions of the province that approach in climatic and other conditions to those which prevail in the North-West Provinces. The unimportance of *juár*, indeed of all millets, may be learned from the opinion arrived at by the Famine Commission, namely, that famine in Bengal meant essentially the loss of the rice crop sown in April to June and reaped from November to January. The loss of the *bhadoi* or intermediate crops, which consist very largely of Indian-corn, millets, etc., would not produce famine, nor even scarcity.

The following passages are extracted from the Agricultural Department reports on the subject of *juár* as a fodder in Bengal:—

"The following fodder-grasses were sown on the Farm:—**Sorghum**, *yellow cholam, white cholam,* and **Reana (Euchlæna) luxurians.** The **Sorghum** was manured with bone-meal. The increase in yield thus obtained more than covered the enhanced cost of cultivation due to the purchase of the meal. The results showed that fodder-grasses can be raised at a nominal cost of two to three pice per maund, and are likely to prove profitable crops in Bengal" (*1888-89*).

FODDER. **2467**

Conf. with pp. 287, 294, 304, 306, 309, 311.

Sorghum.—The following statement shows that **Sorghum** can be profitably cultivated in Bengal as a fodder crop. The price is taken at 4 annas per maund of green **Sorghum**, and has been arrived at by a comparison with paddy straw. It appears that a bullock of more than ordinary size can be kept in perfect health during the height of the working season, on half a maund of **Sorghum** (well matured and with the seed and stalk together), and a seer of mustard cake per diem. The outturn may be given:—

Field.	Row.	Plot.	Whether manured or unmanured.	Cost of cultivation per acre.		Outturn per acre in lb.	Price of the outturn.
				R a. p.			R a. p.
1	11	3	Unmanured .	51 6 0	1st cutting.	18,696	58 6 6
...	...	...		25 11 0	2nd do. .	8,364	26 1 9
...	...	...		2 13 0	3rd do. .	7,440	23 4 0

SORGHUM vulgare.

Provincial Account of the

CULTIVATION in Bengal.

It has hitherto been said, that **Sorghum** as a fodder crop will not pay in Lower Bengal (*1889-90*).

"*Sorghum.*—Experiments made with this crop last year confirmed the opinion expressed in the Annual Report of the previous year that **Sorghum** can be profitably grown as a fodder crop in Bengal. The produce was sold at 2 annas 6 pies per maund to a Calcutta dairy, the manager of which, **Mr. G. C. Bose**, a Cirencester scholar, reported very favourably on it. The yield of milk of a number of cows was increased from 20 to 25 seers by the substitution of **Sorghum** for straw; while the cost of feeding was at the same time reduced. **Mr. Bose** has been so far encouraged by the results, that he proposes to grow **Sorghum** as a fodder for his milch cows. The cultivation of **Sorghum** realized a net profit of R12-12-3 per bigha" (*1890-91*).

Fodder, *Conf. with pp. 287, 294, 304, 305, 309, 311.*

ASSAM. 2468

Cultivation in Assam.—In the report of the Agricultural Department of Assam (1886-87, p. 23), the *Jowár* and *Bájra* crops are dealt with as follows:—"These two admirable fodder crops are never grown in Assam, except, perhaps, occasionally by tea-garden coolies, and yet, in almost all parts of the Province, there are seasons when the cattle suffer from want of food. It was thought that possibly the people might be induced to grow fodder, and in 1886 a small consignment of one maund of *bájra* and one maund of *jowár* was tried. It was distributed to the Deputy Commissioners of Sibságar, Darrang, Kámrúp, Nowgong, and Lakhimpur, and to the Sub-divisional officer of Mangaldai. At Lakhimpur, Jorhát, Golághát, Nowgong, and Kámrúp, the experiments failed. At Sibságar a poor crop was obtained; at Tezpur the produce was good; at Mangaldai a part of the seed germinated well, and the Sub-divisional Officer (**Mr. Gait**) reports that the experiment has resulted in much good, as it has started the cultivation of these crops, and much is being grown by the coolies this year on their own account."

N.-W. PROVINCES & OUDH. 2469

Cultivation in North-West Provinces & Oudh.—*Judr* is a very important crop. According to **Messrs. Duthie & Fuller** there were in 1882-83 1,356,961 acres devoted to the pure crop; 1,543,486 acres of *juár-arhar* (that is, *juár* mixed with pulses); and 247,076 acres under *juár* fodder (called *chari*). Some idea may also be learned of the importance of the crop from the review of opinions collected by the Famine Commission regarding the staple food of these provinces:—"The poorer classes, that is to say, the great majority of the people, live on the cheapest *kharif* grains from September or October till the spring harvest comes in, and on barley, gram, and peas from that time till the autumn. Where rice is much grown the coarser kinds are mainly consumed by the lower classes. Wheat and the finer kinds of rice are almost entirely reserved for the consumption of the rich, who also live on the millets, *juár*, and *bájra*, to some extent in the cold weather. Both the millets and cereals are usually ground into meal and baked in round thin cakes; they are eaten either with *dál, i.e.*, pulses mashed and boiled into porridge, or with *ghi*, or both together; salt is always added, but there is not much demand for pepper. *Juár* and *bájra* are sown shortly after the commencement of the rains, and lateness in the fall of rain is a disadvantage. *Bájra* can, however, be put in later than *juár*. If no rain falls before the 15th of August, they can hardly be sown, or if sown and no rain falls for a month after the crops come up, they are burned up unless protected by irrigation, destroyed by heat and drought. Rain, therefore, in June or the beginning of July, and again in August or early in September is most required. *Arhar* cannot be sown later than *juár* and *bájra*, but it stands a subsequent failure of rain better, and lasts when they do not."

For the purpose of this article it is perhaps undesirable to attempt a

CULTIVATION in the N.-W. Provinces & Oudh.

review of the numerous district reports which have appeared on the *juár* of these provinces. This has been so recently done by **Messrs. Duthie & Fuller** (*Field & Garden Crops*) that it may be said nothing new has transpired, and from that work the following information has been derived:—

"The dry stalks and leaves of *juár*, chopped into small pieces, form the ordinary cattle fodder of the country, for some months in the year, being known by the name of *karbi*. Occasionally *juar* is grown solely for cattle fodder, and not for its grain at all, in which case the stalks are cut while green before the seed has had time to mature. It is almost the only green fodder crop grown as such in the provinces, and hence, when grown for this purpose, has no more distinctive name than *chari*, which simply means fodder. *Chari* cultivation is, however, almost entirely restricted to the districts of the Meerut Division, where the cattle are mostly purchased from the outside, and are of a far better quality than those in other parts of the provinces.

"Its value as a green fodder may be inferred from the following analysis made by **Professor Voelcker**, in which its nutritive qualities are compared with those of turnips:—

	Chari.	Turnips.
Water	85·17	90·43
Flesh-forming matters	2·55	1·04
Fatty and heat-producing matters	11·14	7·89
Inorganic matters	1·14	·64
	100·00	100·00

The area under *juár* is larger than that under any other *kharif* crop, with the exception of rice, and amounts to nearly 36,98,000 acres, 31½ lakhs acres of which are in the 30 temporarily settled North-West Provinces districts, forming 13 per cent. of their total cropped area, and 25 per cent. of the area under *kharif* crops. It is, however, almost entirely confined to the districts of Rohilkhand, and Doáb and Bundelkhand, and comparatively rare in the east of Oudh and in the districts of the Benares Division, where its place is taken by rice."

"*Juár* is a *kharif* crop, being sown at the commencement of the rains and cut during November. When grown for fodder and irrigation is available, it is often sown in the hot weather, before the commencement of the rains, that it may be got off the ground as soon as possible, since it is generally followed by a crop in the succeeding *rabi*, this rarely, if ever, happens with *juár* when grown for its grain. No particular rotation appears to be followed, but it frequently alternates with rice on clay or loamy soils not subject to flooding. It is comparatively rarely sown alone, being, as a rule, mixed with several other crops, of which *arhar* (**Cajanus indicus**) is the chief. The oil seed called *til* or *gingelly* (**Sesamum indicum**) and the low-growing pulses *múng*, *urd* or *mash* (**Phaseolus mungo and radiatus**) and *lobia* or *rawás* (**Vigna Catiang**) form an undergrowth in most *juár* fields, yieding but a small return if the *juár* prospers and overshadows them, but occasionally forming the principal part of the crop if the *juár* suffers from failure of rain, which it feels more keenly than its deeper-rooted associates. Loamy or clayey soils are preferred, where possible, and perhaps the best crops of *juár* in the provinces are borne by the heavy black soil of Bundelkhand. So far indeed as soil is concerned, *juár* in the *kharif* answers to wheat in the *rabi*, the place of barley and its mixtures being taken by the bulrush millet (*bajra*—**Pennisetum typhoideum**). Manure is but rarely given, unless the crop be grown for fodder, when it is generally succeeded by a *rabi* crop, and the land requires therefore artificial stimulation. The number of ploughings varies from one to four, land which has borne a crop in the preceding *rabi* not being held to require so much tillage as land which has lain fallow since the end of the *kharif* preceding (Bareilly). Clods are usually broken before sowing by the use of the log clod-crusher. The first sowings of the *kharif* are those of cotton, and as soon as these are finished *juár* is commenced with. The seed is sown broadcast and ploughed in, being used at the rate of 3 to 6 seers per acre if for a grain crop, and 12 seers per acre if for fodder, when thickness is the chief thing looked to. The seed of the minor crops (*arhar*, *múng*, etc., known collectively as *utara*) is mixed with the *juár* seed and scattered with it; *lobia* alone being sown by hand in lines across the field (Cawnpore). In some parts of the provinces the finest heads are picked out at each harvest and set aside for sowing in the succeeding year

SORGHUM vulgare.

Provincial Account of the

CULTIVATION in the Panjab.

(Cawnpore). Irrigation is very seldom used unless the crop has been sown before the commencement of the rains, or the season is peculiarly unpropitious. The crop is generally weeded at least once, sometimes by hand, but often by merely driving a plough in lines through the field when the plants are about a foot high, so as to open out the soil round the plant roots which has a very beneficial effect."

PANJAB. 2470

Cultivation in the Panjáb.—Mr. Baden Powell (*Panjáb Products*) has very little to say regarding *juár* proper, from which circumstance it might be inferred to be of little value. In the Agricultural Returns of the Province, however, 2,369,400 acres are shown to have been devoted to the crop in 1889-90. In the Famine Commissioner's Report it is stated that from the replies of the District Officers it may be inferred that barley and the *kharif* millets and maize, with a little rice, constitute 47 per cent. or nearly half the food consumed in the province. Wheat contributes 37 per cent. and pulses 15 per cent. But in certain districts no wheat is consumed, while in others, such as Lahore, Rawal Pindi, and Multan wheat forms half of the food annually consumed by the population. It will thus be seen that millets are fairly important and *juár* is the chief one. By the railway returns, the Panjáb is also shown to export millets (*juár* and *bájra*) very largely to Bombay and the North-West Provinces. There are said to be two varieties, one with a reddish grain and the other white. It is, however, fairly extensively cultivated as a fodder crop known as *chári*, and for this purpose it is sown much thicker than when grown for food. The following extracts from district reports may be found to furnish the chief ideas regarding the Panjáb cultivation of this millet:—In the Settlement Report of Jhang district it is stated that—

"In this district there are 38,268 acres under cultivation. *Jowar* and cotton are the two *kharif* staples. *Jowar* is grown largely on wells and *sailab* lands.. On the Barani lands of the Chiniot *tahsil* its place is taken by *bájra*. It is not grown to any large extent on the wells on the northern villages of the *tahsil* where *makai* takes its place. A recent accretion of good soil, land well manured and soil that is clayey, and has lain fallow for some years are the three best soils for *jowar*. On the river lands the best soil for *jowar* is a light sandy loam of recent formation, well moistened by percolation. There is not very much preparation in the way of ploughing. Twice is considered ample. The seed is then sown broadcast and ploughed in. The ground is not rolled unless it is cloddy. If the soil is not very moist the seed is sown with a drill, in order to get it as deep down into the soil as possible. Sowings commence at the beginning of *Sáwan*, and go on to the beginning of *Bhadron*. The earlier the *jowar* is sown the better. It ripens before the frost and the stocks are sweetest. *Jowar* is only sown late for fear of floods. If there has been rain, and the soil is sufficiently moist, the land is prepared and sown. If there has been no rain the land is first irrigated, then ploughed twice and rolled. The seed is sown broadcast and ploughed in. *Jowar* seed is always steeped in water the night before it is sown. *Jowar* is watered about every eighth day, but it is hardly safe to lay down any rate other than that it is watered whenever it begins to dry up. *Jowar* when needing water is a sure index to the quality of the soil. Where the soil is poor the *jowar* leaves shrivel up very soon, while the rest of the crop, if the soil is good, may show scarcely any signs of distress. The amount of seed sown is ½ ℔ an acre. There are numerous kinds of *jowar*. That grown near Khíva and Khánínvána has the highest reputation. The varieties usually denote little more than grades of flavour in the grain when parched or scorched. Of one kind of *jowar* the ear is compact and the grains close together, of another the ear is made up of a number of small branched stems, each carrying grain. The first is called *gumma*,* the second *tilya*.* *Jowar* is often manured. The kachi *jowar* ripens earliest in the latter end of October, that grown in the Kichand next, and, that on the Chenab last, in the beginning of February. The two best crops of *jowar*, the writer has seen, were one in Ladkana Mirati, a *bela* in the river Chenab, *tahsil* Shorkot, and the other at Chakk Bandi, on a well largely assisted by surface drainage from the Bár.

Washing seed. 2471 *Conf. with p. 302.*

Varieties. 2472 *Conf. with pp. 277—9, 282, 290, 295, 305, 309, 310, 311.*

Jowar is rather a delicate plant. Besides the maladies to which it is subject

* *Conf.* with the distinction established in the chapter on varieties: *gumma* is very probably **S. vulgare**, *tilya* one of the forms of **S. halepense.**—*Ed. Dict. Econ. Prod.*

before it comes to ear, early frost and late rain greatly diminish the yield and render the stalks tasteless and dry. It is also liable to '*toka*' and '*tela*'." The writer of the above passage enumerates 29 forms of **Sorghum** as known in the district.

CULTIVATION in the Panjab.

In Dera Ghazi Khan.—"*Jowar* is the staple food of the district and the chief crop grown in the Pachád. It is grown between June and August, and the later sown crops are considered the best. *Jowar* takes less water than most other crops. The average outturn is 10 maunds in the Pachád. In the Sind circle *jowar* is grown for fodder and is not a good grain-bearing crop. The stalks are eaten like those of sugar-cane" (*Report of the Settlement of the Dera Ghazi District*).

Fodder. 2473 *Conf. with pp. 287, 294, 304, 305, 306, 311.*

Of Dera Ismail Khan.—It is said: "About three-fourths of the *jowar* grown Cis-Indus is in the Thal; all through the district a great deal of the *jowar* is cut green for fodder; any grain that ripens goes to feed the cattle. The zemindars themselves do not eat *jowar* grain unless driven to it by necessity. In the Thal however, a certain amount is commonly eaten. Near towns green *jowar* is a very valuable crop, and often sells at R4 or R5 per *kaal*" (*Dera Ismail Khan District Report p. 344*).

CENTRAL PROVINCES. 2474

Cultivation in the Central Provinces.—Some idea of the immense importance of this millet may be learned from the remarks made by the Famine Commission regarding the staple articles of diet with the people of these provinces:—"*Juár* is the staple food in Nagpur, Wardha, Nimar, Western Chanda, the south of Chhindwara and of Betul. *Juár* ordinarily is sown from the first fall of rain on the poorer and higher grounds till the end of July in lower lands. The critical time is through August, when a total cessation of rain will destroy the plant on the high lands, and an excessive fall will rot the seed and young plant in the lower grounds. Again, after full rains in August, a total cessation in September will destroy the plant on the high grounds, but heavy rain will benefit high lands, and will not injure the low lands to any considerable extent. The rains which fall at the end of September, or early in October, will determine whether the outturn be well above or far below the average period. The inferior variety of *juár*, cultivated on the high grounds, is harvested in November, the bulk of the crop on rich soil in December. The cultivation of the lighter sort is on the increase, as it is almost independent of the uncertain October rains." Speaking of Sironcha the Famine Commissioners specially allude to the existence of a *rabí juar* crop.

The following extract from the Report on the Land Revenue Settlement of the Nimar district will indicate the mode and extent of the Jowar industry in that part of the country:—

Varieties suitable to different climates. 2475 *Conf. with pp. 277—9, 282, 290, 305, 310, 311.*

Jowar forms the "staple crop of the autumn harvest. A great number of varieties of it are recognised as possessing distinct qualities in the matters of weight of crop, time of ripening, hardiness in season of drought, and flavour when eaten. The best of these grown only on the best *mál* land, are *honri, amneri, silwa, gúngei, andlei,* and *dúdmogar*; the poorest, which are always sown on *kukrah* or inferior *mál* are *hoglei, santei, agya-kondal,* and *sat-páni* (which will ripen with only severe showers). There are about twenty more besides these, and it is believed that these distinctions are not imaginary, but really form a part of the agricultural knowledge of the people. *Jowar* is sown with the *tífun*, or drill, as soon as the first rains have sufficiently moistened the land, and the appearance of the weather indicates continued rain. This period ranges between the 15th June and the 15th July. It requires to be weeded once or twice, according to the strength of the soil in breeding weeds, and at the same time the plants are thinned out till they stand about 1½ feet apart, as they shoot up very rapidly, they soon top the weeds, and require no further attention. *Jowar* is held to be an exhausting crop and is seldom sown in successive seasons on the same land; the alternating crops being cotton, *túr* or some of the pulses, if the land is good enough. The different species ripen at different times, but cutting usually commences about the 15th November, and all are generally reaped by the end of December. The quantity of seed sown is from 6 to 10fb, and the yield varies from about 150fb only on the poorest *kukrah* soil to 700fb or more on the strong *mál* land of the Taptee Valley. On average *mál* the yield may be about 400fb, but the writer is inclined to estimate the average yield for the whole district at about 350fb per acre, which is a return about forty-fold. Besides this the stalks (*kurbi*) form a valuable cattle fodder, fetching when sold, which is rarely the case, about R4 or 5

CULTIVATION in the Central Provinces.

per acre. Forty-five per cent. of the whole cultivated area is devoted to this crop, which forms the staple food of all classes of the people. The heads are eaten largely at harvest time, simply roasted in the field, but it is usually made into unleavened cakes and eaten with pulse or gram flour and *ghi* (clarified butter). It is much lighter than wheat and extremely digestible, so that a good deal greater weight of it than of wheat forms the average ration of a working man."

In the Settlement Report of the Chhindwara district, referring to this plant it is stated that it forms the chief food of the cultivating classes, rice and wheat being alike but little used by them. It is sown at the beginning of the rains in June and ripens towards the end of November. The Upper Godavery district possesses both a *kharif* and *rabi* crop of this millet, the latter is said to have a yellow grain. The white *juár* constitutes the chief food of the people.

BERAR. 2476

Striga. 2477 *Conf. with p. 303.*

Cultivation in Berar and Hyderabad.—Very little of a definite nature can be learned regarding the *juár* crop of this province. It often suffers badly from the parasite **Striga lutea**, though the Natives appear to have made the same observation as in the Central Provinces and Bombay, that the ravages of this pest are greatly mitigated by alternating cotton and *juár*. In the Famine Commission Report the following passage occurs regarding this millet:—"*Juári* is the staple food of the province. In 1876-77 it occupied 35·10 per cent. of the total cultivated area of the province, and 68·31 per cent. of the area cultivated with food-grains. It is sown in the end of June or beginning of July, as soon as the cotton sowings, which first receive attention, are completed. It is reaped in the end of November or beginning of December. From the time it is sown, until the end of September, any prolonged period of dry weather is hurtful to *juári*. It specially requires good rain in August. The best crops of *juári* are raised when the monsoon rains are equitably distributed over the months from June to September, and aggregate 28 to 35 inches. Prolonged wet weather, such as we have had this season, is damaging to the *juári* in low lands. The *juári* crop is again sometimes damaged in autumn by unseasonable rain falling at a time when the heads of grain are ripening, the effect of which is to blacken and rot the grain."

HYDERABAD. 2478

Two crops. 2479 *Conf. with pp. 279, 282, 283, 287, 291, 293, 306, 310, 311-12, 313-14.*

Of Hyderabad it is stated there are two crops of *juár*: the one is sown from the 6th June to 17th July, and reaped from the 22nd October to the 30th November. Rain is essential at the time of sowing; up to a little before the full development of grain, a period of two months, after which at the reaping of the crop and a little before rain is injurious. Excessive rainfall a month after sowing, that is, when the seed has germinated, blights the crop. From this cause proceeded the chief injury in 1878. The second crop known as white *juár* is sown between the 25th September and 3rd November. It is reaped between the 17th February and 15th March. Rain is necessary from 5 to 6 weeks after sowing, but failure of rains does not injure the crop when grown on *regar* soil.

MADRAS. 2480

Cultivation in Madras.—In the Famine Commission Report a table is given of the food-stuffs grown in the *Ryotwar* and *Inam* lands of the Presidency. Cholam appears as occupying in the following districts the percentages to the total food area indicated by the figures shown:—Kurnool 50, Bellary 40, Cuddapah 31, Coimbatore 30, Madura 22, Trichinopoly 17, Tinnevelly 9, Godaveri 9 and Salem 5. These seem to be the only districts in which it is grown, but in some of the districts not named its place is taken by *cumbú* (**Pennisetum typhoideum**—*bájra*), in others by *rágí* (**Eleusine Coracana**), and two or all three of these millets may be grown in certain districts. Thus where rice is either not suitable or less popular, an abundant harvest is obtained by the people of Madras from millets. In the table furnished at page 296, Madras is shown to have

SORGHUM vulgare.

CULTIVATION in Madras.

had last year 4,276,5,09 acres under *judr* (*cholam*) and 2,596,947 acres of *bájra* (*cumbú*).

A full account of the cultivation of *cholam* (*juár*), as carried on in the Coimbatore district, is given in Mr. Nicholson's Manual.

Two crops. 2481 *Conf. with pp. 279, 282, 283, 291, 305, 308, 309, 311—12, 313—14.*

"*Cholam* (**Sorghum vulgare**), he says, is grown either as a garden or dry crop, also frequently on tank-fed lands, or as a first crop on poorly-irrigated channel lands. This is a very useful crop, giving a good outturn with a minimum of rainfall or water and labour, as it is said *cholam* is the crop for a lazy man. Its grain is abundant and nutritious, while its stem is excellent fodder for cattle, which eat it readily whether green or dry. In garden cultivation there are two great seasons and kinds of crop: (1) that cultivated in March and reaped in June-July (hot weather crop); (2) that sown in October-November and reaped in January-February. It is called a four-month crop. The chief varieties are *periya-vellei, tovaram, siru-vellei, uppam, karuvettu, sen, ennei-kittan,* and *kakayrettu.* The first five are usually sown in October-November and reaped in February-March, the latter grow from March-April to July. There are variations, however, in each *taluk* as to the kinds grown in the two seasons. The soil is manured by cattle and sheep-penning and by village refuse, the manure being well ploughed in when rain falls; seed is then sown broadcast, and ploughed in, after which the plots for watering are formed, but if the soil is fairly moist, it is not irrigated for two or three weeks. After the first watering itis hoed and thereafter watered about once a week till harvest. Seed per acre is 4 measures, or 14℔; and the outturn up to 768 measures, or 2,660 ℔. This crop, as grown in gardens, is a marked contrast to that on dry lands; even in fairly good years the latter, on soils other than black or new lands, is short (5 to 6 feet high) and with but poor heads, while in less favourable years it is not even 3 or 4 feet high, with stalks as thick as quills and heads as big only as an egg. On gardens it is 8 to 9 feet high, stout, healthy, and with splendid heads full of grain. Such is the effect of regular water and good soil and cultivation. On the higher classed red soils and on black *cholam* is, however, of excellent growth, almost equal to that in gardens, though not grown so densely; in the red soils of Anamalai and the neighbourhood, which have been recently reclaimed from the forest and get the south-west monsoon, the crops are of the finest description. The dry land *cholam* embraces most of the above varieties and is grown at two seasons, *viz.*, *Kár,* sown in May, June, July, and reaped from three to four months later; and *paruvam*, sown in October-November and cut in February. Seasons differ widely in the various *taluks, Kár cholam,* for instance, being rare in Erode, and common in Karúr and North Dhárápuram. The *Kár* crop is usually mixed with cotton, gingelly, dholl, castor-oil, and pulses; the *paruvam* crop is not mixed. Castor-oil and dholl are sown in lines; other seeds broadcast with the *cholam.* After sowing, the crop is inter-ploughed when from four to six weeks old. The *Kár* crop is reaped in July or later, according to sowing; the dholl is reaped in January and castor beans twice, *viz.*, in November-December and February-March."

The same author elsewhere, speaking of *cholam* when grown as a fodder crop, says:—

Fodder. 2482 *Conf. with pp. 287, 304, 305 306, 309.*

"In the Kángyam division of Dhárápuran, where the best cattle are still reared, there is a regular practice in February-March of growing either *cholam* or *kambu* (chiefly the former) under well irrigation, this is called *adar* (= close, crowded) *cholam* from its being sown closely so as to yield heavily, and is grown at any time that fodder may be wanted. It is cut down before earing and affords considerable provision during the hot weather. Fodder crops are not grown on dry lands; there is considerable pasture except in the hot weather, and as it is unusual to get rain sufficient even for ploughing from the end of December to 15th April, no such crops are possible except on garden lands. *Cholam* straw is a favourite fodder and is carefully stalked for use; the numerous stalks that dot the black cotton soil of Udamalpet and all gardens are an agreeable feature in the landscape."

On the subject of outturn Mr. Nicholson furnishes useful fmormation. Space cannot, however, be afforded to continue the quotation of the original passage. Mr. Nicholson furnishes a table in which he shows the average yield from good land to be 374 Madras measures,* of middling soils 261 measures, of bad soils 129 measures, and the average of all these 161 Madras measures, or, say, 563℔. Commenting on the district as a

* A Madras measure appears from the figures given above to be 3½℔.

SORGHUM vulgare.

Provincial Account of the

CULTIVATION in Madras.

whole he remarks :—"The outturn of *cholam* on dry land varies immensely from field to field and from year to year; that on the Anamalai red soil is equal to any in the Presidency, and that on the black soils is probably nearly equally good, while that on thousands of acres of poor land in Erode and Dhárápuram is hardly worth calling a crop."

Godavery. 2483

Of the GODAVERY district an account has been furnished of the cultivation of *cholam* which agrees in many points with the above as may be seen from the following extract :—

"Black *regada*, or moderately *regada* soil, suits this plant. It does not grow in very sandy or in swampy soils. The preparation of the ground commences in June, when it is ploughed twice to get rid of the weeds, and after that it is ploughed once every fifteen days till September. *Cholam* cannot bear too much moisture. If water lies on the ground the plant rots. The ground should be slightly moistened before the first ploughing in June, again in September, and again in October. In black *regada* scarcely any moisture is required. The seeds of the previous year's crop are the best, taken from heads of *cholam* well dried and preserved in bags for this special purpose. Four seers of seed suffice for an acre, one seed being sown every eight inches. A small quantity of *pesalu* seed, or *alasanda*, or both are frequently mixed with the *cholam* seed and sown together with it. The ground is again ploughed after the seed is sown and fresh-harrowed. The shoots appear in six days, and by the end of a month will be half a yard high. After three months more, the stalk will have attained a height of two yards, and the head will begin to form, and after another month the crop ripens. The ground requires weeding about a month after the shoots appear, and a little later it is usual to run a plough between the lines of stalks, taking care not to injure them. A head of *cholam* is very often twelve inches in circumference. In the harvest the heads are cut off with twelve inches of stalk attached to them. They are allowed to dry for two days and then left in heaps for three weeks after which they are stacked. To separate the grain, the heads are cut off and thrashed with sticks or palmyra leaf stalks."

BOMBAY & SIND. 2484

Cultivation in Bombay and Sind.—In the table furnished at p. 297, it is shown that there were last year 8,282,876 acres devoted to this millet, and only 2,311,558 acres under wheat and very nearly the same amount of rice. The *juár* crop thus covered twice the area devoted to wheat and rice conjointly. The passages quoted below will convey the chief ideas regarding the methods, etc., of cultivation, but special attention may be drawn to the fact that in this Presidency the *rabí*, or cold weather crop of *juár*, attains far greater importance than in any other province. It will, in fact, be noted that in certain districts a *rabí* crop is the chief, if not the only, *juár* crop raised. What is also very significant is the further fact that the *juár* of Bombay is chiefly the forms which the writer has referred to **S. halepense**, *var.* **Roxburghii**, a series of forms known collectively in Western India by the name *shállú*. Many of these are not only rich in saccharine matter, the stems being accordingly largely eaten as sweetmeats, but they are even expressed for their juice and sugar made therefrom like the sugar-candy of Bikanir. When Imphee was first talked of as a desirable crop to introduce into India, the idea seems to have been currently circulated that none of the Indian forms of this millet had been known to be used for the purpose of sugar, and hence the supposed value of the Kaffir **Sorghum.** **Mr. Wray**, formerly a sugar-planter in India and author of a work on that product, says in an essay on Imphee: "So much has this been the case, that although I have been a sugar-planter so many years of my life, and have, as an author, had to look closely into all information connected with the production of sugar, in both ancient and modern times, yet I can truly say that I knew nothing of this plant until I resided in the colony of Natal, on the south-east coast of Africa, where it grows in abundance; nor was it until I reached England this year" (1857?) "that I could obtain the works of scientific authors, from which to learn its botanical character, and what efforts have been made, if any to introduce it to notice" (*Sorgho and Imphee by Olcott, 201*). The

Rabi crop most important. 2485

Conf. with pp. 279, 282, 283, 291, 293, 296, 305, 308, 309, 310, 311, 313-14.

Sorghum sugar. 2486

Conf. with pp. 283, 296.

Cultivation of the Juar. (*G. Watt.*)

SORGHUM vulgare.

CULTIVATION in Bombay.

Sugar.

reader might consult the opinions which the Editor of this work has expressed regarding the failure to establish sugar-cane planting in India. One of the chief contentions there advanced is that ignorance of the resources of India had much to say to the failure. Had **Mr. Wray** extended his studies while resident in India to the indigenous forms of cane, he would assuredly have learned much of great value and could not have failed to find that sugar from time immemorial had in certain parts of India, been prepared from a race of **Sorghum** quite as good as the Imphee of his special studies in Natal. This illustration is, however, given here in order to enforce the writers firm conviction that much needless waste of money, both public and private, has been occasioned in the experiments to introduce to India exotic forms of Imphee and Sorgho at the expense of a total neglect to foster and develope the natural resources of the country. No better field of research could be afforded than is available in Bombay. The forms of *juár* in that Presidency not only constitute the chief article of food to a large proportion of the population* but, judging from the reports that have appeared, there are doubtless many races of this crop annually cultivated that are both more valuable as sources of food and fodder and even richer in sugar than are any of the exotic forms as yet made known.

The following passages will give the reader some idea of the character and value of the *juár* crop of Bombay.

Mr. Ozanne (in the *Statistical Atlas of Bombay*, published in 1888) gives the following brief review of the **Sorghum** crops of the Presidency:—

Three crops of Sorghum.
2487
Conf. with pp. *279, 283, 291, 293, 296, 305, 308, 310, 311, 312.*

"*Jowári* is the most important cereal, and at the same time the largest and most widely grown fodder crop of the Presidency. There are many varieties adapted to varying conditions of soil and climate. The grain is not exported, but it is stored for many years in underground pits with small damage by weevil, but with occasional serious loss by the flooding of the pits. The fodder is also stored with great care in the tracts of uncertain rainfall. Jowári fodder is superior to bájri. The grain is eaten chiefly as a bread stuff in unleavened cakes, but the custom of parching unripe ears as the crop is maturing is everywhere prevalent and is one of the reasons why estimates of yield are so intricate. Agriculturally speaking, three main divisions of the crop may be made: (1) the early sown or kharíf varieties, red and white; (2) the late sown or rabi, all white; and (3) the irrigated hot season crop grown for fodder alone. In the Karnátak, the early kinds are red and the late white-grained. They are also very important in Khándesh and Gujarát where they are white. Early jowari is not common in the Deccan, being replaced by bájri, and may be said to be grown only as a fodder crop for consumption in the early monsoon. It is sown in the Karnatak, as bájri is sown in the Deccan, with lines of pulses—generally a mixture of five pulses. In Khándesh it is a later crop, sown after cotton has been put into the ground, and generally alone, though often as a row crop in cotton fields. It matures in four months, but when only grown for fodder is cut in three. The stem is tall and leafy. The white kharíf jowari of Khándesh and Gujarát gives better fodder than the red of the Karnátak. Late or rabi jowári is essentially the crop of the Deccan. It is a five-month crop sown about September. Its growth is not usually large. The fodder is comparatively light in weight but in quality is preferred, containing more sugar and less woody fibre than the stem of the early red jowári. (This must not be confused with sorgho, **Sorghum saccharatum,** which has a black seed and is not yet grown except experimentally.) Nevertheless rabi jowári is rather a grain than a fodder crop, and in good seasons yields very heavily. In the Karnátak the rabi jowári is also white and is sown alone or with rows of safflower or linseed as in the Deccan. In Khándesh and Gujarát rabi jowári is little sown. It prefers a black soil and cannot thrive in the cold season without both depth and water-retaining power which good black soil possesses. It grows with great luxuriance on river-side land enriched by silt deposits during floods. Hot season jowári is known all over. It must be irrigated, and is almost universally cut before maturity and fed to the cattle at once in the hot weather."

* In the Ain-i-Akbari (*Gladwin's Translation, Ed. 1800, Vol. II, 62,*) "*Jewary and Bajera*" it is said were the grains chiefly cultivated in the Subah of Gujerat during the 16th century. They were the chief food of the inhabitants.

SORGHUM vulgare.

Provincial Account of the

CULTIVATION in Bombay.

The following passages from the Gazetteers may help to amplify the information already furnished :—

Ahmednagar. 2488

AHMADNAGAR.—"Indian millet, *juvari* or *jondhla*, with, in 1881-82, a tillage area of 679,879 acres, is the staple grain of the open country, and is largely exported. Except two early varieties, *kondya* or *kundya* and *kálbondi*, Indian millet forms the chief cold weather crop, and, without either water or manure, yields plentifully, especially in black soils. It requires little outlay and is grown by all the poorer land-holders. The most esteemed variety is *shálu*. It is grown in black soils and is seldom watered or manured. The grain is white and the stalk thin, three to five feet high and with sweet juice. Other varieties are *dudhmogra* which is sown with *shálu* either mixed or in separate furrows. The grain is full and milky and is much valued when parched and made into *láhi*. The stalk is inferior to *shálu* as fodder, being straight and hard. The head is so thin and feathery that birds cannot rest on it or harm it. The stem of a dark-husked variety of *dudhmogra* is sometimes used as a hand-rod by weavers. *Támbdi*, or red *juari*, is sown earlier and in lighter soils than *shálu*, and ripens more rapidly. The stem is three or four feet high and makes poor fodder. Of the two early varieties, *kondya* or *kundya* is grown and cut for fodder before the head appears, and *kálbondi*, so called from its dark husk, gives the husbandman food in bad years before the regular crop ripens. The stem is six or eight feet high and the head is large. In black soils in March after the wheat is harvested the land is as hard as brick, except two or three inches of the surface soil. But by April, the cracks and seams become two or three inches wide and often two feet deep, and the surface soil becomes pulverised. The cultivator then harrows it with the two-bullock harrow or *aut*. The pulverised soil is driven into the cracks and a new layer is brought up to the immediate weathering influences of the hot sun. The value of this change of soil is fully appreciated. The soil weathers till the rains in June. As soon after this as it is workable, it is harrowed with a four-bullock *aut* in a direction opposite to that of the former harrowing. By this means the sprouting of annual weeds is hastened, and the surface soil is loosened. In the next break in the monsoon the two-bullock *aut* is again brought on the land. It works in the same direction as the hot weather harrowing. It cuts down and kills the annual weeds, and levels the land. The seed—bed is now ready, though seed is not sown till the *uttara nakshatra* (22nd September—6th October). Meantime, the more harrowings and cross-harrowings the land receives the better. The seed is sown with the three-tined seed-drill or *tiphan* which is followed by the *rakhia* or beam-harrow. Except that late sowings require five or six pounds, the seed is put in at the rate of about four pounds the acre. *Kardai*, or safflower, is mixed with the seed at the rate of about half a pound to a pound the acre. The mixing does no harm. Generally, when the *juári* crop is poor, the safflower more than suffices to meet the assessment on the land. The two early varieties of Indian millet are sown thick and broadcast in June and July taking twice as much seed as by the drill. In clayey loam or *khalga* land when the early rains are not favourable, *juári* is sown as in black soil in the cold weather. When the *juári* is about a foot high, it must be weeded with the bullock-hoe. Two hoes, or *kulpás*, are placed side by side each in charge of a man, but drawn by only one pair of bullocks. With the bullock-hoeing, hand-weeding along the rows by women is necessary. Till the crop has grown so as to shade the land and prevent weeds from coming up, one or two hand-weedings by women are usual. This work has to be done quickly, both because the breaks in the rain do not last and because the weeds grow apace. It is usual to put at least ten women on a field, though as many as twenty and twenty-five are set to work by good cultivators, as supervision is not then so costly. Watching is a heavy item in the cost of growing *juári*. One man to about ten acres of land where there are no trees, and double the number, if there are trees, are required. These sit on raised platforms in the field armed with slings. Watching begins when the crop begins to ear and lasts 1½ to 2 months; when ripe the crop is pulled up and tied into sheaves. Five sheaves form a *páchundá*. The sheaves are laid in *páchundás* to dry. The size of the sheaves varies with the length of the stalk which is used as a binder. Occasionally when the husbandman finds it inconvenient to carry the produce home at once, he builds it into stacks or *kátrás*. On the thrashing floor women are employed to break off the ears and throw them on the floor. When this is done, muzzled bullocks tread out the corn which is then winnowed by three men. One stands on a raised platform and another hands up baskets of the grain mixed with the outer coverings of the grain and the small stalks. When there is wind enough, the man on the platform slowly empties the basket. A third man below keeps the pile of good grain separate from the chaff. Two practices materially affect the outturn, if it is judged by the yield

Seed Drilled. 2489

Cultivation of the Juar. (*G. Watt.*) SORGHUM vulgare.

CULTIVATION in Bombay.

on the thrashing floor. First, that of eating parched unripe ears while the crop is standing. Not only are the watchmen allowed to eat as much as they like, but the owner and his family and his invited friends mainly live on the unripe ears of *hurda*, in a good season for six or eight weeks. Secondly, that of pulling up the standing crop for fodder. This is more usual in a poor season when many of the stalks are earless or so behind in growth that they are not likely to be ready for reaping with the rest of the crop, and when other fodder is scarce. The acre yield varies from 150 to 11,000 and averages 500 pounds. *Juari* is chiefly in use as a bread grain, but is also eaten parched as *láhi*. The parched unripe heads called *horda* are a leading article of food with the labouring classes a short time before and after harvest season. Indian millet is the only cereal whose straw is used as fodder in its natural state. The fodder, though not abundant, is superior. In parts of the west it is stacked and thatched; in the east, where the rainfall is lighter, it is stowed in long grave-like ridges covered with clods of black soil."

Broach. 2490

BROACH.—"*Jowari* is the principal food grain. It is partly sown in June and harvested in October, but the main crop is sown in August and harvested in February. Wheat is sown in September or October, and ripens in March. *Jowari* is the staple food in the black soil and *bajra* in the light soil villages. The failure of the early rain would injure the *bajra*, and of the later rain, the *jowari*. The later rain is supplemented by heavy dews from October to December. Cold following on an excessive rainfall (month not given) destroyed the crops (cotton and *jowari* (?)) in A.D. 1835."

Dharwar. 2491

DHARWAR.—"The staple food is rice in the west border of the district, and *jowari* in the rest. Rice is sown in May and reaped in November. Early *jowari* is sown in July and reaped in November or December. Late *jowari* is sown in September and reaped in February. For the rice and early *jowari* rain is needed in June and July. For the late *jowari* a few heavy falls in October. Heavy rain in October causes mildew in the early *jowari*. Heavy rain in December causes blight in the wheat."

Kaladgi. 2492

KALADGI.—"*Jowari* and *bajra* are the staple food. *Bajri* is grown as a *kharif* crop, and *jowari* as both *kharif* and *rabi*. The times of sowing and harvesting are those given above. The *kharif* requires rain in June, July, and August, and is injured by heavy rain after it is in ear. The *rabi* requires rain in August and September."

Satara. 2493

SATARA.—"*Jowari* and *bajra*, and, in the western *talukas*, rice and *nachni*, are the food grains. *Jowari* and *bajra* are sown in June, *jowari* also as a rain crop in October. Rice and *nachni* in June. Those require heavy rain at sowing and transplanting. The *rabi jowari* requires rain at the end of September, and showers in December."

Sholapur. 2494

SHOLAPUR.—"*Bajra* and *jowari* are the staple food. *Bajri* is sown in June or July, and *jowari* as a *rabi* crop (September or October), and harvested in March. If the ground is well saturated when *jowari* is sown in October it does not require much more moisture."

Surat. 2495

SURAT.—In 1874-75 there were 72,521 acres under *júwár*, out of 247,224 under actual cultivation. *Júwár* holds the second place with 18·52 per cent. of the total area under cultivation. "*Júwár* is very extensively grown north of the Tápti, but is less grown towards the south until in the sub-divisions of Balsár and Pardi, it is almost entirely superseded by rice, *nágli*, and *kodra*. Along with rice, *júwár* forms the common food of the people of the district."

TRADE IN JUAR.

TRADE. 2496

The returns of this trade are unfortunately given conjointly with those of *Bájra*, so that the exact share taken by each cannot be made out. The reader may have noted, however, how very much more important *Juár* is than *Bájra*, so that he may be enabled to arrive at some conception of the proportion of the trade statistics that should be accepted as relating to *Juar*. The subject may be discussed under two sections.—Foreign and Internal.

Foreign. 2797

FOREIGN TRADE IN JUAR AND BAJRA.—It may at once be stated that India does not appear to import any millets from foreign countries. It is therefore more than self-supporting in the matter of these grains, since it exports annually considerable amounts. The following were theexports during the past five years:

Years.	Cwt.	R	Yeas.	Cwt.	R
1885-86.	275,747	7,71,432			
1886-87.	641,406	18,77,476	1888-89.	611,960	19,84,220
1887-88.	640,705	18,47,620	1889-90.	932,572	31,38,226

SORGHUM vulgare.

Trade in Juar.

TRADE.

It will thus be seen that the exports of these two millets have manifested a remarkable expansion, being at the present moment in round figures 1,000,000 cwt., valued at £300,000. The estimate of production of *Juar* alone, given above, was 107,000,000 cwt., so that the exports, very considerable though they be, are, comparatively speaking, of no moment. To the estimate just stated would have to be added the outturn of *Bajra* (which must be very large), so that it seems likely these exports do not amount to much over one-half per cent. of the production.

Bombay is the chief exporting port and is followed by Karáchi. Of the exports of last year, Bombay supplied 740,624 cwt. and Karáchi 190,321 cwt. The most important consuming countries appear to be Aden, which took last year 428,280 cwt., Abyssinia 210,267 cwt., Arabia 191,453 cwt., Mekran and Sonmiani 40,864 cwt., Egypt 38,579 cwt., Belgium 10,016 cwt, and France 4,300 cwt. The total of all other countries does not, as a rule, amount to 4,000 cwt., so that Great Britain consumes absolutely none of the Indian *Juar* and *Bajra*. A large import trade, however, exists in England and to some extent in Scotland and Ireland in the Egyptian *Dari* (as it is called), the millet here dealt with. It is used as one of the ingredients in poultry mixtures of grain and as a cattle food. Apparently it is not at present eaten as an article of human diet in Europe. The Egyptian traffic is a yearly increasing one and may comprise some re-exports from Egypt of Indian *juár*. It seems desirable that an effort be put forth to secure for India some share in the yearly expanding demand for this millet. The Italian Millet (**Setaria italica**) seems to be that which is in Europe preferred as a bird seed.

Internal, 2498

The Internal trade of India in these millets is, as might be inferred, far more important than the external. The statistical information available on this subject is not, however, sufficiently complete to allow of a detailed statement being furnished. But it may be judged of, to some extent, from the returns of the rail-borne trade and coast wise shipments. Of the rail-borne transactions it may suffice to take the figures of 1888-89. One striking peculiarity may at once be set forth, namely, that Calcutta imported only 7,496 maunds and Bengal Presidency 895 maunds, so that the statement already made that Bengal does not consume these millets is abundantly confirmed. The total rail-borne imports came to 39,21,067 maunds, of which Bombay Presidency took 9,92,583 maunds, the North-West Provinces 3,21,765 maunds, Bombay port town 11,61,566 maunds, Karáchi 11,25,553 maunds, and the other provinces the balance. Of the exporting provinces Sind heads the list, having furnished 11,10,614 maunds mostly to Karáchi, the Panjáb followed with 6,18,055 maunds mostly to Bombay Presidency (2,30,901 maunds), to the North-West Provinces (3,00,858 maunds). The Nizam's Territory also exports largely—to Bombay town 2,34,318 maunds and to Bombay Presidency 3,64,127 maunds. These figures thus demonstrate the sources from which the *juar* and *bajra* exported from India are mainly drawn. Roughly speaking, the Panjab, Sind, Bombay, and the Nizam's Territory grow for exportation a surplus over local consumption. The North-West Provinces (one of the chief consuming provinces) cannot be said to be self-supporting in this respect since it draws largely on the Panjáb, and it would appear that Bombay Presidency, though perhaps the most important producing province, does not yield enough to meet its home and foreign markets. It accordingly draws largely from the Panjáb and the Nizam's Territory. The effect, therefore,

Effect on Wheat Trade. 2499

of the foreign demand for wheat on Indian agriculture (as it would appear) might be conveniently studied in reference to the expansion or contraction of the demands for these millets. Before a foreign market for

TRADE.

Indian wheatarose, it very likely could be demonstrated that most provinces (as Madras at the present day) produced their own amounts of these millets. Wheat requires a better soil than *juár*, and, moreover, must be grown as a winter crop. It is, however, more profitable than millet and the provinces with suitable soil and a sufficiently long winter season have found it to their advantage to discontinue a *rabi* crop of *juár* and to substitute wheat. This has had the effect of not only raising the value of the land now under wheat, but of enhancing and expanding the cultivation of millets in tracts of country suitable to these crops but not good enough for wheat. The sale of surplus wheat and the purchase of the millet supply has accordingly taken place. That is to say, from regions above the level of rice inundation, but with sufficient *kharif* rains to allow of *juár* and *bajra* being grown a market has arisen in India itself for these millets. The area of millet cultivation, therefore, has expanded into such regions, and the value of millet land has been enhanced by the demand for Indian wheat in foreign countries. The remarkable prosperity of the Panjáb in its being both an exporting province for wheat and *juár* is due doubtless to the immense tracts of rich land which have been rendered productive through the extension of canal irrigation. The Panjáb thus produces a very considerably larger amount of food than it requires. In other parts of India the balance may almost be said to be struck between linseed and wheat or other food crops. Thus Bengal, the North-West Provinces, and the Central Provinces are the chief regions concerned in that crop. The Central Provinces are the most important in the production of *til* (**Sesamum**). The Panjáb may almost be said to produce no linseed, whereas in the Central Provinces that crop is regarded as more profitable than wheat. In this connection, therefore, the reader might consult the remarks which have already been offered in Vol. V., pp. 60-68. In Kathiawar and Cutch the place of *juár* may be said to be taken by *til* (see **Sesamum**). The extension of wheat, oilseed crops, or millet cultivation is a matter of personal advantage in which the Indian agriculturist is by no means backward in appreciating the facilities afforded him by the modern aspects of internal and external trade.

Coastwise Trade. **2500**

The coastwise trade may almost be said to amount to an export from Sind into Cutch, Kathiawar, and Bombay. The interchange between the other provinces is unimportant, except that Bombay port also sends large quantities to Cutch and Kathiawar. This traffic, therefore, adds to the supply from which Bombay port draws its exports to foreign countries. The Karáchi coastwise exports are to a certain extent drained from the Panjáb by rail, but the province of Sind is by no means unimportant in the supply of *juár*. But it may safely be said that the chief items of the coastwise trade are the quantities taken by Cutch and Kathiawar from Karáchi and Bombay. The former obtained last year from these ports 193,856 cwt. and the latter 236,955 cwt. The exports coastwise from Karáchi were 415,238 cwt. and from Bombay 222,596 cwt., out of a total from all Indian ports of 639,649 cwt. This fact will confirm the statement made above that Madras is self-supporting in its production of these millets, since it neither imports nor exports any to speak of. Bengal, as already remarked, practically does not consume millets, and its rail and coastwise as also its foreign transactions in these grains are unimportant in the extreme. One feature of this trade may, in conclusion, be added, namely, that so far as we know the foreign and internal exports from the producing areas are to tracts of country inhabited by simple agricultural communities, or to regions where modern civilization with its concomitant luxury has not penetrated to any appreciable extent.

SOY MDA febrifuga.

The Indian Red-wood.

SOYMIDA, *A. Juss.; Gen. Pl., I., 338.*

[*Sylv., t. 8;* MELIACEÆ.

2501

Soymida febrifuga, *Adr. Juss.; Fl. Br. Ind., I., 567; Beddome, Fl.*

INDIAN RED-WOOD, BASTARD CEDAR.

Syn.—SWIETENIA FEBRIFUGA, *Roxb.;* S. SOYMIDA, *Duncan;* S. RUBRA, *Rottl.*

Vern.—*Rohun, rohunna, rakat rohan,* HIND.; *Rohan, rohina, rohra,* BENG.; *Rakat rohen,* KOL; *Ruhen,* SANTAL; *Sohan, suam,* MAL (S. P.); *Rohni, bugut rori, rohun,* C.P.; *Soimi,* GOND; *Royta,* BHIL; *Rohan,* MERWARA; *Rohun, rohunna, rouen, ruhin,* DECCAN; *Rohan, rohing,* BOMB.; *Rorna,* KATHIAWAR; *Rohina,* GUZ.; *Shem, wond, wundmarum, shemmarum,* TAM.; *Súmi, sómida manu, chéva manu,* TEL.; *Suámi, sime-mara, some,* KAN.; *Patránga, rohuna,* SANS.

References.—*Roxb., Fl. Ind., Ed. C.B.C., 370; Voigt, Hort. Sub. Cal., 137; Brandis, For. Fl., 71; Kurz, For. Fl. Burm., I., 228; Gamble, Man. Timb., 76; Dalz. & Gibs., Bomb. Fl., 38; Rev. A Campbell, Rept. Econ. Pl., Chutia Nagpur, No. 8443; Graham, Cat. Bomb. Pl., 32; Sir W. Elliot, Fl. Andhr., 169, 170; Pharm. Ind., 55, 444; Flück. & Hanb., Pharmacog., 156; Fleming, Med. Pl. & Drugs (Asiatic Reser., XI.), 179; Ainslie, Mat. Ind., I., 123; O'Shaughnessy, Beng. Dispens., 247; Irvine, Mat. Med. Patna, 93; Murray, Pl. & Drugs, Sind, 83; Dymock, Mat. Med. W. Ind., 2nd Ed., 175; Dymock, Warden & Hooper, Pharmacog. Ind., I., 337; Bidwood, Bomb. Prod., 15, 326; Useful Pl. Bomb. (XXV., Bomb. Gaz.), 45, 258, 397; Royle, Ill Him. Bot., 142, 275; Liotard, Dyes, vii.; Wardle, Dye Report, 52; Man. Madras Adm., I., 313; Boswell, Man., Nellore, 100; Gribble, Man. of Cuddapah, 263; Settlement Reports:—Central Provinces, Upper Godavery, 37; Chhindwara, 110; Nimar, 305; Seoni, 10; Baitúl, 127; Raipore, 75; Chanda, app. VI.; Gazetteers:—Bombay, V., 285; VI., 13; VII., 39; VIII., 11; XV., 75; Mysore & Coorg, I., 47; Agri.-Hort. Soc. Ind., Journ. (N. S.) I., (Sel.) 62; Ind Forester:—III., 201; VI., 332; VIII., 417; X., 543; XI., 230; XII., 3183; app. 9; XIII., 120; Balfour, Cyclop. Ind., III., 715.*

Habitat.—A lofty, glabrous tree of North-Western, Central, and Southern India, extending southward to Travancore, and of Ceylon.

GUM & RESIN. Bark. 2502

Gum and Resin.—The deep red BARK contains a gum which is said by **Dymock** to form a good adhesive mucilage. The bitter principle of the bark was ascertained by **Broughton** to be a nearly colourless resinous substance, sparingly soluble in water, but more so in alcohol, ether, or benzol. It does not unite with acids or bases, and is less soluble in water containing them than in pure water. It has a very bitter taste and refuses to crystallise either from benzol or ether. It contains no nitrogen (*Mat. Med. W. Ind.*).

DYE & TAN. Bark. 2503

Dye and Tan.—The BARK is astringent and has been used for tanning. It is employed in Mysore as one of the second-rate dye-substances producing the dirty browns in which cotton fabrics are often dyed. It is also used in combination with more valuable dyes such as arnatto (**Bixa orellana**) (*Liotard*).

FIBRE. Bark. 2504

Fibre.—The BARK yields a strong red fibre, which is used in Chutia Nagpur for making ropes.

MEDICINE. Bark. 2505

Medicine.—The astringent and antiperiodic properties of the BARK of this tree have long been known to the inhabitants of India, but **Roxburgh** was the first to bring it to the notice of Europeans as a substitute for Peruvian bark. In 1791 he sent specimens home to Edinburgh, where **Duncan** made it the subject of a thesis which subsequently led to its introduction into the Pharmacopæias of Edinburgh and Dublin. **Ainslie** in describing it says:—"The bark given to the extent of 4 or 5 drachms, in the twenty-four hours, I have found to be a useful medicine, but beyond that

MEDICINE.

quantity it, in every instance in which I tried it, appeared to me to derange the nervous system, occasioning vertigo and subsequent stupor."

In 1868 **Soymida** bark was made officinal in the Pharmacopæia of India, but in the practice of Europeans it is now seldom if ever used as an antiperiodic, and only occasionally as an astringent and tonic in cases of dysentery and diarrhœa. The decoction forms a good substitute for oak-bark and is well adapted for gargles, vaginal injections, and enemata. In Native practice, too, it is very little used, but among the Santals a decoction of the bark is sometimes given for rheumatic swellings, and the powdered bark is applied as a poultice (*Dymock: Revd. A. Campbell*).

Fluckiger & Hanbury describe the bark from a young tree as occurring "in straight or somewhat curved half tubular quills, an inch or more in diameter and about ⅛ of an inch in thickness. Externally it is of a rusty grey or brown with a smoothish surface, exhibiting no considerable furrows or cracks, but numerous small corky warts. These form little elliptic scars or rings, brown in the centre and but slightly raised from the surface The inner side and edges of the quills are of a bright reddish colour. A transverse section exhibits a thin outer layer coloured by chlorophyll and a middle layer of a bright rusty hue traversed by large medullary rays and darker wedge-shaped rays of liber. The whole bark, when comminuted, is of a rusty colour, becoming reddish by exposure to air and moisture. It has a bitter astringent taste with no distinctive odour." **Dymock** adds that the old bark has a ragged dry tuber a quarter of an inch thick, of a rusty blackish brown colour, deeply fissured longitudinally and minutely cracked transversely. Old bark is generally in half quills, the total thickness being about half an inch; its colour is a rich reddish brown, its substance when soaked in water becomes very compact (*Pharmacog. Ind.*).

TIMBER. 2506

Structure of the Wood.—Sapwood small, whitish; heartwood very hard and close-grained, reddish black, very durable, not attacked by white-ants. Weight about 76lb per cubic foot.

DOMESTIC Wood. 2507

Domestic.—The WOOD is heavier and stronger than the better known American mahogany and is used for almost every purpose. It is much prized for house-building, and the wood-carving in many of the temples in Southern India is made from this timber. It is formed into pestles and pounders for oil-seeds, and is well adapted for ornamental furniture. Sleepers made from it are very durable, but the price it commands for other purposes precludes this use.

Spanish juice, see **Glycryrhiza glabra,** *Boiss.*; LEGUMINOSÆ; Vol. III., [512

Spathodea Rheedii, *Wall.*; see **Dolichandrone Rheedii,** *Seem.*; BIGNONIACEÆ; Vol. III, 174.

SPATHOLOBUS, *Hassk.*; *Gen. Pl., I., 534.*

2508

Spatholobus Roxburghii, *Benth.*; *Fl. Br. Ind., II., 193*; *Wight, Ic., t. 210*; LEGUMINOSÆ.

Syn.—BUTEA PARVIFLORA, *Roxb.*; B. SERICOPHYLLA, *Wall.*

Vern.—*Mala, mula, maula, chihut lar,* (fruit=) *bando,* HIND.; *Moru, mûrrd,* KOL; *Chihut lar,* (fruit=) *bando,* SANTAL; *Debrelara,* NEPAL; *Tetrobrik,* LEPCHA; *Bamdu, durang,* KHARWAR; *Pouk-nwav,* BURM.

References.—*DC., Prodr., II., 415; Roxb., Fl. Ind., Ed. C.B.C., 541; Voigt, Hort. Sub. Cal., 239; Brandis, For. Fl., 143; Kurz, For. Fl. Burm., I., 365; Gamble, Man. Timb., 122; Cat., Trees, Shrubs, &c., Darjiling, 27; Dalz & Gibs., Bomb. Fl., 71; Rev. A. Campbell, Rept. Econ. Pl., Chutia Nagpur, No. 7884; Graham, Cat. Bomb. Pl., 54; Atkinson, Him. Dist. (X., N.-W. P. Gaz.), 309; Gazetteer N.-W. P., IV., lxxi.*

Habitat.—A common gigantic climber of the forests of the Sub.-Himálayan tract, from the Jumna eastward to Bengal and Burma. It occurs also in the forests of the Konkan, and in Ceylon.

GUM. 2509

Gum.—A red gum resembling "Kino" exudes from this plant.

OIL. Seeds. 2510

Oil.—The SEEDS yield an oil, which is used for cooking and anointing purposes.

FIBRE. Bark. 2511

Fibre.—A fibre obtained from the BARK is twisted into ropes and bowstrings.

Spearmint, see **Mentha viridis**, *Linn.*; LABIATÆ; Vol. V., 231.

SPERGULA, *Linn.*; *Gen. Pl., I., 152.*

2512

Spergula arvensis, *Linn.*; *Fl. Br. Ind., I., 243*; CARYOPHYLLEÆ.

SPERGULA OR CORN SPURRY.

References.—*Boiss., Fl. Orient., I., 731; Roxb., Fl. Ind., Ed. C. B. C., 385; DC., Orig. Cult. Pl., 114; Atkinson, Him. Dist. (N.-W. Prov. Gaz., X) 306; Gazetteer, N.-W. P., IV., lxviii.; Ind. Forester., XII., app., 6; Journ. Agri.-Hort. Soc., II., Sel., 171, 172.*

Habitat.—An annual herb, which appears in cultivated fields in various cool parts of India, and throughout the Northern Hemisphere.

FODDER. 2513

Fodder.—Cultivated in Europe as a fodder plant.

DOMESTIC 2514

Domestic.—The cultivation of this plant as a green manure, in the dry and sandy soils of Upper India, has been recommended (*Journ. Agri.-Horti. Soc. Ind.*).

Spermaceti, see **Whales**; Vol. VI., Part III.

SPERMACOCE, *Linn.*; *Gen. Pl., II., 145.*

2515

Spermacoce hispida, *Linn.*; *Fl. Br. Ind., III., 200*; RUBIACEÆ.

Syn.—S. ARTICULARIS, *Linn. f.*; S. SCABRA, *Willd.*; S. HIRTA, *Rottl.*; S. RONGICAULIS, AVANA, and FAMOSISSIMA, *Wall.*; S. TUBULARIS, *Br.*

Vern.—*Madanaghanti*, HIND.; *Pitua arak'*, SANTAL; *Ghánta-chi-baji, dhoti*, BOMB.; *Nattai-chúri*, TAM.; *Madana, modina, madana budata káda, madana grandhi*, TEL.; *Madanaghanti*, SANS.

References.—*DC., Prodr., IV., 555; Roxb., Fl. Ind., Ed. C.B.C., 125; Voigt, Hort. Sub. Cal., 396; Thwaites, En. Ceyl. Pl., 151; Trimen, Sys. Cat. Cey. Pl., 46; Dalz. & Gibs., Bomb. Fl., 111; Aitchinson, Cat. Pb. & Sind Pl., 70; Graham, Cat. Bomb. Pl., 93; Sir W. Elliot, Fl. Andhr. 108; Rheede, Hort. Mal., IX., t. 76; Thesaueus, Zey., t. 20, f. 3; Ainslie, Mat. Ind., II., 259; Bidie, Cat. Raw Pr., Paris Exh., 55; Dymock, Mat. Med. W. Ind., 2nd Ed., 408; Atkinson, Him. Dist. (X., N.-W. P. Gaz.), 311; Gazetteer:—Bombay, V., 26; N.-W. P., I., 81; IV., lxxiii.; Mysore & Coorg, I., 70.*

Habitat.—A common herbaceous weed, found throughout India from the Western Himálaya eastwards to Assam and southwards to Ceylon. It is distributed to Southern China and the Malayan Archipelago.

MEDICINE. Root. 2516

Medicine.—The ROOT, which is a little thicker than the officinal Sarsaparilla root, is, like that, used as an alterative. It may be given in decoction to the extent of 4 ounces or more daily (*Ainslie*).

FOOD. Leaves. 2517

Food.—The LEAVES are eaten as a vegetable, but only in times of scarcity.

SPHÆRANTHUS, *Linn.*; *Gen. Pl., II., 294.*

2518

Sphæranthus indicus, *Linn.*; *Fl. Br. Ind., III., 275*; *Wight, Ic.*, [*t. 1094*; COMPOSITÆ.

Syn.—S. INDICUS and HIRTUS, *Willd.*; S. AFRICANUS, *Vall.*; S. MOLLIS, *Roxb., Fl. Ind., Ed. C.B.C., 608.*

Vern.—*Mundi, gorak mundi*, HIND.; *Chagul-nadi, murmuriá, ghork, múndi*, BENG.; *Belaunja*, SANTAL; *Múndi, múndi búti, guruk mundi,*

(Bazár names=) *sakhmihaiyát, khamádrús, ghúndi*, PB.; *Mundi, múndhri*, DECCAN; *Mundi, gorakh-mundi*, BOMB.; *Kóttak, karandai*, TAM.; *Bóda-tarapu*, TEL.; *Miran-gani, attaká manni*, MALAY.; *Mun-ditiká, múndí*, SANS.; *Kamás ariyús*, ARAB.; *Kamáduriyús*, PERS.

References.—*DC., Prodr., V., 369; Voigt, Hort. Sub. Cal., 409; Thwaites, En. Ceyl. Pl., 162; Trimen, Sys. Cat. Cey. Pl., 47; Dalz. & Gibs., Bomb. Fl., 123; Stewart, Pb. Pl., 130; Aitchison, Cat. Pb. & Sind Pl., 73; Rev. A. Campbell, Rept. Econ. Pl., Chutia Nagpur, No. 9203; Graham, Cat. Bomb. Pl., 96; Sir W. Elliot, Fl. Andhr., 28; Rheede, Hort. Mal., X., t. 43; Thesaurus, Zey., t. 94, f. 3; Irvine, Mat. Med. Patna, 30; Medical Topog., Oudh & Sultanpore, 42; U. C. Dutt, Mat. Med. Hind., 310; Murray, Pl. & Drugs, Sind, 182; Bidie, Cat. Raw Pr., Paris Exh., 55; Dymock, Mat. Med. W. Ind., 2nd Ed., 426; Year-Book Pharm., 1878, 288; Drury, U. Pl. Ind., 402; Gazetteers:—Bombay, V., 26; VI., 15; XV., 436; N.-W. P., I., 81; IV., lxxiii.; X., 312; Mysore & Coorg, I., 62; Ind. Forester, IV., 233; XII., App., 15; Balfour, Cyclop. Ind., III., 718.*

Habitat.—A low annual of the Sub-tropical Himálaya, from Kumáon to Sikkim, found up to altitudes of 5,000 feet. It occurs also in Assam, Sylhet, and southward to Ceylon. It is distributed to the Malay Islands, Australia, and Africa.

MEDICINE. Seeds. 2519 Root. 2520 Flowers. 2521 Bark. 2522

Medicine.—The small oblong SEEDS and the ROOT are considered anthelmintic and are prescribed in powder; the latter also is given as a stomachic in the Malabar coast. The FLOWERS are highly esteemed in the Pánjab as alteratives, depuratives, and tonics. The BARK, ground small and mixed with whey, is said to be a valuable remedy in piles (*Ainslie; Stewart; Dymock*). An OIL prepared from the root, by steeping it in water and then boiling it in Sesamum oil until all the water is expelled, is said by Muhammadan medical writers to be a valuable aphrodisiac.

CHEMISTRY. Oil. 2523 2524

Chemical Composition.—"One hundred and fifty pounds of the fresh herb distilled with water in the usual way yielded a very deep shiny coloured, viscid essential oil, very soluble in water and having much the odour of oil of Lavender. The oil does not appear to have any rotatory power, but is difficult to examine on account of its opacity" (*Dymock, Mat. Med. W. Ind.*).

SPICES & CONDIMENTS.

2525

These terms are applied to a class of substances which possess aromatic and pungent properties, or are mixed with food for the purpose of exciting the appetite. Many articles of this class not only assist digestion, but by tempting the palate increase the amount of food consumed, and thus stimulate a flagging appetite. To the Native of India, Spices and Condiments are indispensable, and in their production India has always been pre-eminent, her ports having been resorted to by the nations of the West, from pre-historic times, in search of these commodities. India possesses a very large export trade in Spices, the details of which will be found fully given under the various articles treated of in this work, and most of which will be found briefly alluded to in the present chapter. Here it may suffice to say that in the statement of the foreign trade of India published by the Finance and Commerce Department of the Government of India, during the official year 1889-90, a total of 22,694,986lb of spices of Indian produce, valued at R46,41,236, are shown to have been exported from the country. This, of course, does not include the very large exports of Cutch and Turmeric which, although used in India as spices, are exported mainly for industrial purposes, nor does it include 53,164 cwt. of essential seeds, valued at R4,40,697, all of which in India and many in Europe are employed as Spices and Condiments. During the same period a large quantity of foreign spices, *viz.*, 52,830,819lb, valued at R85,23,501, was imported into the country, of which the bulk was betel-nut. A total, amounting to 3,503,336lb, valued at R11,67,790, of these foreign spices was again re-exported, so that, besides the spices of Indian origin consumed by the Natives, India imported 49,327,483lb, valued at R73,55,711, of foreign spices for consumption in the country.

2526

Spices & Condiments.

References.—*Dr. Watt's Special Catalogue of Exhibits by the Government of India (Col. and Ind. Exhib.), 80; Royle, Prod. Res., 13, 74; Linschoten, Voyage to East Indies (Ed. Burnell, Tiele, & Yule, I., 13, 44, 47, 109, 112, 113, 130; II., 72; Madras Man. Adm., I., 363; Bomb. Adm. Rep. (1871-72), 372, 378; Ind. Agric., May 11, 1889; Times of India, April 27, 1889; Atkinson, Him. Dist. (X., N. W. P. Gaz.), 704; Stewart, Food of Bijnour Dist. (N.-W. P. Select.), (1866), II., 479.*

The following list of the spice-producing plants of India is chiefly a reprint of the article on the subject in **Dr. Watt's** Special Catalogue of Exhibits of this nature at the Colonial and Indian Exhibition—more recent statistics of those in the earlier volumes of this work only being added :—

CUTCH. 2527

1. Acacia Catechu, *Willd.*; LEGUMINOSÆ.

THE CATECHU, CUTCH, or KATH.

A common tree, in most parts of India and Burma.

The extract obtained by boiling down chips of the heartwood forms the Cutch of Commerce. This substance is chiefly employed in dyeing and tanning, but it is also largely used as a drug, especially in America.

In the preparation of *káth*, twigs are placed in the boiling fluid and upon these, crystals of this substance are deposited. It is generally regarded as purer than cutch and is largely used as an ingredient in the *pán* or betel-leaf preparations which the Natives of India are so fond of chewing (*see* Vol. I., 27).

Trade. 2528

TRADE.—In continuation of the trade returns of Cutch given in Volume I. of this work, the total exports, of this important substance, from India, during the six years ending April 1890 are shown in the following tables :—

YEAR.	Weight in Cwt.	Value in Rupees.
1884-85	246,122	28,20,785
1885-86	205,355	25,28,394
1886-87	199,397	29,52,491
1887-88	273,068	43,38,466
1888-89	290,896	44,23,219
1889-90	221,986	33,34,004

The following analysis of the Cutch returns shows the Provinces from which exported and the countries to which consigned, for the year ending 1889-90 :—

Province from which exported.	Weight in Cwt.	Value in Rupees.	Country to which exported.	Weight in Cwt.	Value in Rupees.
Bengal	65,749	9,67,077	United Kingdom	108,853	16,47,330
			France	1,811	28,914
Bombay	2,108	26,162	Germany	1,219	13,780
			Holland	3,190	50,980
Madras	132	2,400	Italy	2,845	44,290
			Egypt	27,018	4,31,567
Burma	153,997	23,38,365	St. Helena	7,000	84,000
			United States	62,546	9,12,353
			Ceylon	1,308	26,319
			Straits Settlements	6,096	91,925
			Other Countries	100	2,546
TOTAL	221,986	33,34,004	TOTAL	221,986	33,34,004

2. Allium; LILIACEÆ.

ONIONS & GARLIC. 2529

The Onion and the Garlic are extensively cultivated all over India. The Onion is supposed to be a native of Western Asia, but it exists in India only under cultivation. The Onions of Patna and Bombay are, however, of very fine quality and are now being exported. It is surprising how powerfully the climate of India seems to affect the Onion. In one province excellent Onions are to be had, while in another they are very inferior. The Garlic is eaten to a much greater extent than the Onion by the Natives of India, the aggregated white bulb being offered for sale in every bázar (*see* Vol. I., 168.)

3. Amomum subulatum, *Roxb.*; SCITAMINEÆ.

THE GREATER CARDAMOM.

GREATER CARDAMOM. 2530

A native of Nepal. The Greater Cardamom is much used in the preparation of sweetmeats on account of its cheapness. This, as well as the Lesser Cardamom, also forms ingredients of the *pán* or betel-leaf preparation (*see* Vol. I., 222).

4. Apium graveolens, *Linn.*; UMBELLIFERÆ.

WILD AND CULTIVATED CELERY.

CELERY. 2531

A native of Europe, Egypt, Abyssinia, and of Asia from the Caucasus to the Himálaya and Baluchistán. Cultivated in different parts of the plains of India. The small fruits are eaten as a spice by the Natives, who never cultivate the plant for any other purpose; consequently it has quite a different appearance from the garden celery cultivated by Europeans (*see* Vol.. I, 271).

5. Areca Catechu, *Linn.*; PALMÆ.

THE ARECA or BETEL-NUT.

BETEL-NUT. 2532

A native of Cochin China, Malayan Peninsula and Islands. Cultivated throughout Tropical India. The nut is one of the indispensable ingredients which enter into the preparation of the *pán* or betel-leaf which is chewed so universally by Natives of all classes (*see* Vol. I., 291)

Trade. 2533

TRADE.—In continuation of the trade figures given in Volume I. of this work, the following table shows the imports and exports of betel-nut brought down to the official year 1889-90:—

YEAR.	IMPORTS.		EXPORTS.	
	Quantity.	Value.	Quantity.	Value.
	℔	R	℔	R
1884-85	30,720,424	33,44,551	218,204	32,099
1885-86	44,876,057	49,53,928	477,178	64,007
1886-87	31,062,619	34,59,436	373,623	50,223
1887-88	30,743,990	36,97,452	749,455	84,387
1888-89	40,371,229	38,20,142	328,056	42,675
1889-90	39,520,007	37,16,965	275,102	37,158

SPICES & Condiments.

List of the Spice-producing

BETEL-NUT.

Details of Imports (1889-90).

Province into which imported.	Quantity.	Value.	Countries whence imported.	Quantity.	Value.
	℔	R		℔	R
Bengal . .	18,215,931	13,66,939	East Coast Africa—Zanzibar .	84,196	9,389
Bombay . .	2,324,368	2,29,768	Mauritius . .	50,288	6,139
Sindh . . .	44,247	4,730	Ceylon . .	10,995,900	15,79,493
Madras . .	15,681,558	18,58,253	Straits Settlements . .	28,381,291	21,21,361
Burma . .	3,253,903	2,57,275	Other Countries .	8,332	583
TOTAL .	39,520,007	37,16,965	TOTAL .	39,520,007	37,16,965

Details of Exports (1889-90).

Province whence exported.	Quantity.	Value.	Countries to which exported.	Quantity.	Value.
	℔	R		℔	R
Bombay . .	2,548	224	Straits Settlements	21,280	2,850
Madras . .	812	44	Other Countries .	3,360	268
Burma . .	21,280	2,850			
TOTAL .	24,640	3,118	TOTAL .	24,640	3,118

MUSTARD. 2534

6. **Brassica**; CRUCIFERÆ.

The seeds of the various forms of Mustard, Rape, and Cole are used as condiments. They are, however, chiefly of interest as oil-seeds (*see* Vol. I., 520).

CHILLIES. 2535

7. **Capsicum**; SOLANACEÆ.

Two or three species yield the various forms of Chillies, Red-pepper, and Cayenne pepper. They are all natives of America, although now extensively cultivated in India (*see* Vol. II, 134).

CARAWAY. 2536

8. **Carum Carui,** *Linn.*; UMBELLIFERÆ.

CARAWAY SEED.

Wild in Kashmír and the North-West Himálaya, cultivated as a cold season crop throughout the plains.

The seed is used, entire or powdered, in curries, cakes, and confectionery (*see* Vol. II., 196).

BISHOP'S WEED. 2537

9. **C. copticum,** *Benth.*

THE TRUE BISHOP'S WEED.

Cultivated from the Panjáb and Bengal to the South Deccan. The aromatic seed forms an ingredient of the preparation, *pán* (*see* Vol. II., 198).

2538

10. **C. Roxburghianum,** *Benth.*

Extensively cultivated and eaten like parsley; the seed is also used in flavouring curry (*see* Vol. II., 201).

CLOVES. 2539

11. **Caryophyllus aromaticus,** *Linn.*; MYRTACEÆ.

THE CLOVE.

Indigenous in the Moluccas; cultivated in South India. The unexpanded and dried flowers form the cloves of commerce. They contain a pungent, aromatic oil, to which their peculiar property is due. They

CLOVES.

appear to have been known to the Sanskrit writers but not to the Romans. One of the earliest Sanskrit Medical writers (Oharaka) gives them the name of *Lavanga*, a word which exists to this day in many parts of India. In Bombay cloves are known as *Lavang;* in Bengal as *Lavanga;* and to the Hindustanis as *Laung* or *Lang*. The early Arabian writers called them *Karanful*. Arabian and Persian writers, however, speak of cloves as coming from Batavia. Cloves were known in Europe after the discovery of the Molluccas by the Portuguese (*see* Vol. II., 202).

Trade. 2540

TRADE.—In continuation of the trade figures for Cloves given in the second volume of this work, the imports and re-exports up to the end of the official year 1889-90 were as follows:—

YEAR.	IMPORTS.		RE-EXPORTS.	
	Quantity.	Value.	Quantity.	Value.
	lb	R	lb	R
1885-86	4,974,918	11,85,354	1,244,252	3,06,295
1886-87	4,445,180	19,59,535	1,789,826	8,01,233
1887-88	8,374,302	37,67,545	4,038,110	20,25,315
1888-89	7,240,043	28,78,898	3,384,638	1,54,433
1889-90	6,355,839	19,36,283	3,230,111	9,80,867

It will be observed that during the year 1887-88 the trade in Cloves almost doubled itself, due, it is said, to a very abundant crop in Zanzibar whence this spice is brought to India, the bulk to Bombay. If this be so, it does not seem to have again decreased to its former figures, and the prices obtained have in consequence diminished.

With regard to Cloves of Indian produce, the cultivation of this valuable spice in India does not seem as yet to have made much headway, the exports during the past five years having been as follows:—

YEARS.	Quantity.	Value.
	lb	R
1885-86	455	350
1886-87	776	335
1887-88	784	840
1888-89	336	300
1889-90	924	740

12. Cinnamomum; LAURINEÆ.

CINNAMON. 2541

Two or three species afford the various forms of Cinnamon. **C. Zeylanicum**, *Breyn.*, or Ceylon Cinnamon, is the true Cinnamon of modern commerce; **C. Tamala** and **C. obtusifolium** yield part of the so-called Cassia Lignea—the Indian Cinnamon. The Ceylon Cinnamon of European commerce does not appear to have been the Cinnamon of the ancients. Everything points to the probability that the Cassia Lignea of China and India was the much-prized Cinnamon of antiquity. Ceylon Cinnamon appears to have been first discovered about the thirteenth century, and it was not cultivated until the eighteenth. Cassia is mentioned by one of the earliest Chinese herbal writers in 2700 B.C (*see* Vol. II. 317).

SPICES & Condiments. List of the Spice-producing

CINNAMON. Trade. 2542

TRADE.—In continuation of the figures given in Volume II. of this work, the following table will suffice to describe the progress of the Cinnamon trade in India since 1883-84 :—

Foreign Trade in Cinnamon.

YEAR.	IMPORTS.		EXPORTS AND RE-EXPORTS.	
	Quantity.	Value.	Quantity.	Value.
	℔	R	℔	R
1884-85	34,786	5,105	25,739	8,895
1885-86	45,449	7,797	30,666	11,263
1886-87	9,071	1,581	18,956	5,899
1887-88	18,764	3,070	32,831	6,894
1888-89	11,352	1,798	17,339	4,711
1889-90	41,876	6,9:8	29,990	5,823

COCOA-NUT. 2543

13. **Cocos nucifera,** *Linn.;* PALMÆ.

THE COCOA-NUT.

A native of the Indian Archipelago, most probably having been brought to India, Ceylon, and China about 4,000 years ago and conveyed to America and Africa at even an earlier date. It is chiefly abundant on the coast, disappearing altogether about 150 miles inland. There are two or three varieties. They flower in the hot season and the nuts ripen in September to November. The albuminous layer from the interior of the shell is largely eaten as a condiment. It is preserved in sugar and made into various sweetmeats (*see* Vol. II., 415).

CORIANDER. 2544

14. **Coriandrum sativum,** *Linn.;* UMBELLIFERÆ.

THE CORIANDER.

Widely cultivated throughout India; indigenous to the Mediterranean and Caucasian regions As a condiment this seed forms one of the indispensable ingredients of Native curry. During the year 1884-85 the exports of Coriander were valued at R1,56,505 (*see* Vol. II., 567).

Trade. 2545

TRADE.—The following table exhibits the quantities and values of the exports of this substance from India during the past five years :—

YEAR.	Weight.	Value.
	Cwt.	R
1885-86	28,328	96,606
1886-87	31,405	1,14,540
1887-88	33,056	1,88,635
1888-89	38,488	2,38,076
1889-90	35,678	1,93,964

Details of Exports (1889-90).

CORIANDER. Trade.

Province from which exported.	Quantity.	Value.	Countries to which exported.	Quantity.	Value.
	℔	R		℔	R
Bengal . .	5,504	32,036	East Coast of Africa (Mozambique and Zanzibar).	165 358	992 2,599
Bombay . .	2,332	14,535	Mauritius . .	2,164	12,077
Sind . . .	20	166	Natal . .	262	1,631
Madras . .	27,226	1,42,905	Réunion . .	277	1,430
Burma . .	596	4,312	Aden . . .	330	2,165
			Arabia . .	322	1,922
			Ceylon . .	13,729	70,674
			Straits Settlements	17,345	95,996
			Australia . .	340	2,129
			Other countries .	386	2,339
TOTAL .	35,678	1,93,954	TOTAL .	35,678	1,93,954

15. Crocus sativus, *Linn.*; IRIDEÆ.

SAFFRON. 2546

SAFFRON.

The Indian supply of Saffron comes from France, China, and Kashmír, and a small quantity from Persia, in the form of cakes known as *kesar-ki-roti*. It is used by the Hindus in various religious rites and for colouring and flavouring their food (*see* Vol. II., 592).

16. Cuminum Cyminum, *Linn.*; UMBELLIFERÆ.

CUMIN. 2547

THE CUMIN; JIRA, *Beng.*

Widely cultivated in the Panjáb plains, Rájputana, and in the Deccan. Indigenous to the upper regions of the Nile, but carried at an early age to Arabia, India, and China. It was an important spice in the middle ages, and even in the fifteenth century was heavily taxed in Europe. The seed is used by the Natives to flavour curry (*see* Vol. II., 642).

17. Curcuma Amada, *Roxb.*; SCITAMINEÆ.

MANGO GINGER. 2548

MANGO GINGER.

Found wild in Bengal and on the hills. Used as a condiment and vegetable (*see* Vol. II., 652).

18. C. longa, *Roxb.*

TURMERIC. 2549

THE TURMERIC.

The Turmeric is a native of Southern Asia and is cultivated all over India for its rhizome or root-stocks, which is the well-known *Haldi*, the powder of which constitutes the chief ingredient in curry stuffs (*see* Vol. II., 659).

19. Elettaria Cardamomom, *Maton*; SCITAMINEÆ.

LESSER CARDAMOM. 2550

THE LESSER CARDAMOM.

A native of the mountain tracts of South India. This is the most valuable of all the Indian condiments. It is extensively used by the Natives of India for flavouring purposes, and is also eaten in *pán* (*see Vol. III., 227.*).

20. Eruca sativa, *Lam.*; CRUCIFERÆ.

2551

Cultivated in North and Central India, and on the Western Himálaya, ascending to 10,000 feet. The seed is used in the same way as the forms of Brassica (*see Vol. III., 266*).

21. Fœniculum vulgare, *Gærtn.;* UMBELLIFERÆ.

FENNEL.
2552

THE FENNEL.

Commonly cultivated throughout India; a native of Europe. The seed is used as a condiment (*see* Vol. III., 495).

22. Humulus Lupulus, *L.;* URTICACEÆ.

HOPS.
2553

HOPS.

A native of Europe. The cultivation of this plant is now being tried experimentally in Kashmír and Chamba, and seems likely to succeed (*see* Vol. IV., 303).

23. Mentha arvensis, *Linn.;* LABIATÆ.

MINT.
2554

THE MARSH MINT.

An herb of the Western Himálaya, the leaves of which are eaten as a condiment (*see* Vol. V., 228).

24. M. piperita, *Linn.*

PEPPERMINT.
2555

THE PEPPERMINT.

A native of Europe; it occurs in Indian gardens and as an escape is almost naturalised in some parts of the country (*see* Vol. V., 229).

25. M viridis, *Linn.*

SPEARMINT.
2556

THE SPEARMINT.

Commonly grown in Native gardens all over the plains of India and much used as a condiment in curries and other forms of food (*see* Vol. V., 231).

26. Murraya Kœnigii, *Spr.;* RUTACEÆ.

CURRY-LEAF.
2557

THE CURRY-LEAF TREE.

A small tree of the outer Himálaya, the leaves of which are used either dry or fresh to flavour curries (*see* Vol. V., 288).

27. Myristica fragrans, *Houtt.;* MYRISTICEÆ.

NUTMEG.
2558

THE NUTMEG and MACE.

Cultivated in South India. The fruits and nuts of **Myristica malabarica, M. laurifolia,** and of **M. corticosa,** are often to be seen in the drug-sellers' shops, where they are apparently sold as substitutes for the true nutmeg. In the Cawnpore bazár a linear oblong nutmeg was offered for sale under the name of the true nutmeg. This has not as yet been identified, but it seems very different from the ordinary nut. It is $1\frac{1}{2}$ inches long by about $\frac{1}{2}$ inch thick; it has got only an abortive mace (*see* Vol. V., 311).

28. Nigella sativa, *Linn.;* RANUNCULACEÆ.

BLACK CUMIN.
2559

THE BLACK CUMIN.

A native of South Europe and of the Levant; extensively cultivated, the seeds constituting a favourite spice with the Natives of India (*see* Vol. V., 428).

29. Ocimum Basilicum, *Linn.;* LABIATÆ.

SWEET BASIL.
2560

THE COMMON or SWEET BASIL.

The seeds and also the leaves are eaten as a cooling condiment (*see* Vol. V., 440).

30. Peucedanum graveolens, *Benth.;* UMBELLIFERÆ.

DILL.
2561

THE DILL.

A native of tropical and sub-tropical India.

Often cultivated in the plains of India. The seed is eaten in curry (*see* Vol. VI., Pt. I., 181).

31. **Pimpinella Anisum,** *Linn.*; UMBELLIFERÆ.

THE ANISE.

ANISE. 2562

Introduced into India by the Muhammadans from Persia. It is now grown in North India. The seed is used in confectionery (*see* Vol. VI., Pt. I., 236).

32. **Piper Betle,** *Linn.*; PIPERACEÆ.

THE PAN-LEAF PEPPER.

PAN-LEAF. 2563

The leaves of this plant, mixed with a little catechu, areca-nut, and lime, and flavoured with spices, constitute the preparation known as *pán*. This is chewed by the Natives of India as a mild stimulant, especially after meals. It stains the mouth and saliva a deep red colour. The trade in this leaf is entirely for Indian consumption (*see* Vol. VI., Pt. I., 247-256).

33. **P. Cubeba,** *Linn.*

CUBEBS.

CUBEBS. 2564

A native of Java and Sumatra; cultivated in India (*see* Vol. VI., Pt. I., 257).

34. **P. longum,** *Linn.* (**Chavica Roxburghii,** *Miq.*).

LONG PEPPER. 2565

One of the forms of LONG PEPPER. A perennial shrub of Eastern India, Nepál, and East Bengal; of Java, Ceylon, and the Philippines. The dried unripe fruit constitutes the long pepper of India (*see* Vol. VI., Pt. I., 258).

35. **P. nigrum,** *Linn.*

THE BLACK PEPPER.

BLACK PEPPER. 2566

A climber, extensively cultivated in South India, where it is indigenous. Introduced from India into the Straits. The berries are largely eaten as a condiment in curry, and when reduced to a powder constitute the black pepper of Europe (*see* Vol. VI., Pt. I., 260).

36. **Trigonella Fœnum-grœcum,** *Linn.*; LEGUMINOSÆ.

THE FENUGREEK.

FENUGREEK. 2567

Cultivated in some parts of India, wild in Kashmír and the Panjáb. The seed is used by the Natives, as a condiment in curries (*see* Vol. VI., Pt. IV., 46).

37. **Zingiber Cassumunar,** *Roxb.*; SCITAMINEÆ

2568

A native of various parts of India, used as a spice instead of **Z. officinale** (*see* Vol. VI., Pt. IV., 357).

38. **Z. officinale,** *Roscoe.*

THE GINGER.

GINGER. 2569

A native of the warmer parts of Asia, but not known in its wild state. Cultivated in many parts of India for its rhizome (*see* Vol. VI., Pt. IV., 358).

39. **Z. zerumbet,** *Roscoe.*

ZERUMBET. 2570

Found throughout both Peninsulas and used as a substitute for the preceding species (*see* Vol. VI., Pt. IV., 366).

Spikenard, *see* **Nardostachys Jatamansi,** *DC.*; Vol. V., 338; VALERIANEÆ.

SPILANTHES, *Linn.*; *Gen. Pl., II., 380.*

2571

Spilanthes Acmella, *Linn.*; *Fl. Br. Ind., III., 307*; COMPOSITÆ.

Syn.—VERBESINA ACMELLA and PSEUDO-ACMELLA, *Linn.* Several varieties are described in the *Flora of British India:*

Var. 1, Acmella proper, *Clarke*=S. ACMELLA, *DC.*; S. CALVA, *Wight, Ic., t. 1109.*

Var. 2, calva, *Clarke*=S. CALVA, *DC.*; S. PSEUDO-ACMELLA, *Linn.*; COTULA CONICA, *Wall.*

SPINACIA oleracea.

The Garden Spinach.

Var. 3, oleracea, *Clarke*=S. OLERACEA, *Jacq.*; BIDENS FERVIDA and FIXA, *Lank.*
Var. 4, paniculata, *Clarke*=S. PANICULATA, *DC.*
Vern.—*A'karkara*, BOMB.; *A'karkarhá, pokarmúl*, PB.; *Maráti mogga, maráti tige*, TEL.; *Akmalla*, SINGH.; *Hen-ka-la*, BURM.
References.—*Roxb., Fl. Ind., Ed. C.B.C., 595; Dalz. & Gibs., Bomb. Fl., 129; Sir W. Elliot, Fl. Andhr., 112; Mason, Burma & Its People, 495, 789; Dymock, Mat. Med. W. Ind., 2nd Ed., 434; S. Arjun, Bomb., Drugs, 95; Year-Book Pharm., 1880, 248; Baden Powell, Pb. Pr., 357; Lisboa, U. Pl. Bomb., 163; Gazetteer, Mysore & Coorg, I., 56.*

Habitat.—An annual herb, found throughout India, both cultivated and wild, and ascending the Himálaya to an altitude of 5,000 feet. It is distributed to all warm countries.

MEDICINE. Plant. 2572 Flower heads. 2573

Medicine.—The whole PLANT, and more especially the FLOWER-HEADS are very acrid and have a hot burning taste, which causes profuse salivation. It is considered by Natives a powerful stimulant and sialogogue, and is used in headaches, paralysis of the tongue, affections of the throat and gums, and for toothache. It is employed by Europeans in India as a remedy for toothache, instead of Spanish Pellitory (**Pyrethrum** root).

SPINACIA, *Linn.; Gen. Pl., III., 53.*

2574

Spinacia oleracea, *Linn.; Fl. Br. Ind., V., 6; Wight, Ic., t. 818*
GARDEN SPINACH. [CHENOPODIACEÆ.

Syn.—SPINACIA TETRANDRA, *Roxb., Fl. Ind., Ed. C.B.C., 718.*
Vern.—*Pálak, ság, ság-pálak, palki, isfanáj, pinnis*, HIND.; *Pálang, pinnis*, BENG.; *Baji*, C.P.; *Pálak, isfanáj*, N.-W. P.; *Pálak, isfanák* (bazar name=) *bij-palak*, PB.; *Spinaj*, AFG.; *Pálak*, (bazar name=) *pálak-bij*, SIND; *Pálak, isfanaj*, BOMB.; *Vusayley-kiray*, TAM.; *Dum-pa-bachchali, mattur bachchali*, TEL.; *Is.-panaj*, ARAB.; *Is finaj*, PERS.
References.—*Boiss., Fl. Orient., IV., 906; Roxb., Fl. Ind., Ed. C.B.C., 718; Voigt, Hort. Sub. Cal., 320; Dalz. & Gibs., Bomb. Fl., Suppl., 73; Stewart, Pb. Pl., 180; Rept. Pl. Coll. Afgh. Del. Com., 101; DC., Orig. Cult. Pl., 98; Graham, Cat. Bomb. Pl., 171; Sir W. Elliot, Fl. Andhr., 48, 113; Firminger, Man. Gard. Ind., 143; S. Arjun, Cat. Bomb. Drugs, 113; Birdwood, Bomb. Prod., 69, 173; Baden Powell. Pb Pr., 372; Useful Pl. Bomb, (XXV., Bomb. Gaz.), 169; Econ. Prod., N.W.P., Pt. V. (Vegetables, Spices, & Fruits), 21; Royle, Ill. Him. Bot., 319; Ain-i-Akbari (Blochmann's Trans.), I., 63; Settlement Reports:—Panjáb, Dera Ismail Khan, 349; Central Provinces, Bailúl, 77; Gazetteers, Mysore & Coorg. I., 65; Agri.-Horti. Soc.: Ind., Trans., III., 69, 199 (Pro.) 229; IV., 145; Journ; V., 18; X., 32, 91; Journ. (New Series), IV., 22; V. 35, 44; Balfour, Cyclop. Ind., III., 719; Smith, Dict. Econ. Pl., 388.*

Habitat.—An annual herb, cultivated throughout India. The native country of the 'garden spinach' is unknown, but **M. DeCandolle** suggests that S. **oleracea**, *Linn.*, is a derivative from S. **tetrandra**, *Stev.*, which is indigenous to the Caucasus, and occurs sometimes cultivated and sometimes to all appearance wild.

Cultivation. 2575

CULTIVATION IN INDIA.—This plant, although at one time chiefly cultivated in the gardens of Europeans, seems now to be generally adopted by Natives as an article of diet. The seed is sown in October broad cast, or, better still, in drills. The distance between each drill should be a foot, and between each plant in the drills four inches. Spinach loves a rich soil and a shady situation well watered. The young plants, if not protected, are very liable to be devoured by sparrows (*Firminger*).

OIL. Seeds. 2576

Oil.—The SEEDS are reported to yield a fatty oil.

MEDICINE. Seed. 2577

Medicine.—The SEED is held by Muhammadan medical writers to have cooling and laxative properties, and to be efficacious in difficulty of breathing and biliary derangements (*Tal̃if Sherif*). The GREEN PLANT is believed to act as a solvent for urinary calculi (*S. Arjun*).

MEDICINE. Green Plant. 2578

Food.—The LEAVES are used as a favourite vegetable, both by Europeans and Natives, during the early spring and summer months.

FOOD. Leaves. 2579

SPINIFEX, *Linn.; Gen. Pl., III., 1109.*

Spinifex squarrosus, *Linn.*; GRAMINEÆ.

2580

Vern.—*Rávaná suruni misálu*, TEL.

References.—*Mueller, Select Extra Tropical Pl., 402; Sir W. Elliot, Fl. Andhr., 163; Mason, Burma & Its People, 479; Agri.-Horti. Soc. Ind. Jour., IX., 175-177; Ind. Forester, IX., 238; Balfour, Cyclop. Ind., III., 719.*

Habitat.—A grass which grows abundantly on various parts of the Madras coast.

Domestic.—It is extensively cultivated on the sandy reaches near Madras for sand-binding purposes. The fishermen of the Madras coasts collect its ROOTS and SHOOTS and dry them for fuel (*see* **Sand-binding Plants**, Vol. VII., Pt. II., 455).

DOMESTIC. Roots. 2581 Shoots. 2582

SPIRÆA, *Linn.; Gen. Pl., I., 611.*

2583

A genus of Rosaceous plants, the species of which are widely distributed over the temperate and cold regions of the Northern Hemisphere. Eleven species are indigenous to India, none of which are of economic value. Several of them (**S. canescens**, *Don.*, and **S. sorbifolia**, *Linn.*) are shrubs of considerable size, the wood from which is described as "hard, compact, and even-grained," but it does not appear to be used economically for any purpose whatsoever. Some Indian species of **Spiræa**, *e.g.*, **S. bella**, *Sims.*, are very handsome, and are sometimes cultivated as ornamental shrubs in the gardens of Europeans.

TIMBER. 2584

SPIRITS.

2585

[The knowledge of Spirits of various kinds, by the Natives of Hindustan, is very wide-spread, and dates from an early period. At the present day there are few tracts in India where locally-prepared spirits are not largely consumed by, at any rate, the lower castes of Natives. In the Ordinances of **Manu**, the text of which, as it now stands, dates, by the latest estimate, between 100 and 500 A.D., there are frequent references to the drinking of spirituous liquors to excess, so that it is evident that this must have been a common offence, and the twice-born are often urged to avoid the temptation. Three kinds of spirituous liquor are described, *viz.*, that made "of sugar (*molasses*), of ground rice, and of the flowers of the honey tree" (**Bassia latifolia**),—liquors which, down to the present day, are those most commonly consumed by the Natives of India. **U. C. Dutt**, in his *Materia Medica of the Hindus*, says that the later Sanskrit writers describe thirteen kinds of distilled liquors, one or other of which was widely used in their time.

Coming down to more modern dates, the *Ain-i-Akbari* describes fully an intoxicating liquor made from the sugar-cane or from brown sugar by simple fermentation, and says that this is sometimes drunk as a beverage, but is mostly employed for the distillation of *arak.* "This latter," continues **Abul Fazl**, "they have several methods of accomplishing: *first*, they put the above liquor into brass vessels, in the interior of which a cup is put so as not to shake, nor must the liquor flow into it. The vessels are then covered with inverted lids, which are fastened with clay. After pouring cold water on the lids, they kindle the fire, changing the water on the lids as often as it gets warm. As soon as the vapour inside reaches the cold

DISTILLATION of Arak. 2586 *Conf. with Vol. V., 332.*

DISTILLATION of Arak.

lid, it condenses and falls as arak into the cup. *Secondly,* they close the same vessel with an earthen pot, fastened in the same manner with clay, and fix to it two pipes, the free ends of which have each a jar attached to them, which stands in cold water. The vapour through the pipes will enter the jars, and condense. *Thirdly,* they fill an earthen vessel with the above-mentioned liquor and fasten to it a large spoon with a hollow handle. The end of the handle they attach to a pipe, which leads into a jar. The vessel is covered with a lid, which is kept full with cold water. The arak, when condensed, flows through the spoon into the jar. Some distil the arak twice, when it is called *duátashah,* or twice burned. It is very strong. If you wet your hands with it, and hold them near the fire, the spirit will burn in flames of different colours, without injuring the hands. It is remarkable that when a vessel, containing arak, is set on fire, you cannot put it out by any means; but if you cover the vessel, the fire gets at once extinguished." The same author describes a spirit distilled from the *Mahúa,* and records the fact that excessive spirit-drinking prevailed among the grandees at the Court of **Akbar. Linschoten,** in the sixteenth century, deplored the fact that the Portuguese soldiers were learning from the Natives of India the pernicious practice of drinking spirits in place of the wine imported from their own country. **Tavernier,** in his *Travels in India* (1670-1689), mentions a spirit distilled from palm wine which was largely drunk by the idolators of India at certain feasts; he describes the method in which it was prepared.

The above facts may serve to show the reader that the custom of spirit-drinking in India is by no means new, while the details hereafter to be given of the multitudinous substances employed in various parts of the country for the preparation of the liquors, will show that the custom is wide-spread and certainly not an adaptation of a European habit. The means of procuring fermentation in a saccharine or malted liquor are and have been for long much more extensively understood in India than in Europe generally.

Numerous other publications might be quoted in support of this view, but those cited would seem sufficient. For the details of the recent liquor traffic in Bengal, and the statistics relating to the importation of foreign spirits, the reader is referred to the article under **Narcotics,** Vol. V., 332, and to **Vitis vinifera,** Vol. VI., Pt. IV.--*Ed., Dict. Econ. Prod.*]

References.—*Stewart, Pb. Pl. (Index of Uses), 100; U. C. Dutt, Mat. Med. Hind., 272-274; Grierson, Behar Peasant life, 77, 78; Atkinson, Him. Dist. (X., N.-W. P. Gaz.), 768; Ordinances of Manu (Burnell & Hopkins, Trans.), VII., 47, 50; XI., 51 ff., 147 ff.; XI., 95, 154 Hove, Tour in Bombay, 99; Ain-i-Akbari (Blockmann's Trans.), I., 69, 70; Linschoten, Voyage to East Indies (Ed. Burnell, Tiele & Yule,) II., 48; Strettell, The Ficus Elastica in Burma proper, 118; Westland, Rep. on Jessore, 372-390; Bombay Adm. Rep. (1871-72), 372-374, 387, 390, 403, 419; Gazetteers:—Bombay, X., 36; XI., 27-29; XIII., 395, etc., etc.*

The following are the principal substances used in India at the present day, in the preparation of Spirits, arranged in their alphabetical order:—

MATERIALS. 2587

I.—SUBSTANCES FROM WHICH SPIRITS ARE MANUFACTURED.

1. **Agave americana,** *Linn.;* AMARYLLIDEÆ.

THE CENTURY PLANT.

Mezcal. 2588

The juice is used in Mexico in the preparation of a spirit called *Mezcal.* Although the plant is naturalised in many parts of India, it does not as yet appear to have been utilised in India for that purpose, probably on account of the abundance of other spirit-producing substances.

Spirits are Manufactured. (*W. R. Clark.*) SPIRITS.

MATERIALS.

2. **Anacardium occidentale,** *Linn.;* ANACARDIACEÆ.

THE CASHEW NUT.

Cashew Nut. 25

The people of Goa distil a spirit from the succulent fruit-stalk.

3. **Andropogon laniger,** *Desf.;* GRAMINEÆ.

2590

This grass is mixed with spices, and a spirit or arak distilled from it (*Stewart*).

4. **Anthocephalus Cadamba,** *Bth. & Hook. f.;* RUBIACEÆ.

2591

According to **U. C. Dutt**, a spirit distilled from the flowers is mentioned by the later Sanskrit writers.

5. **Arenga saccharifera,** *Labill;* PALMÆ.

Sago Palm. 2592

The Sago Palm of Malacca and the Malaya, the juice of which is used in Batavia for the production of the celebrated Batavian *arak*, but it is apparently not so employed in India.

6. **Bassia latifolia,** *Roxb.;* SAPOTACEÆ.

MAHÚA.

Mahua. 2593

The spirit resulting from the fermentation and distillation of the flowers of this tree is very largely consumed by Natives of Central India, Chutia Nagpur, the Central Provinces, and those parts of the Bombay Presidency where the tree occurs. For a description of the ordinary Native method of obtaining *Mahúa* spirits, the reader is referred to the account given under **Narcotics** (*see Vol. V.*, 323). Under Government restrictions a very large quantity of *Mahúa* spirit is manufactured at Uran on the island of Karanja in the south-west corner of the Bombay harbour, for use in the town of Bombay. As the methods employed there are somewhat different to the rude Native ones, they may be described in detail. "There are about twenty distilleries on the island, all of which are owned by Pársis. The Collector of Salt Revenue issues yearly licenses for working the distilleries. Provided they mix nothing with the spirit, the holders of licenses are free to make liquor in whatever way they choose. The *mahúa* flowers are brought to Bombay by rail from Jabalpur, and from Kaira, the Panch Mahâls and Rewa Kántha in Gujarát. Much of the Gujarát *mahúa* comes by sea direct to Uran. Most of the Jabalpur *mahúa* comes by rail to Bombay, and from Bombay is sent to Uran in small boats by Pársis, who are the chief *mahúa* merchants. When set apart for making spirits, *mahúa* flowers are allowed to dry, and are then soaked in water. Fermentation is started by adding some of the dregs of a former distillation, and the flowers are generally free to ferment for eight or nine days.

Methods employed. 2594

"The Native stills formerly in use have given place to stills of European fashion, consisting of a larger copper boiler and a proper condenser. The cover of the boiler has a retort-shaped neck which is put in connection with the winding tube or warm in the condenser, and the condenser is kept full of sea water, all the distilleries having wells connected by pipes with the sea. Even in these stills the first distillation, technically called *rasi*, is very weak and would find no market in Bombay. It is therefore re-distilled, and becomes *bonda* or twice distilled, which is nearly as strong as ordinary brandy, and on being poured from one glass into another, gives a proper 'head' or froth without which Bombay topers will not have it. Spirit is sometimes scented or spiced by putting rose leaves, imported dry from Persia, cinnamon or cardamoms into the stills with the *mahua*. This is generally weak; it is often made to order for the cellars of wealthy Pársis in Bombay, or for wedding parties. Date rum is manufactured in the same manner as plain double-distilled *mahua* spirit, and, though colourless at first, it acquires the colour of rum after stand-

Rose Leaves. 2595 *Conf. with Vol. VI., Pt. I., 561-566.*

SPIRITS. | **Substances from which**

MATERIALS.

Mahua.

ing in wood for a few months, as *mahua* spirit also does. Small quantities of spirit are sometimes made from raisins or from molasses. Palm spirit is not allowed to be manufactured in the Uran distilleries. It is made in a single distillery in the town of Uran. Since 1880 two of the distilleries have held licenses for the manufacture of spirits of wine, which is sold in Bombay to chemists. This is made from weak *mahua* spirit, in English or French stills of superior construction.

"Each distillery has a strong room in which the outturn of the day's distilling is every morning stored. Each strong room is kept under a double lock, the key of one lock remaining with the owner, and the key of the second lock with the Government officer in charge of the distilleries. All liquor intended for transport to Bombay or the Thána and Kolába ports is brought every morning from the distilleries into a large gauging-house near the wharf. The liquor is, therefore, gauged by the Government officers in charge, and, on payment of the duty, permits are granted for its removal and transport. The liquor is sent in boats belonging to, or hired by, the liquor owners, which start with the ebbtide and cross the harbour to the Carnac wharf in Bombay. At the Carnac wharf the liquor is examined and occasionally tested by customs officers, who also compare each consignment with the permit covering it."

2596

7. **Bassia longifolia**, *Willd.*

The Mowa or Mahúa tree of South India. A spirit is prepared from the flowers of this species.

Palmyra Palm. 2597

8. **Borassus flabelliformis**, *Linn.*; PALMÆ.

THE PALMYRA PALM.

The distillation of the toddy or fermented juice yields *arak.*

Akunda. 2598

9. **Calotropis gigantea**, *R. Br.*; ASCLEPIADACEÆ.

An intoxicating liquor called *bar* is said to be prepared from the milky juice of this plant (*Birdwood*). Other authors say it is only used as an adjunct in the fermentation of an alcoholic liquor.

Sago Palm. 2599

10. **Caryota urens**, *Linn.*; PALMÆ.

THE HILL PALM or SAGO PALM.

The toddy when distilled is made into *arak.*

Pereira. 2600

11. **Cissampelos Pereira**, *Linn.*; MENISPERMACEÆ.

In Garhwál a spirit is said to be distilled from the root.

COCOA-NUT. 2601

12. **Cocos nucifera**, *Linn.*; PALMÆ.

COCOA-NUT PALM.

The toddy is largely made into native spirits. Five *paras* (or measures) of good *arak* may be made from a single tree devoted to the purpose during a single year, but some very good trees will give, though rarely, eight to ten *paras* (*Simmonds' Tropical Agriculture*).

Coffee. 2602

13. **Coffea arabica**, *Linn.*; RUBIACEÆ.

The ripe pulp of the coffee berry contains a quantity of sugar which might be converted into alcohol. In some experiments made by Dr. Shortt it was found that 8 oz. of the dried husk when steeped in water, fermented and distilled, yielded one ounce of spirits. This is not, however, used by the Natives of India.

2603

14. **Cordia Myxa**, *Linn.*; BORAGINEÆ.

The fruit which is known to Anglo-Indians as Sebesten is used in the preparation of spirits.

2604

15. **Daphne oleoides**, *Schrel.*; THYMELÆACEÆ.

Brandis says that on the Sutlej a spirit is distilled from the berries.

MATERIALS.

16. **Diospyros Lotus,** *Linn.*; EBENACEÆ.

THE EUROPEAN DATE PLUM.

Date Plum. 2605

According to Irvine spirits are in the Panjáb distilled from the fruit Stewart is, however, of opinion that no such use is made of them.

17. **Eleagnus latifolia,** *Linn.*; ELEAGINEÆ.

2606

A brandy is made in Yarkand from the fruit of this tree.

18. **Eleusine Corocana,** *Gærtn.*; GRAMINEÆ.

Ragi. 2607

A beer and a spirit are made from the fermented infusion of this grain in the Sikkim Himálaya, in Madras, etc., etc.

19. **Ephedra vulgaris,** *Rich.*; GNETACEÆ.

Soma. 2608

[The reader will find in Vol. III, 246-252, an account of the "Soma" of Sanskrit authors. Since the appearance of that chapter, the Editor has received a specimen of this plant from G. G. Minniken, Esq., Forest Department, Bashrh, which bears the information that it is used on the Upper Himálayan ranges to flavour spirits and to assist fermentation. In Afghánistan it is also employed in the preparation of a preservative fluid used in the manufacture of raisins. See **Vitis vinifera**].—*Ed., Dict. Econ. Prod.*

20. **Eugenia Jambolana,** *Lam.*; MYRTACEÆ.

2609

A spirit called *jambúa* is distilled from the juice of the ripe fruit.

21. **Grewia asiatica,** *Linn.*; TILIACEÆ.

2610

A spirit is distilled from the fruit.

22. **Hordeum vulgare,** *Linn.*; GRAMINEÆ.

Beer, Whisky, etc. 2611

The grain is much employed in some parts of India in the preparation of a kind of spirituous liquor. In Spiti a liquor is distilled from it which is called *chang*. "It is sold at 30 *puttahs* for the rupee. A *puttah* is a liquid measure of 2 seers= $\frac{3}{4}$ of a *pucka* seer. The people of these regions consume large quantities of *chang*, and on occasions of festivity one man is said to consume as much as four puttahs" (*Baden Powell*).

23. **Melia Azadirachta,** *Linn.*; MELIACEÆ.

THE NIM OR MARGOSA TREE.

Nim. 2612

The fermented toddy is occasionally distilled.

24. **Morus alba,** *Linn.*; URTICACEÆ.

Mulberry. 2613

A spirit is distilled from the fruit in Kashmir (*Lowther*).

25. **Opuntia Dillenii,** *Haw.*; CACTEÆ.

THE PRICKLY PEAR.

Prickly Pear. 2614

Proposals have been made in Spain to utilise the better varieties of prickly pears for the preparation of alcohol.

26. **Oryza sativa,** *Linn.*; GRAMINEÆ.

Rice. 2615

Rice-beer, or *pachwai*, is often distilled and a spirit obtained from it. In Burma rice spirit, *Sham-shao*, is largely used. It is very simply prepared. The rice is first steeped in water, to which herbs have been added to promote fermentation; when thoroughly fermented, the liquor is transferred to an iron cauldron covered with an inverted pail, the two being lightly secured by a paste of flour and water and allowed to boil on a slow fire. In the lower part of the pail a hollow bamboo, 4 feet long, is inserted; this connects the apparatus with a double walled vessel, the inner compartment of which is constantly kept cool by fresh supplies of water. The liquor passes into this and condenses. The first quality sells for R2-8 per bottle; the second, which is only the old material with an addition of water re-distilled, at R1-8, and the third at R1. The first and second burn with a light blue flame, and ignite immediately, but not so with the third, for which, indeed, there is hardly any sale. It is principally used to adulterate the first qualities (*Strettel*). (*Conf. with III., 249; V., 330.*)

MATERIALS.

Date Palm. 2616

27. **Phœnix dactylifera,** *Linn.*; PALMÆ.

THE DATE PALM.

It yields a saccharine juice, from which sugar, and a fermented and distilled spirit, may be made, but it is little used for these purposes, since the fruit is more valuable.

2617

28. **P. sylvestris,** *Roxb.*

THE WILD DATE PALM.

The fermented toddy is distilled and made into *arak* and a spirit resembling rum is obtained from the scum which oozes out from the *gúr*, while in the process of being refined to form *dhulua* sugar.

2618

29. **Rhizophora mucronata,** *Lamk.*; RHIZOPHOREÆ.

The fermented juice of the fruit is said to be sometimes used as a source of a spirit.

Sugar-cane. 2619

30. **Saccharum officinarum,** *Linn.*; GRAMINEÆ.

Rum is obtained chiefly by the distillation of the uncrystallisable portion of the expressed juice of this plant. A coarse rum obtained in this way is largely drunk in India. [Indeed, so ancient is this cutsom that many forms of cane are chiefly valued because of the large quantity and peculiar flavour of the molasses. Rum distilled direct from the juice of these canes is said to be much superior to that distilled from the juice that has been boiled down to *rab* or *gur* and the molasses separated by filtration or straining. The Natives thus recognise the fact that a mixture of crystallizable with uncrystallizable sugar yields the best quality of rum. But the peculiar flavour of rum by direct distillation is highly extolled by some of the more ancient authors, and the subject was accordingly thought worthy of special enquiry by the Honourable the East India Company.—*Ed., Dict. Econ. Prod.*]

Juar. 2620

31. **Sorghum vulgare,** *Pers.*; GRAMINEÆ.

A spirit is distilled from the grain.

Vine. 2621

32. **Vitis vinifera,** *Linn.*; AMPELIDEÆ.

In some parts of North-West India, as Peshawar, a kind of coarse brandy is obtained from grape juice, but it is not common nor used further than locally.

2622

33. **Woodfordia floribunda,** *Salisb.*; LYTHRACEÆ.

"In Kangra part of the plant is used in the preparation of spirits" (*Stewart*).

SUBSTANCES USED TO FLAVOUR SPIRITS.

FLAVOURING SUBSTANCES.

Acacia Bark. 2623

34. **Acacia arabica,** *Willd.*; LEGUMINOSÆ.

The root-bark of this species, as well as of **A. ferruginea, A. Jacquemontii,** and **A leucophlœa** is widely used in India for flavouring native spirits and to arrest the further stage of fermentation.

Berberry. 2624

35. **Berberis aristata,** *DC.*; BERBERIDEÆ.

The fruits of this and of **B. asiatica** and **B. Lycium,** are used in the Himálaya to flavour *arak.*

Gentian. 2625

36. **Gentiana tenella,** *Fries.*; and **G. Kurroo,** *Royle*; GENTIANACEÆ.

In Ladak the root of the former species is put in spirits to flavour them; the latter is similarly used in other parts of the Himálaya.

Anise. 2626

37. **Illicium verum,** *Hook. f.*; MAGNOLIACEÆ.

THE STAR ANISE OF CHINA.

The fruits are largely used throughout the East for flavouring spirits.

Substances to produce the narcotic properties of Spirits. (*W. R. Clark.*) SPIRITS.

FLAVOURING SUBSTANCES. Juniper. 2627

38. Juniperus communis, *Linn.*; CONIFERÆ.

From the berries, together with barley meal, a spirit is distilled. The berries are used only to impart a gin-like flavour (*Stewart*).

Spices. 2628

39. Spices. Spices of various sorts are also added as flavouring materials to the fermented liquors before they are distilled. Those most commonly employed are betel-nuts, cloves, sandal-wood, cumin seeds, black pepper, ginger, nutmegs, cardamoms, cinnamon, and the tubers of fragrant grasses belonging to the genus **Andropogon.**

SUBSTANCES USED TO EITHER CAUSE THE FORMATION OF, OR TO INCREASE THE NARCOTIC PROPERTY OF, SPIRITS.

STRENGTH-IMPARTING SUBSTANCES. 2629

40. Acacia leucophlœa, *Willd.*

Acacia Bark. 2630

[Bark is used in manufacture of Native spirits. It is supposed to increase the quantity of alcohol by arresting (as hops do) the secondary fermentation.—*Ed., Dic., Econ. Prod.*]

41. Anamirta Cocculus, *W. & A.*; MENISPERMACEÆ.

Cocculus Indicus. 2631

The seeds are used to increase intoxicating effects of country spirits sold in retail.

42. Cannabis sativa, *Linn.*; URTICACEÆ.

INDIAN HEMP.

Indian Hemp. 2632

The leaves are employed in the preparation of the intoxicating liquor *hashish*. *Bhang*—the young leaves—is used to make Native beer or spirits more narcotic.

Yeast. 2633

43. Cerevisiæ Fermentum—Yeast; see Vol. II., 257-260.

2634

44. Clerodendron serratum, *Spreng.*; VERBENACEÆ.

The root is used by the Santals to cause fermentation.

Datura, 2635

45. Datura fastuosa, *Linn.*, and other species; SOLANACEÆ.

The smoke from the seeds burnt on charcoal, or a powder of the seeds themselves, is sometimes mixed with Native spirits to render it more intoxicating.

Soma. 2636

46. Ephedra vulgaris, *Rich.*; GNETACEÆ.

See the remarks under 19 above.

47. Humulus Lupulus, *Linn.*; URTICACEÆ.

HOPS.

Hops. 2637

Used in India by European brewers only, the supply being imported.

2638

48. Ligustrum Roxburghii, *Clarke.*; OLEACEÆ.

In South India the bark of this tree is put into toddy of **Caryota urens** to accelerate fermentation.

49. Phyllanthus Emblica, *Linn.*; EUPHORBIACEÆ.

Emblic Myrobalan. 2639

The fruit is mixed with the substances used in the preparation of some Native spirits. It is supposed to increase their strength.

50. Stychnos Nux-vomica, *Linn.*; LOGANIACEÆ.

Nux-vomica. 2640

In many parts of India the seeds are eaten to produce intoxication, or are mixed with beverages for that purpose.

51. Terminalia belerica, *Roxb.*; COMBRETACEÆ.

Myrobalans. 2641

The fruit of this or of **T. Chebula,** *Retz.*, is employed in the Panjáb to increase the strength of spirits.

Spodiopogon angustifolius, *Trin.*; see **Ischœmum angustifolium**, [*Hack.*; Vol. IV., 526; GRAMINEÆ.

SPONDIAS, *Linn.*; *Gen. Pl.*, *I.*, *426*, *1001*.

2642 **Spondias acuminata**, *Roxb.*; *Fl. Br. Ind.*, *II.*, *42*; ANACARDIACEÆ.

Vern.—*Ambat, ambadah*, BOMB.

References.—*Roxb., Fl. Ind., Ed. C.B.C., 387; Graham, Cat. Bomb. Pl., 42; Lisboa, U. Pl. Bomb., 56, 57; Gazetteer, Mysore & Coorg, I., 71.*

Habitat.—An elegant, middle-sized tree of the Western Peninsula, found in Malabar and the Konkan hills.

FOOD Fruit 2643

Food.—The FRUIT, an ovoid, globose drupe, about the size of a small hen's egg; has been eaten by the Natives during famine seasons.

2644 **S. dulcis**, *Willd.*; *Fl. Br. Ind.*, *II.*, *42*.

THE OTAHEITE APPLE OR VI.

Vern.—*Amara, umra*, HIND.

References.—*Roxb., Fl. Ind., Ed. C.B.C., 387; Voigt, Hort. Sub. Cal., 144; DC., Orig. Cult. Pl., 202; Ind. Forester, VI., 240; Smith, Econ. Dic., 304; Firminger, Man. Gard. Ind., 234.*

Habitat.—A tree of from 50 to 60 feet high, indigenous in the Society, Friendly, and Fiji Islands; introduced thence into India, and cultivated there and in many other parts of the tropics.

In India it does not appear to germinate freely; indeed, **Firminger** says the stones never germinate, and that young plants are usually obtained by grafting upon seedlings of **S. mangifera**, the common country *amara* or Hog Plum.

FOOD. Fruit. 2645 Rind. 2646 Pulp. 2647

Food.—The FRIUT, which weighs sometimes as much as 1℔ 2 ozs., and measures a foot in circumference, is shaped like a hen's egg, and is of a deep amber colour, often blotched with deep russet patches. It has a large fibre-covered stone in the centre. The RIND tastes of turpentine, but the PULP has an apple-like smell, and its flavour is, in the best varieties, very agreeable. Not much can be done with it in the way of cooking, either as a preserve or as a pudding.

DOMESTIC. Wood. 2648

Domestic.—In Otaheite, the WOOD is much valued for making canoes.

2649 **S. mangifera**, *Willd.*; *Fl. Br. Ind.*, *II.*, *422*; *Wight*, *Ill.*, *I.*, *186*, [*t. 76*.

THE HOG PLUM.

Syn.—S. AMARA, *Lamk.*; MANGIFERA PINNATA, *Kœn* (not *Lamk.*); EVIA AMARA, *Comm.*

Vern.—*Amrá, amara, ambodha, ámbrá*, HIND.; *Amna, ámrá, ambra*, BENG.; *Amburri;* KOL.; *Amara*, ASSAM; *Tongrong, adai*, GARO; *Amara*, NEPAL; *Ronchiling*, LEPCHA; *Kat ambolam*, MAL (S. P.); *Ambulá*, URIYA; *Ambera*, KURKU; *Hamara*, GOND; *Amra, amúr, bahamb, amara, amabára*, KUMAON; *Báhamb, ambárá*, PB.; *Ran-amb, jungli-am*, DECCAN; *Ambáda, jangliám, ambara, amra, rhan-amb, amarah*, BOMB.; *Rán-amba, amb, ambada*, MAR.; *Kat-máa, mari-manchedi, kat-mara*, TAM.; *Puliille*, KADERS; *Ambála chettu, pita vrikshamu, ivuru mámidi, amatum, adivio-mamadie, toura-mamidi*, TEL.; *Amte, ambatte mara, amate, pundi*, KAN.; *Gwe, kywae*, BURM.; *Æmbærælla*, SING.; *Amrátaka*, SANS.; *Darakhte-moryam*, PERS.

References.—*DC., Prodr., II., 75; Roxb., Fl. Ind., Ed. C.B.C., 387; Voigt, Hort. Sub. Cal., 143; Brandis, For. Fl., 128; Kurz, For. Fl. Burm., I., 321, 322; Beddome, Fl. Sylv., t. 169; Gamble, Man. Timb., 112; Dalz. & Gibs., Bomb. Fl., Supp., 19; Stewart, Pb. Pl., 50; Graham, Cat. Bomb. Pl., 42; Mason, Burma & Its People, 461, 489, 774; Sir W. Elliot, Fl. Andhr., 14, 71, 154; Sir W. Jones, Treat. Pl. Ind., V., 125, No. 46; Rheede, Hort. Mal., I., t. 50; O'Shaughnessy, Beng. Dispens., 270; U. C. Dutt, Mat. Med. Hind., 291; Dymock, Mat. Med. W. Ind., 2nd Ed., 205; Dymock, Warden & Hooper, Pharmacog. Ind.,*

395, 396, 549; Birdwood, Bomb. Prod., 147, 219; Baden Powell, Pb. Pr., 597; Atkinson, Him. Dist. (X., N.-W. P. Gaz.), 751; Useful Pl. Bomb. (XXV., Bomb. Gaz.), 56, 250; Econ. Prod. N.-W. Prov., Pt. V. (Vegetables, Spices, & Fruits), 59; Aplin, Rep. on Shan States (1887-88); Gazetteers:—Bombay, V., 285; XV., 76; XVIII., 41; N.-W. Prov., III., 239; IV., lxx.; X., 308; Mysore & Coorg, I., 71; III., 17; Hunter, Orissa, II., 158, App. IV.; 180, App. VI.; Agri.-Hort. Soc. Ind.:—Journ., IV., 128; VI., 41; IX., Sel., 53, 57; X., 2; XIII., 351, Sel., 63 (New Series), V., 74; Ind. Forester:—III., 201, 238; VII., 250; VIII., 410; XI., 2; XIII., 120.

Habitat.—A small, deciduous tree, found wild or cultivated throughout India, from the Indus eastwards and southward to Malacca and Ceylon, ascending to 5,000 feet in the Himálaya. It is distributed to Tropical Asia.

GUM. Bark. 2650

Gum.—A mild insipid gum, somewhat resembling gum arabic, but darker in colour, exudes from the BARK. It occurs in stalactiform pieces of a yellowish or reddish brown colour, and with a smooth shining surface. With a large volume of water it forms a gelatinous mucilage, which is precipitated by acetate of lead, gelatinized by the basic acetate and by ferric chloride, but not by borax (*Dymock*).

MEDICINE. Fruit. 2651

Leaves. 2652

Bark. 2653

Gum. 2654

Juice. 2655

Medicine.—The pulp of the FRUIT is described by Sanskrit writers as acid and astringent, and useful in bilious dyspepsia, for which reason the name of *Pittavriksha* or "bile tree" is sometimes applied to it. The fruit is much used by the Hindus as an acid vegetable, and is made into a preparation resembling gooseberry which they call *ráyeté*. The LEAVES and BARK are astringent and aromatic, and are administered in dysentery. The GUM is used as a demulcent (*Dymock*). The JUICE of the leaves is applied locally in earache (*Atkinson*). By some of the Shan tribes in Burma the frūit is considered an antidote for wounds from poisoned arrows, and for this purpose is eaten either green or dry. When that cannot be obtained, alum is considered the next best remedy, a fact which would seem to indicate that astringents counteract the effects of these wounds on the system (*Mason, Burma & Its People*).

SPECIAL OPINION.—§ "'*Amra*' is a useful antiscorbutic. I use it both in its green and ripe state in curries for the prisoners" (*Surgeon R. L. Dutt, M. D., Pabna*).

FOOD & FODDER. Fruit. 2656

Food and Fodder.—The FRUIT, which ripens in October, when largest, is of the size of a goose's egg, of a rich olive green colour, mottled with yellow and black. It has but little scent. The part nearest the rind is extremely acid, but, that being removed, the part near the stone is sweet and eatable. It is sometimes eaten raw when ripe, but more commonly it is put, while green, in fish or vegetable curries, or in lentils, to give these dishes the acid taste so much appreciated by Natives. It is also made into a pickle with mustard oil, salt, and chillies (*Firminger; Liotard*). Cattle and deer eagerly feed on the fruit (*Atkinson*).

TIMBER. 2657

Structure of the Wood.—Soft, light-grey. Weight about 43℔ per cubic foot (*Gamble*).

DOMESTIC. 2658

Domestic.—The timber is used only for fuel.

2659

SPONGES, *Encyclop. Brit., XXII., 412.*

The PARAZOA or SPONGIÆ form one great phylum of the Invertebrata, which comprises many sorts of Sponges besides those which are of economic value. These latter belong to the Order **Ceratosa**, in which the skeleton consists of horny fibres which never include 'proper' spicules or of introduced foreign bodies or of both in conjunction. It is this skeleton or network of elastic horny fibres, which constitutes the Sponge of Commerce.

SPONIA orientalis.

Species of Sponges.

SPONGES. 2660

Sponges.

References.—*Simmonds, Commercial Products of the Sea, 155-197; Balfour, Cyclop. Ind., III., 725; Fisheries Exibition Literature, IV., 422; V., 369, and many other passages; Tro. Agriculturist, June 1st, 1889, 828.*

The species of Sponge in common use are three :—

2661

1. **Euspongia officinalis,** *Linn.*

THE FINE TURKEY or LEVANT SPONGE.

Habitat.—It occurs in the North Pacific, South Atlantic, and Indian Oceans, but its most frequent habitat is in the shallow waters off the coasts of the Mediterranean.

2662

2. **E. zimocca,** *O. Schmidt.*

THE HARD ZIMOCCA SPONGE.

Habitat.—In similar localities and often in conjunction with the preceding on the shores of the Mediterranean, and off the Bahama Islands.

2663

3. **Hippospongia equina,** *O. Schmidt.*

THE HORSE SPONGE or COMMON BATH SPONGE.

Habitat.—Also found, often along with the preceding, on the Mediterranean shores, at the Bahama Islands, and on the north coast of Cuba.

Collection. 2664

COLLECTION.—The methods employed to obtain sponges from the bottom of the sea, where they grow attached to rocks and stones, depend on the depths from which they have to be brought. In shallow water they are hooked up by a harpoon, at greater depths they are dived for, and at still greater depths they are dredged up with nets.

In harpooning one of the chief difficulties is to see the bottom through the troubled surface of the water and to obviate this the Dalmatian fishers throw a stone dipped in oil a yard or so in front of the boat, the stone scatters drops of oil as it flies, and so makes a smooth track for the look-out. The Greeks use a zinc plate cylinder, about 1½ feet long and 1 foot wide, closed at the lower end by a plate of glass, which is immersed below the surface of the sea; on looking through this the bottom may be clearly seen even at 30 fathoms. This plan is also adopted in the Bahamas.

Balfour, in his *Cyclopædia*, says, that sponges are gathered from the rocks of Vizagapatam at about 12 feet below the sea, but no other reference to this fact occurs among the books available to the writer, and therefore no further information on the subject of Indian Sponges can be given.

Preparation. 2665

Preparation for the market.—After the Sponge has been taken from the sea it is exposed to the air till signs of decomposition set in, and then without delay is either beaten with a thick stick or trodden by the feet in a stream of flowing water till the soft parts are removed. After cleaning it is hung up in the air to dry, and then with others finally pressed into bales.

DOMESTIC. 2666

Domestic.—Besides being used for cleansing purposes in the bath and by the Surgeon, the coarser sorts of American Sponges have recently been employed in America for stuffing in upholstery. For this purpose it is first thoroughly cleansed and cut into pieces. These are then placed in a solution of glycerine and water, and after passing through heavy rollers they are dried. The water evaporates and leaves the Sponges so permeated with glycerine that a permanent elasticity is maintained.

Sponia amboinensis, *Dcne.*, see **Trema amboinensis,** *Blume;* URTICACEÆ; Vol. VI., Pt. IV., 35.

S. orientalis, *Planch,* see **T. orientalis,** *Blume;* Vol. VI., Pt. IV., 35.

SPOROBOLUS, *R. Br.; Gen. Pl., III., 1148.*

2667

A genus of grasses comprising upwards of eighty species, six at least of which are found in the plains of Northern India. Three of these (**Sporobolus commutatus,** *Boiss.*, **S. coromandeliana,** *Roxb.*, and **S. pallidus,** *Nees.*) are comparatively unimportant, since they do not occur in sufficient abundance to be considered valuable as fodder grasses. The grain of the last-mentioned is said to be occasionally eaten in the Panjab in times of scarcity.

Sporobolus diander, *Beauv.; Duthie, Fodder Grasses of Northern India, 48;* GRAMINEÆ.

2668

Syn.—AGROSTIS DIANDRA, *Linn.*

Vern.—*Bena-joni,* BENG.; *Chiriya-ka-dána, galphula,* N.-W. P.; *Nonak,* PB.

References.—*Roxb., Fl. Ind., Ed. C.B.C., 106; Rev. A. Campbell, Econ. Pl., Chutia Nagpur, Nos. 8728, 9844; Gazetteers, N.-W. P., IV., lxx.; X., 320.*

Habitat.—Common on the plains of India and at moderate elevations on the hills.

Fodder.—It is described by Mr. **Duthie** as a favourite fodder grass in the Panjáb for cattle and horses.

FODDER 2669

S. indicus, *R. Br.; Duthie, Fodder Grasses of Northern India, 49.*

2670

Syn.—S. TENACISSIMUS, *Beauv.;* VILFA TENACISSIMA, *Trin.*

Vern.—*Ratua,* N.-W. P.; *Ghorla,* C. P.; *Khir,* PB.; *Tomagarika,* TEL.

References.—*Sir W. Elliot, Fl. Andhr., 183; Gazetteer, N.-W. P., IV., lxxx.; Ind. Forester, XII., App., 2, 25.*

Habitat.—Found on the plains of India, ascending to moderate elevations on the hills and generally distributed over the tropical and sub-tropical parts of the world.

Fodder.—It is considered a good fodder grass, especially when young. "In the United States this grass is of considerable value for grazing purposes if frequently cut or grazed down, but if allowed to remain untouched long, cattle and horses will not eat it unless they are very hungry, as it becomes tough and unpalatable. It is not used to any considerable extent for hay, but makes splendid feed if cut while young and as a pasture plant" (*Duthie*).

FODDER. 2671

S. orientalis, *Kunth; Duthie, Fodder Grasses of Northern India, 49.*

2672

Syn.—VILFA ORIENTALIS, *Kunth.*

Vern.—*Usar-ki-ghás, kar-usura-ghás, kálusura,* N.-W. P.; *Tandua, kheo,* PB.

Reference.—*Report, Agric. Dept. (1882), 45, 46.*

Habitat.—This grass is strictly confined to saline soils, and is found on all the *usar* tracts in Northern India, Sind, and Madras, often constituting the entire vegetation. It does not, however, appear to hold its own on ground which is capable of supporting other grasses, and gradually disappears as the reclamation of the *reh*-infected tracts proceeds.

Fodder.—" It is a capital fodder grass, both horses and cattle eating it readily when it is green. If it is cut and dried at the end of the rains it makes good hay, and cattle will eat it in this state if it is cut and mixed with some green food " (*Rep. Agri. Dept.*).

FODDER. 2673

Spruce Fir, see **Abies excelsa,** *DC.;* Vol. I., 2; CONIFERÆ.

Squill, see **Scilla indica,** *Linn.;* Vol. VI., Pt. II., 489; also **Urginea indica,** *Kunth.;* and **U. scilla,** *Steinheil;* Vol. VI., Pt. IV., 173-174; LILIACEÆ.

Squirrels, see **Furs,** Vol. III., 458; **Skins,** Vol. VI., Pt. III., 244-250; also **Rats, Mice, & Marmots,** Vol. VI., Pt. I., 395.

STACHYS, *Linn.; Gen. Pl., II., 1208.*

2674

Stachys parviflora, *Benth.; Fl. Br. Ind., IV., 677;* LABIATÆ.

Vern.—*Kirimar, baggi búti,* PB.; *Speraghunai,* PUSHTU.

References.—*DC., Prodr., XII., 490; Boiss., Fl. Orient., IV., 740; Stewart, Pb. Pl., 173; Lace, Flora of Quetta, MSS.*

Habitat.—An erect undershrub, common on the Panjáb plains and hills, from the Jhelum eastwards and northwards to Marri, and distributed to Afghánistán.

MEDICINE. Stems. 2675

Medicine.—"In the Salt Range the bruised STEMS are applied to the guinea-worm" (*Dr. Stewart*).

FODDER. Leaves. 2676 Shoots. 2677

Fodder.—In Baluchistán the LEAVES and tender SHOOTS are much browsed by cattle (*J. H. Lace*).

STAPHYLEA, *Linn.; Gen. Pl., I., 412.*

2678

Staphylea Emodi, *Wall.; Fl. Br. Ind., I., 698;* SAPINDACEÆ.

Vern.—*Márchol, chúal, ban shágali, ban bakhwiú, thanári, gúldar, nágdaun, kaghania,* HIND. & PB.; *Chitra, kúrkni,* KASHMIR; *Márchob,* PUSHTU & AFG.

References.—*Voigt, Hort. Sub. Cal., 169; Brandis, For. Fl., 114; Gamble, Man. Timb., 101; Stewart, Pb. Pl., 40; Aitchison, Kuram Valley Rept., Pt., II. 156; Baden Powell, Pb. Pr., 597; Atkinson, Him. Dist. (X., N.-W. P. Gaz.), 308; Gazetteers:—Panjáb, Dera Ismail Khan, 19; Rawalpindi, 15; Hoshiarpur, 12; Bannu, 23; Agri.-Horti. Soc., Ind., XIV., 13.*

Habitat.—A large shrub or small tree of the Western Temperate Himálaya, from Marri to Kumáon, occurring at altitudes between 6,000 and 7,000 feet. It is distributed to Afghánistán.

TIMBER. 2679

Structure of the Wood.—Soft grey. Weight 44lb per cubic foot.

DOMESTIC. Bark. 2680

Domestic and Sacred.—The BARK on the young stems is marked with splashes of white on a dark olive-green ground, resembling the markings on the skin of a snake; hence sticks of this shrub are carried by the Natives as a protection against snakes. This same protective power is ascribed to them along the whole frontier through Hazara to Kashmír, where they are similarly employed—a practice which no doubt took its origin from the "doctrine of signatures," in this case applied to the resemblance borne by the colour of the bark of **Staphylea** to a serpent's skin. As walking-sticks they are useless, since they are hollow and the wood is very brittle (*Stewart, Baden Powell, Aitchison*).

(*G. Watt.*)

2681

STARCHES, *Special Catalogue of Col. & Ind. Exhib., 74, 90.*

The term "Starch" is applied to the fecula or amylaceous matter contained in the fruit, roots, and cellular tissue of the majority of plants, but which is extracted, on a commercial scale, from a few which possess an exceptionally large supply. Starch occurs in minute granules of various sizes, having, when pure, a slightly yellowish colour, and a form or structure characteristic of each kind. Under the microscope these granules are seen to consist of a nucleus or hilum surrounded by layers arranged concentrically or eccentrically; the relations of hilum and layers are the most distinctive features of individual starches. The sp. gr. varies with the kind of starch, and with its dryness; the amount of water reaches 30 per cent. in some instances, while in others it descends to 7, when air-dry. It is only slightly acted on by cold water, but under the influence of boiling water it swells up, forming a cloudy opalescent paste or fluid

Iodine acts on it in water by producing a brilliant blue coloration, a reaction which forms a very delicate and characteristic test. Some starches are prepared for use as alimentary substances, while others are only used for industrial purposes. The best qualities of starches serve for sizing fancy papers, as well as for finishing textiles, for making white dextrine, and as an aliment. Inferior sorts are used for weavers' dressing, for thickening mordants and colouring substances used in cloth printing. Another important application is the "dusting" of forms in metal foundries, in lieu of charcoal dust. The use of starch for stiffening linen and washed clothes is also general. The weighting of cotton goods with starch is mentioned in the Institutes of Manu. (*Conf. Vol. IV., 43.*)

Maize, wheat, and rice starches are principally employed for the direct application, while for the manufacture of dextrine or starch-sugar, potato starch is almost exclusively selected.

Starch.

2682

Vern.—*Ganji,* HIND.; *Kalap, kalaph,* BENG.; *Kanji, gams,* MAL.; *Godambe mas,* TAM.; *Nishashta,* PERS.

References.—*Royle, Prod. Res., 12, 230; Birdwood, Bomb. Prod., 240-245; Simmonds, Waste Products and Undeveloped Resources, 191; Smith, Econ. Dict., 392; Balfour, Cyclop. Ind., III., 733; Spons, Encyclop., II., 1821; Encyclop. Brit., XXII., 455.*

The following account of the principal plants used in India as sources of starch is chiefly derived from the articles on the subject in the Special Catalogue of Exhibits shown at the Colonial and Indian Exhibition of London in 1886:—

STARCHES. Horse-Chestnut. 2683

1. **Æsculus indica,** *Colebr.*; SAPINDACEÆ.

THE INDIAN HORSE-CHESTNUT.

Found abundantly in the North-West Himálaya. The fruits of the European horse-chestnut are used in Southern France in the manufacture of starch, which is readily extracted by the methods adopted with starch from the cereals. One hundred pounds of dry starch are obtained from 240 to 250℔ of the nuts. The bitterness is removed by treating with water containing carbonate of soda (*see* Vol. I., 126).

Mankachu. 2684

2. **Alocasia indica,** *Schott.*; AROIDEÆ.

Generally cultivated around the huts of the poorer classes in Bengal. Its esculent stems and root-stocks contain a large quantity of starch, and are important articles of diet (*see* Vol. I., 178).

Oil. 2685

3. **Amorphophallus campanulatus,** *Blume*; AROIDEÆ.

A native of India and Ceylon; cultivated throughout the peninsula in rich moist soils. The starchy corms are in common use as an article of food (*see* Vol. I., 225).

Peruvian Carrot. 2686

4. **Arracacia esculenta,** *DC.*; UMBELLIFERÆ.

THE PERUVIAN CARROT.

A native, it is believed, of the elevated regions of equatorial America. Experimentally introduced into India. Its tubers yield a large quantity of easily digested starch (*see* Vol. I., 318).

Oats. 2687

5. **Avena sativa,** *Linn.*; GRAMINEÆ.

OATS.

Of recent and not extensive cultivation in India. The grains contain a large percentage of starch (*see* Vol. I., 356).

Shot. 2688

6. **Canna indica,** *Linn.*; SCITAMINEÆ.

THE INDIAN SHOT.

A herbaceous plant, common all over India, chiefly in gardens where it is grown as an ornamental plant. The root-stock contains starch and may be cooked for food (*see* Vol. II., 102).

STARCH. **Plants used as sources of Starch.**

STARCHES. Batard Sago. 2689

7. **Caryota urens,** *Linn.;* PALMÆ.

THE BASTARD SAGO.

A beautiful palm of the Western and Eastern moist zones of India. It affords good sago (*see* Vol. II., 206; also **Sago** Vol. VI., Pt. II., 383).

Arum. 2690

8. **Colocasia antiquorum,** *Schott.;* AROIDEÆ.

THE TARO or EGYPTIAN ARUM.

Cultivated all over India on account of its corms, which constitute an important article of diet (*see* Vol. II., 509).

Kanchura. 2691

9. **Commelina bengalensis,** *Linn.;* COMMELINACEÆ.

Generally distributed throughout India. The fleshy rhizomes contain much starch (*see* Vol. II., 515).

Indian. arrowroot. 2692

10. **Curcuma angustifolia,** *Roxb.;* SCITAMINEÆ.

THE WILD or EAST INDIAN ARROWROOT.

Grows in abundance in the forests of the Deccan and in Malabar. An arrowroot is prepared from its tubers (*see* Vol. II., 652).

Yam. 2693

11. **Dioscorea,** various species; DIOSCOREACEÆ.

THE YAM.

There are several species (wild and cultivated) common all over India. The tubers contain large quantities of a rich and delicate starch, and are much used as articles of diet (*see* Vol. III., 115).

Salep. 2694

12. **Eulophia campestris,** *Lindl.;* ORCHIDACEÆ.

Found in North-Western India. The tubers of this are used as Salep, which is chiefly valuable on account of the starch which it contains (*see* Vol. III., 291, also Vol. VI., Pt. II., 385-386).

Buckwheat. 2695

13. **Fagopyrum esculentum,** *Mœnch.;* POLYGONACÆ.

THE BUCKWHEAT or BRANK.

Cultivated, along with another species, **F. tataricum,** throughout the Himálaya, and distributed to Central Europe and North Asia. A few English firms prepare starch from buckwheat. It is a fine powder of nearly pure white colour (*see* Vol. III., 310).

Jerusalem Artichoke. 2696

14. **Helianthus tuberosus,** *Linn.;* COMPOSITÆ.

THE JERUSALEM ARTICHOKE.

Cultivated as a vegetable in India. Its tuberous roots yield starch (*see* Vol. IV., 211).

Barley. 2697

15. **Hordeum vulgare,** *Linn.;* GRAMINEÆ.

Cultivated in India. The grains contain a large percentage of starch (*see* Vol. IV., 274).

Sweet Potato. 2698

16. **Ipomœa Batatas,** *Lamk.;* CONVOLVULACEÆ.

THE SWEET POTATO.

Most probably a native of America, its cultivation in India became general in the eighteenth century, and at the present day it is grown to a limited extent in almost every part of the country. It requires little care and grows in any soil. There are two kinds—the one with red, and the other with white, tubers. They yield a large quantity of starch and are eaten by all classes of Natives and by Europeans (*see* Vol. IV., 478).

Tapioca. 2699

17. **Manihot utilissima,** *Pohl.;* EUPHORBIACEÆ.

THE TAPIOCA.

Commonly cultivated in equatorial or tropical regions, especially in America, from Brazil to the West Indies, introduced in South India (*see* Vol. V., 157).

STARCHES.

18. Maranta arundinacea, *Linn.*; SCITAMINEÆ.

THE ARROWROOT.

Arrowroot. 2700

Introduced from America, and now extensively cultivated in India. Yields the true arrowroot of commerce.

19. Musa sapientum, *Linn.*; SCITAMINEÆ.

THE BANANA OR PLANTAIN.

Banana. 2701

A native of Southern Asia, extensively cultivated in India. The green fruits contain about 12 per cent. of starch which disappears as they mature. In South America, where the plantain is largely grown, the starch is extracted by slicing, sun-drying, powdering, sifting, and washing with water. The article is exported to Europe in the form of meal, and the starch is there manufactured. The flour is said to contain 66 per cent. of starch (*Spons' Encycl.*) (*see* Vol. V., 180).

20. Nelumbium speciosum, *Willd.*; NYMPHÆACEÆ.

THE SACRED LOTUS OR PADMA.

Lotus. 2702

An aquatic herb, common throughout India, extending as far to the North-West as Kashmír. The starchy tubers are eaten (*see* Vol. V., 343).

21. Oryza sativa, *Linn.*; GRAMINEÆ.

THE RICE.

Rice. 2703

The common rice is largely used as a source of starch, both in India and Europe. In India the *kanji* used by every *dhobi* throughout the country is in almost every instance rudely prepared by himself from rice grains, while in England a considerable amount of the starches prepared by Colman and other large starch manufacturers are derived from the same material (*see* Vol. V., 649).

In Europe it has been found that mere steeping and bruising do not suffice to separate the starch from the other components of the grain, and recourse is had to caustic alkalis for the purpose.

22. Pueraria tuberosa, *DC.*; LEGUMINOSÆ.

Tirra. 2704

A climbing shrub of the sub-tropical tracts of the Western Himálaya, the Western Gháts, Chutia Nagpur, and Orissa. The roots, which are large and tuberous, are eaten by the hill tribes and are exported to the plains, where they are used as a source of starch (*see* Vol. VI., Pt. I., 363).

23. Pulses.

Pulses. 2705

A large proportion of the laundry starches, now in use, are derived from the pulses, and it seems probable that some portion of the pulses exported from India to Europe are utilised in this manufacture, but no separate figures as to the amounts employed for this purpose seem to be available (*see* Vol. VI., Pt. I., 364-368).

24. Salep Misri.

Salep. 2706

The tubers of several species of orchid are all sold under the name of Salep. The commercial article comes from Persia, Afghánistán, and the Panjáb Himálaya. There are three forms—the Panjáb or palmate tuber, the Persian, and Afghan, which are long, ovoid, and small round tubers. Part of these at least is supposed to be derived from **Eulophia campestris** and **E. herbacea.** Salep in all its forms yields a large quantity of easily digested starch, and is supposed by the Natives to be an excellent food for invalids and children (*see* Vol. VI., Pt. II., 385-386).

25. Solanum tuberosum, *Linn.*; SOLANACEÆ.

THE POTATO.

Potato. 2707

Introduced into Europe in the sixteenth and into India in the eighteenth century. It is now cultivated throughout India in the plains and on the hills up to 9,000 feet. The potato is eaten by all classes. In Europe it is

 Plants used as sources of Starch.

STARCHES.

largely employed as a source of starch, it is made into British gum, an altered form of that product, and into fictitious arrowroots and sago. For the manufacture of starch the potatoes are first cleaned and grated. They are then sifted to separate the starch granules from the cellular and fibrous matters, and the sifted starch is then washed and allowed to deposit itself in water. It is then bleached by the application usually of sulphuric acid, afterwards neutralised, and last of all dried (*see* Vol. VI., Pt. III., 264).

South Sea arrowroot. 2708

26. Tacca pinnatifida, *Forsk.;* TACCEACEÆ.

THE SOUTH SEA ARROWROOT.

Found in Chutia Nagpur, Central India, the Konkan and southwards, also on the Parell hills near Bombay, cultivated at Travancore. Its large round tuberous roots yield a quantity of white nutritious fecula which resembles arrowroot (*see* Vol. VI., Pt. III., 401).

Wheat. 2709

27. Triticum sativum, *Lamk.;* GRAMINEÆ.

WHEAT.

The ripe grain of the wheat plant contains 50-75 per cent. of starch. There are three chief methods for preparing wheat-starch based on different principles, *viz.*, (1) by acetous fermentation, (2) without fermentation, (3) from flour. The first plan, which is the one employed in India, where only small quantities of wheat-starch are occasionally prepared for local use, has the disadvantage that the gluten is, for all practical purposes, destroyed, and that noxious vapours and foul liquids are largely produced. The wheat is first soaked in large *ghurrahs* containing sufficient water to thoroughly immerse the grain, clean water being supplied daily. When the grain is soft and pulpy it is rubbed between the hands, and the milky-looking fluid containing the starch-grains sinks to the bottom, while the husks float. The latter are then skimmed off, the supernatant fluid removed, and fresh water is again and again supplied till fermentation is complete and the starch is free from any disagreeable odour. It is then dried in the sun on papers or clothes, and after powdering in a mortar is fit for use. Modifications of this simple method are largely employed in Europe in the manufacture of wheat-starch, as well as the more scientific methods above mentioned, and which need not here be further dealt with (*see* Vol. VI., Pt. IV., 49-162).

Maize. 2710

28. Zea Mays, *Linn.;* GRAMINEÆ.

INDIAN CORN OR MAIZE.

After the millets and rice, this is now, perhaps, the most important food-stuff of modern India, although in the time of **Roxburgh** it was only cultivated in gardens as an ornamental plant. There are now over 2¼ million acres annually under this crop. While cultivated extensively in every district of India, it is not exported, but is either eaten green as a vegetable or matured as a grain crop; it is not made into flour to any extent. To the hill tribes it is an important article of diet, and in Southern India it may be grown all the year round. There are many varieties, some with white, others with yellow, and still others with dark red orange or mottled grains. Maize is much used as a source of starch. It is obtained in an analogous manner to that from the cereals, but the proportion of gluten in the grain being smaller and less tenacious in nature, the operations present fewer difficulties. Under one method, the separation of maize-starch is facilitated by steeping, swelling, and softening the grain in a weak solution of caustic soda, and good results are obtained also by a process in which the pulp from the crushing mill is treated with water acidulated with sulphurous acid (*Encycl. Brit.*).

The finer qualities of maize-starch are largely used as a substitute for arrowroot and in biscuit making; inferior qualities serve for laundry pur-

purposes. The most extensive factories where it is produced are **Brown & Polson's** in Scotland, **Erkenbrecker's** in Cincinnati, and the **Glen Cove** Company's in New York; it is also made on a smaller scale in Brazil, New South Wales, France, and Hungary (*Spons' Encyclop.*) (*See* Vol. VI., Pt. IV., 286-313).

(*John Watt.*)

STEATITE, *Ball, Man. Geol. Ind. (Economic), III., 426, 439-445.*

2711

Steatite or Talc is a soft magnesian or talcose mineral, which has received its name from its soapy smooth feel. It occurs in subordinate beds of serpentine and chlorite-schist, is usually of a greyish or yellowish-green colour, and has a laminated texture. Soapstone, Potstone, and other Talcose rocks (silicates of magnesia) are in fact only varieties. In popular *parlance*, MICA (*see Vol. V., 239-240*) is often (though incorrectly) called Talc. There is usually, says **Murray**, no difficulty in distinguishing these two articles, for, while both are flexible, mica alone is elastic (*See Ainslie, Mat. Ind., I., 421*). The foliated varieties of talc are inapplicable to the purposes for which mica is principally applied. Serpentine or Ophite, when pure, is a hydrous magnesium silicate, that contains more water but less silica than talc. Some of the magnesian potstones of India have been described as serpentines, so that the soapstone plates, bowls, etc., sold all over the country may sometimes be made of serpentine, at other times of a transitional mineral between steatite and serpentine (*See Vol. VI., pt. II., 501*), as well as of pure soapstone. Verd antique marble is limestone with included serpentine.

It seems desirable to deal with the manufactures of all these stones in this place, since they are scarcely separable.

Steatite or Talc, *Mallet, Records, Geol. Surv., Ind., Vol. XXII.,* SOAPSTONE, POTSTONE. *[Pl. 2, 1889.*

2712

Vern.—*Abrak* (? Mica), HIND.; *Silkhari*, PB.; *Appracum*, TAM. and TEL.; *Kokubulurs, minirum*, SING.; *Tulk*, PERS.; *Abraka*, SANS.; *Sang-i-palaun* (French Chalk), HIND.; *Bulpum* (French Chalk), TAM.; *Zahr-muhra* (Serpentine), HIND.

References.—*Ball, l.c., gives complete list of the papers that have appeared in the Mem. & Rec., Geol. Surv., India; also in Asiatic Researches, Journ. & Proc. Asiatic Soc., Bengal, etc., etc.; Ainslie, Mat. Ind., I., 421; U. C. Dutt, Mat. Med. Hind., 76-80; Baden Powell, Pb. Prod., I., 42-43, 100.*

OCCURRENCE 2713

Occurrence.—**Ball** remarks that "owing to the wide distribution of the varieties of Talc throughout the metamorphic rocks which occupy so extended an area in India, it would be impossible, except with the expenditure of a very considerable amount of space, to give a detailed account of all that is known on the subject." Some short time ago interest was taken in the possibility of India being able to meet some of the modern demands for this stone. **Mr. J. R. Royle, C.I.E.**, issued from the India Office a Note on the subject, and this called for the collection of specimens and particulars regarding these, from the chief known Indian sources of supply. **Mr. Mallet**, Superintendent of the Geological Survey, reported on these samples, and as his report is not only the most recent paper on the subject but the most instructive that has as yet appeared, it may usefully be republished here :—

"Altogether about 50 specimens (mostly 6-inch cubes) have been received from 38 different localities, scattered over 19 districts. Concerning all of these, more or less full information was sent at the same time. The samples and papers having been forwarded to the Geological Survy Office, the work has been assigned to me of making a preliminary examination of the former and systematizing the information obtained.

"As it would be useless to send to England stones that are unfit for employment in any of the uses to which steatite can be applied, all those have been rejected which are obviously worthless. Of these a few are sandstones, the rest being potstones, and steatites so impure as to be valueless, except for such coarse purposes

STEATITE or Talc.

Localities in which Soapstone

OCCURRENCE

as potstones can be applied to. These rejections include about 30 specimens, leaving 22 specimens from 15 localities in 12 districts. Of these districts, 7 are in the Madras Presidency; the remainder being in the Central Provinces, Rajputana, and Burma.

"The specimens last mentioned vary greatly in quality, but none are so bad as to have been deemed worthy of summary rejection. Although the final decision as to which of them are suitable for the purposes of gas engineers must be left to the engineers themselves, an attempt has been made to form some idea on this point in the laboratory here. It appears, from **Mr. Royle's** memorandum, that the latter stages in the manufacture of gas-burners are performed with circular saws, running at 1,700 revolutions per minute, and that the most minute particles of grit would suffice to destroy both the burners and the saws. As one of the most necessary qualities in the steatite, then, is the capability of being cut into sufficiently small pieces without injury to the saw, it appeared that it would be a fair test to reproduce the actual conditions of manufacture as nearly as could be. A saw two inches in diameter was not obtainable, but one of four inches was used, which was run at about 1,300 revolutions per minute, giving therefore a circumferential speed even greater than that of the two-inch. But this, in as far as injury to the saw itself was concerned, was probably more than balanced by the fact that the larger saw was probably of thicker steel and had stronger teeth. A straight-edge was placed parallel to the saw on one side, and adjustable in its distance, so that slices of steatite of any required thickness could be cut. As a matter of fact, however, all the specimens were tested as to their capacity of being cut into slices ·03 inch thick. Slabs measuring about 4″ × 1½″ × ½″ were taken, and slices cut from these of 4″ × ½″. From some specimens, slice after slice of this kind could be cut without breaking. Others, again, gave slices which generally broke in two, while the most brittle samples gave slices which broke up into three or four pieces. These grades are distinguished roughly in the notes below as—

Cut easily in slices.
Cut in slices.
Cut with difficulty in slices.

At the same time the presence of grit could be detected by the peculiar noise made by the saw, and the feel of the slab in the hand as it was pushed forward.

"Previously to their being sent to Calcutta the Madras specimens were tested by **Mr. Bosworth-Smith**, Mineralogist to the Government of that Presidency, in as far as the means at his disposal allowed.

USES. Gas-burners. 2714 Marking chalk. 2715 Fire-proof paint. 2716 Lubricant. 2717

"'It appears,' he writes, that the stone is principally wanted for cutting into caps for GAS-BURNERS, but as the demand for stone for such a purpose would of necessity be very small, the economic value of the mineral for other purposes should be given. There are several minor uses to which steatite is put, such as 'MARKING-CHALK' for tailors, etc. It has been proposed to use the mineral in a FIRE-PROOF PAINT, but with what success I am not aware. Its most general use, however, is as a LUBRICANT, and as this quality can be fairly well tested by rubbing the powdered mineral between the finger and thumb, I have given, in the remarks upon each specimen, a note upon its quality as a lubricant. In determining its capabilities of being cut, but little can be done away from the cutting machine; but as it is clear that grit and a heterogenous structure would be absolutely fatal to its use, I have classed all those specimens that contain grit, and which are not homogeneous, as unfit for cutting. In determining the grit, the method used was to take an ordinary office penknife and cut a small plane on one of the corners of the specimen; then, if on drawing the edge of the knife *backwards* up this streak there were particles of grit loosened, these will scratch the surface of the cut. Such a specimen I have called 'gritty.' If the grit particles could be felt with the edge of the knife when cutting *forwards*, I have termed such a specimen 'very gritty.' On cutting one specimen the grating effect was so great that only the term 'sandy' would apply.' **Mr. Bosworth-Smith's** remarks on the various specimens are quoted in the following notes.

MADRAS. 2718

"MADRAS PRESIDENCY.—A1.—*From Maddawaram Village, Nandyal Taluk, Kurnool District.*—The specimen from this locality is white, and faint reddish, in colour, with a compact structure. Cuts very freely: cuts easily in slices. **Mr. Bosworth-Smith** says—'This specimen seems the best of those sent in for the required purpose. It does not seem too hard, neither is it over soft. It appears to be remarkably free from grit.' The present Director of the Geological Survey, **Dr. W. King**, who has examined the rock *in situ*, writes—'Further north still, between Moodwaram (Maddwaram) and Yenkatgerry, the shales of this series* are very magnesian, some of the layers being nothing else but fine grey and greenish steatite.

* *i.e.*, the Paupugnee beds of the Kadapah formation.

OCCURRENCE in Madras.

There are also seams of the finer form, or French chalk, which is here called, and known over South India, as 'Bulpum.' This Bulpum is largely used by the people as a chalk for writing on their blackened boards, or small folding books of blackened paper or canvas. The associated bands of steatite and steatitic shale, which are of various shades of brown, green, and purple, are carved at Kurnool into paper-weights, etc., * * *. This is the quarrying place for steatite in the district.' †

"In the preliminary note alluded to above, Dr. King remarks that 'any amount of it (the steatite) can be easily obtained, as it occurs in well-marked bands or thin beds.' 'This locality is 22 miles due south of the town of Kurnool.' It is four miles distant from the Bethumcherla Railway Station. The Collector of the district gives the 'dimensions of the quarries' as one square mile (doubtless meaning that they are scattered over that area), and estimates the cost of delivery at Madras per ton at R20 to R30, the equivalent of which in sterling is about £1-7 to £2-0.‡ To this must be added the sea-freight. This varies greatly, but at the time of writing, would be, by steamer, about 40 shillings per ton (by weight), making the cost of the stone in London about £3-7 to £4-0.‡

2719 "A2.—*Same Locality as* A1.—Colour pale green; slightly crystalline in structure. Contains occasional dark red, highly gritty specks of some size, which are minute garnets. Harder to cut than A1, but cuts tolerably freely except where specks are met with, which are *very* gritty: cuts in slices. Mr. Bosworth-Smith remarks that this specimen 'does not seem so suitable for cutting as the above; for, on cutting it, the powder seems to clog somewhat. It makes a good lubricant and 'French Chalk' powder.'

2720 "B1.—*From Pendakallu Village, Ramallakot Taluk, Kurnool District.*—Colour pale green; very slightly crystalline in structure. Cuts freely: cuts in slices. 'This seems free from grit, is soft and easily cut; does not clog much. Its powder is very greasy and would make a good lubricant (*Bosworth-Smith*). The locality is seven miles from the Railway Station. The quarries are stated by the Collector to have the dimensions of (to be scattered over?) 10 acres, and the cost of delivering the stone at Madras is estimated at R20 to R30 (£1-7 to £2-0) a ton, which would be equivalent to about £3-7 to £4-0 in London.

2721 "B2.—*Same locality as* B1.—Pale green; somewhat crystalline in structure. Cuts tolerably freely; cuts with difficulty into slices. 'Very much like the above, but not quite so homogeneous, and therefore not likely to cut so well' (*Bosworth- Smith*).

2722 "G1.—*Somalapuram, Bellary District.*—Pale green; somewhat crystalline in structure. Cuts freely; cuts in slices. 'A soft stone; fairly well free from grit; giving a very greasy powder. Seems homogeneous and free from included crystals' (*Bosworth-Smith*). The Collector of the district remarks that the place is 37 miles from Bellary, and that there are five quarries, of which the smallest measures 8′ × 12′ × 7′, and the largest 12′ × 24′ × 10′. This, and the two following specimens, were probably obtained from different quarries, but it is not so stated. The Collector estimates the cost of delivering the stone at Madras at R25-1 (£1-13) per ton, to which must be added about £2 for freight to London.

2723 "G2.—*Same locality as* G1.—Similar to G1 in colour and structure Cuts freely: cuts in slices 'Similar to above, but in two places tried there was a fairly large piece of grit. Gives a greasy powder' (*Bosworth-Smith*).

2724 G3.—*Same locality as* G1.—Pale green. Somewhat crystalline in structure, and slightly schistose. Cuts freely: cuts in slices. 'A good soft stone free from grit: will make an excellent lubricant and should cut wel (*Bosworth-Smith*).

2725 "*Narjampalli of Gulumarri Village, Tadpat Taluk, Anantapur District.*—Yellowish-white and greenish: compact in structure. Cuts very freely: cuts easily in slices. 'This is very similar to A1 * *. If not too hard for the cutting machine, it will probably make a serviceable stone, as it is compact and free from grit. It gives a fair lubricating powder, but it is inferior to the softer stones in this respect' (*Bosworth-Smith*). The quarry is situated in a hill called Balapapurangi Gutta. The stone, according to the Collector, may be had in abundance, but it is not regularly worked. It may, apparently, be inferred from this that pieces are to be obtained much larger than that sent, which only measures 6″ × 1″ × 1″. The Collector estimates the cost of delivering it at Madras at R28-12-4 (£1-18) a ton, but adds that this is 'merely an approximate estimate.' It would make the cost in London about £3-18.

2726 "L1. *From Pathur Village, Chittoor Taluk, North Arcot.*—Pale green: nearly compact in structure, but traversed by thin veins of crystallized talc, which constitute lines of weakness along which the stone breaks more easily than elsewhere. Rather

† Memoirs, Geological Survey of India, Vol. VIII, p. 166.
‡ Here, and elsewhere, the rupee is taken as equal to 1*s*.4*d*.

Localities in which Soapstone

OCCURRENCE in Madras.

hard to cut, but free from grit: cuts easily in slices. 'This seems a fairly good stone for cutting, as it is free from grit, and seems compact and homogeneous. Its lubricating qualities are only moderate' (*Bosworth-Smith*). There is one quarry at present. No estimate is given of the cost of delivering the stone at Madras, but the cost of delivering stone from Gangadaranellur, in the same taluk, is put down by the Collecter at R7-8-0 (10*s.*) a ton, to which must be added the sea-freight.

2727

"C1.—*From Eswaramalai Hills, Atur Taluk, Salem District.*—Pale green finely crystalline in structure: contains disseminated crystals of dolomite, which are occasionally as much as ¼ inch long. Gritty in cutting: cuts with difficulty in slices. 'This specimen is found slightly gritty on cutting. It contains some calcite. Its powder, when free from grit, makes an excellent lubricant' (*Bosworth-Smith*). The Collector states that there are three quarries, the smallest of which measures 9′×9′×9′, and the largest 33′×21′×42′, besides several smaller pits at the foot of the hills. He estimates the cost of delivery at Madras at R20 to 25 per 50 Madras maunds, or a little over half a ton—say R40 to 50 (£2-13 to £3-6) a ton, which would make the cost in Loodon about £4-13 to £5-6.

2728

"C2.—*Same locality as* C1.—Pale green: finely crystalline. Contains numerous, small disseminated crystals of a chloritic mineral, and minute acicular colourless crystals (tremolite?). Gritty: cuts with some difficulty in slices. 'Similar to above but rather more gritty' (*Bosworth-Smith*).

2729

"E2.—*Edamaranahalli Village, Kollegal Taluk, Coimbatore District.*—Pal green: finely crystalline: contains minute chloritic crystals. Rather gritty in cutting cuts in slices. 'Distinctly *sandy* to the cut. Powder greasy, but with gritty particles' (*Bosworth-Smith*). The stone is obtained in the hills near the village. No estimate of cost is given.

2730

"E3.—*Kollegal Taluk, Coimbatore District.*—Reddish-white and greyish: finely crystalline. Cuts tolerably freely: cuts easily in slices. 'Gritty to cut. The powder is very greasy, and would make a good lubricant if grit were removed. The specimen is thin, and not up to the required size' (*Bosworth-Smith*). It is not clear whether this specimen is from the same village as E2, or from a different one, and it is not stated whether pieces thicker than that sent (1½ inch) can be obtained.

2731

"K4.—*Manavalike Village, Nerankimagane, Uppinangadi Taluk, South Canara District.*—Pale buff, with reddish specks in places: schistose. Cuts very freely: cuts easily in slices parallel to the foliation, but not across it. 'A rather hard stone, free from grit and may do for cutting, but will not do for lubricant' (*Bosworth-Smith*). The specimen sent is a small one, about 2 inches thick, and it is not stated whether larger ones can be obtained. The Collector puts down the dimensions of the quarry or quarries as 600 square yards, and estimates the cost of delivery at Madras at R20-2-0 (£1-7) a ton, which would be equivalent to about £3-7 in London.'

CENTRAL PROVINCES 2732

"CENTRAL PROVINCES.—No. 6.—*Kanheri Village, Sakoli Tashil, Bhandara District.*—Colour buff: crystalline in structure, and intersected by occasional thin veins of crystallized talc, along which the stone breaks easily. Cuts very freely: cuts easily in slices if free from veins: powder very greasy. The local authorities say that there is a large quarry which is extensively worked, the stone being largely used for making vessels. The cost of quarrying is roughly estimated at R2 a ton, and that of the cartage to the railway (27 miles) at R6. The railway charge to Bombay would be R25-12, giving a total at the port of R33-12 or £2-5. To this must be added the sea-freight, which, at the present time, would be, by steamer, about 30 shillings per ton (by weight), making the cost in London about £3-15.

2733

"No. 1.—*Marble Rocks, Jabalpur District.*—White, with pale reddish blotches here and there: somewhat schistose. Cuts very freely: cuts easily in slices with the foliation, but with difficulty across it: powder very greasy. Dr. King writes—'I know of this steatite myself, and have seen the people grubbing it from the pockets in the Marble Rock dolomite and schists. The rocks about there are much crushed and twisted, the steatite having, in this way, been stretched and squeezed into irregular pockets, which of course, in case of more extended exploitation in depth, will be difficult to get at. At present there seems quite enough of the material, either at the surface or close to it, for the purpose required.' According to the local authorities—'The extent of supply cannot be given with certainty. The steatite is found in irregular pockets, imbedded in limestones and schists. As the local demand is inconsiderable, the quarries have not been as yet worked on any large scale, but it is said that some thousands of maunds can be made readily available. If the specimens now sent are up to commercial standard, the only doubt which can arise is, not whether the local supply is sufficient, but whether (having regard to the peculiar formation) it will be feasible to extract blocks of the size required.' The cost of quarrying is roughly estimated at R2 a ton, and that of cartage to the railway

OCCURRENCE in RAJPUTANA. 2734

(3 miles) at R1-12-0. Adding to this R28-7-0, for railway carriage, gives a total at Bombay of R32-3-0 (£2-3), or, with sea-freight, about £3-13 in London.

"RAJPUTANA.—*Mora Village, 15 miles north-west of Hindaun, State of Jaipur.*—Pale green: very finely crystalline, and somewhat schistose, in structure. Cuts very freely: cuts easily in slices parallel to the foliation; more difficulty across it: powder very greasy. This is the material so much used at Agra for manufacturing elaborately carved ornamental articles. Mr. C. A. Hacket describes the stone as occurring in a bed (intercalated with quartzites of the Arvali series) which varies in thickness, but averages 2 feet, and which dips at about 30°, the outcrop being on the side of a hill, 150 feet above the plain. The quality of the material varies, but cubes of pure stone of 12 to 18 inches can be obtained. It is not excavated from open quarries, but from rude mines, the entrances to which are inclines following the dip of the bed. The mines are only worked in the dry season, and then intermittently, when the merchants from Agra arrive with orders, the total of which amounts on an average to 1,500 maunds* (55 tons) per annum. The cost per maund of delivering the stone at Mora (1 mile from the mines) was given by the head villagers to Mr. Hacket as follows†:—

	a.	*p.*
To the khatis or miners	1	0
Carriers from mine to Mora	2	0
Zamindar of Mora	2	0
Sonar, or broker (financier)	0	6
Chowkidar of village	0	6
Chowkidar on guard	0	6
Putwari for weighing	0	6
Village charities	0	3
Maharaja of Jaipur (royalty)	3	0
Total cost delivered at Mora	10	3

2735

"According to information given to the Executive Engineer of Jaipur, the cost at the mine is 2 annas a maund, but this apparently refers to the cost of extraction only. The rate for cartage to Hindaun Road railway station is estimated at 6 annas, the railway charge to Bombay being R1-7-1. Adopting Mr. Hacket's figures this gives—

	R	*a.*	*p.*
Cost at Mora	0	10	3
Cartage to Hindaun Road Station	0	6	0
Railway carriage to Bombay	1	7	1
Cost per maund at Bombay	2	7	4

which is equivalent to R66-15-0 or £4-9 per ton, or to about £5-19 in London.

2736

"*Raiwala (or Raialo) Village, 15 miles north of Jatwara Railway Station, Jaipur State.*—White, with occasional reddish markings: nearly compact (slightly crystalline) in structure. Cuts very freely: cuts easily in slices: powder very greasy. According to information obtained for the Executive Engineer of Jaipur, there are 3 quarries, with the following dimensions—

Length.	Breadth.	Depth.
800′	30′	15′
300	30	?
450	50	?

It is added, however, that 'these quarries are apparently of dimensions stated, but, not having been opened out, it is impossible to speak with certainty as to extent or quality.' 'We have never worked these quarries; the information is therefore necessarily imperfect.' It would appear from this that the figures given must refer to the supposed extent of the soapstone deposit, not to that of existent quarries. As the place is some 7 or 8 miles nearer the railway than Mora, and involves about 30 miles less railway carriage, the charges for delivering the stone at Bombay may probably be taken as about the same, or slightly less.

2737

"*Gisgarh Village, Jaipur State.*—Green: highly schistose, in structure. Rather gritty in cutting, but cuts easily in slices, except across the direction of foliation.

* 27·2 maunds = 1 ton.

† Manual of the Geology of India, Part III, p. 443.

STEATITE or Talc.

Localities in which the Soapstone

OCCURRENCE in Rajputana.

According to information obtained for the Executive Engineer of Jaipur, the quarry is 200 feet long, 3 feet wide, and 2 feet deep, from which it may be inferred that it extends along the outcrop of a bed having a rather high dip. Slabs can be obtained of large size, but not more than 2½ or 3 inches thick. The cost per maund of delivering the stone at Bombay is estimated as follows:—

2738

	R	a.	p.
Cost at Gisgarh	0	4	0
Cartage to Bandakari Station	0	4	0
Railway carriage to Bombay	1	6	6
Cost at Bombay	1	14	6

This is equivalent to R51-14 (£3-9) per ton, or to about £4-19 in London.

BURMA. 2739

"BURMA.—*Myingudé Mountain, Yoma Range, Kyaukpyu District,*—Pale green, but contaminated a good deal with ferruginous impurity: compact in structure. The purer pieces cut freely: cuts easily in slices. The Native Extra Assistant Commissioner at An writes—'I have the honour to report my visit to the soapstone mines at the Myingudé Mountain, which (the mines) are evidently on our side of the Yoma Range, and on a small hill joining the Myingudé Mountain. Last year the Burmese from Upper Burma dug six mines on our side and two on theirs. As these soapstones are to be found between other stones, it is very difficult to know the approximate yield of the mines. I am, however, informed that the Burmese who dug the mines last year received about 5,000 *viss*.* A hundred *viss* of first quality will fetch at An R70 or R80 at least; of the second R50 or R60. The greatest depth of the mines dug last year is 8 cubits. Having discovered soapstones all over the hill, I hope to find more mines, but the discovery, and the production of them, entirely depend on the amount of labour employed.' The specimens sent to Calcutta are only about an inch square by 4 or 5 inches long. It is not stated whether larger can be obtained, but as **Mr. Theobald**, speaking of the steatite of the Arakan Hills generally, says that the veins are usually of small dimensions,† probably the pieces in question are fair samples of what can be procured. The rates given above are equivalent to—

	R		R		£	s.		£	s.
First quality, per ton,	429	to	491	or	28	12	to	32	15
Second ,, ,,	307	,,	368	,,	20	9	,,	24	11

"To this has to be added carriage from An, by river, creek, and sea to Akyab (estimate for which is not given), and freight from Akyab to London, so that the stone would be far more expensive in England than that from Germany.

"It may be suspected that the very high prices quoted are due in part to much of the stone being of inferior quality, so that only a small portion of that extracted is saleable. The remote position of the mines or quarries, and the high rates paid for labour in Burma, also tend to raise the cost.

2740

"*Hills behind Pa-aing, Sidóktaya Township. Minbu District.*—Pale green: compact in structure: cuts freely. The pieces sent are only ½ or ⅝ inch square by 2 or 3 inches long, and it is not stated if larger can be obtained. Concerning this stone the Deputy Commissioner writes—'It is not exactly known how much there is, but the revenue paid for the quarry license this year was R2,300, so probably there is a good deal. We have been such a short time in the country that we have not been able to investigate the quarries, which are right up in the hills away from villages I understand. It is procurable at Sidóktaya at R80 to R85 a 100 *viss* (365lb), and has then to be carted about 50 miles to the river. This would cost another R5 or so. The steamer charges to Rangoon (about R15 a ton of 50 cubic feet) have then to be borne.' The price at Sidoktaya given above is equal to R491 to R521 (£32-15 to £34-15) a ton, a rate which would be quite prohibitive, in as far as export to England is concerned.

"Although, as previously remarked, the final selection of the most suitable material must be left to the gas engineers themselves, it will, I think, be found that the variety like A1, from Maddawaram in Kurnool, and the stones from Gulumarri in the Anantapur District, and Raiwala in Jaipur, are amongst the best, while several other samples appear to be very promising. But it has been pointed out by **Dr. King** in his preliminary note that a difficulty is likely to arise in the first instance with reference to the due selection of the best stone. 'It is this difficulty of selection

* About 8 tons.

† Memoirs, Geol. Survey of India, Vol. X, p. 336.

OCCURRENCE in Burma.

which must introduce a considerable factor in the cost of the stone as placed in the English market, for I fear that for some time the native quarrymen and contractors cannot be depended on for sending well-selected stone.' **Dr. King** adds that a European, and preferably one from the German quarries, in charge, would be highly desirable, were it not that owing to the comparatively small demand now existing for steatite in England, the nascent industry would not be able to support the expense. As a practicable alternative, therefore, **Dr. King** suggests that natives might be obtained from some of the existing quarries, who, after some training, would be competent to make a proper choice. Mora, in Jaipur, would probably be the most likely place to seek such overseers. The steatite there varies in quality, but the demand at Agra for the best material has led to the acquirement of the necessary skill in selection on the part of those engaged in the work.

"It is not clear whether the estimates for quarrying given by the various district authorities include manual labour only, or whether provision is made for supervision. Royalty, also, does not seem to be allowed for. An addition ought further to be made to the estimates, on account of expenses connected with breaking bulk at the railway, and at the ports of shipment and delivery, as well as, perhaps, on account of some incidental charges. But even if a liberal allowance be made under such heads, it appears clear that the steatite from every locality mentioned, except those in Burma, can be delivered in London at prices far below that now paid for continental stone. When the gas engineers have made their final selection, the export of a trial consignment will lead to more closely accurate information than is available at present, as to the cost of delivering the material in England."

MANUFACTURES. 2742

Manufactures of Steatite.—In consequence of the receipt of the samples reported on by **Mr. Mallet**, an English firm offered to purchase a small experimental consignment of Indian steatite. The correspondence that took place between **Mr. J. R. Royle** and various merchants on the subject of Indian steatite may be here published, however, as it furnishes many additional facts regarding the nature and extent of the European trade in soapstone.

Mr. Royle in a Memorandum (*dated 6th November 1889*) wrote as follows :—

"In accordance with your instructions, on receipt of the specimens of steatite referred to in the letter from the Government of India, No. 2 (Museums and Exhi bitions), dated 2nd April 1889, and fully described in the Memorandum by **Mr. F. R. Mallet**, I made the necessary arrangements with the Stores Department, and personally superintended the cutting up of the specimens at the Stores Depôt, Belvedere Road."

Gas-burners. 2743

"The whole of the samples were then arranged for inspection at the India Office, and **Mr. Sugg** (of the firm of **W. Sugg & Co.**, Gas Engineers, at whose request the specimens were originally asked for from India), was then invited to call and see them. He selected seven (out of the 22 sent from India) as being worth experimenting on, with a view to testing their suitability for the manufacture of gas-burner tops. The remaining samples were pronounced not suitable for the purpose required, either through their being manifestly too full of hard and gritty specks or crystals, or through being too flaky to stand manipulation.

"The samples selected for trial were—

New Mark.	Old Mark.	Locality.	Remarks.
A	A 1	Maddawaram, Kurnool .	This sample was finally approved of by **Messrs. Sugg.**
B	A 2	Maddawaram, Kurnool.	
C	B 1	Pendakallu, Kurnool.	
D	B 2	Pendakallu, Kurnool.	
I	L 1	Pathur, N. Arcot.	
R	None	Mora, Jaipur.	
S	None	Mora, Jaipur.	

"**Messrs. Sugg's** report on the samples experimented with is appended, and

STEATITE or Talc.

Manufactures of Steatite.

MANUFACTURES. Gash-burner Tops. 2744

a specimen of a perfect burner top made from sample A, as well as a damaged burner top from sample R, showing how a slight flaw in the material makes it unsuitable for working, are also forwarded herewith.

"As a result of their examination, Messrs. Sugg ask that a trial consignment of five tons of the steatite, sample A, (A 1 in Mr. Mallet's memorandum) may be procured for them in blocks averaging six inches cube. The probable cost of this variety, which comes from Maddawaram village, Nandyal taluk, Kurnool district, is estimated in Mr. Mallet's memorandum to be about 3*l.* 7*s.* 4*d.* per ton delivered in London, but I explained to Messrs. Sugg that it was possible to quote a firm price until a regular trade in the mineral was established, and that, seeing that the first order was so small, the cost of a trial consignment would almost certainly be increased owing to the initial expenses of selecting suitable blocks of stone. They have consequently agreed to pay up to 10*l.* or 11*l.* per ton for the five tons ordered, to include delivery at their works, but they ask that the cost may be kept as low as possible, and that, if the material can be supplied for less than their limit, they may have the benefit of the lower price. So far as it goes this is satisfactory, but it is evident that so small an order as five tons, or even the possible 10 tons a year which Messrs. Sugg might eventually consume for gas-burner tops, is of little use alone for establishing a trade with India in the material, and if this is to be done a more extended use for it must be found. Messrs. Sugg are confident that there must be many uses for it if it can be procured at a reasonable price, but it has not been possible as yet to arouse much interest in the subject, although numerous letters with invitations to inspect the samples have been addressed to makers of gas fittings, gas stoves, gas engines, acid pumps, fire-proof paints, lubricating materials, makers of toilet powders, and others, while attention has also been drawn to the subject by the editor of a paper devoted to the gas and water interests.

Toilet-powder 2745

Soaps. 2746

White Fuller's earth. 2747

"Since the receipt of Messrs. Sugg's order, endeavours have been made to obtain further orders for the same variety as that they have selected, but as yet without success, although a maker of toilet powders and soaps, and of a preparation used as a desiccant by medical men, under the name of *Terra Cimolia lævigata*, or 'White Fuller's Earth,' has pronounced this variety of steatite to be good enough to replace a variety which now costs him about 30*s.* per cwt. But the total quantity required for this purpose annually is very trifling, and the same may be said with regard to the requirements of the various branches of the gas engeneering trade.

"It has been ascertained that about 90 per cent. of the steatite at present imported into England from the Continent is in the ground state, in which condition it is used by various trades. It is estimated that about 3,000 tons of different varieties of steatite are at present imported annually, and of this quantity probably about 100 tons only are in lumps or slices. Good crushed steatite may generally be purchased at from, say, 50*s.* to 80*s.* per ton. Of course higher prices than the above are paid for small quantities for special purposes, but the trade in these special varieties is very limited. It would appear, therefore, that there is really an ample supply of steatite of good quality already imported into England, and that the price of small lots for retail use,—in which would be included the small quantities at present used for gas-burner tops,—is, to a great extent, artificially kept up. One large importer of steatite, whose letter is appended,* remarks, that, 'They (*i.e.*, makers of gas fittings and toilet preparations) sell their articles at very high figures, and can afford it, therefore why should not the producer and merchant have a share of it?' If we judge solely by this letter, it would seem hopeless for India to compete with other countries in supplying England with the crushed steatite, which is nearly all ground by water power at the place of production, but it is clear that the importer in question is interested in keeping matters as they are, and is opposed to the introduction of any new supplies. It is, however, manifest that the average prices fetched by the ground material are so low that it could not pay to import from India solely for grinding, though, if once there were a sufficient demand for the mineral in blocks, considerable quantities of small pieces and the *débris* from quarrying would be procurable at a very trifling cost, and could be utilized for packing around the blocks when shipping them to England.

* See No. 5 below.

"In connection with the employment of crushed or ground steatite, it has been ascertained that shipments of this material have been made from this country to Bombay; there is, therefore, probably some demand for it there, and it might be better to grind steatite of local production in Bombay than to continue to import it ready ground.

"But it is evidently only in blocks or slices that its importation into this country from India can be really remunerative, and at present its use in this form seems to be very limited. Meanwhile, Messrs. Sugg & Co. are anxious to receive the trial consignment for which they have asked as early as possible, and I would suggest that, if the Government of India see fit to comply with the request for five tons of the Madras steatite, an additional supply of one or two cwt. of the same should be sent to this office, and that on its arrival here the attention of all trades interested in the matter should be drawn, through a paper such as the Journal of the Society of Arts, to the opportunity afforded them of procuring good samples free, for experimenting with a view to a more extended use of this material."

Messrs. William Sugg & Co. wrote as follows :—

Vincent Works, Regency Street, Westminister, 2nd August 1889.

"In reply to your favour of the 1st instant, we enclose report showing result of our tests of the samples of steatite received from you, from which you will see that samples A and C are very good, particularly A.

"Would it be possible for us to get 5 or 10 tons of A for trial, and in what time could we have it after ordering?"

Report of Tests made with Samples of Indian Steatite received from Mr. Royle.

Sample A.—Cut eight blocks for 5 feet T tops; burned well inside and outside; no waste, all good in slotting and screwing; eight good burners, all baked well.

Sample C.—Cut six blocks for 5 feet T tops; worked well, but rather gritty; slotted and screwed well; none bad; six good burners, all baked well.

Sample D.—Cut nine blocks for 5 feet T tops; rather flaky, but worked well; three blocks split and broke; six good burners; slotted and screwed well, not so good as samples A and C; baked fairly well, not so good as A and C.

Sample R.—Cut 11 blocks for 5 feet T tops; two broke in first process, one chipped in slotting; rather flaky, but worked well, no grit; slotted and screwed well, not so good as samples A and C; eight good burners; baked fairly well, similar to D.

Sample S.—Cut seven blocks for 5 feet T tops; three spoilt in first process, two chipped in slotting; two good burners; too flaky, but worked well; not so good as any of the other samples; baked fairly well.

Sample B.—Soft, but full of iron pyrites; too dangerous for tools.

Sample I.—Hard, and very dangerous for the tools; almost too hard for the saws.

Messrs. W. Sugg & Co. in their next letter wrote :—

Vincent Works, Regency Street, Westminister, 8th August 1889.

"We thank you for your favour of the 7th instant, and in reply beg to say that we should not mind giving 10*l*. or 11*l*. per ton delivered here for a sample five tons of this steatite, for trial. We cannot offer any higher price, as there is always a certain amount of waste which cannot be utilized, and which is not so good as the sample. Of course, if it could be procured for less, we should be pleased to have benefit of the reduction. When our **Mr. David Sugg** had the pleasure of calling upon you, he understood you to say that this steatite was easily procurable, and, so far as selecting it was concerned, this did not appear to be very necessary, inasmuch as he understood that the quarry yielded a large stone similar to what he saw in your sample room.

"We trust there will now be no difficulty in obtaining this sample five tons, and, with regard to payment, we should be willing to pay for the material on delivery."

Messrs. W. Sugg & Co. again wrote :—

Vincent Works, Regency Street, Westminister, 13th August 1889.

"In reply to yours of the 12th instant, we beg to thank you for the trouble you have taken to procure steatite for us. We trust that the cost will be kept as low as possible, and, if the material can be supplied for less, we may have the benefit of the reduction. There would be no hesitation in paying 10*s*. per ton if one could be sure of obtaining the quality approved of. There should be a big field for this mineral, providing the price can be kept low. There must be numbers of articles which could be made out of steatite, such as ornaments for instance. The burner trade alone would not be sufficient to warrant a big trade with this mineral."

Mr. George G. Blackwell communicated the following :—

Liverpool, 29th October 1889.

"I have your favour of yesterday, and shall only be too happy to give you any information that shall lead to business of mutual advantage. I note that you mention that the cost price laid down in London for the quality for grinding is 4*l*. per ton, but

Manufactures of Steatite

MANUFACTURES.

even so it is practically out of the market in comparison with other productions. For your information I send you herewith sample of steatite marked No. 54, which, including bags and everything, can be done here at your lowest point of 80s., and this is as pure as you can possibly have the article. Of course it sells for something more than that figure, but at the figure named it can be laid down here, and in large quantities.

"With regard to slices, you are quite right. Engineers have been accustomed to pay the figure you name, and why should they not? They only use a very small quantity. I suppose that 10 tons per year will cover the whole of the consumption for that purpose in England. That you see will never pay the minor, especially at the figure you put it down at, and if you think of encouraging shipments of the Indian article at the price you name, you will spoil a trade, and to no purpose. There are other shapes and sizes which you have not touched upon, which also bring fair prices, but practically this is only a retail trade, and at present not likely to be of any size.

"As to the toilet powder manufacturers, this is similar. The most insignificant part of the trade altogether, and what small quantity they do use, it must be the very finest, and then they must pay a good price for it. They sell their articles at very high figures, and can afford it; therefore, why should not the producer and merchant have a share of it?

"I can bring in steatite here of a very good quality, ground, bagged, and ready for consumption, at something like 50s. per ton. I can see, therefore, that it is quite impossible for India to try to supply this market in competition with other productions. You are not perhaps aware of the fact that there, in our own country, we have a very fair deposit of steatite, but it has not been workable to any extent, because the article could not be transported to the mill and ground to compete with foreign production. I am going to make a further trial of it in my own mill; still, although I may do as well, I am not in hopes of doing better than I can do from abroad.

"I may inform you that this is a trade in which I have been engaged for the last 15 to 20 years, and I am the largest importer of the article. This year alone I shall have run through something like 20,000 bags."

Powdered Steatite. 2748

The above record of official proceedings may be concluded by the statement that it is understood samples more recently to hand from Madras, are not regarded equally satisfactory with the former parcels. It is thus probable that, with the extensive supply already on the European market, little progress is likely to be made by India in exporting this article. It is, perhaps, therefore, unnecessary to repeat that all the European uses of steatite have not as yet been made fully known. The chief demand is for powdered steatite, but the price paid for that article is too low to justify any anticipation of a large traffic from India. Powdered steatite of a very pure quality has for many years been used in the preparation of certain cosmetics. This subject is alluded to so long ago as by **Ainslie** (1823) and in the correspondence quoted above, this use of the substance is incidentally mentioned as of insignificant proportions. The employment of steatite in fire-proof paints, French Chalk, lubricating-materials, pencils, etc, etc., is probably of much greater importance.

Pencils. 2749
Plates. 2750
Bowls. 2751
Fancy work. 2752

But even were India to obtain one-half the world's present demand for steatite, the traffic could not be regarded as of anything like equal importance to the Indian local demand. Steatite (soapstone, potstone, and even serpentine and other such magnesium silicates or talcose schists) are extensively employed all over the country in the form of plates, bowls, fancy boxes in graceful shapes and plain or elegantly carved. Many ornamental articles, such as paper weights, pen-holders, etc., are made of soapstone. Some of the plates or vessels being able to withstand the action of heat, may be used as cooking pots, and to the Hindu this is a great advantage as they can be purified by fire. Stone platters and vessels are, in fact, much appreciated, since they do not communicate any flavour to the food cooked in them such as is given by unglazed earthenware. The following notes, chiefly from **Ball's** account of these stone manufactures, may be found useful, arranged province by province:—

South India. 2753

South India.—The employment of soapstone in the preparation of idols and ornamental stones in temples and palaces is very extensive. A special form found at Mysore has, owing to its suitability for this purpose, received the name of *Pratima kaller* or image stone. According to the *Mysore and Coorg Gazetteer*, the material for the famous carvings of the temple at Halebid was obtained from potstones. The so-called black-stone of Mysore is also of this nature; it was used for the pillars of the Mausoleum of Hyder at Seringapatam. Large quantities of plates, bowls, etc., are manufactured at Salem district from a local supply of potstone. The manufactures of Kurnool have already been dealt with. Ball alludes to extensive manufactures in this district, as also in Cudapah which he classes under serpentine.

Steatite is largely used as a sort of white ink for writing on black wooden slates in many parts of India and in connection with Hyderabad. Heyne alludes to the employment of steatite to give a gloss to Chunam work. The refractory property of potstone has long been appreciated in India: hence its use in the floors of hearth's. Ainslie alludes to "a kind of apple-green coloured talc, called by the Tamals *mungil appracum*, and in Dukhani *píla talk*" as being by inaccurate observers mistaken for golden-coloured orpiment. Its beautiful translucent flakes are used by the Natives for ornamenting many of the baubles employed in their ceremonies."

BENGAL.—Many of the temples of Orissa contain sculptured steatite, and small idols of slaty black steatite are sold at Puri and carried all over India as mementos of Jagannath. The potstones of Midnapur furnish one of the chief sources of the stone platters sold in Calcutta, but from several localities of Chutia Nagpur similar articles are largely manufactured, some being even made of serpentine. A dark blue stone is made at Gya into cups, plates, vases, figures of animals, etc., which are largely purchased, by the pilgrims to that city. Manbhum and Singbhum also furnish large quantities of potstone, small articles, and even large basins. Each plate or curry platter, says Ball passes through four hands: (1) the man who quarries (earns 1 anna); (2) the rough shapers; (3) the clean shaper; and (4) the turner, who uses a rude lathe in which the vessel is finished off—each earns one peiss. The employer receives from the merchants about 2½ or 3 annas, and the plate sells in Calcutta for, perhaps, 8 annas.

CENTRAL PROVINCES.—Talcose schists with bands of steatile are frequent in the metamorphic rocks of these provinces. A dark-coloured potstone found in Chanda is said to have been reserved by the Maratha authorities for the manufacture of idols; the lighter-coloured stone found at Dini and at Biroli have long been used for making vessels.

RAJPUTANA (Raipur State).—A beautiful bluish-grey soapstone occurs, which is in much demand for the construction of art-objects, especially by the manufacturers of Agra. The stone is found at Mora, a village 16 miles north-west of Hinduan. Mr. Mallet has furnished particulars of these mines, so that these need not be repeated.

BOMBAY.—"In the Southern Mahratta country", says Ball, "talcose or steatitic rocks have been extensively used in many places for manufacturing vessels and in architecture." At Ratnagiri potstones, cropping out from underneath the laterite of the Konkan, are quarried and made into vessels for which there is a steady demand in Bombay and Goa. Similar articles are also made in the south-eastern portion of Dharwar.

NORTH-WEST PROVINCES.—The chief industry in stone articles is in Agra, which has already been alluded to as drawing its supplies from the Mora mine in Raipur. But in Garhwal a brittle soapstone is turned into cups and vessels which, when polished, look like marble.

PANJÁB.—A dark-green massive serpentine occurs in association with chloritic schists in Puga valley, as also in the Hanle valley. According

MANUFACTURES. South India. Images. 2754
Ornamental carvings. 2755
Domestic Utensils. 2756
Ink. 2757
Chunam. 2758
Hearth stones. 2759
BENGAL. 2760
Ornamental work. 2761
Domestic utensils. 2762
Idols. 2763
Gya cups, etc. 2764
CENTRAL PROVINCES. 2765
Idols. 2766
Ornamental work. 2767
Domestic vessels. 2768
RAJPUTANA. 2769
Art ware of Agra. 2770
Plates, etc. 2771
BOMBAY. 2772
Art manufactures. 2773
Domestic vessels. 2774
N.-W. PROVINCES. 2775
Agra ware. 2776
Garhwal cups 2777
PANJÁB. 2778

Ornamental work. 2779 Domestic vessels. 2780

to Mr. Calvert, there is also a serpentine quarry on the Rangal mountain in Kulu. These, as also a beautiful stone found beyond the Kali river and the Shigri mines of Ludak, afford stones that are made into cups and other ornamental objects. The Shigri stone is a verd antique, which is incorrectly called *Yessham* or jade. Cups of this substance are said to split if poison be put into them. (*Conf. with Jade IV., 540*).

BURMA. 2781 Pencils. 2782

BURMA.—The subject of the soap-stones of Burma have already been alluded to, but it may be added "that pencils made of steatite or French chalk are largely used throughout the province for writing on blackened paper" (*Ball*).

MEDICINE. Dark coloured. 2783 Gney. 2784 White and yellow 2785 Rouge. 2786

Medicine.—"Several varieties of talc and mica," says Ainslie, "are found in India and Ceylon ...... the most esteemed by the Natives is a dark-coloured sort, *koushno-abrak*. The common grey mica is in Tamul called *vullay appracum* and in Hindi *suffiad talk*; this, and another dark species of mica, termed by the Tamuls *istna kappracum*, are prescribed by the Vytians, in small doses, in flux cases! they are also employed for ornamenting fans, pictures etc." "The white and yellow micas, in powder, are used for sanding writing while wet; by the names of gold and silver sand. In Europe, talc enters into the composittion of cosmetic, called *rouge*. The Romans prepared with it a beautiful blue, by combining it with the colouring fluid of particular kinds of testaceous animals." The Indian medicinal uses of talc are incidentally alluded to by many writers. Thus, for example, in 1843, Mr. Wilkinson stated that steatite was largely so used in Nagpur. It was sold at the rate of 10 seers for a rupee and was supposed to be obtained from the Jabalpur District. He however, furnishes no particulars as to its reputed actions, nor the diseases for which it was prescribed. U. C. Dutt tells us that according to the Hindu Sanskrit writers the black variety called *vajrá-bhra*, after being purified and reduced to a powder, was regarded as tonic and aphrodisiac, and was prescribed in combination with iron in anæmia, jaundice, chronic diacœhrra and dysentery, chronic fever, enlarged spleen, urinary diseases, impotence, etc. Its efficacy was held to be increased by combination with iron. He then furnishes several very elaborate prescriptions, in most of which mercury and sulphur were combined with many vegetable drugs and talc.

DOMESTIC & SACRED. Chunam. 2787 Wax-cloth. 2788

Domestic and Sacred.—In addition to the use as a material from which to construct platters, cups, bowls, basins, etc., talc, like mica, is largely employed for ornamental purposes. Ground into a powder it is employed as white ink, or is added to plaster (Chunam) to make it shining. In the preparation of wax-clothes, talc is often powdered on to the moist designs, and for this purpose may be coloured prior to use. The wax, on hardening, fixes the greater portion of the talc or mica powder (for both substances appear to be so used). For further information see the article **Mica**, Vol. V, 239-240.

Steel, see **Iron**, Vol. IV., 503.

(*W. R. Clerk.*)

STELLARIA, *Linn.; Gen. Pl., I., 149.*

[CARYOPYHLLEÆ.

2789

Stellaria media, *Linn.; Fl. Br. Ind., I., 230; Wight, Ic., t. 947;* CHICKWEED.

Syn.—S. MONOGYNA, *Don.;* ALSINELLA WALLICHIANA, *Benth.*

Vern.—*Morolia*, ASSAM.

References.—*Boiss., Fl. Orient., I., 707; Voigt, Hort. Sub. Cal., 178; Thwaites, En. Ceyl. Pl., 24; Trimen, Sys. Cat. Cey. Pl., 6; Aitchison, Cat. Pb. & Sind Pl., 14; Atkinson, Him. Dist. (X., N.-W. P. Gaz.) 306; Note on the "Condition of the People of Assam" (Agri. File No. 6* *of 1888), app. "D"; Gazetteers:—Bombay, XV., 427; N.-W. P., IV., lxviii.; Mysore & Coorg, I., 57.*

Habitat.—A very common and most variable weed, found throughout the Panjáb and temperate regions of India, ascending in the Himálaya to 12,000 feet and in Western Tibet to 14,500 feet. It is distributed to all Arctic and North temperate regions, and is elsewhere a doubtful native.

Food.—The Natives of Assam eat the LEAVES and tender STALKS, boiled in *khar* water, either by themselves or with fish. It is used also as a vegetable by the Natives of the Nilghiris.

FOOD. Leaves. 2790 Stalks. 2791

STEMODIA, *Linn.; Gen. Pl., II., 950.*

[SCROPHULARINÆ; Vol. IV., 642.

Stemodia ruderalis, *Vahl.;* see **Lindenbergia urticæfolia,** *Lehm.;*

S. viscosa, *Roxb.; Fl. Br. Ind., IV., 265; Wight, Ic., t. 1408.* 2792

Syn.—S. MARITIMA, *Heyne;* S. ARVENSIS, *Steud.*

Vern.—*Nukachúni,* BENG.; *Bóda-sarum, gunta kaminam,* TEL.

References.—*Roxb., Fl. Ind., Ed. C.B.C., 489; Dalz. & Gibs., Bomb. Fl., 176; Sir W. Elliot, Fl. Andhr., 28, 66; Irvine, Mat. Med. Patna, 78; Gazetteer, N.-W. P., I., 83.*

Habitat.—A small annual plant, found from Central India throughout the Deccan, and distributed to Afghánistán.

Medicine.—The dried PLANT, which is slightly fragrant and mucilaginous, is used by the Natives of Bengal in infusion as a demulcent (*Irvine*).

MEDICINE. Plant. 2793

STEPHANIA, *Lour.; Gen. Pl., I., 37, 962.*

[*Ic., t. 939;* MENISPERMACEÆ. 2794

Stephania hernandifolia, *Walp.; Fl. Br. Ind., I., 103; Wight,*

Syn.—CISSAMPELOS HERNANDIFOLIA, *Willd.; Wall. Cat. No. 4977 D. E. F. G. H. & K.;* C. DISCOLOR, *DC.;* C. HEXANDRA, *Roxb.;* CLYPEA HERNANDIFOLIA, *W. & A.*

Vern.—*A'kanádi, ágnád nemuka,* BENG.; *Sha-ma-say-nway,* BURM.; *Vanatik-tika, ambashthá, páthá,* SANS.

References.—*Roxb., Fl. Ind., Ed. C.B.C., 742; Pharm. Ind., 7, 11; U. C. Dutt, Mat. Med. Hind., 103, 290, 313; O'Shaughnessy, Beng. Dispens., 200, 201; Dymock, Warden, & Hooper, Pharmacog. Ind., I., 54; Gazet. Bombay, XV., 427.*

Habitat.—A climbing shrub, found in the forests from Nepál to Chittagong, and in Ceylon and Singapore. It is distributed to the Malay Islands, Tropical Australia, and Africa.

Medicine.—In the Materia Medica of the Hindus the ROOT of this plant is described as light, bitter, astringent, and useful in fever, diarrhœa, urinary diseases, and dyspepsia. **Sir W. O'Shaughnessy,** speaking of it under the native name *neemooka,* strongly recommends it as a substitute for **Pereira brava.** The author of the *Indian Pharmacopœia* also describes it as a substitute for **Pereira,** but seems to be in some confusion as to the identification of the plant known under the name of *nemuka.* This confusion is perhaps due to the fact that the name **Cissampelos hernandifolia,** *Wall.,* has to appear both under **C. Pereira,** *Linn.,* and **Stephania hernandifolia,** *Walp.* (*Conf. with Flora of British India, I., 103, 104*). The matter, however, clears up when we look to the other synonym **C. hexandra** figured by **Roxburgh** as the *nemuka* of Bengal. The *nemuka* root is then seen to be perfectly distinct from **Pereira brava. Dymock, Hooper & Warden** say, however, that it has the same properties, and that the same vernacular names are given to both plants (*Watt, Calc. Exhib. Cat.*).

MEDICINE. Root. 2795

STEPHEGYNE, *Korth.; Gen. Pl., II., 31.*

[RUBIACEÆ.

2796

Stephegyne diversifolia, *Hook. f.; Fl. Br. Ind., III., 26;*

Syn.—NAUCLEA DIVERSIFOLIA, *Wall.*; N. PARVIFOLIA, *var. 2, Kurz*; N. ROTUNDIFOLIA, *Roxb., Fl. Ind., Ed. C.B.C., 173*; N. BRUNONIS, *Wall.*

Vern.—*Bingah,* BURM.; *Taingthe,* SHAN.

References.—*Kurz, For. Fl. Burm., II., 67; Aplin, Report on Shan States (1887-88).*

Habitat.—A small tree of Chittagong and Burma, distributed to the Philippines.

TIMBER. 2797

Structure of the Wood.—Weight 45℔ per cubic foot.

DOMESTIC. 2798

Domestic Use.—Wood used for similar purposes to that of **S. parvifolia** (*q.v.*).

2799

S. parvifolia, *Korth; Fl. Br. Ind., III, 25; Wight, Ill., t. 123.*

Syn.—NAUCLEA PARVIFOLIA, *Willd.*; N. PARVIFLORA, *Pers.*; CEPHALANTHUS PILULIFER, *Lamk.*

Vern.—*Kaddam, kallam, keim, kangi,* HIND.; *Gui, kómba,* KOL; *Góré,* SANTAL; *Kutebi,* KURKU; *Kalam,* C. P.; *Mundi,* GOND.; *Khem,* BUNDEL.; *Kaim, kangai, phaldu,* N.-W. P.; *Phaldu,* KUMAON; *Kalam, kám, kalkam, keim,* PB.; *Guri, gurikaram, kumra,* RAJ.; *Tamák,* BHIL; *Kaddam, kangei, kalam, kadamb,* BOMB.; *Kadamb, karamb, kalam,* MAR.; *Buta-kadambe,* TAM.; *Mir-kadambe, bata ganapu, karmi,* TEL.; *Congú, hedu, yetega, kadwar, kadani,* KAN.; *Htein thay,* BURM.; *Helembé,* SING.

References.—*Roxb., Fl. Ind., Ed. C.B.C., 172; Brandis, For. Fl., 262; Kurz, For. Fl. Burm., II., 67; Beddome, Fl. Sylv., t. 34; Gamble, Man. Timb., 222; Dalz. & Gibs., Bomb. Fl., 118; Stewart, Pb. Pl., 116; Rev. A Campbell, Rept. Econ. Pl., Chutia Nagpur, No. 7803; Atkinson, Him. Dist. (X., N.-W. P. Gaz.), 311, 817; Useful Pl. Bomb. (XXV., Bomb. Gaz.), 83, 84, 278, 393; Gribble, Man. Cuddapah, 262; Gazetteers:—Bombay, V., 285; VI., 13; VII., 32, 36; XIII., 25; Panjáb, Kangra, 166; N.-W. P., I., 81; IV., lxxiii.; Agri-Horti. Soc. Ind.:—Journ. (New Series), VII., 135; Ind. Forester, II., 19; III., 203; IV., 292; VIII., 29, 115, 117, 126, 128; X., 31, 325; XII., app., 14; XIII., 121.*

Habitat.—A large, deciduous tree, found in the dry forests of the Tropical Himálaya up to an altitude of 4,000 feet, and throughout the drier parts of India, Burma, and Ceylon.

FIBRE. Bark. 2800

Fibre.—The BARK yields a cordage fibre.

MEDICINE. Bark. 2801 Root. 2802

Medicine.—Among the Santals the BARK and ROOT are given in fever and colic, and the former, ground and made into a paste, is applied for muscular pains (*Revd. A. Campbell*).

FODDER. Leaves. 2803

Fodder.—In Rájputana, the LEAVES are used as fodder (*J. F. Duthie*).

TIMBER. 2804

Structure of the Wood.—Light pinkish-brown, moderately hard, devoid of heartwood. It is easily worked, polishes well, and is durable if not exposed to wet. Weight about 47℔ per cubic foot.

DOMESTIC. Timber. 2805

Domestic.—The TIMBER is used for planking and furniture, for making agricultural implements, gun-stocks, combs, cups, spoons and platters, and for turned and carved articles.

STERCULIA, *Linn.; Gen. Pl., I., 217, 982.*

2806

Sterculia alata, *Roxb.; Fl. Br. Ind., I., 360;* STERCULIACEÆ.

Syn.—PTERYGOTA ROXBURGHII, *Schott. & Endl.*; STERCULIA COCCINEA, *Wall.*; S. HEYNII, *Bedd.*

Vern.—*Buddha narikella,* CHITTAGONG; *Túla,* ASSAM; *Jaynkatala, bekaro,* KAN.; *Let-kope,* BURM.

References.—*Roxb., Fl. Ind., Ed. C.B.C., 509; Kurz, For. Fl. Burm., I., 134, 135; Beddome, Fl. Sylv., t. 230; Mason, Burma & Its People, 457-754; Gazetteer, Bombay, XV., 76.*

Habitat.—A large tree of the Western Peninsula, Sylhet, Chittagong, Pegu, and Martaban down to Tenasserim. It is found also on the Andaman Islands.

FOOD. Seeds. 2807

Food.—The winged SEEDS are sometimes eaten by the Natives of Burma. According to Roxburgh they are used in Sylhet as a cheap substitute for opium.

TIMBER. 2808

Structure of the Wood.—Light, coarsely fibrous, yellowish white, perishable.

2809

Sterculia Balanghas, *Linn.; Fl. Br. Ind., I., 358.*

Syn.—In the *Fl. Br. Ind.*, three varieties are given, which correspond to three species of many writers:—

Var. (1) **mollis**=S. MOLLIS, *Wall.*

Var. (2) **angustifolia**=S. ANGUSTIFOLIA, *Roxb.*

Var. (3) **glabrescens.**

Vern.—*Cavalum*, MAL.

References.—*Roxb., Fl. Ind., Ed. C.B.C., 508; Kurz, For. Fl. Br. Burm., I., 138; Graham, Cat. Bomb. Pl., 17; Lisboa, U. Pl. Bomb., 21; Gazetteers:—N.-W. P., IV., lxix.; Mysore & Coorg, I., 58.*

Habitat.—A tree, found throughout the hotter parts of India, on the coasts of Tenasserim, in Ceylon, and the Andaman Islands.

DYE. Capsules. 2810

Dye.—In Amboyna the CAPSULES are burnt for the preparation of the colouring matter called by the Natives *kussumbha.*

FOOD. Seeds. 2811

Food.—The SEEDS are sometimes roasted and eaten by the Natives of India.

TIMBER. 2812

Structure of the Wood.—Soft and open-grained It does not appear to be used for any economic purpose.

2813

S. campanulata, *Wall.; Fl. Br. Ind., I., 362.*

Syn.—PTEROCYMBIUM JAVANICUM, *Br.*

Vern.—The gum=*kothila*, BENG.; *Tshaw*, BURM.

References.—*Kurz, For. Fl. Burm., 139; Mason, Burma & Its People, 487, 754; Journ. Agri.-Horti. Soc., IX., Sel., 50.*

Habitat.—A tree, 50 to 60 feet high, frequently found in the tropical forests of Pegu, and distributed to Java.

GUM. 2814

Gum.—It yields a gum resembling African tragacanth.

TIMBER. 2815

Structure of the Wood.—Soft, white, coarsely fibrous and rather loose, but straight-grained. It is very light and perishable, but polishes well.

2816

S. coccinea, *Roxb.; Fl. Br. Ind., I., 357.*

Syn.—S. LANCEOLATA, *Ham.*

Vern.—*Sitto udal*, NEPAL; *Katior*, LEPCHA.

References.—*Roxb., Fl. Ind., Ed. C.B.C., 509; Kurz, For. Fl. Br. Burm., I., 137; Gamble, Man. Timb., 47.*

Habitat.—A small tree of the Tropical Eastern Himálaya, found in Sikkim and Bhután at altitudes between 3,000 and 6,000 feet. It occurs also in Assam and the Khásia mountains.

FIBRE. Bark. 2817

Fibre.—Its BARK yields a strong but coarse fibre, which is sometimes made into ropes, but is less frequently used for this purpose than is that of **S. villosa.**

TIMBER. 2818

Structure of the Wood.—Grey, spongy, extremely soft. Weight 17℔ per cubic foot.

2819

S. colorata, *Roxb.; Fl. Br. Ind., I., 359.*

Syn.—S. RUBICUNDA, *Wall.*; ERYTHROPSIS ROXBURGHIANA, *Schott. & Endl.*

Vern.—*Bodula, walena, samarri*, HIND.; *Múla*, BENG.; *Khowsey, pinj*, BERAR; *Pisi, sisi*, KOL; *Bolazong*, GARO; *Sitto udal, phirphiri, omra*, NEPAL; *Kanhlyem*, LEPCHA; *Bodála, bodál*, KUMAON; *Mutruk*, MER-

The Jangli-badam.

WARA; *Lersima*, KHARWAR; *Bhái-koi, khowsey, bheckhol, samarri, walena*, BOMB.; *Karaka, karu boppayi*, TEL.; *Wet-shaw, yaseng-shaw*, BURM.; *Berdá*, ANDAMAN.

References.—*Roxb., Fl. Ind., Ed. C.B.C., 507; Kurz, For. Fl. Burm., I., 139; Gamble, Man. Timb., 47; Thwaites, En. Ceyl. Pl., 29; Dalz. & Gibs., Bomb. Fl., 23; Sir W. Elliot, Fl. Andhr., 86; Atkinson, Him. Dist. (X., N.-W. P. Gaz.), 792; Useful Pl. Bomb. (XXV., Bomb. Gaz.), 21, 229; Kew Reports, 1879, 34; Gazetteers:—Bombay, Kanara, XV., 76; Agri.-Horti. Soc. Ind.:—Jour., IX., Sel., 50; Pro., 140, 145, 194; X., 60, 61; Ind. Forester, XII., app. 7.*

Habitat.—A large tree of Eastern Bengal, the Western Peninsula, Pegu, and Ceylon.

FIBRE. Bark. 2820

Fibre.—The BARK yields an inferior fibre, which is used like that of the preceding. In 1856 it was reported on by the Hemp and Flax Committee of the Agri.-Horticultural Society of India, as a worthless fibre which, if freed from the gum which it contains, would be of no possible use. Fine specimens of it were, however, sent from Berar to the Paris Exhibition of 1878, but no information seems to exist as to how these were reported on. Specimens sent to Kew in 1878 were pronounced by the late Mr. Routledge, Ford Works, Sunderland, as a harsh and wiry fibre, which, he considered, would not pay if imported from India for paper-making.

FODDER. Twigs. 2821 Leaves. 2822

Fodder.—The TWIGS and LEAVES are used in the Western Peninsula as cattle fodder.

TIMBER. 2823

Structure of the Wood.—Dingy, greyish white in colour, very soft, marked with conspicuous medullary rays.

2824

Sterculia fœtida, *Linn.; Fl. Br. Ind., I., 354; Wight, Ic., t. 181, 364.*

Vern.—*Jangli-badam*, HIND.; *Jungli-badam, pún*, BOMB.; *Jangli-baddam, kuo-mhad, virhoi*, GOA; *Jangali-badam, goldarú, nágalkuda*, MAR.; *Pinári, kuddurai-pudduki, kudra-plukku, pinari-marum*, TAM.; *Gurapu-badam*, TEL.; *Bhatala penari*, KAN.; *Shawbyu, hlyanpyoo, showbju, let-khok*, BURM.; *Kaditeni, telemboo*, SING.

References.—*DC., Prodr., I., 483; Roxb., Fl. Ind., Ed. C.B.C., 510; Voigt, Hort. Sub. Cal., 103; Kurz, For. Fl. Burm., I., 135; Gamble, Man. Timb., 43; Thwaites, En. Ceyl. Pl., 29; Dalz. & Gibs., Bomb. Fl., 10; Graham, Cat. Bomb. Pl., 18; Mason & Its People, 457, 487, 753; Sir W. Elliot, Fl. Andhr., 66; Rumphius, Amb., III., t. 107; Ainslie, Mat. Ind., II., 119; O'Shaughnessy, Beng. Dispens., 226; Moodeen Sheriff, Mat. Med. S. Ind. (in MSS.), 67; S. Arjun, Cat. Bomb. Drugs, 213; Bidie, Cat. Raw Pr., Paris Exhib., 61, 77; Dymock, Warden, & Hooper, Pharmacog. Ind., I., 231; Birdwood, Bomb. Prod., 324; Drury, U. Pl. Ind., 404; Useful Pl. Bomb. (XXV., Bomb. Gaz.), 18, 229; Man. Madras Adm., II., 135; Gazetteers:—Bombay, XIII., 26; Mysore & Coorg, I., 58; Agri.-Horti. Soc.:—Ind., Trans., VII., 81; Jour., VIII., 39, 40; Pro., 24; IX., Sel., 50; XIV., Sel., 166; New Series, VII., 362; Ind. Forester:—I., 367; II., 18, 176; III., 238; VIII., 102; X., 31; XI., 277, 572.*

Habitat.—A large evergreen tree of Western and South India, Burma, and Ceylon. It is distributed to East Tropical Africa, the Moluccas, and North Australia.

GUM. 2825

Gum.—It exudes a gum resembling tragacanth.

FIBRE. 2826

Fibre.—Lisboa includes it in his list of fibre-yielding trees.

OIL. 2827 Seeds. 2828 Kernels. 2829

Oil.—An oil is extracted by boiling the SEEDS in water. The KERNELS contain about 40 per cent. of this. It is a fixed oil and is thick, pale yellow, bland, and non-drying. "It commences to deposit crystalline solid fats at 18°C, and the whole congeals at about 8°. The specific gravity at 15·5° is ·9277. Saponification equivalent 266·2. The crystalline fatty acids melt at 29° to 30°. With sulphuric acid, the oil forms a thick orange red mixture. With cold nitric acid it becomes opaque and slightly deepens in colour; when heated with the acid it changes to a deep coffee brown. The portion of the lead soap of the fatty acids insoluble in ether amounted to 68·9 per cent., and the liberated acid without any purification had a melting point approximating that of stearic acid. The fatty acids from the lead soap soluble in ether consisted of oleic with a small quantity of lausic acid" (*Pharm. Ind.*).

OIL.

Medicine.—The flowers have a most offensive odour. "The LEAVES are considered as repellent and aperient. **Loureiro** informs us that the SEEDS are oily, and that when swallowed incautiously they bring on nausea and vertigo. **Horsfield** adds that the LEGUME is mucilaginous and astringent" (*Ainslie, Mat., Ind.*).

MEDICINE Leaves. 2830 Seeds. 2831 Legume. 2832

Food.—Especially in times of scarcity the SEEDS are roasted and eaten like chestnuts.

FOOD. Seeds. 2833

Structure of the Wood.—Grey, soft, spongy, takes an indifferent polish. Weight 29lb per cubic foot.

TIMBER. 2834

Domestic.—The WOOD is used for house-building and for the construction of masts and canoes; it is good for making packing cases (*Lisboa*).

DOMESTIC. 2835 Wood. 2836

Sterculia guttata, *Roxb.; Fl. Br. Ind., I., 355; Wight, Ic., t. 487.*

2837

Syn.—S. CUNEATA, *Heyne*; S. ALATA, *Wall.*

Vern.—*Kukar, goldar, koketi,* BOMB.; *Goladára,* MAR.; *Kawili,* TAM.; *Bikro, happusavaga,* KAN.

References.—*Roxb., Fl. Ind., Ed. C.B.C., 508; Beddome, Fl. Sylv., t. 105; Dalz. & Gibs., Bomb. Fl., 23; Thwaites, Enum. Pl. Zeyl., 29; Rheede, Hort. Malab., IV., t. 61; Lisboa, U. Pl. Bomb., 20, 229; Royle, Fibrous Pl., 266; Gazetteers:—Bombay, Thana, XIII., 25; Kanara, XV., 76; Ind. Forester:—III., 200; V., 184; Agri.-Horti. Soc., Ind., Trans., VII., 81, 83.*

Habitat.—A tree found in the Eastern and Western Peninsulas, Ceylon, and the Andaman Islands.

Fibre.—The BARK of the younger parts of the tree abounds in a very strong white flaxen fibre. This was first brought into notice in 1802 by **Captain Dickenson** of the Bombay Military establishment, who says that in his time the Natives of the lower coasts of the Wynaad contrived to make a sort of clothing from it. It was not customary to manufacture the bark until the tenth year of the life of the tree, when its size would be equal to that of most forest trees. The tree was then felled, the branches lopped off, and the trunk cut into pieces of 6 feet long; a perpendicular incision was made in each piece; the bark opened and taken off whole, chopped, washed, and dried in the sun. By these means and without any further process it became fit for clothing purposes (*Trans. Agri.-Horti. Soc., Ind.*). The fibre is said to be well adapted for cordage and the making of coarse paper.

FIBRE. 2838 Bark. 2839

Food.—The SEEDS are roasted and eaten, especially in times of scarcity (*Lisboa*).

FOOD. Seeds. 2840

S. Roxburghii, *Wall.; Fl. Br. Ind., I., 356.*

2841

Syn.—S. LANCEÆFOLIA, *Roxb.,* not of *Cav.*; S. OVALIFOLIA, *Wall.*; S. ALATA, *Wall.* in part.

Vern.—*Gód-gadala,* PB.

References.—*Roxb., Fl. Ind., Ed. C.B.C., 508; Baden Powell, Pb. Pr., 598; Gazetteer, N.-W. P., X., 306.*

Habitat.—A tree of the Temperate Himálaya, common to the west of the Jamna, and found as far east as Sikkim. It occurs also in Sylhet and Assam.

Fibre.—In the Panjáb a cordage fibre is made from its BARK.

FIBRE. 2842 Bark. 2843

S. scaphigera, *Wall.; Fl. Br. Ind., I., 361.*

2844

Syn.—SCAPHIUM WALLICHII, *Br.*

A Substitute for Tragacanth.

References.—*Kurz, For. Fl. Burm., 140; Dymock, Warden, & Hooper, Pharmacog. Ind., I., 230.*

Habitat.—A lofty tree, frequent in the tropical forests along the eastern and central slopes of the Pegu Yomah and Martaban. According to Kurz it is found also in Chittagong and Tenasserim. It is distributed to Malacca.

GUM. 2845 Fruits. 2846

Gum.—According to the authors of the *Pharmacographia Indica,* the FRUITS of this species yield a large quantity of gum, which, in China, is used medicinally.

FIBRE. 2847 Liber. 2848

Fibre.—The LIBER yields a fibre (*Kurz*).

MEDICINE Fruit. 2849

Medicine.—The FRUIT, which in China receives the name of *Tahai-tsze,* is used as a remedy for dysentery. It was introduced into France under the name of *Boa-tampaijang,* but was found to act simply as a demulcent. Guibert found in the pericarp green oil, 1·06; bassorin, 59·04; brown astringent matter and mucilage, 1·60; woody fibre and epidermis, 3·20; and in the nucleus, fatty matter, 2·98; saline and bitter extractive, 0·21; starch and cellular tissue, 31·91 per cent.

2850

Sterculia urens, *Roxb.; Fl. Br. Ind., I., 355.*

Syn.—CAVALLIUM URENS, *Schott. & Endl.*

Vern.—*Gúhú, kúlú, gúlar, gulu, tabsi, karrai, bali, tanuku, kalru,* HIND.; *Keonge,* MANBHUM; *Kanaunji, mogul, karaunji,* MONGHYR; *Feley, kaunji, teley,* KOL; *Telhec',* SANTAL; *Odla, hatchanda,* ASSAM; *Kavili,* URIYA; *Takli,* KURKU; *Goorloo, gooloo,* C. P.; *Hittúm, pinoh,* GOND; *Kuli, gulli,* N.-W. P.; *Kulu,* BANDA; *Kalru, katila,* AJMIR; *Kurdu,* DECCAN; *Kalauri,* PANCH MEHALS; *Kavalee, kandol, gwira, gulu, kulu, karái, pándrúk, kándo, kullin, gular, kadai,* BOMB.; *Pándrúka, kávalí, kándúla, karai, kandol, gwira,* MAR.; *Karái, kada,* GUZ.; *Vellay pútali, vellay bootali,* TAM.; *Tabsu, talbsu, kevalee, erra puniki chettu, tansi, kavile, tabasi, tanuku mánu, taosee, thubisee,* TEL.; *Penári,* KAN.

References.—*Roxb., Fl. Ind., Ed. C.B.C., 507; Brandis, For. Fl., 33; Kurz, For. Fl. Burm., I., 135, 136; Gamble, Man. Timb., 46; Dalz. & Gibs., Bomb. Fl., 23; Rev. A. Campbell, Rept. Econ. Pl., Chutia Nagpur, No. 9243; Sir W. Elliot, Fl. Andhr., 53, 89, 172, 173; O'Shaughnessy, Beng. Dispens., 225; Irvine, Mat. Med. Patna, 46; Dymock, Mat. Med. W. Ind., 2nd Ed., 112; Dymock, Warden, & Hooper, Pharmacog. Ind., I., 228; Year-Book Pharm., 1878, 288; Birdwood, Bomb. Prod., 11, 257; Atkinson, Him. Dist. (X., N.-W. P. Gaz.), 751; Useful Pl. Bomb. (XXV., Bomb. Gaz.), 19, 229, 250, 398; Econ. Prod., N.-W. P., Pt. I. (Gums and Resins), 3, 4; Gums and Resinous Prod. (P. W. Dept. Rept.), 19, 48; Kew Reports, 1879, 34; Man. Madras Adm., II., 88; Selections from the Records of the Madras Govt., 1856, 8-9; Settlement Reports:—Central Provinces, Upper Godavery, 38; Raipore Dist., 76, 77; Chhindwara, 111; Nimar, 306; Gazetteers:—Bombay, V., 285; VII., 39; XIII., 25; N.-W. P., Bundelkhand, I., 79; Agra, IV., lxix; Agri.-Horti. Soc.:—Ind., Trans., VII., 81; Pro., 88; Jour., New Series, I., Sel., 58; VII., 148; Ind. Forester:—III., 200; IV., 227, 232, 322; VI., 101, 108; VIII., 377, 378, 411; IX., 15; X., 549; XII., app. 7; XIII., 119.*

Habitat.—A soft-wooded tree of North-Western India, Assam, Behar, the Eastern and Western Peninsulas, and Ceylon.

GUM. 2851

Gum.—It yields a gum called *katíla* or *katíra,* which belongs to the tragacantha series, and which, although inferior to the genuine article, has been issued as a substitute for it to the Government hospitals in Bombay. It has been repeatedly valued in Europe and has been pronouned worth only about 20 shillings a cwt.

CHEMISTRY. 2852

CHEMICAL COMPOSITION.—It is completely soluble in cold water, forming an almost colourless solution. Seen in volume it is slightly opalescent. Thirty grains dissolved in twenty ounces of water forms a thick tasteless mucilage, whlch entirely passes through a paper filter. A solution of this strength is neutral to litmus and not precipitated by alcohol, although a very thick solution is precipitated. It is gelatinized by basic acetate and gives a faint precipitate with neutral acetate of lead, but is unaffected by ferric chloride or borax and not coloured bleaue by iodine, By boiling with an alkaline solution of cupric tartrate it is precipitated but the copper is not reduced. The gum treated with nitric acid yields abundant crystals of mucic acid. The mucilage possesses little or no adhesive power. From some comparative experiments made with codliver and castor oils it appears to be about equal to tragacanth as ah emulsiying agent (*Pharm. Ind.*).

CHEMISTRY.

Fibre.—The LIBER yields a good fibre, samples of which were seut from Berar to the Paris Exhibition. Specimens sent to Kew in 1878 were tested by Mr. Routledge of the Ford Paper Works, Sunderland, and as regards their yield when converted into rough paper-stock, he reported that, the bark of **S. urens** yielded 59·3 per cent. of green fibre, 47 per cent. when bleached, and that the fibre was a good strong one, but it would not pay to import it from India. The Natives of India use it for rope-making.

FIBRE. 2853 Liber. 2854

Medicine.—The GUM is used medicinally by the Natives as a substitute for tragacanth. Among the Santals it is considered a useful medicine in throat affections. The LEAVES and tender BRANCHES steeped in water yield a mucilaginous extract which is largely employed in the pleuro-pneumonia of cattle.

MEDICINE. Gum. 2855 Leaves. 2856 Branches. 2857

Food.—The SEEDS, which are oblong and chestnut coloured, are roasted and eaten by the poorer tribes of Natives such as the Gonds and Kurkis of the Central Provinces, but are said to possess cathartic propetties. In some parts of the North-West Provinces they are roasted, ground, and made into a sort of coffee. The gum, under the name of *karai-gond*, is largely used in Bombay in the manufacture of native sweetmeats (*Dymock*).

FOOD. Seeds. 2858

Structure of the Wood.—Very soft, reddish brown in colour, with lighter coloured sapwood. Weight about 42lb per cubic foot. It has a most unpleasant smell.

TIMBER. 2859

Domestic.—The WOOD is used for making native guitars and children's toys, and also as fuel.

DOMESTIC. Wood. 2860

Sterculia villosa, *Roxb.; Fl. Br. Ind., I., 355.*

2861

Vern.—*Udal, udar,* HIND.; *Lisi, walkóm, pironja, sisi, piro,* KOL.; *Ganjher,* SANTAL; *Omak, odela, salua,* ASSAM; *Udal,* CACHAR; *Udare,* GARO; *Kanhlyem,* LEPCHA; *Sambeing,* MAGH; *Buti,* KURKU; *Pironja,* MUNDARI; *Kudar, baringa,* GOND; *Sisír,* ORAON; *Poshwa, gulbodla, osha, gul-kandar, massú, kúri, gódgúdála,* PB.; *Gulkhandar, anni-nar, udal, udar,* BOMB.; *Sarda, saldhol,* MAR.; *Vake-nar, arni, ani-nar,* TAM.; *Erra-pulike,* TEL.; *Savaga, shi-anvige, sangana mara,* KAN.; *Shawni,* BURM.; *Bájada,* ANDAMAN.

References.—*Roxb., Fl. Ind., Ed. C.B.C., 510; Brandis, For. Fl., 32; Kurz, For. Fl. Burm., I., 136; Gamble, Man. Timb., 46; Dalz. & Gibs., Bomb. Fl., 22; Stewart, Pb. Pl., 25; Baden Powell, Pb. Pr., 598; Drury, U. Pl. Ind., 405; Atkinson, Him. Dist. (X., N.-W. P. Gaz.), 792; Useful Pl. Bomb. (XXV., Bomb. Gaz.), 19, 229, 250; Royle, Fibrous Pl., 266; Liotard, Mem. Paper-making Mat., 33; Kew Reports, 1879, 34; Nicholson, Man. Coimbatore, 40; Adm. Rep. Chutia Nagpur, 1885, 28; Gazetteers:—Bombay, Kanara, XV., 96; N.-W. P., Him. Dist., X., 306; Agra, IV., 69; Mysore & Coorg, I., 48; Agri.-Horti. Soc., Ind.:—Trans., VII., 81; VIII., 274; Jour., IV., 206; VI., 135-141; Pro., 108; Sel., 177; X., Sel., 5-25; XIII., 317, 347; Ind. Forester, I., 367; II., 176; VIII., 414; IX., 377; X., 325; XI., 272, 381; XII., 73; XIII., 119.*

Habitat.—A tree of North-Western India, Bengal, and Malabar, and the Tropical Himálaya from Kumáon eastwards.

Gum.—It yields a light pellucid gum, which exudes freely from scars on the BARK. It is only slightly soluble and has no adhesive properties.

GUM. Bark. 2862

FIBRE. Liber. 2863

Fibre.—A valuable fibre is obtained from the LIBER, which is made into ropes and bags. It is very strong, and in Southern India and Burma is much esteemed for the purpose of making elephant ropes. In Northern India the ropes from this fibre are chiefly used in making cattle halters. The rope is said to become stronger for a time from being frequently wetted, but if constantly exposed to moisture it seldom lasts more than eighteen months. A good paper is said to have been made from it in India, but the samples of fibre sent to Europe were not favourably reported on as paper-making materials (*see Kew Bulletin, 1879*).

FOOD. Root. 2864

Food.—" The ROOT of the tree is eaten on the hills " (*Atkinson*).

STEREOSPERMUM, *Cham.; Gen. Pl., II., 1047.*

2865

Stereospermum chelonoides, *DC.; Fl. Br. Ind., IV., 382;* [*Wight, Ic., t. 1341;* BIGNONIACEÆ.

Syn.—BIGNONIA CHELONOIDES, *Linn.;* HETEROPHRAGMA CHELONOIDES, *Dalz. & Gibs.*

Vern.—*Pader, padri, parral,* HIND.; *Pandair,* LOHARDAGA; *Pandri,* KHARWAR; *Dharmar, atcapali,* BENG.; *Kandior, pondair,* KOL; *Parolli,* ASSAM; *Pareya-auwal,* CACHAR; *Bolsel,* GARO; *Parari,* NEPAL; *Singyen,* LEPCHA; *Sirpang,* MICHI; *Pamphunia,* URIYA; *Tsaingtsa,* MAGH; *Taitu,* BERAR; *Padurni,* BHIL; *Pádal, padri, paral, kirsel, tuatuka,* BOMB.; *Kirsel, tuatuka, pádul, padvale, pádhri,* MAR.; *Pádri, pon-padira, pathiri, vela-padri,* TAM.; *Tagada, thágu, kala gorú, moka-yapa, pisúl,* TEL.; *Kalíhútrú, kall-udi, bondh-vála, bile padri, maradakarji,* KAN.; *Nai-udi, mallali,* COORG; *Thakúppo, tha-khwot-hpo,* BURM.; *Lúnú-madala, ela-palol,* SING.

References.—*DC., Prodr., IX., 210; Roxb., Fl. Ind., Ed. C.B.C., 493; Brandis, For. Fl., 352; Kurz, For. Fl. Burm., II., 230; Beddome, Fl. Sylv., t. 72; Gamble, Man. Timb., 278; Dalz. & Gibs., Bomb. Fl., 160; Rheede, Hort. Mal., VI., t. 25; Dymock, Mat. Med. W. Ind., 2nd Ed. 545; Birdwood, Bomb. Prod., 333; Useful Pl. Bomb. (XXV., Bomb. Gaz.), 106, 290; Gazetteers:—Bombay, XV., 76; N.-W. P., IV., lxxiv., X., 313; Mysore & Coorg, I.. 52, 53; Agri-Horti. Soc.:—Ind., Jour., IV., 128, 135; VI., 49; Ind. Forester, III., 204; IX., 358.*

Habitat.—A large, deciduous tree, met with throughout the moister parts of India, from the Tarai of Oudh and Assam to Ceylon and Burma.

MEDICINE. Roots. 2866 Leaves. 2867 Flowers. 2868 Juice. 2869

Medicine.—The ROOTS, LEAVES, and FLOWERS are used medicinally. The roots and flowers are said by **Ainslie** to be prescribed by the Vytians in infusion as a cooling drink in fevers, and **Rheede**, who speaks of the tree under the name of *Padrie*, says that the JUICE of the leaves, mixed with lime juice, is of use in maniacal cases. The dose of the infusion is about half a teacupful twice daily.

TIMBER. 2870

Structure of the Wood.—Reddish brown or orange coloured, close and even-grained, elastic and durable, but soft. It takes a good polish. Weight 45 to 48℔ per cubic foot.

DOMESTIC & SACRED. Flowers. 2871

Domestic and Sacred.—The wood is used for canoe and house-building, and in Assam for making tea-boxes. It is also employed in various kinds of fancy work. The highly-scented FLOWERS are much used in Bombay as an offering to the gods.

2872

S. fimbriatum, *DC.; Fl. Br. Ind., IV., 383.*

Syn.—BIGNONIA FIMBRIATA, *Wall.*

Vern.—*Than-that,* BURM.

References.—*DC., Prodr., IX., 211; Kurz, For. Fl. Burm., II., 231; Gamble, Man. Timb., 279.*

Habitat.—A tall, deciduous tree, found in the tropical forests of Martaban and Upper Tenasserim, up to an elevation of 3,000 feet. It is distributed to Malayana.

Structure of the Wood.—Heartwood small, dark brown; sapwood light brown. It is very hard. Weight 54℔ per cubic foot.

TIMBER. 2873

Stereospermum neuranthum, *Kurz; Fl. Br. Ind., IV., 382.*

2874

Vern.—*Than-day*, BURM.

References.—*Gamble, Man. Timb., 277; Kurz, For. Fl. Br. Burm., II.,* [230.

Habitat.—A tree, 40 to 60 feet high, common in the forests of Pegu and Moulmein.

Structure of the Wood.—Pale greyish or reddish brown, very close-grained, fibrous, rather heavy, tolerably soft. Weight 33 to 36℔ per cubic foot.

TIMBER. 2875

S. suaveolens, *DC.; Fl. Br. Ind., IV., 382; Wight, Ic., t. 1342.*

2876

Syn.—BIGNONIA SUAVEOLENS, *Roxb.*; TECOMA SUAVEOLENS, *G. Don*; HETEROPHRAGMA SUAVEOLENS, *Dalz. & Gibs.*

Vern.—*Páral, padal, padiála, pád, padaria, parur, purula, pár*, HIND.; *Parlú, párul, ghunta, múg*, BENG.; *Pandri*, KHARWAR; *Kandior*, KOL.; *Pader*, SANTAL; *Parari*, NEPAL; *Singyen*, LEPCHA; *Patúli*, URIYA; *Padar*, KURKU; *Pandri*, C. P.; *Phalgataitu*, MELGHAT; *Unt katar, padar*, GOND; *Pádal*, N.-W. P.; *Pádal, kaltháun, summe*, PB.; *Pan, dan*, BHIL; *Paral, paddal, pahad*, BOMB.; *Padal, padialú, parúl, kala-gori*, MAR.; *Padiri, goddatipalusu*, NELLORE; *Padri*, TAM.; *Kala-goru, kuberakashi, padari, patali*, TEL.; *Húday, billa, vulunantri marada, kavi*, KAN.; *Pátalá*, SANS.

References.—*DC., Prodr., IX., 211; Roxb., Fl. Ind., Ed. C.B.C., 493; Brandis. For. Fl., 351; Kurz, For. Fl. Burm., II., 231; Beddome, For. Man., 169; Gamble, Man. Timb., 278; Dalz. & Gibs., Bomb. Fl., 161; Stewart, Pb. Pl., 148; Sir W. Jones, Treat. Pl. Ind., V., 131; U. C. Dutt, Mat. Med. Hind., 203, 313; Dymock, Mat. Med. W. Ind., 2nd Ed., 546; Birdwood, Bomb. Prod., 333, 334; Useful Pl. Bomb. (XXV., Bomb. Gaz.), 106; Boswell, Man., Nellore, 99; Gazetteers:—Bombay, XV., 76; N.-W. P. I., 82; IV., lxxiv.; X., 313; Mysore and Coorg, I, 63; II., 7; Agri.-Horti. Soc.:—Ind., Jour., II., 356; VI., 49; XIII., 295; New Series, VII., 132; Ind. Forester:—III., 204; IV., 323; VIII., 126; X., 326; XII., 311; XIII., 121.*

Habitat.—A tree, 36 to 60 feet high, found throughout moister India, from the Himálayan Terai to Travancore and Tenasserim, and in Ceylon. In the Himálaya it ascends to altitudes of 4,000 feet.

Gum.—The BARK yields a gum, one of the dark-coloured Hog or Tragacanth series (*Watt*).

GUM. 2877 Bark. 2878

Medicine.—The FLOWERS and ROOT-BARK are used medicinally by the Natives of India. The former are given by the Hindus rubbed up with honey to check hiccup, while the latter is an ingredient of the preparation *dasamula* (*see* **Desmodium gangeticum**, *DC.*; Vol. III., 82), and is thus largely used in Native medicine. It is considered cooling, diuretic, and tonic, and is generally given in combination with other medicines. The ASHES are used in the preparation of alkaline water and of caustic pastes (*Hindu Mat. Med.*).

MEDICINE. Flowers. 2879 Root-bark. 2880 Ashes. 2881

Structure of the Wood.—Sapwood large, grey, hard; heartwood small, yellowish brown, beautifully mottled with darker streaks, very hard, seasons and polishes well. It is fairly durable and easy to work. Weight about 46℔ per cubic foot (*Gamble*).

TIMBER. 2882

Domestic.—The WOOD is much valued for building and makes excellent charcoal.

DOMESTIC. Wood. 2883

S. xylocarpum, *Wight, Ic., t. 1335-36; Fl. Br. Ind., IV., 383.*

2884

Syn.—BIGNONIA XYLOCARPA, *Roxb.*; SPATHODEA XYLOCARPA, *Brandis*; TECOMA XYLOCARPA, *G. Don.*

Vern.—*Kursingh*, BOMB.; *Kharsing, bersinge*, MAR.; *Jai-mangal, sondar-padal*, MANDLA; *Dhótamara, dhotte*, GOND; *Teto*, KURKU; *Vadencarni*, TAM.; *Ghansing*, KAN.

References.—*DC., Prodr., IX, 169; Roxb., Fl. Ind., Ed. C.B.C., 494; Brandis, For. Fl., 349; Beddome, Fl. Sylv., t. 70; Gamble, Man. Timb., 279; Dalz. & Gibs., Bomb. Fl., 159; Dymock, Mat. Med., W. Ind., 544; Lisboa, U. Pl. Bomb., 105, 167; Ind. Forester, III., 204; X., 222.*

Habitat.—A tree, 30 to 60 feet high, common in the Deccan Peninsula, and extending north to the Satpura Range.

RESIN. 2885

Resin.—The wood contains a resin which is extracted by the Natives of Western India. For this purpose they use two earthen pots, the upper of which has a perforated bottom, is fitted with a cover, and is luted to the mouth of the lower pot. The wood in small pieces is placed in the upper pot, and the whole apparatus is heated with cowdung cakes. The resin drops from the wood into the lower pot and is there collected. It is a thick fluid of the colour and consistence of Stockholm tar (*Dymock*).

MEDICINE. Resin. 2886

Medicine.—The RESIN is used as a remedy for scaly erruptions of the skin. Its properties seem to be similar to those of Pine Tar (*Dymock*).

FOOD. Seed capsules. 2887

Food.—The young SEED CAPSULES are eaten as a vegetable (*Lisboa*).

TIMBER. 2888

Structure of the Wood.—Sapwood large, grey; heartwood brown and very hard; tough and elastic, close grained. Weight 47℔ per cubic foot.

DOMESTIC. Wood. 2889

Domestic.—The WOOD is used for cabinet-work.

Stillingia sebifera, *Michaux.*, see **Sapium sebiferum,** *Roxb.*; [EUPHORBIACEÆ; Vol. VI., Pt. I, p. 472.

STONES, BUILDING; *Ball, in Man. Geology [of India, III., 532.*

2890

Stones, Building.

The subject of the occurrence of stones suitable for building has been treated of at various other parts of this work, see **Gneiss,** Vol. III., 517; **Granite,** Vol. IV., 176; **Marble,** Vol. V., 185-186; and **Trap,** Vol. VI., Pt. II.; but it has been thought necessary to give here a short collective article on the subject as a whole, in order to furnish details of the stones which have been found in India best adapted for this purpose, the methods of working them pursued by the Natives, and the amount of stones of this class produced in India.

References.—*Mason, Burma & Its People, 587, 735; Manual, Geology of India, Pt. III, (Ball, Econ. Geol.), 446-448, 456-473, 532 555; Baden Powell, Pb. Pr., 36, 37; Buchanan, Journey through Mysore and Canara, &c., Vol. II., 441; Moore, Man., Trichinopoly, 66; Bomb., Admin. Rep., 1871-72, 364, 365; 1872-73, 376; Settlement Reports:—Central Provinces, Chanda, 106; Nellore, para. 4; Nagpore, 276 of Suppl.; Gazetteers:—Bombay, II., 38; V., 22, 285; VI., 11, 200, 201, 243, 246, 254, 258; VIII., 91, 92, 262; X., 31; Panjáb, Delhi, 130; Central Provinces, 59; Mysore & Coorg, II., 3, 841; W. W. Hunter, Orissa II., 36, App. II.; Tropical Agriculturist, July 1889, 6; Balfour, Cyclop. Ind., III., 240.*

SUITABLE STONES FOR BUILDING.

VARIETIES. 2891

The following account of the stones which have been found most suitable for building purposes is an abstract of **Mr. Ball's** article on the same subject in the *Manual of the Geology of India*:—

Gneiss & Granite. 2892

Gneiss and Granite.—Most of the so-called granite of India is a granitoid gneiss resulting from the excessive metamorphism of sedimentary rocks. The varieties of these which are suitable for building purposes are very numerous. As building stones, the dense crystalline unpolished varieties are the most durable. In the alluvial tracts of Bengal ancient buildings of stone are of rare occurrence, but in Behar many temples are found in the construction of which granite was employed. On the East Coast, again, from Midnapur throughout Orissa, the use of granite in the construction of

VARIETIES.

Gneiss & Granite.

temples seems to have been common, and in Orissa Hindu temples and deities are very frequently made of granitiferous gneiss. In Madras, in the Ganjam district, on the Nilghiris and in Chingleput, granites have been extensively used in the construction of temples; and in the last-mentioned district there is a wonderful series of temples which have been carved with an incredible amount of labour out of solid bosses of granite *in situ*. In Mysore a variety of gneiss is obtained which can be split into posts 20 feet high, and these have been used for the support of telegraph wires. Some of the gneisses of Southern India are wonderfully susceptible of fine carving, an instance of which is the rings appended to the drooping corners of some pagoda buildings. These rings, the links of which are moxable, and the projecting corners, are carved out of single blocks of gneiss. It will thus be seen that the use of gneiss and granite in the construction of temples was very common throughout Southern India, but except for purely local purposes, *viz.*, the construction of bridges, etc., where, upon economical grounds, the rock nearest to hand has been made use of, the varieties of these stones have not, on account of their hardness, commended themselves as building materials to English engineers. There are throughout the country no English buildings of importance, in the construction of which these materials have been used except for rough work.'

Limestone & Marble. 2893

Limestone and Marble.—A detailed account of the occurrence of stones of this class will be found in another portion of this work. It will be, therefore, sufficient to add here that the abundant supplies of Indian marble have been but little utilised for public buildings by the English in India, as it has been found cheaper either through the at present inaccessible position of the principal marble quarries, or the difficulty of obtaining skilled labour at a low rate, to import Italian marble than to use the Indian article. It is extensively employed locally both in the building of temples and for internal decoration, but there is not much general demand for any of the Indian marbles. Limestones occur abundantly throughout India, and are in some places extensively used for building purposes. Thus in Guzerât a more or less calcareous rock, for which Dr. Carter proposed the name "Milcolite," has a very wide distribution. Its greatest development is in the Gir hills, where it rests on an arenaceous clay; it is largely made up of foraminifera, and is supposed to be of the pleiocene age. As a building stone it is admirably suited for some purposes, but is said to be incapable of sustaining great pressure. It is largely quarried about 12 miles from Porebunder, whence it is shipped to Bombay. It has been freely utilized in the erection of many of the public buildings of that city. The limestones near Simla have been utilized in the construction of most of the larger buildings of that town, such as Viceregal Lodge, the Town Hall, the Clubs, the Court, etc., etc.

Sandstones 2894

Sandstones.—Several of the recognised formations in India, *viz.*, the Vindhyan, Gondwana, Cretaceous and Tertiary, afford sandstones suitable for building, and some of them have, from very early periods, been largely used for this purpose. Thus the Vindhyan series have furnished the great monoliths or *lats*, some of which are said to have been erected about 250 years B.C., and which afford striking evidence of the size of the stones obtainable from the Vindhyan sandstones. The quarries at Dehri on the Son are the most eastern of all those which have been opened in the Vindhyan rocks. These are largely worked in connection with the Son irrigation and canal works. The stone is a compact, whitish sandstone, susceptible of artistic treatment, and very strong and durable. The next locality of importance where there are quarries, is Chunar, where the Ganges affords a ready means of transport for the building stones. Benares and other cities and towns of less importance have largely used

Building Stones produced in India.

VARIETIES.

Sandstones.

the Chunar sandstone, and even to Calcutta a certain quantity has been conveyed. The next quarries are those of Mirzapur, which, with those of Partabpur and Seorajpur, have supplied both Mirzapur and Allahabad with material for the construction of their buildings, both ancient and modern. Sandstones of this formation have also been extensively worked at Gwalior, in the construction of forts and temples; but perhaps the most important quarries in India are those in the Upper Bhaurers to the south of Bhurtpur at Fatipur Sikri and Rupas, which have furnished building materials, since the commencement of the Christian era, to the cities of the adjoining plains. Portions of the Taj at Agra, Akbar's palace at Fatipur Sikri, the Jamma Masjid at Delhi, and buildings generally in Agra, Delhi, and Muttra, have drawn upon these quarries for their materials.

Among the sandstones, too, of the Gondwana series, there are several varieties which are suited for building purposes, and which have already been made use of to a small extent. Thus those at Barakar have been quarried largely for local use in the construction of the Barakar bridge, and for various purposes in connection with the East Indian Railway. A considerable portion of the new High Court in Calcutta is also built of this material, and as it is readily accessible at the terminus of the Barakar branch of the railway, this rock will probably be always more or less used for purposes for which brick is not suited. Some of the beds in the Jabalpur group yield a useful building material, and a very hard indurated variety, which occurs in the station itself, has been largely quarried for local purposes, and is the stone of which the viaduct over the Narbada below Jabalpur was built.

The sandstones of the eocene beds in the Himálaya have, at Dagshai, Kasauli, Subathu, and Dharmsala, furnished excellent building stones.

Laterite. **2895**

Laterite.—This name has been applied to a group of tertiary rocks which are very widely spread throughout India. By the Natives they are largely used for building purposes, since they are easily worked, harden on exposure, and some varieties are fairly durable. In the coastal districts many temples, some of considerable antiquity, have been built of this material and appear to have stood well. Recently, it has been much employed in Orissa in connection with irrigation operations, and has given engineers much satisfaction. **Dr. Balfour** gives the Arcade Inquisition at Goa, St. Mary's Church, Madras, and the old fortress at Malacca as examples of its use in the construction of buildings by Europeans. Where of poor quality laterite soon crumbles away when exposed to the influences of weather and moisture, but when well selected, continued exposure to atmospheric influences or wet, as in the case of tanks and *bouries*, only tends to improve the stone. Most of the religious edifices and tanks constructed of laterite show the lines and angles of the carvings as sharply as though fresh from the builders' hands.

Slate. **2896**

Slate.—The great majority of the Indian slates are so imperfectly fissile that even tolerably thin roofing slates cannot be produced from them, and in consequence of this they are not generally well adapted for sloping roofs. Even where locally an exceptional degree of fineness of texture and capability of sub-division is present, it cannot be counted on to extend through a large mass. Thick slates, which are not suitable for sloping roofs, may, however, be employed for flat roofs and for paving instead of tiles. In this way, owing to their strength, they may be used of a much larger size than it would be safe to employ tiles, and stretching from beam to beam they may actually cause a great economy in timbering, and the concrete which is laid upon tiles might, where slate is used, be very considerably diminished. Another advantage of the use of large flag-like slates would be that as the number of joints are diminished the chances of leakage are lessened, and so also the consequent cost of repairs.

STONE-CUTTING and POLISHING. 2897

STONE-CUTTING AND POLISHING.

From the very earliest periods the Natives of Hindustan have been notable workers in stone, and some of the structures raised by the ancient Hindus are conspicuous for the exquisite polish they have contrived to give to even the hardest rocks. Numerous quarries exist throughout India from which, as above mentioned, large supplies of good building stones are obtained. The methods of working these employed by the Natives, differ largely according to the hardness of the material. Where this is soft, as in the case of laterite, the soil is first removed and the rock smoothed on its surface. A space about 12 feet each way is next divided into slabs 1 foot square, the grooves between them being cut with a light flat-pointed single-bladed pick. These slabs are raised successively by a tool something between an adze and a mattock, a single stroke of which is in general sufficient to detach each slab from its bed. The blocks thus cut and raised are then put aside, the bed once more smoothed, and the operation resumed till the pit reaches a depth of 6 or 8 feet, when, as it is no longer convenient to remove the stones by hand or basket, a new pit is cut. Where the stone is harder, as in the case of granite, it is split by means of wedges. A number of small square holes, about 1½ inches in diameter, and 4 inches deep, are cut in the line where the stone is meant to split. When the rock or stone is very long or deep, these holes must be almost contiguous, but when the surface to be split is small, they may be put at considerable distances. Blunt wedges of steel are then put into the holes and each is struck in its turn until the stone splits, which it does in almost a straight line down to the bottom of the mass. These purely Native methods of stone-working have now, however, been almost wholly superseded by the use of explosives, although they are still sometimes employed in rural districts for the stones used in building Native houses.

Cutting. 2898

For dressing stones, preparatory to polishing, the Natives of India use a small steel chisel and an iron mallet. The chisel in length is not more than 6 inches and tapers to a round point like a pencil. The iron mallet does not weigh more than a few pounds. Its head is fixed on at right angles to the handle, and has only one striking face which is formed into a tolerably deep hollow and lined with lead. The stone having been dressed with these instruments is next smoothed with water in the usual way and is then polished. This is accomplished in a very rude but efficient fashion. A block of granite of considerable size is rudely fashioned into a shape like the end of a rough pestle. The lower face of this is hollowed out into a cavity and this is filled with a mass composed of pounded corundum stone mixed with melted lac. By means of two sticks or pieces of bamboo placed on each side of its neck and bound together by cords which are twisted and tightened by sticks the block is moved. The weight of the whole is such that two workmen can easily manage it. They sit themselves upon, or close to, the stone which they are to polish, and by moving the block backwards and forwards between them the polish is imparted by means of the mass of corundum and lac.

Dressing. 2899

Polishing. 2900

PRODUCTION. 2901

PRODUCTION OF BUILDING STONES.

From the statement published by the Revenue and Agricultural Department of the Government of India, showing the quantities and values of minerals produced in each British Province and Native State of India during the calendar year of 1889, it appears that a total of 146,959,054·07 tons quarry stones, valued at R54,84,644, was obtained during that period.

PRODUCTION.

Of these 47,966,802 tons,valued at R13,84,732, were granite; 9,505,127 tons, valued at R15,35,247, were limestones; 749,558 tons, valued at R4,36,914, were laterite; 559,835·07, valued at R8,02,522, were sandstones; 58,007 tons, valued at R1,19,843, were slates; 791,992 tons, valued at R4,51,867, were trap; 1,031,251 tons, valued at R3,90,484, were clays, and 296,482 tons, valued at R3,63,035, were stones of other kinds. Bengal, with an outturn of 139,309,772 tons, valued at R10,31,122, was the largest producing province; next came Madras, with an outturn of 2,357,266 tons, valued at R6,96,625; then Bombay, with an outturn of 1,391,075 tons, valued at R10,22,552; then Rajputana, with an outturn of 1,054,161 tons, valued at R3,43,584, and then Mysore with an outturn of 442,544 tons, valued at R8,10,429. The other provinces produced smaller amounts. The quantities produced in the Central India States are not shown in the return.

Storax, see **Liquidambar orientalis,** *Miller*; HAMAMELIDEÆ.; Vol V., 78.

STORKS.

2902

The family of CICONIDÆ or Storks, comprising three genera, *viz.*, Ciconia, Leptoptilos, and Xenorhynchus, contains the adjutant birds as well as the true storks. Six species of CICONIDÆ are found in the Indian Peninsula:—

Storks; *Murray, Avifauna of Br. Ind., II., 647.*

Species. 2903

1. **Ciconia alba,** *Belon; Jerdon, Birds Ind., III., 736.*
THE WHITE STORK.

Vern.—*Lag-lag, ujli, haji-lag-lag,* HIND.; *Dhák,* N.-W. P.; *Wadume, konga,* TEL.

Habitat—Found all over Northern, Central, and Eastern India. It is a cold weather visitant, and is not present all the year.

2904

2. **C. leucocephala,** *Gm.; Jerdon, Birds Ind., III., 737.*
THE WHITE-NECKED STORK.

Habitat.—Found, but with less frequency, all over India, British Burma, and Ceylon.

2905

3. **C. nigra,** *Linn.; Jerdon, Birds Ind., III., 735.*
THE BLACK STORK.

Habitat.—A winter visitant, found all over Northern, Central, and Eastern India.

2906

4. **Leptoptilos argala,** *Lath.; Jerdon, Birds Ind., III., 730.*
THE ADJUTANT.

Vern.—*Hargila, dusta,* HIND.; *Chaniari-dhauk,* BENG.; *Garur,* N.-W. P.; *Pini-gala-konga,* TEL.

Habitat.—Found throughout the greater part of India, Burma, and the Malayan Peninsula, rare in the south of India, and absent in Malabar, where it is replaced by the next species.

2907

5. **L. javanicus,** *Horsf.; Jerdon, Birds Ind., III., 732.*
THE LESSER ADJUTANT.

Vern.—*Chinjara, chandana, chandiari,* HIND.; *Madanchur, moduntiki,* BENG.; *Bang-gor,* PURNEAH; *Dodal-konga, dodal-gatti-gadu,* TEL.

Habitat.—Found in small numbers over nearly the whole of India, frequenting marshes, paddy-fields, and the edges of lakes and rivers. Common in Southern India. In distribution it extends over the Malayan Peninsula.

DOMESTIC. 2908

Domestic.—The undertail covers of birds of this family, composed of long, lax, feathers, are known as *marabou*, and were formerly much esteemed as ornaments and for ladies' head-dresses. The down of the young adjutant bird is also made into ladies' boas and victorines; See

STORKS.

Feathers, Vol. III., 321. The adjutants and true storks are all more or less foul feeders, and in the East do much service as scavengers.

2909

6. Xenorhynchus asiaticus, *Lath.; Jerdon, Birds Ind., III.,* 734.
THE BLACK-NECKED STORK.

Habitat.—Found over the greater part of India, Burma, and Ceylon, and distributed also to Malayana.

Stramonium, see **Datura Stramonium,** *Linn.*; SOLANACEÆ; Vol. III., 40.

STRANVÆSIA, *Lindl.; Gen. Pl., I.,* 627.

2910

Stranvæsia glaucescens, *Lindl.; Fl. Br. Ind., II.,* 382; ROSACEÆ.

Syn.—CRATÆGUS GLAUCA, *Wall.*; PYRUS NUSSIA, *Ham.*; COTONEASTER AFFINIS, *Lindl.*

Vern.—*Garmekal, sánd,* KUMAON.

References.—*Brandis, For. Fl., 210; Gamble, Man Timb., 170.*

Habitat.—A small, evergreen tree, of the Central Himálaya, found from Kumáon to Nepál at altitudes between 3,000 and 7,500 feet, and in the Khásia hills between 4,000 and 5,000 feet.

TIMBER. 2911

Structure of the Wood.—Light-coloured when fresh cut, turning reddish-brown on exposure, fine and even-grained; annual rings marked by a thin line. Weight 48lb per cubic foot.

STREBLUS, *Lour.; Gen. Pl., III.,* 359.

[URTICACEÆ.

2912

Streblus asper, *Fl. Br. Ind., V., 489; Wight, Ic., t. 1961;*

Syn.—TROPHIS ASPERA, *Retz.*; EPICARPURUS ORIENTALIS, *Bl.*; E. ASPER, *Steud.*

Vern.—*Siorá, karchanua, rúsa, daheyá,* HIND.; *Sheora, syáora,* BENG.; *Sehora,* PATNA; *Hara saijung,* KOL.; *Sahra,* SANTAL; *Sahuda,* URIYA; *Ungnai,* MAGH; *Karasni,* GOND; *Sihora, rúsa,* N.-W. P.; *Jindi, dahya,* PB.; *Karvati, karera, karaoli, karchanna, rúsa,* BOMB.; *Karera, kharaoli, poi,* MAR.; *Prayám, palpirai,* TAM.; *Bariniki, bari venka, barranki, pakki,* TEL.; *Mitli, punje,* KAN.; *Op-nai,* BURM.; *Geta-netul,* SING.; *Sákhotaka,* SANS.

References.—*Roxb., Fl. Ind., Ed. C.B.C., 714; Brandis, For. Fl., 410; Kurz, For. Fl. Burm., II., 464, 465; Beddome, For. Man., 221, t. 26, f. 1; Gamble, Man. Timb., 326; Trimen, Sys. Cat. Cey. Pl., 83; Dalz. & Gibs., Bomb. Fl., 240; Rev. A Campbell, Rept. Econ. Pl., Chutia Nagpur, No. 8417; Sir W. Elliot, Fl. Andhr., 24; Sir W. Jones, Treat. Pl. Ind., V., 152; Rheede, Hort. Mal., I., t. 48; Irvine, Mat. Med. Patna, 95; U. C. Dutt, Mat. Med. Hind., 316; Birdwood, Bomb. Prod., 600; Atkinson, Him. Dist. (X., N.-W. P. Gaz.), 317, 751; Useful Pl. Bomb. (XXV., Bomb. Gaz.), 127; Rept. Agri. Dept. & Exper. Farms, Madras (1882-83), 10; (1883-84), 10, 46, 47; Strettel, New Sources of Revenue for India, 8, 81; Gazetteers:—Bombay, XIII., 25; XV., 78; N.-W. P., IV., lxxvii.; Ind. Forester, III., 205; X., 325.*

Habitat.—A rigid shrub or gnarled tree, found in the drier parts of India, from Rohilkhand eastward, and southward to Travancore, Penang, and the Andaman Islands. It is distributed to the Malay Islands, Cochin China, China, and Siam.

RESIN. 2913

Resin.—A white resin is exuded by the tree (*Kurz*).

MEDICINE. Juice. 2914 Bark. 2915

Medicine.—The milky JUICE has astringent and antiseptic qualities, and is applied to sore heels and chapped hands. The BARK in decoction is given in fevers, dysentery, and diarrhœa (*Atkinson; Revd. A Campbell*).

Roots. 2916

SPECIAL OPINIONS.—§ "The ROOTS are used as an application to unhealthy ulcers and sinuses. It is said to be an antidote to snake poison" (*Civil Surgeon J. H. Thornton, B.A., M.B., Monghyr*). "The powder of the dry root in 5 to 10 grain doses is used with good effect in dysentery in certain parts of East Bengal" (*Assistant Surgeon Nobin Chunder Dutt, Durbhunga*). "Juice is used externally to remove glandular swellings" (*Surgeon Anund Chunder Mukerji, Noakhally*).

TIMBER. 2917

Structure of the Wood.—White, moderately hard, devoid of heartwood, tough, and elastic. There are no annual rings. Weight about 75lb per cubic foot.

DOMESTIC and SACRED 2918 Timber. 2919 Twigs. 2920 Leaves. 2921

Domestic and Sacred.—In Southern India the TIMBER is sometimes used for cart wheels and furnishes excellent fuel. The tree is good for hedges, and coppices well. The TWIGS are used as toothbrushes, and the rough LEAVES to polish wood and ivory. In 1883, some experiments were made at the Madras Experimental Farms with reference to the value as milk-coagulating substances of various kinds of vegetable rennet, and it was found that the milky juice of this plant produced rapid coagulation of milk and a very firm curd from which a crude cheese was made. In 1884 these experiments were continued, and it was found that it yielded better results than did the berries of **Withania coagulans** (*q.v.*). "To 5 *ollocks* of milk (drawn at 6-15 A.M.), heated to the boiling point and cooled, 25 grains of **Withania coagulans** seed, powdered, was added. To another 5 *ollocks* dried *prayam* juice was added. To another the fresh juice of 32 *prayam* leaves. The coagulants were all added at 10-15 A.M. The milk treated with the fresh *prayam* juice was well curdled at 12-45 P.M., whereas that to which the **Withania coagulans** seed had been added was only viscid, and that to which the dry *prayam* juice had been added showed no material change even at 1-30 P.M., and by that time the coagulation of the milk to which the fresh *prayam* juice had been added was complete. Dry *prayam* juice, in fact, caused no coagulation. The milk treated with it was not fully curdled till next morning, and this was probably due to natural fermentation.

"To 4½ *ollocks* of raw milk the juice of 16 *prayam* leaves was added at 10 A.M., and to another 4½ *ollocks* five drops of rennet extract. Both the samples of milk were curdled at 12-30 P.M. It was not ascertained which of the two had coagulated most quickly. Three ounces of cheese were obtained from each lot of milk," but this it was found impossible to preserve through the ripening process even with the aid of boro glyceride. The cheese was not marketable, and special means would have to be adopted in a climate like that of Southern India to preserve it until it had ripened (*Rep. Exper. Farms, Madras*). (*Conf.* with **Withania**, Vol. VI., Pt. IV., 269, also **Rennet**, Vol. VI., Pt. I., 427.)

In Southern India, in April and May, the twigs of this tree are stuck in and around the thatched roofs of houses to ward off lightning.

2922

Striga, *Lour.*; *Fl. Br. Ind. IV., 298*; SCROPHULARINEÆ.

A genus of parasitic plants. They often do much harm to crops in India. See **Saccharum** (Sugar), paragraph on Diseases, *Vol. VI., Pt. II., 126*; as also **Sorghum vulgare**, paragraph on Diseases.—(*Ed., Dict. Econ. Prod.*)

STROBILANTHES, *Blume.; Gen. Pl., II., 1086.*

[ACANTHACEÆ.

2923

Strobilanthes auriculatus, *Nees; Fl. Br. Ind., IV., 453;*

Syn.—S. AMPLECTENS, *Nees*; RUELLIA AURICULATA, *Wall.*

Vern.—*Gada-kalha, harna pakor,* SANTALI.

References.—*Beddome, Ic. Pl. Ind. Or., t. 210; Rev. A. Campbell, Econ. Prod., Chutia Nagpur, Nos. 8149, 8150; Gazetteer, N.-W. P., X., 315; Ind. Forester, XIV., 153.*

Habitat.—A shrub, 2 to 6 feet high, common in Central India, from Jubbulpore to Chutia Nagpur, at altitudes between 1,000 and 4,000 feet.

Medicine.—The pounded LEAVES are rubbed on the body during the cold stage of intermittent fever.

MEDICINE. Leaves. 2924

2925

Strobilanthes ciliatus, *Nees.; Fl. Br. Ind., IV., 439.*

Syn.—RUELLIA CILIATA, *Wall.;* GOLDFUSSIA ZENKERIANA, *Wight, Ic., t. 1517 (?), not of Nees.*

Vern.—*Kárvi, kara,* BOMB.

References.—*Beddome, Ic. Pl. Ind. Or., t. 211; Dymock, Mat. Med. W. Ind., 591; Gasetteer, Bombay XV., 440; Ind. Forester, XIV., 158.*

Habitat.—A very common, erect, shrub, found in the South Deccan Peninsula. It occurs on the Ghâts up to 4,000 feet; on the Nilghiris and at Mangalore.

Medicine.—The PLANT has a strong aromatic odour, and is much used in domestic medicine by the country-people of the regions where it occurs. The BARK, with an equal proportion of that of **Calophyllum inophyllum**, is applied as a fomentation in tenesmus. The JUICE of the bark, with an equal quantity of that of **Eclipta alba**, boiled down to one-half and mixed with old sesamum oil, a few pepper-corns and ginger, is heated and used as an external application in parotitis, and equal quantities of the juice of the flowers and of those of **Randia dumetorum** are smeared over bruises (*Dymock*).

MEDICINE. Plant. 2926 Bark. 2927 Juice. 2928

Domestic.—The STEMS are much used like those of the bamboo in the construction of mud walls and fences (*Dymock*).

DOMESTIC. Stems. 2929

2930

S. flaccidifolius, *Nees; Fl. Br. Ind., IV., 468.*

THE RÚM OR ASSAM INDIGO PLANT.

Syn.—S. FLACCIDUS, *Mann;* RUELLIA INDIGOFERA, *Griff.;* R. INDIGOTICA, *Fortune;* GOLDFUSSIA CUSIA, *Nees.*

Vern.—*Rúm, rámpát, rampat,* ASSAM; *Khuma, khum,* MANIPUR; *Khom, hom.* PHAKIAL; *Sapro,* ANGAMI NAGA; *Chimohu,* LHOTA NAGA; *Tonham, ton kham, rom gas,* KHAMPTI; *Mai-gyee, man-kyee,* BURM.

References.—*DC., Prod., XI., 194; Gamble, Man. Timb., 280; Mason, Burma & Its People, 510, 795; Watt, Cal. Exhib. Cat., II., 59, 60; Darrah, Note on Cotton in Assam, 29; Trotter, Report on Manipur Dyes; Report on Raw Prod. Col. & Ind. Exhib., 1886, No. 2481; Strettell, The Ficus Elasticus in Burma Proper, 151; Mann, Rep. on Assam Forest Admin. (1876-77), 135; Agri.-Horti. Soc.:—Ind., Jour., II., 249; III., 232; VI., 49, 68, 69, 142; Pro., 2; IX., 51; X., Sel. 18; XI., 155, Sel., 4; (New Series), VII., 361; Ind. Forester, XIV., 153, 394; Balfour, Cyclop. Ind, III., 452; Smith, Dict. Econ. Pl., 353.*

Habitat.—A shrub, found in North and East Bengal, Assam, and Manipur, and distributed to Northern Burma and Southern China. It is often cultivated, and usually occurs on the lower hills of these regions, at altitudes between 1,000 and 4,000 feet.

Dye.—The account of the *rúm* dye of Assam, given by **Dr. Watt** in the Calcutta Exhibition Catalogue, may be here reproduced:—

DYE. 2931

"Both **Mann** and **Kurz** speak of a plant yielding a blue dye, the former in Assam and the latter in the Karén country, under the name of **S. flaccidus**. This is probably a mistake for **S. flaccidifolius**, *Nees.* This exceedingly valuable dye was first made known by **Griffith**, who met with it during one of his Assam explorations. It is pretty generally cultivated by the hill tribes of the eastern frontier, and extends into North-Western China. The plant was called **Ruellia indigotica** by **Balfour**, as he explains, in the absence of any better name. It grows freely on the plains of Manipur in a climate not very different from that of many parts of Bengal, Behar, or the North-West Provinces, and might be extensively cultivated in Assam. It does not require the flooding, which is necessary for the early growth of the Bengal indigo plant, and is therefore not exposed to the danger of having its colour extracted during an exception-

DYE.

ally rainy season. In fact, in many respects, it possesses properties eminently suited for a profitable indigo crop, and in China at least the dye is pronounced finer than that obtained from any other plant. It is propagated freely by cuttings, yields prunings twice or three times a year, and is perennial. It would give little or no anxiety to the planter, and if not sufficiently remunerative to take the place of the Bengal indigo plant, it seems natural to expect that the two plants might, with great advantage, be cultivated together. The *rúm* would flourish on the higher dry lands in the plantation, and yield its crop probably in the cold and the hot season, while the ordinary indigo might be grown in the low flooded lands and occupy the attention of the planter during the rest of the year. At present an indigo factory is idle for more than half the year, but with **Strobilanthes flaccidifolius** this need not be so.

"In Manipur the *khúma* is largely cultivated, and the dye is extracted for home use; nearly every owner of a farm cultivates a small plot of it and prepares his own dye. The twigs, about a foot long, are twice or three times a year plucked and deposited in large earthen pots filled with water. In these primitive vats they are left for the required time, and when ready the decoction of a greenish colour is poured into another pot and violently shaken or stirred by a few twigs. A little lime is generally added, and when the transformation of green into blue indigo has been effected, the liquid is poured into a small earthen vessel and boiled down, more and more being added until from the evaporation of the water the vessel is filled with the dye-stuff. A little lime is placed in the mouth of the vessel, which is thereafter placed in the sun to complete the drying of the dye. In this form it is stored for family use or sold in the market.

They use the dye in combination with turmeric to produce shades of green; with lime and turmeric, browns and almost reds; with lime alone, deep blue black; with safflower, purple; and so on as in the ordinary combinations."

Method of dyeing. 2932

The method of dyeing with *rúm* pursued in Manipur has been subsequently to the above fully described by **Mr. Darrah** as follows:—

"Put as large a quantity of leaves of the *khúm* plant into a pot as it will hold, and then fill it with water and let stand until the leaves become partially decomposed; then wring out the leaves gently in the pot, and throw them away; next put a *chittak* or two of shell-lime into the pot, and let stand for 24 hours; this lime has the effect of precipitating all the organic matter in the water to the bottom of the pot apparently; then pour off the water gently and scrape out the sediment; this sediment is *khumbang*. In from 10 to 12 quarts of water mix one seer of *khumbang*, and let stand for two or three hours; then mix 8 chittaks of shell-lime in the pot and let stand for 24 hours; then add a pint and a half of *heibung** water; stir up the mixture thoroughly, and let stand for 24 hours; then add 2 quarts of water that uncooked rice has been washed in, stir up thoroughly, and let stand for another 24 hours; now add 3 tolas of shell-lime and 2 quarts of *ootee* water (a lye prepared from plantain or other ashes); mix thoroughly and again let stand for 24 hours, when the dye ought to be ready for use. If it is not, then more of some of the above ingredients will have to be added to it, but only experience teaches what particular ingredients are not in sufficient quantity in the dye. Wash the cloth to be dyed perfectly clean and steep thoroughly in the vessel containing the dye; squeezing and pressing it about for 8 or 10 minutes; then wring out and wash with soap and water [if soap is not

* *Heibung* water, prepared by soaking ½ seer of the fruit of **Garcinia pedunculata** cut in slices in a pint of water for 22-24 hours.

DYE. Method of Dyeing.

at hand, rice water (*i.e.*, water that uncooked rice has been washed in) to be used; this ought to be stale, that is, two or three days old]. If it is found that the cloth has not taken the dye well, repeat the above process and hang out to dry in the sun."

The *Khámpti* process, as described by Mr. Darrah, is somewhat similar to the above:—

"The tops of the plants are cut twice a year, in May and October tied into bundles and immersed in a large earthen vessel containing water, where they are left to steep about three days in May and about six days in October. Then the vegetable matter is taken out and thrown away, and the liquid, after being thoroughly stirred up by means of *khaloi* (a jug-shaped wooden basket in which fish are put when caught), is allowed to settle for the night. In the morning the liquid at the top is poured off and the sediment put aside in an earthen vessel for use as required. In this state the dye (now called *nám-ham* from *nám*=water and *ham*=, *rum*) can be retained for six months or so without deterioration. When required for use a solution of ashes in water (*khárpani*) is added in equal quantities to the *nám-ham*, and into this is put a small quantity of a mixture composed of almost equal parts of the following ingredients:—(1) the juice of a small esculent plant called *Jya hamkhia*; (2) juice expressed from the bark of the *bhat-ghila* tree; (3) a fluid obtained by squeezing a number of *amruli* (**Cricula trifenestrata**) worms into pulp. The solution thus prepared is exposed to the sun, and the shade of the dye is tested from time to time by dipping in the finger nail, alterations being made in the quantity of the ingredients according to the depth of the colour required. When the right shade is obtained the thread or cloth is steeped in the liquid for half an hour, and then exposed to the sun. The oftener is it steeped and dried, the darker is the blue produced. To fix the colour, the dyed material is placed over a fire in a closed basket till quite hot. It is then allowed to cool and finally exposed to the sun. This method of fixing the colour is only adopted in the case of cloths highly valued by the *Khámptis*."

"Shades of green and black also are produced by the use of other ingredients. Thus to produce a green shade, Major Trotter (*Report on Manipur Dyes*) says that the Manipuris dip a cloth which has been recently dyed blue by the process above described, in a solution, which is prepared by soaking, in as much water as it will absorb, a seer of fresh turmeric root cut and pounded, and squeezing out the juice. The cloth is then washed clean in fresh water, wrung out and steeped thoroughly in a pint of *heibung* water, again wrung out and dried in the shade. In Assam the leaves of **Vigna Catiang**, *Endl.* (*see* Vol VI., Pt. IV.), are employed along with *rúm* to give green shades to cotton cloths.

"For shades of black "an earthen pot is half filled with the leaves of the *rúm* plant and filled up with water. When the leaves are well soaked, the pot is put on the fire. Before the mixture comes up to the boiling point the leaves are thrown out and a fresh supply of equal quantity of leaves is put into it and warmed. Then the water is taken out of the pot and poured into a wooden vessel. The leaves also are taken off, and before the water becomes cool, cotton threads which have previously for three or four days been kept in water with *rum* plants are well soaked into it and dryied. Under this process the threads become black. In order to make the colour fast, ash water is mixed with the warm *rúm* water" (*Watt. Cal. Ind. Exhib. Rep.*).

Strobilanthes Simonsii, *T. Anders.; Fl. Br. Ind., IV., 447.* 2933

Reference.—*Kurz, For. Fl. Br. Burm., II., 244.*

Habitat.—A shrub found in Assam, Martaban, and Tenasserim.

TIMBER. 2934

Structure of the Wood.—White, rather light, very soft but close-grained, of a fine silky fibre.

2935

STROPHANTHUS, *DC.; Gen. Pl., II., 714.*

A genus of Apocynaceous plants, comprising about eighteen species, natives of Tropical Asia and Africa. The members of this group are small trees or shrubs, or are sometimes climbers. At least five species are indigenous to the tropical regions of India, Burma, or the adjacent Malayan Peninsula. Several of them are very handsome plants and well worthy of cultivation, but the chief interest in the genus centres in the fact that an active principle separated from a closely allied African species, is one of the chief ingredients of the arrow poisons of some African tribes, and has lately been discovered to be a most valuable remedy in fatty degeneration of the heart and other forms of cardiac disease. This active principle, which has received the name of *Strophanthin*, occurs most abundantly in the seeds and is easily separated from a watery solution of the alcoholic extract by agitating repeatedly with ether to remove the fat and colouring matter, and then evaporating the water solution at a low temperature.

No experiments appear to have as yet been made to see whether *Strophanthin* exists in any of the Indian species.

STRYCHNOS, *Linn.; Gen. Pl., II., 797.*

2936

Strychnos colubrina, *Linn.; Fl. Br. Ind., IV., 87*; LOGANIACEÆ.

Syn.—S. BICIRRHOSA, *Lesch.*; S. MINOR, *Blume.*

Vern.—*Kuchila lata*, HIND. & BENG.; *Módira-kaniram*, MAL (S.P.); *Goagari-lakei*, BOMB.; *Kájar-wel*, MAR.; *Nága-musadi, kousu kandira, tansu-paum*, TEL.

References.—*Roxb., Fl. Ind., Ed. C.B.C., 194; Thwaites, En. Ceyl. Pl., 201; Dalz. & Gibs., Bomb. Fl., 155; Sir W. Elliot, Fl. Andhr., 100; Rheede, Hort. Mal., VII., t. 5; Pharm. Ind., 145; Ainslie, Mat. Ind., II., 202; O'Shaughnessy, Beng. Dispens., 442; Moodeen Sheriff, Supp. Pharm. Ind., 234; Dymock, Mat. Med. W. Ind., 2nd Ed., 533-535; Year-Book Pharm., 1880, 249; Drury, U. Pl. Ind., 406; Useful Pl. Bomb. (XXV., Bomb. Gaz.), 267, 275; Linschoten, Voyage to East Indies (Ed. Burnell, Tiele & Yule), II., 104; Agri.-Horti. Soc. Ind., Jour., VI., 13.*

Habitat.—A scandent plant of the Western Deccan Peninsula, frequently met with from the Konkan to Cochin.

MEDICINE. Root. 2937

Medicine.—Linschoten, in his *Voyage to the East Indies*, describes it under the name of "Snake-wood," and says that in Ceylon the ROOT, bruised in water and wine, is "very good and well proved against all burning fevers," and "against all poison and sickness, as the collick, wormes, and all filthie humours and coldness in the body, and specially against the stinging of snakes whereof it hath the name." He then goes on to state that this is the root which the mongoose eats when bitten in its encounters with snakes (*see* **Ophiorrhiza Mungos** (*Vol. V., 488*) and **Rauwolfia serpentina** (*Vol. VI. Pt. I., 398*).

Wood. 2938

In India, the WOOD is much used among the Hindus as a tonic in dyspepsia and malarious affections. Its claims as an antiperiodic have been examined by Dr. Bardenis van Berkelow (*Brit. & For. Chir. Rev., April 1867, 527*), who reports favourably of its action in quartan and tertian fevers, and considers that from its cheapness it might be advantageously used as a febrifuge. It contains strychnia in considerable quantities, and must, therefore, be employed very cautiously (*Waring, Pharm. Ind.*).

It is commonly supposed to be the **Arbor lignicolubrini** of Rumphius, who says that a preparation from its root was, in Amboina, a common domestic remedy for quartan fevers, and that in Batavia an arrack flavoured with it was drunk daily by many persons as a prophylactic against malarial affections. It was used by the Sinhalese of his time as an anthelmintic, and was considered by them one of the most excellent of all specifics for the bites of poisonous reptiles.

MEDICINE. Arak. 2939

2940

Strychnos Ignatii, *Bergius.*

Syn.—IGNATIA AMARA, *Linn.*

By an error the younger **Linnæus**, confusing two plants, formed the genus **Ignatia**, hence the synonym (*Watt, Cal. Exhib. Cat.*).

Vern.—(The seeds=) *Pipita* (a corruption of the Spanish term, *Pepita*), HIND., BENG., & BOMB.; *Kayap-pankottai*, TAM.

References.—*DC., Prodr., IX., 19; Bentley & Trimen, Med. Pl., t. 179; Pharm. Ind., 146; Flückiger & Hanbury, Pharmacographia, 431; Irvine, Mat. Med. Patna, 84; S. Arjun, Bomb. Drugs, 89; Dymock, Mat. Med. W. Ind., 535; Baden Powell, Pb. Pr., 368; Birdwood, Bomb. Prod., 55.*

Habitat.—"A large, climbing shrub, growing in Bohol, Samar, and Cebu, islands of the Bisaya group of the Phillipines, and, according to **Loureiro**, in Cochin China where it has been introduced" (*Pharmacographia*).

Medicine.—The SEEDS are met with in Indian bazárs, and are described by Native physicians as alexipharmic and a useful remedy in cholera; they also give them in doses of from 1 to 2 grains in asthma, dropsy, rheumatism, and piles. The bean contains about 1·5 per cent. of Strychnine and ·5 per cent. of brucine.

MEDICINE, 2941 Seeds. 2942

S. Nux-vomica, *Linn.; Fl. Br. Ind., IV., 90; Wight, Ic., t. 434.*

2943

THE NUX-VOMICA OR STRYCHNINE TREE.

Syn.—S. LUCIDA, *Wall*, partly; S. COLUBRINA, *Wight* (non other authors).

Vern.—*Kuchlá, kajra, nirmal, chilbinge, bailewa,* HIND.; *Kuchilá, thalkesur,* BENG.; *Nirmali,* NEPAL; *Kanni-rak-kuru, kariram, tettamperel marum,* MAL (S.P.); *Kuchla, kerra, korra,* URIYA; *Kuchlá,* C.P.; *Kuchlá,* OUDH; *Kúchila, hub-ul-jarāb, kágphala, kajra,* PB.; *Kajra, kara, jhar-katchura,* BOMB.; *Kájra, kara, jhar katchura,* MAR.; *Kuchlá,* GUZ.; *Yetti, yetti-maram, ettik-kottai,* TAM.; *Mushti, musidi, indupu,* TEL.; *Kasaraka, kujarra, khasca, kasaragadde, mushti,* KAN.; *Khaboung,* BURM.; *Goda kadúra, goda-kadura-atta, kanchura,* SING.; *Kupilu, kulaka, víshamúshti,* SANS.; *Izaragi, khanek-ul-kella,* ARAB.; *Fulúzmáhi, izaraki,* PERS.

References.—*A. D.C., Prodr., IX., 15; Roxb., Fl. Ind., Ed. C.B.C., 193, 194; Voigt, Hort. Sub. Cal., 530; Brandis, For. Fl., 317; Kurz, For. Fl. Burm., II., 166, 167; Beddome, Fl. Sylv., t. 243; Gamble, Man. Timb.. 269; Dalz. & Gibs., Bomb. Fl., 155; Mason, Burma & Its People, 489, 802; Sir W. Elliot, Fl. Andhr., 120; Rheede, Hort. Mal., I., t. 37; Pharm. Ind., 143; British Pharm., 280, 386; Flück. & Hanb., Pharmacog., 428; U. S. Dispens., 15th Ed., 575, 1415; Fleming, Med. Pl. & Drugs (Asiatic Reser., XI.), 178; Ainslie, Mat. Ind., I., 318; O'Shaughnessy, Beng. Dispens., 436; Irvine, Mat. Med. Patna, 47, 115; Moodeen Sheriff, Supp. Pharm. Ind., 234; U. C. Dutt, Mat. Med. Hind., 198, 199, 306, 324; K. L. De, Indig. Drugs Ind., 113; Murray, Pl. & Drugs, Sind, 151; Bent. & Trim., Med. Pl., III., 178; Dymock, Mat. Med. W. Ind., 2nd Ed., 526, 532; Year-Book Pharm., 1874, 91; 1875, 13; 1877, 223; 1881, 23; Trans. Med. & Phy. Soc. Bomb. (New Series), IV., 85, 152; XII., 174; Birdwood, Bomb. Prod., 55, 168; Baden Powell, Pb. Pr., 360; Useful Pl. Bomb. XXV., Bomb. Gaz.), 101, 166, 267; McCann, Dyes & Tans, Beng., 137; Wardle, Dye Report, 44; Man. Madras Adm., I., 314; Nicholson, Man. Coimbatore, 6; Boswell, Man., Nellore, 99, 129; Moore, Man., Trichinopoly, 80; Aplin, Report on Shan States (1887-88); Gribble, Man. Cuddapah Dist., 200; Settlement Reports:—Central Provinces, Upper Godavery, 38; Raepore, 76; Mundlah, 89; Bundara, 20; Chanda, App. VI.; Gazetteers:—Bombay, XV., 76; XVI., 323; N.-W. P., IV., lxxiv.; Mysore & Coorg., III., 17; Hunter, Orissa, II.,*

27 (App. I.) 121; (App. IV.); 159 (App. IV.); 182 (App. VI.), Agri.-Horti. Soc.:—Ind., Trans., VI., 121, 124; VII. (Pro.), 129; Jour., IX. (Sel.). 41; New Series, V., 75; Ind. Forester, I., 114; III., 203; IV., 292; VIII, 29, 416; X., 31, 547; XII., 188; (xxii) XIV., 199; Balfour, Cyclop. Ind., III., 746; Smith, Ec. Dict., 290.

Habitat.—A tree which attains a height of 40 feet; found throughout Tropical India up to an altitude of 4,000 feet. It is rare in Bengal, but common in Madras and Tenasserim.

DYE. Seeds. 2944

Dye.—The SEEDS yield a dye, which is used, in Balasore, for producing lightish brown shades on cotton cloth. Boiled with proto-sulphate of iron and lime they give darker shades of brown (*McCann*). **Wardle** (*Dye Report*) says that they give lightish drab shades on silk, but are not well adapted for wool, on which only a very faint colour can be obtained.

OIL. 2945 Seeds. 2946

Oil.—An empyreumatic oil, prepared from the fresh SEEDS, is used medicinally by Native practitioners.

MEDICINE. 2947

Medicine.—Nux-vomica does not appear to have been used in early Sanskrit medicine, although it is quite possible that some part of the tree may have been used by the aboriginal tribes of India from a very early date, since we find the wood used now as a common tonic over very extensive tracts of country. In the more recent Sanskrit compilations we find it mentioned under its vernacular name *Kuchila* (*Dutt*). Among the Hindu practitioners of the present day, the SEEDS are, in various combinations, used as a medicine for dyspepsia and diseases of the nervous system. On the Malabar coast the ROOT is given in snake-bite, and in Bombay the WOOD is a popular remedy, like that of **S. colubrina**, in the dyspepsia of vegetarians. In the Konkan small doses of the seeds are given in colic combined with aromatics, and the JUICE of the fresh wood (obtained by applying heat to the middle of a straight stick to both ends of which a small pot has been tied, is given in doses of a few drops in cholera and acute dysentery (*Dymock*). The OIL from the fresh seeds is used as an external application in chronic rheumatism. "Nux-vomica seeds produce a sort of intoxication, for which they are habitually taken by some Natives as an aphrodisiac. Those who do so gradually become so far accustomed to this poison that they often come to take one seed daily which is cut into small pieces and chewed with a packet of betel leaf (*U. C. Dutt*). The Muhammadans' knowledge of the uses of Nux-vomica seems to have been derived from the Hindus, as **Makhzan-el-Adwiya** concludes his description of the drug by saying that much information will be found about the drug in Hindu works. He recommends great caution in its use, but says it is a valuable medicine in palsy, relaxation of the muscle and tendons, debility and chronic rheumatism (*Dymock*).

Seeds. 2948 Root. 2949 Wood 2950 Juice. 2951 Oil. 2952

In European medicine it was first known about the middle of the sixteenth century when **Valerius Cordus** who wrote in Germany gave a very accurate description of the appearance of the seeds, but for, at any rate, a century later it was not used in medicine, since **Parkinson** (1640) remarks that its chief use was for poisoning cats, dogs, crows and ravens, and it was not till the beginning of the present century that its value as a nervine tonic was recognised by European practitioners. An extract and a tincture are now, however, officinal in all the principal European Pharmacopœias as well as preparations from the alkaloid strychnine, which is extracted from the seeds. These preparations are largely used in various nervous disorders, as nervine tonics and stimulants, in bronchitis, emphysema, and phthisis as respiratory stimulants, and in chronic constipation from atony of the bowels to increase the peristaltic action of the intestine.

COLLECTION. 2953

COLLECTION.—The seeds are collected either together with the fruit or from the ground, where they have been thrown by birds while eating the pulp. They are then washed to free them from fragments of pulp, dried and exported to Europe.

COLLECTION.

Characters. 2954

CHARACTERS.—As met with in the bazárs of India or in European medicine they are rounded, ⅞ to 1 inch in diameter, about ¼ inch thick, flattish or concavo-convex, rounded at the margin. They are marked on one side by a central scar whence a projecting line passes to the margin, where it ends in a slight prominence. Externally they are of an ash grey colour, glistening with short satiny hairs, internally horny, somewhat translucent. They have no odour, but an extremely bitter taste.

Composition. 2955

COMPOSITION.—Nux-vomica seeds contain two alkaloids, ·2 to ·5 per cent. of strychnine and ·12 to 1 per cent. of brucine, united with an acid, strychnic or egasuric acid. The characters of these alkaloids will be found so fully described in every work on Materia Medica that it is needless to repeat them here. These substances seem to exist not only in the seeds but in the wood and root, and the leaves of parasitic plants even, growing on the tree, seem to acquire its poisonous properties and to contain the same alkaloids.

SPECIAL OPINIONS.—§ "In muscular and chronic rheumatism a paste made by rubbing the seeds of Nux-vomica, dry ginger, and the horn of the antelope on a stone is used with benefit; the paste made by rubbing the seeds of Nux-vomica only is also used in rat-bites" (*Surgeon Major A. S. G. Jayakar, Muskat*). "Useful in treatment of tobacco amaurosis, in paralysis following on exhausting diseases, such as diphtheria and gastric catarrh. It is antagonistic to calabar-bean and markedly so to chloral" (*Surgeon-Major E. G. Russell, Superintendent, Asylums, Calcutta*). "The leaves, when applied as poultice, promote healthy action in sloughing wounds or ulcers, more specially in those cases when maggots have formed. It arrests any further formation of them, and those in the deeper parts perish immediately when the poultice is applied. The root-bark is ground up into a fine paste with lime-juice, and made into pills which are said to be effectual in cholera" (*Surgeon-Major D. R. Thomson, M.D., C.I.E., Madras*). "I have found strychnine very useful in malarious fevers of a low type" (*Surgeon-Major H. F. Hazlitt, Ootacamund*). "Valuable nervine tonic, useful in intermittent fevers, enlargement of spleen, &c., also in impotence" (*Assistant Surgeon S. C. Bhuttacharji, Chanda, Central Provinces*). "The bark, ground with lemon juice and aromatics, such as ginger and made into pilts, &c., checks cholera" (*V. Ummegudien, Mettapolian, Madras*). "Used in paralysis and impotence" (*Assistant Surgeon Nehal Sing, Saharunpore*). "A good nervine tonic, obtainable in all bazárs and much used in dispensary practice" (*Surgeon G. Price, Shahabad*). "The preparations of the Strychnos Nux-vomica are highly prized by the Natives on account of their efficacy in many cases of impotency. Strychnine is a valuable drug in the bronchitis of the debilitated. Its action as an expectorant appears to be considerable" (*Surgeon S. H. Browne, M.D., Hoshangabad, Central Provinces*).

FOOD. Fruit. 2956

Food.—Although the pulp of the FRUIT also contains strychnine (*Hanbury*), yet it appears to be eaten voraciously by birds, and **Birdwood** (*Bombay Products*) says: "The fruit is commonly eaten in the Koncans for the sake of the pulp enclosing its deadly seeds." Several African travellers also state that the pulp of the fruit of a variety of the Nux-vomica is eaten by many native tribes in that country, is of a pleasant juicy nature, and has a sweet, acidulous taste.

Structure of the Wood.—Brownish-grey; hard, close-grained, splits and warps in seasoning. It is said not to be attacked by white ants. Weight from 52 to 62℔ per cubic foot.

STRYCHNOS potatorum. **The Clearing Nut Tree.**

DOMESTIC

Seeds. 2957

Domestic.—The SEEDS are used by the hill tribes of the Nilghiris as a fish poison. They are also employed by Native distillers, who sometimes add a small quantity of them to arrack so as to render it more intoxicating. In consequence of the large importations to Europe far over what can possibly be consumed for medicinal purposes, it was thought that they were sometimes used by brewers to give a bitter taste to ales. This is said to have been disproved, but it is still unknown to what use the bulk of the imports to European countries are put.

Wood. 2958

In Burma the WOOD is used for making carts and agricultural implements and for fancy cabinet work.

Trade. 2959

Trade.—"Large quantities of Nux-vomica are exported from India. The annual exports from Bombay amount to about 4,000 cwt., all shipped to the United Kingdom. Madras and Cochin export still larger quantities, and Calcutta rather less. Value R2 per maund of 37$\frac{1}{2}$ cwt. Bombay Nux-vomica is preferred by the manufacturers of strychnia, as it yields the largest quantity of the alkaloid" (*Dymock, Mat. Med. W. Ind.*).

2960

Strychnos potatorum, *Linn. f.; Fl. Br. Ind., IV., 90; Wight, Ill., II., t. 156.*

THE CLEARING NUT TREE.

Syn.—S TETTANKOTTA, *Retz.*

Vern.—*Nirmali, nelmal, neimal,* HIND.; *Nirmali,* BENG.; *Kuchla,* SANTAL; *Titrán-parala, tettam-parel,* MAL (S. P.); *Kotaku,* URIYA; *kuwi,* C. P.; *Ustumri,* GOND; *Nirmali,* PB.; *Chil-binj* (=the nuts), DECCAN; *Nirmali, gajrah,* BOMB.; *Niwali, chilbing* (=the nuts), MAR.; *Tetan-kottai, tettian, tétta,* TAM.; *Induga, katakami, indupa, judapa, chilla chettu, indupu, chettu,* TEL.; *Chilu,* KAN.; *Kamon-yeki, khaboung-yac-kyie,* BURM.; *Ingini,* SING.; *Kátaka, ambu-prasáda,* SANS.

References.—*A. D.C., Prodr., IX., 15; Roxb., Fl. Ind., Ed. C.B.C., 104; Voigt, Hort. Sub. Cal., 530; Brandis, For. Fl., 317; Kurz. For. Fl. Burm., II., 167; Beddome, For. Man., 163; Gamble, Man. Timb., 268; Dalz. & Gibs., Bomb. Fl., 156; Mason, Burma & Its People, 497, 802; Sir W. Elliot, Fl. Andhr., 38, 70, 88; Pharm. Ind., 146; U. S Dispens., 15th Ed., 1755; Fleming, Med. Pl. & Drugs (Asiatic Reser., XI.), 178; O'Shaughnessy, Beng. Dispens., 443; Irvine, Mat. Med. Patna, 75; Medical Topog., Ajmir, 147; Moodeen Sheriff, Supp. Pharm. Ind., 235, 359, 360; U. C. Dutt, Mat. Med. Hind., 200, 304; S. Arjun, Cat. Bomb. Drugs, 88; K. L. De, Indig. Drugs, Ind., 114; Dymock, Mat. Med. W. Ind., 2nd Ed., 532, 533; Year-Book Pharm., 1880, 249; Birdwood, Bomb. Prod., 56, 345; Baden Powell, Pb. Pr., 360; Drury, U. Pl. Ind. 408; Useful Pl. Bomb. (XXV., Bomb. Gaz.), 101, 166, 252; Ordinances of Manu, Sect. VI, para. 67; Boswell, Man., Nellore, 96, 120; Moore, Man., Trichinopoly, 80; Butter, Topog. States of Oudh & Sultanpore, 43; Settlement Reports:—Central Provinces, Chanda, App. vi.; Upper Godavery, 38; Hunter, Orissa, II., 181, App. VI.; Gazetteers:—Bombay, XV., 76; Central Provinces, 503; Ind. Forester:—III., 203, 238; X., 315; XIV., 199.*

Habitat.—A small tree plentiful in the Deccan, Central and South India, and Burma.

MEDICINE. Seeds. 2961

Medicine.—Medicinally the SEEDS are chiefly used by Hindu practitioners in eye diseases: they are rubbed up with honey and a little camphor and the mixture applied to the eyes in cases of too copious lachrymation. Rubbed up with water and rock salt, they are applied to chemoses in the conjunctiva. They enter also into the composition of several complex applications for corneal ulcer (*U. C. Dutt*). In Muhammadan works of medicine it is considered a cold and dry remedy, useful in dysentery, and as an external application in colic, and to the eyes for the purpose of strengthening the eyesight. It is also given in irritation of the urinary organs and gonorrhœa (*Baden Powell*). In Southern India, Ainslie says that the FRUIT in its mature state is regarded by the Tamil doctors as an emetic when given as a powder in doses of about half a teaspoonful,

Fruit. 2962

The Benzoin Tree. (*W. R. Clark.*)

MEDICINE.

With reference to this use the late **Moodeen Sheriff**, in his unfinished *Materia Medica of Southern India*, says that the seeds have no such property, but that the PULP is an excellent emetic and a good substitute for Ipecacuanha in the treatment of dysentery and bronchitis. In European practice this drug is not used, but it has a secondary place in the Indian Pharmacopœia, where it is mentioned, on the authority of **Kirkpatrick**, as a remedy for diabetes. No details are available as to its chemical composition.

Pulp. 2963

SPECIAL OPINIONS.—§ "In long standing and chronic diarrhœa which resists all treatment one-half or a full seed, rubbed up into a fine paste with some butter-milk and given internally for one week, is effectual" (*Surgeon-Major D. R. Thomson, M.D., C.I.E., Madras*). "The powder of the seed is internally given in milk for gonorrhœa" (*V. Ummegudien Mettapolian, Madras*). "The seed rubbed in a little water has the property of throwing down all suspended impurities from muddy water as a precipitate" (*Assistant Surgeon S. C. Bhattacharji, Chanda, Central Provinces*). "From experiments made some years ago I came to the conclusion that the construction of the nut to be clearly vegetable albumen, and that this, when rubbed down with water, acted mechanically as a precipitant of suspended matter in water" (*Surgeon-General W. R. Cornish, F.R.C.S., C.I.E., Madras*).

FOOD. Fruit. 2964

Food.—The pulp of the FRUIT, which is black, one-seeded, and of the size of a cherry, is eaten by the Natives. When young it is made into a preserve by the people of Southern India (*Ainslie*).

TIMBER. 2965

Structrue of the Wood.—White when fresh cut, turning yellowish grey on exposure, hard, close-grained, seasons well. It has no heartwood or annual rings. Weight 57lb per cubic foot.

DOMESTIC. Wood. 2966 Seeds. 2967

Domestic.—The WOOD is used for buildings, carts, and agricultural implements. The ripe SEEDS are employed by the Natives of India, and, before filters were in use, by Europeans also, for clearing muddy water. The cut seeds are rubbed on the inside of a rough earthen vessel in which the water is kept, so that the expressed juice mixes with the fluid. On allowing the water to stand, most of the impurities subside and the water becomes quite drinkable. **Dr. Pareira** suggests that this property depends upon the albumen and casein which they contain. If the seeds be sliced and digested in water, they yield a thick mucilaginous liquid which, when boiled, yields a coagulum (albumen), and by subsequent addition of acetic acid it furnishes a further coagulum casein (*Pharm. Ind.*).

Sturgeon, see **Acipenser huso,** *Linn.*; PISCES; Vol. I., 83.

STYRAX, *Linn.*; *Gen. Pl., II., 669.*

2968

Styrax Benzoin, *Dryand*; *Fl. Br. Ind., III., 589*; STYRACEÆ.

THE BENZOIN TREE.

Vern.—The vernacular names all apply to the resin and not to the tree itself. *Lúban, úd, lobaní úd, hussí,* HIND.; *Lubán, úd,* BENG.; *Lúbán,* PB.; *Ud, lúbani-úd,* DECCAN; *Lubán, úd,* BOMB.; *Loban,* GUZ.; *Shambiráni,* TAM.; *Kaminian,* MALAY.; *Loban, loban jáwí,* ARAB.; *Hussí-lúban, hussí-úl-jawí, durí-haskhak arisa, kalangúra, kamkam,* PERS.

References.—*Roxb., Fl. Ind., Ed. C.B.C., 375; Voigt, Hort Sub. Cal., 347; Gamble, Man. Timb., 253; Mason, Burma & Its People, 486, 783; Pharm. Ind., 132; Flück. & Hanb., Pharmacog, 403; Fleming, Med. Pl. & Drugs (Asiatic Reser., XI.), 188; Ainslie, Mat. Ind., II., 33; Irvine, Mat. Med. Patna, 59; Moodeen Sheriff, Supp. Pharm. Ind., 66; S. Arjun, Cat. Bomb. Drugs, 82; Murray, Pl. & Drugs, Sind, 148; Bent. & Trim., Med. Pl., III., 169; Dymock, Mat. Med. W. Ind., 2nd Ed., 485-487; Year-Book Pharm., 1873, 83, 191; 1874, 85; Trans*

STYRAX Benzoin.

The Gum Benzoin of Commerce.

Med. & Phy. Soc. Bomb. (New Series), 12, 174; Birdwood, Bomb. Prod., 50; Baden Powell, Pb. Pr., 359, 412; Linschoten, Voyage to East Indies (Ed. Burnell, Tiele, & Yule), II., 96, 98; Balfour, Cyclop. Ind., III. 748; Smith, Econ. Dict., 47.

Habitat.—A small tree of the Malayan Archipelago.

RESIN. 2969

Resin.—It yields the true Benzoin or Gum Benjamin of commerce. Two kinds of Benzoin are met with in the market known by the names of Siam and Sumatra Benjamin. The following account of this important resin is an abstract of what is contained in **Flückiger & Hanbury's** *Pharmacographia*:—

Collection. 2970

METHOD OF COLLECTION.—The trees, which are of quick growth, are raised from seeds and are grown on the edges of fields. When they are six or seven years old and have trunks 6 to 8 inches in diameter, they are judged capable of yielding the resin, and incisions are then made in the stems from which exudes a thick whitish resinous juice. This soon hardens by exposure to the air and is then carefully scraped off with a knife. The tree continues to yield at the rate of about 3℔ per annum for ten or twelve years, at the end of which time it is cut down.

Characters. 2971

CHARACTERS.—Siam Benjamin imported in cubic blocks, which take their shape from the cases in which the resin is packed while still soft. It consists most frequently of a compact mass of rich, amber-brown, translucent resin, containing a number of white tears of the size of an almond. Occasionally the resin predominates, and the white tears are almost wanting. In some cases the tears of white resin are very small, and the whole mass has the appearance of reddish-brown granite. The most esteemed sort consists entirely of flattened tears or drops loosely agglutinated into a mass. There is always a certain amount of admixture of pieces of wood, bark, and other accidental impurities. The white tears when broken display a stratified structure, with layers of greater or less translucency, and by keeping, they become brown and transparent on the surface. Siam Benzoin is very brittle. It easily softens in the mouth and may be kneaded with the teeth like mastich. It has a delicate balsamic vanilla-like fragrance, but very little taste. When heated it evolves a more powerful fragrance together with the fumes of benzoic acid; its fusing point is 75°C. The Sumatra drug is imported in cubic blocks like the preceding, but differs from that in its generally greyer tint. When of good quality the mass contains numerous tears set in translucent greyish-brown resin mixed with bits of wood and bark. The tears may, however, be wanting, and, as a rule, the amount of the impurities is greater than in the Siam resin In odour the Sumatran drug is weaker and less agreeable, and generally commands a much lower price. The greyish portion melts at 95°, the tears at 85°C.

Chemistry. 2972

CHEMICAL COMPOSITION.—Benzoin consists mainly of amorphous resins perfectly soluble in alcohol and potash, having slightly acid properties and differing in their behaviour to solvents. Subjected to dry distillation benzoin affords as its chief product benzoic acid which exists to the extent of from 14 to 18 per cent. ready formed in the drug. The benzoic acid may be readily extracted by the aid of an alkali, and the amorphous resins are then left. About one-third of those will be found soluble in ether; the prevailing portion dissolves in alcohol, and a small portion remains undisolved (*Flückiger & Hanbury, Pharmacographia, 407*).

MEDICINE. 2973

Medicine.—Benzoin does not appear to have been known to the early inhabitants of India, nor to the Greeks and Romans. It is used by the Hindus and Muhammadans of the present day as an internal remedy in phthisis and asthma, and Muhammadan physicians recommend the inhalation of its fumes in diseases of this class. In Bengal it is given as an aphrodisiac in doses of from grs. v to xx. In European medicine Benzoin and its preparations are used externally as disinfectants and stimulants; internally Benzoin and Benzoic acid enter the blood in the form of Bensoate of sodium, and there, as well as in the kidneys, the acid is partly converted into hippuric acid. They are thus valuable in inflammation of the bladder with alkalinity of the secretions and phosphatic deposits in the urine, as they acidulate the urine and stimulate and disinfect the mucous surfaces of the bladder.

MEDICINE.

Domestic and Sacred.—It is burnt as an incense by the Buddhists and Hindus in their worship, and in Europe by the Roman Catholics and members of the Greek Church. On account of its disinfectant properties, the smoke it gives out is sometimes employed in the East to drive away mosquitos and sand-flies.

DOMESTIC & SACRED. 2974

Trade.—The total imports of Gum Benjamin into Indian ports during the official year 1889-90 amounted to 12,498 cwt., valued at R4,00,322. All of this, with the exception of one cwt., came from the Straits Settlements. The largest amount imported was by Bombay (7,921 cwt.); Madras imported 4,130 cwt., Bengal 384 cwt., and Burma 63 cwt. During the period under review, 1,125 cwt., valued at R37,477, was again exported, of which over 1,000 cwt. went from Bombay. The largest importing country from India was Arabia, which took 400 cwt., the United Kingdom, the East Coast of Africa, and Aden took, respectively, 117, 109, and 103 cwt., and smaller quantities were sent to other European and Asiatic countries.

Trade. 2975

Styrax Hookeri, *Clarke; Fl. Br. Ind., III., 589.*

2976

Syn.—STYRAX SP., *Gamble, Darjiling List, 54.*

Vern.—*Chamokung*, LEPCHA.

Reference.—*Watt, Cal. Exhib. Cat., VII., 238.*

Habitat.—A small tree, frequently met with in Sikkim and Bhután at altitudes between 6,000 and 7,000 feet.

Structure of the Wood.—White, close-grained, moderately hard.

TIMBER. 2977

S. officinale, *Linn.*

2978

THE TRUE STORAX TREE.

Vern.—*Silajit*, BENG.; *Boe, usturak*, BOMB.; *Usturak*, ARAB.

References.—*Gamble, Man. Timb., 253; Flück. & Hanb., Pharmacog, 276; Birdwood, Bomb. Prod., 51; Irvine, Mat. Med Patna, 107; Ayeen Akbary, Gladwin's Trans., I., 92; Agri.-Horti. Soc. Ind., Trans. III., 41.*

Habitat.—A native of the Levant, Asia Minor, and Syria.

Resin.—This was a solid resin somewhat resembling benzoin, which was held in great estimation from the time of Dioscorides and Pliny down to the close of last century. In most localities, the tree which yielded it has been reduced by ruthless lopping to a mere bush, the young stems of which yield not a trace of exudation, and thus true Storax has almost entirely disappeared.

RESIN. 2979

Medicine.—In the East Storax was given internally as a stimulant in doses of from ½ to 10 grs.

MEDICINE. 2980

S. serrulatum, *Roxb.; Fl. Ind., III., 588.*

2981

Syn.—S. PORTERIANUM, *Wall.*; S. FLORIBUNDUM, *Griff.*

Vern.—*Kum-jameva*, BENG.; *Chamo*, LEPCHA.

References.—*Roxb., Fl. Ind., Ed. C.B.C., 375; Kurz, For. Fl. Br. Burm., II., 142; Agri.-Horti. Soc. Ind., Jour., VI., 47.*

Habitat.—A tree, or more often shrub, frequently met with in Eastern India, from Nepál to Pegu, at altitudes between 3,000 and 7,000 feet.

SUÆDA nudiflora.

A Plant from which crude Soda is prepared.

RESIN. 2982

Resin.—It yields a resin somewhat like Gum Benjamin, but of inferior quality.

TIMBER. 2983

Structure of the Wood.—Whitish, rather soft, but not liable to warp or split.

SACRED. Wood. 2984

Sacred.—The WOOD is used by the Bhutias for prayer poles.

SUÆDA, *Forsk.; Gen. Pl., III., 66.*

2985

Suæda fruticosa, *Forsk.; Fl. Br. Ind., V., 13;* CHENOPODIACEÆ.

Syn.—SALSOLA FRUTICOSA, *Linn.;* S. INDICA, *Wall.;* S. LANA, *Edgew.*

Vern.—*Lúnak, choti láni, usak lani, khar-khusa, khaskhasa, phesak láne, baggi lána, dána, samái,* PB.; *Zimeh,* PUSHTU; *Shorag,* AFG.; *Ushuk lani,* SIND; *Morasa,* MAR.

References.—*DC., Prodr., XIII., 2; Boiss, Fl. Orient., IV., 939; Stewart, Pb. Pl., 180; Aitchison, Cat. Pb. & Sind Pl., 126; Rept. Pl. Coll. Afgh. Del. Com., 102; Graham, Cat. Bomb. Pl., 170; Murray, Pl. & Drugs, Sind, 104; Baden Powell, Pb. Pr., 372; Gazetteer, Panjáb, Muzaffargarh, 27.*

Habitat.—A sub-erect, perennial plant, common on the plains of North-Western India from Delhi throughout the Panjáb to the Indus, and distributed westward to the Atlantic, and in Africa and America.

MEDICINE. Branches. 2986 Leaves. 2987

Medicine.—The plant is subject to have woolly excrescences on the tips of the BRANCHES, which are mixed with an empyreumatic oil and used as an application to the sores on the backs of camels (*Murray*). The officinal *kaskasa* of the bazárs appears to be the dried LEAVES of this plant, and from the Pushtu name these would appear in the Peshawar Valley to be mashed when fresh and applied as a poultice to the eyes for ophthalmia (*Stewart*).

FODDER. 2988

Fodder.—It is eaten by camels.

DOMESTIC. 2989

Domestic.—In Sind crude soda (*Sajji-khar*) is prepared from this plant, together with species of **Salsola** and **Chenopodium**; that prepared from this plant alone is considered inferior in quality (*Murray*).

2990

S. maritima, *Dumort.; Fl. Br. Ind., V., 14.*

Syn.—S. NUDIFLORA, *Moq.*; CHENOPODIUM MARITIMUM, *Linn.;* CHENOPODINA MARITINA, *Moq.;* SALSOLA SALSA, *Jacq.;* S. INDICA, *Willd.;* S. NUDIFLORA, *Wall.;* S. SATIVA, *Wight.*

Vern.—*Lani,* PB.; *Kharri lani,* SIND; *Lana,* MAR.; *Yella kiray,* TAM.; *Ila kura,* TEL.

References.—*DC., Prodr., XIII., ii., 155; Boiss., Fl. Orient., IV., 941; Roxb., Fl. Ind., Ed. C.B.C., 261; Graham, Cat. Bomb. Pl., 170; Sir W. Elliot, Fl. Andhr., 70; Irvine, Mat. Med. Patna, 47; Murray, Pl. & Drugs, Sind, 104; Ind. Forester, III., 238.*

Habitat.—An erect, perennial, shrubby plant, found on the Upper Gangetic plains, the sea coasts of Bengal, Bombay, the Deccan, and Ceylon, and distributed to Siam, Europe, North Africa, North and West Asia, and North America.

FOOD. 2991 Leaves. 2992

Food.—The green LEAVES are universally eaten by Natives wherever the plant occurs. It is considered a very wholesome vegetable, and during famine times an essential article of food among the poor (*Roxburgh*).

DOMESTIC. 2993

Domestic.—This species also is incinerated to produce an impure carbonate of soda.

2994

S. nudiflora, *Moq.; Fl. Br. Ind., V., 14; Wight, Ic., t. 1796.*

Syn.—S. INDICA, *Moq.;* SALSOLA NUDIFLORA, *Willd.;* S. FRUTICOSA, *Wall.;* S. ELATA, *Wight;* CHENOPODIUM PROSTRATUM, *Roxb.*

Vern.—*Geriá,* URIYA *Khari-láni,* SIND; *Morasa,* BOMB.; *Kiray,* TAM.; *Ráva kada, reyi káda,* TEL.

References.—*Roxb., Fl. Ind., Ed. C.B.C., 261; Thwaites, En. Ceyl. Pl., 246; Trimen, Sys. Cat. Cey. Pl., 72; Dalz. & Gibs., Bomb., Fl., 213;*

Aitchison, Cat. Pb. & Sind Pl., 126; Graham, Cat. Bomb. Pl., 270; Sir W. Elliot, Fl. Andhr., 163, 164; Pharm. Ind., 183, 323; Irvine, Mat. Med. Patna, 47; Dymock, Mat. Med. W. Ind., 2nd Ed., 654; Useful Pl. Bomb. (XXV., Bomb. Gaz.), 203; Gazetteers:—Bombay, V., 28; Orissa, II., 178, App. VI.; Agri.-Horti. Soc. Ind, New Series VII., Pro. (1883), 124; Ind. Forester, III., 238; XII., App., 20.

Habitat.—An abundant shrub on the sea-coasts of India.

Food.—The green LEAVES are eaten by the Natives.

FOOD. Leaves. 2995

Domestic.—This species also is used as a source of *sajji.*

DOMESTIC. 2996

Sugar, see **Saccharum officinarum,** Vol. VI., Pt. II., p *et seq.*

2997

SULPHATES.

All the Sulphates which are of any economic importance will be found treated of under their respective bases or commercial names. Thus, for example, Sulphate of Aluminia, see **Alum,** Vol. I., 201; Sulphate of Copper, see **Cuprum,** Vol. II., 649; Sulphate of Iron, see **Iron,** Vol. IV., 523; Sulphate of Lime, see **Gypsum,** Vol. IV., 195; and **Manures,** Vol, V., 175; and for Sulphates of Potash and Soda, see **Barilla,** Vol. I., 304; **Reh,** Vol. VI., Pt. I, 400.

SULPHUR, *Ball in Man. Geol. of Ind., III., 155.*

2998

Elementary Sulphur occurs as a mineral, chiefly in the upper miocene deposits and in the Flötz, associated, in general, with gypsum, massive limestone, and marl. In India, as in other parts of the world, the principal natural sources of the mineral are either deposits formed in connection with hot springs, or deposits which have originated from active or extinct volcanoes. Native Sulphur may occur either in the massive or crystalline condition. In the latter case the crystals are acute octohedra or secondaries to that form and belong to the trianetric system.

In combination with several metals, sulphur forms sulphides or sulphurates which, by atmospheric influence alone, may, to a small extent, become decomposed and form deposits of sulphur. By artificial treatment of these ores, a considerable proportion of the sulphur of commerce is prepared.

Sulphur.

2999

Vern.—*Gundhak,* HIND.; *Gandrok,* BENG.; *Gandak, kibrit, ánwlásár, hassan dhúp, gogird,* PB.; *Gandakam,* TAM.; *Gandhakam,* TEL.; *Avalasara-gandhka* (= the mineral), *sadha-gandhaka* (= roll sulphur), *gandhaka-ka-phula* (= sublimed sulphur), MAR.; *Blerong,* MAL.; *Kan,* BURM.; *Gandhaka,* SANS.; *Kábrit,* ARAB.; *Gangird,* PERS.

References.—*Mason, Burma & Its People, 573, 731; Pharm. Ind., 292; Fleming, Med. Pl. & Drugs (Asiatic Reser., XI.), 193; Ainslie, Mat. Ind., I., 411, 412, 635; O'Shaughnessy, Beng. Dispens., 61; Irvine, Mat. Med. Patna, 32; Medical Topog., Ajmir, 131; Moodeen Sheriff, Supp. Pharm. Ind., 235; U. C. Dutt, Mat. Med. Hindus, 26; Sakharam Arjun, Cat. Bomb. Drugs, 159; Trans. Med. & Phys. Soc. Bombay (New Series), XII., 174; Manual Geology of India, Pt. IV. (Mineralogy), 7; Baden Powell, Pb. Pr., 18-28, 57, 96, 109; Atkinson, Him. Dist. (X., N.-W.-P. Gaz.), 293; Royle, Prod. Res., 383; Davies, Trade and Resources, N.-W. Boundary, India, xl., 69; Buchanan-Hamilton, Account of the Kingdom of Nepal, 78; Man. Madras Adm., II., 33; Bombay, Adm. Rep. (1872-73), 366; Settlement Report, Panjáb, Kohat, 32; Gazetteers:—Panjáb, Rawal Pindi, 11; Gurgaon, 14-16; Indian Agri., Feb. 20, 1886; Encyclop. Brit., XXII., 634; Balfour, Cyclop. Ind., III.*

SOURCES.

SOURCES 3000

The following account of the sources of Sulphur in India is an abstract of that given by Ball in the *Manual of the Geology of India:*—

Bengal.—Barren Island, a volcano in the Bay of Bengal, is the only source known to exist in this province, and the quantity of Sulphur there

Bengal. 3001

SOURCES. Bengal.

is so limited that it is not likely ever to become an article of commerce. Both in and near the crater of the central one, and at a point 250 feet lower down, where a recent lava stream has broken out, crusts varying from 2 or 3 to 6 or 8 inches or even a foot in thickness were found, but the total amount was estimated not to exceed a few dozen tons, and there is reason to believe that the deposition of the mineral has taken place very slowly during the last quarter of a century at least.

Madras. 3002

Madras.—In the Godaveri district between the two mouths of the river of that name, **Dr. Heyne** examined a deposit of Sulphur, the results apparently of the decomposition of sulphate of lime. The Sulphur was collected on the dried-up margin of a tidal creek, but since no information as to whether it was ever obtained in quantities sufficient to render it of commercial importance, it probably was not so. A sulphurous earth from the same district, and perhaps the same substance as described by **Dr. Heyne**, was quite recently examined by **Dr. McNally**, Chemical Examiner, who reported that the earth contained 28·32 per cent. of a free sulphur which was easily extracted. The commercial importance of the discovery must, however, depend on the extent of the deposit and the local supply of fuel, and on these points no information is available.

Bombay. 3003

Bombay.—In 1843, a deposit of sulphur was discovered in the Ghizri creek near Karachi, which, on analysis, was found to contain 60 per cent. of the mineral, but was only of trifling extent and therefore valueless commercially.

Panjab. 3004

Panjab.—In the Kohat and Bannu districts Sulphur occurs in several localities, and was worked during the sovereignty of the Sikhs. The sulphur used to be extracted by roasting the loose earth, and in the Gunjully Hills near Kohat it is reported that as much as 1,000 tons used to be annually manufactured. The deposition of sulphur is evidently still in progress as the whole place emit sulphurous fumes, and native sulphur occurs on the sides of small cracks in the shales. Although on the spots where the former workings are situated, the best part of the deposit may have been exhausted, it is probable that it extends beneath the neighbouring debris.

Kashmir. 3005

Kashmir.—Sulphur mines belonging to the Maharaja of Kashmír are worked in a valley in Rupshu at an elevation of 14,500 feet. They were visited by **General Cunningham** and **Mr. Mallet**, who describe them as vertical holes, about 8 feet deep, from the bottom of which the rock (a quartz schist in the lamina of which the sulphur occurs) is excavated laterally, and the mines are then deserted for new sides. The outturn is said to be from 500 to 600 maunds per annum, but the process of manufacture is rude and wasteful in the extreme.

N.-W. Provinces. 3006

North-West Provinces.—Unimportant deposits of sulphur have been described in the Kumaon and Garhwál districts.

Nepal.—Sulphur is known to exist in Nepál and was at one time worked, but are said to have been now abandoned, as owing to want of skill it could not be done with profit.

Upper Burma. 3007

Upper Burma.—Sulphur is said to exist in unlimited quantity in the Shan States. The ore is called "a hard metallatic pyrites, but not iron pyrites." The method of extraction is said to be by means of a simple retort consisting of two *gharas*, one of which is placed on the fire and the other inverted over it so as to catch the fumes. A total of 28,000 *viss* of sulphur is the annual outturn of these workings.

Baluchistan. 3008

Baluchistan and Afghanistan.—Sulphur occurs in several localities in these countries, and in considerable quantities. It is worked in a rude fashion, but a considerable amount of sulphur is said to be obtained. The mines at Sunni in Cutchi and in Gurmsael and Balkh used to supply Kandahar with the Sulphur required for the manufacture of gunpowder, but no information seems available as to whether this is still the case.

MEDICINE. 3009

Medicine.—Four varieties of Sulphur are mentioned by Sanskrit authors, *viz.*, red, yellow, white, and black. The yellow variety is called *ámlá-sar*, because of its resemblance to the translucent ripe fruits of the *ámalaki* (**Phyllanthus Emblia**). For internal use this is preferred, and isgiven in combination with mercury. The white variety is said to be inferior to the yellow, and is employed as an external application in skin diseases. In Hindu medicine, Sulphur, purified by being dissolved in an iron ladle smeared with butter, and then washed in milk, is given internally to increase the bile, act as a laxative and purgative and as an alterative in skin diseases, rheumatism, consumption, and enlarged spleen. Externally it is used in various combinations as a remedy in most skin diseases. Thus :—Rubbed up with an impure carbonate of potash and mustard oil it is applied in pityriasis and psoriasis, and an oil composed of equal proportions of sesamum oil, madder, myrabolans, lac, turmeric, orpiment, realgar and sulphur mixed and exposed in the sun is said to be a specific in eczema and scabies (*U. C. Dutt*). Hindu and Muhammadan physicians employ it extensively as an application for itch and other cutaneous disorders, and direct it to be taken in cases of leprosy, venereal diseases, and chronic contractions of the limbs (*Ainslie*).

Sulphur was a therapeutic agent well known to, and much used by, the ancients. It was spoken of by **Hippocrates** under the name of Θειον and was prescribed by him and his followers in asthma and cutaneous complaints. **Celsus** says of it "It matures and removes pus, cleans wounds, and purifies the body." In modern works on therapeutics, Sulphur is described as alterative and diaphoretic in small, and mildly aperient in larger, doses. It is valuable in scabies and some other cutaneous diseases both as an external and an internal remedy. By itself it has probably no local action, but is partially converted by the acids of the skin into sulphuretted hydrogen and sulphides which are energetic substances. Its internal action in all probability is similar; it is not the Sulphur but its hydrogen compound which possesses the therapeutical properties.

DOMESTIC. 3010

Domestic.—By far the greatest proportion of the sulphur obtained in India is used in the manufacture of gunpowder. Smaller quantities are employed to make sulphuric acid.

TRADE. 3011

Trade.—In the "Statement showing the quantity of minerals produced in each British Province and Native State during the calendar year 1889," published by the Revenue and Agricultural Department, 29 cwt. of sulphur valued at R200, are shown to have been obtained in the Panjáb. Other provinces are not shown to have produced the mineral.

The import trade in Sulphur is pretty considerable, 28,167 cwt. valued at R1,47,426, having been brought into the country during the official year 1889-90. The great proportion of this (15,964 cwt.) was brought from the United Kingdom, 7,057 cwt. came from France, 4,396 cwt. from Italy, and 750 cwt. from Abyssinia. Bombay took the largest share in the import trade, *viz.*, (13,398 cwt.) 11,586 cwt. was taken by Bengal, 2,513 cwt. by Madras, and smaller quantities by Burma and Sind. During the same period 99 cwt. of sulphur was re-exported from India to foreign countries. The fluctuations both in the import and export trades are insignificant.

Sumbul, see **Ferula Sumbul,** *Hook. f.*; UMBELLIFERÆ; Vol. III., 339.

Sumach, see **Cæsalpinia coriaria,** *Willd.*; LEGUMINOSÆ; Vol. II., 6; and [**Rhus coriaria,** *Linn.*: ANACARDIACEÆ; Vol. VI., Part I., 496.

SWERTIA Chirata.

The Chireta - a bitter, alterative medicine.

SWERTIA, *Linn.; Gen. Pl., II., 816.*

3012

Swertia affinis, *Clarke; Fl. Br. Ind., IV., 126;* GENTIANACEÆ.

Syn.—OPHELIA ELEGANS, *Wight, Ic., t. 1331;* O. AFFINIS, *Arn. in Wight, Ill., II., 175, t. 157, bis. fig. 3b.*

Reference.—*Year-Book Pharm., 1875, 228.*

Habitat.—An annual herb of the Deccan Peninsula, occurring at altitudes between 2,000 and 4,000 feet.

MEDICINE. 3013

Medicine.—An infusion of this plant is employed locally as a substitute for the true *chiretta* (**Swertia Chirata,** *q.v.*).

3014

S. alata, *Royle; Fl. Br. Ind., IV., 125.*

Syn.—OPHELIA ALATA, *Griseb.;* AGATHOTES ALATA, *Don.*

Vern.—*Búi,* KASHMIR; *Chiretta, kasb ul zarira, hátmúl, harúntútia,* [PB.

References.—*DC., Prodr., IX., 127; Stewart, Pb. Pl., 147.*

Habitat.—An annual herb, closely resembling **S. Chirata,** found on the Temperate Western Himálaya, from Kashmír to Kumáon, at altitudes between 4,000 and 6,000 feet.

MEDICINE. 3015

Medicine.—An infusion of this plant also is largely used by the Natives of the Panjáb, and is believed to have tonic and febrifuge properties.

3016

S. angustifolia, *Ham.; Fl. Br. Ind., IV., 125.*

Syn.—OPHELIA ANGUSTIFOLIA, *D. Don.*

Vern.—*Pahari kiretta,* HIND.; *Pahádi kiraita,* MAR.

References.—*DC., Prodr., IX., 126; Year-Book Pharm., 1875, 12, 228; O'Shaughnessy, Beng. Dispens., 460; Royle, Ill. Him. Bot., 277.*

Habitat.—An annual herb of the Sub-tropical Himálaya from the Chenab to Bhután, occurring at altitudes between 1,000 and 6,000 feet.

MEDICINE. 3017

Medicine.—The dried plant is used locally as a substitute for true *Chiretta,* and is exported to Europe as such. It is very inferior in its bitter tonic properties to the genuine article.

3018

S. Chirata, *Ham.; Fl. Br. Ind., IV., 124; Bentley & Trimen, Med. Pl., III., t. 183.*

Syn.—GENTIANA CHIRAYITA, and CHIRAYTA, *Roxb.;* G. CHIRATA, *Wall.;* AGATHOTES CHIRATA, *D. Don.;* OPHELIA CHIRATA, *Griseb.*

Vern.—*Charayatah,* HIND.; *Chirétá,* BENG.; *Cherayta,* PATNA; *Cherata,* NEPAL; *Nila-véppa,* MAL (S.P.); *Charayatah,* DECCAN; *Chiráita, kiráita,* BOMB.; *Chiráyitá,* MAR.; *Chirayata,* GUZ.; *Shirat-kuch-chi, nila-vémbu,* TAM.; *Níla vém,* TEL.; *Nelabevu,* KAN.; *Sekhági,* BURM.; *Kirata-tikta, bhunimba,* SANS.; *Qasabuzzarirah,* ARAB. & PERS.

References.—*DC., Prodr., IX., 127; Roxb., Fl. Ind., Ed. C.B.C., 264; Pharm. Ind., 57; British Pharm., 106, 206, 413; Flück. & Hanb., Pharmacog., 436; U. S. Dispens., 15th Ed., 405; Fleming, Med. Pl. & Drugs (Asiatic Reser., XI.), 167, 446, 536, 540; Ainslie, Mat. Ind., II., 373; O'Shaughnessy, Beng. Dispens., 459; Irvine, Mat. Med. Patna, 19; Medical Topog. Ajmir, 130; Moodeen Sheriff, Supp. Pharm. Ind., 189; U. C. Dutt, Mat. Med. Hind., 200; S. Arjun, Cat. Bomb. Drugs, 90; K. L. De, Indig. Drugs Ind., 8; Dymock, Mat. Med. W. Ind., 2nd Ed., 446; Year-Book Pharm., 1874, 14; 1875, 228; W. W. Hunter, Orissa, II., 159; Buchanan-Hamilton, Kingdom of Nepal, 85; Birdwood, Bomb. Prod., 56; Agri.-Horti. Soc., Ind., Trans., IV., 182; VI., 241; Jour., XII., 347; XIII., 389; Balfour, Cyclop. Ind., III., 25.*

Habitat.—A small, erect, herbaceous plant, 2 to 5 feet in height, met with in the Temperate Himálaya at altitudes between 4,000 and 10,000 feet from Kashmír to Bhután, and in the Khásia mountains between 4,000 and 5,000 feet. It is nowhere gregarious nor so plentiful as other allied species, still curiously enough the true *Chiretta* is always to be met with in the bazárs of Bengal (*C. B. Clarke in Fl. Br. Ind.*).

MEDICINE. Plant. 3019

Medicine.—The drug obtained from the dried PLANT has long been held in high esteem by the Natives of India. It was mentioned by Sanskrit writers under the name of *kirat-tika, i.e.*, the bitter plant of the Kiritas, an outcast race of mountaineers in the north of India, and was described by them as possessing tonic, febrifuge, and laxative properties (*Flückiger & Hanbury*). By Hindu medical writers it is much esteemed on account of its tonic, anthelmintic, and febrifuge properties, and it is prescribed in fevers of all sorts in a variety of forms, and in combination with other medicines of its class. An oil of *chiretta*, prepared from the decoction, combined with mustard oil, whey, and a number of other drugs, is used by Hindu physicians for rubbing on the body in chronic fever with emaciation and anæmia (*U. C. Dutt*). Muhammadan writers upon Indian drugs have identified *chirayata* with the *kasab-el-daríra* of the Arabs, but this is now doubted. *Chiretta* is used by Muhammadan practitioners, however, as a remedy for colds and bilious affections, burning of the body, and the fever arising from derangement of the three humours. In Western India it enjoys a high repute as a remedy for bronchial asthma, and Dymock remarks that he has known it used with success (*Mat. Met. W. Ind.*).

In England *chiretta* began to attract attention about the year 1829, and in 1839 it was introduced into the Edinburgh Pharmacopæia. It is now officinal in the British Pharmacopæia, and is administered usually in the form of infusion or tincture. It is not, however, used extensively in England, and very little on the Continent or in America. By European practitioners in India it is very frequently prescribed in place of gentian, and was described by Fleming as "possessing all the stomachic, tonic, febrifuge, and anti-diarrhœtic virtues which are ascribed to that drug and in a greater degree than they are generally found in it in the state in which it comes to us from Europe." It is a pure bitter tonic without aroma or astringency, and is even more bitter than gentian. It is administered for cases similar to those in which that drug is used, and is said to be specially serviceable in the dyspepsia of gouty subjects. Like gentian it is sometimes employed to impart flavour to the compositions known as *Cattle Foods* (*Bentley & Trimen*).

Description. 3020

DESCRIPTION.—The entire plant collected, most commonly, when the capsules are fully formed and dried, is tied up into flattish bundles with a slip of bamboo. Each of these bundles is about 3 feet long, and from 1½ to 2℔ in weight. The stems have an orange brown or purplish colour, and an average thickness of that of a goose quill. They are rounded below, faintly quadrangular above, and consist in their lower portion of a large woody column coated with a very thin rind and enclosing a large pith. With the exception of the wood of the stronger stems, the whole plant is intensely bitter (*Flückiger & Hanbury*).

Chemistry. 3021

CHEMICAL COMPOSITION.—The analysis made by M. Hohn of Jena, at the request of the authors of the *Pharmacographia*, show that *chiretta* contains two bitter principles—ophelic acid and chiratin, and also a tasteless yellow crystalline substance, which was obtained in too small a quantity for investigation.

SPECIAL OPINIONS.—§ "Among the Natives *chiretta*, in combination with *gulancha* in equal parts, is used as a good alterative. The ordinary process is to macerate in cold water for a whole night, and, after straining, the liquid is taken in the morning" (*Assistant Surgeon N. R. Banerji, Etawah*). "Tonic, stomachic. Dose of infusion ℥ii. Its infusion is used in cases of weakness after fever and atonic dyspepsia, and

MEDICINE.

in urticaria with pulv. rhei and soda bicarb." (*3rd Class Hospital Assistant Abdulla, Civil Dispensary, Jubbulpur*). "Largely used in charitable hospitals. It is a most useful medicine, and easily obtained at a small cost. I agree with those who are of opinion that boiling injures the strength of the drug. I always recommend an infusion in cold water" (*Civil Surgeon S. M. Shircore, Moorshedabad*). "Is tonic and febrifuge. It is also used in atonic dyspepsia to give tone to the stomach and thereby improve digestion" (*Assistant Surgeon R. C. Gupta, Bankipore*). "An extremely useful bitter tonic, much employed in dispensary practice in the shape of infusion either alone or combined with other tonics" (*Surgeon G. Price, Shahabad*). "Have used the Bhután and Nepál *chirata* for years as an emollient, tonic, febrifuge, and laxative" (*Surgeon D. Picachy, Purneah*). "A bitter tonic and thought to be cholagogue by some. A grain or two of quinine and a few drops of diluted sulphuric, added to each dose of the infusion, increase its efficacy" (*Civil Surgeon C. M. Russell, Sarun*). "Valuable bitter tonic, cold infusion preferable". (*Assistant Surgeon S. C. Bhuttacharji, Chanda, Central Provinces*).

Trade. 3022

Trade.—Most of the *Chiretta* of commerce is said to be collected in Nepál. It is packed in large bales, each of which contains about 1 cwt., and exported to Calcutta, where it is distributed.

3023

Swertia decussata, *Nimmo; Fl. Br. Ind., IV., 127.*

Syn.—OPHELIA DENSIFOLIA, *Griseb.*; O. ALBA, *Arn. in Wight, Ill., t. 157, bis. fig, 3 f.*; O. MULTIFLORA, *Dalz.*

Vern.—*Silájit*, DEC.; *Kadú*, MAHABLESHWAR.

References.—*DC., Prodr., IX., 125; Graham, Cat. Bomb. Pl., 249; Dalz. & Gibs., Bomb. Fl., 156; Dymock, Mat. Med. W. Ind., 540; Trans. Med. & Phy. Soc. Bomb. (N. S.), VI., 1860, App., 58, 59.*

Habitat.—A herbaceous plant, found in the Western Deccan Peninsula from the Konkan to Travancore. It is met with at altitudes between 3,000 and 6,000 feet.

MEDICINE. Plant. 3024

Medicine.—The whole PLANT is bitter, but the ROOT is preferred, and is said by Dr. Broughton (*Trans. Med. & Phys. Soc. Bomb.*) to be an excellent substitute for gentian or *chiretta*. It is sold in the bazár at Mahableshwar under the name of *Kadu*, which simply means "bitter" (*Dymock*).

3025

S. paniculata, *Wall.; Fl. Br. Ind., IV., 122.*

Syn.—OPHELIA PANICULATA, *D. Don.*; O. WALLICHII, *G. Don.*

Vern.—*Kadavi*, MAR.

References.—*DC., Prodr., IX., 124; Gazetteer, Panjáb, Simla, Dist., 13.*

Habitat.—A herbaceous plant of the Temperate Western Himálaya from Kashmir to Nepál, found at altitudes between 5,000 and 8,000 feet.

MEDICINE. 3026

Medicine.—Collected as a substitute for the true *chiretta* and exported to the plains.

3027

S. purpurascens, *Wall.; Fl. Br. Ind., IV., 121.*

Syn.—OPHELIA PURPURASCENS, *D. Don.*; O. DALHOUSIANA, *Griseb.*; O. CILIATA, *G. Don.*

Vern.—*Cheretta*, HIND.

References.—*DC., Prodr., IX., 123, 124; Gazetteers:—Simla Dist., 13; N.-W. P., Him. Dist., 313.*

Habitat.—A herbaceous plant of the Temperate North-Western Himálaya from Kashmír to Kumáon; found at altitudes between 5,000 and 12,000 feet.

MEDICINE. 3028

Medicine.—Used in a similar manner to the preceding species.

SWIETENIA, *Linn.; Gen., Pl., I., 338.*

Swietenia febrifuga, *Roxb.;* see **Soymida febrifuga,** *Adr. Juss.,* [MELIACEÆ

S. Mahagoni, *Linn.; Fl. Br. Ind., I., 540.*

3029

THE MAHOGANY TREE.

References.—*Brandis, For. Fl., 70; Gamble, Man. Timb., 74; Stewart, Pb. Pl., 34; O'Shaughnessy, Beng. Dispens., 247; Dymock, Warden, & Hooper, Pharmacog. Ind., I., 548; Birdwood, Bomb. Prod., 260; Useful Pl. Bomb. (XXV., Bomb. Gaz.) 45; Royle, Prod. Res., 198, 218; Gums and Resinous Prod. (P. W. Dept. Rept.), 21; Kew Reports, 1377, 33; 1879, 22; 1880, 19; 1881, 17; 1882, 25; Gazetteer, Mysore & Coorg I., 71; Agri.-Horti. Soc. Ind., Trans., III., 39; Jour. V., Sel., 13-15; Pro., 10, 16; IX., 290; XIII., 299; XIV., Sel., 193-203; New Series, I., Sel. 55; IV., 179-182; Pro., 1873, 6, 21; V., Pro., 20 (1875), 4, 13-15 (1876), 46; VII., Pro., 23; Ind. Forester:—III., 359; IV., 411; VIII., 85; IX., 513, 516; X., 156, iii., 171.*

Habitat.—A large evergreen tree indigenous to the West Indies and Central Africa.

CULTIVATION IN INDIA.

CULTIVATION in Bengal. 3030

Bengal.—In 1795, several Mahogany trees were introduced as seedlings from Jamaica into the Botanic Gardens at Calcutta, and in 1796 Dr. Roxburgh, in a letter to the Sub-Secretary to the Government of Bengal, mentions among other things that "the Mahogany plants sent out by the Court of Directors in 1794-95 thrive very well." In 1799 he again wrote: "They thrive exceedingly and have multiplied to some hundreds. This useful tree is perfectly at home here, and I may venture to say it is also fairly established in India." These plants were the first introduced into India, and until August 1865 there seems to have been no fresh introductions. The trees, however, appear to have continued to flourish, for several, which were destroyed in the great cyclone of 1864, when they were probably about 71 years of age, averaged about 12 feet in girth at 4 feet above the ground, and a log taken from one of them, and consisting merely of the trunk to the commencement of the first branches, gave, after squaring and removal of the sapwood, 169 cubic feet of timber.

In 1865, 183 pods, containing 8,235 seeds, were received from Jamaica by the Superintendent of the Government Botanical Gardens, Calcutta. From these only 460 plants were produced, 338 of which were sent to Darjiling to be planted in the Taraï there, and the remaining 112 were kept in the Botanic Gardens. The plantation in the former locality proved a failure, but at other places in Bengal the trees throve very satisfactorily. The numbers of the original trees increased vastly, for from them a total of 5,548 plants were distributed in the Bengal Presidency between the years 1802—1866, sixty-seven were sent to other presidencies and provinces, four were sent to Europe, and nine to Africa.

Some difficulty in propagating Mahogany was experienced in Bengal and other parts of India, as the trees, although they flowered freely, did not produce seeds, but in 1876, the Vice-President of the Agri.-Horticultural Society of India, distributed 43 capsules grown in the Society's gardens, containing on an average each 40 seeds, to the members of the Society, while trees in other parts of Bengal have seeded,* although not freely; and large numbers of plants have been produced by layering. Fresh introductions, too, of seeds from Jamaica have been made from time to time by the authorities of Kew, so that, altogether, the successful cultivation of Mahogany in Bengal seems not improbable.

* An avenue of large Mahogany trees, near the Hugli railway station, seeded freely in 1878.—*Ed., Dict. Econ. Prod.*

SWIETENIA
Mahogny.

Cultivation of the

CULTIVATION in N.-W. Provinces. 3031

North-West Provinces.—Several trees from Calcutta have been introduced into the Saharunpore Gardens, and in 1878 we find **Mr. J. F. Duthie** reporting:—"The tree appears to thrive well in these gardens. There are some fine large specimens, about 60 years old, one of which measures 8 feet 9 inches in girth at 4 feet above the ground. The tree flowers regularly every year, but as yet no seed has been produced." Further accounts are not given with regard to mahogany cultivation at Saharunpore.

Bombay. 3032

Bombay.—In 1815, four plants, propagated from those introduced into the Calcutta Botanic Gardens in 1795, were sent to the Bombay Presidency, but as to the ultimate fate of these no information, save the passing allusion in **Lisboa's** *Useful Plants*, quoted below, seems to be available, and it is not till 1879 that we find any account of mahogany cultivation in Bombay. In that year **Colonel Peyton** reports:—"A small parcel of seed was received from the Superintendent of the Botanical Gardens. It was not understood how to treat them and only 18 germinated. They were put out at Yellapur and are now splendid specimens varying from 1 foot 8 inches to 4 feet 7½ inches. Subsequently another parcel was received. They are now understood and are germinating famously. They require much watering, good air and light, and a light covering only of earth." In 1881 **Mr. Shuttleworth** reports that in the Northern Division of the Bombay Presidency "the mahogany trees, if carefully looked after, thrive and have a quick growth," and in the Southern division **Colonel Peyton** says:—"The mahogany, of which there are many plants, is attacked by a grub which destroys the terminal shoots and distorts the trees; otherwise the plants are green and appear fairly healthy" (*Kew Reports*). In **Lisboa's** *Useful Plants of Bombay*, published in 1886, a letter is quoted from **Mr. Woodrow,** Superintendent of the Botanical Gardens, Ganishkhind, in which he says:—"I have measured many mahogany trees; I find the average of eight years' growth is 20 feet high and 15 inches circumference at 3 feet from the ground. Much larger trees 40 years old are at Hewra."

Madras. 3033

Madras.—Early in the present century eight of the trees raised from the original stock were sent to the Madras Agri.-Horticultural Society. These were a few years afterwards reported to be thriving well, and some of the specimens now in the Society's garden are over 40 years of age and continue to grow. They have flowered regularly, but have not borne fruit. A few years ago, two small bags of fresh seed were obtained by the Agri.-Horticultural Society of Madras direct from the Government Botanist in Jamaica Out of 1,280 seeds put down, about 500 germinated and 423 young trees were raised, the majority of which were placed with District officers, Forest officers, and others to be planted on Government land.

Burma. 3034

Burma.—In Burma also no information seems to exist as to the fate of the original trees sent by **Roxburgh** from the Calcutta Botanic Gardens, but in 1871 **Dr.** (now) **Sir D. Brandis** informed the authorities at Kew that the mahogany trees which had been introduced, at what period he does not state, throve well and produced good marketable furniture wood. In the Kew Report for 1879, a letter is quoted from **Lieutenant-Colonel Hawkes,** Honorary Superintendent of the Government Gardens at Rangoon, from which it appears that fresh experiments in the cultivation of Mahogany had been undertaken. Seeds were sown in September 1878, and were very successful, for when the plants raised from these seeds were measured, exactly a year after they were planted, three of them were found to average 7 feet 3 inches in height with a girth of from 3⅛ to 3⅝ inches and "in addition to this rapid growth they appear to be very hardy, and promise to bear with impunity the extremes of heat and moisture characteristic of the Burmese climate." All the last plants raised from seed had succeeded, a large number had been planted out and they seemed to thrive in the poorest soils where it was difficult to keep other plants alive.

CULTIVATION in Burma.

Of the Tenasserim Circle **Major Seaton** in the same year reports:—"The four mahogany trees planted out in 1875 continue to thrive, the maximum and minimum heights of the trees being 17 feet and 7 feet, respectively." In 1879, **Lieutenant-Colonel Hawkes** again reports that "the mahoganies are still growing well under every variety of soil and disadvantage of situation; some planted in a swamp survived the whole monsoon and do not seem behind others more favourably placed. At least 75 per cent. of the plants were this year attacked by a species of borer which drilled holes near the points of the leading shoots and bored down through the pith. This disfigured the plants somewhat, but they rapidly threw out side shoots, the strongest of which, being allowed to remain, soon took the place of the original leading shoot. In addition to this, a large caterpillar was very destructive, but the great vitality of the plant enables it readily to overcome these insect pests. Every wound thus inflicted causes the plant to throw out a fresh shoot, thus disturbing the parallelism of the fibres and producing the mottled appearance of the grain which is, by dealers in cabinet work, called the curl. As the value of the best kinds of wood depends on the amount of this curling of the grain, it seems not unlikely that to the liability of the mahogany to the attacks of these insect enemies, the wood owes much of its value as an article of commerce. "Three young mahogany trees measured on the 7th September (when they were exactly two years old) were found to have a height of 16 feet 6 inches, 14 feet 4 inches and 12 feet, with a girth of 7, 5½, and 7 inches respectively."

GUM. 3035

Gum.—"The mahogany tree yields a gum which, at first liquid, soon dries up into brittle, white, shining fragments. These become yellow on keeping. The gum dissolves readily in water, forming a weak dark-coloured mucilage which freely reduces Fehling's solution; is precipitated by acetate of lead and gelatinised by the basic acetate and by ferric cloride but not by borax" (*Dymock, Pharmacog. Ind.*).

MEDICINE. Bark. 3036

Medicine.—In the West Indies, the BARK is sometimes used by the Natives as a substitute for cinchona. It is very astringent, but is said to contain "no alkaline principle" (*O'Shaughnessy*).

TIMBER. 3037

Structure of the Wood.—Heartwood hard, reddish-brown, seasons and works well. Weight about 53℔ per cubic foot (*Gamble*). In the trade several varieties of wood are distinguished. Of these the dark dense timber known as Spanish mahogany chiefly imported from San Domingo, Central Mexico, and Cuba, is regarded as of best quality; while the light open grained kind termed "Bay wood" or Honduras, obtained from regions further south, although in its best forms approaching the quality of the former, is usually more nearly of the texture of cedar and is now much used for joiners' work. Indian mahogany has been found equal in quality to some of the best varieties of "Honduras" although inferior in fineness of grain and curl to the best qualities of "Spanish" wood (*Royle*).

DOMESTIC 3038

Domestic.—In Europe the wood is used perhaps more extensively than any other for furniture; it is also sometimes employed in ship-building. **Mr. May** in an interesting paper on "Timber for ship-building" read before the Society of Arts, states, however, that mahogany is only classed in the second rate at Lloyds among timber for ship-building purposes, the reason being that fine dense mahogany is too costly for the purpose, while the cheaper bay wood, the price of which would enable it to be used, comes only in the second division. The timber of a Spanish line of battle ship built at Havannah of the finest picked wood, when captured by the

English, and broken up more than 100 years after she had been launched, was found to have every timber sound (*Ind. Forester*).

TRADE. 3039

Trade.—About 40,000 tons are annually imported into Great Britain from Honduras, Jamaica, and San Domingo, and in the London market mahogany fetches from 4*d.* to 1*s.* 6*d.* per superficial foot of planking one inch thick, so that if Bengal and Burma are to prove, as would seem likely, favourable fields for the growth of this tree, another most valuable timber may soon be added to the exports from these provinces.

SWINTONIA, *Griff.*; *Gen. Pl., I.*, 421.

[ANACARDIACEÆ.

3040

Swintonia Schwenkii, *Teysm. & Binnend.*; *Fl. Br. Ind., II.*, 26;

Syn.—ANAUXANOPETALUM SCHWENKII, *Teysm. & Binnend.*; ASTROPETALUM 2, *Griff.*

Vern.—*Boilam, boilsur,* BENG.; *Sambúng, sanginphrú,* MAGH.; *Shibiku,* CHAKMA; *Thayet San,* BURM.

References.—*Kurz, For. Fl. Br. Burm., I.*, 316; *Gamble, Man. Timb.*, [104.

Habitat.—A large tree, very frequent in the tropical forest of Martaban down to Tenasserim, and distributed to Malacca and Sumatra.

DOMESTIC. Wood. 3041

Domestic.—The WOOD is sometimes used for boats and is said by Major Lewin to last better in salt water than other woods.

3042

SYMPHOREMA, *Roxb.*; *Gen. Pl., II.*, 1159.

[*Wight, Ic., t.* 362; VERBENACEÆ.

Symphorema involucratum, *Roxb.*; *Fl. Br. Ind., IV.*, 569;

Syn.—CONGEA PANICULATA, *Wall.*

Vern.—*Surúdú, konda tekkali, gubba dára,* TEL.; *Nway-sat,* BURM.

References.—*Roxb., Fl. Ind., Ed. C.B.C.*, 326; *Kurz, For. Fl. Br. Burm., II.*, 254; *Gamble, Man. Timb.*, 282; *Dalz. & Gibs., Bomb. Fl.*, 199; *Sir W. Elliot, Fl. Andhr.*, 63, 97; *Gazetteer, Bombay, XV.*, 440; *Ind. Forester, VIII.*, 411.

Habitat.—A large deciduous scandent shrub, frequent in the Western Deccan Peninsula from the Konkan southwards; also in Burma and Ceylon.

TIMBER. 3043

Structure of the Wood.—Grey, close-grained, rather heavy. Used for fuel.

SYMPHYTUM, *Linn.*; *Gen. Pl., II.*, 854.

3044

Symphytum asperrimum, *Sims.*; *DC., Prod., X.*, 38; BORAGINEÆ.

THE PRICKLY COMFREY.

References.—*Report of Experimental Farms, Madras* (1876), 21; (1878) 24.

Habitat.—A native of the Caucasus; introduced into England as an ornamental plant.

FODDER. 3045

Fodder.—In England it is cultivated as a fodder plant and is said to be particularly useful for dairy-cows. It was recommended for experiment to the Agricultural Department of India, but trial showed that it "had little if anything to recommend it for cultivation on the plains of Southern India," although it was suggested that on the Nilghiris and other hill tracts it might be expected to produce good results. Further trials however, do not appear to have been made.

SYMPLOCOS, *Linn.; Gen. Pl., II., 668.*

[STYRACEÆ.

Symplocos cratægoides, *Hamilton; Fl. Br. Ind., III., 573;* 3046

Syn.—S. PANICULATA, *Wall.*; LHODRA CRATÆGOIDES, *Dcne.*

Vern.—*Lodh,* KUMAON; *Lodar, lú, laudar, loj, losh,* (Bazár, bark=) *lodh patháni, loja,* PB.; *Lodur, pathani lodh,* SIND.

References.—*A. DC., Prodr., VIII., 258; Brandis, For. Fl., 299; Kurz, For. Fl. Burm., II., 147; Gamble, Man. Timb., 253; Stewart, Pb. Pl., 138; Murray, Pl. & Drugs, Sind., 149; Baden Powell, Pb. Pr., 359, 598; Atkinson, Him. Dist. (X., N.-W. P. Gaz.), 313, 598, 751, 776; Liotard, Dyes, 88; Settlement Report, Panjáb, Simla Dist., xliv., App. II., A; Gazetteer, Panjáb, Simla Dist., 12; Agri.-Horti. Soc. Ind., Jour., XIV., 43; Balfour, Cyclop. Ind., III., 794.*

Habitat.—A large shrub or small tree of the Himálaya, from the Indus to Assam, occurring at altitudes between 3,000 and 8,000 feet. It is found also on the Khásia mountains and the hills of Martaban, and is distributed to Japan.

DYE. Bark. 3047 Leaves. 3048

Dye.—The BARK and LEAVES are used in dyeing and yield a yellow colour. They are principally employed in combination with madder, and are said by Liotard to act probably more as a mordant than as a colouring material. The average annual export from the tract between the Ganges and the Sárda is stated by **Atkinson** to be about twenty tons, of which about nine tons come from the Kumáon forest division.

OIL. 3049 Seeds. 3050

Oil.—An oil is said to be extracted from the SEEDS (*Stewart*).

MEDICINE. Bark. 3051

Medicine.—The BARK is considered by Hindu practitioners to have tonic properties. It is also used in ophthalmia.

FODDER. Leaves. 3052

Fodder.—The LEAVES are lopped in the Himálaya to feed sheep and goats.

TIMBER. 3053

Structure of the Wood.—White, hard, close-grained, splits and twists in seasoning. It is, however, durable, and has been recommended for turning. If properly seasoned, it would do for carving. Weight 45 to 54℔ per cubic foot (*Gamble*).

S. ferruginea, *Roxb.; Fl. Br. Ind., III., 574.* 3054

Syn.—S. MOLLIS, *Wall.*; S. JAVANICA, *Kurz.*

Vern.—*Fúlinasur,* GARO.

References.—*Roxb., Fl. Ind., Ed. C.B.C., 416; Kurz, For. Fl. Burm., II., 145; DC., Prodr., VIII., 257.*

Habitat.—A small tree of the Khásia and Mikir hills and the Malay Peninsula from Mergui to Malacca. It is distributed to the Malay Archipelago.

DOMESTIC. 3055

Domestic.—In the Garo country the WOOD is used for house-posts.

S. grandiflora, *Wall.; Fl. Br. Ind., III., 578.* 3056

Vern.—*Bumroti,* ASS.; *Moitsúm,* PHAKIAL.

References.—*DC., Prodr., VIII., 257; Report on Dyes of Assam; Liotard Mem. on Indian Dyes, 75, 95, 125; Ind. Forester, V., 212; VIII., 405.*

Habitat.—A tree of the Khásia mountains.

DYE. Leaves. 3057

Dye.—In Assam the LEAVES are used as a mordant in dyeing with **Garcinia Xanthochymus** (*see* Vol. III., 478), and with **Bixa Orellana** (Vol. I, 457).

DOMESTIC. Leaves. 3058

Domestic.—The *múnga* or *múga* silkworm is sometimes fed on the LEAVES (*Ind. Forester*).

S. phyllocalyx, *Clarke; Fl. Br. Ind., III., 575.* 3059

Vern.—*Chandan, lal-chandan,* HIND., & BENG.

References.—*Gamble, Darjiling List, 54; McCann, Dyes & Tans, Bengal, 89; Hooker, Himálayan Journals, II., 41.*

Habitat.—A small tree or shrub, almost wholly glabrous, which occurs frequently in Sikkim and Bhutan, at altitudes between 8,000 and 12,000 feet.

DYE. Root. 3060 Leaves. 3061 3062

Dye.—The wood is grey with streaks of red. "This red part which is darkest in the ROOT is ground into a paste by Paharias and used in their religious ceremonies and for caste marks" (*Gamble*). **Sir J. D. Hooker** in his *Himálayan Journals*, Vol. II., 41, describes the women in Sikkim as preparing a yellow dye from the LEAVES of a **Symplocos** for export to Tibet.

Symplocos racemosa, *Roxb.; Fl. Br. Ind., III., 576.*

THE LODE or LODH TREE.

Syn.—S. HAMILTONIANA & RIGIDA, *Wall.*; S. NERVOSA, *A. DC.*; S. PROPINQUA, *Hance.*

Vern.—*Lodh*, HIND.; *Khoidai, kaidai, singen, súngen*, DARJILING; *Lodh* BENG.; *Lodh, ludam*, KOL.; *Lodam*, SANTAL; *Kaviang, bhom roti*, ASSAM; *Lapongdong*, KHASIA; *Singyan*, BHUTIA; *Chamlani*, NEPAL, *Palyok*, LEPCHA; *Kaiday*, MICHI; *Ludhu, nidhu*, URIYA; *Lodh, tinsah* C. P.; *Lodh*, N.-W. P., OUDH, & KUMAON; *Lodhra, lodh, hura*, BOMB.; *Lodar*, GUZ.; *Ludduga, erra lodduga*, TEL.; *Lodhra, márjana, tillaka*, SANS.

References.—*A. DC., Prodr., VIII., 254; Roxb., Fl. Ind., Ed. C.B.C., 415; Brandis, For. Fl., 300; Kurz, For. Fl. Burm., II., 144; Gamble, Man. Timb., 253, xxv; Cat. Trees, Shrubs, &c., Darjiling, 53; Rev. A. Campbell, Rept. Econ. Pl. Chutia Nagpur, No. 7887; Sir. W. Elliot, Fl. Andhr., 52, 107; Irvine, Mat. Med. Patna, 59; Medical Topog., Ajmir, 144; U. C. Dutt, Mat. Med. Hind., 189, 308; S. Arjun, Cat. Bomb. Drugs, 82; Dymock, Mat. Med. W. Ind., 2nd Ed., 489; Year-Book Pharm., 1879, 163; Darrah, Note on Cotton in Assam, 33; Baden Powell, Pb. Pr., 598; Drury, U. Pl. Ind., 409; Atkinson, Him. Dist. (X., N.-W. P. Gaz.), 313; Useful Pl. Bomb. (XXV., Bomb. Gaz.), 96, 247; Econ. Prod., N.-W. Prov., Pt. III. (Dyes and Tans), 23, 33; Liotard, Dyes, 71, 88, 136, 138; McCann, Dyes & Tans, Beng., 3; 4, 31-36, 54, 55, 74, 86-90, 124, 160; Trotter, Report on Manipur Dyes, 1883; Wardle, Dye Report, 88; For. Admn. Rep. Ch. Nagpore (1885), 32; Gazetteers:—Bombay, XVIII., 46; Mysore & Coorg, II., 7, W. W. Hunter, Orissa, II., 160, App. IV.; Agri.-Horti. Soc., Ind.; Trans., VI., 240; Jour., XIII., 359, 391; New Series, VII., 139, 276; Ind. Forester, VIII., 416; XI., 370.*

Habitat.—An abundant small tree, of the plains and lower hills of Bengal, Assam, and Burma, chiefly met with in dry forests, up to altitudes of 2,500 feet.

DYE & TAN. Bark. 3063 Leaves. 3064

Dyes and Tans.—The BARK and also LEAVES of this species are used in dyeing, and a yellow dye is said to be extracted from both. By itself the *lodh* bark yields a yellow dye, which is obtained by simply steeping it in hot water; but this seems rarely, if ever, employed, and Balasore is the only district in Bengal in which mention is made of *lodh* being thus employed. McCann states that there "the bark is boiled in water for eight hours, and cotton cloths are steeped in this infusion for half an hour. *Lodh* bark appears to be chiefly employed as an auxiliary or mordant in dyeing with *ál* (**Morinda tinctoria**), *bac, bakam* (**Cæsalpinia Sappan**), and *paras* (**Butea frondosa**). The Michis of Darjiling *terai* use it along with turmeric and *gumbengfong* (**Plecospermum spinosum**) in dyeing silk a yellow colour. *Lodh* bark is also an ingredient of the *abir* or red powder used during the *Holi* festival (*McCann*). **Roxburgh** states that this bark is in request among the dyers of red in Calcutta as a mordant in *manjít* dyeing. He describes the process as follows:—"To dye with *manjít* (East India Madder) in which the bark called *lodh* is an ingredient: For three yards of cloth take *lodh, bura hur* (**Terminalia Chebula**, *Roxb.*), of each one chittak, pound and rub them with water on a

DYE & TAN.

stone, mix them up with water and steep the cloth in it; then dry it. Take one chittak of alum, dissolve it in water, and boil it, put the cloth into this solution, and let it boil for an hour, then wash it and dry it. Then take *ál, viz.*, **Morinda tinctoria,** *Roxb.*, one chittak, *dhawra* flowers (**Woodfordia floribunda**) one chittak, *munjit* (**Rubia Munjeet,** *Roxb.*) half a seer, separately, mix them with lukewarm water, and let it boil. Then put in the cloth and let it remain boiling for forty minutes." In the North-West Provinces, Sir E. C. Buck (*Dyes and Tans of the North-West Provinces*) reports that—" The *lodh* tree in these provinces is confined to the forests of Bijnor, Kumáon and Garhwal districts. The bark is imported in some quantity from Calcutta, and is used in calico-printing and dyeing leather as an auxiliary to other dyes, being usually pounded up and mixed with them."

In the Central Provinces the bark is used locally as a dye-stuff, but there does not appear to be any distinct trade in it. In these provinces it is also regarded as one of the most valuable tans (*Liotard*). In the Bombay Presidency, Colonel Drury says that the bark is used locally in the Kotah district, where the tree occurs, to dye a red colour, and that it is exported for that purpose.

With reference to its use as a tan Professor Hummell reported, in a communication addressed to the Editor, that it contained no traces whatever of tannic acid, and that he was, therefore, at a loss to account for its reputed value for tanning purposes.

MEDICINE. Bark. 3065 Decoction. 3066 Wood. 3067

Medicine.—*Lodh* BARK is largely used in Hindu medicine. It is "considered cooling, astringent, and useful in bowel complaints, eye diseases, and ulcers. A decoction of the WOOD is used as a gargle for giving firmness to spongy and bleeding gums" (*U. C. Dutt*). In Bombay the bark is often employed in the preparation of plasters, and is supposed to promote the maturation and resolution of stagnant tumours. In Europe it was formerly looked upon as a cinchona bark and has been known at various times as 'Ecorce de latour,' 'China nova,' 'China calafornica,' 'China Brasilensis,' and 'China Paragua tan.' Drs. Charles and Kanai Lal De recommend the bark in 20 grain doses, mixed with sugar, as a medicinal agent in menorrhagia, due to relaxation of the uterine tissue; it should be given two or three times a day for three or four days. It is considered that the drug has a specific action on relaxed mucous membranes (*Dymock*).

CHEMISTRY. 3068

CHEMICAL COMPOSITION.—According to the account in the *Materia Medica of Western India*, Dr. Hesse obtained from this bark three alkaloids which he named 'Loturine,' 'Coiloturine, and 'Loturidine.' " Loturine is present in largest quantity, it is crystalline, and forms crystalline salts. Colloturine also is crystalline, while Loturidine is amorphous. All three alkaloids in dilute acid solutions show an intense blue-violet fluorescence.

TIMBER. 3069

Structure of the Wood.—White, hard, close-grained, durable when properly seasoned, but very apt to warp and split in seasoning. Weight 54lb per cubic foot. It is sometimes used for making furniture.

Symplocos ramosissima, *Wall.; Fl. Br. Ind., III., 577.*

3070

Vern.—*Kala kharani, silingi,* NEPAL; *Tungchong,* LEPCHA.

References.—*Brandis, For. Fl., 299; Gamble, Man. Timb., 254; Gazetteer, N.-W. P., X., 313.*

Habitat.—A shrub or small tree of the Temperate Himálaya, from Garhwál to Bhután, occurring at altitudes between 4,000 and 8,000 feet. It is found also in the Khásia hills.

TIMBER. 3071

Structure of the Wood.— White, soft, even-grained. Weight 37lb per cubic foot.

DOMESTIC Leaves 3072

Domestic.—" In Sikkim, the yellow silkworm is raised on its LEAVES" (*Brandis*).

3073

Symplocos spicata, *Roxb.; Fl. Br. Ind., III., 573.*

Syn.—S. POLYCARPA, *Wall.;* S. LOHA, *Don.;* S. RACEMOSA, *Wall. (in part, not of Roxb.).*

Vern.—*Bholia, lodh,* HIND.; *Buri,* BENG.; *Gyong,* LEPCHA; *Bhúmrati,* ASSAMESE; *Moit-súm,* PHAKIAL.

References.—*Roxb., Fl. Ind., Ed. C.B.C., 416; Brandis, For. Fl. Burm., II., 146; Lisboa, U. Pl. Bomb., 96, 286; McCann, Dyes & Tans, Beng., 89, 154; H. Z. Darrah, Note on Cotton in Assam, 80; Gazetteers:—Bomb., XV., 437; N.-W. P., X., 313; Ind. Forester, III., 203; XI., 276; Agri.-Horti. Soc. Ind. Trans., IV., 103; V., 64.*

Habitat.—A small tree, found in North and East India up to altitudes of 4,000 feet. It is common near the base of the hills from Kumaon to Bhután and in Assam and Martaban, and is distributed to China and Japan.

DYE. Leaves. 3074

Dye.—In the Eastern Himáláya the LEAVES are dried, pounded, mixed with the fruit of a plant called *kauda,* which is also pounded, and the article to be dyed is steeped in the infusion; a yellow colour is thus obtained.

DOMESTIC. Seeds. 3075

Domestic.—The fluted SEEDS are strung as beads and hung round the necks of children to avert evil spirits (*Lisboa*).

3076

S. theæfolia, *Ham.; Fl. Br. Ind., III., 575.*

Syn.—S. LUCIDA & RACEMOSA, *Wall.*

Vern.—*Bhauri,* BENG.; *Kharani,* NEPAL; *Cashing,* BHUTAN.

References.—*Brandis, For. Fl., 300; Kurz, For. Fl. Burm., II., 143; Gamble, Man. Timb., 254; Watt, Calc. Exhib. Cat., II., 61; McCann, Dyes & Tans, Beng., 88, 89.*

Habitat.—An erect tree of the Eastern Himálaya, from Nepál to Bhután, occurring at altitudes between 4,000 and 6,000 feet. It is common also in the Khásia hills and in Martaban.

DYE. Leaves. 3077

Dye.—"Dr. McCann gives the vernacular name of *bhauri* to this species, and says its LEAVES are used in Durágepur as an auxiliary in dyeing with **Morinda tinctoria** and lac. The name *bhauri,* however, is very near to *buri*—the Bengali name for **S. spicata,** and it is probable that they both bear the same name; Dr. McCann, however, spells them differently.

TIMBER. 3078

Structure of the Wood.—White, soft. Weight 36lb per cubic foot. It is used for fuel and rough house-posts.

SYRINGA, *Linn.; Gen. Pl., II., 675.*

[*Bot., t. 65, f. 2;* OLEACEÆ.

3079

Syringa Emodi, *Wall.; Fl. Br. Ind., III., 605; Royle, Ill. Him.*

Vern.—*Ghia, tworsing,* KUMAON; *Bán phúnt, ban dákhúr, bánchír, razli-júari, ranikrún, kehimu, lolti, leila, sháfri, shapri duden, chilanghati kármar, chinú, dúdla, rang chul,* PB.

References.—*DC., Prodr., VIII., 283; Brandis, For. Fl., 306; Gamble Man. Timb., 256; Stewart, Pb. Pl., 140; O'Shaughnessy, Beng. Dispens, 435; Atkinson, Him. Dist. (X., N.-W. P. Gaz.), 313; Royle, Ill. Him. Bot., 267, t. 65; Gazetteers, Bombay, V., 285; Ind. Forester, XI., 4; Balfour, Cyclop. Ind., III., 795.*

Habitat.—A large shrub met with in the Sub-alpine Himálaya from Kashmir to Kumáon at altitudes between 9,000 and 12,000 feet.

MEDICINE. Seeds. 3080

Medicine.—The SEEDS are said to be astringent and also to contain a bitter principle (*O'Shaughnessy*).

FODDER. Leaves. 3081

Fodder.—The LEAVES are eaten by goats.

TIMBER. 3082

Structure of the Wood.—Smooth, hard, with a small, dark-coloured heartwood. Weight 59lb per cubic foot.

Syzygium Jambolanum, *DC.;* see **Eugenia Jambolana,** *Lamk.;* MYRTACEÆ; Vol. III., 284.

(*J. Murray.*)

TABASHIR.

1

Tabáshír, see Vol. I., 383-386.

TABERNÆMONTANA, *Linn.; Gen. Pl., II., 706.*

2

A genus of trees or shrubs, which comprises about one hundred and ten species (natives of the Tropics of both Hemispheres), of which fourteen are Indian. Several shrubs which belong to the genus are cultivated as ornamental plants. In addition to the species described below, two are of interest—**T. recurva,** *Roxb.* (*Fl. Br. Ind., III., 648*), *Tau-sa-lap,* BURM., a shrub of Chittagong and Burma, which has handsome white flowers, and **T. crispa,** *Roxb.* (*Fl. Br. Ind., III., 648*), frequently cultivated as a garden shrub.

3

Tabernæmontana coronaria, *Br.; Fl. Br. Ind., III., 646; Wight, Ic., t. 477;* APOCYNACEÆ.

Syn.—T. DIVARICATA, *Bl.;* NERIUM DIVARICATUM, *Linn.;* N. CORONARIUM, *Jacq.;* JASMINUM ZEYLANICUM, *Burm.*

Vern.—*Chándui, taggai, taggar,* HIND.; *Tagar, tagur, tugur, chameli,* BENG.; *Asuru,* NEPAL; *Krim,* LEPCHA; *Tagar,* BOMB.; *Sagar, tagar,* MAR. & GUZ.; *Grandi tagarapu, nandivardhana,* TEL.; *Nágin-kada,* KAN.; *Tagara,* SANS.

References.—*Roxb., Fl. Ind., Ed. C.B.C., 249, 250; Brandis, For. Fl., 322; Kurz, For. Fl. Burm., II., 174; Beddome, Fl. Sylv. Anal. Gen., 159; Dalz. & Gibs., Bomb. Fl., 144; Mason, Burma & Its People, 414, 798; Elliot, Fl. Andhr., 63, 129; Irvine, Mat. Med. Patna, 111; Med. Top. Ajmir, 152; Baden Powell, Pb. Pr., 360; Atkinson, Him. Dist., 313; Lisboa, U. Pl. Bomb., 391; Gazetteers:—N.-W. P., I., 82; IV., lxxiv.; Mysore & Coorg, I., 62; Ind. Forester, XIV., 298; Agri.-Horti. Soc. Ind., Journal (Old Series), X., 17.*

Habitat.—A small, evergreen shrub, with silvery bark and glossy leaves, cultivated in gardens throughout India; native country unknown.

DYE. Pulp. 4

Dye.—The red PULP obtained from the aril, or extra coat of the seed, gives a red colour, which is occasionally used as a dye by the hill-people.

MEDICINE. Wood. 5

Medicine.—The WOOD is employed medicinally as a refrigerant (*Irvine*).

TIMBER. 6

Structure of the Wood.—White, moderately hard, close-grained; weight 47lb per cubic foot. It is hard and tough, but too small to be of much value as a timber.

DOMESTIC. Wood. 7 Flowers. 8

Domestic.—Irvine states that the WOOD is used as incense, and for purposes of perfumery. The sweet-scented FLOWERS are esteemed for decorative purposes.

9

T. dichotoma, *Roxb.; Fl. Br. Ind., III., 645; Wight, Ic., t. 433.*

Syn.—CERBERA DICHOTOMA, *Lodd.;* C. MANGHAS, *Linn.;* TANGHINIA DICHOTOMA, *G. Don.*

Vern.—*Píli karbír, kaner zard,* PB.; *Kát-aralie,* TAM.; *Dirvi-kaduru,* SING.

References.—*Roxb., Fl. Ind., Ed. C.B.C., 248; Gamble, Man. Timb., 262; Thwaites, En. Ceyl. Pl., 192; Ainslie, Mat. Ind., II., 260, 262; O'Shaughnessy, Beng. Dispens., 447; Baden Powell, Pb. Pr., 360; Gazetteer, Mysore & Coorg, I., 62.*

Habitat.—A small tree, common in the Western Gháts and the warmer parts of Ceylon.

OIL. Seed. 10

Oil.—An OIL is said to be prepared from the SEED.

The French and African Marigold.

MEDICINE. Seeds. 11 Leaves. 12 Bark. 13 Sap. 14

Medicine.—Ainslie informs us that the SEEDS are said to be powerfully narcotic and poisonous, producing delirium and other symptoms similar to those caused by **Datura.** They are said by Lindley to be purgative. The LEAVES and BARK act as purgatives and are believed to be used in Java as substitutes for senna; the milky SAP is also described as cathartic.

TACCA, *Forst.; Gen. Pl., III., 741.*

15

Tacca pinnatifida, *Forsk.; Fl. Br. Ind., VI., 287;* TACCACEÆ.

Syn.—T. SITOREA, *Rumph.;* T. PINNATIFOLIA, *Gærtn.*

Vern.—*Dhai,* SANTAL; *Bará-kandá,* DEC.; *Diva, diva kanda,* BOMB.; *Periya karunaik-kizhangu, karachunai,* TAM.; *Pedda-kanda-gadda, kanda, chanda,* TEL.; *Chane-kizhanna,* MALAY.; *Pánkhadé, pembwaú, touk-ta, touta,* BURM.

References.—*Roxb., Fl. Ind., Ed. C.B.C., 297; Voigt, Hort. Sub. Cal., 600; Mason, Burma & Its People, 507, 807; Rev. A. Campbell, Rept. Ec. Pl., Chutia Nagpur, No. 9425; Moodeen Sheriff, Supp. Pharm. Ind., 238; Grah., Cat. Bomb. Pl., 230; Dalz. and Gibs., Bomb. Fl., 276; Dymock, Mat. Med. W. Ind., 2nd Ed., 889; Drury, U. Pl., 411; Lisboa, U. Pl. Bomb., 178; Birdwood, Bomb. Prod., 238; Smith, Ec. Dict., 319; Gazetteer, Bombay, XV., 445; Ind. Forester, I., 261; Agri.-Horti. Soc. Ind., Journals (Old Series), IV., Sel., 24, 25; Pro., 8; X., Pro., 91, 111, 115.*

Habitat.—A native of the hilly tracts of India generally, more especially of Chutia Nagpur, Central India, the Konkan, Malabar, Burma, and the Malay Peninsula.

FOOD. Root. 16

Food.—It has a large, round, tuberous ROOT, which yields a considerable quantity of white nutritious fecula. This substance resembles arrowroot, and is much eaten by the Natives, especially in Travancore, where the plant is cultivated and forms an important article of trade (*Drury*), and in Mergui (*Mason*). It is said to be equal to the best arrowroot. The tubers, dug up after the leaves have died down, are rasped and macerated for four or five days in water, when the fecula separates in the same manner as that of sago. When in the crude and raw state it is intensely bitter and acrid, but these objectionable qualities are removed by frequent washing in cold water. In Travancore it is generally dressed with some agreeable acid, which improves its flavour (*Drury*).

TAGETES, *Linn.; Gen. Pl., II., 411.*

17

Tagetes erecta, *Linn.; Clarke, 142;* COMPOSITÆ.

THE FRENCH MARIGOLD AND (**T. Patula**) THE AFRICAN MARIGOLD.

ÆILLET D' INDE, ROSE INDE, *Fr.*

Vern.—*Genda,* HIND. & BENG.; *Gendu,* URIYA; *Tangla, mentok, genda sadbargi,* PB.; *Makhmal, gul-jáfari,* BOMB.; *Guljháro,* GUZ.; *Rojia chaphúl,* MAR.; *Gulgoto,* KATHIAWAR; *Banti,* TEL.

References.—*Roxb., Fl. Ind., Ed. C. B. C., 604; Stewart, Pb. Pl., 130; Elliot, Fl. Andhr., 23; Dymock, Mat. Med. W. Ind., 2nd Ed., 466; Pharmacog. Ind., II., 321; Baden Powell, Pb. Pr., 358; Atkinson, Him. Dist., 778; Lisboa, U. Pl. Bomb., 247, 390; McCann, Dyes & Tans Beng., 140; Buck, Dyes & Tans, N.-W. P., 29; Liotard, Dyes, App., ii., v.; Wardle, Dye Report, 46; Aplin, Rept. Shan States; Gazetteer:—Mysore & Coorg, I., 62, 82; N.-W. P., I., 82; IV., lxxiii.; Orissa, II., 179; Agri.-Horti. Soc., Ind. Jour. (Old Series), VIII., 242; X., 12.*

Habitat.—African and French Marigolds are quite naturalized in India, and are also extensively cultivated as garden plants.

DYE. Flower. 18

Dye.—The FLOWER yields a yellow dye, which, though little used by dyers, is not unfrequently employed by the poorer classes for colouring their clothes. The shade of yellow known as *gendia* is produced by it; it is also occasionally used in place of *harsinghár* or turmeric in producing the colour *champai* (*Buck*). McCann informs us that it is employed in Lohárdágá to produce a dull green dye. "The flowers are first freed from the calyx, &c., the petals alone being retained. These are dried in the shade. To one seer of dried flowers, four seers of water are added, and the whole boiled till two seers remain; four tolas of alum are then added. The cloth is steeped in this solution and then dried in the shade; the colour resulting is a dull green" (*Dyes and Tans of Bengal*). Wardle reports that the flowers contain a moderate amount of colouring matter, and produce, by the various processes he employed, several shades, all more or less yellow. With unbleached Indian *tasar* he obtained colours from drab to brownish-yellow; with bleached Indian *tasar*, light brownish-yellow; with *corah* silk, yellowish grey; and with wool a deep brownish-yellow. Of the last, he remarks: "This is much the deepest and yellowest colour I have obtained with this sample of dye-stuff."

MEDICINE. Flowers. 19 Juice. 20 Leaves. 21

Medicine.—The FLOWERS are used in diseases of the eyes, and, when given internally, are supposed to purify the blood. "One tola of the JUICE of the petals, heated with an equal quantity of *ghi*, is given daily for three days, as a remedy for bleeding piles" (*Dymock*).

SPECIAL OPINIONS.—§ "The LEAVES, formed into a paste, are used as an application to boils and carbuncles. Beaten up with nitre and indigo, this paste is employed as a remedy for suppression and for retention of urine, when due to spasm. It is also applied externally with common salt and indigo in cases of tympanitis" (*Civil Surgeon J. H. Thornton, B.A., M.B., Monghyr*). "The juice is useful in earache, acting as a sedative" (*Narain Misser, Hoshangabad*). "The flowers, made into ointment, are used for unhealthy ulcers" (*Surgeon Anund Chunder Mukerji, Noakhally*).

DOMESTIC & SACRED. Flowers. 22

Domestic and Sacred.—The FLOWERS are much admired all over India, but perhaps especially so with the hill-people of the Himálaya. They are largely employed for decorative purposes, and are strung into garlands and hang round idols, or in front of shrines, temples, etc. *Rojia*, the name current in Western India, perhaps denotes the introduction of the plant by the Portuguese with whom it appears to represent the *Rosa de ouro* or golden rose, which the Pope usually blesses at mass on a Sunday in Lent (*Dymock*).

TALAUMA, *Juss.; Gen. Pl., I., 18, 955.*

23

A genus of trees or shrubs which belongs to the Natural Order MAGNOLIACEÆ. It comprises some fifteen species, which are distributed through the Tropics of Eastern Asia, Japan, and South America. Of these four are natives of India.

[MAGNOLIACEÆ.

Talauma Hodgsoni, *H.f. & T.; Fl. Br. Ind., I., 40;*

24

Vern.—*Siffú*, LEPCHA; *Laigongron*, MICHI; *Harré*, NEPAL; *Patpatta*, PAHARIA; *Pankakro*, GARO.

Habitat.—An evergreen tree, found in the forests of the Sikkim Himálaya and the Khásia hills from 4,000 to 5,000 feet.

TIMBER 25

Structure of the Wood.—Very soft, even-grained; weighing 21lb per cubic foot (*Gamble, Man. Timb.*, 4).

26

Talauma Rabaniana, *H.f. & T.; Fl. Br. Ind., I., 40.*

Vern.—*Sappa,* ASSAM.

Habitat.—A large tree of the Khásia hills and Burma.

TIMBER. 27

Structure of the Wood.—Sometimes used in Assam for furniture and planking.

Talc, see **Steatite.**

Talipot Palm, see **Corypha umbraculifera,** *Linn.;* Vol. II., 575.

Talispatra, see **Abies Webbiana,** *Lindl.;* Vol. I., 4.

Tallow, see **Oils,** Vol. V., p. 459; see **Wool** (article SHEEP & GOATS), [Vol. VI., Pt. II., 583.

TAMARINDUS, *Linn.; Gen. Pl., I., 581.*

28

Tamarindus indica, *Linn.; Fl. Br. Ind., II., 273;* LEGUMINOSÆ.

THE TAMARIND TREE.

Syn.—T. OCCIDENTALIS, *Gærtn.;* T. OFFICINALIS, *Hook.*

Vern.—*Amli, anbli, imli, teter, amlicá, tamru'lhindi, nuli,* HIND.; *Téntúl, tintiri, tintúri, tintil, tétai, nuli, ambli, ámbli,* BENG.; *Joj, jojo,* KOL; *Jojo,* SANTAL; *Teteli,* ASSAM; *Titri,* NEPAL; *Tentúli, koyam, koyan, asok,* URIYA; *Chicha,* KURKU; *Imlí, chinch,* C. P.; *Sitta, hítta, chíta,* GOND; *Imlí,* N.-W. P.; *Imlí,* OUDH; *Imlí,* (dried fruit=*tamar hindí,* seeds=*imlí-ká-bíj,* leaves=*chínchá,*) PB.; *Amli,* MERWARA; *Gidamrí, amri,* SIND; *Amlí, amli-ká-bót, ambli,* DECCAN; *Chints, amlí, ámbli, chinch, chincha,* BOMB.; *Chincha, amlí, chits, ámbali, chinch, chicha,* MAR.; *Amli, ámblí,* GUZ.; *Púli, puliyam-pasham, púlia, pollium,* TAM.; *Chinta-pandu, chinta, asek,* TEL.; *Karangi, kamal, asam, hunase,* MYSORE; *Hunashé-hannu, hunase,* KAN.; *Puli, balam, polli, puliyam-pasham,* MALAY.; *Magyí, magi,* BURM.; *Siyembela, maha siyambala, siyambula,* SING.; *Tintidi, tintiri, tintili, téntrání, ambia, amliká,* SANS.; *Tamare-hindí, humar, sabárá, umblí, dár-al-sída,* ARAB.; *Anbalah, tamar-i-hindí,* PERS.

References.—*Roxb., Fl. Ind., Ed. C.B.C., 530; Voigt, Hort. Sub. Cal., 247; Brandis, For. Fl., 163; Kurz, For. Fl. Burm., I., 414; Beddome, Fl. Sylv., t. 184; Gamble, Man. Timb., 142; also Cat., Trees, Shrubs, &c., Darjiling, 32; Thwaites, En. Ceyl. Pl., 95; Dalz. & Gibs., Bomb. Fl., 82; Stewart, Pb. Pl., 76; Mason, Burma & Its People, 458; Sir W. Elliot, Fl. Andhr., 41; Sir W. Jones, Treat. Pl. Ind., V., 75, No. 9; Rheede, Hort. Mal., I., 23; Pharm. Ind., 64, 445; Flück. & Hanb., Pharmacog., 224; Ainslie, Mat. Ind., I., 425; II., 327; O'Shaughnessy, Beng. Dispens., 309; Moodeen Sheriff, Supp. Pharm. Ind., 238; U. C. Dutt, Mat. Med. Hind., 157, 321; Murray, Pl. & Drugs, Sind, 134; Bent. & Trim., Med. Pl., t. 92; Dymock, Mat. Med. W. Ind., 2nd Ed., 270, 272, 887; Dymock, Warden & Hooper, Pharmacog. Ind., I., 532; Cat., Baroda Durbar, Col. Ind. Exhib., No. 169; Year-Book Pharm., 1883, 236; Trans. Med. & Phys. Soc., Bomb., New Series, No. I., 1851 52, 115; No. 4, 156; Birdwood, Bomb. Prod., 30, 148, 219, 329; Baden Powell, Pb. Pr., 270, 344, 598; Drury, U. Pl. Ind., 411; Useful Pl. Bomb. (XXV., Bomb. Gaz.), 65, 153, 198, 254, 387; Econ. Prod. N.-W. Prov., Pt. V. (Vegetables, Spices and Fruits), 44, 61; Gums and Resinous Prod. (P. W. Dept. Rept.), 16; Cooke, Oils & Oilseeds, 76; also Gums & Resins, 25; Bidie, Prod. S. Ind., 6, 85, 111; Stocks, Rept. on Sind; Simmonds, Science and Commerce, their Infl. on Manuf., 520; Ayeen Akbary, Gladwin's Trans., I., 86; II., 46; Linschoten, Voyage to East Indies (Ed. Burnell, Tiele & Yule), II., 119, 121; Milburn, Oriental Commerce (1813), II., 276; also 307; (1825), Buchanan-Hamilton, Statistics Dinajpur, 158; Taylor, Topography of Dacca, 61; Bomb., Man.*

Rev. Accts., 102; Man. Madras Adm., II., 123; Gribble, Man. Cuddapah, 199; Boswell, Man. Nellore, 96; Moore, Man. Trichinopoly, 80; Appendix of a Note on the Condition of the People of Assam, vide Agri. File No. 6 of 1888; Aplin, Rep. on the Shan States, 1887-88; Settlement Reports:—Port Blair, 1870-71, 33; Panjáb, Delhi, cclxi.; N.-W. P., Shahjehánpur, ix; Allahabad, 38; C. P., Mandla Dist., 89; Upper Godavery, 39; Chánda, App. VI.; Bhundára, 20; Beláspore, 77; Nimár, 306; Chindwára, 110, 111; Survey and Settlement of the Chellumbrum & Manangoody Taluks of South Arcot, 34; Gazetteers:—Bombay, II., 39; V., 23, 25, 2`5; VI., 12; VII., 37, 41; VIII., 183; XIII., 24; XV., 77; XVII., 18; XVIII., 45; Panjáb, Delhi, 19; Rohták, 14; N.-W. P., I., 80; II., 56; III., 33; IV., lxxi.; Orissa, II., 160, 180; Mysore & Coorg, I., 49, 59; II., 8, 11; III., 25; Agri.-Horti. Soc., Ind.:—Trans., III., 59; VI., 241; Jour., Old Series, VIII., Sel., 176; IX., 419, Sel., 56; XIII., 318; Sel., 59, 61; New Series, II, 235; VII., 148; Tropical Agriculturist, VIII., 1888-89, 72, 144, 216, 288, 360, 432, 504, 648, 794, 868; Ind. Forester:—III., 202, 238; IV., 322; VI., 239, 298, 301; VIII., 301, 401; XII., 188 (xxii.), app. 27; XIII., 120; Spons' Encycl., I., 826, 1028; II., 1415, 1694; Encyclop. Brit., XXIII., 40; Smith, Econ. Dict., 401.

Habitat.—A large, evergreen tree, which grows to a height of 80 feet, with a circumference of 25 feet. It is cultivated throughout India and Burma, as far north as the Jhelum. Is probably indigenous in Africa, possibly also in some parts of South India. It is one of the most beautiful of the common trees of India, and is frequently planted in avenues and topes. In the Central Provinces, Central India, and many parts of Southern India it is found self-sown in waste and forest lands.

Gum.—It yields a dirty black gum of little value (*Cooke*).

GUM. 29

Dye.—An infusion of the LEAVES is said to yield a red dye and to impart a yellow shade to cloth previously dyed with indigo (*Atkinson*). The leaves, FLOWERS, and FRUIT contain a large proportion of acid and are much employed as auxiliaries in dyeing, especially along with safflower. They are believed to act as mordants. *Conf.* with account of Silk (Tasar) dyeing.

DYE. Leaves. 30 Flowers. 31 Fruit. 32

Oil.—An oil of an amber colour, free of smell and sweet to the taste, is prepared from the SEEDS by expression. This oil appears to have been brought to notice for the first time in 1856, when a **Captain Davies** sent a sample to the Agri.-Horticultural Society of India, with the remark that it was, in his opinion, suitable for culinary purposes. The Society's Sub-Committee reported favourably on the oil, and suggested that it might be found useful in the preparation of varnishes and paints, as well as for burning in lamps. A member of the Committee remarked that it was occasionally employed in Bengal for making a varnish to paint idols, and for finishing *kurpa* cloth, but that it was very little appreciated. The Sub-Committee further noticed that the sample submitted to them had an odour of linseed-oil, but **Captain Davies** explained that this was not a property of the oil itself, but was due to the mill in which it had been expressed, having been one ordinarily employed for making linseed-oil (*Agri.-Horti. Soc. Ind., Jour., 1885*). The authors of the *Pharmacographia Indica* have examined it, and write, "**Braunt** states that the seeds contain 20 per cent. of a thickly fluid oil with an odour of linseed, and classes it with the non-drying oils. By expression from the dry seeds, we were unable to obtain any oil, and by solvents the yield was only 3·9 per cent. The oil possessed greater siccative properties than boiled linseed oil." The subject appears to be well worthy of further investigation, the more so from the contradictory nature of the literature regarding it, for the seeds might be obtained in any quantity and cheaply, should the oil prove of commercial value.

OIL. Seeds. 33

Medicine.—The tamarind has been valued from a very remote period in Sanskrit medicine, and, through the Hindus, appears to have become known

MEDICINE.

Medicinal uses of the Tamarind.

MEDICINE.

to the Arabians. Through the writings of the Arabs, it attracted attention in Europe, where it became known during the middle ages. The Arabs and Persians called it *Tamare-hindi* or "Indian Date," thus adopting for its name that of their own best known fruit. In European writings of the middle ages tamarinds are called *'οξυφοίνικα*, and "dactyli acetosi," both evidently translations of the Arabian term, the idea of resemblance to the date being accepted. **Saladinus** and other writers of the period considered the fruit to be obtained from a wild palm growing in India (*Flückiger & Hanbury*). The tree was first accurately described in the sixteenth century by **Garcia D'Orta** and by **Linschoten.** Sanskrit writers consider the FRUIT "refrigerant, digestive, carminative, and laxative, and useful in diseases supposed to be caused by deranged bile, such as burning of the body, costiveness, intoxication from spirituous liquors, or **Datura.** The SHELLS of the ripe fruit are burnt and their ashes used in medicine as an alkaline substance, along with other medicines of the sort. The pulp of the ripe fruit, as well as a poultice of the LEAVES, is recommended to be applied to inflammatory swellings" (*U. C. Dutt*). **Dymock**, quoting from the *Makhzan-el-Adwiya*, writes: "Mahometan physicians consider the pulp to be cardiac, astringent, and aperient, useful for checking bilious vomiting, and for purging the system of bile and adust humours; when used as an aperient it should be given with a very small quantity of fluid. A gargle of Tamarind water is recommended in sore-throat. The SEEDS are said to be astringent; they are used as a poultice to boils. The leaves crushed with water and expressed yield an acid fluid, useful in bilious fever, and scalding of the urine; made into a poultice they are applied to reduce inflammatory swellings and to relieve pain. A poultice of the FLOWERS is used in inflammatory affections of the conjunctiva; the JUICE expressed from them is given internally for bleeding piles. The BARK is considered to have astringent and tonic properties. Natives believe the acid exhalations of the Tamarind tree to be injurious to health, and it is stated that the cloth of tents allowed to remain long under the trees becomes rotten. Plants also are said not to grow under them, but this is not universally the case, as we have often seen fine crops of **Andrographis paniculata** and other shade-loving plants growing under Tamarind trees." **Ainslie** and other writers comment on this belief in the baneful influence of the neighbourhood of the tree. There is no doubt that Natives in all parts of the country are at one in considering its influence unwholesome, especially during the rainy months.

Fruit. 34 Shells. 35 Leaves. 36 Seeds. 37 Flowers. 38 Juice. 39 Bark. 40

During the middle ages the fruit appears to have enjoyed a reputation in Europe for all the virtues ascribed to it by Arabian and Persian writers. **Linschoten** described it as a very good purgative;—"For the poore that are of small habilite and are not able to be at charges of Rhubarbo, Manna, and such like costlie apothecarie's ware, doe only use Tamarinio pressed out into a little water, which water being drunk fasting in a morning, is the best purgative in the world." **Paludanus**, his learned commentator, informs us that "the Turkes and Egyptians use this tamarinde muche in hotte diseases and feavers, they put it into faire water, and so drinke it. I healed myselfe therewith of a pestilent fever being in Siria." He adds that it is a valuable prophylactic against fever and useful in all diseases attended by heat, and that the leaves are vermifuge.

In India at the present day it is employed for all the purposes for which it was considered valuable in ancient times.

The fruit is officinal in modern Pharmacopœias—West Indian tamarinds in England, East Indian in most Continental countries. It is, however, valued only as a laxative and refrigerant for preparing drinks

MEDICINE.

in febrile and inflammatory affections, and as an adjunct to other preparations. In the *Pharmacopœia of India* **Dr. Shortt** is stated to recommend the seed, deprived of its testa, as a valuable remedy in diarrhœa and dysentery. Many of the minor uses of different parts of the tree will be found detailed in the Special Opinions quoted below.

Chemistry. 41

CHEMICAL COMPOSITION.—**Flückiger & Hanbury** write that water extracts sugar, together with acetic, tartaric, and citric acids, combined chiefly with potash, from unsweetened tamarinds. East Indian tamarinds differ from West Indian, in being prepared for the market without the use of sugar or syrup, and contain less citric acid. No peculiar principles, to which the laxative action of the fruit can be attributed, are known. The fruit pulp diffused in water forms a thick tremulous, somewhat glutinous, and turbid liquid, owing to the presence of pectin. The hard seed has a testa which abounds in tannin, and after long boiling is easily separated, leaving the soft cotyledons, which have a bland mucilaginous taste (*Pharma-cographia*).

C. Mueller found from an analysis of nine samples of East Indian tamarinds that the percentage of seeds averaged 13·9. The pulp, free from seeds, contained, on an average, 27 per cent. water, 16·2 insoluble matter, 5·27 potassic bitartrate, 6·63 tartaric acid, and 2·20 citric acid. The dry pulp had 7·20 per cent. potassic bitartrate and 9·09 per cent. tartaric acid. Very small quantities of malic acid, calculated as citric acid in the above figures, were also found (*Pharmacog. Ind.*).

SPECIAL OPINIONS.—§ "Tamarinds are especially useful in preventing or curing scurvy" (*Surgeon-Major C. W. Calthrop, M.D., Morar*). "Commonly used as an antiscorbutic" (*Assistant Surgeon Nehal Sing, Saharunpore*). "I have found 'Tamarind sherbet' (made by boiling tamarind pulp in water and sweetening with sugar) a good laxative for children. It sometimes causes the motions to be very offensive" (*Surgeon-Major H. DeTatham, M.D., M.R.C.P. Lond., Bombay Army, Ahmednagar*). "Sherbet prepared with the pulp of fresh ripe fruit is a refrigerant and laxative in fevers" (*Thomas Ward, Apothecary, Madanapallè, Allahabad*). "The tender leaves and flowers cooked and eaten are said to be cooling and antibilious. A decoction of the leaves used as a wash for indolent ulcers promotes healthy action. The seeds coarsely bruised, soaked in water over night and given to pigs, are fattening. The red outer covering of the seeds is a soothing and mild astringent, very useful in chronic dysentery if administered as below:—To half a drachm of the above, powdered, add one and a half drachm of cumin seeds powder and palmyra sugar-candy to taste, mix and divide into three powders, one for a dose every three or four hours. The ASH of the bark is given internally as a digestive" (*Surgeon-Major D. R. Thomson, M.D., C.I.E., Madras*). "It is useful as an antiscorbutic in place of lime juice, and is a common domestic, cooling laxative, useful in febrile and billiary affections" (*Assistant Surgeon Bhagwan Das, Rawalpindi*). "The juice of the leaves heated by dipping a red hot iron is useful in dysentery. The seed rubbed on a rough stone with water is a specific for Delhi boil" (*Assistant Surgeon T. N. Ghose, Meerut*). "Natives dislike the shade of the tamarind tree, and say it is hot. There is probably some reason in this. As an antiscorbutic in food, tamarind is a most useful adjunct to the dietary of a grain-feeding people" (*Surgeon-General W. R. Cornish, F.R.C.S., C.I.E., Madras*). "In upwards of fifty cases of dysentery of all kinds I have used the ROOT of one-year plants with black pepper ground up into a paste with butter-milk, and made into pills, given three times a day. Average number of days for recovery six; longest twenty days and shortest two days. In one case the patient had been

Ash. 42

Root. 43

 Uses of Tamarind.

MEDICINE.

treated with Ipecacuanha for several days. This preparation is certainly most useful in milder cases" (*Civil Surgeon D. Basu, Faridpur, Bengal*). "Is a mild laxative and refrigerant. It is used in fever attended with costiveness" (*Assistant Surgeon R. C. Gupta, Bankipur*). "The juice of the leaves if mixed with a little sugar may be given in cases of dysentery" (*Surgeon William Wilson, Bogra*). "The pulp of the fruit made into a sherbet with sugar and water is a pleasant laxative. Useful in bilious nausea" (*Civil Surgeon S. M. Shircore, Murshedabad*). "Two tablespoonfuls of the fruit mixed with an equal quantity of dates, boiled with a quart of milk and strained, forms an excellent cooling drink in feverish states and also a good antiscorbutic. The pulp of the fruit mixed with cold water and applied to the shaved head is useful in apoplexy and sunstroke" (*Surgeon-Major G. Y. Hunter, Civil Surgeon, Karáchi*). "Decoction of the leaves is used in acute dysentery" (*Surgeon A. Crombie, Dacca*). "Syrup of tamarind is cooling, refrigerant, and laxative" (*Honorary Surgeon A. E. Morris, Tranquebar*). "Cooling, digestive, and laxative. Older pulps are preferred for medicinal purposes. Prescribed as a drink in certain bilious fevers" (*Assistant Surgeon S. C. Bhattacherji, Chanda*). "Air blowing from tamarind trees is considered as very injurious to health" (*Surgeon A. C. Mukerji, Noakhally*). "Fruit is laxative, the seed ground into a powder is used as an astringent. The leaves made into a decoction are used also as an astringent" (*Surgeon-Major Lionel Beech, Cocanada*). "The pulp (preserved for fifteen or twenty years) is used in the form of sherbet in cases of habitual constipation. The expressed juice of the leaves, a little warmed, is used internally in dysentery. The juice of an old tree taken internally promotes the secretion of milk" (*Civil Surgeon J. H. Thornton, B.A., M.B., Monghyr*).

FOOD. Fruit. 44

Food.—The FRUIT, a large flat pod, from 4 to 6 inches in length, filled with acid pulp, is largely eaten, being a favourite ingredient of curries and *chatnis*. It is also employed in making a pleasant cooling drink or sherbet, with sugar and water. Two kinds are recognised: the small-seeded red Guzerát tamarind, and the common red dish-brown. In India the fruit is not prepared for the market by adding sugar or syrup as in the West Indies. The seeds and epicarp are more or less removed by hand, and the pulp is then generally mixed with about 10 per cent. of salt and trodden into a mass with the naked feet. The best qualities are free from fibre and husk, the worst contain both, as well as a large proportion of seed. Careful house-keepers prepare their own pulp by exposing it for about a week to the sun and dew (*Dymock*).

Seeds. 45

The SEEDS are largely eaten by Natives. After the outer skin is removed by roasting and soaking, they are then boiled or fried, and are said, when thus prepared, to be tolerably palatable. They are also used as a flour after being dried and ground. The SEEDLING or tender plant when about a foot high is eaten as a vegetable. The LEAVES are employed to make curries, especially in times of scarcity, and the FLOWERS, made into a dish called *chingar*, are also eaten.

Leaves. 46 Flowers. 47

The tamarind, nearly every part of which is useful, is necessarily of considerable value. In Government Forests it is farmed annually to the highest bidder, the product constituting an important item of forest revenue. A good tree will give about 350℔ of fruit, which varies considerably in value according to quality. In Bangalore the average selling price is said to be about 4 pie per ℔ (*Liotard*); in Bombay the best quality fetches about R50 per kandy of 7 cwt., while some of the inferior kinds are not worth more than R20 (*Dymock*). Large quantities are sent to the seaport towns from the interior, and are thence shipped to Europe, Persia, and other countries. No exact information can be given as to the quantity annually exported.

Structure of the Wood.—Yellowish white, sometimes with red streaks, hard and close-grained; heartwood small, near the centre of old trees only dark purplish-brown, with an irregular outline and radiating ramifications It is regarded as very durable; weight from 60 to 83lb per cubic foot. This timber is highly prized though extremely difficult to work, and is used chiefly for wheels, mallets, planes, furniture, rice-pounders, oil and sugar mills. It is also excellent for turning purposes, and is one of the woods preferred for making gun-powder charcoal. It is also much prized for fuel when great heat is necessary, as in brick-making. From the liability of the tree to become hollow in the centre, it is extremely difficult to get a tamarind-plank of any width. Partly for this reason, partly because it is liable to the attacks of insects, it is not used for house building.

TIMBER. 48

Domestic and Sacred.—The pulverised SEED, boiled into a paste with thin glue, form one of the strongest of wood-cements. A size made from the seed is used to dress country-made blankets. In Southern India a strong infusion of the FRUIT mixed with sea-salt is used by silversmiths in preparing a mixture for cleaning and brightening silver (*Drury*). In certain localities the tree is considered to be haunted by spirits, and is worshipped in Jámbughoda, on a day called *Ámlí Agiáras* (*Lisboa*).

DOMESTIC & SACRED. Seed. 49 Fruit. 50

TAMARIX, *Linn.; Gen. Pl., I., 160.*

Tamarix articulata, *Vahl.; Fl. Br. Ind., I., 249;* TAMARISCINEÆ.

51

Syn.—T. ORIENTALIS, *Forsk.;* T. PHARAS, *Ham.;* THUYA APHYLLA, *Linn.*

Vern.—*Lál-jháv* (galls=*chhóti-máyín, nahnui-máyín*) HIND.; *Rakta-jháv*, BENG.; *Farás, farásh, farwá, úkhán, rúkh, kharlei, narlei, khagal, pharwán, ghwá, ghuz* (galls=*mái vari, mái chhóti*), (manna=*gazanj-bín, misrí lei*), (flowers=*búr*), PB.; *Ghwá, ghwhaz*, PUSHTU; *Kirri*, BALUCH.; *Gaz, gaz-lan, asri, asrelei* (galls=*sakem*), (SIND; *Lál-jháv*), (galls=*chhóti-máyí, nahní-máyí*) DEC.; Galls=*Magiya-máin*, BOMB.; *Lál-jháv-nu-jháda*, GUZ.; *Shivappu-átru-shavukku, shivappu-kóta-shavukku, shivappu-shiru-shavukku*, TAM.; *Erra-érusaru, erra-shiri-saru*, TEL.; *Aslul-armar, tarfál-ahmar*, (galls=*habbul asle, samaratul-asl, aazbah*), ARAB.; *Gaze-surkh, kohr-a-gaz*, (galls=*gazmázaje-khurd, may-ínekhurd*), PERS.

References.—*Brandis, For. Fl., 22; Beddome, For. Man., xx.; Gamble, Man. Timb., 20; Stewart, Pb. Pl., 92; Rept. Pl. Coll. Afgh. Del. Com., 42; Pharm. Ind., 29; Flück. & Hanb., Pharmacog., 598; O'Shaughnessy, Beng. Dispens., 332; Irvine, Mat. Med. Patna, 72; Moodeen Sheriff, Supp. Pharm., 239, 240; Ind. Mat. Med. S. Ind. (in MSS.), 40; Murray, Pl. & Drugs, Sind, 46; Dymock, Mat. Med. W. Ind., 2nd Ed., 78; Dymock, Warden, & Hooper, Pharmacog. Ind., I., 162; Baden Powell, Pb. Pr., 331; Drury, U. Pl. Ind., 413; Liotard, Dyes, 13, 14; Stocks, Rept. on Sind; Settlement Reports:—Panjáb, Montgomery, 16; Jhang, 20; Lahore, 15; Gazetteers:—Panjáb, Hoshiárpur, 10; Montgomery, 17; Muzaffargarh, 22; Múltán, 102; Bunnu, 23; Jálandhar, 5; Peshávar, 26; Karnal, 16; Rohtak, 14; Amritsar, 4; Jhang, 16; Ludhiána, 10; N.-W. P., IV., lxviii.; Agri.-Horti. Soc., Panjáb, Select Papers to 1862, 18, 51, 148, 149, Index, 55; Ind. Forester:—II., 171, 407; IV., 233; VIII., 335; XIV., 367; Smith, Econ. Dict., 402.*

Habitat.—A large or moderate-sized tree, common in Sind and the Panjáb, cultivated in Rohilkhand. Beyond India it extends to Afghánistán, Persia, Arabia, North and Central Africa. It grows well on sandy and saline soils, requires little water after it has once taken root, springs up freely from seed, and is readily propagated by cuttings.

TAMARIX dioica.

A Tanning Material.

GUM. 52

Gum.—It yields a small quantity of gum of little value.

DYE & TAN. Bark. 53 Galls. 54 Flowers. 55

Dye and Tan.—The BARK is occasionally employed for tanning; the small irregularly rounded tuberculate GALLS, often abundantly produced on the branches by the punctures of an insect, are used as a mordant in dyeing and also in tanning. They are similar in properties to those of **T. gallica**, from which they differ only in their smaller size and irregular shape. In a few localities of the Panjáb, the FLOWERS are said to be also used in dyeing (*Stewart*).

MEDICINE. Bark. 56 Galls. 57

Medicine.—The BARK is bitter, astringent, and probably tonic (*Pharm. Ind.*). The astringent GALLS are used in the Panjáb for making gargles, &c., and for most of the purposes to which common galls are applied in medicine. Tamarisk manna, produced on the twigs by the puncture of an insect, in parts of the Panjáb and Sind, is collected during the hot weather and employed for medicinal purposes and to adulterate sugar. It will not keep more than a year, and readily deteriorates if exposed to damp (*Brandis*).

SPECIAL OPINION.—§ "The bark powdered, and in combination with oil and *kamela*, is used as an aphrodisiac by the Natives. It is also employed as an application in eczema capitis and other diseases" (*Narain Misser, Hoshangabad*).

TIMBER. 58

Structure of the Wood.—White, moderately hard; weight about 61℔ per cubic foot. This timber is used for many kinds of ordinary work, as for making ploughs, Persian-wheels, the frame-work of native beds, and small ornaments. In Western Rajputána it is said to be employed for making screws of mills and presses, carts, and for house-building. In the dry and more desert tracts of the Panjáb and Sind the rapid growth of the Tamarisk renders it a valuable source of firewood. The wood is also used for making charcoal.

DOMESTIC. Twigs. 59

Domestic.—The TWIGS are frequently covered with a slight efflorescence of salt, which, according to **Edgeworth**, is utilized by poor people near Múltán, who use the water into which they have dipped the branches to season bread.

TRADE. Galls. 60

Trade.—The GALLS are occasionally offered in the drug market of Bombay in large quantities, but are frequently not obtainable. They fetch from R12 to 13 per maund of 37½℔ (*Dymock*).

61

Tamarix dioica, *Roxb.; Fl. Br. Ind., I., 249.*

Syn.—T. ARTICULATA, *Wall, not of Vahl.*

Vern.—*Jau, jhau*, HIND.; *Lal jhau*, BENG.; *Rgelta*, LADAK; *Jhaú, lái, lei, leh, panj, panj-pilchi, farwan, harwan, farás, kachlei, ghazlei, pilchí, rúkh, koán*, (twigs=*hásha*), (galls=*máín*), PB.; *Khwa*, (gum= *chirodheli, vadhál*), PUSHTU; *Gaz, láo, jau, lyi*, (manna=*maki*), ? *turun-jabín*,* SIND; *Pilchí, kachleí, jhau*, MERWARA.

* This name is given on the authority of **Stocks**. According to **Flückiger & Hanbury** *turanjabin* is the Arabic name for the exudation of **Alhagi camelorum**, *Fisch.*

References.—*Roxb., Fl. Ind., Ed. C.B.C., 274; Brandis, For. Fl., 21; Kurz, For. Fl. Burm., I., 83; Beddome, For. Man., xx.; Gamble, Man. Timb., 19; Cat., Trees, Shrubs, &c., Darjiling, 6; Stewart, Pb. Pl., 91; Trans. Med. & Phys. Soc. Bomb., New Series, No. III. for 1855-56, 147; Birdwood, Bomb. Prod., 9, 247, 309; Baden Powell, Pb. Pr., 331, 397, 452, 598; Atkinson, Him. Dist. (X., N.-W. P. Gaz.), 306; Liotard, Dyes, 14; Cooke, Gums & Resins, 25; Stocks' Rep. on Sind; Settlement Report, Panjáb, Delhi, 28; Gazetteers:—Bombay, V., 23; Panjáb, Montgomery, 17; Muzaffargarh, 22; Lahore, 11; Dera*

Ghási Khán, 10; Dera Ismáil Khán, 19; Delhi, 20; Bannu, 23; Karnal, 16; N.-W. P., IV., lxviii.; Agri.-Horti. Soc., Panjáb, Select Papers up to 1862, 40, 50, Index, 55; Ind. Forester: —II., 171, 175, 407; IV., 229; X., 325; XII., App., 6; XIV., 361.

Habitat.—A gregarious shrub, found near rivers and on the sea-coast; throughout India from Sind to Burma; often planted for ornament. Like **T. gallica** it grows freely in soil impregnated with salt, and is easily reproduced from coppice shoots.

GUM. 62

Gum.—Baden Powell states that a gum is obtained from this species or from **T. articulata**, which he describes as follows:—"This occurs in nodules, highly friable, of a granular texture; the nodules appear opaque or a pale yellow; but the little grains of which the nodule consists are individually transparent; the centre of each nodule is more transparent and of a red colour. Its taste is very peculiar, of a bitter combined with sweet, like a mixture of liquirice, aloes, and sugar; it is quite soluble in water."

DYE & TAN. Galls. 63

Dye and Tan.—The GALLS like those of the other species are astringent and are employed as a mordant in dyeing, to produce various shades of grey and black with salts of iron, and in tanning (*Baden Powell*).

MEDICINE. Twigs. 64 Galls. 65

Medicine.—The TWIGS and GALLS are used medicinally as an astringent. Stocks states that this species is the chief source of tamarisk manna, see **T. articulata and T. gallica.**

FOOD. Manna. 66

Food.—*Maki*, the MANNA from this tree, is used in Sind for making confections, &c. (*Stocks*).

TIMBER. 67

Structure of the Wood.—Red, outer portion white, moderately hard; weight 49lb per cubic foot. It is mainly employed for fuel, but, in the Northern Panjáb, it is also used for making Persian-wheels, in turning, &c. In Ladák, where wood is scarce, polo sticks are said to be made of it (*Stewart*). In the Southern Panjáb the twigs are utilized in the manufacture of baskets, rough brooms, and the lining of *kacha* walls (*Settlement Rep., Delhi*).

68

Tamarix ericoides, *Rottl.; Fl. Br. Ind., I., 249; Wight, Ill., t. [24 B; Ic., 22.*

Syn.—T. MUCRONATA, *Smith;* T. TENACISSIMA, *Ham.;* TRICHAURUS ERICOIDES, *W. & A.*

Vern.—*Javra*, MERWARA.

References.—*Beddome, For Man., xx.; Dalz. & Gibs., Bomb. Fl., 14; Lisboa, U. Pl. Bomb., 9; Gaz., N.-W. P., I., 81; IV., lxviii.; Ind. Forester, XII., App. 6.*

Habitat.—A slender shrub or small tree, met with in Bengal, Central and Western India, and Ceylon.

DOMESTIC. Branches. 69

Domestic.—The BRANCHES are used for firewood in Bombay.

70

T. gallica, *Linn.; Fl. Br. Ind., I., 248.*

THE TAMARISK.

Var. indica, *Wild.*=T. INDICA, *Kœn.;* T. GALLICA, *Wight, Ill., t. 24A;* T. ARTICULATA, *Wall. Cat. 3756 a. & d.*

Var. Pallasii, *Desv.,*=T. RAMOSISSIMA, *Ledeb.*

Vern.—*Jháv jháu*, (galls=*barí-máin*), HIND. & BENG.; *Jaulá, jaura*, URIYA; *Telta, rgeeta*, TIBET; *Pilchí, koá, rúkh, lainyá, jhau, laí, ? farásh*,(galls=*mahín, barí-mahín*), PB.; *Gaz-surkh, sura-gaz*, PUSHTU; *Gaz-khera*, BALUCH.; *Lei, lái, jhan*, SIND; *Jháv*, (galls=*barí-mái*) DEC.; Galls=*magiya-máin*, BOMB.; *Jháv-nu-jháda*, GUZ.; *A'tru-sha-*

TAMARIX gallica.

The Tamarisk.

vukku, kóta-shavukku, shiru-shavukku, TAM.; *E'ru-saru, shiri-saru, pakké, pakki, prakke,* TEL.; *Jhávuka, shávaka,* SANS.; *Asl, tarfá,* (galls =*samaratul-asl, samaratul-tarfá, habbul-asl, aazbah, jazmázaj, habbut-tarfá*), (manna=*gazánjabín*), ARAB.; *Gaz, shór-gaz, gaz-shakar,* (galls=*gazmázaj, gazmázak, gazmájú,* (manna=*gazangabín*), PERS.

References.—*Roxb., Fl. Ind., Ed. C.B.C., 274; Voigt, Hort. Sub. Cal., 179; Brandis, For. Fl., 20, t., 5; Kurz, For. Fl. Burm., I, 83; Beddome, Fl. Sylv., t. 20; Gamble, Man. Timb., 19; Rept. Pl. Coll. Afgh. Del. Com., 42; Sir W. Elliot, Fl. Andhr., 141, 157; Pharm. Ind., 29; Flück. & Hanb., Pharmacog., 414; Irvine, Mat. Med. Patna, 71; Moodeen Sheriff, Supp. Pharm. Ind., 238, 239; Mat. Med. S Ind. (in MSS.), 39; U. C. Dutt, Mat. Med. Hind., 301; Murray, Pl. & Drugs, Sind, 46; Dymock. Mat. Med. W. Ind., 2nd Ed., 76, 78; Dymock, Warden, and Hooper, Pharmacog. Ind., I., 150; Trans. Med. & Phys. Soc. Bomb., New Series, No. III. for 1855-56, 153; Notes by Mr. Duthie's collector—Trans-Indus No. 6, 8056; Baden Powell, Pb. Pr., 331, 598; Atkinson, Him. Dist. (X., N.-W. P. Gaz.), 306; Useful Pl. Bomb. (XXV., Bomb. Gaz.), 8; Liotard. Dyes, 13; Bidie, Prod. S. Ind., 20, 108; Settlement Reports: Panjáb, Lahore, 13, 14; Guzrát, 134; Delhi, cclxix.; Montgomery, 16; Gazetteers:—Bombay, V., 288; Panjáb, Múltán, 102; Dera Gházi Khán, 10, 85; Jalandhar, 5; Montgomery, 17; Gujrát, 11; Delhi, 18; N.-W. P., lxviii.; Orissa, II., 178; Agri.-Horti. Soc., Ind:—Jour. (Old Series), IX., 423; XIII., 172, 173; Ind. Forester, II., 171; IV., 228; XII., App. 1, 6; XIII., 93; XIV., 367; Smith, Econ. Dict., 401.*

Habitat.—A shrub or small tree, found, especially along rivers and near the sea-coast, throughout India from the North-West Himálaya to Burma and Ceylon; distributed to Afghánistán, Persia, the Western and Southern shores of Europe, China, and Japan. It favours sandy or gravelly soils, and flourishes on land impregnated with salt. When young it grows moderately fast, but soon reaches maturity and decays rapidly. It is easily propagated from seeds and cuttings.

DYE & TAN. Galls. 71 Twigs. 72

Dye and Tan.—The GALLS, which are similar in properties to those of **T. articulata**, are from two to three times larger, and are extensively employed in dyeing and tanning in the same way as common Galls (see **Quercus**). The bruised TWIGS are similarly used in the Trans-Indus.

MEDICINE. Galls. 73 Bark. 74 Twigs. 75 Ashes. 76

Medicine.—The GALLS have long been known and used in Northern India as astringents. The late Dr. **Stocks** speaks highly of their value, and, from personal experience, recommends a strong infusion as a local application to foul, sloughing ulcers and phagedenic buboes. By Natives the galls and astringent BARK of the TWIGS are principally employed in the treatment of diarrhœa and dysentery. They are noticed in the non-officinal list of the *Indian Pharmacopœia*, and there appears to be no reason why they, together with the galls of **T. articulata**, should not altogether replace the imported officinal oak-gall. The ASHES of the bark are said to contain a large proportion of sodic sulphate (*Baden Powell*).

A variety of this species (*var.* **mannifera**, *Ehrenb.*) produces the greater part of the Tamarisk manna collected in Arabia and Persia which is frequently met with in Indian bazárs under the Arabic or Persian name. It exudes, in consequence of the puncture of an insect (**Coccus manniparus**, *Ehrenb.*), in little honey-like drops, which solidify on exposure (*Pharmacog.*). It is employed medicinally in India, like officinal manna, *i e.*, as a detergent, aperient, and expectorant, and is also said to be recommended by *hakims* in cases of chronic enlargement of the spleen. **Dymock** informs us that it is imported into Bombay from Persia, and sells for about 8 annas per ℔. It is doubtful to what extent the Indian varieties of this species shed manna, if they do so at all. The galls, like those of **T. articulata**, are sometimes abundantly procurable in the Bombay market, while at others they cannot be obtained at all. They fetch the same price as *chhotí máyín*.

CHEMICAL COMPOSITION.—The galls contain as much tannic acid as oak-galls and are readily purchased when offered for sale in Europe (*Pharm. Ind.*). A specimen of the manna from Sinai, examined in 1861 by **Berthelot**, had the appearance of a thick yellow syrup mixed with vegetable fragments. It was found to consist of cane sugar, invert sugar (lævulose and glucose), dextrin and water, the latter constituting one-fifth of the whole. A sample from Persia yielded to **Ludwig**, dextrin, uncrystallizable sugar, and organic acids (*Pharmacog.*).

Chemistry. 77

Structure of the Wood.—Whitish, occasionally with a red tinge, open and coarse-grained, fairly hard and tough but not strong. It is occasionally used in Sind and the Southern Panjáb for making agricultural implements and turned lacquered work, but its chief value is as a plentiful source of fuel and firewood (*Brandis*). The smaller twigs are employed for thatching and basket-making (*Settle. Rep., Lahore*).

TIMBER. 78

Tamarix macrocarpa, *Bunge.; Boiss., Fl. Or., I., 779.*

79

Vern.—*Kirri, gaz-surkh*, BALUCH.; *Gazlei*, SIND.

References.—*Aitchison, Bot. Afgh. Del. Com., 42; Lace, Notes on Quetta Pl. (MSS).*

Habitat.—A very common large shrub throughout Northern Baluchistán and the Sind-Baluchistán Frontier.

Fodder.—Mr. **Lace** informs the writer that the LEAVES and young BRANCHES form one of the chief camel fodders in Peshin. He adds, "A well-known camel contractor tells me that camels only care for this after the branches have been well washed by rain; when covered with dust and salt after drought they don't eat it if they can get anything else."

FODDER. Leaves. 80 Branches. 81

TANACETUM, *Linn.; Gen. Pl., II.. 434.*

Tanacetum senecionis, *Gay; Fl. Br. Ind., III., 319;* COMPOSITÆ.

82

Syn.—T. TOMENTOSUM, *DC.*

Vern.—*Púrkar*, LAD.

Reference.—*Stewart, Pb. Pl., 131.*

Habitat.—A plant of the Western Himálaya, found in Lahoul, Kunawar, and Garhwál, at altitudes of 11,000 to 14,000 feet.

Fodder.—It is browsed by goats.

Domestic.—The thick ROOTS are occasionally used for fuel (*Stewart*).

FODDER. 83 DOMESTIC. Roots. 84

T. tenuifolium, *Jacq.; Fl. Br. Ind., III., 319.*

85

Syn.—ARTEMISIA TENUIFLORA, *Jacq.*

References.—*Stewart, Pb. Pl., 130, 131; Year-Book, Pharm., 1874, 626; Gazetteer, Panjáb, Simla, 12; Agri.-Horti. Soc. Ind.:—Trans., III., 199; Jour. (Old Series), II., Sel., 182.*

Habitat.—A herb, which closely resembles **T. tibeticum**, *Hook. f. & Thoms.*, met with in Kumáon and Western Tibet, at 14,000 feet.

Food.—It is said by **Cleghorn** to be useful for flavouring puddings.

FOOD. 86

Tar, see **Pinus**, Vol. VI., 243.

Taramira Oil, see **Eruca sativa**, *Lam*; CRUCIFERÆ; Vol. III., 266.

TARAXACUM, *Hall; Gen. Pl., II., 522.*

87

Taraxacum officinale, *Wigg.; Fl. Br. Ind., III., 401;* COMPOSITÆ.

THE DANDELION.

Syn.—T. DENS-LEONIS, *Desf.;* LEONTODON TARAXACUM, *Linn.*

Var.—α, typica.

" β, glaucescens,=T. WALLICHII, *DC.*!; LEONTODON GLAUCESCENS, *M. Bieb.*

" γ, riopoda,—T. ERIOPODUM, *DC.;* LEONTODON ERIOPODUM, *Don.;* L. ERIOPUS, *Spreng.*

" δ, parvula,—LEONTODON PARVULUM, *Wall.*

Vern.—*Yamaghi khá, rasúk,* LADAK; *Dúdal, baran, kanphúl, dúdlí, radam, dúdh batthal, shamukei, shamúke,* PB.; *Bathur,* SIND; *Pathri,* DEC.

References.—*Stewart, Pb. Pl., 131; Aitchison, Bot. Afgh. Del. Com., 82; Pharm. Ind., 123; Ainslie, Mat. Ind., I., xxiii.; O'Shaughnessy, Beng. Dispens., 407; Flück. & Hanb., Pharmacog., 392; Bent. & Trim., Med. Pl., t, 159; Murray, Pl. & Drugs, Sind, 186; Bidie, Prod. S. Ind., 59; Atkinson, Him. Dist., 310; Birdwood, Bomb Pr., 49; Smith, Econ. Dict., 150; Rep. Govt. Bot. Gardens, Nilghiris, 1881-82, 1884-85; Gazetteer, Simla, 13; Agri.-Horti. Soc. Ind.:—Jour. (Old Series), XIV., 7; (New Series), III. Pro., 36; IV., 18, 19.*

Habitat.—A common herb, found throughout the Temperate Himálaya and Western Tibet, from 1,000 to 18,000 feet, also found on the Mishmi hills. Sir J. D. Hooker remarks, "It is remarkable that this common Himálayan plant should not be found in the Khásia or Nilghiri mountains, even as a garden escape." Of late years, however, it is said to have begun to spread rapidly in all directions from Ootacamund, and occurs in such quantity in the deeply-tilled soil of the Cinchona plantations that abundance of its roots might be annually collected to supply the wants of all the medical stores in India (*Nilghiri Bot. Garden Reports*).

MEDICINE. Root. 88

Medicine.—The medicinal properties of the ROOT are too well known to require any notice in this work. It appears to be little known to, or appreciated by, Natives.

FOOD & FODDER. Leaves. 89 Root. 90

Food and Fodder.—The LEAVES may be used in the form of salad in place of lettuce, though rather too bitter to be very palatable. They, however, form good food for cattle. The ROOT is sometimes employed instead of chickory for mixing with coffee; in France it is eaten raw as a salad, and in Germany is boiled for use as a vegetable.

Taro, see **Colocasia antiquorum,** *Schott.;* AROIDEÆ, Vol. II., 509.

Tasar, Tusser, Tussur, or Tussah, etc., see **Silk,** Vol. VI., Pt. III., [p. 1 *et seq.*

TAVERNIERA, *DC.; Gen. Pl., I., 511.*

91

Taverniera nummularia, *DC.; Fl. Br. Ind., II., 140; Wight, Ic., t. 1055;* LEGUMINOSÆ.

Syn.—T. SPARTEA, *DC.;* T. CUNEIFOLIA, *Arn.;* HEDYSARUM NUMMULARIFOLIUM, *DC.;* H. SPARTEUM, *Burm.;* H. GIBSONI, *Grah.*

References.—*Dalz. & Gibs., Bomb. Fl., 67; Graham, Cat. Bomb. Pl., 49; Murray, Pl. & Drugs, Sind, 123; Gazetteer, Mysore & Coorg, I., 56; Agri.-Horti. Soc., Ind., Journals (Old Series), XIV., 8.*

Habitat.—A copiously branched under-shrub, met with in the plains of Sind and the Panjáb, distributed to Afghánistán.

MEDICINE. Leaves. 92

Medicine.—The LEAVES are said to be useful, in the form of a poultice, as a cleaning application for sloughing ulcers (*Murray*).

Medicinal Uses of the Yew. (*J. Murray.*)

TAXUS baccata.

TAXUS, *Linn.; Gen. Pl., III., 431.*

93

Taxus baccata, *Linn.; Fl., Br. Ind., V., 648;* CONIFERÆ.
THE YEW.

Syn.—T. NUCIFERA, *Wall.;* T. NEPALENSIS, *Jacq.;* T. CONTORTA, *Griff.;* T. ORIENTALIS, *Bertholini.*

Vern.—*Thúno, birmi, sirnub birmi,* HIND.; *Burmie, bhirmie, sugandh,* BENG.; *Dingsableh,* KHASIA; *Tcheiray sulah, tcheiray gulab,* NEPAL; *Nhare,* TIBET; *Tingschi, tsashing,* BHUTIA; *Cheongbu,* LEPCHA; *Thúner, geli, gallu, lúst,* N.-W. P.; *Thaner, thúner, lúet, nhare, bráhmi, geli, gallu, lúst,* KUMAON; *Pung cha, pung chu, sung cha,* LADAK; *Túng, sungal, postil, postal, barma, birmi, barini, thúnu, chogu, chatúng;* KASHMIR; *Birmi, barma, barini, tung, thúnu, sungal, pastal, chogu, chatúng,* CHAMBA; *Birmi, túng, thúnú, barma, barmi, kautú, dhúnú, chogú, rakhal, rikhái, nyamdal, thona, kadenrú, rikaling,* leaves=*birmi,* PB.; *Badar, sarop, saráp, kharoa,* PUSHTU; *Bhirmi, sugan,* AJM.; *Barmi,* leaves=*tálispatr,* BOMB.

References.—*Brandis, For. Fl., 539; Gamble, Man. Timb., 413; Stewart, Pb. Pl., 227; Irvine, Mat. Med. Patna, 11; Bent. & Trim., Med. Pl., t. 253; Dymock, Mat. Med. W. Ind., 2nd Ed., 758; Birdwood, Bomb. Prod., 85; Baden Powell, Pb. Pr., 373, 599; Atkinson, Him. Dist., 751, 775, 843; Settle. Rep., Simla, xliii, App., ii.; Gazetteers:—Bannu, 23; Rawalpindi, 15; Peshawar, 28; Hazara, 14; Dera Ismail Khan, 19; Simla, 10; Gurdaspur, 54; Agri.-Horti. Soc. Ind.:—Jour. (Old Series), IV., Sel., 259, 261, 267, Pro., 71; VII., 155-159, 161; VIII., Sel., 196; XIII., 381, 383, 391; XIV., 13, 36, 267, 269, 271, 272; Ind. Forester:—IV., 292; V., 184-186; VIII., 38, 404; IX., 321; XI., 284; XIII., 62, 66; Smith, Econ. Dict., 447.*

Habitat.—An evregreen tree, found on the Temperate Himálaya from 6,000 to 11,000 feet, the Khásia hills at 5,000 feet, and in Upper Burma; distributed to North and Temperate East Asia, all Europe, North Africa, and North America. In the Himálaya it attains a considerable size; **Madden** mentions a tree at Gangútri 100 feet high and 15 feet in girth. In the Panjáb hills it is commonly about 5 to 6 feet in girth, in Hazára 8 to 9 feet is not unfrequent (*Brandis*).

GUM. 94

Gum.—The tree yields a gummy exudation, which forms a portion of the incense used in Tibet.

DYE. Bark. 95 Juice. 96

Dye.—According to **Madden** the people of Ladak import yew BARK from Kashmír and Kanáwar, on account of the inner part, which they employ as a red dye. **Hooker** notes that the red JUICE of the bark is used in Nepál as an inferior dye, and by Brahmins for staining the forehead.

MEDICINE. Leaves. 97 Fruits. 98

Medicine.—"The LEAVES contain a volatile oil, tannic and gallic acids, and a resinous substance called *taxin*. Yew leaves and FRUITS have been given for their emmenagogue, sedative and antispasmodic effects. **Pereira** says that therapeutically the yew appears to hold an intermediate position between Savin and Digitalis, being allied to the former by its acrid, diuretic, and emmenagogue properties, and to the latter by the giddiness, irregular and depressed action of the heart, convulsions and insensibility, which it produces. Yew is, however, reported to have one decided advantage over Digitalis by its effects not accumulating in the system, so that it is a much more managable remedy than Digitalis. Besides its use as an emmenagogue and sedative in the same cases in which Savin and Digitalis are administered, it has also been employed as a lithic in calculus complaints; and as an antispasmodic in epilepsy and convulsions. According to **Dr. Taylor,** the yew tree is sometimes used by ignorant persons to cause abortion. At the present time, yew is never used in regular medical practice in Europe, the principal interest attached to it has, therefore, reference to its poisonous properties. Thus, the leaves and young branches act as a narcotico-acrid poison, both to the human subject and to certain animals,

TAXUS baccata.

Economic Uses of the Yew.

MEDICINE. Shoots. 99

but more especially to horses and cows. Fatal cases of poisoning have also occurred from swallowing the fruit. It is frequently stated that animals may feed upon the young growing SHOOTS with impunity, but that when these have been cut off, and left upon the ground for a short time, they are then poisonous. This is an entirely erroneous notion, for yew shoots and leaves are poisonous both in a dried and fresh state. It seems certain, however, that the red, succulent CUP of the fruit is harmless, for a fatal case of poisoning has been recorded of a child from swallowing the entire fruit with its contained seed; whilst other children, who had partaken of the fruit at the same time, but who had rejected all but the fleshy cup, suffered no ill effects" (*Bentley & Trimen*).

Cup. 100

In Northern India the leaves are largely employed for medicinal purposes, under the name of *birm* or *bráhmí*, chiefly as a remedy for indigestion and epilepsy and as an aphrodisiac (*Irvine*). They are exported in considerable quantities from the hills to the plains. According to **Dymock** they constitute at least a portion of the drug known as *talispatr* in Bombay. Considerable confusion exists regarding this drug. In Calcutta it appears to consist of the leaves of **Abies Webbiana**, while **Moodeen Sheriff** assigns the name of *talishpatri* to the leaves of **Cinnamomum Tamala**, *Nees*. It is, therefore, doubtful to which plant the *talispatr* of Sanskrit writers belongs, but a medicine described as "carminative, expectorant, and useful in phthisis, cough, and asthma," can hardly have been the leaves of the Yew. (*Conf.* with **Abies Webbiana**, Vol I., 5; **Cinnamonum Tamala**, Vol. II., 321; and **Flacourtia Cataphracta**, Vol. III., 398.) It is difficult to understand how *baníer* should have come to bear the name, and presumably to be applied to the same uses in Bombay.

FOOD & FODDER. Berries. 101 Bark. 102 Leaves. 103

Food and Fodder.—The BERRIES appear to be eaten by Natives in most parts of the Himálaya, their poisonous effect being probably prevented by rejecting the seeds. The BARK is largely employed as a substitute for, or mixed with, tea, for which purpose a considerable quantity is annually imported from Kulu and Kashmír into Ladak. The LEAVES are eaten by goats and sheep, though stated by **Lindley** and others to be poisonous to horses and cattle. [It appears probable that the Yew is less poisonous on the Himálaya than in Europe.—*Ed.*]

TIMBER. 104

Structure of the Wood.—Sapwood white, heartwood red or orange-red, hard, smooth-grained; average weight 44℔ per cubic foot (*Gamble*). The timber of yew is celebrated for its toughness and elasticity. It works easily, polishes well, and is much valued in Europe for these qualities and for its beautiful colour. It is extensively used by cabinet-makers for making furniture veneers, and was at one time highly prized as the best wood for making bows. In India it is still employed for that purpose in the Sutlej valley (*Cleghorn*), but is more generally used for making native bedsteads, banghy poles, and furniture; in Kashmír it is employed for making clogs (*Vigne*). It is well fitted for purposes of turning.

Furniture. 105

DOMESTIC & SACRED. Branches. 106 Twig. 107 Wood. 108

Domestic and Sacred.—The BRANCHES are used in the Panjáb Himálaya for putting under the earth covering of the roofs of native houses. A TWIG is worn by young unmarried Nága females as a charm to prevent pregnancy, chastity being an exceptional virtue amongst them (*Watt*). The WOOD is burned as incense, the branches are carried in processions in Kumáon, and in Nepál are used to decorate houses at religious festivals. In certain localities of the Himálaya and Khásia hills the tree is held in great veneration, and shares the name of *Deodár* (God's tree) with the Cedar. [On the Himálaya the Yew appears to be greatly subject to some root-parasite or some other disease, since very large numbers are seen decaying.—*Ed.*]

TEA.

109

In Vol. II. (*pp. 65-83*) of this work the reader will find an account of the various species of **Camellia**, more especially **C. theifera,** *Griff.* The historic chapter of that article is believed to be sufficiently complete on the main issues, and to require only the further details necessary to convey an idea of the origin and progression of the industry in the provinces of India. Details of the methods of cultivation and manufacture, on the other hand, are deemed of too technical a character for this work. A brief compilation would convey to the général reader very little of practical value and would be useless to the planter. What is more especially desirable may be said to be an exhaustive review of the principles of manufacture and the adaptability to requirements of the numerous patented appliances now in use. But to furnish such a review would necessitate the allotment of a much larger space than can be assigned to the subject, and could not be satisfactorily accomplished without a personal investigation of processes presently pursued. That this cannot be undertaken at present is the more to be regretted since, in the writer's opinion, every detail in the manufacture of tea, so far as can be discovered, needs to be critically examined by a chemist and by a botanist, working in conjunction, so as to establish the chemical changes in the manipulation and the character of the plants now grown and their possible improvement. The diseases of the tea plant have not as yet assumed serious proportions, but it would be contrary to experience in all other branches of agricultural enterprise to suppose this is always to be so. The greatest possible changes must of necessity be insidiously taking place now, changes that may result in weakness, disease or degeneration in the flavour of the produce. It is perhaps to the greater natural strength of the comparatively wild plant of the majority of the Indian gardens, that is attributable the strength and rich flavour of the Indian as compared with the China teas. Doubtless greater cleanliness, more uniformity, and a carefully prepared article have largely contributed to the success attained by India. But while admitting all this, it has still to be confessed that the manufacture of tea is far too dependant on skill and judgment, and hence would very possibly be immensely improved could experience be reduced to scientific principles. Few industries, at all events, of anything like the importance of tea, have had so small a share of the chemist's attention. It has, accordingly, been thought desirable that in this article the treatment of the subject should be restricted very largely to statistical information and more especially to statistics of a local nature. The foreign exports of tea from India are fairly well known. What seems more especially deficient are particulars of area of cultivation, yield, outturn, prospects, local consumption, capital invested, number of hands employed, etc., etc. The writer's failure to obtain definite information on some of these topics may, it is hoped, serve to impress on those in a position to assist, the advisability of rectifying such defects when the doing so would involve no personal disadvantages or inconvenience.

Tea.

110

References.—A list of works and reports consulted while drawing up the article **Camellia** will be found in Vol. II. A supplementary enumeration may now be furnished, consisting of those that have a special bearing on the information furnished below. But this supplementary enumeration, it may be explained, has been given under two headings—those in this paragraph being of a general character, and those in the further sections, of a more local and specific kind:—*Linschoten, Voyage to E. Indies in 1598, I., 156-158, II., 157; DC., Origin Cult. Pl., 117; Bretschneider, Study and Value of Chinese Bot. Works, 13, 45; Wallich, New Camellia in Nepal, As. Res., XIII. (1820), 428; Royle, Report of Cult. of Tea in the Himálaya from 1835-1837 (Jour. Royal Asiatic Soc., 1849), also Prod. Res. Ind., 257-311 and Ill. Him. Bot., 125; The Tea Plant, Quarterly Journal of Agriculture, 1831-32, III., 560; VI. (1855), 630; Corbyn's Indian Review (Calcutta 1837), I., 311; Fortune, Visit to the Tea Districts of India and China, 1852; also Three Years' Wanderings in China, Ball, Account Culti. and Manuf. of Tea in China, 1848; Martin; Tea Trade of Europe and America, 1832; Bruce, Manuf. of Black Tea (Calcutta), 1838; Houssaye, Monographie, Du Thé; Lees, Tea Cultivation, etc., 1863; also Memo. A Tour through the Tea Districts of Bengal,*

1864-65; Markham, Mission of Bogle & Manning to Thibet, 148; Moorcroft & Trebeck, Travels (Ed. Wilson, 1841), I., 329, 351; Walker, Tea Report (in which Buchanan-Hamilton's MS. regarding Tea in Burma is quoted), 183; Heber, Journal I., 513; Papers on Tea Industry in Bengal (Calcutta, 1873); Hassal, Food & Its Adulteration, London, 1876; Haworth, Information and Advice for Tea Planters, 1865; Money, Culti. and Manuf. of Tea, 4th Ed., 1883; Campbell Brown, Jour. Chem. Soc., Lecture on Agri. Chemistry of the Tea Plantations of India, June 3rd, 1875; Railway, Mining, Banking, and Commercial Almanac, 1866; Cochran, a series of articles in 'Food Journal' in 1871; Report of Select Committee of Parliament on Commercial Relations with China in 1849; Tea in Johnston's Chem. Com. Life, Ed. Church, 115; Tea Cyclop., 1-355; Anderson, Dict., Commerce 1859; Hooker, Him. Jour., I., 5, 144, 408; II., 347; Bell, Chemistry of Food, 1-39; Allen, Pharm. Jour., 1873 (also in Chemical News XXX.); Chemical News, XXVIII., 186; Blyth, A. Wynter, Mico-Chemistry as applied to the detection of foreign leaves in Tea, "Analyst," 1877; also Indian Tea in Jour. Chem. Soc., 1875; U. S. Dispens., 15th Ed., 1762; Pareira, Mat. Med.; Royle, Mat. Med.; Moodeen Sheriff, Supp. Pharm. Ind., 240; Bent. & Trim., Med. Pl., I., 34; Müller, Extra-Tropical Plants; Stanton in Col. & Ind. Exh. Reports, 1885; Calcutta Review (Tea Cultivation in India), LXXX.; Baden Powell, Pb. Prod., 275-285; Atkinson, Him Districts, 887-907; Drury, Useful Plants of India, 423, 477; Simmonds, Trop. Agri., 79; Note on Tea Industry of the N.-W. Prov. and Pb., by L. Liotard, Simla, 1882; Tea Culture as a probable American Industry by Saunders, 1879; Ure, Dict. Arts, Manuf., etc., III., 870; Spons' Encycl., II., 1994; Encyclopædia Britanica; Balfour, Cyclopædia India; Smith's Dic. Econ. Pl.; In connection with the present article the writer has consulted files of most of the leading periodicals and newspapers such as the Society of Arts Journal, The Tropical Agriculturist, The Indian Agriculturist, Indian Forester, The Planters Gazette, The Pioneer, The Englishman, Capital, etc., etc.

AREA & OUTTURN. III

AREA & OUTTURN.

This section of an essay on an Indian agricultural product is one of the most difficult to deal with when India, as a whole, is under consideration. Some provinces have not been surveyed as yet, such as Bengal. The area under tea has, therefore, to be guessed at when the owners of plantations do not choose to furnish particulars. Leaving Bengal out of consideration, the following has been published as the tea area in 1890-91:—

	Acres.
Madras	5,738 *
Bombay	...
Bengal	...
North-West Provinces	7,977
Oudh	...
Panjáb	9,229
Central Provinces	...
Upper Burma	1,001
Lower Burma	78
Assam	230,822
Berar	...
Coorg	...
Ajmír	...
TOTAL	254,845

* Exclusive of Travancore, which has between 4,000 and 5,000 acres: see table, p. 420.

Area under Tea.

AREA.

For the five previous years the areas under the above mentioned provinces were as follows :—

	Acres.
1885-86	219,111
1886-87	227,258
1887-88	234,176
1888-89	241,077
1889-90	251,672

Travancore. 112 *Conf. with pp. 448, 469.*

It will thus be seen that tea cultivation has been steadily expanding; but as Bengal has not been given, the figures are necessarily imperfect. It will also be observed that no return has been furnished of the area under tea in the Native State of Travancore. This is to be regretted on at least two accounts—(*a*) the area has so rapidly increased (and appears to be still expanding) that the] defect is becoming material : (*b*) in no other Native State (with the exception of coffee in Mysore) has any modern industry taken so prominent a position. But authorities in Europe seem agreed that there must be some special merit in the climate and soil of Travancore, since the production is not only vastly improved, within the past few years, but in point of flavour and quality Travancore might be spoken of as the Ceylon of Indian Tea planting. As showing the behaviour of Assam, the most important producing province, the following table may be furnished :—

Assam. 113

	Assam.	Rest of India except Bengal.
	Acres.	Acres.
1885-86	194,480	24,631
1886-87	203,963	23,295
1887-88	211,079	23,097
1888-89	216,676	24,401
1889-90	227,249	24,423
1890-91	230,822	24,023

It will thus be seen that of the surveyed tea area of India the Assam portion has increased 36,342 acres since 1885-86, while that of the other Indian provinces, exhibited above, has fluctuated very considerably, but on the whole shown a decrease.

For two of the years mentioned above, the writer has, on the next page, endeavoured to furnish an estimate of the total area in all India, in which will also be found the approximate yield and total production. From that statement it will be seen that the area in Bengal has been calculated (for the two years in question, *viz.*, 1888-89 and 1889-90) to have been between 79,000 and 80,000 acres.

TEA.

AREA.

Area, Approximate yield,

Statement illustrative of the state of Tea Cultivation in

Provinces and Districts.	Number of gardens or plantations.		Area in acres.					
			Under mature plants.		Under immature plants.		Total area under tea.	
	1888-89.	1889-90.	1888-89.	1889-90.	1888-89.	1889-90.	1888-89.	1889-90.
Assam—	No.	No.	Acres.	Acres.	Acres.	Acres.	Acres.	Acres.
Cachar	181	181	48,493	Details not received.	6,908	Details not received.	55,401	57,598
Sylhet	115	116	34,686		9,459		44,145	44,791
Goalpara	4	4	300		67		367	377
Kamrup	89	89	5,987		240		6,227	6,424
Darrang	86	87	16,863		3,149		20,012	21,085
Nowgong	61	61	10,283		690		10,973	11,551
Sibsagar	176	178	43,925		3,452		47,377	51,861
Lakhimpur	150	151	27,762		4,382		32,144	33,532
Khasi and Jaintia Hills	1	1	30		...		30	30
Total	863	868	188,329	196,689	28,347	30,560	216,676	227,249
Bengal—								
Chittagong	26	25	3,572	3,777	523	187	4,095	3,964
Do. Hill Tracts	1	1	100	100	...	...	100	100
Hazaribagh	6	6	877	978	30	43	907	1,021
Lohardugga	30	30	2,070	2,302	919	564	2,989	2,866
Darjeeling	174	165	35,755	35,978	9,170	5,993	44,925	41,971
Julpigoree	159	166	21,997	23,658	4,919	5,399	26,916	29,057
Dacca	6	6	27	27	2	...	30	27
Total	402	399	64,398	66,820	15,563	12,186	79,962	79,006
North-West Provinces—								
Kumaon	38	40	2,646	2,844	297	173	2,943	3,017
Garhwal	7	7	425	459	35	105	460	564
Dehra Dun	32	33	4,408	4,311	455	422	4,863	4,733
Total	77	80	7,479	7,614	787	700	8,266	8,314
Panjab—								
Simla	1	1	100	50	...	...	100	50
Kangra	2,565	2,193	8,905	8,193	684	583	9,589	8,776
Total	2,566	2,194	9,005	8,243	684	583	9,689	8,826
Madras—								
Nilgiris	95	98	4,741	4,503	506	445	5,247	4,948
Madura	2	2	6	6	·50	·50	6·50	6·50
Malabar	7	7	72	72	340	340	412	412
Travancore	49	57	2,045	2,181	1,968	2,544	4,013	4,725
Vizagapatam	3	3	8·50	8·50	...	...	8·50	8·50
Cochin	2	2	35·50	32	...	3·50	35·50	35·50
Total	158	169	6,908	6,802·50	2,814·50	3,333	9,722·50	10,135·50
Burma—								
Toungoo	1	1	14	14	...	...	14	14
Akyab	...	1	...	150	...	...	...	150
Tavoy	...	...	...	8†	...	...	...	8†
Total	1	2	14	172	...	...	14	172
						Grand Total	324,329	333,702

* Incomplete.
† Particulars for other columns not available.
‡ Exclusive of 129,970lb, which have been shown in the Chittagong return, but not classed as

T. 113

and Total Production of Tea in India. (*G. Watt.*)

TEA.

AREA.

British India during 1888 and 1889.

Taken up for planting but not yet planted.		APPROXIMATE YIELD IN ℔. Black.		Green.		Total.		Average yield in ℔ per acre of mature plants.	
1888-89.	1889-90.	1888-89.	1889-90.	1888-89.	1889-90.	1888-89.	1889-90.	1888-89.	1889-90.
Acres.	Acres.	℔	℔	℔	℔	℔	℔	℔	℔
180,98	Not received.	15,477,096	Not received.	...	Not received.	15,477,096	15,631,692	319	312
137,375		13,456,078		29,260		13,575,338	15,741,409	391	432
1,011		102,400		...		102,400	92,083	341	302
19,281		1,115,128		...		1,115,128	1,163,727	186	192
81,044		8,033,149		...		8,033,149	8,445,916	476	464
48,640		3,763,044		...		3,763,044	3,521,595	366	340
173,290		16,126,800		440		16,127,240	17,615,211	367	378
97,150		14,470,853		10,734		14,481,587	14,701,195	522	514
50		3,000		...		3,000	3,000	100	100
738,823		72,637,548		40,434		72,677,982	76,915,828	386	391
8,749	1,875	948,731	1,123,177	23,871	...	972,602t	1,123,177	272·2	297·37
373	373	23,540	26,407	...	...	23,540	26,407	235·4	264·07
1,847	1,857	114,022	88,860	1,255	13,509	115,277	102,369	131·4	104·6
2,053	1,894	339,383	363,960	...	1,280	339,383	365,240	163·9	158·6
12,256	10,715	10,274,131	10,503,620	...	334,600	10,274,131	10,838,220	287·3	391·2
21,562	20,952	1,375,826	12,632,880	...	...	10,375,826	12,632,880	471·6	533·97
...	...	1,215	1,130	...	...	1,215	1,130	44·1	41
46,840	37,666	22,076,848	24,740,034	25,126	349,389	2 ,101,974	25,089,423	343·2	375·47
393	378	438,887	377,333	84,819	14',329	523,706	518,662	198	182
...	...	58,080	49,314	1,797	4,500	59,877	53,814	41	117
49	1,421	1,001,440	1,137,992	61,030	60,255	1,062,470	1,198,247	241	278
442	1,799	1,498,407	1,564,639	147,646	206,084	1,646,053	1,770,723	220	232
...	...	2,500	2,089	...	...	2,500	2,089	25	42
2,722	2,165	1,024,370	1,080,169	788,568	713,673	1,812,938	1,793,842	204	219
2,722	2,165	1,026,870	1,082,258	788,568	713,673	1,815,438	1,795,931	202	218
1,168	1,257	...	...	...	...	972,836	856,085	205	190
9·50	...	...	...	...	...	93	31	15	10
640	640	...	...	...	...	28,800	28,800	400	400
9,794	11,872	535,212	571,756	10,000	...	545,212	571,756	270	...
...	...	...	...	...	...	...	...	...	...
...	...	2,556	2,048	...	...	2,556	2,048	72	64
1,817·50	...	537,768*	573,804*	10,000*		1,549,497	1,458,720	Not given.	...
...	...	1,600	250	...	...	1,600	250	...	250
...	...	...	...	...	12,000	...	12,000	...	80
...	...	...	...	...	...	...	...	...	...
...	...	1,600	250	...	12,000	1,600	12,250	Not given.	...
GRAND TOTAL OF PRODUCTION .						99,792,544	107,042,875	...	...

either black or green.

TEA. **Financial Results**

AREA.

It will thus be seen that the major portion of Indian tea is prepared in the black form. Although green teas could be prepared at any plantation, they are mainly produced at Sylhet, Chittagong, Dajeeling, the North-West Provinces, and Kangra. The average acre yield in Assam and Cachar is returned as 386℔ in 1888-89, and 391℔ in 1889-90; the corresponding figures in Bengal for these years were 343·2℔, 375 47℔; in the North-West 220℔ and 233℔, and in the Panjáb 202℔ and 218℔. These figures of acreage yield may be said to indicate the greater success of Assam and Bengal as compared with the North-West Provinces and the Panjáb. They correspond approximately to the regions of cultivation of the Assam indigenous or strongly Assam hybrid and to the China or strongly China hybrid.

Yield. 114

The outturn worked out for 1888-89 and 1889-90, *viz.*, 99,792,544℔ in the former year and 107,042,875℔ in the latter, will be seen to approximate very closely with the recorded receipts in Calcutta and the published foreign exports from Calcutta. The remarks below will be found to establish that an allowance of 3 million pounds of Indian tea for the annual Indian consumption would be sufficient. Further, since almost the whole of the foreign trade goes from Bengal, it would involve no serious error to accept the exports of Calcutta as the production less an allowance of the value indicated. The total exports from all India in 1890-91 were 107,014,993℔ and in 1891-92 120,149,407℔. If now we assume the area of Bengal tea to have been at least 80,000 acres, we are in a position to form some sort of opinion as to the total area that afforded last year's outturn. The tea crop of last year may, therefore, be assumed as having been obtained from 334,845 acres. But there is a feature of much interest connected with the acreage and production that had perhaps be better dealt with under.

FINANCIAL RESULTS. Capital Invested. 115

FINANCIAL RESULTS OF TEA-PLANTING.

Capital Invested in Indian Tea.—It may be here explained that there are three classes of tea gardens: (1) Companies Registered in England: (2) Companies registered in India; and (3) Private concerns not Registered at all. To arrrive, therefore, at some idea of the total Capital Invested in Indian Tea, it becomes necessary to bear these three classes of gardens in view. Through the very great kindness of **Messrs. Gow, Wilson, & Stanton**, the author has had the pleasure to receive a series of circular letters, which have been issued by both English and Indian firms, such as those of **Mr. Alex. W. Martin** of 27, Throgmorton Street, London, and of **Messrs. Barry & Co.**, Calcutta. The Tea Association, Calcutta, has also most obligingly furnished its circulars and reports. The author has further consulted the *Home and Colonial Mail; Capital*, etc., etc., so that he is confident the figures upon which he rests the following estimate of the probable total capital invested in Indian tea may be accepted as fairly accurate:—

Statement of the Indian Tea Industry for 1891-92.

		Acres.
Total area under tea in India		334,845
	Acres.	
Companies registered in London	108,277	
Companies registered in Calcutta	55,414	
Deduct total of Registered Companies		163,691
Balance representing private concerns		171,154
		℔
Exports of tea from India (approximate production)		120,149,407

FINANCIAL RESULTS.

Capital Invested

	lb	
Production of Registered London Companies	35,56c,859	
Production of Registered Indian Companies	19,902,249	
Deduct Total Production of Registered Companies . .		55,463,108
Balance representing Production of private concerns .		64,686,399

These statements would thus show a yield of 339lb per acre to the registered Companies and of 378lb for the private concerns.

The total capital of the registered Companies may be put at—

(a) London	£5,062,194
(b) Calcutta, R3,57,75,117, or	£3,577,511
TOTAL .	£8,639,705

Total Capital invested: 17½ million pounds. 116

That capital expressed to the acreage of 163,691 would be £52-16. Now, if it be admitted, from the above calculation, that the acreage of private concerns comes to 171,154, and that the capital invested in that area of cultivation is proportionate to that in the public Companies, we would arrive at the conclusion that the capital invested in these private concerns may be £9,036,931. It is thus possible that the total capital invested in Indian tea may be 17½ million pounds sterling. It may further be stated that ⅘ths of that large capital is British money, or the investments of Englishmen resident in India. But to be safe in allowing for the error involved in the expression of rupees at a nominal pound sterling and in a possible over-statement of the value of private concerns, it may be affirmed that the private or non-registered tea gardens have at least a capital of 5 millions, so that it may, with perfect safety, be accepted that the capital in tea is 13½ to 14 million pounds sterling. The magnitude of the stake Great Britain thus holds in her Indian Tea Plantations will be seen to justify the very strongest statements of the value of India to England. While India pays no tribute to Britain, such as many other nations demand from their Colonies or newly-acquired provinces, it affords an opening for British capital that would otherwise have poured into foreign countries. India has thus its resources developed, its vast expanses of country opened up, and its people enriched by lucrative employment and by the expansion of the cultivated area through the colonization of new regions, such as has followed the expansion of the tea area. The diversion of British capital to foreign countries cannot be too strongly depricated—both in the interests of the speculator and in the light of the claims of his nationality for support and encouragement. The British investments in Assam tea have converted that province into one of the most flourishing and prosperous in India. A few years ago the officers who were sent to Assam regarded themselves as condemned to a penal servitude.

Working of Tea Plantations. 117

The Working of Tea Plantations in India.—From the facts just mentioned—that there are three classes of gardens in India, *viz.*, Companies Registered in London; Companies Registered in India; and Private Concerns, that is, non-registered Companies—it is impossible to furnish acomplete statement of the financial results of tea-growing. The writer has, however, been favoured with particulars of the Registered Companies, and it would seem, from the comparison of the Indian with the London, that the inferences deducible from these are very likely to be fully applicable to private concerns. There does not seem, however, any occasion to publish the results of all the Registered Companies, and the following statement of the London Companies may, therefore, suffice. The statement, it will be noted, deals with the working season of 1890, and is so very exhaustive that further comment is unnecessary:—

TEA.

FINANCIAL RESULTS.

Working of Tea Plantations.

Results of Working of

COMPARATIVE TABLE OF INDIAN TEA

Showing the results of Work

	Assam Company.	Jorehaut Tea Company, Limited.	Jhanzie Tea Association, Limited.	Tiphook Tea Company, Limited.	Noakacharee Tea Company, Limited.	Scottish Assam Tea Company Limited.
CAPITAL— Paid up . . . £	187,160	100,000	55,000	28,000	70,000	79,590
TOTAL— Area of cultivation . acres	9,595	4,715	2,070	1,050	2,300	835
Do., mature do. . ,,	7,521	4,096	1,415	780	2,074	742
PROPORTION OF YOUNG CULTIVATION . per cent.	21·61	13·13	31·64	25·71	9·83	11·14
YIELD (total crop)— Per mature acre . ℔	360	353	364	276	360	536
OUTTURN— (*Account Sales weight*) ,,	2,712,274	1,446,565	514,657	215,333	746,772	397,553
LOSS IN TARING . per cent.	·70	1·38	1·25	·38	·87	1·39
CAPITAL VALUE— Per acre of Total Cultivation . . . £	19-10-2	21-4-5	26-11-5	26-13-4	30-8-8	95-6-4

Results per ℔, etc., on the

	s. d.	s. d.	s. d.	s. d.	s. d.	s. d.
TOTAL COST*— Average per ℔ . . .	0 9·67	0 9·32	0 9·70	0 9·80	0 8·44	0 8·50
SURPLUS . Commission to Managers, average per ℔	0 0·22	0 0·25	0 0·36	0 0·23	0 0·17	—
SURPLUS . Profit for Shareholders, average per ℔ .	0 1·75	0 1·56	0 2·67	0 1·82	0 1·56	0 2·75
GROSS PROCEEDS— Average per ℔ . . .	0 11·64	0 11·13	1 0·73	0 11·85	0 10·17	0 11·25
SHAREHOLDERS' PROFIT— Per mature acre . . £	2-12-8	2-6-0	4-0-11	2-1-11	1-5-11	6-2-11
Per cent. on Total Cost* . . per cent.	17·69	16·79	27·54	18·61	18·52	32·40
Per cent. on Capital ,,	10·57	9·43	10·41	5·84	6·95	5·74
DIVIDEND PAID PROFIT— Per cent. on Capital ,,	10	10	10	5	*Nil.*	5
RESERVE FUND PROFIT— Per cent. on Capital . ,,	19·90	7·44	12·73	8·21	*Nil.*	2·83

NOTE.—Where blanks appear the
* Old non-dividend

COMPANIES (REGISTERED IN LONDON).

ing, etc., in Season 1890.

FINANCIAL RESULTS.

Working of Tea Plantations.

Jokai (Assam) Tea Company, Limited.	Wilton Tea Company of Assam, Limited.	Doom Dooma Tea Company, Limited.	Assam Frontier Tea Company, Limited.	Upper Assam Tea Company, Limited.	British Indian Tea Company, Limited.	Brahmaputra Tea Company, Limited.	Moabund Tea Company, Limited.
200,000	28,000	116,100	220,000	204,224	60,825	114,500	35,007
5,193	974	1,868	3,877	3,170	1,690	2,848	746
4,377	790	1,412	3,326	2,771	1,349	2,458	597
13·78	18·89	24·41	14·21	12·59	20·18	13·69	19·97
520	340	623	718	379	442	595	547
2,275,094	268,406	879,236	2,386,664	1,050,601	596,379	1,462,019	326,726
1·02	1·59	1·64	1·17	1·48	1·56	1·11	·39
38-10-8	28-15-0	62-3-0	56-14-11	64-8-6	35-19-9	40-4-1	46-18-6

Working of Season 1890.

s. d.	s. d.	s. d.	s. d.	s. d.	s. d.	s. d.	s. d.
0 9·03	0 9·44	0 7·99	0 8·24	1 0·19	0 7·69	0 6·53	0 11·82
0 0·43	0 0·22	0 0·50	0 0 19	0 0·14	0 0·11	...	0 0·47
0 3·14	0 1·65	0 4·12	0 3·40	0 1·92	0 2·00	0 4·67	0 4·24
1 0·60	0 11·31	1 0·61	0 11·83	1 2·25	0 9·80	0 11·20	1 4·53
6-16-1	2-6-8	10-13-8	10-3-4	3-0-10	3-13-6	11-11-6	9-13-2
34·80	17 48	51·55	41·24	15·81	25·94	71·46	35·82
14·89	6·59	12·99	15,37	4·13	8·15	24·84	16·49
10	6¼	13¼	10	*Nil.*	3	20	15
15·00	5·75	*Nil.*	4·59	*Nil.*	*Nil.*	13·52	16·17

figures are not shown in the Reports.
paying Companies.

TEA. **Results of Working of**

FINANCIAL RESULTS.

Working of Tea Plantations.

COMPARATIVE TABLE OF INDIAN TEA

Showing the results of Work

	Borelli Tea Company, Limited.	Luckimpore Tea Company of Assam, Limited.	Attaree Khat Tea Company, Limited.	Majuli Tea Company, Limited.	Dejoo Tea Company, Limited.	Dooars Tea Company, Limited.
CAPITAL— Paid up . . . . £	78,170	76,852	57,280	55,970	43,580	161,008
TOTAL— Area of cultivation . acres	993	961	1,399	1,184	841	3,927
Do. mature do. . „	988	906	1,239	1,035	733	2,957
PROPORTION OF YOUNG CULTIVATION . . per cent.	·50	5·72	11·44	12·58	12·84	24·70
YIELD (total crop)— Per mature acre . . lb	550	447	495	459	572	458
OUTTURN— (*Account Sales weight*) „	543,615	405,091	613,701	475,309	419,012	1,355,546
LOSS IN TARING . per cent.	·85	1·40	1·02	·50	1·23	2·17
CAPITAL VALUE— Per acre of Total Cultivation . . . . £	78-14-5	84-16-6	40-18-10	47-5-5	51-16-4	41-0-0

Results per lb, etc., on the

	s. d.	s. d.	s. d.	s. d.	s. d.	s. d.
TOTAL COST*— Average per lb	0 0·80	0 10·53	0 8·84	0 9·62	0 9·15	0 7·98
SURPLUS . Commission to Managers, average per lb	0 0·40	0 0·22	0 0·12	0 0·36	0 0·18	0 0·18
SURPLUS . Profit for Shareholders, average per lb .	0 2·44	0 2·03	0 1·87	0 2·04	0 1·76	0 2·90
GROSS PROCEEDS— Average per lb . . .	0 11·64	1 0·78	0 10·83	1 0·02	0 11·09	0 11·06
SHAREHOLDERS' PROFIT— Per mature acre . . £	5-11-8	3-15-7	3-17-3	3-8-1	4-4-1	5-10-7
Per cent. on Total Cost* . . per cent.	27·73	19·26	21·16	21·13	19·30	36·28
Per cent. on Capital . „	7·04	4·46	8·35	7·20	7·08	10·16
DIVIDEND PAID PROFIT— Per cent. on Capital . „	7	5	7	7	7	10
RESERVE FUND PROFIT— Per cent. on Capital . „	5·57	2·78	*Nil.*	*Nil.*	*Nil.*	2·13

NOTE.—Where blanks appear the
* Old non-dividend

FINANCIAL RESULTS.

Working of Tea Plantations.

COMPANIES (REGISTERED IN LONDON).

ing, etc., in Season 1890.

Lebong Tea Company, Limited.	Land Mortgage Bank of India, Limited.	Darjeeling Company, Limited.	Borokai Tea Company, Limited.	Indian Tea Company of Cachar, Limited.	Langla Tea Company, Limited.	Shumshernugger Tea Company, Limited.	**Totals and Averages of the whole 27 Companies.**
82,070	360,380	135,420	43,560	94,060	36,100	21,100	**2,743,956**
1,544	8,393	2,094	1,055	1,063	955	812	**66,152**
1,254	7,857	1,839	938	813	820	665	**55,752**
18·78	6·39	12·18	12·47	23·52	14·14	18·10	**15·72**
317	297	324	295	495	502	473	**448**
397,962	2,334,793	595,578	276,212	402,628	412,050	314,552	**23,824,328**
1·09	—	1·87	1·54	1·68	1·08	1·02	**1·20**
53-3-1	42-18-8	64-13-6	41-5-10	88-9-9	37-16-0	25-1-98	**47-17-3**

Working of Season 1890.

s. d. 0 10·52	*s. d.* 0 9·79	*s. d.* 0 9·89	*s. d.* 0 10·15	*s. d.* 0 7·96	*s. d.* 0 7·29	*s. d.* 0 7·43	**s. d. 0 9·12**
0 0·32	—	0 0·34	—	—	0 0·17	0 0·9	**0 0·21**
0 3·26	0 1·20	0 2·52	0 2·95	0 4·80	0 2·12	0 2·62	**0 2·58**
1 2·10	0 10·99	1 0·75	1 1·10	1 0·76	0 9·58	0 10·14	**0 11·91**
4-6-3	1-9-9	3-8-2	3-12-6	9-18-3	4-8-11	5-3-6	**5-0-2**
31·02	12·27	25·55	29·10	60·38	29·16	35·34	**29·38**
6·59	3·24	4·63	7·80	8·57	10·10	16·30	**9·39**
6	*Nil.*	6	7½	7½	8	13	*** 3=Nil. 24=8·69**
34·74	*Nil.*	3·56	18·94	9·30	*Nil.*	*Nil.*	**10=Nil. 17=10·77**

figures are not shown in the Reports.
paying Companies.

HENRY EARNSHAW,
14, St. Mary Ax, E.C.

TEA. **Cultivation and Manufacture of Tea**

CULTIVATION
118

CULTIVATION.

Space cannot be afforded to deal province by province with the various systems of Cultivation and Manufacture, but the following brief remarks may be offered as manifesting some of the more interesting features, or, at all events, as denoting the distribution of tea-growing in India. The numerous references quoted under each province will, it is hoped, enable the reader to procure the more useful publications that furnish fuller details and more technical information.

ASSAM & CACHAR.
119

ASSAM & CACHAR.

References.—*Griffith, Report on Tea Plant in Upper Assam, 1836; Manufacture of Black Tea as now prosecuted in Suddiya by C. A. Bruce, 1838; Assam: Sketch of its History, Soil and Productions with the Discovery of the Tea-plant, 1839; Report of Manuf. of Tea and Extent of Produce of the Tea Plantations of Assam by C. A. Bruce, 1839; also Tea Cultivation, Cotton and other Agri. Exper. in India—A Review by W. Nassau Lees, LL.D., 1863; Discovery of Tea Plant in Sylhet (Sel. Rec. Gov. Bengal, XXV., 1857; Memo. written after a Tour Through the Tea Districts (1864-65); Prize Essay on Cultiv. and Manuf. of Tea in Cachar, by H. A. Shipp, Esq., 1866; Report of Commission appointed to enquire into the State and Prospects of Tea Cultivation in Assam, Cachar and Sylhet, 1868; Assam Tea, by R. P. Wingrove, 1870; Report on Tea cultivation by Mr. Edgar (Parliamentary Paper on Tea and Tobacco Industries in India), 1874; Tea in Assam, Origin, Culture and Manufacture of, S. Baildon, 1877; Cultivation and manufacture of Tea by Col. E. Money (1883), 15; Tea operations in Assam, 1873-74; Papers regarding the Tea Industry in Bengal (History of Assam cultivation, pp. 121-128), 1873; Correspondence regarding Tea cultivation in Assam (Sel. Rec. Beng. Gov. XXXVII., 1861, pp. 1-73); Robinson, Account of Assam, 1841; Report on Tea-mite by J. Wood-Mason, 1884; Tea cultivation, evidence received from Gardens in Assam, Cachar, Sylhet, Darjeeling, etc., 1870; Tea Cultivation in Assam and Cachar—Tea Cyclopædia, 236-237; Annual Reports on Tea Culture in Assam by Secretary to Assam Government 1881-91; Administration Reports of Bengal (including Assam) from 1860-61; Tea in the Andaman Islands—Tea Cyclopædia, 262-263; Indian newspapers:—Englishman, Pioneer, Planters Gazette, Indian Agriculturist, Tropical Agriculturist, etc.,—passages too numerous for quotation; Agri.-Hort. Soc. Ind. Transactions, II., 153; III., 35; IV., 1-58; VI., 10; VII., 1-38; (Proc.) 45; VIII., 27-29, 282-301, 361, 380, 389; Journals Vol. I. (1841-42), Proc. 9-40; II., Pt. I., 337-345; III., 1-8, 61-69; (Sel.) 102; V., 79-82; (Proc.) 47; (Sel.) 132-135; IX., 201-207; 207-210; 342-352; (Proc.), 1857, clxxxix; X., 193-204; XII., 113-122; 164-175, 299-310, 364-379; XIII., 31-47; XIV., 282-294, 303-339; (Proc.), 1867, xliii-xliv; Journal, New Series, IV., 126-132; VI., 82-87; VII., 364-365; (Proc.) xxxvii, xlii, lxxvii, lxxxiii, lxxxix, xcv, xcix, clix, clxxxv.*

Climate.
120

Climate, Soil, etc.—Cultivation of tea commenced in 1835, and the first Company in Assam was organised in 1839. This is not only the home of the tea plant of India, but it is the province in which cultivation has been carried to the greatest extent. In official returns the province is generally spoken of under two sections: the Brahmaputra Valley (Assam proper) and the Surma Valley (or Sylhet and Cachar). The northern portions of the province are commonly held to be superior to the southern for tea cultivation, owing to the higher rainfall in spring. **Colonel Money** remarks: "The tea plant yields most abundantly when hot sunshine and showers intervene. For climate, then, I accord the first place to Northern Assam. Southern Assam is, as observed, a little inferior. The soil of this province is decidedly rich. In many places there is a considerable coating of decayed vegetation on the surface, and inasmuch as in all places where tea has been or is likely to be planted, it is strictly virgin soil, considerable nourishment exists. The prevailing soil also is light and friable, and thus, with the exception of the rich oak soil in parts

CULTIVATION in Assam and Cachar.

Climate.

of the Himálayas, Assam in this respect is second to none. As regards labour we must certainly put it last in the list. The Assamese, and they are scanty, won't work, so the planters, with few exceptions, are dependent on imported coolies; and inasmuch as the distance to bring them is enormous, the outlay on this head is large, and a sad drawback to successful tea cultivation." The rivers of Assam and Cachar are at present the channels by which the tea is exported, but it is anticipated that before long the province will be tapped by railways, when the difficulties both of export and of immigration of coolies may be greatly mitigated.

The climate of Cachar differs but little from that of Assam. **Colonel Money** adds "in one respect it is better: more rain falls in spring. The soil is not equal to Assamese soil; it is more sandy, and lacks the power. Again, there is much more flat land fit for tea cultivation in Assam, and there can be no doubt as to the advantage of level surfaces." Cachar is, moreover, not so distant from Calcutta, and this is of value not only in lowering freights but in favouring the supply of coolies.

Discovery by Bruce. 121

Original Discovery of the Tea Plant in Assam by Major Bruce.— Few features of the early historic records of tea cultivation in India have been more hotly contested than that of the person to whom the honour of first discovery should be assigned. **Mr. A. Burrell**, in a lecture delivered before the Society of Arts, London, on the 9th February 1877, announced a discovery made by him that would seem at first sight to leave little room for doubt that that honour should be paid to **Mr. David Scott**,—the first Commissioner in Assam, or rather the Governor General's Agent, as that officer was then called. **Mr. Burrell** says: "Some time after 1819, and certainly not later than 1821, **Mr. Scott** sent down to Calcutta to his friend, **Mr. James Kyd** (son of **Colonel Kyd**—the correspondent of **Sir Joseph Banks**—), a specimen of this tea, writing thus:—'The enclosed leaves are said to be those of the wild tea plant. I have not been able to get the flowers with seeds, but I have some plants that I hope will survive, and plenty more are procurable. The Shans, Burmans, and a Chinaman that is here, say that it is the tea plant, perhaps the species that **Mr. Gardner** sent down from Nepal.'" "That specimen" (continues **Burrell**) "was handed to **Dr. Wallich** by **Mr. Kyd**, and included in the Indian herbarium brought home by him, and presented by the East India Company to the Linnæan Society of London. When examining the catalogue of that collection drawn up by **Dr. Wallich**, I was struck by an entry titled '**Camellia? Scottiana**,' and on referring to the actual plant I found still attached to it the last sheet of **Mr. Scott's** letter, quoted above, in the shape of a P.S., but unfortunately without the date of the year, which also the post mark does not bear. I do not detain you with the evidence that satisfied me of the real time, and that this specimen was in no way to be confounded with others sent from Manipur by **Mr. Scott** in and after 1826, to the Secretary of the Indian Government, and others that have been confounded with it. I found also that **Mr. W. T. Thiselton Dyer** of Kew had recently examined the plant, and in a memoir he read to the Society, 'On the Determination of Three Imperfectly known Indian Tea Plants' in 1873, had described it, stating that, 'After careful examination I feel satisfied that **Mr. Scott's** belongs to the Assam tea plant, and the late **Dr. Anderson** appears, from a MS. note in the Kew Herbarium, to have arrived at the same conclusion.' Desiring to satisfy my surmise (continues **Burrell**) I applied to **Mr. Clark Marsham**, resident of Calcutta, at the time, and he was kind enough to write to **Mr. M'Clelland**—the very highest living authority on the subject—and that gentleman, in a letter I received yesterday, fully confirms the claim, writing thus:—'The circumstances brought to light by **Mr. Burrell** coincide exactly with what I

Scott stated to be the discoverer. 122

CULTIVATION in Assam and Cachar.

Discovery.

have always understood to be the fact, that specimens of the plant and seeds of the indigenous tea plant had been sent by **Mr. Scott** through **Mr. Kyd** to **Dr. Wallich** as early at least as 1821.'"

Now it may be pointed out that the full force of this argument turns on whether **Scott's** specimen in the Wallichian Herbarium was collected in Assam and by himself. **Wallich** says that while stationed in Cooch Behar, **Scott** took an interest in the introduction of tea. So many persons give **Scott** the credit of first discovery of the Assam tea plant that the author would be prepared to regard that discovery as a just tribute to the memory of a truly great man, who, if he was not actually the first discoverer, was one of the earliest discoverers of tea in Assam. The matter might be left thus were it not that **Wallich's** express statements on this subject appear to have been overlooked by **Burrell** and by nearly every other writer. There is also another point of some importance, *viz.*, that it is highly doubtful whether **Scott** had actually been in Assam so early as 1821. At all events, he furnished a report of the weather of a portion of the district of Rungpore in 1823 (*see Trans. Agri.-Horti. Soc., Ind., I., 82*); and it should also be recollected that war was only declared against the Burman invaders of Assam, by the British Government, on the 5th of March 1824. Little more than a year from that date the Burmans had been driven from Assam, and **Mr. Scott,** during the military operations, had the direction of all civil matters. When the conquests were completed he became Governor General's Agent on the North-East Frontier, and **Captain Neufville** was associated with him, as in command of the troops. There is thus the very strongest presumptive evidence that **Scott** had very probably not visited the Assam region, from which wild tea is said to have been sent by him to **Wallich**, before the year 1824. According to **Burrell, Scott** collected wild tea and sent it to **Kyd** in 1821. If he did so, it must have been procured in Rungpore or from the lower portion of the Assam Valley, the portion which was in the possession of the East India Company before the year 1824. But **Dr. Wallich's** statements are very much to the point. In the Proceedings of the Agri.-Horticultural Society of India, Vol. I., for the year 1841-42, *pp. 9-40* (Proceedings which formed a large separate volume), a discussion arose as to the claim made by **Captain Charlton** for the Society's gold medal, on the ground that he was the discoverer of the wild tea of Assam. **Charlton** appears to have thought that certain letters had been withheld in the reprint of the correspondence that had been placed on the tables of the House of Parliament. **Dr. Wallich** was then a member of the "Tea Committee" appointed by order of **Lord Bentinck**, and he, therefore, regarded that charge as especially directed against him. He accordingly furnished a long report and published all the letters that had passed between himself personally or the Tea Committee and **Scott, Bruce, Jenkins, Charlton** and others. There are certain very significant points in the correspondence, (1) there is no mention of specimens sent by **Scott** to **Kyd** and which, by the latter gentleman, are said to have been forwarded to **Wallich,** (2) the three first letters from **Scott** to **Wallich** are dated from the British portion of Assam or from Eastern Bengal, *viz.*, November 30th, 1823 (Gowalpara); 6th January 1824 (Singimari); and 2nd August 1824 (Gowalpara). These either ask for or acknowledge receipt of the tea seed which he desired to obtain from **Wallich.*** Some time before,

Scott was not in Assam before 1824. 123

Captain Charlton's claim. 124

* The writer has examined **Scott's** letter to **Kyd** in the Wallichian Herbarium (quoted by **Burrell**), and it is not oneof the letters given by **Wallich** in the Journal of the Agri-Horticultural Society; nor are **Scott's** own specimens and drawings which he sent to **Wallich** in that Herbarium.—*Ed.*

CULTIVATION in Assam and Cachar.

Discovery.

Sir Joseph Banks had mentioned Kuch Behar and Rungpore as regions likely to prove suitable for tea culture, and **Scott's** first letters with **Wallich** appear to have been written in consequence of a desire to procure seed. Had he discovered tea in Rungpore it might fairly have been expected that he would have made that announcement with some pride, seeing that he had previously been so anxious to introduce the plant from **Colonel Kyd's** Calcutta stock, raised in 1780 from China seed. A year later (1825) **Scott** forwarded a drawing of a wild tea fruit, and by a *P.S.*, added that he had forwarded specimens. He does not claim to have discovered the plant and does not seem to lay much stress on its discovery. The tone of his letter implies that **Wallich** must have previously heard of the discovery of the tea plant in Assam. In 1827 he again wrote to **Wallich**: "I have the pleasure to forward by this day's *dâk* a small box containing seeds, said to be those of the tea plant, and which have lately been received from a chief residing on the borders of Yunnan. About a year ago I had the pleasure of addressing you on the subject of the Assamese tea plant, and at the same time forwarded some seeds *preserved for inspection.*" Had **Scott** sent specimens in 1821 he would surely not have written in 1827 that "*about a year ago*" he had sent other specimens of the *Assamese tea,* and his Assam plant is thus spoken of apart from his later discovery of the tea plant in Manipur.

Scott's letters to Wallich.

China seed sent to Calcutta in 1780.

125

It would seem perfectly clear, however, that **Wallich** made a mistake. He had regarded the plants first sent to him as those of a **Camellia** but not **C. theifera,** *Griff.* It was not until **Captains Jenkins** and **Charlton** rediscoverd the plant in 1834 that **Wallich** admitted he had obtained evidence, sufficiently strong, to justify the announcement that Assam possessed the true tea plant. He had previously given the subject only a passing consideration and now sought to undo the effects of his former opinion. **Wallich,** for some unaccountable reason, would thus seem to have been prejudiced against the possibility of an indigenous tea or of the value of such a discovery. In 1832 he reported that the attempts made to introduce tea cultivation in Penang, Ceylon, and Java had resulted in failure, and he apparently wished the inference to be drawn that it was unlikely the commercial article could be produced anywhere out of China. When, however, **Lord Bentinck's** keen interest in an experiment to introduce into India the tea industry became known, **Wallich** was aroused from his former apathy. He next erred in giving too much prominence to the discovery just then made by **Captain Charlton.** All he meant to say, however, was that evidence had at last been produced sufficient to remove all doubt as to the existence of the tea plant in Assam. This position was irresistible, for he not only obtained from **Charlton** botanical specimens, but fairly good tea made of the Assam indigenous plant. Upon the announcement thus made by **Wallich, Charlton** claimed to be the discoverer of the tea plant, and **Wallich** was then forced to confess his shortcomings in order to do justice to the real discoverers—the brothers **Bruce.** "I will not deny—never—on the contrary, I will proclaim loudly that I have been sceptical as to the solution of the question, 'Is the Upper Assam shrub, a **Camellia** or a real **Thea** or Tea;' but that we could not venture to decide the question until we had seen the seeds." He seems to have forgotten, however, that ten years before **Charlton** had done so, **Scott** had furnished him with seed. But in discussing **Charlton's** claim he wrote to **Jenkins** on the 15th March 1836: "It was **Mr. Bruce,** and his late brother, **Major R. Bruce,** at Jorehath, who originally brought the Assam tea into public notice, many years ago, when no one had the slightest idea of its existence; a fact to which the late **Mr. David Scott** has borne ample testimony." **Mr. C. Bruce,** in a letter to **Captain Jenkins,**

Wallich's mistake.

126

Lord Bentinck's action.

127

Scott bears testimony in favour of Bruce.

128

TEA. **Cultivation and Manufacture of Tea**

CULTIVATION in Assam and Cachar.

Discovery.

says: "My brother was the first person that ever thought of the tea plants in these parts. Before the Burma war, when he was at Rungpore, he offered a musical snuff box for two plants, to the Beesa Gaum. In the course of the war I was at Suddeya. I begged and got from the same man a canoe full of the plants and seeds. Since that **Mr. Scott,** and **Captain Neufville,** and every one in Assam have been in possession of them." In a further letter **Mr. C. Bruce** gives **Captain Jenkins** a brief account of himself. He says, "at the breaking out of the Burma war" (1824) "I offered my services to **Mr. Scott,** then Agent to the Governor General, and was appointed to command gun-boats. As my command was in Suddeya, I was the first who introduced the tea seeds and plants, and sent them to **Mr. Scott** and other officers below. My late brother, who was in Assam before the breaking out of the war, had previously informed me of their existence." It may accordingly be concluded that **Scott's** anxiety to test the value of **Sir Joseph Bank's** suggestion that tea might be grown in Kuch Behar and Rungpore, may have inspired the two **Bruce's** to keep a sharp look-out for the plant. Seeing that crudely prepared tea was often imported into Eastern Bengal from Assam and that it was regularly used by many of the Siam invaders of the valley, it is not difficult to understand how they might have come to be shown the plant. The above and many other letters were published by **Wallich** in Calcutta in 1841, and thus at a time when the subject must have been fresh in the memory of many persons. Neither **Captain Charlton** nor any one else ventured to challenge the accuracy of the statement published by **Wallich.** On the contrary **Major Wilcox** wrote to the Secretary of the Agri.-Horticultural Society that there was no doubt **Mr. Scott** was aware of the existence of the tea plant in 1825 "when I first met him." "During his short visit to Suddeya in 1826, I well recollect his making particular inquiries regarding it of the Singfoh Chiefs who were assembled to meet him." "The Beesa Gaum promised to produce it, and accordingly, on his return home, he immediately sent in five or six plants." "**Bruce** was then with us at Suddeya, and I see no reason to doubt his statements that he sent down plants and seeds." An anonymous writer "Tea-plant, etc., of Assam, 1839" says **Major Bruce** went to Assam in 1823 with an assortment of goods. "He formed the acquaintance of a Singpho Chief, Beesa Gaum, with whom he made a written engagement to be furnished with some tea-plants." On the completion of the war, the Beesa Gaum came to **Mr. C. Bruce** and showed him the agreement he had made with his late brother **Major Bruce.**

Chronological order of the Pioneers of the Tea Industry.

129

Sir J. Banks. Colonel Kyd. Mr. D. Scott. Major R. Bruce. Lord Bentinck. Mr. C. Bruce. Captain Charlton. Captain Jenkins, etc., etc.

Conf. with pp. 499-460.

The reader is now possessed of all the facts over and above those given by **Mr. Burrell** in his lecture to the Society of Arts. He will therefore be in a position to make up his own mind as to whether **Scott** or **Bruce** should be regarded as the real discoverer of the Assam wild tea plant. Personally, the writer thinks the tea industry of India should primarily be attributed to **Sir Joseph Banks**; then to **Colonel Kyd** who procurred and cultivated the first Chinese seed ever grown in India (1780); next to the enlightened action of **Mr. David Scott** who appears to have taken steps to put **Sir Joseph Banks'** suggestion into action and inspired those around him with interest in the subject. But without **Lord Bentinck** it is highly probable that **Jenkins'** and **Charlton's** rediscovery of the plant would have shared the fate of the earlier discovery. To the wisdom, ability, and energy, of **Lord William Bentinck** India is most undoubtedly very largely indebted for this valuable accession to her wealth. In an address to his Council, on January 24th, 1834, he made it clear that he was to leave nothing untried that might help to attain the object he aimed at—the acclimatization of the best Chinese plants in India. But if credit of an exceptional character be

CULTIVATION in Assam and Cachar.

necessary, for any one of the pioneers who may have first seen the indigenous Assam plant, there would seem little doubt that credit must be given to **Major R.** (not to **Mr. C.**) **Bruce.** But it was **Scott** who knew how to utilize that discovery, and but for the apathy shown by **Wallich**, when his attention was first drawn to it, the tea industry of Assam might have been started ten years earlier than it was. It was the anticipation of trouble with China that induced the Board of Directors of the East India Company to recommend the subject of the introduction of tea into India, to the consideration of **Lord Bentinck**, who was then about to proceed to India as Governor General. That nobleman lost no time in arousing interest. **Wallich** and others urged the suitability of the temperate or sub-temperate tracts of the Himálaya and recommended the effort being made to obtain fresh seed from China. The subject became freely discussed and, it may be added, **Mr. Gordon — Dr. Wallich's** Secretary in the Tea Committee—was on his way to China on board the *Water Witch* when **Charlton** and **Jenkins** rediscovered the Assam indigenous plant. The Tea Committee then addressed His Excellency the Governor General in these words:—"It is with feelings of the greatest possible satisfaction that we are enabled to announce to His Lordship in Council that the Tea Shrub is, beyond all doubt, indigenous to Upper Assam, being found there, through an extent of country of one month's march within the Honourable Company's Territories." The Committee then added that they were not altogether unprepared for this discovery. It is remarkable, however, that that Committee had not thought it necessary to mention to **Lord Bentinck** the fact that **Scott** and **Bruce** had previously reported the existence of a plant which these gentlemen held to be the true tea-yielding **Camellia.** Not only so, but a circular letter, issued by the Tea Committee, was sent to **Dr. Falconer** in the North West, to **Mr. Trail** in Kumaon, and to **Captain Jenkins** in Assam, in precisely similar words, telling these officers to look out for situations that might likely prove suitable for the cultivation of the China tea plant. It was not explained to **Captain Jenkins** that **Dr. Wallich** had been unable to make up his mind regarding the reputed tea plant sent to him by **Mr. D. Scott** in 1824—26. These omissions are quite inexplicable, unless it be assumed that **Wallich** had a preconceived notion that it was impossible the tea plant could be found in Assam. That some such idea seems to have been in his mind there would appear to be little doubt, for, as **Royle** had, and as **Falconer** was then, advocating, he believed with **Dr. Abel** (the botanist who accompanied **Lord Ahmerst** to China) that a country that had a winter of at least six weeks or two months frost and snow was essential to the successful cultivation of tea.

Temperate Countries alone recommended for Tea. 130 *Conf. with p.* 438.

Mr. Gordon despatched to China for seed. 131 *Conf. with p.* 450.

A mistaken idea. 132

But the rediscovery of the plant in Assam was the result of the abovementioned letter in which the Chief Commissioner (**Captain Jenkins**) was desired to look out and prepare a suitable place for the China plants which, it might be expected, would shortly be sent to him for experimental cultivation. The Secretary to the "Tea Committee"—**Mr. Gordon**—had been, as stated above, deputed to China for the purpose of procuring seed and plants of the best varieties. On the Assam discovery being made known, it was precipitately assumed there was now no more any occasion to get plants from China. **Gordon** was accordingly recalled, only to be deputed to China a second time to complete the work he had begun. On his second return to Calcutta with the plants and seeds, he at once resigned being Secretary to the Tea Committee, and does not appear, as is customary with officers on such missions, to have published a report of his travels in China. That privilege devolved on **Mr. Robert Fortune,** who, at a later date, and in consequence of a third mission for tea, seed, and plants, visited China.

Gordon was recalled. 133

Mr. Robert Fortune's Mission to China. 134 *Conf. with p.* 442.

Lord Bentinck, however, very properly regarded the Assam discovery

CULTIVATION in Assam and Cachar.

as one that could not be allowed to incubate with the obscure problem of the separation of the tea plant into the genus **Thea** from the other species of **Camellia.** Whether a **Camellia** or a **Thea,** the Assam plant yielded excellent tea, which had been manufactured for centuries in a crude fashion by certain tribes in that province. It was also said to be distributed from the wild hilly tracts of Assam to the tea fields of China. These were statements that called for immediate investigation. A Commission was accordingly formed which consisted of **Drs. Wallich** and **Griffith** as botanists and **Dr. M'Clelland** as geologists. These gentlemen were directed to proceed to Assam and to there study the plant in its wild habitat, as also the soils and climates under which it was found. **Griffith** and **M'Clelland** favoured the opinion that the Assam home of the tea plant was precisely similar to that of two at least of the best tea districts of China. **Wallich** adhered on the main to his former view, that it belonged to a temperate loving family and would be most successfully grown on the hilly slopes of Assam or in the Himálaya. To this controversy may be attributed much of the disagreement that sprang up between two so eminent men as **Wallich** and **Griffith.** In guarded yet unmistakable language **Griffith** told his opinions, even though these were inimical to those held by his superior and colleague—**Dr. Wallich.** Looking back on the events that subsquently transpired, it is difficult to avoid the conviction that some share, of the irreparable loss India sustained in the prematurely early death of **Dr. William Griffith,** may be attributable to the misunderstanding between himself and **Wallich.** Thus true to the object of his deputation, but perhaps unwise in his own interests, he wrote, "I cannot conclude this part of my report without adverting to the desultory manner in which the question of tea culture in India has been treated by every author who has written on the subject, with the exception of **Mr. M'Clelland.** To what conclusion, but one, can we come, when we find an authority, who has been supposed to be acquainted with the question in all its details, stating very gravely that a temperature between 30° and 80° is requisite; and when we find that this is as gravely taken up by a popular and more philosophical author." Can it be doubted that the poignancy of that remark was aimed at **Abel** and **Wallich.** The Tea Committee at all events in its circular letter regarding the selection of suitable localities laid it down "that a decided winter climate of six weeks or two months' duration with frost as well as snow, is essential to ensure final success with really good sorts of tea." **Dr. Wallich** was the scientific officer of the Tea Committee, and as such was responsible for the publication and circulation of that erroneous idea—an idea that did much harm to the tea industry and ruined many of the pioneer planters. **M'Clelland** showed conclusively that **Dr. Abel** had arrived at a perfectly false opinion regarding the tea cultivation of a large portion of China. **Abel's** views were not only adopted by **Wallich** but by **Govan, Royle,** and **Falconer,** with the result that an undue importance was given to the tea plantations of Dehra Dun and Kangra. Speaking on this subject **Mr. W. Nassau Lees** says: "It is a source of regret that the experiment (of tea cultivation) was not placed under the superintendence of one or other of these officers (**Griffith, M'Clelland**) at a time when their services might have been of the utmost value, instead of being jeopardized, and all but abandoned, as was subsequently the case. For I have little doubt, that had the reports of **Drs. M'Clelland** and **Griffith,** received, at the time they were furnished, the attention that experience gained, has proved they were deserving of, India instead of *two,* might now (1866) be exporting annually *ten* million pounds of tea."

Commission appointed to visit Assam. 135

A winter climate regarded as necessary. 136

Results of Gordon's visit to China. 137

Gordon brought to Calcutta several casks of seeds, some plants and

CULTIVATION

Assam and Cachar.

eight or ten Chinamen. From the seeds 42,000 plants were raised which were distributed as follows:—

To Madras	2,000
,, Assam	20,000
,, the North-West Provinces	20,000
	42,000

C ina Tea: Mr. C. Bruce the first Indian Tea Planter, 1836. 138

The plants sent to Madras were six months after (*August 22nd, 1836*) reported to have all died, with the exception of a few that had been sent to the Nilghiri Hills. Those sent to Assam had a like fate. When they reached their destination only 8,000 were alive. **Mr. C. Bruce** was appointed to origanize a nursery and to take charge of the plants. No attempt had been made to keep down the weeds or to improve the soil, and **Dr. Griffith** found on inspecting the plantation that scarcely 500 were alive. Acclimatization of the plants had been entirely neglected, by their being freely exposed to the much stronger sun of Assam than that of China; they had accordingly been suffocated by weeds or scorched by the sun. The industry was thus on the eve of extinction when it was saved by **Griffith**. On his recommendation greater care was bestowed on the China stock, and the experiments begun by **Jenkins** and **White** in the cultivation of the indigenous plant, were prosecuted with more zeal and on more rational lines. **Griffith** urged the absolute necessity of careful cultivation. At last samples of Assam tea were seen in the markets, and the prices obtained were so encouraging that absurdly exalted expectations were entertained. A Joint Stock Company was projected in 1839—the Assam—with a nominal capital of a million sterling. In 1840, the Company commenced operations, and by the most wasteful extravagance had spent £200,000 without any prospect of a return. The Company was on the verge of proposing liquidation, when **Mr de Mornay** visited the garden. He saw the defects of the past efforts, *viz.*, expenditure in useless directions, with a total neglect of the plantation. **Colonel Hannay** had brought a small experimental garden of the *China* plant to a high state of perfection, and had thus demonstrated what was deficient in the Assam Company's Concern. Reform was thus imperatively demanded, and the results soon thereafter obtained were such as to bring into prominence many men not likely to be daunted by difficulties or by past errors of judgment. **Williamson, Warren, Jenkins, Barry, Martin,** and many others were in the field, and gardens sprang up in every direction. While the Himálayan cultivation was struggling to acclimatize an exotic plant, the Assam planters were discovering that it had been better for them at least, had they never seen the Chinese Tea. They were unlearning Chinese experience and developing methods of their own which were destined soon to give to Assam an industry that would raise it from the position of a penal settlement, to that of one of India's most prosperous provinces.

Indigenous Tea: First planted by Jenkins & White. 139

First Joint Stock Company, 1840. 140

Colonel Hannay's Garden. 141

Government ceased to have Tea Plantations in 1865. 142

Numerous reports were issued by Government (from the date of the appointment of **Bruce** in 1836 to the time when it ceased to have any direct interest in Tea (1865)) which contained every discovery that had been made and experience gained. It was freely announced that when the industry no more required the fostering care of Government, it would be handed over entirely to private enterprise. The progress made was such that long before the same position had been arrived at in the Himálaya, Government had ceased to have any direct interest in Assam Tea. The discovery of the Tea plant in Sylhet and Cachar also greatly encouraged the opening out of gardens in that province. In concluding this section, therefore, it may be of interest to add that the writer, while reading the

CULTIVATION in Assam and Cachar.

Expenditure incurred by Government. 143

numerous published and manuscript papers on the subject of Tea, came across certain statements of the expenditure incurred by Government in connection with the Tea Committee. **Mr. Gordon's** mission to China and **Drs. Wallich, Griffith** and **M'Clelland's** deputation to Assam, came to the total of £17,819—a direct gift from the country, in the creation of the now prosperous industry of Tea Planting.

Area, Outturn, etc. 144

Area, Outturn, etc.—The reader who may desire further historic facts, or to trace out the development of the present system of cultivation and manufacture, could not do better than read the numerous articles that appeared in the Transactions and Journals of the Agri-Horticultural Society of India. Some of the more important of these may be here named: Report on Physical condition of the Assam Tea Plant, etc., by **J. M'Clelland** (*Trans. IV.*); Discovery of Wild Tea in Tipperah Hills by **H. Walters**, on behalf of **Mr. P. Wise** (*Trans., VI.*); Report on manufacture of Tea and on the extent and Produce of the Tea Plantations in Assam by **C. A. Bruce**, (*Trans., VII.*); Report of samples of Assam Tea (*Trans., VIII.*); also correspondence with **W. J. Thomson** of Mincing Lane; **Messrs. J. Travers & Sons**, Swithin's Lane; **W. J. Bland** of Fenchurch Street and **Richad Gibbs**, White Hart Court, during 1840 on the samples of Assam Tea submitted to them for report; Memorandum on the Manufacture, etc., of Black Tea as practised in Assam by **J. Owen** (*Journ., II., Pt. I.*); Observations on Tea Culture by **J. W. Masters**, late Superintendent of Tea Plantations in Assam (*Journ. III.*); Assam Tea Plant compared with the Tea Plant of China by **J. W. Masters** (*Journ., III.*); Reports on the Sale of Assam Tea in London during February and March 1846 (*Vol. V.*); Notices Respecting the Culture and Manufacture of Tea at Cachar, Manipur and Darjeeling (*Vol. IX.*); Discovery of the Tea plant in Sylhet by **F. A. Glover** (p. 207), by **T. P. Larkins** (p. 242) (*Vol IX.*); Progress of Tea Cultivation in Cachar by **Lieutenant R. Stewart** (*Vol. X.*); Hints on the Cultivation of Tea by the Nursery method in Cachar by **C. Brownlow**, (*Vol. XII.*); Hints for the formation of Tea Gardens and the Culture of Tea by **Dr. J. B. Barry** (*Vol. XII.*); Notes on the Cultivation of Tea in Assam by **A. C. Campbell**, (*Vol. XII.*); Reports on Tea Cultivation for Season 1861-62 in Assam, Cachar, Sylhet, and Darjeeling, by **Major W. Agnew** (*Vol. XII.*); Observations on the Assam Tea Plant in Upper Assam by **J. W. Masters** (*Vol. XIII.*); Particulars regarding the yield of Tea from Plants of different ages in certain districts of India (*Vol. XIV.*); Prize Essay on the Cultivation and Manufacture of Tea in Cachar by **H. A. Shipp** (*Vol. XIV.*); Journal, New Series; The Tea Bug of Assam (*Vol. IV.*); Results of Trials with certain Manures on Tea Gardens in Assam, Cachar, and Chittagong (*Vol. VI.*); Tea Insect and Blights in Cachar (*Vol. VII. Proc.*), etc.

Publications on Assam Tea. 145

The returns of the crop in 1890-91 showed a total of 230,822 acres, but the actual area held by the planters has for some years past been given as a little under 1,000,000 acres. The detailed report of 1890-91 has not as yet reached the writer's hands, but the figures of 1889-90, which were only published on the 27th April 1891, may be accepted as affording fairly recent particulars. The number of gardens were then shown to have been 867. The report gives the outturn of the province and of the districts of the province as follows:—

Outturn. 146

Total Outturn of the Province.—The total outturn of tea during the year is reported as 82,119,252lb, or an increase of 5,203,424lb, or 6·77 per cent. as compared with 1889. The following table compares the figures according to the Indian Tea Association and according to the

CULTIVATION in Assam and Cachar.

Area, Outturn, etc.

trade returns with those supplied by Deputy Commissioners :—

	1889.			1890.		
	Brahmaputra Valley.	Surma Valley.	Total.	Brahmaputra Valley.	Surma Valley.	Total.
	lb	lb	lb	lb	lb	lb
Outturn according to Indian Tea Association . .	42,030,564	29,750,054	71,780,618	45,416,721	31,472,703	67,889,424
Outturn according to trade returns (for the financial year) . .	44,843,739	30,364,992	75,208,731	44,793,380	31,528,676	76,322,056
Outturn according to Annual Tea Report .	45,539,727	31,376,101	76,915,828	48,144,401	33,974,851	82,119,252

There is thus a difference of over five million pounds between the first and last set of figures, and it is probable that the Deputy Commissioners have over-estimated the crop.

Outturn by Districts.—The following table is interesting, as giving the outturn of each district for the last two years :—

District.	Rate of outturn per acre.		Total yield.		Increase or decrease.	Percentage of increase or decrease.
	1889.	1890.	1889.	1890.		
	lb	lb	lb	lb	lb	lb
Cachar . .	312	336	15,631,692	16,966,008	+1,334,316	+8·54
Sylhet . .	432	454	15,741,409	17,005,843	+1,264,434	+8·03
Khási and Jaintia Hills . .	100	100	3,000	3,000	...	...
Goálpára . .	302	265	92,083	93,464	+1,381	+1·49
Kámrúp . .	192	194	1,163,727	1,152,086	—11,641	—1·00
Darrang . .	464	467	8,445,916	8,433,809	—12,107	—·14
Nowgong . .	340	360	3,521,595	3,823,377	+301,782	+8·57
Sibságar . .	378	413	17,615,211	19,083,484	+1,468,273	+8 34
Lakhimpur . .	514	493	14,701,195	15,558,181	+856,986	+5·83
TOTAL .	391	409	76,915,828	82,119,252	+5,203,424	+6·76

NORTH-WEST PROVINCES AND OUDH.

N.-W. PROVINCES & OUDH. 147

References.—*Report, on the Tea Plantations in the N.-W. Provinces by Fortune (Agra 1851); Papers on the Tea Factories and Plantations in Kumaon and Garhwal, Published by Authority, Agra, 1854; Bell, Remarks on Tea manufacture in the North-West Provinces of India; Tea Plantations in Kumaon by Dr. Geo. King, Sel. Rec. N.-W. P., II. (Second Series), 1869, p. 433; J. H. Batten, Tea Cultivation in Kumaon, (Jour. Royal As. Soc. X., 131); Tea in Kumaon, Saunders Monthly Mag., 1851; Lees, Tea Cultivation, etc., 1863, pp. 35-91; Recollections of Tea Cultivation in Kumaon and Garhwal by J. H. Batten—The Tea Cyclopædia, 245-253; Tea in Dehra Dun—The Tea Cyclopædia, 254-262; Selections from Records Government of India (Home Dept.), No. XXIII, 1857; Jameson, Selections from the Records of the N.-W. Prov. (Part XXXVII.), 1862; Selections from the Records of the Govt. of the N.-W. Provinces (Second Series), 1867-68; Agra Exhib. Cat., 1867; Atkinson, Him. Dist., 887-907; Agri.-Hort. Soc. Ind. Journ., I. (Lord Auckland), 288-*

CULTIVATION in N.-W. P. and Oudh.

289; II. (Jameson), 323-333; IV. (Jameson), 173-197; V. Sel. (Jameson), 146; VI. (Jameson), 81-118 (Sel.), 14-16; (Proc.) XLV.; VIII., (Sel.) (Fortune), Report of 1851, 1-14; XIV., 119-123, (Proc., 1866) XXII; New Series, III. (King on prunning), 82, 98.

Climate 148 *Conf. with p. 433.*

Climate, Soil, etc.—It seems scarcely necessary to traverse the numerous arguments for and against the Sub-Himálayan sites which were selected by the Tea Committee for the experimental cultivation of the plant in these provinces. The chief historic facts will be found traced out in the remarks below under the Chapter on Introduction and early cultivation. The first plantation was organised in 1836, and Chinese planters and manufacturers reached the Himálayan gardens in 1842. The first Company was formed in 1863, when the Government Paoree Plantation was sold for R10,00,000. The degree of success attained has given origin to the hottest possible controversy; some writers may still be found to maintain that these regions possess high claims that will always secure a certain extent of cultivation and a ready market for the produce. It seems sufficient for the present purposes to indicate some of the modern opinions that have been advanced by persons well qualified to the claim of guiding public opinion. **Mr. W. Cochran** has written :—" Dehra Dhun is a tract of country situated in 30 degrees 20 minutes, north latitude, and consequently within the limits of the best tea districts of China. This circumstance, and its moderate elevation of 2,000 feet above the sea level, would seem to indicate a locality not unsuited for tea cultivation. But the advocates for the superiority of Assam say that the moist heat of Eastern Bengal is wanting; consequently the Dhun is not adapted for dividend-paying gardens. On the other hand, it is found that it possesses a tolerably rich soil, consisting of clay and vegetable remains resting on a gravelly substratum of lime, sandstone, clay-slate, and quartz, but destitute of iron; such a soil, in fact, as, arguing from its similarity to that of the Chinese Moyuen district, ought to be devoted to the production of green tea only. Accordingly, the experiment has been tried with considerable success. Fair, and even fine *young hyson*, and *gunpowder* are now prepared, and sell on the spot at R1·2 per ℔, equal to 2*s*. 3*d*. sterling."

Arrival of Chinamen 1842. 149

Colonel Money, in his work on the *Cultivation and Manufacture of Tea*, which is deservedly popular with Planters as a useful manual, advances opinions which, on the whole, are unfavourable to tea cultivation in these provinces, when contrasted with the results that have been obtained in Assam, Cachar, Darjeeling, etc. His views may be here given, therefore, on the subject of soil, climate, etc., in relation to cultivation. He deals with Dehra Dhun, Kumaon, and Garhwal separately :—

Dehra Dhun. 150

DEHRA DHUN.—" The lucky men, two officers, who commenced the plantation, sold it, I believe, in its infancy, to a Company for five lakhs of rupees. What visions did Tea hold forth in those days! In climate the Dehra Dhoon is far from good. The hot dry weather of the North-West is not at all suited to the Tea plant. Hot winds shrivel it up, and though it recovers when the rains come down, it cannot thrive in such a climate. One fact will, I think, prove this. In favourable climates, with good soil and moderate cultivation, 18 flushes or crops may be taken from a plantation in a season. With like advantages, and *heavy* manuring, 22 or even more may be had."

Referring to the subject of yield as given in the report of 1857, **Colonel Money** holds that it was a mistake to suppose that only five flushes could be obtained. He tells us that 10 or 12 flushes may be got with high cultivation, but then adds," what is this as against twenty and twenty-five." " Labour is plentiful and cheap. The great distance from the coast makes transport very expensive."

Kumaon. 151

KUMAON AND GARHWAL.—**Colonel Money's** views on these districts are as follows :—

"It was in this district (a charming climate to live in, with magnificent scenery to gaze at) I first planted Tea in India, and I much wish for my own sake, and that of others, I had not done so. I knew nothing of Tea at the time, and I thought a district selected by Government for inaugurating the cultivation, must necessarily be a good one. . . . . . Yet, there it was, Government made nurseries, distributed seed gratis, recommended the site for Tea, and led many on to their ruin by doing so. The intention of the Government was good, but the officers in charge of the enterprise were much to blame, perhaps not for making the mistake at first (no one *at the first* knew what climate was suitable), but for perpetuating the mistake, when later very little enquiry would have revealed the truth. I believe it was guessed at by Government officials long ago, but it was easier to sing the old tune; and a very expensive song it has proved to many. . . . . I need scarcely, after this, add that I do not approve of Kumaon for Tea. An exhilarating and bracing climate for man is not suited to the Tea plant. The district has one solitary advantage—rich soil. I have never seen richer, more productive, land than exists in some of the Kumaon oak-forests, but even this cannot in the case of Tea counterbalance the climate. Any crop which does not require much heat and moisture will grow to perfection in that soil. Such potatoes as it produces! Were the difficulties of transport not so great, a small fortune might be made by growing them. Could any part of Kumaon answer for Tea, it would be the lower ranges of the hills, but these are precisely the sites that have *not* been chosen. Led, as in my own case, partly by the Government example, partly by the wish to be *out* of sight, of the 'horrid plains,' and *in* sight of that glorious panorama, the snowy range, planters have chosen the interior of Kumaon. Some wisely (I was not one of them) selected low sites, valleys sheltered from the cold winds; but even their choice has not availed much. The frost in winter lingers longest in the valleys, and though doubtless the yield there is larger, owing to the increased heat in summer, the young plants suffer much in the winter. The outer ranges, owing to the heat radiating from the plains, are comparatively free from frost, but there again the soil is not so rich. Still they would unquestionably be preferable to the interior. Labour is plentiful in Kumaon and very cheap—R4 per mensem. Transport is very expensive. It costs not a little to send Tea from the interior over divers ranges of hills to the plains. It has then some days' journey by cart ere it meets the rail, to which 1,000 miles of carriage on the railroad has to be added. Since the above was written, Kumaon has secured a good local market, and I believe sells most of its Tea unpacked to merchants who come from over the border, to buy it. It has also improved its position greatly by making Green Teas, for which, as observed before, the China plant is so well fitted. With those two advantages though the climate is inferior, I suspect that Tea there now pays better than in Darjeeling. Gurhwall is next to Kumaon, and so similar that I have not thought it necessary to discuss it separately. The climate is the same, the soil as a rule not so good. There is one exception though, a plantation near 'Lohba,' the Teas of which (owing, I conceive, to its peculiar soil) command high prices in the London market. The gardens, both in Kumaon and Gurhwall, have been generally much better cared for than those in Eastern Bengal. As a rule they are private properties managed by the owners."

CULTIVATION in N.-W. P. and Oudh.

Kumaon.

The reader will find much useful information regarding the Tea Industry of Kumáon in the chapter on that subject (from the pen of **Mr. J. F. Duthie**) which appeared in **Atkinson's** *Himálayan Districts.*

Introduction. 152

Introduction and Early Cultivation.—In the remarks in the corresponding section to the present under the chapter devoted to the province of Assam, the writer has endeavoured to show that two widely different opinions prevailed regarding the climate best suited for tea cultivation. **M'Clelland** and **Grffith** advocated the claims of Assam, while **Govan, Wallich, Royle, Falconer** and **Jameson** held that the Himálayas afforded a better token of ultimate success. The Tea Committee confessed, however, that their selection of the sub-Himálaya was largely in consequence of a report they had received from **Dr. Falconer.** "For the facts and reasonings which led them to adopt the sub-Himalayan regions as entirely suitable for the projected culture, they relied especially on the able and interesting report that they had received from **Dr. Falconer** on the subject." **Dr. Royle**, in a paper read before the Royal Asiatic Society, showed that he had advocated the exact same arguments as **Falconer**, though quite independently. The results of the past half century have abundantly proved the former opinion to have been on the main the correct one, but experience has not belied entirely the position taken by the other side. On the contrary, it is now admitted that even in China, the dis-

Advocates in favour of the Himalaya. 153

CULTIVATION in N.-W. P. and Oudh. Introduction

tinction into cultivated races suited to cold regions and others to sub-tropical conditions—races that formerly may be said to have been represented by the two forms—**Thea viridis** or plant of the Northern tea districts of China and **Thea Bohea** or plant of the Southern—is of considerable agricultural, though not of botanical, value. The former is a more temperate-loving plant than the latter, and it affords some of the best green teas. Roughly speaking, this distinction exists in India in the temperate-loving plant of the hilly plantations and the sub-tropical stock of the plains. It had not been shown that a tea prepared from the Assam indigenous stock would meet with favour, and, therefore, if it was desired to attempt the cultivation of some of the better Chinese plants, a colder clime than Assam would be preferable. And this view has been abundantly confirmed. In all or at least in most of the hilly tracts of India, the China plant, or a hybrid with a strong strain of the China stock is preferred to the Assam. It is where the purest forms of China stock are grown that the best green teas of India are manufactured. The rich soft flavour of the Kangra tea approaches nearest to that of the old favourites of China. Experience has thus given a new elemant to the tea trade which it took some time to educate the consumers to appreciate, namely, the strength of flavour of the Indian teas, more especially of those of Assam and Cachar. It does not seem necessary to traverse the ground which by so many writers has already been beaten into winding pathways, and pathways through scattered and often conflicting records but which in the end led to one termination, namely, the above conclusion. **Wallich** never gave his opinions in the form of a separate report, but **Royle, Falconer, Fortune, Jameson** and many other writers each tried to vie with the other in laudation of the prospects of the Himálayan plantations. Though these plantations still exist, the student of this subject, after working through a library of records and books of travel, comes unavoidably to the conclusion that it had been better for India if less had been said of Dhera Doon, Kumaon, and Kangra, and more of Assam, Cachar, the Duars, Darjeeling, etc. The contagion of wild anticipations from Himálayan tea affected Europe quite as much as India, and sometimes even assumed ludicrous proportions. In one of the leading London daily papers, dated 6th October 1863, an editorial was published much in the strain of the writings of that time. While confessing ignorance of Indian matters as a national failing in Great Britain, the writer proceeded to mingle truth with error in a ridiculous manner. The article deals, however, with some of the main expectations and figures of the experiments which were then being performed, and may serve the author's purpose as well as any of the more accurate though less readable official productions that are at present bestrewed around him :—

The Contagion of preference for the Himalaya. 154

"The chemists have been at considerable trouble to tell the world why it drinks tea; some of them have gone so far as to declare that the only benefit of the beverage was the milk and sugar put into it; but, with its usual sublime disregard of science when it finds something to its taste, mankind goes on 'making tea' and leaves the *savans* to fight out the question about 'theine' and 'theobromine.'

Conf. with pp 443, 444.

Tea-Caddy. 155

Conf. with p. 465.

Clearly this is not a topic on which we can any longer pretend to be indifferent, and the first item of news from India accordingly concerns everybody, because it concerns tea-drinking. 'The Governor-General will leave Simla in September, and his earliest visit will be to the Kangra Valley to see the progress of tea-planting there.' 'What of that?' we hear *Materfamilias* say; and we reply to her deferentially, 'Madam, put another spoonful in 'for the pot;' for this news is good news for tea-tables.' Conquests do not usually concern the tea-caddy; but when we took the Punjab from the Sikh Singh, a great revolution in grocery was inaugurated. We found on the Kangra slopes and the uplands of the Murree hills just the kind of soil which the tea-plant delights in. What was more, we found a tea-plant growing wild; and the credit of putting two and two together is due to **Dr. Royle**, who urged upon **Lord Dalhousie**'s Government and his splendid lieute-

CULTIVATION in N.-W. P. and Oudh.

Introduction.

nants, the **Lawrences**, to plant tea there. New seeds and new plants were procured, and the long spurs of the Hímalayas grew rapidly green and then white with the foliage of **Thea viridis** and the scattered snow of its flowers. All sorts of obstacles, of course, arose. At Holta, for example, a fine sub-Himálayan plain, thousands of acres lay unused, because they were reported to belong to a 'djin'. We ploughed them to the horror of the valley men; sowed them; and in spite of genii and giants, reaped a crop that made tea cheaper, and the Chinamen anxious. This kind of domestic revolution has been quietly progressing ever since. Kumaon, Gurwhal, Dehra Doon, and a score of other localities in the north and north-west of India, have their tea-gardens, which four years ago covered three thousand acres. In 1861 the report upon the Kangra tea-gardens showed that they produced 30,000℔ weight of fine tea, 1,258 maunds of tea-seed—for a share in the last of which about 400 new growers applied. The greatest supply attainable did not meet a tenth of the demand, and the same was the case in the north-west, at Rawul Pindee, and the Simla Hills. Once embarked in the enterprise, the Governments of the Punjab and North-West Provinces showed a capital faculty for their opportunity. In one season they distributed 75 tons of tea-seed and 2,500,000 seedling plants, while they reserved 42 tons of seed and 4,000,000 young plants to extend their own plantations like those at Kangra which the Governor-General is now about to visit. These things are in the category of 'the not generally known' partly because tea-dealers are not much inclined to talk about them, and partly because of the dense ignorance prevailing among us about India, and languidly tolerated by those who know the magnificent land and wish it well. However, this cultivation, which a Viceroy goes out of his way at last to visit, is creeping over all the eligible uplands of the Himálayas, becoming an immense and established trade; and what is notable, quietly collecting about the cool and healthy seats of tea plantations a colony of Europeans, who seem likely to strike root in the soil.

Conf. with p. 475.

Even now that **Lord Elgin*** has made public the progress of Indian tea by a visit of state to 'young Hyson' we shall probably not feel at once the benefit of the growing culture. Himálayan tea has to fight its way against the brick-tea that comes in from Yarkhand and Lhassa— has to fill all the teapots of the Mussulmans and Hindoos, and to make a cheap and innocent substitute for deadly opium and mad 'bhang' in the bazaars of Hindoostan before it can challenge China. But we shall feel it before long; **Mr. Gladstone's** late remissions are altogether in its favour; so is the shorter distance to Europe, either by sea or through Afghanistan and Russia or Turkey; so is the popularity of Assam tea, which is a cousin of the Himálayan Hyson. Either brought direct, or cheapening Chinese tea by ousting it from many an Oriental market, we may be quite sure that the advantage of this fruit of Punjab conquest will very soon be felt at the British tea-table; and does not that mean by everybody? And if it should indeed turn out, as seems most probable, that a moiety of America's great cotton cultivation, and of China's huge tea monopoly are to be transferred permanently to India, how brilliant a prospect opens for this country and for that!— a prosperity worthy a statesman's energy to achieve; worthy a philanthropist's self-sacrifice to hasten; worth— shall we venture to say it?—worth, consummation even at the price of war, which first gave Peace these new and emerald-green garden-plots.?"

Conf. with p. 444.

It is needless to point out the numerous errors of the above passage, since they are for the most part self evident. The long spurs of the Himálaya are certainly not green through tea plantations, and the colony of Europeans, who seemed likely to strike root in the soil, nowhere exists. Though Himálayan tea has certainly not filled the tea-pots of the Mussulmans and Hindus (for a very good reason—they do not possess and never have possessed tea-pots), yet Indian teas have all the same successfully contested with China the European markets. But there is one misstatement in the above passage that did more harm than the imaginative pictures of future greatness—*viz.*, the oft-repeated assertion that a wild tea had been found on the Himálaya. **Moorcroft**, in 1821, pointed out that the natives of Basáhrh used the leaves of a species of **Osyris** in making a fairly good tea substitute. **Bishop Heber**, while on a visit to Almora, fell into the mistake of regarding that plant as 'a wild tea bush', and this mistake, though repeatedly corrected, has persistently re-appeared since, and been made by independent observers or by persons

Tea substitute.
156

* *Conf. with p. 443.*

CULTIVATION in N.-W. P. and Oudh.

Introduction.

Tea in Nepal in 1816.
157

unfamiliar with **Bishop Heber's** unbotanical remark. No species of **Camellia** occurs wild anywhere west of Nepal, though several are abundant plants in the Eastern Himálaya, more especially on the mountain ranges of Assam and Manipur. While this is so, a historic fact of some interest may be here added. The first specimen of a tea plant grown on Indian soil (other than the plants raised experimentally in Calcutta) was sent from Nepal in 1816 by the **Honourable E. Gardner,** then resident at the Court of that State. The specimen alluded to was ultimately forwarded by **Wallich** to **Mr. Don,** and by him was accordingly described in the *Prodromus Floræ Nepalensis.* It is presumed the plant must have been introduced into Khatmandu during some of the early expeditions from that kingdom into China, and **Wallich** appears to have obtained evidence that the seed of the bush at Khatmandu came from Pekin. But it is not stated that more than one plant was found in the palace garden, and it may be added it has not as yet been recorded as occurring in a wild state on the lower Nepal forest tracts, nor in Sikkim, though it is plentiful in Assam, and in Manipur it almost constitutes forests, the plants attaining to the dimensions of trees.

Fortune's Mission to China.
158
Conf. with p. 433.

In connection with the subject of Himálayan tea, it is desirable to allude here to **Mr. Fortune's** deputation to China. The stock imported by **Mr. Gordon's** two missions had apparently failed to suffice. It became necessary to obtain a further supply of seeds and plants. The writer has consulted the correspondence on this subject, and it may be said **Mr. Fortune's** deputation was mainly in consequence of representations by **Royle. Mr. Fortune's** first letter to **Royle** on this subject is dated 16th February 1848. In 1847 **Jameson** had submitted an able report on the tea experiments of the Himálaya. He dealt with soil, elevation, atmosphere and systems of cultivation and manufacture. In that report he announced an addition of 252,842 seedlings, and the manufacture of 1,023℔ 11 oz. of tea. The report was illustrated, and so much importance was attached to it that it was reprinted in the Selections from the Records of the Government, No. XXIII., and was thus widely distributed. The results attained were considered satisfactory, and, as **Mr. W. Nassau Lees** remarks, they "attracted the attention of Government to the desirability of, at last, taking efficient measures to provide for supplementing the labours of their zealous Superintendent, by obtaining the best information, and further supplies of seed and seedlings of the finest tea plants from China, as successively urged by **Drs. Royle, Griffith,** and **Falconer.** It was a fortunate circumstance, moreover, that the London tea brokers had reported most favourably on the specimens of tea sent home from the Himálayas the previous year." The opinions of the experts will be found in the *Agri.-Hort. Soc*, *Jour. VI.* (*Sel.*), *14-16.* **Mr. W. Hunt** said, "I am quite satisfied that the climate and soil of Kumaon is as suitable to the favourable growth of the shrub, as the finest of the China localities." **Mr. Thompson** said, "The flavour is very strong, and it is so coarse burnt that all richness of flavour is destroyed." So, again, **Messrs. Ewart** said, "In flavour the Dyhrah Dhoon most resembles the better descriptions of Orange Pekoe, having with the brisk flavour of that description, more than its usual strength." Was it to be wondered at therefore that Assam was for the time neglected, and that the scientific advisers of the Government triumphed over those who took a less favourable view of the Himálayan than of the Assam experiments? In 1848, accordingly, **Mr. Fortune** was directed to proceed to China for the purpose of obtaining the finest varieties of the tea plant, as well as native manufacturers and implements, for the Government plantations in the Himálaya. "**Mr Fortune** discharged the duties entrusted to him with

CULTIVATION in N.-W. P. and Oudh.

Introduction.

Fortune reached India in 1851.
159

energy and enterprise" and he reached Calcutta in March 1851. He introduced into India a large quantity of seed and upwards of 20,000 tea plants. **Mr. Fortune** then visited the tea plantations, but his report was not favourable. The system that had been pursued was defective and the whole of Dhera Dhun unsuited. Matters then assumed a different aspect, and tea cultivation in India was almost regarded as a total failure, for at this juncture the Assam Company had spent all its capital and could show nothing. Indeed, but for the courage and resources of **Dr. Jameson**, the enterprise might have been abandoned. He keenly felt the need of more technical aid, both in cultivation and manufacture. But Government had begun to weary of the expenditure on its tea estates, and began to remind **Dr. Jameson**, as they had impressed on **Mr. Bruce** and others in Assam, that the policy of the Government should not be lost sight of, namely, EXPERIMENT. No expenditure was to be incurred that might be regarded as passing beyond the definition of experiment into that of a commercial undertaking. The technical assistance which **Jameson** required was, therefore, refused him, and it was with the greatest difficulty that he succeeded in inducing the Government to depute **Mr. Fortune** a second time to China in order that he might himself study 'all the different processes of black tea manufacture, from the gathering of the leaf to the firing, preparing, and packing, of the teas, including the winnowing, sifting, etc., and thus be enabled to communicate the result of his observations.' In 1852, the Marquis of Dalhousie, then Governor General of India, visited Kangra and authorised the expansion of the experiment by the cultivation of Holta with tea. That particular locality had been selected and highly recommended by **Mr. Fortune**. But owing to the disturbed state of China **Mr. Fortune's** second mission was not so successful as his first. He returned to India, and on visiting the plantations in Dhera Dhun had to admit that **Jameson** had succeeded in demonstrating that some of his (**Mr. Fortune's**) strictures in his former report (1851) were unnecessarily severe. His report in 1855, accordingly, announced that "I have great pleasure in stating that I have never seen finer or more productive plantations in China." But he still found fault with much of what he saw in India. Some of his new suggestions were accordingly acted on (and proved beneficial), while others such as his arguments that tea should not be irrigated, subsequent experience has proved him to have been in error. To **Fortune**, **Dr. Jameson** again replied, in a most exhaustive report, the substance of which may be said to be a just condemnation of the parsimonious treatment of the tea experiment, in that he had not been furnished with assistants of sufficient skill and intelligence to see that his instructions were carried out. A half-hearted experiment, he practically contended, was worse than no experiment at all, since failures, criticised in the spirit **Mr. Fortune** had displayed, were likely to be regarded as proving the impossibility of tea ever being made a commercial success.

Fortune's Second Mission.
160

Conf. with pp. 440-1.

Effect of differences of opinion.
161

Thus in the Himálayas as in Assam the efforts of the Government had to a large extent been frustrated through the conflicting opinions held by the experts and advisers of Government, whose united action might have secured more liberal support. But these differences of opinion were not without their good influence, for every aspect of the experiment was in time put to final test. Experience was thereby gained, and India developed for herself systems of cultivation and manufacture of a higher and more scientific character than might have come into existence through a blindfold adoption of the cruder methods of China. The country reaps now the benefit and, as has been said in connection with Assam, so with the Himálayas, the tea industry of to-day owes much to the pioneers who in many cases were ruined during the evolution of opinion and the rejection

CULTIVATION in N.-W P and Oudh.

Area. 162

of error, that took place. But it may in conclusion be added that no name stands out as deserving of greater gratitude than that of **Dr. W. Jameson.**

Area, Outturn, etc.—It does not seem necessary to discuss these subjects separately in connection with this province, nor that of the Panjáb, nor of Madras. The information furnished regarding India, as a whole, at pp. 420-421 is believed to supply all that is necessary.

PANJAB. 163

PANJAB.

References—*Baden Powell, Pb. Pr., 275; Davies, Trade and Resources, N.-W. Frontier, CXCI., XV., VI., 68; Aitchison, Hand-book of the Trade Products of Leh, etc., 33-40; also Notes on Prod. W. Afgh. and N. E. Persia, 32; Papers connected with the Cultiv. of Tea in the District of Kangra (Sel. Records Pb. Govt., No. XIV.), 1853; Report by Col. E. H. Paske on the Tea Plantations of Kangra in 1869; Jour. Agri-Hort. Soc. Ind., XIV. (Proc. 1866) LXIII.; Tea Plantations in the Panjab (Sel. Rec. Pb. Govt.) IV., 1859; Cultiv. in N.-W. P. and Pb. (Home Dept. Sel., XXIII., 1857); Tea Culti. in Kangra Dist. (Sel. Rec. Pb. Govt. New Series, V., 1-35 1869); Report Settlement of Kangra Dist., 25-28, 80; Notes on Tea Industry of Kangra; The Tea Cyclopædia, 238-245; Koteghur 265; Brick tea and Trade with Ladakh in Allen's Indian Mail March 1865 (reprinted from "Friend of India"); Panjab, Administration Reports; also Resources of Panjab July 1868.*

Climate. 164

Climate, Soil, etc.—So much has been incidentally said regarding this province in connection with the remarks under the headings of Assam and the North-West Provinces, that it does not seem necessary to do more than furnish **Colonel Money's** remarks regarding the physical features of Kangra. The reader who may desire further information should consult the publications and reports mentioned above, as also the chapter below regarding the trade in Kangra Valley Tea. **Colonel Money** says:—

Kangra. 165

KANGRA.—"This is a charming Valley, with a delightful climate, more favourable to Tea than the Dehra Dhun; still it is not a perfect Tea climate. It is too dry and too cold. The soil is good for Tea, better than that of the Dhun but inferior to some rich soils in the Himálayan oak forests. Local labour is obtainable at cheap rates. Distance makes transport for export very difficult; but a good local market now exists in the Punjab, and a good deal of Tea is bought at the fairs, and taken away by the wild tribes over the border. With the limited cultivation there, I should hope planters will find a market for all their produce. Manure must be obtainable (manure had not been thought of for Tea when I visited Kangra), and if liberally applied, it will increase the yield greatly." "Kangra is *not* the best place for a man who wants to make money by Tea, but for one who would be content to settle there, and content to make a livelihood by it, a more desirable spot with a more charming climate could not be found. Land, however, is not easily procured. The Teas produced in Kangra are of a peculiarly delicate flavour, and are consequently highly esteemed in the London market."

Holta Plantations. 166 *Conf. with pp. 440-1, 443.*

Mr. Cochran writes much in the same strain. A charming country, honourable occupation, and a fair return—inducements that have attracted to Kangra the only colony of European planters that can be said to have taken root in India (*Conf. with p. 441*). In addition to tea many of these planters have gone in largely for fruit growing, and the valley may undoubtedly be said to produce apples and pears that are unsurpassed anywhere in the world Kangra tea is finding a large market in India and across the Himálaya into Central Asia.

BENGAL. 167

BENGAL.

References.—*Papers Regarding the Tea Industry in Bengal, 1873, more especially—Chutia Nagpur, 8-14, Cooch Behar, 16-22, Dacca, 129, Chittagong, 160-166; Muller, Observ. on Culti. and Manuf. of Tea in the Darjeeling District (Jour Agri.-Hort. Soc. Ind., XIV., App. XXXVII.-LVI); Tea Cyclopædia—The Terai and Western Doars 264-265, Chutia Nagpur, 265; Col. Money, Cult. and Manuf. Tea (numerous passages); Notes on Tea in Darjeeling by Planter, 1888; Administration Re-*

ports of Bengal; Agri.-Hort.-Soc. Ind. Jour., II. (Sel.), 408-409; VI. (Note on Darjeeling Tea, etc., by Dr. Campbell), 123-126; X. (Report on Tea grown at Tugvor, Darjeeling), 107-109; (also Report by Capt. J. Masson of Tea grown at Tugvor) 227-28; also Progress Tea Cultivation Darjeeling, by Dr. Campbell, 229-237.

CULTIVATION in Bengal.

Climate. 168

Climate, Soil, etc.—Speaking generally it may be said the greater part of the tea plantations of this province are located in the hilly tracts. The physical features of the districts are, however, so very different as to necessitate their being treated separately.

Darjeeling. 169

DARJEELING.—To **Dr. Campbell** is due, it would appear, the honour of having introduced tea cultivation to this district In the Journal of the Agri.-Horticultural Society of India he wrote in 1847: "About six years ago I received a few tea seeds from **Dr. Wallich**; they were of the China stock grown in Kumaon. I planted them in my garden in the month of November 1841, and had a dozen seedlings in the month of May following." **Dr. Campbell** then tells us that in August 1847, his plants were examined by **Mr. Macfarlane**, a gentleman from Assam, and who was acquainted with tea planting. His report was so satisfactory that **Dr. Campbell** was induced to procure seed from Assam and extend his experiment. In several subsequent reports **Dr. Campbell** told of his continued efforts, and in 1858 he informs us of his having examined seven plantations which had by that time come into existence. The above may, therefore, be accepted as denoting the origin of tea cultivation in this district. Coming down to modern opinions **Colonel Money** says of Darjeeling:—

"The elevation of the station, 6,900 feet, is far too great; but plantations lower down do tolerably well (that is, well for hill gardens). The climate, like all hill climates, is too cold." "Like elevations in Darjeeling and Kumaon are in favour of the former, *first* because the latitude is less; *secondly*, because Darjeeling has much more rain in the spring. I believe, therefore, that the hill plantations of Darjeeling have a better chance of paying than the gardens in Kumaon, but, as stated before, no elevated gardens, that is, none in the Himálayas, have any chance in the race against plantations in the plains, always providing the latter are in a good Tea climate. In two respects, however, Darjeeling is behind Kumaon. The soil is not so good, and the land is much steeper. It is more than absurd, some of the steeps on which Tea is planted in the former; and such precipices can, I am sure, never pay. Gardens, barely removed above the Terai (and there are such in Darjeeling) can scarcely be called 'elevated,' and for them the remarks applied to the Terai are more fitting. As a broad rule it should be recognised that the lower Tea is planted in the Himálayas the better chance it has. All the plants in the Darjeeling gardens, with but few exceptions, are China."

In a small book by a practical Tea planter, some useful hints are given. Speaking of the soils and the nature of Darjeeling cultivation he says:—

"Tea will grow in almost any soil, but a black loam with a fair amount of sand in it is to be preferred; if the subsoil is clay, it will not do so well, except when the drainage is exceptionally good, in which case gardens situated on dry ridges often give a very fair first flush (which they would not do if the subsoil was sand), although if the rainfall afterwards is very heavy, they get clogged and blighted badly. Soil, which has many rocks in it, is always good, as the rocks keep it open, and some gardens, which are full of rocks, pay very well." "Some gardens have put out all their good land and are now extending on land that is positively precipitous and really dangerous to cross. This will never pay, as the soil will soon be washed away when cultivation begins, (leaving the roots bare). I saw a piece of new extension, the other day, which was too steep to terrace and grooves about 4 inches deep had been cut."

In a very instructive paper on "A Few Observations on the Cultivation and Manufacture of Tea in Darjeeling" (*Jour. Agri.-Hort. Soc. l. c.*) **Mr. D. C. Muller** offers certain remarks on the soils and climates of Darjeeling that help to elaborate a conception of the physical characteristics of this tea district:—

"The tea plant will *grow* in any soil, but *flourishes* in a light sandy loam. It is essentially necessary, whatever the surface earth may be, that the *subsoil* consist of a

CULTIVATION in Bengal.

Darjeeling.

yellowish earth, composed of clay and sand, sometimes passing into pure sand; the former gives sustenance to the plant, holding rain and moisture in deposit, and the admixture of the latter, renders the soil porous, and of such consistency that the tap root can easily penetrate it without injury to the spongioles. There are two descriptions of soil to be specially avoided,—*1st*, the stiff red, almost crimson clay, impervious to rain, and which cakes, and hardens under the sun,—*2nd*, the loose, friable micaceous, black-looking earth, intensely dry and hot, which although it may in a favourable, that is, in very rainy season give apparently healthy seedlings with *stiff* blackish leaves, will ultimately produce a stunted scant-yielding tree. A sandy loam, though stony, is far preferable to either of the above. I have heard some planters remark that they have seen the best trees on stony soils, and that, consequently, stone, instead of being a disadvantage, was decidedly beneficial to the Tea tree. I am inclined to think that too hasty an inference has been drawn from isolated facts, and that the *merit* due to the soil has been ascribed to the presence of stone. Stone undoubtedly retains heat, in a measure prevents the wash of soil, and in fact often retains rich earth in crevices; but these are, as far as I am aware, all the advantages it possesses. I should give the preference to good soil without it, and consider the first of the above enumerated advantages as decidedly disadvantageous in unfavourable soil."

"Planters are at variance as to the advisability of allowing any trees whatever to remain within that portion of the land intended for tea cultivation. The argument against it is that the shade of such trees must necessarily be injurious to the Tea plant, depriving it also of the quantum of rain it would otherwise receive."

Mr. W. C. Muller calls these reasons in question, while acknowledging that the seedling under the "*immediate drip of a tree*" would be likely to suffer injury, but not otherwise; the partial shade is of great benefit to a transplanted seedling. The reasons given for arriving at that conclusion being derived from a study of the localities and situations in which the indigenous tea tree has been found. The seedlings reared from seed which has been allowed to fall and germinate, being "especially *healthy and numerous* where they are *immediately under the fostering shade of the parent tree.*" A further objection to the wholesale clearing off of the trees is the well known attractive power of forests for rain clouds.

Darjeeling Terai. 170

TERAI BELOW DARJEELING.—"The soil is very good for Tea. The climate is also a good one, but there is not as much rain in the early part of the year as planters could wish. Much difficulty exists about labour, owing to the very unhealthy climate. As the jungle is cleared, however, this last objection will be in a measure got over. As it stands now, it is perhaps the most unhealthy Tea locality in India" (*Money*).

Western Doars. 171

WESTERN DOARS.—This, comparatively speaking, new tea district, seems to have attained a highly encouraging position. **Colonel Money** said of it in his 'Cultivation and Manufacture of Tea' (*Edition 1883*):—

"My attention was directed to it in 1874; I was the second who planted Tea in it; and I have now completed a garden there. As regards climate, soil and lay of land, it is perfect, and I believe it will eventually prove the most paying district in India for Tea. The Northern Bengal Railway, just opened, gives it great advantages for transport."

Chittagong. 172

CHITTAGONG.—It would appear from the Journals of the Agri.-Horticultural Society of India that **Mr. A. Sconce**, Collector, introduced tea into this district somewhere in 1841. He obtained a boat-load of plants from Assam from **Major Jenkins**, and started cultivation in his private garden. In 1843 he forwarded a sample of tea made by him in an earthen pot over the fire. The sample was very favourably reported on by **Mr. C. Terry**, from which date therefore the industry may be regarded as taking its origin. In the papers regarding the tea industry of Bengal, 1873, a long and very instructive report is given (written by **Mr. H. Hankey**, the Commissioner) in which it is stated that, in November 1862, **Dr. J. B. Barry** visited Chittagong, and, being satisfied with what he saw of the prospects of tea, commissioned **Mr. Fuller** to take up on his behalf 20,000 acres of land. After this, applications poured into the Collectorate for waste land. Some of the Companies that were started collapsed with the failure of the Agra

CULTIVATION in Bengal.

Chittagong.

Bank; others suffered through the ignorance of their managers; while of others, the sites were so badly chosen that they never could have paid. The Chittagong tea mania ran its course from 1863—1867, and the gardens that now (1873) remain had either weathered the storm or changed hands during the crisis, the present owners having benefited by the failure of others in being able to purchase at a low figure.

The chief features of the Chittagong tea districts are set forth briefly in the following passage from **Colonel Money's** work:—

"This is a comparatively new locality for Tea. The climate is better than Cachar in one respect, that there is less cold weather, but inferior in the more important fact that much less rain falls in the spring. In this latter respect it is also inferior to Assam, particularly to Northern Assam. There is one part of Chittagong, the Hill Tracts (Tea has scarcely been much tried there yet), which, in the fact of spring rains, is superior to other parts of the province, as also in soil, for it is much richer there. On the whole, however, Chittagong must yield the palm to both Assam and Cachar on the score of climate, and also, I think, of soil. For though good rich tracts are occasionally met with, they are not so plentiful as in the two last-named districts. Always, however, excepting the Hill Tracts of Chittagong, there the soil is, I think, quite equal to either Assam or Cachar. As regards labour (a very essential point to successful Tea cultivation), Chittagong is most fortunate. With few exceptions (and those only partial) all the plantations are carried on with local labour, which—excepting for about two months, the rice-time—is abundant. For transport (being on the coast with a convenient harbour, a continually increasing trade, ships also running direct to and from England), it is very advantageously situated. Chittagong possesses another advantage over all other Tea districts in its large supply of manure. The country is thickly populated, and necessarily large herds of cattle exist. The natives do not use manure for rice (almost the sole cultivation), and, consequently, planters can have it almost for the asking. The enormous advantages of manure in Tea cultivation are not yet generally appreciated; it will certainly double the ordinary yield of a Tea garden."

Chutia Nagpur. 173

CHUTIA NAGPUR.—Tea cultivation seems to be pursued in two districts of this division, *viz.*, Hazaribagh and Ranchi. The reports of some four or five planters for the year 1872 will be found in the papers regarding the Tea industry in Bengal, pp. 8-15. The writer has not been able to procure any more recent reports, but it is believed this locality has not as yet given indications of a future of much importance. The mistake appears to have at first been made to try and cultivate the China plant, but a hybrid Assam is now grown. The following brief note from **Colonel Money's** work manifests the cheif features of this tea district:—

"The climate is too dry, and hot winds are felt there. A great compensation, though, is labour; it is more abundant and cheaper in this district than in any other. The carriage is all by land, and it is some distance to the rail. But the Tea gardens at Hazareebaugh can never vie with those in Eastern Bengal, inasmuch as the climate is very inferior. The soil is very poor."

Area. 174

Area, Outturn, etc.—The province of Bengal not having as yet been surveyed the greatest possible difficulty is experienced in obtaining particulars of the area, outturn, etc., of crops. It is believed that the remarks already given for India collectively (pp. 420-421) on these topics, when read in the light of the returns of the Calcutta trade below, may be accepted as fairly conveying an idea of these subjects.

MADRAS.

MADRAS. 175

References.—*Bidie, Report of Coffee Borer (Chapter on Tea, Nilghiris), 1869; Madras Manual of Administration, I., 292; Robertson, Tea Cultivation in Nilghiri Hills (in Report on Agri. Condition Nilghiri District), 1875; Tea in Mysore & Coorg Gaz., III., 44; Tea Cyclopædia (Nilghiri Hills), 265.*

Climate. 176

Climate, Soil, etc.—It has been remarked by many writers that South India takes a keener interest in Coffee than in Tea. Nevertheless, tea gardens exist here and there on most of the uplands of the presidency and

TEA. **Cultivation of Tea**

CULTIVATION in Madras.

Climate.

in Travancore, Mysore and Coorg. The tea plantations of the Nilghiri hills have, in fact, attained a considerable reputation, and the exports from Travancore are yearly increasing in importance. In a memorandum, published in 1874 by **Mr. O. G. Masters**, some useful facts are told regarding the tea cultivation of the Nilghiri hills. The first tea plants (already alluded to (p. 434) as sent to Madras from **Mr. Gordon's** Chinese stock) reached the presidency in 1835, and six months later they had all perished except a few plants that had been forwarded to the Nilghiri hills. In 1861 **Captain Mann** imported seedlings from China and opened a plantation near Coonoor, which eventually succeeded. To that gentleman is, therefore, due the credit of having started the Madras tea industry. **Mann's** example was gradually followed, until in 1874 there were upwards of thirty estates, and the exports of Madras, produced tea (besides local consumption) for that year came to 80,766lb, valued at R89,357. **Mr. Masters**, in his report of 1874, goes on to say that "The plant grows well on these hills. There is a very large area of land suited to the culture of the different varieties on these and other hilly ranges in the presidency, and there can be no doubt that if the culture be conducted by persons possessing the requisite knowledge and capital in suitable localities, it will prove alike profitable to the cultivator and advantageous to the State." "The Government some years ago* opened a plantation at Ootacamund, with the view of supplying planters with seed and young plants." "The hybrid plant appears to be that generally preferred." "The Government also procured Tea-makers from Bengal, in order to give instructions in the manufacture, but the attempt was not altogether satisfactory, nor, indeed, generally acceptable." "It is generally allowed that 5 cwt. per acre is a fair average of the crop, and 50*s*. per cwt. a fair average price." These and such like passages occur in **Mr. Masters'** memorandum and may be useful as marking the progression and retrogression that has since taken place. There were last year (1890-91) 5,738 acres under the crop. **Colonel Money** makes the following remark regarding the physical features of the Nilghiri plantations :—

"The climate is superior to the Himálayan, for the frost is very slight. Were, however, more heat there in summer, it would be better. Some of the Teas have sold very well in the Indian market, for as regards delicacy of flavour they take a high place. The soil is good, but the temperate climate which holds on these 'blue mountains' is not favourable to a large produce."

Travancore. *Conf. with pp. 419, 469.* **177** **Coorg.** **178** **Nilghiri.** **179**

TRAVANCORE, COORG, NILGHIRI HILLS, ETC.—**Balfour** mentions the successful introduction of tea at Coorg by **Colonel Dyce** in 1843 ; at Nandidrug; at Bababudín hills in 1847 ; at the Pulni hills by **Major Hamilton** ; at Travancore (**Messrs. Bering & Co.'s** plantation) by **Mr. Huxham. Dr. Wight**, in a letter to **Wallich** (*Agri.-Hort. Soc. Jour , 1841*), alludes to his having been on a visit to **Mr. Huxham's** plantation. The success of the experiment, **Wight** tells us, has been established. **Huxham** obtained his supply of plants from Assam in June 1839. In the Agricultural Gazette (*April 15th, 1871*) **Mr. J. Macpherson** furnished a detailed account of the manufacture of tea in the Nilghiri hills, and in many other public papers frequent mention of these gardens occur, but the writer has failed to discover a recent statement of the plantations of Southern India. Their importance may be fairly well judged of by the returns of the trade reviewed below. **Messrs. Gow, Wilson & Stanton**, however, in their Circular of July 18 o, furnish the following instructive particulars regarding the Tea Planting and Tea Trade of Travancore :—" Another feature of the past season

* The plantation in question was under **Mr. Rae** in 1862-63.

CULTIVATION in Madras.

has been the increase which has taken place in the output from the recently opened district of Travancore, which has at the present time an area of about 4,700 acres under Tea cultivation. The following table shows the growth of the industry:—

TEA EXPORTS FROM TRAVANCORE.

1882-83.	1883-84.	1884-85.	1885-86.	1886-87.	1887-88.	1888-89.
lb	lb	lb	lb	lb	lb	lb
3,577	16,660	19,418	87,493	245,952	175,987	678,363

Some forty different estates were represented in the London auctions last season, comprising a total of over 9,000 packages. In this district, Tea can be produced of good quality, well suited for self drinking, as also for blending purposes."

BURMA.

BURMA. 180

References.—*Mason's Burma and its People, 505, 752; Tea in Aracan—The Tea Cyclopædia 264; Tea in the Andaman Islands—Tea Cyclopaedia, 262-264.*

Climate. 181

Climate, Soil, etc.—The writer can discover no separate account of the new tea plantations of this province. In the report of the Agricultural Statistics of British India for 1890-91, the area under the crop in Upper Burma was given as 1,001 acres and of Lower Burma as 78 acres. In 1888-89 the total area was only 14 acres and in 1889-90, 172 acres, so that from these facts it may be assumed prospects are brightening. In one or two publications mention is made of the Chittagong planters having endeavoured to prepare pickled tea for the Burma market. In that undertaking they seem to have met with some degree of success. The pickled tea used in Burma is for the most part manufactured in the Shan States and in Manipur. The prepared leaves are eaten, not used in decoction. The returns of the import trade from China (reviewed below) would seem to justify the opinion that a great future may be in store for this new industry, in competing for the local market alone. Large tracts of Upper Burma are, in climate and soil, remarkably alike to the Manipur regions of wild tea. In them and in Manipur itself it is possible, with the advance of civilisation and the opening up of means of communication, a great tea industry may be organised. There is perhaps no more hopeful country than Manipur anywhere in India. Immense tracts of rich and often flat land, within the valleys, are wholely uncultivated. The difficulty of labour could not be greater than in Assam, and it remains to be seen whether some of the numerous hill tribes might not be willing to engage as tea labourers, on the advantages of British influence beginning to be appreciated. The talked of railway communication would bring Manipur and Upper Burma into the position of highly attractive fields of future tea enterprise.

Bickled Tea. 182

Manipur Tea. 183

TRADE IN INDIAN TEA.

HISTORY of—

TRADE—184

History of Foreign. 185

History of Indian Tea Trade.—The prosperity of the INDIAN TEA TRADE may fairly be mentioned as a striking result of the British Administration of India. This is, however, by no means the only example that could be cited. The present position of the trades and industries in Jute, Indigo, Cotton, Oil-seeds, Silk, Wheat, etc., may be characterised as living monuments of national and commercial prosperity alike traceable to one and the same cause. Prior to the arrival of the pioneers of the Honourable the East India Company, it may safely be said this country had no foreign commerce. Her then rulers would have regarded (and

TEA. **Trade in Indian Tea**

HISTORY of TRADE. Foreign.

did regard) proposals of extended exportation as the pernicious doctrines of self-seeking adventurers. That an interchange in productions could be the surest indication of prosperity, it took even Europe centuries to learn. And now it may be said that India, of all the great commercial countries of the world, leads in the path of free trade. Its commerce is unrestricted, so that very nearly every article may be held to be free to leave its shores or to penetrate to its utmost recesses without let or hindrance in any form. Local manufacturing enterprise, it has been contended, has neither been stayed thereby nor the country's revenue lessened. In the success of its merchants, the nation is held to have been elevated and its people enriched, while the capabilities of the administration to further advancement has in no way been restricted. But, on the other hand, it must not be forgotten that the fertility of India and its vast resources are such that granted facilities of foreign trade, prosperity would have ensued without there having been absolute free trade. This is, however, neither the place nor is it the writers purpose to enter on a political dissertation as to the possibility of greater national prosperity under any other system than that observed in India, and which, for many years, has guided the Government in its conduct of commercial problems. It suffices that Tea and other great agricultural and manufacturing industries have not only prospered but become necessities of life for which the world now looks to India as the natural country of supply.

Conf. with pp. 432-433.

In the year 1788, **Sir Joseph Banks** suggested to the Court of Directors of the East India Company the practicability of cultivating the tea-plant in British India. It was not, however, till 1834 that the subject was submitted by **Lord W. Bentinck**—the Governor General—to his Council for serious consideration. The reader will find most of the subsequent historic facts detailed in another volume of this work (*vol. II., 65-83*), and it may suffice, therefore, to recall the more salient points only. A Committee was appointed under **Lord Bentinck's** supervission to obtain the necessary preliminary information with respect to the soils and situations most favourable for the growth of the plant. The Committee deputed its Secretary, **Mr. G. J. Gordon**, to China for these purposes, and in 1834 plants were raised in the Botanic Gardens, Calcutta, from some of the seeds collected by **Mr. Gordon**. While that gentleman was busily engaged in China the rediscovery was made by **Charlton** and by **Jenkins** of a wild tea plant in Upper Assam. It is believed that this same discovery had been made by **Bruce** and **Scott** some years previously, but that the plants then forwarded to Calcutta for determination had not received even the most casual consideration. Had it been otherwise, **Mr. Gordon** would, very probably, never have been sent to China and what many planters hold as of far greater moment, the China plant might never have been brought to India, or at all events not until a much later date. It is contended that the China plant not only retarded the development of the tea industry in India, through the trouble and expense involved in its acclimatization, but that for the greater part of the tea districts, even the hybrids from it, now to some extent cultivated, are inferior to the pure "indigenous" Assam or Manipur stock.

Introduction of China Retarded the growth of Indian industry. 186

In the early stages of tea cultivation the Government of India owned the gardens. Natives were encouraged to embark in the undertaking and for a time few Europeans seemed to consider the results attained as sufficiently encouraging. When opportunity afforded, however, the Government withdrew from direct ownership of gardens (about 1865), and made over its plant to Natives or Europeans on highly favourable terms. The incubation was at first slow, but when vitality manifested, the expansion became rapid, indeed all too rapid. Reckless and even criminal

Government withdraw from Ownership. 187

HISTORY of TRADE. Foreign.

speculation led to a crash in which many persons were ruined. From this reversion the industry very slowly recovered. When it did, all unnecessary expenditure was curtailed. It was realized that economy was essential to commercial prosperity, and inventions, to reduce labour and to guarantee results, became the planter's most certain and satisfactory assistants. It would be superfluous to attempt to indicate the reforms in planting and the improvements in manufacture that year by year were adopted. The industry now developed rapidly, and India was soon recognized as possessing advantages over China in the supply of tea. These in **Col. Money's** words may be briefly summed up thus:—

Criminal Speculation. 188

1. "Grown on large estates, the cultivation and manufacture being superintended by educated and skilled Englishmen."

Advantages of Indian over China Teas. 189

2. "Manufactured in a clean way by machinery, as opposed to hand manufacture which is the reverse of cleanly."
3. "Unadulterated."
4. "Stronger, thus going further, and in consequence more economical. On this point," **Colonel Money** continues, "I have shown that a superior class of plant, and a hotter and therefore more suitable climate, necessarily give this superior strength and body to the Teas."

Capital invested. 190

With such advantages it became a matter of time only for India to establish itself in the European market, and as it began to obtain a footing China lost ground. Year by year the relation of the two countries to each other changed, until India not only led but is now yearly giving indications of its possibly supplanting China altogether. Such then are the salient historic features of the Indian tea trade. We are thus in a position to review the statistical returns of this very striking industry which, within a very few years has caused the investment of Indian and English capital, in Joint-Stock Companies alone, to the amount of over 8½ million pounds sterling, and of private Companies at least other 5 million pounds sterling, while it has given to India an export trade valued annually at close on 6 million pounds sterling. Even these figures, however, convey but a partial conception of the industry, since many gardens, not being Joint-Stock Companies, are under no obligation to publish their capital or outturn. But even did we possess full details of all the money invested in Indian tea plantations, we should be in a position to judge thereby but of the pulsation of the industry. The thousands of persons who now obtain a highly-paid market for their labour, have gradually become colonists in tracts of country formerly almost uninhabited. It has been estimated that the Tea Industry of India now gives employment to close on half a million Natives of India. These have been drained for the most part from the over-populated tracts and from a condition of poverty to one of opulence. Valuable to India though such a satisfactory achievement undoubtedly is, we have mentioned but a few out of the many advantages which the Tea Planters have conferred on India. With the establishment of tea plantations it was soon recognized by other capitalists that it would pay to extend railway communication where success had been attained by the planter. The great rivers also rapidly carried on their watery road-ways immense fleets of native crafts, and in time resounded with the throb of the steam engine. Facilities of import and export create trade, so that the tea plantations in time became the nuclei around which a new civilization and a new commerce gravitated. India could ill afford to loose her tea planters now, and were any calamity to suddenly ruin the present flourishing trade, the effects of such a disaster would be felt from one end of the country to the other, from the rice grower of Bengal to the mill worker of Bombay. The prosperity and stability of tea planting, of indigo growing, and of the numerous other branches of trade in which Bri-

Annual Exports £6,000,000. 191

Tea gives Employment to half a million Natives of India. 192

HISTORY of TRADE: Foreign.

tish capital is invested, are of the gravest possible moment to the labouring classes of India; they certainly do not alone concern the capitalists. It is not to be wondered at, therefore, that apart from all other considerations an industry like tea that may be said to have secured an investment in India of perhaps little short of 20 million pounds sterling, mostly of foreign money, and which gives an annual turn over of close on 6 million sterling, must extend its influence to the uttermost corners of the land and be of importance to every subject of the Empire. That a trade of such magnitude could have been developed in so short a time betokens the value to India of British rule, and gives a foretaste of still further prosperity.

Earliest Advertisement of Indian Tea. 193

One of the earliest announcements of the sale of Indian teas appeared in 1841. Messrs. Mackenzie, Lyall & Co. of Calcutta issued one of their usual circulars thus:—

"NOVEL AND INTERESTING SALE
OF
ASSAM TEAS
AT THE EXCHANGE HALL.

To be sold by Public Auction this day, Wednesday, the 26th May 1841, at one o'clock P.M., precisely—*By Order of Government*—the First Importation for the Calcutta Market
OF
ASSAM TEAS.

These Teas were manufactured by the Singfo Chief, Ningroolla, of the Province (aided by the Government establishment) with the greatest possible care, and will be disposed of by Auction for his benefit.

At the same time, on account of Government,
The Entire Consignment
Consisting of Ninety-five (95) Chests
OF
ASSAM TEAS,

The produce of the Government Tea Plantations in Assam, for the Season of 1840."

The above (a fac simile as near as possible of the original) would, therefore, appear to have been the earliest public sale of Indian tea, though, doubtless, samples were seen in Europe slightly before this date (12 small chests were reported on in 1838), but only as test samples of the work done. From 1841 the commercial course of the Indian tea industry may be traced. It soon began to be noticed in the chief newspapers of Europe. *The Times* had many brief notices as, for example, 1st April 1847; 4th May 1848; and many others down to recent dates, such as 2nd April 1883; 16th August 1883; 22nd September 1888, etc., etc. In the article of the 4th May 1848, *The Times* reviews the success that had attended the undertaking on the Himálaya, in which it is stated that the Government of India had authorised an outlay of £10,000 per annum for a series of years. The imported Chinese labourers are stated to have regarded the plant grown in Kumaon as the true tea plant, and as far superior to the wild plant of Assam. But subsequent discoveries have proved that opinion to have differed in no material respect from the vain expectations of the European sugar-planters, who wasted their fortunes and ruined sugar-planting in India, through the mistaken notion that the superior stock of the West India plantations, which gave such splendid results in these Islands, were the only canes worth growing, and that hence upon their acclimatization depended the prospects of sugar-cane growing in this country. Had they turned from the exotic to the indigenous stock, as the tea-planters fortunately did, the results of sugar production might have been as great as those of tea.

HISTORY of TRADE: Foreign.

Growth of the Consumption of Tea in Great Britain.

CONSUMPTION in Great Britain. 194

Before proceeding to demonstrate the growth and present position of the Indian tea trade, it seems desirable that some general idea should be given of the China trade. Long anterior to the records of this modern cultivation of tea in India, mention was frequently made of Indian tea. The East India Company had dealings with China as well as with India, and it would appear that tea often reached England as a re-export from India. The first supplies of tea came to Europe through the Dutch East India Company and were apparently obtained from Japan. Very shortly after tea was brought from China, but from the beginning until near the close of the seventeenth century the whole of the European demand was met by the Dutch East India Company's sales. The total receipts in Europe up to 1671 amounted to only 516lb 10 oz. In 1678, 4,717lb are said to have been imported by the East India Company from Ganjam and Bantam; in 1680, 143lb were obtained from Surat; in 1685, 12,070lb came from Madras and Surat; in 1687, 4,995lb, from Surat; in 1689, 25,300lb, from Amoy and Madras; and in 1690, 41,471lb, were recorded as imported by England through the East India Company from Surat. These and such like records of traffic in tea must be regarded as China tea imported into India and re-exported from India. Soon, however, direct dealings took place between China and England, and the growing importance of the tea industry brought into birth swift sailing vessels for the sole purpose of conveying the China tea to Europe. The Dutch lost their hold in the market; and Europe and America then looked mainly to England for its supplies. The following table, taken from Milburn's *Oriental Commerce*, manifests the averages of the imports and exports by Great Britain for each ten years from 1711 to 1810:—

Early Re-exports from India. 195

Early British Traffic. 196

	Quantity sold.	Sale amount.	Exported.	Remaining for Home use.
	lb	£	lb	lb
1711 to 1720 inclusive	2,645,337	1,769,649	823,580	1,821,757
1721 to 1730 „	7,467,874	2,731,078	2,401,733	5,066,141
1731 to 1740 „	13,263,164	2,901,324	2,787,214	10,475,950
1741 to 1750 „	18,069,606	4,399,556	3,465,972	14,603,634
1751 to 1760 „	25,869,753	7,236,421	3,012,132	22,857,621
1761 to 1770 „	58,587,416	11,519,985	7,627,610	50,959,806
1771 to 1780 „	60,689,183	10,548,302	12,989,112	47,700,071
1781 to 1790 „	123,171,196	19,808,196	19,882,481	103,288,715
1791 to 1800 „	200,017,212	30,617,781	27,387,772	172,629,440
1801 to 1810 „	240,438,275	38,272,303	36,093,069	204,345,206

TEA.

Growth of the Consumption

CONSUMPTION in Great Britain.

The above therefore shows in one hundred years a total of 750,219,016℔ sold at the Company's sales, the value of which was £129,804,595. But of that amount 116,470,675℔ were re-exported, and the remaining 633,748,341℔ retained for home consumption. And this traffic paid to the British nation in Customs and Excise, during the one hundred years, £77,017,480, but even that large sum would not fully denote the revenue obtained through tea; sugar also paid duty, for example, and it has been estimated that one-half the sugar used in Great Britain during the one hundred years named was consumed up in connection with tea.

But it may be useful to carry the record of the quantities of China tea retained in Great Britain for home consumption down to about the period of the first importation of Indian tea:—

Quantity of Tea Retained for Home Consumption in the United Kingdom.

Years.	℔	Years.	℔	Years.	℔
1811 . . .	22,454,532	1823 .	27,093,015	1835 .	36,574,004
1812 . . .	24,584,402	1824 .	27,648,295	1836 .	49,142,236
1813 . . .	25,409,855	1825 .	29,232,174	1837 .	30,625,206
1814 . . .	24,389,501	1826 .	29,045,852	1838 .	32,351,593
1815 . . .	25,917,853	1827 .	29,931,178	1839 .	35,127,287
1816 . . .	22,693,992	1828 .	29,305,757	1840 .	32,252,628
1817 . . .	24,605,794	1829 .	29,495,205	1841 .	36,675,667
1818 . . .	26,527,531	1830 .	30,046,935	1842 .	37,355,911
1819 . . .	25,241,693	1831 .	29,997,055	1843 .	40,293,393
1820 . . .	25,712,935	1832 .	31,548,381	1844 .	41,363,770
1821 . . .	26,754,587	1833 .	31,829,620		
1822 . . .	27,574,025	1834 .	34,969,651		

CHINA TRADE. 197

EFFECT OF THE PRODUCTION OF TEA IN INDIA ON CHINA TRADE.—Messrs. Gow, Wilson & Stanton, in the Circular dated July 1890, carry the record of the British trade in China tea down to a recent date, while showing at the same time the growth of the consumption of Indian and Ceylon teas.

"The Home Consumption (in ℔) may be exemplified as follows:—

	1849.	1854.	1859.	1864.	1869.	1874.	1879.	1884.	1889.
China, etc.	50,021,576	61,953,041	76,303,661	85,799,235	101,080,491	118,751,000	126,340,000	110,843,000	61,100,000
Indian	...	...	...	2,800,000	10,716,000	18,528,000	34,092,000	62,717,000	96,028,491
Ceylon	...	...	...	...	...	...	...	1,500,000	28,500,000
TOTAL	50,021,576	61,953,041	76,303,661	88,599,235	111,796,491	137,279,000	160,432,000	175,060,000	185,628,491

"In 1889 the quantity of China Tea used in Great Britain was *less* than the Home Consumption in 1854—or 35 years previously.

"In 1889 the quantity of British Grown Tea used was *more than double* the entire Home Consumption in the same year, 1854.

"In 1889 the quantity of British Grown Tea used was in excess of the entire Home Consumption in 1871—18 years previously."

In their further report, dated April 1890, **Messrs. Gow, Wilson & Stanton** furnish instructive particulars regarding the relative positions of the Indian and Chinese Tea Trade:—

CHINA TRADE.

" Tea drinking in Great Britain has largely increased during recent years. The additional weight of Tea annually used clearly proves this, but inadequately gauges the increased volume of ***actual Tea drinking***, or the use of Tea in the form of a beverage.

" The stronger Teas from India and Ceylon have, for many years, been gradually displacing the weaker Teas of China. Indian and Ceylon Teas are capable of producing a far greater quantity of liquid Tea, owing to their superior strength and quality. Thus, as the use of these stronger Teas progressed, a given weight of Tea would yield a larger volume of liquid.

" This property has long been recognised by the trade, and was considered of such importance by the Board of Customs, that they took steps to obtain reliable information as to the relative strengths of Indian and China Teas.

In a Report subsequently written, the following passages occur :—

"From the information which has been afforded us on the subject, we believe that we make a moderate estimate in assuming that Indian Tea goes half as far again as Chinese Tea, so far as depth of colour and fullness (not delicacy) of flavour are concerned.

"Thus, if 1 lb of Chinese Tea produces 5 gallons of Tea of a certain depth of colour and fullness of flavour, 1 lb of Indian Tea will produce 7½ gallons of a similar beverage."

" Where we have to deal with Teas of such varied strength as those from India and Ceylon, and China, the only reliable standard by which to measure the extent of Tea drinking, consists in ascertaining the volume of Tea consumed in a liquid form—or, in other words, the number of gallons annually used for drinking. To determine this absolutely is manifestly impossible, but an estimate may be made which, although perhaps not indicating the exact volume of liquid Tea used, will be of value as showing the rate of progress in Tea drinking. This progress is known to have been considerable during many years past, in some measure traceable to the spread of the temperance movement, and partly due to numerous other causes.

"We have already stated that the consumption, when measured by weight, cannot give an accurate idea of the growth in the use of Tea in the form of a beverage. The Customs' Report, to which we have alluded, affords valuable information for calculating the increased *liquid* consumption.

Home Consumption of Indian, Ceylon and China Teas. 198

" Taking this estimate as a moderate one, which it appears to us to be and basing our calculations thereupon, we arrive at the figures given in the Diagram on the preceding page. The regular annual increase in the number of gallons of Tea used, is somewhat remarkable owing to its constancy. The sluggish advance which took place in the weight of Tea, consumed during the years 1886 to 1889 is thus clearly seen to have been caused by the ***great displacement of China Tea*** which occurred in those years—the rate at which the gallon, of liquid, consumption progressed, continuing uninterrupted; and, although in those years the lbs weight used per head of population hardly increased perceptibly, the ratio in which the gallon consumption per head of population increased continued about equal to previous years. The displacement of China Tea is shown in the following table of Home Consumption.

	1885.	1886.	1887.	1888.	1889.	1890.
INDIAN . .	65,678,000	68,420,000	83,112,000	86,210,000	96,000,000	101,961,686
CEYLON. .	3,217,000	6,245,000	9,941,000	38,503,900	28,500,000	34,516,469
CHINA, etc. .	113,514,000	104,226,000	90,508,000	80,653,000	61,100,000	57,530,337
lb .	182,409,000	178,891,000	183,561,000	185,416,000	185,600,000	194,008,492

TEA.

Growth of the Consumption

CONSUMPTION increased in 1890.
199

	1885.	1886.	1887.	1888.	1889.	1890.
lbs. per head of population.	5 02	4'87	4'95	4'95	4'91	5'07
Gallons per head of population	29'75	29'28	0'75	31'26	32'01	33'40

"Turning to the year 1890, we notice a sudden increase in the weight of Tea used. This is to a great extent accounted for by the reduction of the duty from *6d.* to *4d.* per ℔. on 1st May last, but it is not due solely to this cause. From the above figures it will be seen that during the years 1886-1889, large quantities of China Tea were regularly being displaced by Indian and Ceylon Tea, but that, in the year 1890, this displacement almost entirely ceased. As so little displacement of the weak Teas of China by the stronger Indian and Ceylon Teas had occurred, a *greater weight* of Tea became necessary, in order to supply the regularly expanding demands of the country.

"These two causes amply suffice to account for the greatly increased weight of Tea used in 1889, and in the early months of the present year.

Probable future increase in the Home Consumption.
200

"The inference thus to be drawn from these calculations is that we have now reached a point when a greater weight of Tea will be annually required to supply the expanding gallon consumption of the country, owing to the small quantity of China Tea now left for displacement; also that any further expansion in the gallon consumption must immediately cause an increase in the weight of dry Tea used, a condition which did not exist so long as a weak Tea was being displaced by a stronger one.

"These inferences, taken in conjunction with the late reduction in the duty, point to a larger Home Consumption of dry Tea in the future.

"The diagram also strongly illustrates the greater economy effected by the use of British Grown Tea, and its favour with the public is thus readily understood, especially when quality and flavour are taken into consideration.

Present Season.

201

The season just closing has been remarkable for many events—

(1) Reduction of duty from *6d.* to *4d.*, and the sudden increase in Home Consumption.

(2) The high rate of Exchange ruling and the consequent decrease in supplies of Tea from China.

(3) The unexpected shortness of the Indian Crop which proves to be about 10,000,000 ℔ below the original estimate.

(4) The quotation of Indian Tea on the London Produce Clearing House.

(5) The increased use of Indian and Ceylon Teas in Australia.

(6) The endeavour to increase the sale of Ceylon Tea in North America by the formation of the Ceylon Planters Tea Company; and efforts to open Russia to Ceylon Tea.

(7) The growing importance of the Tea industry in Travancore.

"We have already stated that a great increase had taken place in the weight of Tea consumed. The high rate of Exchange having tended to check supplies from China; this, together with the unexpected shortage in the Indian Crop, caused a somewhat unlooked for curtailment of supplies, and a consequent advance in quotations of the lower grades towards the close of the season. This advance naturally became emphasized, and probably intensified by Clearing House Operaters, and considerable tension in the market ensued during the earlier months of 1891."

The immense expansion within recent years of the Indian and Ceylon tea trade and the decrease of the demand for China and Japan teas, in certain markets, suggests the enquiry whether this state of affairs necessarily denotes a decline in the production of tea in China and Japan. It will be found in further page that the imports of China tea by India

CONSUMPTION in Great Britain.

have been considerably augmented and, therefore (while Persia has become a valuable market for Indian teas), it might be the case that the China (home), Central Asiatic, and Russian, American, etc., demands for China teas had compensated for the loss of the British trade. So, again, the success of India and Ceylon might be but the expression of the increased demand for tea in the world generally, and not an indication of the decline of the production of tea in China. Unfortunately statistics cannot be obtained to allow of a complete solution of this enquiry, but such details as exist justify the affirmation that China and Japan have suffered very greatly through the prosperity of the Indian and Ceylon planters. The particulars published by **Captain Gill** and other travellers in China, of the inferior teas procurable in the towns of China itself, and of the filthy materials used in the preparation of the Central Asiatic brick teas, show conclusively that the teas of India have only to be placed in the markets hitherto met by China, in order to repeat in other parts of the world what has been accomplished in Great Britain, namely, the complete displacement of China tea. When the time arrives for the Indian and Ceylon planters to seek in earnest an extension of their trade into further countries, they need not fear competition even in China itself. Hitherto the success attained has been such as to justify the rejection of all other considerations than the production of an honestly prepared article of superior quality. The requirements of each individual market have as yet scarcely entered into the planter's mind but the day is perhaps not far distant when the manufacture of special teas will be taken in hand.

New Markets. 202 *Conf. with p. 478.*

China and Japan. 203

As exemplifying some of the leading features of the China and Japan export trade in tea, the following particulars (furnished by **Messrs. Gow, Wilson & Stanton**) may be here given:—

Exports. 204

EXPORTS (IN lbs.) OF TEA FROM CHINA AND JAPAN TO

Season.	Great Britain.	United States.†	Australia New Zealand.	Continent of Europe.*
	I.	II.	III.	IV.
1870-71 . . .	131,753,470	48,029,184	11,305,598	2,386,282
1871-72 . . .	146,687,870	56,492,000	12,639,962	5,108,429
1872-73 . . .	149,151,046	60,787,830	16,799,738	4,269,766
1873-74 . . .	139,204,428	46,584,806	14,838,899	8,347,035
1874-75 . . .	161,964,407	56,030,888	15,281,862	6,357,143
1875-76 . . .	155,837,183	53,989,070	16,287,325	10,481,194
1876-77 . . .	167,335,130	40,221,446	16,822,955	5,626,586
1877-78 . . .	156,785,618	56,691,166	16,245,264	882,601
1878-79 . . .	164,435,363	57,273,240	15,805,511	1,238,415
1879-80 . . .	160,686,307	69,197,900	15,588,240	2,714,673
1880-81 . . .	174,514,982	83,977,682	24,824,218	7,189,093
1881-82 . . .	163,845,306	80,227,182	22,699,080	10,022,465
1882-83 . . .	149,077,980	68,173,885	19,452,062	9,338,800
1883-84 . . .	154,102,122	61,017,946	15,950,289	12,433,238
1884-85 . . .	145,530,654	72,245,553	19,078,900	12,216,970
1885-86 . . .	147,258,663	82,004,577	21,455,921	9,455,600
1886-87 . . .	149,516,600	90,761,972	19,961,614	13,891,818
1887-88 . . .	123,180,744	87,471,755	22,599,220	14,209,420
1888-89 . . .	101,809,467	81,338,801	26,085,467	20,740,732
1889-90 . . .	94,420,961	83,244,744	20,755,467	21,432,547
1890-91 . . .	70,243,662	89,996,584	15,250,738	25,256,663
1891-92 . . .	67,256,263	82,617,050	14,936,037	25,299,373

It may now be instructive to exemplify the share taken in the above by Japan as distinct from China. Since Japan tea goes, however, mainly

* Exclusive of the over-land traffic to Russia. † Conf. with p. 479.

TEA.

China and Japan Tea Trade.

to the United Kingdom and the United States, the figures of the above table (columns I and II) only need be analysed :—

Japan Exports. 205

Exports of tea (in lbs.) from Japan to

Great Britain.				United States.			
1870-71 .	43,400	1881-82 .	565,306	1870-71 .	13,529,184	1881-82 .	35,027,182
1871-72 .	10,000	1882-83 .	137,980	1871-72 .	15,492,000	1882-83 .	30,773,885
1872-73 .	...	1883-84 .	222,122	1872-73 .	17,287,830	1883-84 .	35,517,946
1873-74 .	...	1884-85 .	97,654	1873-74 .	19,584,806	1884-85 .	35,245,553
1874-75 .	...	1885-86 .	78,663	1874-75 .	20,330,888	1885-86 .	40,604,577
1875-76 .	...	1886-87 .	32,600	1875-76 .	24,389,070	1886-87 .	47,531,972
1876-77 .	6,004,380	1887-88 .	34,244	1876-77 .	11,891,446	1887-88 .	44,915,755
1877-78 .	103,268	1888-89 .	46,000	1877-78 .	22,091,166	1888-89 .	38,733,581
1878-79 .	672,063	1889-90 .	101,491	1878-79 .	29,673,240	1889-90 .	40,113,084
1879-80 .	579,757	1890-91 .	52,162	1879-80 .	35,697,900	1890-91 .	46,887,434
1880-81 .	149,982	1891-92 .	536,663	1880-81 .	40,577,682	1891-92 .	49,397,690

There are several very striking facts brought out by the above tables :—(1) The United States of America now afford the chief markets for China and Japan teas: (2) within the past 22 years the exports to these States have increased from 48 to 92 million pounds: (3) of these exports Japan furnishes more than half and seems to be steadily gaining favour: (4) the consumption of Chinese tea in the Continent of Europe is giving distinct indications of expansion: and (5) the trade in Chinese tea with the Australian Colonies seems if any thing to be declining. In the efforts that are presently being made to introduce Indian and Ceylon tea to the United States, it would seem desirable that the conditions of and recommendations for the flourishing state of the Japanese supply should receive careful consideration.

Having thus shown very briefly the chief items of the trade in China and Japan teas, we are in a position to more fully elaborate the actual share of the imports of China Tea that is consumed in Great Britain. **Mr. A. W. Martin** of 27, Throgmorton Street, London, gives the following instructive figures for the twelve months, July 1891 to June 1892 :—

Re-exports from United Kingdom. 206

Consumption of Tea in the United Kingdom.

	Indian.	Ceylon.	Total Indian and Ceylon.	China, etc.	GRAND TOTAL.
	lb	lb	lb	lb	lb
United Kingdom	105,000,000	58,000,000	163,000,000	40,000,000	203,000,000
Exported . .	4,000,000	3,000,000	7,000,000	27,000,000	34,000,000
TOTAL IMPORTS	109,000,000	61,000,000	170,000,000	67,000,000	237,000,000

These are admittedly only approximate figures, and since the year which they embrace does not correspond with the official years, comparisons are in some cases difficult. It will be noted, however, how very large a share of the China tea is re-exported from England as compared with the re-exports of Indian and Ceylon teas. The total imports of Chinese and Japanese teas by Great Britain cannot be regarded as corresponding in any way to the Home Consumption, though this might almost be said to be the case with the teas of India and Ceylon.

FOREIGN TRADE.

Exports.

207

I.—INDIAN FOREIGN TRADE.

Exports.—But turning now to the more immediate Indian side of this subject, it may be said that the total exports of Indian-grown tea from India up to 1838 were 488℔; thirty years later (1868-69) they were 11,480,213℔, or as 10 to 90 Chinese. Ten years later (1878-79), they were 34,432,573℔, or at 22 to 78, and five years still later (1883-84) the Indian exports were 59,911,703℔, valued at £4,083,880, and bore the relation of 37 to 63 with the Chinese exports. The exports from Ceylon first exceeded one million pounds in 1882 and ten years later they were close on 70 million pounds. The following table in continuation of the figures given in Vol. II., page 83 of this work, may now be furnished as illustrative of the Freign Trade:—

Years.	Chief Items of the Foreign Tea Trade of India: Exports, Indian Produce.		Exports, Foreign Produce.		Imports, Foreign Produce.		Of Ceylon: Exports—Figures taken from Colonial Statistical Abstract.	
	℔	R*	℔	R*	℔	R*	℔	£
1871-72[illegible]	17,187,328	1,45,49,846	272,810	2,72,011	2,025,129	20,25,129	...	...
1872-73	17,789,911	1,57,76,907	130,528	1,32,350	2,465,761	24,65,761	...	...
1873-74	19,324,235	1,74,29,256	118,044	1,16,926	1,828,571	18,28,588	...	...
1874-75	21,137,087	1,93,74,292	255,673	2,61,211	1,701,475	16,99,824	...	...
1875-76	24,361,599	2,16,64,168	200,227	1,74,638	2,771,204	24,75,663	784	180
1876-77	27,784,124	2,60,74,251	141,276	1,27,150	1,755,300	14,01,096	2,105	310
1877-78	33,459,075	3,04,45,713	197,640	1,72,960	2,323,033	19,06,107	19,607	1,743
1878-79	34,432,573	3,13,84,235	367,454	3,16,948	1,822,345	13,05,185	95,969	10,018
1879-80	38,173,521	3,05,10,200	231,111	2,12,236	2,534,518	21,20,624	228,680	17,878
1880-81	46,413,510	3,05,42,400	505,029	4,56,466	3,322,407	27,13,094	348,798	26,938
1881-82	48,691,725	3,60,91,363	563,617	4,59,226	2,845,212	19,96,906	697,268	49,317
1882-83	57,766,225	3,69,94,965	467,120	3,93,458	2,751,085	19,30,515	1,666,423	74,566
1883-84	59,911,703	4,08,38,805	561,410	5,03,411	3,065,170	23,76,141	2,[illegible]93,023	116,661
1884-85	64,162,055	4,04,47,592	986,842	9,25,919	3,874,412	32,55,477	4,373,754	213,322
1885-86	68,784,249	4,30,61,335	881,867	9,10,437	4,005,637	30,42,585	7,851,562	38[illegible],996
1886-87	78,702,857	4,72,79,917	1,854,472	15,51,509	4,214,342	32,42,604	13,824,701	587,967
1887-88	87,514,505	5,17,44,400	1,467,841	12,80,064	3,623,872	26,04,185	23,820,724	894,270
1888-89	97,011,112	5,26,73,149	2,328,756	20,58,226	4,767,004	31,79,373	34,346,432	1,227,885
1889-90	103,760,104	5,27,76,496	1,849,429	16,78,385	5,382,851	36,36,806	45,799,518	1,717,482
1890-91	107,014,993	5,21,92,335	3,179,826	28,50,600	4,770,008	32,51,408	46,901,554	...
1891-92	120,149,407	5,96,81,294	3,368,662	31,57,410	6,353,017	44,31,610	67,720,544	2,222,080

* To express nominally the valuation of rupees to pounds sterling strike out the last figure.

TEA.

Foreign Trade in Indian Tea.

FOREIGN TRADE.

Exports.

The above table thus refers the Indian Foreign trade in TEA to three sections, *viz.*, (*a*) Exports to Foreign Countries of Indian tea; (*b*) Exports to Foreign Countries of Imported (that is, Foreign) tea; and (*c*) the Imports of Foreign tea from which the exports of (*b*) are made. It will be observed that the exports of Indian tea since 1871-72 have never manifested the slightest fluctuation. Year by year they have steadily expanded—from 17,187,328℔ in 1871-72 to 120,149,407℔ in 1891-92. In these twenty-one years the exports have therefore expanded by fully sevenfold. But no indication has been given that the trade has even as yet reached its highest level. In 1873, Ceylon began in earnest to substitute the cultivation of tea for coffee, and shortly after the good results of this departure became visible. In 1875-76 the Ceylon exports of tea were 784℔; five years later they stood at 348,788℔; and five years, later still (1885-86) they were 7,851,562℔. During that year, in connection with the Colonial and Indian Exhibition at London, a great effort was made to advertise Ceylon tea. A very enlightened spirit has been shown by the Ceylon planters, not only at that Exhibition but since. In consequence the exports of tea from the colony became 13,824,701℔ in 1886-87; 23,820,724℔ in 1887-88; 34,346,432℔ in 1888-89; 45,799,518℔ in 1889-90; 46,901,554℔ in 1890-91; and it is estimated that the exports of 1891-92 may be close on 70 million pounds. The present Ceylon exports are, therefore, a little more than one-third of those of India. In passing, however, it may be remarked that the official and commercial returns of Indian and Ceylon teas often seem to differ seriously. This is due to the years of tabulation of returns being different. Exports are carried forward into 1891-92 by the trade which by Government would appear in 1892-93. The direct effect of the Ceylon and Indian prosperity has been the contraction of the demand for China and Japan teas, at least in Great Britain—a not unnatural result of the superior and cheap article which both Ceylon and India have been able to place on the market. In the heated controversies which have been thrust on the public recently, on the subject of the depreciation of the value of silver, it has been customary to read that India had reaped thereby a benefit. The expansion of the export traffic, we have been told, was due very largely to the fall in silver. The fact that the import trade had increased to a greater extent than the export— a trade which should have been inversely affected—was apparently overlooked. It was not, moreover, deemed necessary to investigate the influence of the appreciation of gold on each and every individual article of India's export traffic. It was held to be enough that the total exports had shown a steady expansion with the downward tendency of silver. How utterly false all this becomes when we endeavour to work out the grand total of expansion by a balance sheet of all the items of Indian trade. It is then seen that the supposed talisman of India's prosperity operates not only erratically on the different items of export, but that its influence is not even constant with any single article. Witness the decline of the exports of wheat from 1886-87 to 1890-91 and their sudden expansion in 1891-92—a term of years during which the value of silver steadily declined. So, again, were we to neglect the commercial and administrative reforms of India, we might fall into the error of regarding the prosperity of the Indian and Ceylon tea Trade as due to a talisman which must have had the same influence on China as on India and Ceylon.

Distribution. 208

Distribution of the Indian Exports.—It may now serve a useful purpose to analyse the returns of the exports of Indian tea during certain arbitrarily selected years of the series shown above but it must be added that the full returns for 1891-92 have not as yet been published:—

FOREIGN TRADE.

Distribution of the Exports.

Analysis of the Exports (of Indian Grown Tea) from India to Foreign Countries, during each fifth year since 1871-72, as also the approximate Returns for 1891-92.

Countries to which Exported.	1875-76.		1880-81.		1885-86.		1890-91.		1891-92.	
	lb	R	lb	R	lb	R	lb	R	lb	R
United Kingdom	24,287,488	2,15,96,155	45,416,582	2,99,28,827	66,640,947	4,18,39,403	100,208,625	4,92,62,170	111,168,895	5,52,47,129
Austria	...	...	1,050	1,403	1,647	1,695	7,423	6,317	Shown under "other countries."	
France	...	...	2,543	2,426	10,329	8,052	8,862	6,007		
Germany	...	...	...	...	...	...	29,449	14,672		
Greece	...	...	...	...	...	...	2,343	1,876		
Holland	...	...	...	...	...	...	13,972	7,194		
Italy	...	...	1,384	1,182	...	...	4,207	3,899		
Malta	...	...	...	...	...	...	3,143	1,628		
Spain	...	...	...	...	...	...	180	110		
Turkey in Europe	...	...	...	...	6,634	5,088	...	...		
Total of Europe	24,287,488	2,15,96,155	45,421,559	2,99,33,838	66,659,557	4,18,54,238	100,278,204	4,93,03,873	111,168,895	5,52,47,129
Aden	...	...	17,124	8,994	3,291	2,286	4,659	3,147	Shown under "other countries."	
Arabia	...	...	...	...	...	...	33,786	18,894		
Persia	1,126	1,249	10,828	9,892	31,404	20,505	1,221,478	7,11,329	2,789,175	19,38,486
Turkey in Asia	...	...	6,340	5,564	18,875	16,143	89,160	81,311	Shown in other countries.	
Total	1,126	1,249	34,300	24,450	53,570	38,934	1,349,083	8,14,681	2,789,175	...
China { Hong-Kong	...	...	...	...	19,364	16,101	60,542	26,499	Shown under "other countries."	
China { Treaty Ports	...	...	...	...	...	...	530	377		
Straits Settlements	7,900	1,237	69,262	17,725	169,289	31,819	25,999	17,326		
Total	7,900	1,237	69,262	17,725	188,653	47,920	87,071	44,202	...	...
Cape Colony	...	...	...	...	2,447	1,259	27,334	14,806	Shown under "other countries."	
Canada	...	...	...	...	...	...	61,040	26,965		
United States	5,715	5,653	68,597	39,568	98,343	52,320	79,457	37,638	83,405	49,486
Australia	44,836	45,374	807,608	5,15,909	1,766,447	10,53,292	5,118,714	19,40,037	5,203,995	18,88,344
Total	50,551	51,027	876,205	5,55,477	1,867,237	11,06,871	5,286,545	20,19,446	5,287,400	19,37,830
Other Countries with each less than 10,000lb	14,534	14,497	12,184	16,910	15,232	13,372	14,090	10,133	903,937	5,57,849
GRAND TOTAL	24,361,599	2,16,64,168	46,413,510	3,05,42,400	68,784,249	4,30,61,335	107,014,993	5,21,92,335	120,149,407	5,99,81,294
Provinces from which Exported.										
Bengal	24,220,458	2,15,00,366	45,797,823	2,99,67,217	67,857,088	4,23,79,187	104,545,622	5,04,10,901	Not as yet published.	
Bombay	20,232	24,478	68,924	60,570	73,578	62,296	788,176	6,75,628		
Sind	622	843	207,612	2,06,110	345,373	2,42,133	763,405	2,98,634		
Madras	112,562	1,37,303	263,940	2,84,262	344,617	3,50,203	912,704	8,05,347		
Burma	7,725	1,178	75,211	24,241	163,593	27,516	5,086	1,825		
	24,361,599	2,16,64,168	46,413,510	3,05,42,400	68,784,249	4,30,61,335	107,014,993	5,21,92,335	120,149,407	5,96,81,294

T. 208

Foreign Imports of Tea.

FOREIGN TRADE.

Distribution of the Exports.

It will thus be observed that by far the major portion of the exports of tea from India go to the United Kingdom and from Bengal. In other parts of this review particulars will be found regarding the imports of tea from China and their re-shipment from India or conveyance by land to Persia, Afghánistán, Tibet, etc. This subject has been repeatedly brought to the attention of those interested in the tea trade. The above table shows, however, that in the traffic from India to Persia the tide has begun to turn in favour of Indian teas in preference to those of China. The exports from India to Persia in 1885-86 were only 31,404lb, in 1890-91 they were 1,221,478lb, and last year 2,789,175lb. Similarly the exports to Turkey in Asia, and it will be seen (from the transfrontier traffic below) those to Afghánistán and Kashmír also, have recently manifested a tendency to expansion. But passing to the other side of the globe India may be said to have made immense strides in the Australian trade since the date of the Melbourne International Exhibition, when **Sir E. C. Buck** and the other Indian Commissioners made strenuous efforts to bring Indian tea to the notice of the Colonies. The Indian exports to Australia in 1880-81 were only 807,608lb; in 1885-86 they stood at 1,766,447lb; in 1890-91 at 5,118,714lb, and in 1891-92 at 5,203,995lb. The direct trade from India to the United States has not as yet given any very striking tokens of expansion, but it may confidently be anticipated that with a great tea consuming nation, like the Americans and one in which this commodity is free of duty, the future is likely to manifest an unprecedented expansion.

Persia. 209 *Conf. with p. 479.* *Conf. with p 478.*

Duty. 210

Duty on Tea.—It may here be remarked that among the tea drinking countries of the world the United States, the Straits Settlements, and India are those which admit this necessity of life free of duty. In Canada Tea brought from the United States has to bear 10 per cent. duty, other teas are free. In the Colonies the import duty on tea is remarkably low, being for the most part only 3*d.* or 4*d.* on the pound. Of the countries that levy an almost prohibitive duty the following may be mentioned. Portugal with 1-7½*d.* a pound; Greece about 1-6*d.* a pound; Spain 10*d.* to 1-1½*d.* a pound; France 9*d.* to 11½ *d.* a pound; Russia, on teas brought across its European frontier 1-10¾ or others 2*d.* to 11½*d.* Jamaica and the Bahamas charge 1*s.* a pound. The lowest duty among European nations appears to be Switzerland 1¾*d.*, and Holland 2¼*d.* a pound. As remarked above, the duty in England has recently been lowered from 6*d.* to 4*d.* India charges neither an import nor export duty, so that China tea comes free of any restrictions. It may be added that since the date of **Messrs. Gow, Wilson & Stanton's** Circular of November 1889 (from which the above facts have been derived) several of the duties charged on tea have been changed. They may, in fact, be said to be modified almost from year to year.

Imports. 211

Imports of Tea from Foreign Countries.—The imports are not very large doubtless, but they are none the less instructive.

Analysis of the Imports of Tea from Foreign Countries during each fifth year.

Countries whence imported.	1875-76.		1880-81.		1885-86.		1890-91.	
	lb	R	lb	R	lb	R	lb	R
United Kingdom	236,287	2,05,385	113,922	87,026	7,843	6,205	11,426	10,226
Ceylon . .	66,798	81,518	6,293	3,803	11,925	11,041	167,177	98,045
Persia . .	7,402	1,876	...	...	...	...	610	350
Mozambique .	...	...	448	154	...	...	...	...
Zanzibar . .	...	...	2,590	1,052	...	...	...	...
Arabia . .	...	...	109,943	55,235	...	...	...	...
Aden . .	...	...	...	...	28,436	26,267	2,000	1,500
Straits Settlements . .	276,413	1,76,669	554,305	2,61,662	707,519	3,59,581	543,847	2,58,957

FOREIGN TRADE.

Imports.

	1875-76.		1880-81.		1885-86.		1890-91.	
	lb	R	lb	R	lb	R	lb	R
China { Hong-Kong	1,986,125	17,50,677	952,740	5,63,015	1,254,452	7,54,608	748,441	3,85,949
China { Treaty Ports	195,666	2,57,992	1,581,350	17,40,471	1,972,401	18,71,713	3,192,143	23,91,525
Japan . .	...	...	...	...	...	...	4,090	1,239
Java . .	...	...	...	...	20,940	18,985	97,640	1,03,074
Australia. .	...	...	...	...	...	...	2,634	543
Others . .	2,513	1,546	816	676	2,120	1,185	...	...
TOTAL .	2,771,204	24,75,663	3,322,407	27,13,094	4,005,637	30,42,585	4,770,008	32,51,408

The above table shows the countries from which the Foreign Imports are usually derived by India. It will be observed that the supplies drawn from China have within the past twenty years increased from 2 to 4 million pounds, and that this increase has been mainly in the teas obtained from the Treaty Ports, the supply from Hong-Kong having decreased from 1,986,125lb to 748,441lb.

In this connection it may be as well to exhibit the shares taken in these imports by the provinces of India.

Shares taken by the Provinces. 212

Analysis of the Imports of Tea from Foreign Countries during each fifth year since 1871-72.

	1875-76.		1880-81.		1885-86.		1890-91.	
	lb	R	lb	R	lb	R	lb	R
Bengal . . .	538,823	3,73,418	343,459	1,75,094	159,495	52,123	58,008	23,474
Bombay . . .	2,089,521	20,00,073	2,748,181	24,29,217	3,498,554	28,49,821	4,350,025	31,01,317
Sind	...	...	58	52	2,771	2,445	191	183
Madras . . .	28,790	23,979	26,525	17,780	14,361	9,014	14,409	7,925
Burma	114,070	78,193	204,184	90,951	330,456	1,29,182	347,375	1,18,509
TOTAL .	2,771,204	24,75,663	3,322,407	27,13,094	4,005,637	30,42,585	4,770,008	32,51,408

It will thus be seen that the increase of the imports noted above has been mainly in the supply to Bombay. The Bengal production of tea has checked materially the import of foreign supply to that province, but it has as yet failed to contest the market of foreign supply in Western India. Doubtless this is largely due to the same cause that has created a Bombay demand for China silk, namely, the low freight of the return (opium) ships.

Re-exports. 213

Re-exports of Foreign Tea.—From the tables of the exports of Foreign Tea (the Re-exports as they are often called), which may now be given, it will be seen that two-thirds at least of the tea brought to India is again exported from it and that by far the major portion of this re-export traffic is with Persia. The possibility of an enhancement of the traffic with Persia, in Indian teas, is therefore a subject that deserves the careful consideration of all parties interested in the Indian Tea trade. That the

TEA.

Foreign Trade in Tea.

FOREIGN TRADE.

Persian market is capable of considerable expansion would seem justified by past results.

Re-exports. 214

Analysis of the Exports of Foreign Tea from India each fifth year during the past twenty years.

Countries to which Exported—	1875-76.		1880-81.		1885-86.		1890-91.	
	lb	R	lb	R	lb	R	lb	R
United Kingdom	41,785	33,373	21,380	14,237	...	...	3,809	2,323
France	...	...	...	...	...	...	220	220
Russia	...	...	18,159	22,698	...	...	...	...
Greece	...	...	...	...	...	...	2	2
Malta	...	...	...	...	...	...	471	409
Turkey in Europe	3,902	2,897	...	...	5,146	7,297	1,225	918
East Coast Africa — Mozambique	11,838	14,694	13,978	14,183	12,835	10,028	20,594	17,141
East Coast Africa — Zanzibar	...	...	8,857	6,317	12,193	6,504	18,966	9,788
Egypt	...	...	2,370	2,277	1,817	3,123	3,569	3,144
Mauritius	1,436	1,426	...	...	2,837	1,155	1,780	785
Abyssinia	...	...	...	...	...	...	1,378	699
Aden	16,675	12,069	7,540	5,070	20,707	10,939	20,145	9,859
Arabia	5,376	5,300	16,980	17,178	60,028	61,680	96,373	87,342
Mekran and Sonmiani	...	...	...	...	...	...	260	180
Persia	77,292	66,603	354,022	3,32,864	702,677	7,57,095	2,973,817	26,78,582
Turkey in Asia	29,425	27,047	54,760	37,903	62,788	52,108	36,408	38,712
Ceylon	10,060	8,569	1,562	1,431	...	...	196	140
Maldives	...	...	...	...	...	...	40	10
China—Hong-Kong	...	...	...	...	...	...	98	98
Straits Settlements	...	...	...	...	...	...	427	210
Australia	...	...	...	...	...	...	48	38
Other Countries	2,438	2,660	5,421	2,308	839	508	...	...
Total	200,227	1,74,638	505,029	4,56,466	881,867	9,10,437	3,179,826	28,50,600

The shares taken in these exports of foreign tea by the various provinces may now be exhibited.

215

Shares taken by the Provinces.

	1875-76.		1880-81.		1885-86.		1890-91.	
	lb	R	lb	R	lb	R	lb	R
Bengal	4,926	4,842	3,447	940	50	37	164	80
Bombay	193,586	1,67,928	501,181	4,55,219	881,501	9,10,234	3,177,671	28,49,214
Sind	206	314	204	184	36	30	1,476	1,014
Madras	240	287	85	75	...	...	88	82
Burma	1,269	1,267	112	48	280	136	427	210
Total	200,227	1,74,638	505,029	4,56,466	881,867	9,10,437	3,179,826	28,50,600

It will thus be seen that the re-exports shown by the above table (a period of twenty years) have expanded from 200,227lb (the highest return) up to 1875-76 to 3,179,826lb in 1890-91. These re-exports are made almost entirely from Bombay and are consigned mainly to Persia, Arabia, Aden, and the East Coast of Africa. By the tables of imports of foreign tea it has been shown that the Bombay supply comes from the Treaty Ports of China, Hong-Kong, and the Straits Settlements. This feature of India's traffic has been repeatedly urged to the consideration of the planting interests, and it is believed the effort has been made, and is being made, to secure a larger participation in the supply of the teas consumed by

FOREIGN TRADE.

the Natives of India and in the teas consigned to Persia, but the figures given show how much remains still to be done. The Natives of India generally have not as yet taken appreciably to consume tea, but the market is one of yearly expansion and possible future magnitude. This subject will, however, be found fully dealt with in the chapter below on the Indian Internal Trade in Tea, pp. 467 to 477. A special chapter will also be seen (pp. 478 *et seq*) to have been set apart for the subject of new or not sufficiently exploited markets. That subject derives special interest in connection with the present consideration. That India should not only import foreign tea but that it should largely re-export that tea to markets, such as Persia, which have shown a decided tendency to increase their consumption of Indian and Ceylon teas, is very much to be deplored. It will be recollected, however, that India first figures in the tea trade as a re-exporting country (*Conf. with p. 453*). In the 17th Century the East India Company appear to have appreciated the full advantage of Western India as a distributing depôt for Chinese produce. The return opium ships brought among other articles tea, and we accordingly read of shipments of tea from India to Europe at a period when of course India produced no tea. It might almost be said that England secured her Eastern possessions through a discussion over "the pepper box" and lost America through the "tea-caddy" (*Conf. with p. 440*). "It was the imposition of a prohibitive charge for pepper, made by the Dutch, that led to the formation, in 1600, of the "Old" or "London" East India Company, and it was the absurd demands of the Hon'ble East India Company and of England, in connection with the American supply of tea, that originated the War of Independence (1775-78 A.D.) It was the difficulties of trade with China that ultimately suggested to the East India Company the desirability of ascertaining whether or not tea could be successfully grown in India. As has abundantly been shown, the effort was crowned with complete success and already indications are not wanting of India and Ceylon being able in a not very distant future of completely supplanting China from this, one of her most ancient and most highly prized articles of foreign commerce. Few articles of trade have indeed been so closely connected with the destinies of nations nor afforded anything like so largely the sinews of war than tea.

II.—TRANS-FRONTIER LAND TRADE.

TRANS-FRONTIER. 216

There is very little to be said regarding this trade which the reader will not find alluded to below in connection with the Internal Trade. The table which may, however, be here furnished, of the returns during the past three years will be seen to be referred, as is customary, to two sections, *viz.*, Indian and Foreign teas. The table manifests the imports and exports to and from each country, so that the balance between these or between the totals would show the net transactions. The exports of Indian tea appear to be growing in importance. They were during the past three years as follows:—837,648℔ in 1888-89, 887,824℔ in 1889-90, and 1,004,640℔ in 1890-91. The corresponding re-exports of Foreign teas were 1,066,800, 1,050,602, and 648,704℔, so that the re-exports by land routes may be regarded as shrinking. As to the imports it will be seen that within the past few years Burma has drawn considerable supplies of China tea across its land frontier, a circumstance that may be viewed as due to the peace that has been established since the annexation of Upper Burma. It will be observed that taking Indian and foreign teas conjointly, the net result is an exportation of over 1½ million pounds to countries across the land frontier.

TEA

FOREIGN TRADE.

Trans-frontier.

Trans-frontier Land Trade.

Statement of the Transfrontier Land Trade of India in Tea during the past three years.

	Imports from						Exports to					
	1888-89.		1889-90.		1890-91.		1888-89.		1889-90.		1890-91.	
	Cwt.	R	Cwt.	R	Cwt.	R	Cwt.	R	Cwt.	R	Cwt.	R
INDIAN, From or to—												
Kandahar	...	...	...	...	...	...	59	3,200	...	...	...	...
Sewestan	...	...	...	...	...	...	10	72	...	...	...	...
Tirah	...	...	...	...	...	...	...	...	...	...	1	75
Kabul	...	...	...	...	...	...	2,333	1,06,615	2,655	1,64,685	4,429	2,29,004
Bajaur	...	...	...	...	...	...	6	264	50	2,297	28	1,254
Kashmir	...	...	...	...	...	...	4,036	1,82,122	4,539	2,07,050	3,529	1,98,995
Ladakh	...	...	...	...	...	...	242	10,290	612	25,065	673	27,570
Thibet	21	1,660	61	5,180	113	9,820	...	...	...	...	6	640
Nepal	208	11,878	3	240	385	20,935	693	42,362	12	960	254	14,119
Sikkim	...	...	...	...	...	...	...	...	1	45	2	121
Bhutan	...	...	6	359	...	...	...	...	...	...	...	...
Naga and Mishmi Hills	...	...	...	...	...	...	3	87	4	150	1	50
Manipur	...	...	...	...	...	...	...	3	1	36	...	...
Transfrontier, Sind-Pishin Railway	2	120	1	80	...	...	97	7,860	53	4,440	47	3,960
TOTAL	231	13,658	71	5,859	498	30,755	7,479	3,52,875	7,927	4,04,728	8,970	4,75,788
TOTAL IN lb	25,872	...	7,952	...	55,776	...	837,648	...	887,524	...	1,004,640	...
FOREIGN, From or to—												
Khelat	...	...	...	...	...	...	5	528	...	...	...	...
Kabul	...	...	...	...	...	...	8,662	8,94,154	8,424	8,80,525	4,479	4,43,040
Bajaur	...	...	...	...	...	...	...	...	36	3,750	15	1,400
Kashmir	...	...	...	...	...	...	104	9,850	160	16,150	533	65,130
Ladakh	2	420	...	...	...	...	128	6,270	182	12,320	136	8,850
Thibet	5	271	70	4,393	54	3,023	...	...	...	...	2	240
Sikkim	9	580	2	124	...	...	7	385	...	...	...	...
Nepal	290	15,645	387	24,816	107	5,873	...	...	...	...	...	...
Bhutan	33	1,776	14	821	7	406	...	...	...	...	...	...
West China	...	...	...	...	111	4,455	...	...	...	...	...	...
Siam	...	...	8	193	3	80	...	...	...	...	...	...
Northern Shan States	...	...	3,390	92,562	2,350	86,949	...	...	...	...	...	...
Southern Shan States	...	...	2	90	18	490	...	...	...	...	...	...
Karennee	...	...	3	158	...	...	...	...	...	...	...	...
Zimmé	...	...	...	...	52	3,200	...	...	...	...	...	...
Transfrontier, Sind-Pishin Railway	...	...	...	...	...	...	619	67,440	579	62,960	627	68,160
TOTAL	339	18,692	3,876	1,23,157	2,702	1,04,476	9,525	9,78,627	9,381	9,75,705	5,792	5,86,820
TOTAL IN lb	37,968	...	434,112	...	302,624	...	1,068,800	...	1,050,672	...	648,704	...
GRAND TOTAL { In R & CWT.	570	32,350	3,947	1,29,016	3,200	1,35,231	17,004	13,31,502	17,308	13,80,433	14,762	10,62,608
GRAND TOTAL { In lb	63,840	...	442,064	...	358,400	...	1,904,448	...	1,938,496	...	1,653,344	...

III.—INTERNAL TRAFFIC IN TEA.

INTERNAL TRADE. 217

In order to arrive at some sort of conception of the Internal Traffic in, and Consumption of, Tea in India it becomes necessary to bear several important considerations in view: (1) The producing provinces may consume locally manufactured tea, which may largely escape registration. The railways publish their returns under two main sections—(*a*) traffic between the internal blocks (as they are called), that is to say, in the case of Bengal the traffic, for example, from Darjeeling to Calcutta (two internal blocks), and (*b*) between external blocks, *e.g.*, from Calcutta to the North-West Provinces. So far this is very satisfactory, but it must not be forgotten that the railways of India may not inaptly be compared with the main arterial system, and that the ultimate distributing currents of the traffic, which passes along the roads and rivers, to a large extent escape all registration. While, therefore, the returns of the railways furnish particulars of the chief transactions, the producing provinces have to be admitted as possibly possessing means of distribution and consumption of which we have no record. The non-producing, internal provinces, on the other hand, may safely be assumed to drain their chief supplies along the railways or rivers, so that we can more correctly judge of their consumption. But (2) tea is imported into India from Foreign Countries, both by sea and across the land frontier, so that the majority of the provinces, especially those with a sea board, have three sources of possible supply: (*a*) Foreign tea brought by the maritime trade, and Foreign tea carried across the land frontier; (*b*) Foreign and Indian teas conveyed by coastwise interchange; and (*c*) railway supplies of both Foreign and Indian teas. The teas brought to India across the land frontier are not very important, but to some extent these appear in the railway returns, and it is significant that India not only exports her own teas by these routes, but even imports small amounts of them. These imports of Indian tea may be supposed, for example, to have been exported from Sikkim and imported through Nepal.

The tables already furnished of the FOREIGN IMPORTS by sea and land have supplied the data necessary to arrive at an opinion as to the amount that has to be credited to local production, in arriving at the total supply. The foreign imports by sea may be said to be very largely obtained by the Port Town of Bombay and to be either consumed in that city or distributed by railway to the interior, or sent by the coastwise shipping agencies to provinces on the sea board. In order to arrive at a knowledge of the internal consumption, it seems necessary, therefore, to furnish particulars regarding

1st—COASTWISE TRANSACTIONS.

Coastwise. 218

The following table shows the total trade by this route during the past five years:—

TEA.

INTERNAL TRADE.

Coastwise.

Internal Trade in Tea.

Statement of the Indian Coastwise Traffic in Tea showing the Imports and Exports of the Provinces.

	Quantities.						Values.					
	Bengal.	Bombay.	Sind.	Madras.	Burma.	Total.	Bengal.	Bombay.	Sind.	Madras.	Burma.	Total.
	lb	lb	lb	lb	lb	lb	R	R	R	R	R	R
INDIAN PRODUCE. Imported to—												
1886-87	1,032,728	498,693	1,210	78,527	148,430	1,759,588	6,39,681	2,43,645	806	77,211	1,14,529	10,75,872
1887-88	1,115,724	492,699	1,184	73,178	154,148	1,836,933	7,09,108	2,42,806	534	40,456	1,36,503	11,29,407
1888-89	1,145,959	761,205	2,581	21,594	77,556	2,008,895	5,82,198	4,17,137	1,177	13,531	73,633	10,87,676
1889-90	1,175,096	1,157,970	1,598	174,105	123,872	2,632,641	5,61,695	5,89,209	1,074	90,057	84,250	13,26,285
1890-91	1,111,899	877,029	1,773	232,334	119,338	2,342,373	4,85,703	4,32,035	740	1,19,049	87,555	11,25,082
FOREIGN PRODUCE. Imported to—												
1886-87	499	32,377	191,533	11,927	12,036	248,372	472	12,160	1,82,158	16,511	6,217	2,07,518
1887-88	195	38,169	182,518	87,561	12,814	321,257	272	14,458	1,63,181	43,603	6,729	2,28,243
1888-89	62	36,656	135,555	20,582	15,200	208,055	42	12,601	1,29,121	12,790	6,841	1,61,395
1889-90	266	37,817	171,640	14,665	10,231	234,619	147	12,452	1,61,219	8,676	4,092	1,89,586
1890-91	823	48,928	111,129	16,343	11,578	188,801	337	19,406	1,08,543	7,539	94,75	1,40,584
TOTAL IMPORTS IN 1890-91	1,112,722	925,957	112,902	248,677	130,916	2,531,174	86,040	4,51,441	1,09,283	1,26,588	92,314	12,65,666
INDIAN PRODUCE. Exported from—												
1886-87	1,996,800	1,926	1,834	22,020	2,841	2,025,421	11,14,752	1,678	1,560	19,875	1,369	11,39,234
1887-88	1,983,841	4,155	30	14,727	2,485	2,005,238	10,58,757	1,985	25	13,030	1,617	10,75,414
1888-89	2,192,267	3,363	1,050	8,937	7,618	2,213,235	10,51,032	1,759	541	8,148	5,267	10,66,747
1889-90	2,743,731	714	...	26,211	16,168	2,786,824	12,47,669	468	...	17,943	9,270	12,75,350
1890-91	2,215,211	4,219	1,111	19,545	14,340	2,254,426	10,15,647	1,780	481	12,872	9,296	10,40,076
FOREIGN PRODUCE. Exported from—												
1886-87	70,050	326,977	31	412	16,789	414,259	33,531	2,59,020	10	316	6,489	2,99,366
1887-88	847	241,949	100	992	14,408	258,296	386	1,63,737	75	1,793	5,743	1,71,734
1888-89	50,712	251,843	3,955	433	14,392	321,335	9,353	1,83,385	2,375	421	5,783	2,01,317
1889-90	2,833	271,586	531	1,185	14,569	290,704	1,045	1,90,668	596	1,301	5,341	1,98,861
1890-91	1,466	233,488	7,164	499	15,549	258,166	636	1,52,820	4,290	555	6,153	1,64,454
TOTAL EXPORTS IN 1890-91	2,216,677	237,707	8,275	20,044	29,889	2,512,592	10,16,283	1,54,600	4,771	13,427	15,449	12,04,530
Net trade (+ being import — Export) in 1890-91	—1,103,955	+688,250	+ 104,627	+ 228,633	+ 101,029	+ 18,582	...	...	...	...	...	...

INTERNAL TRADE.

Coastwise.

The final result of the above table may be said to be that Bengal is the only exporting province, the net balances of the two kinds of tea being an export from Bengal of 1,103,955℔ and an import by Bombay of 688,250℔, by Sind of 104,627℔, by Madras of 228,633℔ and by Burma of 101,629℔. But if the two classes be treated separately we learn that of (a) INDIAN TEAS, Bengal showed a net export of 1,103,312℔ and the other provinces net exports, *viz.*, Bombay 872,810℔, Sind 662℔, Madras 212,789℔, and Burma 104,998℔. A large share of the Bengal imports and exports are usually to other ports within the presidency, *viz.*, a little over one million pounds each way. It may be noted that the net imports of other provinces are a little short of the net exports from Bengal. It does not follow, however, that these coastwise exports and imports are directly drawn from each other. The supplies exported may have been derived from that brought to the sea-board by rail and road, and need not to any appreciable extent, at all events, be coastwise re-shipments. The total (gross) exports of Bengal tea sent coastwise to the provinces of the rest of India may be put at one and a half million pounds, of which Bombay alone takes fully one million pounds. Of the trade in (b) FOREIGN TEAS coastwise it may be said this amounts to above one-fourth of a million pounds. Sind and Madras are the only provinces that show a net import. Last year the consignments were 103,965℔ for Sind and 15,844℔ for Madras. The other provinces had net exports, *viz.*, Bengal 643℔, Bombay 184,560℔, and Burma 3,971℔. The excess of these exports from British provinces (*viz.*, 69,365℔) over the imports by British provinces, represents approximately the consumption in the Native States and Foreign Territory within India.

It may be added, however, that there are one or two defects in all the records of coastwise trade that preclude to a large extent a final balance sheet being framed. Of these may be mentioned the fact that the exports from a province need not actually appear in the imports of the province for which they were exported, since the destination of a shipment may be changed while at sea. So many ships have usually left the ports but not reached their destination when the year of record closes. The Native States, such as Kattywar, Cutch, Cambay, Travancore, Goa, etc., while they are given as regions to which exports and from which imports are made with British provinces, do not obtain an independent place, and hence entries appear on one side of the accounts of the British provinces, *viz.*, exports, which are not adjusted on the other. As already explained, these exports to Native States represent the consumption in these regions, which, if a total balance sheet were prepared, would again appear as imports by these States and thus balance the exports. From these, and such like considerations, it becomes necessary to take the transactions of the coastwise trade as they stand, and to make little attempt to critically analyse them. Each separate item of the coastwise trade has, however, an important bearing when viewed in relation to the port against which it is entered. Thus, for example, we learn that since Bengal, Bombay, and Burma usually show a net coastwise export of foreign teas, we must look to the imports from foreign countries or to the railway traffic in bringing the stocks from which these exports are made. Sind and Madras, on the other hand, do not obtain, by the routes just named, enough of foreign teas, and accordingly they indent on the other provinces coastwise. The trade in tea at Travancore manifests many peculiarities. The State imports Madras tea (shown in the official returns under the section of Indian Produce and Manufacture) and it also exports to Bombay tea also recorded under that section. But being a Native State, it next is shown as exporting to Bengal (very possibly its own tea) under

Native States. 219

Travancore Trade. 220

Conf. with pp. 419, 448.

TEA. | **Internal Traffic in Tea.**

INTERNAL TRADE.
Native States.

the section of Foreign Merchandise and as importing from Bombay China tea. The consumption of tea in Native States appears to be very largely that obtained from China.

In order to illustrate the adjustment of the coastwise trade on the total transactions by all routes, the following figures may be exhibited. The total imports of foreign tea in 1890-91 came to 4,770,008℔ and the re-exports were 3,179 826℔, thus showing a net balance of 1,600,000℔ of foreign tea, as having been retained in the whole of India to meet local demands. This excludes the transactions by land routes, but as these gave in 1890-91 a net import of only 346,080℔, they may, for the present, be left out of consideration, the more so since when Indian and Foreign teas are considered conjointly the transfrontier traffic shows a net export of 1,294,944℔. Of the imports of foreign teas by sea in 1890-91, Bombay Port Town took 4,350,025℔ and re-exported 3,117,671℔, it thus retained a net surplus of 1,232,354℔. From that stock Bombay Port gave by coastwise 237,707℔ net to the provinces and Native States, etc., 880,270℔ by railway to Upper India. There thus remained in 1890-91 114,377℔ of foreign tea for the consumption of the people of the city of Bombay. But of the exports by rail in that year, it may be added that 815,982℔ (=9,951 mds.) went to the Panjáb, 42,230℔ to Bombay presidency, 3,774℔ Rajputana and Central India, and the balance to the North-West Provinces and Oudh, the Central Provinces, Berar, Calcutta, etc. It will thus be seen that from this point of view the Panjáb so far stands out as the largest consuming province for China tea. But the exports by land routes to Kabul, etc., of foreign teas were 4,479 cwt., to Bajaur 15 cwt., to Kashmir 533 cwt., to Ladakh 136 cwt., and to Tibet 2 cwt., or say in all, 578,480℔. It seems probable that a very large share of these exports took place from the Panjáb although doubtless Sind contributed, especially in the further export of 627 cwt. (=70,224℔) which went by the Sind-Pishin Railway. It may be added that the net coastwise import of foreign tea by Sind came to 103,965℔, so that it was not likely to have contributed much more than the amount named towards the transfrontier land exports. But, on the other hand, India, more especially Burma, imported in 1890-91 2,702 cwt. (=302,624℔) foreign tea, *viz.*, 2,350 cwt. from the Northern Shan States, 110 cwt. from West China, 107 cwt. from Nepal, 52 cwt. from Zimme, and 54 cwt. from Tibet. These imports did not, however, affect the Panjáb trade but mainly furnished the source from which the exports from Burma were made. We thus learn that perhaps half the annual drain of China tea made by the Panjáb on Bombay is to meet the transfrontier land traffic which in foreign tea was last year 648,704℔. This trade seems to have given some indication of shrinking in the face of competition with the supply of Indian tea. The exports of foreign teas averaged in the two previous years 1,058,736℔. The exports of Indian teas across the Panjáb frontier have, on the other hand, been steadily increasing. They were last year 1,004,640℔.

Having thus shown some of the main features of the coastwise traffic and the destination of the tea drawn from foreign countries, the details of the INTERNAL Trade may be still further exemplified by

Rail, Road & River.
221

2nd—RAIL, ROAD AND RIVER TRANSACTIONS.

With no other article of Indian trade does a statement of the internal transactions, carried by rail, road and river, result so much in a table of blank columns, as that which will be found on a further page on the tea trade. The only figures of value are those expressive of the transactions between Calcutta and Assam, Cachar, and Eastern Bengal, *viz.*, the totals

INTERNAL TRADE.

Rail, Road & River.

shown horizontally against Nos. 24 and 29. But these figures denote the amounts carried to the port towns to meet the foreign demand. They should, therefore, be practically excluded from consideration, in connection with internal traffic, very much in the same way as the Indian goods that pass through the Suez Canal are not given by the Egyptian Government as representing the internal trade of that country. Speaking on general principles it might be said that the totals shown vertically on columns Nos. 10 and 14 represent the amounts of Indian and Foreign teas that are consumed in the Provinces and Native States, analysed on columns 1 to 9 and 11 to 13, both inclusive. On this estimate the consumption of tea in the provinces of India (for the year 1889) would have been 6,140,570℔ and in the Native States 279,046℔. So, in a like manner, the imports drained from the provinces by the port towns might be accepted as representing the amount required to meet the foreign trade plus the local demands of the port towns themselves. The grand total vertically in column No. 20 will be seen to be 98,448,298℔. Now, in order to arrive at an approximate estimate of local consumption in these towns, it is necessary to deduct from the figure just mentioned the recorded foreign exports for the year in question. The table given on page 459 shows these to have been 97,011,112℔ and the table at page 421 gives the estimated total production as 99,792,544℔. There are, on the face of these quotations, certain points of agreement and others of disagreement. The sufficiency of the estimated production to meet the subsequently ascertained foreign exports, and the confirmation of the accuracy of these figures as given by the railway returns, are features of agreement that are highly instructive. But, on the other hand, the balances shown above as retained for local consumption would be wholly unintelligible did we not recall that besides production India has two other sources of supply—(*a*) foreign imports by sea (pages 462 to 465) and (*b*) foreign transfrontier imports (pages 465—466). Thus, for example, Bombay is shown by the table (pages 472—473) to import by rail and road only 1 maund of tea but to export 10,819 maunds — an amount almost wholly drawn from its foreign imports by sea. It seems probable that after making full allowances for the balance between imports and exports in every branch, the total consumption to tea in India cannot possibly materially exceed 5 to 6 million pounds.

The following table of the returns of 1888-89 may be here given in order to show a complete statement for one of the years dealt with in this review:—

TEA. | **Railand River Traffic**

Weights of articles carried by Rail and River† in British India*

ARTICLES, AND WHENCE EXPORTED.	IMPORTED									
	BRITISH PROVINCES (EXCLUDING CHIEF TOWNS).									
	Madras.	Bombay.	Sind.	Bengal.	N.-W. P. & Oudh.	Panjab.	C. P.	Berar.	Assam.	Total.
	1	2	3	4	5	6	7	8	9	10
TEA.	Mds.	Mds.	Mds.	Mds.	Mds.	Mds.	Mds.	Mds.	Mds.	Mds.
(1) INDIAN.										
British Provinces (excluding Chief Towns).										
Madras	...	226	...	...	12	4	103	28	...	373
Bombay	...	...	...	...	...	...	3	1	...	4
Sind	...	...	...	...	...	3	...	...	...	3
Bengal	...	2	...	...	84	12	5	...	5	108
N.-W. P. & Oudh	...	55	9	48	...	2,833	74	...	...	3,019
Panjab	24	495	160	158	794	...	156	32	...	1,819
Central Provinces	...	1	...	1	...	...	...	2	...	4
Berar	...	...	...	...	...	1	...	...	...	1
Assam	...	...	...	51,610	...	...	...	...	...	51,610
	24	779	169	51,817	890	2,853	341	63	5	56,941
Native States.										
Raj. and Central India	...	9	...	...	2	5	3	...	...	19
Nizam's Territory	2	...	4	...	...	...	...	...	...	6
Mysore	16	...	...	...	...	...	...	...	...	16
	18	9	4	...	2	5	3	...	...	41
Chief Towns.										
Madras seaports ‡	391	208	...	...	...	1	227	...	...	827
Bombay	3	1,422	...	1	220	28	144	35	...	1,853
Karachi	...	...	50	...	...	...	...	...	...	50
Calcutta	...	5	...	1,360	1,938	70	33	1	15	3,422
	394	1,635	50	1,361	2,158	99	404	36	15	6,152
GRAND TOTAL	436	2,423	223	53,178	3,050	2,957	748	99	20	63,134
Ditto in lb	35,752	198,686	18,286	4,360,596	250,100	242,474	61,336	8,118	1,640	5,176,988
(2) FOREIGN.										
British Provinces (excluding Chief Towns).										
Madras	...	...	...	...	...	...	...	...	...	...
Bombay	...	...	...	...	...	12	...	...	...	12
Sind	...	...	...	...	...	7	...	...	...	7
Bengal	...	...	...	...	...	...	...	...	...	...
N.-W. P. & Oudh	...	...	...	1	...	...	...	...	...	1
Panjab	...	27	...	...	2	...	...	...	...	29
Central Provinces	...	...	...	...	...	...	...	...	...	...
Berar	...	...	...	...	...	...	...	...	...	...
Assam	...	...	...	...	...	...	...	...	...	...
	...	27	...	1	2	19	...	...	...	4
Native States.										
Raj. & Central India	...	...	...	...	...	...	...	...	...	...
Nizam's Territory	...	...	...	...	...	...	...	...	...	...
Mysore	...	...	...	...	...	...	...	...	...	...
	...	...	...	...	...	...	...	...	...	...
Chief Towns.										
Madras seaports ‡	18	...	...	...	2	...	...	...	...	20
Bombay	1	1,401	...	3	128	8,975	60	13	...	10,581
Karachi	...	...	246	...	...	818	...	...	...	1,064
Calcutta	...	...	...	5	5	27	...	...	...	37
	19	1,401	246	8	135	9,820	60	13	...	11,702
GRAND TOTAL	19	1,428	246	9	137	9,839	60	13	...	11,751
Ditto in lb	1,558	117,096	20,176	738	11,234	806,798	4,920	1,066	...	963,582
TOTAL (INDIAN & FOREIGN) TEA	455	3,851	469	53,187	3,187	12,796	808	112	20	74,885
Ditto in lb	37,310	315,782	38,458	4,361,334	261,334	1,049,272	66,256	9,184	1,640	6,140,570

* To reduce maunds to lb multiply by 82. The totals vertically of the columns 1 to 20 are the im-
† *i.e.*, (1) Trade between Sind and the Panjab by the River Indus; (2) Trade between Bengal and
‡ Madras seaports are those touched by railways, *viz.*, Madras, Pondicherry, Negapatam, Tuticorin

during 1888-89, i.e., between 1st April 1888 and 31st March 1889.

INTO										
NATIVE STATES.				CHIEF TOWNS.					GRAND TOTAL.	ARTICLES, AND WHENCE EXPORTED.
Raj.& C. I.	Nizam's Terry.	Mysore.	Total.	Madras sea-ports ‡	Bombay.	Karachi.	Calcutta.	Total.		
11	12	13	14	15	16	17	18	19	20	
Mds.	Mds.	Mds.	Mds.	Mds.	Mds.	Mds.	Mds.	Mds.	Mds.	TEA.
										(1) INDIAN.
										British Provinces (excluding Chief Towns.
7	337	515	859	6,164	174	...	1	6,339	7,571	Madras, 21.
1	2	...	3	...	54	...	...	54	61	Bombay, 22.
...	...	...	...	...	...	97	...	97	100	Sind, 23.
2	7	...	9	...	12	...	3,13,798	3,13,810	3,13,927	Bengal, 24.
436	26	...	463	...	1,184	415	10,548	12,147	15,628	N.-W. P. & Oudh, 25.
264	39	4	307	58	2,679	2,082	7,096	11,915	14,041	Panjab, 26.
26	...	...	26	...	4	...	...	4	34	Central Provinces, 27.
...	...	...	...	...	1	...	...	1	2	Berar, 28.
...	...	...	...	...	...	...	7,77,886	7,77,886	8,29,496	Assam, 29.
736	411	519	1,666	6,222	4,108	2,594	11,09,329	11,22,253	11,80,860	
										Native States.
...	...	...	...	...	...	...	...	...	19	Raj. & C. India, 30.
...	...	...	...	...	1	...	...	1	7	Nizam's Territory, 31.
...	...	...	...	2	6	...	...	8	24	Mysore, 32.
...	...	...	...	2	7	...	...	9	50	
										Chief Towns.
1	562	339	902	...	5	...	1	6	1,735	Madras seaports, ‡ 33
369	72	...	441	4	...	1	3	8	2,302	Bombay, 34.
...	...	...	...	...	...	...	...	...	50	Karachi, 35.
15	3	...	18	...	11	...	...	11	3,451	Calcutta, 36.
385	637	339	1,361	4	16	1	4	25	7,538	
1,121	1,048	858	3,027	6,228	4,131	2,595	11,09,333	11,22,287	11,88,448	GRAND TOTAL, 37.
91,922	85,936	70,356	30,832	510,696	338,742	212,790	90,965,306	92,027,534	97,452,736	Ditto in ℔
										(2) FOREIGN.
										British Provinces (excluding Chief Towns).
...	...	...	...	...	...	...	...	...	...	Madras, 38.
5	...	...	5	...	9	...	...	9	26	Bombay, 39.
...	...	...	...	...	...	...	...	...	7	Sind, 40.
...	...	...	...	...	...	...	...	...	...	Bengal, 41.
1	...	...	1	...	...	...	...	...	2	N.-W. P. & Oudh, 42.
2	...	...	2	...	...	...	...	...	31	Panjab, 43.
...	...	...	...	...	...	...	...	...	...	Central Provinces, 44.
...	...	...	...	...	...	...	...	...	...	Berar, 45.
...	...	...	...	...	...	...	...	...	...	Assam, 46.
8	...	...	8	...	9	...	...	9	66	
										Native States.
...	...	...	...	...	...	...	...	...	...	Raj. & C. India, 47.
...	...	...	...	...	...	...	...	...	...	Nizam's Territory, 48.
...	...	...	...	...	...	...	...	...	...	Mysore, 49.
...	...	...	...	...	...	...	...	...	...	
										Chief Towns.
...	...	134	134	...	1	...	...	1	155	Madras seaports,‡ 50.
184	49	1	234	1	...	...	3	4	10,819	Bombay, 51.
...	...	...	...	...	...	...	...	...	1,064	Karachi, 52.
...	...	...	...	...	...	...	...	...	37	Calcutta, 53.
184	49	135	368	1	1	...	3	5	12,075	
192	49	135	376	1	10	...	3	14	12,141	GRAND TOTAL, 54.
15,744	4,018	11,070	30,832	82	820	...	246	1,148	995,562	Ditto in ℔
1,313	1,097	993	3,403	6,229	4,141	2,595	11,09,336	11,22,301	12,00,589	TOTAL (INDIAN & FOREIGN) TEA, 55
107,666	89,954	81,426	279,046	510,778	339,562	212,790	90,965,552	92,028,682	98,448,298	Ditto in ℔

ports; those horizontally against 21 to 55 are the exports.
Assam by the Rivers Megna and Bhahmaputra; and (3) Trade to and from Calcutta by river.
and Calicut.

TEA. **Internal Traffic in Tea.**

INTERNAL TRADE.
Local Consumption.
222

Local Consumption.—The remarks hitherto offered have shown the distribution of the Foreign Tea, and manifested some of the more striking facts in the coastwise supply of both Foreign and Indian tea. It may, therefore, serve a useful purpose to still further criticise the features of the trade, brought out by the above detailed tabular statement of the Rail and River tea transactions. The figures are shown in maunds, but these may be reduced to pounds readily enough since the maund of all railway statistics is 82℔. This reduction has been effected with the grand totals, but any of the other entries may be similarly expressed in pounds. The totals of the vertical columns are the imports and the totals extended horizontally are the exports. It will also be observed that the table has been referred to three main sections both vertically and horizontally, namely, the imports into (1) the Provinces, into (2) the Native States, and into (3) the Port Towns—as distinct from the provinces. The traffic in foreign tea may be accepted as sufficiently dealt with in the remarks that have already been made. The total imports of Indian tea by all Provinces, Native States, and Port Towns came (in 1888-89) to 97,452,736℔*; of that amount Calcutta alone took 90,965,306℔ and drew that quantity almost exclusively from Assam and Bengal. The exports from all India to foreign countries were in that year 97,011,112℔, so that it will be seen very nearly the whole of that amount must have been carried on the railways and rivers, in other words, along registered routes. It will be noted that Bengal is shown to take by far the largest amount of tea of all the Indian Provinces. In the year in question it obtained 4,360,596℔ of Indian tea, by far the major portion of which was obtained from Assam. It would seem, however, a very considerable share of these imports may find its way, by routes not registered, to Calcutta or into the North-West Provinces. But assuming that this is not the case, then Bengal is by a long way the most important tea-consuming Province of India, but even were it granted that the $4\frac{1}{3}$ million pounds, credited to the Province, may actually have been consumed in it, this, to a population of over 71 millions, would come to 0·982 oz. per head per annum. Next to Bengal stands the Panjáb, with, assigned to it in 1888-89, 1,049,237℔ of tea, which expressed to the population of 20 million persons, would be 0·807 oz. of tea per annum to each individual. But this calculation ignores two very important considerations, (*a*) the amount of locally produced tea consumed and (*b*) the exports from the province across its land frontier. These exports were, in 1888-89, 7,479 cwt. Indian, and 9,525 cwt. foreign, or say in all 1,904,448℔, and in 1890-91, 1,653,344℔.

Calcutta Trade.
223
Conf. with p. 478.

Consumption to head of Population.
224

After the Panjáb, Bombay stands as next important—the Province received by rail in 1888-89, 315,782℔ of tea in nearly equal proportions of Foreign and Indian. It obtained its supplies from the Port Town which, it may be added, drew *its* supplies of Indian tea mainly from the Bengal contribution coastwise, and the Panjáb and the North-West Provinces by rail. Very little Bengal or Assam tea seems to be consigned by rail to Bombay or the Panjáb After Bombay the North-West Provinces come next in importance; these provinces took in 1888-89, 261,334℔ of tea, of which 250,100℔ was Indian, derived mainly from Calcutta with a fair amount also from the Panjáb. It will thus be seen that the Panjáb tea plantations find a very large outlet for their teas in the Provinces of Upper and Western India and in the transfrontier land traffic. The value of the last mentioned market to the Panjáb plantations may be judged from the following considerations: The imports of Indian tea into the Panjáb from other Provinces in 1888-89 came to 242,474℔, while in that year the exports across the frontier were 837,648℔. The difference

* According to the estimate of production, the outturn of 1888-89 came to 99,792,544℔.

INTERNAL TRADE.

Local Consumption.

between these two figures must be the quantity drawn from local production. But as showing the consumption of Panjáb tea within India, or exported from India (from Calcutta), the following further facts may be given: Of the 14,041 maunds of Indian tea exported from the Panjáb, in 1888-89, to other Indian provinces, 7,096 maunds went to Calcutta, 2,082 maunds to Karachi, 2,679 maunds to Bombay Port, 264 maunds to Rájputana, 156 maunds to the Central Provinces, 794 maunds to the North-West Provinces and Oudh, and 495 maunds to Bombay Presidency. Half these exports from the Panjáb may, in fact, be said to be used up in India.

It may thus, in perfect fairness, be said that Bengal and the North-West Provinces use Bengal and Assam tea and take very little foreign tea. That the Panjáb obtains a large foreign consignment and a considerable Indian supply in addition to its local production. That Bombay, the Central Provinces, Berar, Rájputana, and Sind use about equal proportions of foreign and Panjáb tea with very much smaller quantities of Bengal and Assam tea. That Madras obtains a certain quantity of foreign tea, but is mainly dependent on its local plantations. Madras tea seems also to penetrate to the Central Provinces and even to Bombay just as the Panjáb teas go very nearly all over India, and little more than half the production is thus available for foreign exportation. The Native States (with the exception of Travancore) seem to prefer Foreign to Indian teas.

It will thus be seen that much still remains to be accomplished before the people of India can be classed among the tea-consuming nations of the world. Hitherto the planter has disregarded the possibilities of the Indian market and he could afford to do so, when the field was open to contest with China the supply of Europe. The writer (quoted on page 441) who prophesied in 1863, that the "Himalayan tea has to fight its way against the brick-tea that comes in from Yarkhand and Lhassa, has to fill all the tea-pots of the Mussulmans and Hindoos, and to make a cheap and innocent substitute for deadly opium and mad 'bhang' in the bazaars of Hindoostan, before it can challenge China," shot wide indeed of the mark. The whole statistical information furnished in the foregoing pages shows that while India has not only challenged but beaten China, during the past 30 years, no progress has been made in teaching the native population of India the value of tea. It would therefore seem highly desirable that this subject should receive careful consideration. And it is probable this would best be accomplished by studying the flavour and other pecularities of the imported teas that find a ready market and if possible (for a time at least) imitating these; and next by making up (at the gardens) tea in sufficiently small packages to be purchasable by even the poorest consumer. The Chinese tea sold in India comes to this country (to a large extent) made up in small packages, each of which bears a recognisable trade mark. This suggestion in the writer's opinion should commend itself to the owners of small private tea concerns.

Bombay 225

Bombay Trade.—In order to illustrate more fully the facts shown above, while at the same time furnishing more recent returns, the following table may be given of the internal tea traffic of the Bombay Presidency and of its port town during the past three years:—

TEA.

INTERNAL TRADE.
Local Consumption.
Bombay.

Internal Traffic in Tea.

Statement of the Bombay Presidency and its Port Town Transactions in Tea during the past three years.

	Imports into						Exports from					
	1888-89.		1889-90.		1890-91.		1888-89.		1889-90.		1890-91.	
	1 Bombay.	2 Port Town.	3 Bombay.	4 Port Town.	5 Bombay.	6 Port Town.	7 Bombay.	8 Port Town.	9 Bombay.	10 Port Town.	11 Bombay.	12 Port Town.
INDIAN.	Mds.	Mds.	Mds.	Mds.	Mds.	Mds.	Mds.	Mds.	Mds.	Mds.	Mds.	Mds.
Provinces. Madras	226	174	207	139	192	150	...	3	3	2	2	3
Provinces. Bombay and Sind	...	54	...	76	...	50	...	1,422	...	1,933	...	2,747
Provinces. Bengal	2	12	...	16	2	2	...	1	...	8	1	3
Provinces. N.-W. Provinces and Oudh	55	1,184	104	2,910	139	1,886	...	220	4	79	1	113
Provinces. Panjab	495	2,679	649	1,610	619	1,255	...	28	1	216	1	42
Provinces. Central Provinces	1	4	...	...	...	2	3	144	4	200	1	610
Provinces. Berar	...	1	...	1	...	1	1	35	2	53	1	82
Total of Provinces	779	4,108	960	4,752	952	3,346	4	1,858	14	2,491	7	3,600
Native States. Rajputana and Central India	9	...	3	2	2	2	1	369	4	898	15	1,095
Native States. Nizam's Territory	...	1	...	...	...	2	2	72	1	108	...	159
Native States. Mysore	...	6	2	...	2	...	...	...	...	1	...	2
Total of Native States	9	7	5	2	4	4	3	441	5	1,007	15	1,256
Port Towns. Madras	208	5	255	4	13	8	...	4	...	...	...	1
Port Towns. Bombay	1,422	...	1,933	...	2,746	...	54	...	76	...	50	...
Port Towns. Karachi	...	...	...	...	1	...	...	1	...	...	...	...
Port Towns. Calcutta	5	11	4	18	6	9	...	3	...	1	...	1
Total of Port Towns	1,635	16	2,192	22	2,766	17	54	8	76	1	50	2
FOREIGN.												
Provinces. Madras	...	...	...	...	...	...	...	1	...	2	...	1
Provinces. Bombay	...	9	...	3	...	...	...	1,401	...	742	...	515
Provinces. Bengal	...	...	...	...	...	...	...	3	...	2	...	1
Provinces. N.-W. Provinces and Oudh	...	...	...	...	...	...	...	128	...	148	...	95
Provinces. Panjab	27	...	1	...	...	3	12	8,975	...	10,463	...	9,951
Provinces. Central Provinces	...	...	...	...	...	...	...	60	...	41	...	33
Provinces. Berar	...	...	...	...	...	...	...	13	...	17	...	29
Total of Provinces	27	...	1	3	...	3	12	10,581	...	11,415	...	10,625
Native States. Rajputana and Central India	...	...	...	...	...	33	5	184	...	51	...	58
Native States. Nizam's Territory	...	...	...	...	...	...	...	49	...	53	...	42
Native States. Mysore	...	...	...	...	...	...	...	1	...	2	...	7
Total of Native States	...	...	...	...	...	33	5	234	...	106	...	107
Port Towns. Madras	...	1	...	...	...	...	...	1	...	...	...	...
Port Towns. Bombay	1,401	...	742	...	515	...	9	...	3	...	...	...
Port Towns. Calcutta	...	...	...	...	...	...	...	3	...	4	...	3
Total of Port Towns	1,401	1	742	3	515	...	9	4	3	4	...	3
Total of Indian and Foreign	3,851	4,141	3,900	4,776	4,237	3,403	87	13,121	98	15,024	72	15,593

Bengal Trade. (*G. Watt.*)

TEA.

INTERNAL TRAD.

Local Consumption.

Bombay may not inappropriately be called the importing province for tea, and Bengal the exporting. The table above demonstrates fairly conclusively that the Bombay trade has been practically stationary. The exports from the Port Town would appear to, if anything, have increased, while the imports have decreased slightly. The amount of Bengal tea drawn to Western India is remarkably small indeed, and the imports, as already demonstrated, are from the North-West Provinces and the Panjáb, while the exports, to the Panjáb, of foreign tea continue very large.

Bengal 226

Bengal Trade.—The following facts may similarly be given regarding the Bengal, or what may practically be called the Calcutta, traffic :—

Specification of Routes.	1887-88.		1888-89.		1889-90.		1890-91.	
IMPORTS	Mds.	℔	Mds.	℔	Mds.	℔	Mds.	℔
By Boat . .	...	...	115	9,463	...	...	...	...
,, Inland Steamer . .	6,69,897	55,122,953	7,77,886	64,008,918	8,91,598	73,365,778	8,88,982	73,150,519
,, Eastern Bengal State Railway .	3,17,877	26,156,936	3,07,358	25,291,173	3,23,571	26,625,271	3,26,332	26,852,461
,, East Indian Railway .	17,386	1,430,619	23,974	1,972,719	20,103	1,654,190	15,812	1,301,102
,, Sea . .	13,544	1,114,475	13,919	1,145,321	14,252	1,172,704	13,499	1,110,802
TOTAL .	10,18,704	83,824,783	11,23,252	92,427,594	12,49,524	102,817,943	12,44,625	102,414,884
EXPORTS								
By Boat . .	60	4,937	2	164	...	...	...	...
,, Inland Steamer . .	20	1,646	25	2,057	22	1,810	16	1,316
,, Eastern Bengal State Railway .	92	7,570	870	71,588	108	8,887	630	51,840
,, East Indian Railway .	2,086	191,648	2,554	210,156	1,736	142,847	1,901	156,425
,, Sea . .	10,56,225	86,912,229	11,75,400	98,718,664	12,55,648	103,321,907	12,83,076	105,578,480
TOTAL .	10,58,483	87,098,030	11,78,851	97,002,629	12,57,514	103,475,451	12,85,623	105,788,066

But as showing the Sources of the Calcutta supply, the following may be given as the estimated outturn :—

	1888.	1889.	1890.
	℔	℔	℔
Assam	41,865,499	42,030,564	45,001,072
Cachar and Sylhet . . .	27,343,505	29,750,054	34,477,770
Darjeeling, Terai, and Duars .	18,950,822	20,497,857	21,070,453
Chittagong and Chutia Nagpur	1,148,458	1,378,920	1,768,716
Dehra Dun, Kumaon, and Kangra	4,000,000	4,500,000	4,500,000
Private and Native gardens .	3,000,000	4,500,000	4,000,000
TOTAL .	96,308,284	102,657,395	110,818,011

TEA.

Internal Traffic in Tea.

INTERNAL TRADE.

Local Consumption.

Bengal.

The actual imports into Calcutta were—

	1888-89.	1889-90.	1890-91.
	℔	℔	℔
Assam	66,456,247	75,241,810	76,320,329
Bengal	24,226,231	26,147,273	24,811,063
N.-W. P. and Oudh	867,950	566,537	400,320
Panjáb	583,899	544,484	399,003
Chutia Nagpur	230,318	259,282	448,622
Behar	60,480	48,466	33,161
Other places	2,469	10,091	2,386
TOTAL	92,427,594	102,817,943	102,414,884

By comparing these imports with the figures of exports in the tables above, pp. 459, 461, 474, 477, it will be seen that almost the whole goes to foreign countries. Under 200,000℔ in all is the amount annually sent from Bengal up-country.

IV.—NEW MARKETS FOR INDIAN TEA.

Messrs. Gow, Wilson & Stanton in their Circular of May 1891 furnish some useful information on the subject of future markets for Indian tea. Their remarks may, therefore, be furnished here: "One thing is evident, climate is no barrier to the free use of tea. When we find Australia, Persia, and Turkey contributing so largely to its consumption, it is absurd to say that its use is debarred in semi-tropical regions, while the fact of its adaptability to colder latitudes is too widely recognised to need advocating. The following statistics show, as nearly as can be ascertained, the quantities of Indian and Ceylon tea taken in the undernoted countries during 1891, compared with their approximate *total* annual consumption:—

	Indian.	Ceylon.	Total Annual Consumption.
	℔	℔	℔
Australian Colonies	4,440,000	3,211,000	30,000,000
Persia	2,400,000	500,000	...
United States	990,000	744,000	80,000,000
Turkey	1,104,000	18,000	...
Canada	680,000	410,000	20,000,000
* Germany	192,000	604,000	4,000,000
Holland	407,000	156,000	5,000,000
South Africa	114,000	111,000	2,000,000
South America	94,000	84,000	...
* Austria	14,000	156,000	1,200,000
Arabia	130,000	...	...
France	43,000	65,000	1,200,000
Mauritius	2,000	69,000	...
Russia	2,000	66,000	70,000,000

* Probably most of the tea sent to Germany and Austria was for Russia.

Conf. with pp. 457, 461, 462.

"AUSTRALIAN COLONIES.—It is not surprising that Australasia should be so large a consumer. Not only are its people our own kinsfolk and countrymen, and have thus inherited similar tastes, but it was here that India made her earliest efforts at establishing a new market; and she can now look back with grateful pride upon the work of those early days in the history of her tea industry. Here, too, Ceylon was eager, in later years, to find not only a near but an important market—for the Australians consume annually nearly 30,000,000℔ of tea—and her efforts have also reaped a rich reward. These colonies have now become the largest of all markets for British-grown tea outside the United Kingdom, although the demand appears still to be only in infancy.

Internal Traffic in Tea.

TEA.

INTERNAL TRADE.

"PERSIA.—This market is a source of surprise to many. It has grown and developed until it has attained its present dimensions. Its nearness to India may be one of the chief causes, but the favour in which Indian tea is there regarded is attributed by some to a preference in the locality towards the use of an article supplied by co-religionists;—and there may be some truth in the suggestion.

Conf. with pp. 462, 464, 465.

"UNITED STATES OF AMERICA.—This is so vast a territory, and the quantity of tea consumed is so large, amounting annually to some 80,000,000lb, that it is curious so small a percentage of our teas should be used. Decided headway has been recently made, and prolonged low rates of the past few months have done much to popularize Indian and Ceylon teas. A very large amount of advertising has of late been done by the Ceylon Planters' Tea Company, who continue steadily pushing the sale of Ceylon tea;—and the present demand may be partially due to their perseverance.

"The kinds of tea used in different parts of the States—separated by so many thousands of miles—are so varied that it is folly to argue that the produce of India and Ceylon is unsuited to the American taste, because Japans, Oolongs, and Greens, as well as Black China teas, are so largely consumed. If the Americans knew where to buy *good* tea, there is little doubt but that they would soon buy it,—although, to commence with, as a rule, they prefer light, flavoury kinds to strong, heavy teas. Happily, both India and Ceylon can supply them with abundance of tea, both light and flavoury, and of really good quality.

"TURKEY has become an important consumer of Indian tea, and it is possible that the religion of her people may induce her to take the teas of a country which contains perhaps the largest Mahomedan population in the world. Ceylon tea is being gradually introduced and appears to be received with some favour.

Conf. with pp. 461, 464.

"CANADA is perhaps one of the most promising outlets. The consumption of all tea is nearly 4lb per head of population and the percentage of British-grown tea is already considerable—even though little systematic attempt has been made to open up this market. Recent low rates have perhaps given the greatest impetus to the trade.

Conf. with p. 461.

"RUSSIA, although a market of considerable magnitude, and taking some 70,000,000lb of China tea annually, takes but little Indian tea, and until recently Ceylons were almost unknown. During the last two or three years a distinct enquiry for Ceylon tea has sprung up, and considerable quantities are now disposed of in this country. Figures showing actual consumption of Ceylon tea are impossible to obtain, but probably the bulk of that which goes to Germany has its final destination in Russia. The work of Ceylon in pioneering this market appears to have created some demand for the finest and most flavoury descriptions of her tea. Russia should prove most valuable as an outlet for high class teas, and of much eventual assistance in maintaining, if not increasing, the value of teas with flavour and quality combined.

Conf. with p. 457.

"Of other European markets Holland appears the most active, but there are reasons why Holland should be a tea-drinking country. She has herself fostered tea culture in her own colony of Java, and having thus acquired a taste for tea, is now one of the chief tea consumers in Europe.

GERMANY also is likely to prove by degrees a useful outlet.

"OTHER MARKETS.—Amongst those countries which as yet take but little of our tea, perhaps that of South Africa, with its rapidly increasing British population, offers the greatest encouragement for prospecting. In time this locality should naturally consume British-grown tea.

SOUTH AMERICA is a consumer of British-grown tea, but the unsettled state of this continent is adverse to immediate development of the trade."

Printed in the United States
By Bookmasters